Trophoblast Research
Volume 9

PLACENTAL MOLECULES IN HEMODYNAMICS, TRANSPORT AND CELLULAR REGULATION

Trophoblast Research
Volume 9

PLACENTAL MOLECULES IN HEMODYNAMICS, TRANSPORT AND CELLULAR REGULATION

Edited by

Toshio Hata
Saitama Medical School
Saitama, Japan

Masaomi Takayama
Tokyo Medical College
Tokyo, Japan

Ichiro Taki
Osaka Police Hospital
Osaka, Japan

UNIVERSITY OF ROCHESTER PRESS

First published 1997

University of Rochester Press
668 Mount Hope Avenue
Rochester, New York 14620 USA
and at PO Box 9, Woodbridge, Suffolk IP12 3DF, UK

ISBN 1-58046-016-X

Library of Congress Cataloging-in-Publication Data

Placental molecules in hemodynamics, transport and cellular regulation / edited by Toshio Hata, Masaomi Takayama and Ichiro Taki
p. cm. -- (Trophoblast research ; v.9)
Derived from the First Meeting of the Japanese Placenta Group, held November 4-5, 1993, in Saitama, Japan" -- P. viii,
Includes index
ISBN 1-58046-016-X (alk. paper)
1. Placenta--Physiology--Congresses. 2. Growth Factors -Congresses. 3. Trophoblast--Congresses, I. Hata, Toshio, 1940-. II. Takayama, Masaomi, 1938- . III. Taki, Ichiro, 1919- . IV. Japanese Placenta Group. MeetingSeries (Ist : 1993 : Saitama-ken, Japan) V. Series..
QP281.P543 1997 97-28672
812.6'3--dc21 CIP

British Library Cataloguing-in-Publication Data

A catalogue record for this book is available from the British Library

This publication is printed on acid-free paper

Printed in the United States of America

Trophoblast Research

Series Editor

Richard K. Miller and Henry A. Thiede
University of Rochester Medical Center
Rochester, New York

Volume 1	FETAL NUTRITION, METABOLISM AND IMMUNOLOGY The Role of the Placenta Edited by Richard K. Miller and Henry A. Thiede Published: 1983
Volume 2	CELLULAR BIOLOGY AND PHARMACOLOGY OF THE PLACENTA Techniques and Applications Edited by Richard K. Miller and Henry A. Thiede Published: 1987
Volume 3	PLACENTAL VASCULARIZATION AND BLOOD FLOW Basic Research and Clinical Applications Edited by Peter Kaufmann and Richard K. Miller Published: 1988
Volume 4	TROPHOBLAST INVASION AND ENDOMETRIAL RECEPTIVITY - Novel Aspects of the Cell Biology of Embryo Implantation Edited by Hans-Werner Denker and John D. Aplin
Volume 5	THE MOLECULAR BIOLOGY AND CELL REGULATION OF THE PLACENTA Edited by Richard K. Miller and Henry A. Thiede Published: 1991
Volume 6	PLACENTAL SIGNALS Endocrine and Paracrine Control of Pregnancy Edited by Lise Cedard and J. Anthony Firth Published: 1992
Volume 7	FETAL GROWTH AND THE PLACENTA - From Implantation to Delivery Edited by Henning Schneider, Paul Bischof, and Rudolf Leiser Published: 1993
Volume 8	HIV, PERINATAL INFECTIONS AND THERAPY The Role of the Placenta Edited by Richard K. Miller and Henry A. Thiede

Published: 1994

Volume 9	PLACENTAL MOLECULES IN HEMODYNAMICS, TRANSPORT AND CELLULAR REGULATION Edited by Toshio Hata, Masaomi Takayama and Ichiro Taki Published: 1997
Volume 10	EARLY PREGNANCY Edited by Jean-Michel Foidart, John Aplin, Peter Kaufmann and Jean-Pierre Schaaps Published: 1997

TROPHOBLAST RESEARCH

Trophoblast Research publishes contributions concerning the placenta and the extraembryonic membranes as they relate to embryonic and fetal development and to trophoblastic neoplasia. Original articles, reviews, and reports are published in single bound volumes. All articles are peer-reviewed.

The Editorial Office for
Department of Obstetrics and Gynecology
University of Rochester School of Medicine and Dentistry
601 Elmwood Avenue, Rochester, New York 14642-8668 USA
716-275-3638 and trophores@obgyn.rochester.edu

Derived from the
First Meeting of the
Japanese Placenta Group
held November 4-5, 1993 in
Saitama, Japan

PREFACE

Evolution of Placental Research in Japan : Organization of the Japanese Placenta Study Group (JPG)

Ichiro Taki, M.D.,
Manager in Chief, Japanese Placenta Study Group Committee
Emeritus Director, Osaka Police Hospital
10-31 Kitayamacho, Tennojiku, Osaka 542, Japan.

It was an epoch making event - the First Academic Meeting of Japanese Placenta Study Group was held on 4-5th November 1993. In commemoration of this meeting, I. Taki presented a lecture relating the historical events leading up to the organization of JPG and the recent trend in the placenta research in Japan. as compared with that, understood by a review of the papers appeared in Placenta, Volume 1-13, 1980-92. There had been stepwise progress concerning the placenta research in Japan as follows.

Since Boyd and Hughs in 1954, Wislocky and Dempsey in 1955 published the initial papers on the findings in the human placenta, several papers appeared in the Japanese literature during the successive period from 1955 to 1963. This research dealt with the electronmicroscopy of the human placenta examining its development, differentiation, function, hormone production, and tumorgensis (H. Isomura, 1955; C. Sawazaki and T. Mori, 1956; I. Yoneyama, 1957; T. Takeyama, 1958; T. Inoue, 1959; A. Ikawa, 1959; H. Watanabe, 1959; Y. Ashitaka and I. Taki, 1958; Y. Okudaira, 1963).

These studies certainly developed the fine structures of the placental cellular components, which could not be differentiated traditional light microscopy and significantly opened wide avenues for further investigation of placental function not only from morphological but biochemical, endocrinological and oncological aspects.

During next ten years, remarkable progress and improvement had been achieved as for the capacity and quality of electronmicroscope, and preparation technique of placental specimens to be observed by means of newly developed methods such as the histochemicals, immunohistochemicals and autoradiographics. In 1970, I. Taki and H. Soma successfully developed the Electron Microscopic Placental Study Group following negotiations with the other researchers in this field. The principal purpose of this study group was to have intimate communications among the researchers and to promote their research, learning from each others' ideas and techniques.

The first scientific meeting of the Electron Microscopic Placental Study Group was held on 4th April 1979 at Tokyo in conjunction with the annual meeting of the Japan Society of Obstetrics and Gynecology. Thereafter the meeting was held annually in the same manner in April, i.e., at Tokyo in 1980, at Niigata in 1981, at Kobe in 1982, at Osaka in 1983, at Sendai in 1984, and at Fukuoka in 1985.

At every meeting, a small number of the selected researchers presented the latest results of their studies. Lively discussions among the speakers and participants concerning each presentation lead to enhanced appreciation and rapid progress in the field.

Obstetrics and Gynecology

Through negotiations between the Electron Microscopic Placenta Study Group during the previous scientific meetings, it had been decided to open the scientific meeting in 1986 as well as in successive years under the modified title, "Morphological and Functional Study in Gynecology and Obstetrics" with the purpose of introducing research and discussion, not limited to only placental morphology, but correlating function with morphology in the placenta. The focus of the first meeting was "Structure and Function of the Placenta".

Since 1986, there have been a total of eight meetings held annually in December or in January until 1993. The date, place, focus, and sponsor of the meetings were as follows:

I. December 6, 1986, Tokyo, Structure and Function of the Placenta, H. Soma.
II. December 7, 1987, Osaka, Structure and Function of the Endometrium, O.Tanizawa.
III. January 30, 1988, Tokyo, Structure and Function of the Placenta, T. Hata.
IV. January 21, 1989, Tokyo, Structure and Function of Endometrium, S. Takagi.
V. January 27, 1990, Placenta and Endometrium, I. Sawaragi.
VI. January 19, 1991, Function and Morphology of the Ovary, H. Ito.
VII. January 8, 1992, Kyoto, Morphological Approach to Ovarian Function, Y. Yoshida.
VIII. January 23, 1993, Tokyo, Morphological Approach to Endometrial Function, M. Takayama.

Distinguished foreign researchers were often invited and presented special lectures. G.E. Wood at the first meeting in 1986, G. Grudzinskas in 1987, M. Panigel in 1988, S.G. Silverberg in 1989, and P. Kaufmann in 1990.

All presentations at every meeting were published as full papers in an issue of "The World of Obstetrics and Gynecology" (in Japanese) in the corresponding year: volume 38(7), 3-29, 1986; volume 39(8), 1-39, 1987; volume 40(7), 1-41, 1988; volume 41(10), 1-50, 1989; volume 43(1), 1-47, 1991; volume 43(9), 1-79, 1991; volume 44(7), 1-43, 1992; volume 45(7), 1-67, 1993.

These papers, including the special lectures given by the invited foreign researchers were actually the publications of the latest achievements of every year and are usually the leading articles.

While the annual meetings of the Morphological and Functional Placental Study Group were held one after another since 1986, the International Conference on Placenta was held on 1-3 October 1990 at Tokyo Medical College, organized by Professor H. Soma, Professor M. Takayama, and colleagues, the content of which was briefly reported by J. Dancis on the Placenta (1991) 12, No. 1. This international event not only stimulated much scientific interest among placenta researchers in Japan, but effectively emphasized the necessity of the international exchange of the recent knowledge.

At the 12th Rochester Trophoblast Conference, an evolution of International Federation of Placental Associations was initiated. Upon returning to Japan, the Japanese attendants at the conferences, H. Soma, M. Takayama, and T. Hata, strongly proposed the

Japanese Placenta Study Group to be an official member in order to participate in future activity of IFPA, if this will be realized. To answer the proposal, preliminary negotiations occurred by voluntary members on 13th April 1993 in Osaka, and at the next meeting in Tokyo on 23th April, the general rules for establishment and activity of an official organization of Placenta Study Group were drafted. Following the draft, the first meeting of the elected managers for the newly organized Japanese Placenta Study Group (JPG) were held, and planning of annual scientific conference and collaboration with the evolving IFPA were negotiated, and an agreement was to have a scientific meeting in November 1993. and to make every effort for collaborating with IFPA.

Thus, the first scientific meeting was held in Osaka on 4-5 November 1993 under the auspices of Professor I. Sawaragi and colleagues (Department of Obstetrics and Gynecology, Kansai Medical College). The details on the meeting was presented by H. Soma as Conference Report on Placenta (1994), 15, 447-449.

The second scientific meeting was held in Omiya, on 10-11th November 1994, organized by Professor Toshio Hata and colleagues (Department of Obstetrics and Gynecology, Saitama Medical College). At this time number of registered members of JPG reached almost four hundreds. The third meeting was held in Osaka on 9-10th November 1995, organized by Associate Professor Yoshio Okudaira and colleagues (Department of Obstetrics and Gynecology, Osaka University Medical School).

In *Placenta*, from volumes 1-13 (1980-92), approximately 579 original papers were published. Four hundred and twenty-four were studies involving the human, and 155 involving the animal placenta. Following classification by I. Taki, seven studies were on the microbials, 22 the physiology, 157(3) the morphology, 162(8) the biochemistry, 39(4) the endocrinology, 24 the immunochemistry, 13(6) the microbiology and genetics. The numbers within the parentheses indicate the contributions of Japanese authors. It appears that the morphological and biochemical studies, were approximately equal in number. There were increasing numbers of studies on placenta function involving microbiology, immunochemistry, and genetic more than ever. Microbial study on viral infections also increased.

In *Trophoblast Research* from volumes 1-9 (1983 - 1997), approximately 228 papers have been published. Twenty- two of these contributions have been from Japanese authors. There were several factors which have promoted the placenta research. The obstetricians and pediatricians who had been engaging in the perinatal problem were eager to lower the perinatal morbidity and mortality. In the clinical field, numerous diagnostic as well as therapeutic procedures have been playing an ever increasingly important role. However, inspite of the improvements in these procedures, it remains difficult to have the complete picture of fetal distress. Therefore, enthusiastic support for continued fundamental research concerning the fetomaternal correlations is mandatory. This must be achieved as the most essential factor for the continuation of placental research. Accordingly, the recent biochemical, immunochemical, microbiological knowledge base and techniques have been actively introduced. The observation of enzyme activity and transportation of such a variety of substances utilizing isolated BBM, BBMV, BM, BMV are replacing the classic methods with brilliant results.

Cytokine have revolutionized our concepts of both the physiology and pathophysiology of the placenta, e.g., infection-induced preterm labor. Flow cytometry has contributed to the differentiation of chorionic tumors. Electronmicroscopes with higher performance than ever has enabled more detailed and precise morphological observations of the placenta, being associated with recent histochemical and immunohistochemical techniques. Highly advanced methods of *in situ* hybridization, *in situ* PCR, tissue culture, transfusion, cytometry have been introduced, and detailed observations of the cellular components and functions of the placenta have been achieved. New pathogenic viruses and bacteria have been found one after another. Finally the medicolegal imperative for the management of delivered placenta give further significance to the pathomorphological examination of the placenta in cases of perinatal difficulties. All demonstrate a fruitful future for placenta research.

At the 45th annual meeting of JSOG, Japan Society of Obstetrics and Gynecology, 1993, there were 446 oral presentations and 295 poster demonstrations. Among them 367 (49%) concerned pregnancy, delivery, puerperium, and correlating infections. And, 37 (10)%) out of the presentations were studies on placenta. The trend for the fundamentals of placenta research, though small in number, are growing and stimulated by our collective efforts. However, many more contributions from researchers in the field of fundamental medicine and zoology are required to intensify the research on the placenta. Such also reflects the urgency for the Japanese Placenta Study Group, now known as the Japan Placenta Association.

CONTENTS

Trophoblast Research 9:1-11, 1997

MOLECULAR DIVERSITY OF RAT PLACENTAL LACTOGENS
-A Review -

Kunio Shiota, Kwan-Sik Min, Ryuichi Miura, Mitsuko Hirosawa, Naka Hattori, Ken Noda and Tomoya Ogawa

Laboratory of Cellular Biochemistry
Animal Resource Sciences/Veterinary Medical Sciences
The University of Tokyo
1-1-1 Yayoi, Bunkyo-ku
Tokyo 113, Japan

INTRODUCTION

The placenta plays an essential role in the maintenance of pregnancy and fetal growth. Many kinds of molecules are produced from the placenta in a pregnancy stage-specific manner. One of the main endocrine functions of the placenta is the production of placental lactogens (PLs), which are prolactin (PRL)- or growth hormone (GH)-like hormones found in several mammals, including the rat (Robertson et al., 1982), cow (Arima and Bremel, 1983), sheep (Chan et al., 1976), human (Hunt et al., 1981), monkey (Shome and Friesen, 1971), and hamster (Souhard et al., 1986).

Placental PRL-like molecules (PRL family) are found in a number of species, and their structures and functions have been described in previous reviews (Brash et al., 1983; Southard and Talamantes, 1991; Soares et al., 1991; Shiota et al , 1994). Many members of the PRL family are found in mouse, bovine, and hamster placentae (Yamakawa et al., 1990; Southard and Talamantes, 1991), but GH-like proteins have not been identified in these species. Rodent PRL family members show structural homology with PRL through cysteine residue positioning within these proteins (Southard and Talamantes, 1991; Soares et al., 1991). Primate PLs are structurally more similar to pituitary GH than PRL (Brash et al., 1983; Shome and Friesen, 1971). Thus, the rat is one of several species which have developed various members of the PRL family in the placenta. In this review, we describe the molecular diversity of rat PLs from their structural and functional aspects, and consider why and how they developed in this species.

The Rat Has Developed Mid-Pregnancy-Specific PLs Which Are Structurally Related To PRL

Rat PLs can be broadly classified into two categories, based on the stage of pregnancy when they are secreted. Mid and late-pregnancy-specific PLs, such as placental lactogen-I (PL-I) and PL-II (Robertson et al., 1982) were originally identified in pregnant rat serum. Rat PL-II has been purified completely and the nucleotide sequence of its cDNA has been determined (Duckworth et al., 1986a). Rat PL-I cDNA has also been cloned by screening a cDNA library prepared from placentae during mid-pregnancy with an antiserum against mouse PL-I (Robertson et al., 1990). During the cloning of rat PL-I and PL-II cDNAs, placental PRL family members were identified and were termed prolactin-like proteins (PLP)-A, PLP-B and PLP-C (Duckworth et al., 1986b, 1988; Deb et al., 1991). Another member of the rat placental PRL family, which is expressed during late pregnancy (Robertson et al., 1991), is known as the placental lactogen-I variant (PL-Iv). With the exception of PL-I, all these members of the rat placental PRL family (PL-II, PL-Iv, PLP-A, PLP-B, PLP-C) are known to be expressed during late pregnancy (Duckworth et al., 1986a, 1986b, 1988; Robertson et al., 1990, 1991; Deb et al., 1991). Recently, we found the new member of the family, PLP-D, which is expressed during late pregnancy (Iwatsuki et al., 1996).

PL activity in the serum during mid-pregnancy does not appear to depend upon a single molecular type. A molecule with a molecular mass of 55-60 kDa has been found in mid-pregnancy and tentatively designated placental lactogen-α (PL-α) (Furuyama et al., 1991; Shiota et al., 1991; Hattori et al.; 1993). Although the cDNA corresponding to PL-α has not been cloned, during the course of investigations into mid-pregnancy-specific PLs, we isolated the cDNA encoding a novel PL from the rat placenta and called PL-I mosaic (PL-Im) (Hirosawa et al., 1994) (Figure 1).

The cDNA of PL-Im comprises an open reading frame of 687 bp encoding 229 amino acids with two putative N-glycosylation sites, and PL-Im contains the twenty amino acids which are conserved in the six other previously reported PLs (Figure 1). The positions of these amino acids are; Cys (80, 199, 216 and 224); Leu (14, 16, 23, 103, 181, 196 and 214); Ser (19, 29 and 183); Trp (24 and 175); Glu (25); His (119); Asp (203) and Arg (217) (Duckworth et al., 1986a, 1986b, 1988; Robertson et al., 1990; 1991; Deb et al., 1991). Of these, the four Cys residues, which form the disulfide bonds that determine the tertiary structures of proteins (Southard and Talamantes, 1991), are conserved in the rat PL family without exception. PL-Im is highly homologous with the other six members of the rat PL family (PL-I, PL-Iv, PL-II, PLP-A, PLP-B and PLP-C) (Hirosawa et al., 1994) (Figure 2). In particular, PL-Im cDNA is over 90% homologous with those of PL-I and PL-Iv. Furthermore, detailed comparisons of their nucleotide and amino acid sequences revealed that PL-Im cDNA does not possess its own sequence but is a mosaic structure of PL-I (mid-pregnancy-specific) and PL-Iv (late-pregnancy-specific) (Figure 3). Thus, PL-Im cDNA does have a unique structure as described above, and is produced in the placenta at mid-pregnancy.

The Importance Of PL-IM In The Evolution Of Rat PLs

The possible molecular evolution process of PLs was deduced by constructing a dendrogram based on the nucleotide sequence homologies of PL family members (Figure 4). Based on the dendrogram, the mid-pregnancy-specific PLs (PL-Im and PL-I) appear to be derived from PL-II (late-pregnancy-specific), and PL-Im is located between PL-Iv and PL-I during the process of their molecular evolution. The replacement to silent substitute ratio of PL-Iv to PL-Im is 1.21 (54.8%/45.2%), whereas that of PL-Im to PL-I is quite high (3.50, 77.8%/22.2%). Therefore, it is probable that positive selective influences caused rapid fixation of replacement substitutes in the genes.

PL-II is not a glycoprotein, and therefore, the evolution of PRL genes appears to have been accompanied by a process that produced molecular diversity with respect to glycoresidues. Alternatively, during the molecular evolution of the rat placental PRL family from the late to mid-pregnancy, PLs may be linked with the evolutionary process that resulted in the acquisition of glycoresidues on protein molecules.

```
       Met Gln Leu Thr Leu Thr Leu Ser Gly Ser Gly Met Gln Leu Leu Leu Leu Val
PL-Im  ATG CAG CTG ACT TTG ACT CTT TCG GGC TCT GGT ATG CAA CTG TTG CTG CTG GTG
                                       27          36          45          54

Ser Ser Leu Leu Leu Trp Glu Asn Val Ala Ser|Lys Pro Thr Ala Ile Val Ser Thr Asp
TCA AGC TTG CTC CTT TGG GAA AAC GTG GCC TCC|AAA CCA ACT GCC ATT GTG TCC ACT GAT
                                81          ▼ 90          99          108

Asp Leu Tyr His Arg Leu Val Glu Gln Ser His Asn Thr Phe Ile Met Ala Ala Asp Val
GAC CTA TAT CAT CGT TTG GTT GAA CAG TCT CAT AAT ACA TTT ATC ATG GCT GCA GAT GTA
        123         132         141         150         159         168

Tyr Arg Glu Phe Asp Ile Asn Phe Ala Lys Arg Ser Trp Met Lys Asp Arg Ile Leu Pro
TAC CGT GAA TTT GAT ATA AAT TTT GCC AAG AGA AGT TGG ATG AAA GAC AGG ATA CTT CCC
        183         192         201         210         219         228

Leu Cys His Thr Ala Ser Ile His Thr Pro Glu Asn Leu Glu Glu Val His Glu Met Lys
CTG TGT CAC ACT GCT TCC ATC CAT ACT CCA GAG AAT CTA GAG GAA GTC CAT GAA ATG AAA
        243         252         261         270         279         288

Thr Glu Asp Phe Leu Asn Ser Ile Ile Asn Val Ser Val Ser Trp Lys Glu Pro Leu Lys
ACT GAA GAC TTC CTG AAC TCA ATC ATC AAT GTT TCA GTT TCC TGG AAA GAA CCT CTG AAA
        303         312         321         330         339         348

His Leu Val Ser Ala Val Thr Asp Leu Pro Gly Ala Ser Val Ser Met Gly Lys Lys Ala
CAC TTG GTG TCT GCA GTG ACT GAT CTT CCG GGA GCT TCT GTT AGT ATG GGG AAA AAA GCT
        363         372         381         390         399         408

Val Asp Met Lys Asp Lys Asn Leu Ile Ile Leu Glu Gly Leu Gln Thr Leu Tyr Asn Arg
GTT GAT ATG AAG GAC AAA AAT CTT ATA ATT CTG GAG GGA CTT CAG ACC TTA TAC AAC AGG
        423         432         441         450         459         468

Thr Gln Ala Lys Val Glu Glu Asn Phe Glu Asn Phe Asp Tyr Pro Ala Trp Ser Gly Leu
ACT CAG GCT AAA GTT GAA GAA AAT TTT GAA AAT TTT GAC TAC CCT GCC TGG TCT GGA CTC
        483         492         501         510         519         528

Lys Asp Leu Gln Ser Ser Asp Glu Asp Thr His Leu Phe Ala Ile Tyr Asn Leu Cys Arg
AAA GAC TTG CAG TCA TCT GAT GAA GAC ACT CAT CTT TTT GCC ATT TAT AAC CTG TGC CGC
        543         552         561         570         579         588

Cys Phe Lys Arg Asp Ile His Lys Ile Asp Thr Tyr Leu Lys Val Leu Arg Cys Arg Val
TGC TTT AAA AGG GAC ATC CAT AAG ATT GAC ACT TAT CTC AAA GTC TTG AGG TGC CGA GTT
        603         612         621         630

Val Phe Lys Asn Glu Cys Gly Val Ser Thr Phe TER
GTC TTT AAG AAT GAG TGT GGA GTG TCC ACC TTT TGA AGTCTTGCACCCAATGTTGAACCAGACTTTT
        663

GTAATGCTTTTTCGCCTCTCGGTGTATTCAGAGCTGTAATGGAATTCTTTTCATAAAATAAAATGGAATTATTTAGAAA

AAAAAAAAAAAAAAAAA
```

Figure 1. Nucleotide and deduced amino acid sequence of PL-Im cDNA. The PL-Im cDNA contains an open reading frame with 687 bp nucleotides encoding 229 amino acids. The nucleotides were numbered from the initiation codon, ATG. The position of the signal cleavage site was predicted according to its homology with other members of the PL family, and is marked with an arrow. The positions of the two potential glycosylation sites, Asn-Val-Ser at position 108-110 and Asn-Arg-Thr at position 157-159, are underlined. The four Cys residues, which are conserved in the rat PL family and are considered to form disulfide bonds, are shown enclosed in boxes. (Reprinted with permission, Hirosawa et al., 1994).

Homology for Identical Nucleotides (%) (upper right) / **Homology for Identical Amino Acids (%)** (lower left)

	PL-Im	PL-I	PL-Iv	PL-II	PRL	PLP-A	PLP-B	PLP-C
PL-Im		97	90	50	45	34	26	22
PL-I	92		88	55	45	34	26	22
PL-Iv	87	80		51	53	29	36	19
PL-II	42	39	42		47	35	39	33
PRL	30	27	31	36		47	28	24
PLP-A	20	18	32	29	19		38	27
PLP-B	11	10	20	30	17	29		50
PLP-C	15	14	12	16	16	14	19	

Figure 2. Homology of PL-Im with other members of the rat PL family. The percentage homologies with identical amino acids and nucleotides are shown. (Reprinted with permission, Hirosawa et al., 1994).

```
                 10        20        30        40
PL-I   MQLTLTLSRAS_GMQLFLLVSSLLLWEKVASKPTAIVSTDDLYHCLVEQS
PL-Im  MQLTLTLS_GS_GMQLLLLVSSLLLWENVASKPTAIVSTDDLYHRLVEQS
PL-Iv  MQLTLTLS_GSTGMQLLLLVSNLLLWENMASKPTVLVSTEDLYHRLVEQS

        50        60        70        80        90
PL-I   HNTFIMAADVYREFDINFAKRSWMKDRILPLCHTASIHTPENLEEVHEMK
PL-Im  HNTFIMAADVYREFDINFAKRSWMKDRILPLCHTASIHTPENLEEVHEMK
PL-Iv  HNTFIKAADVYREFDINFAKRSWMKDRIL LCHTASIHVPENREEVHEIK

       100       110       120       130       140
PL-I   TEDFLNSIINVSVSWKEPLKHLVVCSDCSSGASVSMGKKAVDMKDKNLII
PL-Im  TEDFLNSIINVSVSWKEPLKHLVSAVTDLPGASVSMGKKAVDMKDKNLII
PL-Iv  TEDILSSIINISVSWKEPLKHLVSAVTDLPGASASMRKKAVDMKDKNLII

       150       160       170       180       190
PL-I   LEGLQTLYNRTQAKVEENFENFDYPAWSGLKDLDSSDEEHHLFAICNLCR
PL-Im  LEGLQTLYNRTQAKVEENFENFDYPAWSGLKDLQSSDEDTHLFAIYNLCR
PL-Iv  LEGLQKIFNRTQTKVEEN ENFDYPAWSGLKDLQSSDEDTHLFAIYNLCL

       200       210       220
PL-I   CVKRDIHKIDTYLKVLRCRVVFKNECGVSTF
PL-Im  CFKRDIHKIDTYLKVLRCRVVFKNECGVSTF
PL-Iv  CFKSDIHKIDTYLKVLRCRVVFKNEC
```

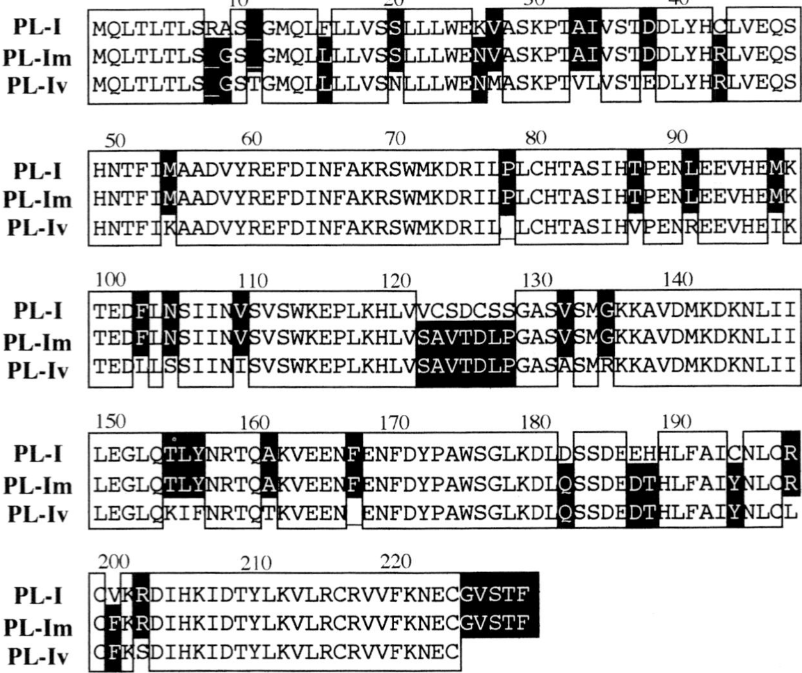

Figure 3. Comparison of the amino acid structure of PL-Im with those of PL-I and PL-Iv. The nucleotide and deduced amino acid sequences of PL-Im show that its structure is a mosaic of PL-I and PL-Iv and that it does not possess its own sequence. (Reprinted with permission, Shiota et al., 1994).

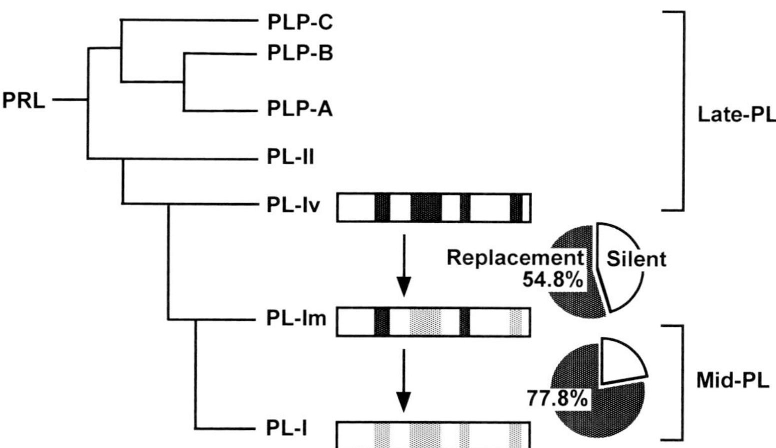

Figure 4. Dendrogram of the PL family. The dendrogram was constructed according to the nucleotide sequence homologies of seven members of the PL family. Replacement substitution (one base change in a codon results in an amino acid replacement) and silent substitution (one base change does not affect the amino acid sequence as it codes for the same amino acid) are shown. (Reprinted with permission, Shiota et al., 1994).

Diversity Of Rat Placental Lactogens

In a study of point mutations in mouse PLs, three other amino acids, Arg at 43 and 202 and Lys at 212 in PL-Im, were postulated to be essential for reactions with the lactogenic hormone receptor (Davis and Linzer, 1989). These three amino acids were conserved in PL-Im and two of them were conserved in PL-Iv (Robertson et al., 1991; Hirosawa et al., 1994). PL-Im as well as PL-Iv have ability to bind PRL-receptor and stimulate proliferation of Nb2 cells (Hirosawa et al., 1995; Cohick et al., 1996). Therefore, PL-Im and PL-Iv might have evolved, maintaining the ability to react with the PRL-receptor during the process. Since only PL-I and PL-Im are expressed in mid-pregnancy, this evolution appears to have been accompanied by the regulation element of mRNA transcription. Genomic analysis including the 5'-upstream region of PL-Im will be necessary to confirm this hypothesis.

Functional Diversity Of PLs

The discovery of these various PLs and PLPs prompted us to speculate on their possible functions. The PRL family members PL-I, PL-II and PL-Im appear to share the PRL receptor (Robertson et al., 1982; Hirosawa et al, 1994), which is distributed in various tissues. In the liver, the number of PRL receptors increases during pregnancy (Kelly et al., 1991), and ovine PL stimulates glycogen synthesis in fetal rat hepatocytes (Freemark and Handwerger, 1984), suggesting that the PLs participate in the maternal metabolic function. PRL affects the growth and proliferation of neonatal and adult rat islet b cells (Brelje and Sorenson, 1991). PLs also make an important contribution to the development of mammary glands in various animals, and the role of the placenta in mammary gland development and function have been reviewed (Thordarson and Talamantes, 1987). Administration of PRL stimulates maternal behavior in steroid-treated rats (Bridges et al., 1990), and the direct action of PRL and PLs on the central nervous system has been studied (Voogt and Greef, 1989). These putative multiple functions may account for the diversity of the PRL-like molecules. Therefore, PLs would be expected to have various functions in maternal and fetal tissues.

Possible Role Of PL As A Luteotropic Factor

Why are most PLs expressed during late-pregnancy and only a few during mid-pregnancy in the rat? The dominant expression of placental PRL- or GH-like molecules in the later half of pregnancy is common in various animals including cattle and humans. Therefore, the expression of late pregnancy-specific PLs in the rat placenta may have a physiological role common among animal species, such as the development of mammary glands as described above. Conversely, development of mid-pregnancy-specific PLs may be required by the rodent for a particular reason.

Hormone production by the placenta is critical for the initiation and maintenance the pregnancy. Progesterone is essential for maintaining pregnancy in all mammals and is secreted by the ovary or placenta, depending on the animal species and stage of pregnancy (Heap et al., 1973). In some species, including humans, the placenta is the main organ of progesterone secretion in middle and late pregnancy (Heap et al., 1973), while in the rat, the main source of progesterone throughout pregnancy is the ovary. In rodents, PRL-like hormones are critical for sustaining the production of progesterone from the corpus luteum in the ovary throughout pregnancy. In view of their structural similarities and the observation that twice-daily surges of PRL ceased when secretion of PLs started in the rat (Tonkowicz et al., 1983), PLs have been proposed to replace the functions of pituitary PRL. Indeed, removal of the anterior pituitary after mid-gestation has been shown not to affect luteal function in the rat, and pregnancy was able to be continued (Heap et al., 1973). The presence of pituitary PRL dose not appear to be necessary after day six of gestation. Thus, mid-pregnancy-specific PLs may act as placental luteotropic factors after the development of the placenta is completed. This differ from the situation in primates and horses, in which the placenta-derived luteotropic factors are the chorionic gonadotropins such as human chorionic gonadotropin, which is highly homologous with lutenizing hormone (LH), and equine chorionic gonadotropin, of which the cDNA is identical to that of LH (Min et al., 1994). In other animals including the sheep, pig and cow, another mechanism involving the inhibition of luteolytic substances is employed to maintain progesterone secretion. Therefore, in contrast to other animals, acquisition of PLs secreted during mid-pregnancy seems to be essential in establishing reproduction in the rat.

Co-Evolution Of Rat PL With A Steroid-Metabolizing Enzyme

20α-hydroxysteroid dehydrogenase (20α-HSD, EC.1.1.1.149) is a NADPH-dependent oxidoreductase that catalyzes the conversion of progesterone to 20α-dihydroprogesterone, a biologically inactive steroid (Mori and Wiest, 1979; Noda et al., 1991). An increase in its activity results in reduction of progesterone secretion (Shiota and Wiest, 1979; Mori and Wiest, 1979), therefore, this enzyme is a key intracellular regulator of ovarian progesterone secretion in rats. During pseudopregnancy and pregnancy until term in the rat, ovarian 20α-HSD activity is considerably suppressed by PRL (Heap et al., 1973; Shiota and Wiest, 1979). This is not the case in other species including man and domestic animals, and adoption of the enzyme into the reproductive system is not common in these animals, where progesterone secretion from the corpus luteum continues for about two weeks without PRL action.

We purified rat ovarian 20α-HSD completely and determined its N-terminal amino acid sequence, which was used to screen for its cDNA (Noda et al., 1991; Miura et al., 1994). The cDNA encodes a polyadenylation signal and an open reading frame encoding 323 amino acids (Miura et al., 1994). Sequence analysis revealed that 20α-HSD was similar to other cytosolic proteins of the aldoketo reductase family, which include rat liver 3α-hydroxysteroid dehydrogenase (3α-HSD) (Cheng et al., 1991), bovine lung prostaglandin F synthase (PGFS) (Watanabe et al., 1988), human liver chlordecone reductase (CDR) (Winters et al., 1990), frog lens γ-crystalline (Fujii et al., 1990) and mammalian aldose reductase (Chung and Lamendola, 1989; Schade et al., 1990). The members of this family may have evolved as detoxification agents to remove steroids from various tissues, and the rat may have developed 20α-HSD in the ovary to metabolize excess progesterone.

Natural molecular evolution is a process of trial and error whereby successful molecules are selected and their genetic heritability is maintained. Communication between the ovary and placenta appears to be elemental in determining the direction of evolution of the various molecules involved in the reproductive system. Thus, the functional relationship between the ovary and placenta seems to have prompted co-evolution of placental PRL-like hormones and aldoketo reductase family members in the rat.

SUMMARY

A major function of the placenta is the production of placental lactogens (Pls). Progesterone is essential for pregnancy in all mammals and is secreted by the ovary and placenta, depending on the animal species. In the rat, the main source of progesterone throughout pregnancy is the ovary, and 20α-hydroxysteroid dehydrogenase (20α-HSD) is a key enzyme controlling ovarian progesterone secretion. The primary action of prolactin (PRL) in the maintenance of ovarian progesterone secretion is suppression of the activity of ovarian 20α-HSD. In this review, the sequence homologies between cDNAs for PLs and PRL and the intimate functional relationship between the ovary and placenta are discussed in order to speculate how and why the molecular diversity of rat PLs has developed.

REFERENCES

Arima, Y. and Bremel, R.D. (1983) Purification and characterization of bovine placental lactogen. *Endocrinology* 113, 2186-2194.

Brash, G.S., Seeburg, P.H. and Gelinus, R.E. (1983) The human growth hormone family gene family: structure and evolution of the chromosomal locus. *Nucleic Acids Res.* 11, 3939-3958.

Brelje, T.C. and Sorenson, R.L. (1991) Role of prolactin versus growth hormone on islet B-cell proliferation in *vitro*: Implication for pregnancy. *Endocrinology* 128, 45-57.

Bridges, R.S. and Paul, M. (1990) Prolactin (PRL) regulation of maternal behavior in rats: Bromocriptine treatment delays and PRL promotes the rapid onset of behavior. *Endocrinology* 126, 837-848.

Chan, J.S.D., Robertoson, H.A. and Friesen H.G. (1976) The purification and characterization of ovine placental lactogen. *Endocrinology* 98, 65-76.

Cheng, K.C., White, P.C. and Qin, K.N. (1991) Molecular cloning and expression of rat liver 3α-hydroxysteroid dehydrogenase. *Mol. Endocrinol.* 5, 823-828.

Chung, S. and LaMendola, J. (1989) Cloning and sequence determination of human placental aldose reductase gene. *J. Biol. Chem.* 264, 14775-14777.

Cohick, C.B., Dai, G., Xu, L., Deb, S., Kamei, T., Leven, G. Szpirer, C., Szpirer, J., Kwok, S.M.C. and Soares, M.J. (1996) Placental lactogen-I variant utilizes the prolactin receptor signaling pathway. *Mol. Cell. Endocrinol.* 116, 49-58.

Davis, J.A. and Linzer, D.I.H. (1989) Mutational analysis of a lactogenic hormone reveals a role for lactogen-specific amino acid residues in receptor binding and mitogenic activity. *Mol. Endocrinol.* 3, 1987-1995.

Deb, S., Roby, K.F., Faria, T.N., Szpirer, C.G., Levan, C., Kwok, S.C.M. and Soares, M.J. (1991) Molecular cloning and characterization of prolactin-like protein C complementary deoxyribonucleic acid. *J. Biol. Chem.* 266, 23027-23032.

Duckworth, M.L., Kirk, K.L. and Friesen, H.G. (1986a) Isolation and Identification of a cDNA clone of rat placental lactogen II. *J. Biol. Chem.* 261, 10871-10878.

Duckworth, M.L., Peden, L.M. and Friesen, H.G. (1986b) Isolation of a novel prolactin-like cDNA clone from developing rat placenta. *J. Biol. Chem.* 261, 10879-10884.

Duckworth, M.L., Peden, L.M. and Friesen, H.G. (1988) A third prolactin-like protein expressed by the developing rat placenta: Complementary deoxyribonucleic acid sequence and partial structure of the gene. *Mol. Endocrinol.* 2, 912-920.

Freemark, M. and Handwerger, S. (1984) Ovine placental lactogen stimulates glycogen synthesis in fetal rat hepatocytes. *Am. J. Physiol.* 246, E21-24.

Fujii, Y., Watanabe, K., Hayashi, H., Urade, Y., Kuramitsu, S., Kagamiyama, H. and Hayaishi, O. (1990) Purification and characterization of r-crystallin from Japanese common bull frog lens. *J. Biol. Chem* 265, 9914-9923.

Furuyama, N., Shiota, K. and Takahashi, M. (1991) Two distinct placental lactogen-like substances in serum during mid-pregnancy in the rat. *Endocrinol. Japon.* 38, 533-540.

Hattori, N., Hirosawa, M., Wakimasu, M., Takahashi, M, Shiota, K and Ogawa, T. (1993) Characterization of rat placental lactogen-a (PL-a) with an antipeptide antibody directed against rat PLs. *Endocrine. J* 40, 673-681.

Heap, R.B., Perry, J.S. and Challis, J.R.G. (1973) Hormonal maintenance of pregnancy. In: *Handbook of Physiology*, (eds.) R.O. Greep, E.B. Astwood and S.R. Geiger SR Section 7: Endocrinology, Vol. II, Part 2, Williams and Wilkins Co., Baltimore, pp. 217-260.

Hirosawa, M., Miura, R., Min, K.S., Hattori, N., Shiota, K. and Ogawa, T. (1994) A cDNA encoding a new member of the rat placental lactogen family, PL-I mosaic (PL-Im). *Endocrine. J.* 41, 387-397.

Hirosawa M., Hattori N., Soares M., Shiota K. and Ogawa T. (1995) Production of recombinant placental lactogen-I mosaic (rPL-Im) by a baculovirus system. *Biol. Reprod.* 52 (Suppl. 1, 28th Annual Meeting Abstract): p. 162

Hunt, R.E., Moffat, K. and Chung, D. (1981) Purification and crystallization of the polypeptide hormone human chorionic somatomammotropin. *J. Biol. Chem.* 256, 7042-7045.

Iwatsuki, K., Shinozaki, M., Hattori, N., Hirasawa, K., Itagaki, S., Shiota, K. and Ogawa, T. (1996) Molecular cloning and characterization of a new member of the rat placental prolactin (PRL) family, PRL-like protein D (PLP-D). *Endocrinology* 137: 3849-3855.

Kelly, P.A., Djane, J., P-Vinay, M.C. and Edery, M. (1991) The prolactin/growth hormone receptor family. *Endocrine. Rev.* 12, 235-251.

Min, K.S., Shinozaki, M., Miyazawa, K., Nishimura, R., Sasaki, N., Shiota, K. and Ogawa, T. (1994) Nucleotide sequence of eCG α-subunit cDNA and its expression in the equine placenta. *J. Reprod. Dev.* 40, 301-305.

Miura, R., Shiota, K., Noda, K., Yagi, S., Ogawa, T. and Takahasi, M. (1994) Molecular cloning of cDNA for rat ovarian 20α-hydroxysteroid dehydrogenase (HSD1). *Biochem. J.* 299, 561-567.

Mori, M. and Wiest, W.G. (1979) Purification of 20α-hydroxysteroid dehydrogenase by affinity chromatography. *J. Steroid. Biochem.* 11, 1443-1449.

Noda, K., Shiota, K. and Takahashi, M. (1991) Purification and characterization of rat ovarian 20α-hydroxysteroid dehydrogenase. *Biochim. Biophys. Acta.* 1079, 112-118.

Robertson, M.C., Gillespie, B. and Friesen, H.G. (1982) Characterization of the two forms of rat placental lactogen (rPL): rPL-I and rPL-II. *Endocrinology* 111, 1862-1866.

Robertson, M.C., Croze, F., Schroedter, I.C. and Friesen, H.G. (1990) Molecular cloning and expression of rat placental lactogen-I complementary deoxyribonucleic acid. *Endocrinology* 127, 702-710.

Robertson, M.C., Schroedter, I.C. and Friesen, H.G. (1991) Molecular cloning and expression of rat placental lactogen, a variant of rPL-I present in late pregnant rat placenta. *Endocrinology* 129, 2746-2756.

Schade, S.Z., Early, S.L., Williams, T.R., Kézdy, F. J., Heinrikson, R.L., Grimshaw, C.E. and Doughty, C.C. (1990) Sequence analysis of bovine lens aldose reductase. *J. Biol. Chem.* 265, 3628-3635.

Shiota, K. and Wiest, W.G. (1979) On the mechanism of prolactin stimulation of steroidogenesis. In: Ovarian Follicular and Corpus Luteum Function, (eds.) C.P. Channing, J.M. Marsh, and W.A. Sadler, Plenum Press, New York, 169-178.

Shiota, K., Furuyama, N. and Takahashi, M. (1991) Two distinct placental lactogen-like substances in serum during mid-pregnancy in the rat. *Endocrinol. Japon.* 38, 541-549.

Shiota, K., Hirosawa, M., Hattori, N., Itonori, S., Miura, R., Noda, K., Takahashi, M. and Ogawa, T. (1994) Structural and function aspects of placental lactogens (PLs) and ovarian 20α-hydroxysteroid dehydrogenase (20α-HSD) in the rat. *Endocrine. J.* 41 (Suppl.), S43-S56.

Shome, B. and Friesen, H.G. (1971) Purification and characterization of monkey placental lactogen. *Endocrinology* 89, 631-641.

Soares, M.J., Faria, T.N., Roby, K.F. and Deb, S. (1991) Pregnancy and prolactin family of hormones: Coordination of anterior pituitary, uterine, and placental expression. *Endocrine Reviews* 12, 402-423.

Southard, J.N., Thordarson, G. and Talamantes, F. (1986) Purification and partial characterization of hamster placental lactogen. *Endocrinology* 119, 508-514.

Southard, J.N. and Talamantes, F. (1991) Placental prolactin-like proteins in rodents: Variations on a structural theme. *Mol. Endocrinol.* 79, C133-140.

Thordarson, G. and Talamantes, F. (1987) Role of the placenta in mammary gland development and function. In: The Mammary Gland, (eds.) C. Neville and C.W Daniel, Plenum Press, New York, pp. 459-498.

Tonkowicz, P., Robertson, M. and Voogt, J. (1983) Secretion of rat placental lactogen by the fetal placenta and its inhibitory effect on prolactin surges. *Biol. Reprod.* 28, 707-716.

Voogt, J.L. and de Greef, W.J. (1989) Inhibition of nocturnal prolactin surges in the pregnant rat by incubation medium containing placental lactogen. *Proc. Soc. Exp. Biol. Med.* 191, 403-407.

Watanabe, K., Fujii, Y., Nakayama, K., Ohkubo, H., Kuramitsu, S., Kagamiyama, H., Nakanishi, S. and Hayaishi, O. (1988) Structural similarity of bovine lung prostaglandin F synthase to lens e-crystallin of the European common frog. *Proc. Natl. Acad. Sci. USA.* 85, 11-15.

Winters, C.J., Molowa, D.T. and Guzelian, P.S. (1990) Isolation and characterization of cloned cDNAs encoding human liver chlordecone reductase. *Biochemistry* 29, 1080-1087.

Yamakawa, M., Tanaka, M., Koyama, M., Kagesato, Y., Watahiki, M., Yamamoto, M. and Nakashima, K. (1990) Expression of new members of the prolactin growth hormone gene family in bovine placenta. *J. Biol. Chem* 265, 8915-8920.

Trophoblast Research 9:13-25, 1997

ISOLATION OF PREGNANCY-ASSOCIATED PLASMA PROTEIN A

Yoshichika Suzuki[1], Junko Takada[1], Keiichi Isaka[1], Masaomi Takayama[1] and J. Gedis Grudzinskas[2]

[1]Department of Obstetrics and Gynecology
Tokyo Medical College
Tokyo, Japan

[2]Academic Unit of Obstetrics
Gynaecology and Reproductive Physiology
The Royal London Hospital
London, United Kingdom

INTRODUCTION

Pregnancy-associated plasma protein A(PAPP-A) is a high molecular weight glycoprotein which was extracted from serum obtained from pregnant women by Lin et al. (1974). PAPP-A is also a heterotetramer which has two characteristic subunits of 200 kDa in molecular weight respectively linking to a 50-90 kDa proform of eosinophil major basic protein (pro MBP). It is known that PAPP-A, which is a heparin binding protein of iso-electric point of 4.4-4.6, has an α2-globulin electrophoretic mobility and contains 19% carbohydrate (Lin et al., 1974; Bischof, 1979; Torsten et al., 1994). Although PAPP-A monomer was reported to have 1547 amino acid sequences in 1994 (Torsten et al., 1994), there still remain many unsettled questions as to its biological activity. Serum PAPP-A level increases during the course of pregnancy and reaches its peak of 50 mg/l in week 39. It is an immunohistologically confirmed fact that PAPP-A is mainly produced and secreted in the syncytiotrophoblast of placental villi (Folkersen et al., 1981; Lin et al., 1976). Also PAPP-A is confirmed to exist not only in the follicular fluid and tubal mucosa of non pregnant women but also in male seminal plasma (Sinosich et al., 1984; Bischof et al., 1983). The measurement of the maternal serum PAPP-A level has been reported to be effective in the detection of abnormal pregnancy during the early gestation period such as an ectopic pregnancy or spontaneous abortion (Grudzinskas et al., 1984). The reports that emphasized the relation between the low serum PAPP-A level during early gestation period and a chromosomal abnormality, especially Down's syndrome, have recently been in the limelight (Wald et al., 1992; Brambati et al., 1993; Bersinger et al., 1994). PAPP-A level has conventionally been measured by radioimmunoassay (RIA). Considering the present condition, we expect elucidation of the biological activity of PAPP-A as well as development of a more effective measurement system. In this study, a new method for PAPP-A purification was developed by three step chromatography with molecular sieve chromatography, heparin-Sepharose affinity chromatography and DEAE-Sephacel chromatography.

MATERIALS AND METHODS

Materials

The serum samples (30 ml) were collected from 6 women having a normal pregnancy (38-40 weeks). The collected samples were centrifuged within one hour of delivery and stored at -20°C.

Methods

PAPP-A Purification

Molecular Sieve Chromatography

Using Tris-HCl buffer (0.05 M, pH 7.8) containing NaCl (0.15 M), a 5.0 x 100 cm column (Pharmacia, Australia) was filled with degassed Ultrogel ACA 34 (L'Industrie Biologique Francaise, France) at a flow rate of 60 ml/hour at 10°C. The column was run using the same buffer at a flow rate of 52 ml/hour. The total bed volume(Vt) was 1.5 l and the void volume(Vo) was 0.6 l. The 30 ml serum sample was applied to the column and the protein contents of the respectively eluted fractions (10 ml) were assessed via the UV method (absorption:280 nm) and fused rocket immuno-electrophoresis using anti human serum antibody. PAPP-A content was determined by conventional RIA (Sinosich et al., 1982). Three hundred milliliter fractions were collected from a high molecular weight zone containing PAPP-A.

Heparin-Sepharose Affinity Chromatography

Using Tris-HCl buffer (0.02M, pH 7.2) containing NaCl (0.15M), a 2.6 x 25 cm column was filled with Heparin-Sepharose(CL-6B, Pharmacia, Sweden). At a flow rate of 35 ml/hour, the column was washed and equilibrated with the same buffer. Three hundred milliliter fractions containing PAPP-A separated by molecular sieve chromatography were applied to the column at a flow rate of 35 ml/hour. The heparin binding protein in the column was eluted by Tris-HCl buffer (0.02M,pH 7.2) containing NaCl (0.15, 0.30, 0.60, 0.90 M) in the order from lower to higher NaCl concentrations. The protein contents of the respectively eluted fractions (2 ml) were assessed by the UV method and fused rocket immunoelectrophoresis. The PAPP-A content was determined by RIA and the fractions(20 ml) containing PAPP-A were collected.

DEAE-Sephacel Chromatography

Using Tris-HCl buffer (0.01 M, pH 7.2), a 2.6 x 30 cm column was filled with DEAE-Sephacel (Pharmacia, Sweden), and the column was washed and equilibrated. The sample (20 ml) containing PAPP-A separated by heparin-Sepharose chromatography was diluted 10 times and a total of 200 ml of the sample was applied to the column. The proteins were eluted in turn by an acetate buffer (pH 5.5) containing NaCl (0.15, 0.30, 0.45 M) at a flow rate of 42 ml/hour. The protein contents of the respectively eluted fractions(2 ml) and PAPP-A values were then determined.

Dialysis and Concentration

The PAPP-A containing fractions collected after DEAE-Sephacel chromatography were pooled and dialyzed against phosphate buffered saline (pH 7.4), and then concentrated 10 times using a centrifugal condenser(Centripep 100, Amicon, USA).

Immunoelectrophoresis

Rocket immunoelectrophoresis and crossed immunoelectrophoresis were performed using a 1% agarose gel 1.5 mm in thickness and 0.02 m Tris-barbital buffer (pH 8.6). After treatment, the agarose gel was pressed and dried for Coomassie brilliant blue staining.

Rocket Immunoelectrophoresis

In a ratio of 50 to 1, the agarose gel was mixed with anti-human serum antibody (Dako Immunoglobulins, Denmark), and in a ratio of 100 to 1, the agarose gel was mixed with anti α2-macroglobulin antibody (Chemicon International, Inc., USA), anti antithrombin III (ATIII) antibody (Chemicon International, Inc., USA),and anti PAPP-A antibody (Dako Immunoglobulins, Denmark). The gel was run at a voltage of 2.5 V/cm for 18 hours.

Crossed Immunoelectrophoresis

In one-dimensional electrophoresis, the gel was run at 10 V/cm for 40 minutes and migration was monitored by a bromophenol blue stained albumin marker. In the two-dimensional electrophoresis, the agarose gel was mixed with anti PAPP-A antibody in a ratio of 100 to 1 and the separated proteins by one-dimensional electrophoresis were electrophoresed at 2.5 V/cm for 18 hour, 10°C.

Polyacrylamide Gel Electrophoresis

The PAPP-A containing samples eluted by the respective chromatography steps were electrophoresed by using a 10% polyacrylamide gel.

SDS-Polyacrylamide Gel Electrophoresis

The technique developed by Laemmli et al. (1970) was adopted for electrophoresis with an 8% polyacrylamide gel. Isolated PAPP-A samples(5 μl) were electrophoresed under nonreducing conditions and reducing conditions with 1% mercaptoethanol.

Immunoblotting Method

After SDS-polyacrylamide gel electrophoresis, the protein transcribed on the nitrocellulose membrane by the semi-dry-blotting technique was placed on immunoblotting treatment with anti PAPP-A antibody of 1/100 concentration.

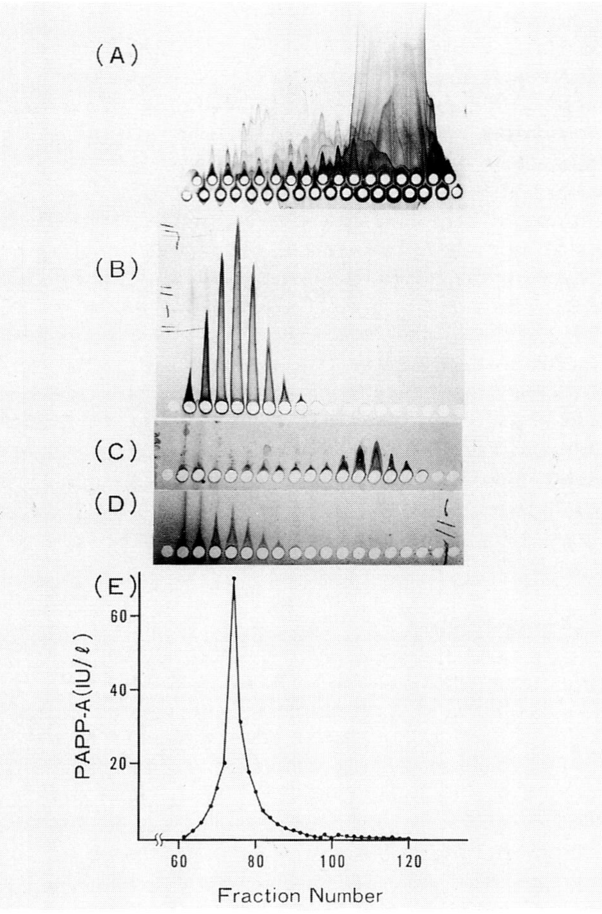

Figure 1. Molecular sieve chromatography. The elution profile of 30 ml of late pregnancy serum applied as described under Materials and Methods. Fractions were assayed by rocket immunoelectrophoresis for total protein content (A), anti-thrombin III (B), α2-macroglobulin (C) and by RIA for PAPP-A (D).

RESULTS

The elution profile of late pregnancy serum (30 ml) containing 5120 mIU PAPP-A on molecular sieve chromatography is shown in Figure 1. The elution of PAPP-A was observed in a high molecular weight zone(Fraction No. 65-95) and 91%(4650 mIU) of total PAPP-A was collected. Figure 2 gives the results of the elution by heparin-Sepharose affinity chromatography. Ninety-three percent(4310 mIU) of PAPP-A which was applied on heparin-Sepharose column was recovered by elution buffer containing 0.6 M NaCl.

Most of the proteins without heparin binding potency were eluted and removed by elution buffer containing 0.15 M NaCl. The process of elution using DEAE-Sephacel chromatography is demonstrated in Figure 3. Ninety-one percent (3920 mIU) of the

PAPP-A applied to the column was eluted and collected by acetate buffer (pH 5.5) containing 0.30 M NaCl. This sample (20 ml) containing PAPP-A was desalted by dialysis and condensed 10 times with a centrifugal condenser, and 3800 mIU (1.7 mg) of PAPP-A (2 ml) was separately extracted and purified. In this procedure, 99% of maternal serum protein was removed and 74% of total PAPP-A was recovered. The purification factor (fold), which was calculated as the increase in specific activity (mIU:PAPP-A/mg:protein) compared to that in the starting materials, was 526 (Table 1). The polyacrylamide gel electrophoresis of the elute obtained in the various stage of PAPP-A purification is demonstrated in Figure 4. After DEAE-Sephacel chromatography, a single band was recognized. The existence of a single band consisting of the protein of 200 kDa in molecular weight was recognized by SDS-polyacrylamide gel electrophoresis of purified protein (Figure 5) and the western blot analysis (Figure 6). Crossed immunoelectrophoresis was employed to compare the PAPP-A purified in our experiment with the PAPP-A purified by Teisner et al. (1983). The morphology of the precipitate and electrophoretic mobility of both PAPP-A were similar (Figure 7a). According to the results of tandem crossed immunoelectrophoresis, there existed some immunological identity between the two PAPP-A's (Figure 7b).

Table 1

Purification of Pregnancy-Associated Plasma Protein A

	Volume (ml)	Total PAPP-A (mIU)[a]	Total protein (mg)	Specific activity (mIU/mg)[b]	Purification factor (fold)[c]
Serum sample	30	5120	1770	2.89	-
Sieve chromatography	300	4650	150	31.0	10.7
Heparin-sepharose affinity chromatography	20	4310	22.4	192.4	66.6
DEAE-Sephacel chromatography	20	3920	8.9	440.4	152.4
Centripep 100	2	3800	2.5	1520	526.0

[a] Assuming 45 μg/ml = 100 mIU/ml.

[b] Total PAPP-A divided by total protein.

[c] Calculated as the increase in specific activity compared to that in the starting materials.

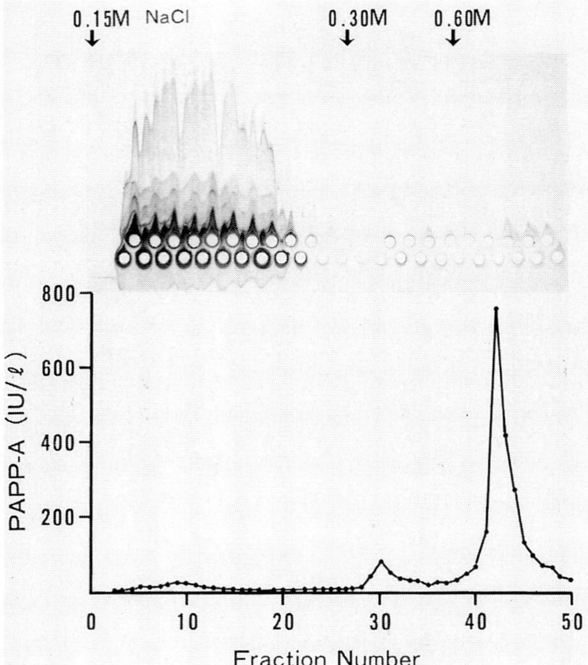

Figure 2. Heparin-Sepharose chromatography. The sample pool from the molecular sieve chromatography was applied on the column of heparin- Sepharose as described under Materials and Methods. Fractions were assayed for PAPP-A by RIA and for total protein by rocket immunoelectrophoresis.

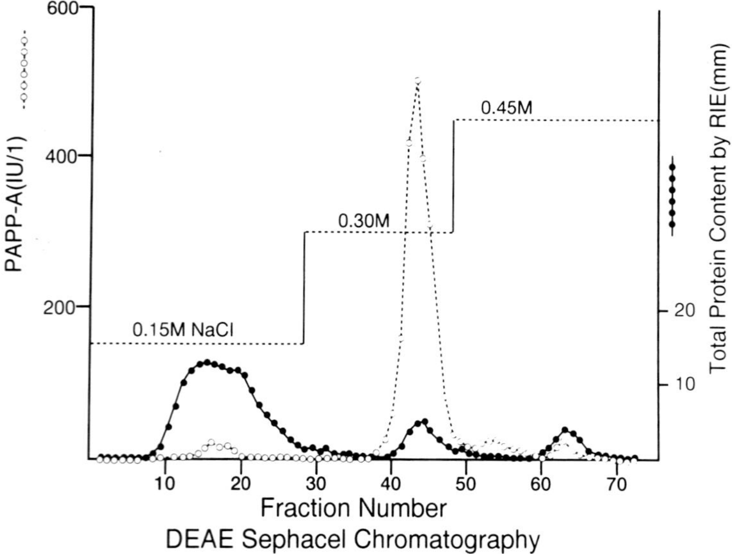

Figure 3. DEAE-Sephacel chromatography. The sample pool from heparin-Sepharose chromatography was applied on the column of DEAE-Sephacel as described under Materials and Methods. Fractions were assayed for PAPP-A by RIA and for total protein by rocket immunoelectrophoresis.

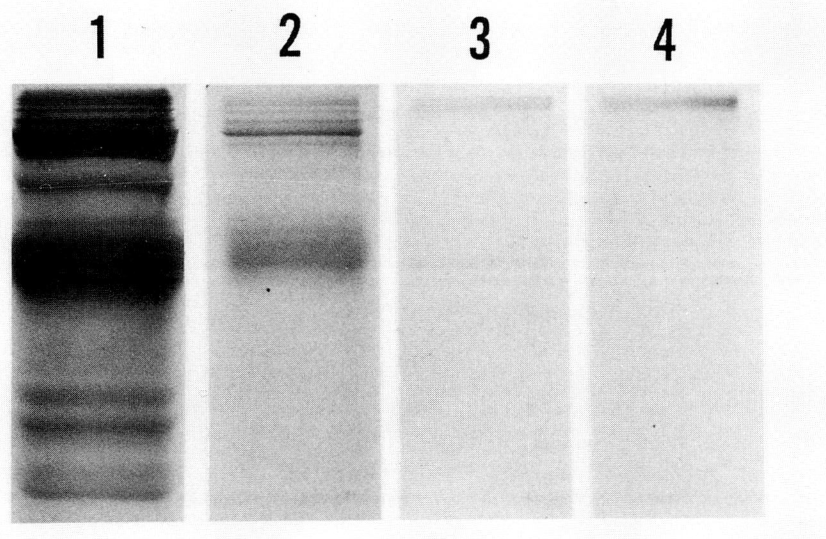

Figure 4. PAGE of each step in the purification schedule. Lane 1: Starting material. Lane 2: After molecular sieve chromatography. Lane 3: After heparin-Sepharose chromatography. Lane 4: After DEAE-Sephacel chromatography.

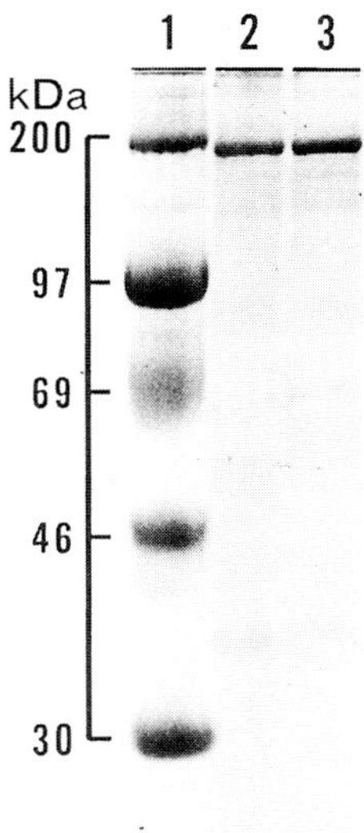

Figure 5. SDS-PAGE analysis of purified PAPP-A. Lane 1: Molecular weight markers. Lane 2: Non-reducing condition. Lane 3: Reducing condition.

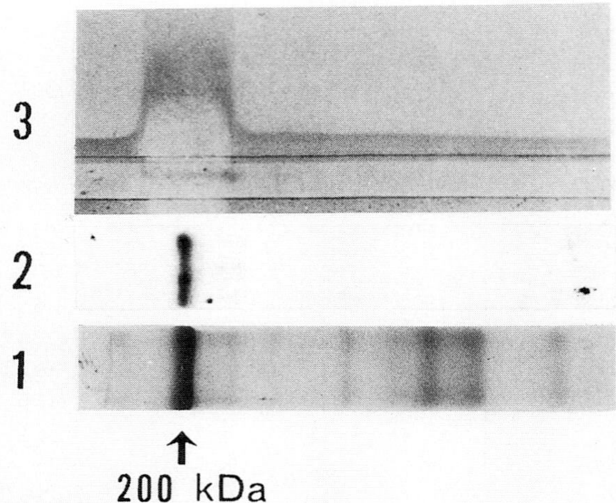

Figure 6. Western blot analysis and rocket immunoelectrophoresis after SDS-PAGE for purified PAPP-A. Lane 1: SDS-PAGE for purified PAPP-A. Lane 2: Western blot analysis. Lane 3: Rocket immunoelectrophoresis.

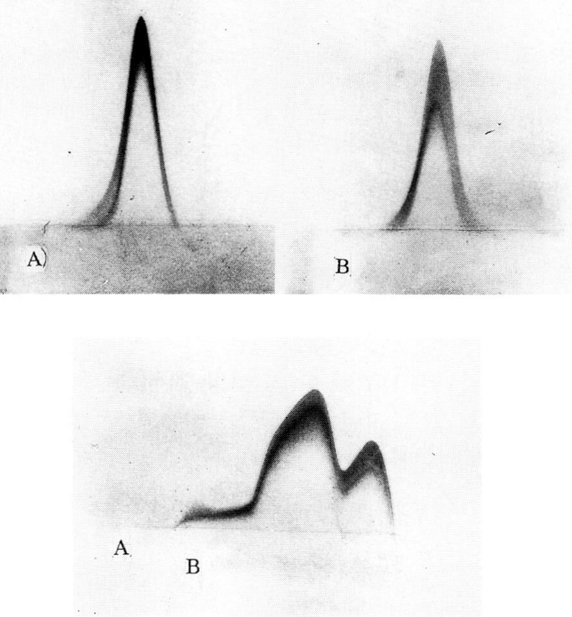

Figure 7. (a): Crossed immunoelectrophoresis of (A) 10 µl of pure PAPP-A donated by B. Teisner and (B) 4 µl of isolated PAPP-A in this study against the agarose gel containing anti PAPP-A antibody (1:100). (b): Tandem crossed immunoelectrophoresis of (A) 9 µl of pure PAPP-A donated by B. Teisner and (B) 3 µl of isolated PAPP-A in this study against the agarose gel containing anti PAPP-A antibody (1:100).

DISCUSSION

Many researchers have developed PAPP-A purification techniques (Bischof, 1979; Folkersen et al., 1981; Davey et al., 1983; Sinosich et al., 1987). In the present study, we used DEAE-Sephacel chromatography to perform PAPP-A purification. In the process of PAPP-A purification, α2-macroglobulin whose biochemical properties are similar to those of PAPP-A and anti-thrombin III which is one of heparin binding protein, were used as the motif of the protein to be separated. At first, serum proteins were eluted by gel filtration chromatography in the order of higher to lower molecular weights. Because PAPP-A has a high molecular weight of over 200 kDa, Ultrogel AcA 34 was used as a gel and 30 ml serum sample was applied to the column so that the sample volume might occupy 2% of total bed volume of the column (1.5 1). As in the case of PAPP-A, α2-macroglobulin was eluted in the high molecular zone while albumin, an important serum protein, and ATIII were eluted in much lower molecular zones (Figure 1). The high molecular zone fraction which contained PAPP-A eluted by molecular sieve chromatography was applied to heparin-Sepharose affinity chromatography (Figure 2). Ninety-three percent of PAPP-A was eluted from the sample using elution buffer containing 0.6 M NaCl. Most of the proteins without heparin binding potency such as α2-macroglobulin were eluted by a buffer containing 0.15 M NaCl. Furthermore ATIII was eluted with a buffer containing 0.9 M NaCl in the same way that Teisner et al. reported (Teisner et al., 1983). The PAPP-A separation by ion exchange chromatography was conducted as follows. Because PAPP-A had an iso-electric point of 4.4-4.6 (Lin et al., 1974; Bischof et al., 1979), acetate buffer (pH 5.5) containing 0.15 M NaCl was used so that the DEAE ion exchanger could absorb PAPP-A in the sample. As shown in Figure 3, 91% of PAPP-A was collected from the sample using an acetate buffer (pH 5.5) containing 0.3 M NaCl. α2-macroglobulin has an iso-electric point of 5.2 and stabilizes in the buffer at a pH value of 5.5. Therefore α2-macroglobulin might be eluted by a highly concentrated NaCl.

The ratio of total PAPP-A amount (IU) to total protein (mg) content of the final product was 526 times as much as the same ratio calculated for the maternal serum. According to the reduction formula defined by Bohn et al., 100 mIU equated to 45 μg (Bohn et al., 1980). On the basis of this formula, the purity of the PAPP-A purified in this study was 68.4%. A total of 1.7 mg PAPP-A was purified from 30 ml serum collected from women at term. In comparison with conventional techniques reported by other researchers, the PAPP-A purification technique which uses three steps of chromatography, successfully demonstrated its efficiency in extracting more PAPP-A of a higher purity in a single process of purification.

SDS-PAGE and western blot analysis showed that the protein purified in this study, which molecular weight is 200 kDa, is PAPP-A. The results of immunoelectrophoresis showed that PAPP-A purified by Teisner and associates was immunologically similar to our PAPP-A. These two PAPP-A's were iodinated by the chloramine-T method and antibody dilution curves were described. The fact that the maximum binding rate of our PAPP-A(95%) was higher than that of Teisner et al.(86%) indicated our success in the development of an ultra-pure PAPP-A. Furthermore the establishment of a more specific radioimmunoassay is expected.

Many researchers have studied and clarified PAPP-A *in vivo* kinetics by immunohistochemistry and measuring the humoral and tissue concentrations of PAPP-A by radioimmunoassay. PAPP-A is produced and secreted by placental syncytiotrophoblast throughout the gestation period. As in the cases of many other pregnancy-associated proteins [human placental lactogen (hPL), pregnancy-specific β1 glycoprotein (SP1), placental protein 5 (PP5), PP10, PP19, PP21, fetal antigen 1 (FA1), etc.], maternal serum PAPP-A level increases over the course of pregnancy and reaches 50 mg/l at term. Considering these findings, many researchers insist on the importance of PAPP-A as an index of fetal development. J.G. Grudzinskas and associates reported low maternal serum PAPP-A level in abnormal pregnancy cases including fetal death occurring during early pregnancy after the confirmation of heart movement by ultrasonography, and ectopic pregnancy (Grudzinskas et al., 1984; Westergaad et al., 1983; Sinosich et al., 1985). Recently, it has been reported repeatedly that the maternal serum PAPP-A level is remarkably low in the cases of early pregnancy complicated with Down's syndrome. On the basis of this finding, PAPP-A has been regarded as an effective marker for screening of Down's syndrome (Brambati et al., 1993; Bersinger et al., 1994). In view of the downward trend of child birth and upward trend of late child bearing, we strongly feel it is necessary to examine and diagnose embryonic abnormalities, especially chromosomal abnormalities including Down's syndrome. In advanced countries, obstetricians often conduct the triple marker test to measure serum levels of human chorionic gonadotrophin (hCG), alpha fetoprotein (AFP) and unconjugated estriol (uE3) during mid pregnancy (Palomaki et al., 1993). PAPP-A serves an effective index to detect Down's syndrome during early pregnancy, the period from week 10 to 13. Also, it is said that the detection rate of Down's syndrome is 60% in the single PAPP-A measurement if false positive rate is 5%, and we can improve the detection rate up to 85% if we additionally measure free β-hCG which is known to show high levels in Down's syndrome (Brambati et al., 1994). This examination method might be more frequently used as an excellent screening technique because it enables us to make an early diagnosis with high accuracy. PAPP-A plays an important role in the embryonic development and possibly serves as a useful marker to detect embryonic abnormalities. Although the biological activity of PAPP-A has not been elucidated, PAPP-A contains Zn^{2+} like most of metalloproteinases and there probably exists certain enzymatic activity (Torsten et al., 1994). Some other researchers pointed out that PAPP-A had immuno-suppressive effects and acted as a granulocyte elastase inhibitor (Martin-du-Pan et al., 1983; Sinosich et al., 1982). Therefore, PAPP-A is supposed to play a key role in the placenta, an important organ which connects the mother with the fetus.

Furthermore PAPP-A exists in the follicular fluid and seminal plasma and probably has some biological effects on the process of human development consisting of ovulation, fertilization and implantation. We hope that new PAPP-A purified in this study would be used for the development of a more sensitive and convenient assay for PAPP-A and its standards, as well as for unraveling the biological function of PAPP-A in human body.

SUMMARY

Pregnancy-associated plasma protein A (PAPP-A) was purified from normal term maternal serum in this study using a three step chromatographic procedure with molecular sieve chromatography, heparin-Sepharose affinity chromatography and DEAE-Sephacel chromatography. Thirty milliliter of the maternal serum samples were applied to the column for sieve chromatography and eluted according to molecular weight. PAPP-A was detected at the high molecular weight fraction area via radioimmunoassay for PAPP-A, and 91% of the total amount of PAPP-A in the maternal samples was recovered. These PAPP-A containing fractions were applied to heparin-Sepharose column and eluted in a stepwise increase of NaCl 0.15, 0.30 and 0.60 M in 0.15 M Tris-HCl buffer, pH 7.8. Ninety-five percent of PAPP-A in the applied samples was recovered after 0.6 M NaCl. The pooled fractions, which were containing PAPP-A after heparin-Sepharose affinity chromatography, were then applied on a DEAE-Sephacel Chromatography and eluted in a stepwise increase of NaCl 0.15, 0.30 and 0.45 M in 0.01 M acetate buffer, pH 5.5. Ninety-two percent of PAPP-A in applied samples was recovered after 0.30 M NaCl. Finally PAPP-A containing fractions were concentrated 10 times and 3.8 IU (1.7 mg) of PAPP-A was isolated. The purification schedule removed approximately 99% of total protein in the maternal serum while 74% of PAPP-A was recovered. The purification factor (fold), which was calculated as the increase in specific activity (mIU:PAPP-A/mg:protein) in comparison with the starting value in maternal serum, was 526. And the purity(mg:PAPP-A/mg:protein) of the final product was 68.4%. Analysis of the final purified material using SDS-PAGE demonstrated a single band of 200 kDa, and western blot analysis showed that purified main protein was PAPP-A. The immunological identity between purified PAPP-A in this study and PAPP-A donated by Teisner et al. was recognized by crossed immunoelectrophoresis and tandem crossed immunoelectrophoresis. The PAPP-A purified in this study by the three step chromatographic procedure will hopefully be utilized for the development of a sensitive and convenient assay for PAPP-A and its standard material, furthermore for the basic study of clarifying the biological function of PAPP-A in human body.

REFERENCES

Bersinger, N.A., Brizot, M.L., Johnson, A., Snijders, R.J.M., Abbott, J., Schneider, H., and Nicolaides, K.H. (1994) First trimester maternal serum pregnancy-associated plasma protein A and pregnancy-specific b1-glycoprotein in fetal trisomies. *Br. J. Obstet. Gynaecol.* 101, 970-974.

Bischof P. (1979) Purification and characterisation of pregnancy-associated plasma protein A(PAPP-A). *Arch. Gynaek.* 227, 315-326.

Bischof, P., Martin-Du-Pan, R., Lauber, K., Girard, J.P., Herrmann, W.L. and Sizonenko, P.C. (1983) Human seminal plasma contains a protein which shares physicochemical, immunochemical and immunosuppressive properties with pregnancy-associated plasma protein A. *J. Clin. Endocr. Metab.* 56, 359 -362.

Bohn, H., Grudzinskas, J.G., Rosen, S., Seppala, M., Tatarinov, Y.S., Chard, T., Klopper, A., Schultz-Larsen, P., Sizaret, P. and Teisner B. (1980) Reference preparation for assay of some pregnancy and cancer associated proteins. *Lancet* 2, 796.

Brambati, B., Macintosh, M.C.M., Teisner, B., Maguiness, S., Shrimanker, K., Lanzani, A., Bonacchi, I., Tului, L., Chard, T. and Grudzinskas, J.G. (1993) Low maternal serum level of pregnancy associated plasma protein (PAPP-A) in the first trimester in association with abnormal fetal karyotype. *Br. J. Obstet. Gynecol.* 100, 323-326.

Brambati, B., Tului, L., Bonacchi, I., Suzuki, Y., Shrimanker, K. and Grudzinskas, J.G. (1994) Biochemical screening for Down's syndrome in the first trimester. In: *Screening for Down's Syndrome,* (eds.), J.G. Grudzinskas, T. Chard, M. Chapman and H. Cuckle Cambridge University Press, pp. 285-294.

Davey, M.W., Teisner, B., Sinosich, M. and Grudzinskas, J.G. (1983) Interaction between heparin and human Pregnancy-Associated Plasma Protein A (PAPP-A): A Simple Purification Procedure. *Analyt. Biochem.* 131, 18-24

Folkersen, J., Grudzinskas, J.G., Hindersson, P., Teisner, B. and Westergaard, G. (1981) Pregnancy associated plasma protein A: Circulating levels during normal pregnancy. *Am. J. Obstet. Gynecol.* 139, 910-914.

Folkersen, J., Grudzinskas, J.G., Hindersson, P., Teisner, B. and Westergaard, G. (1981) Purification of Pregnancy-associated Plasma Protein-A by a two step affinity chromatographic procedure. *Placenta* 2, 11-18.

Grudzinskas, J.G., Westergaard, J.G. and Teisner, B. (1984) Pregnancy-Associated Plasma Protein A in normal and abnormal pregnancy. In: *Protein of the Placenta,* (eds.) P. Bischof and A. Klopper, 5th Int. Congr. on Placental Proteins, pp. 184-197

Laemmli, U.K. (1970) Cleavage of structure proteins during the assembly of the head of bacteriophage T4. Nature 227, 680-685.

Lin, T.M., Halbert, S.P., Kiefer, D., Spellacy, W.N. and Gall, S. (1974) Characterization of four pregnancy-associated plasma proteins. *Am. J. Obstet. Gynecol.* 118, 223-236.

Lin, T.M. and Halbert, S.P. (1976) Placental localization of human pregnancy-associated plasma proteins. *Science* 193, 1249-1252

Martin-du-Pan, R.C., Bischof, P., Bourrit, B., Lauber, K., Girard, J.P. and Herrmann, W.L. (1983) Immuno-suppressive activity of seminal plasma and pregnancy-associated plasma protein-A (PAPP-A) in men. *Arch. Androl.* 10, 185-188.

Palomaki, G.E., Knight, G.J., McCarthy, J., Haddow, J.E. and Eckfeldt, J.H. (1993) Maternal serum screening for fetal Down's syndrome in the United States: A 1992 survey. *Am. J. Obstet. Gynecol.* 169, 1558-1562.

Sinosich, M.J., Porter, R., Sloss, P., Bonafacio, M.D., Saunder D.M. (1984) Pregnancy-Associated plasma protein A in human ovarian follicular fluid. *J. Clin. Endo. Metab.* 58, 500-504.

Sinosich, M.J., Teisner, B., Folkersen, J., Saunders, D.M. and Grudzinskas, J.G. (1982) Radioimmunoassay for pregnancy associated plasma protein A. *Clin. Chem.* 28, 50-53.

Sinosich, M.J., Bonifacio, M.D., and Hodgen, G.D. (1987) Affinity immunoelectrophoresis and chromatography for isolation and characterization of a placental granulocyte elastase inhibitor. In: *Protein Purification: Micro to Macro,* (ed.) R. Burgess, Alan R Liss:New York.

Sinosich, M.J., Ferrier, A., Teisner, B., Porter, R., Westergaard, J.G., Saunders, D.M. and Grudzinskas, J.G.. (1985) Circulating pregnancy-associated plasma protein A and its tissue concentrations in tubal ectopic gestation. *Clin. Reprod. Fertil.* 3, 311-317.

Sinosich, M.J., Davey, M.W., Ghosh, P. and Grudzinskas, J.G. (1982) Specific inhibition of human granulocyte elastase by human pregnancy-associated plasma protein-A. Biochem. Int. 5, 777-786.

Teisner, B., Davey, M.W. and Grudzinskas, J.G. (1983) Interaction between heparin and plasma proteins analysed by crossed immunoelectrophoresis and affinity chromatography. *Clinica Chemica Acta* 127, 413-417.

Torsten, K., Claus, O., Ole, S., Niels, P.H.M. and Lars, S.J. (1994) Amino acid sequence of human pregnancy-associated plasma protein-A Derived from Cloned cDNA. *Biochemistry* 33, 1592-1598

Wald, N.J., Stone, R., Cuckle, H.S., Grudzinskas, J.G., Barkai, G., Brambati, B., Teisner B, Fuhrmann, W. (1992) First trimester concentrations of pregnancy associated plasma protein A and placental protein 14 in Down's syndrome. *Br. Med. J.* 305, 28.

Westergaard, J.G., Sinosich, M.J., Bugge, M., Madsen, L.T., Teisner, B. and Grudzinskas, J.G. (1983) Pregnancy- associated plasma protein A in the prediction of early pregnancy failure. *Am. J. Obstet. Gynecol.* 145, 67- 69.

Trophoblast Research 9:27-39,1997

EPIDERMAL GROWTH FACTOR (EGF) REGULATES TROPHOBLAST PROLIFERATION AND ENDOCRINE FUNCTION IN SYNERGY WITH THYROID HORMONE
- A Review -

Takeshi Maruo[1], Hiroya Matsuo, Tetsuo Otani and Matsuto Mochizuki

Department of Obstetrics and Gynecology
Kobe University School of Medicine
Kobe 650, Japan

INTRODUCTION

The development of the placenta involves a coordinate process of trophoblast proliferation and differentiation. The placenta is known to be an exceedingly rich source of receptors for several growth factors, including thyroid hormone, epidermal growth factor (EGF), insulin and insulin-like growth factor-I (IGF-I). Human placental trophoblast possesses specific nuclear receptors for L-triiodothyronine (T3) and the binding capacity of trophoblastic nuclear T3 receptors in early placenta is much greater than that in term placenta (Ashitaka et al., 1988; Nishii et al., 1989). Consistent with the presence of abundant nuclear T3 receptors in early placental trophoblast, thyroid hormone exerts a stimulatory effect on trophoblast endocine function (Mochizuki, 1988; Maruo et al., 1991). Furthermore, EGF, that is primarily mitogenic for a variety of cells and also affects differentiated cellular function in the absence of the mitogenic effect, is presumed to take a role in feto placental development. On the other hand, abundant evidence is now available that a number of the oncogenes of transforming viruses are related to growth factors and their receptors. A close similarity has been noted between thyroid hormone receptor and erb.A oncogene protein and between the cytoplasmic domain of EGF receptor and the translation product of erb B (Sap et al., 1986; Downward et al., 1984). Thus, in the present communication, attention will be focused on the possible roles of EGF, EGF receptor and c-erb B mRNA expression in human placental growth and function.

MATERIALS AND METHODS

Placental Materials

First and second trimester placentae were obtained from patients who underwent therapeutic abortions at 4-12 weeks of gestation and at 16-21 weeks of gestation, respectively, while third trimester placentae were obtained from patients who had a cesarean section at 38-40 weeks of gestation. Informed consent was obtained from the patients before operation for the use of placental tissues.

[1]To Whom Correspondence Should Be Addressed.

Immunohistochemical Staining

Immunohistochemical staining was performed by avidin/biotin immunoperoxidase techniques with the use of a polyvalent immunoperoxidase kit (Omnitags, Lipshaw) as previously described (Maruo and Mochizuki, 1987). A rabbit polyclonal antibody against EGF (Collaborated Research, Inc.) and a mouse monoclonal antibody against human EGF receptor (Transformation Research, Inc.) were used as the primary antibodies in a dilution of 1:50 and 1:10, respectively. Immunohistochemical staining with a monoclonal antibody to proliferating cell nuclear antigen (PCNA) which can label the nuclei of proliferating cells was performed using frozen cryostat sections. The PCNA antibody was diluted 1:100 before use.

Tissue Culture

An *in vitro* culture system similar to that previously described was used (Maruo et al., 1986). Briefly, the explants of early placental villous tissues (total wet weight, 50 mg) were placed on filter papers (0.45 μm) within multiwell plates, 24 mm in diameter, to which 2 ml McCoy's 5a medium containing 100 x 103 U/l penicillin and 50 mg/l streptomycin were added. The placental explants were cultured at 37°C in an atmosphere of 95% air-5% CO_2 in the presence or absence of EGF (100 mg/l), with or without 10-8 mol/l T3 for five days. The medium was changed everyday.

Radioimmunoassays (RIAs)

Immunoreactive hCG and hPL in the media were determined by homologous double antibody RIAs as previously described (Maruo et al., 1979). EGF immunoreactivity was measured using human EGF reagent pack for RIA (Amersham). The anti-human EGF antibody used had 88% cross-reactivity for mouse EGF, but did not cross-react with transforming growth factor α (TGFα).

Gel filtration

Serum-free conditioned medium (total volume 40 ml) was dialyzed, lyophilized and acidified in 1 mol/l acetic acid. The acetic acid soluble material was applied to a 0.9 x 46 cm Sephadex G-75 column equilibrated with 1 mol/l acetic acid in order to dissociate and resolve EGF-like substance from its binding proteins. The elutant was collected in 0.65 ml fractions.

Northern Blot Hybridization

RNA was prepared from freshly obtained human placental cells according to the method of Chirgwin et al (1979). Each 10 mg of RNA obtained from placental cells at the first, second and third trimester was applied to Northern blot hybridization as previously described (Otani et al., 1988). C-erb B-1 cDNA was generously provided by American Type Culture Collection (Rockville, MD). The cDNA was labeled with 32P-dCTP by random primed DNA labeling kit produced by Boehringer Mannheim (Mannheim, Germany). The labeled cDNA was used as the probe. The filters were also hybridized with the chicken b-actin cDNA (770 base pairs; Oncor, Inc.) as control for the differences in the amount of RNA loaded on the gel and the transfer efficiency. Autoradiographic densities were quantitated by densitometric analysis with a scanning densitometer.

In Situ Hybridization

Dissected placental tissues were rapidly frozen, cut into 5-µm sections in a cryostat at -20°C, and the sections were hybridized with ^{35}S-labeled c-erb B-1 probe as previously described (Kitagawa et al., 1992).

Statistical Analysis

Each experiment was repeated three times with similar results, and the results reported are representative. Data were analyzed by Student's t-test, and a statistically significant difference was considered to be present at p<0.5.

RESULTS

Expression of EGF, EGF Receptor and c-erb B mRNA in the Human Placenta

Cytologic localization of EGF and EGF receptor was immunohistochemically analyzed in developing human early placenta using the avidin-biotin immunoperoxidase procedure. In 4- to 5-week placentae, EGF and EGF receptor were found to be almost exclusively localized to cytotrophoblast cells (Figure 1), whereas in 6- to 12-week placentae EGF and EGF receptor were predominantly localized to syncytiotrophoblast cells. Immunohistochemical measurement of cellular EGF receptor levels in the syncytiotrophoblast revealed that cellular levels of EGF receptor in the syncytiotrophoblast was highest in the first trimester and declined toward the end of gestation (Figure 2).

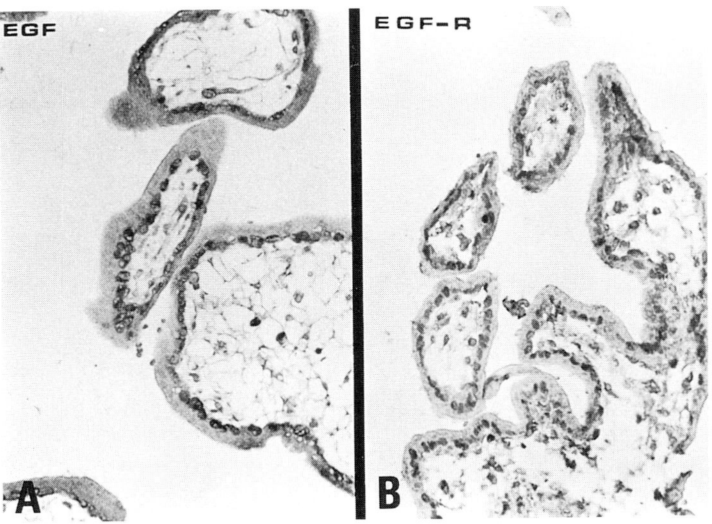

Figure 1. Immunohistochemical localization of EGF and EGF receptor in very early placenta. Tissue sections of a 4-week placenta were stained with antibodies directed against EGF (A) and EGF receptor (B).

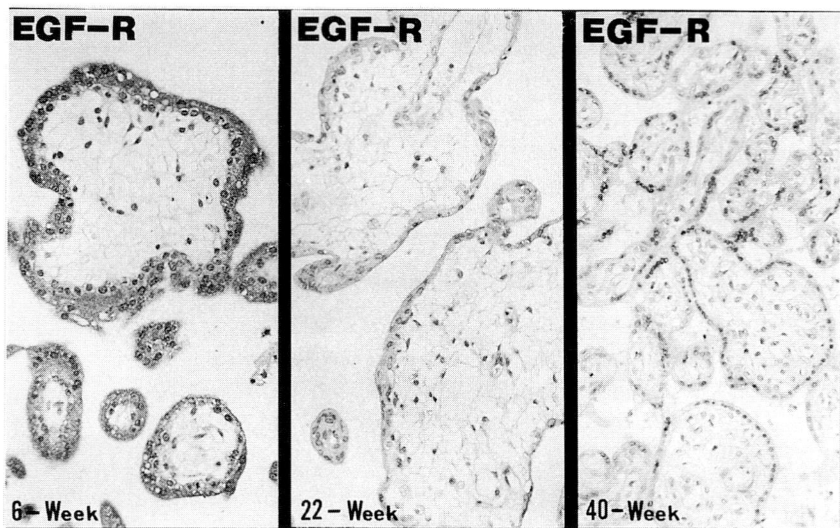

Figure 2. Immunohistochemical localization of EGF receptor in placental tissue sections over the course of pregnancy from 6th week to term.

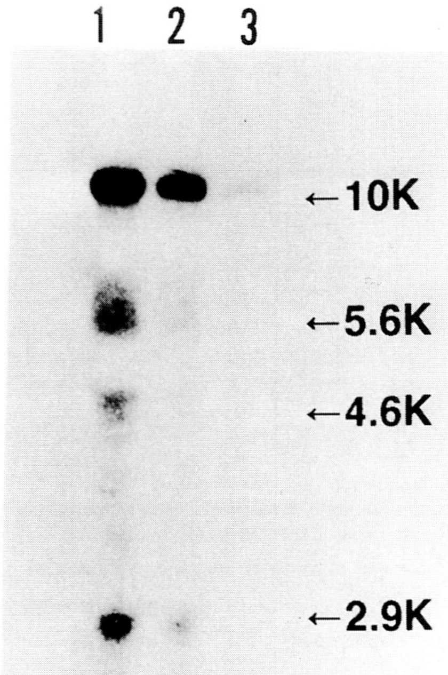

Figure 3. Analysis of total RNA prepared from human placentae at different trimesters during gestation by Northern blot hybridization with a c-erb B cDNA probe. Aliquots of total RNA (10 μg/lane) prepared from 6-week (lane 1), 20-week (lane 2) or 37-week (lane 3) placentae were subjected to Northern blot analysis using a ^{32}P-labeled c-erb B cDNA probe.

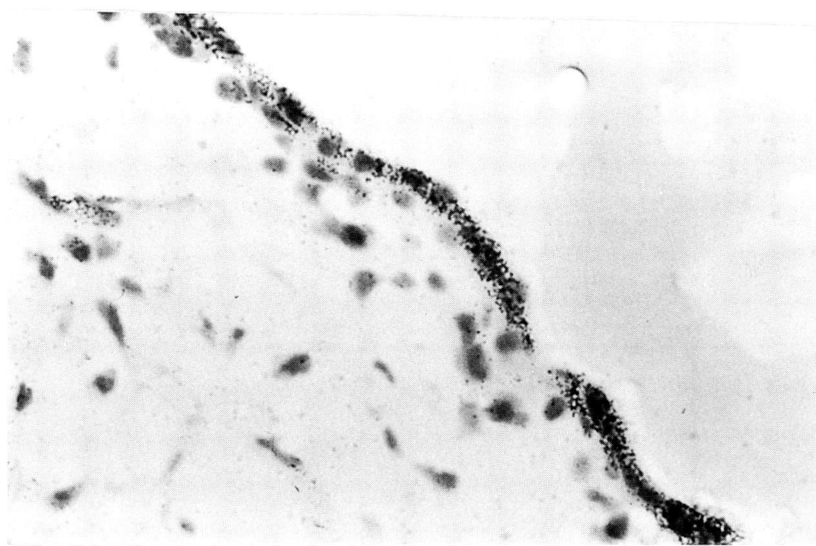

Figure 4. *In situ* hybridization of c-erb B mRNA in a 7-week human placental sample. Sections were hybridized with a 35S-labeled c-erb B probe.

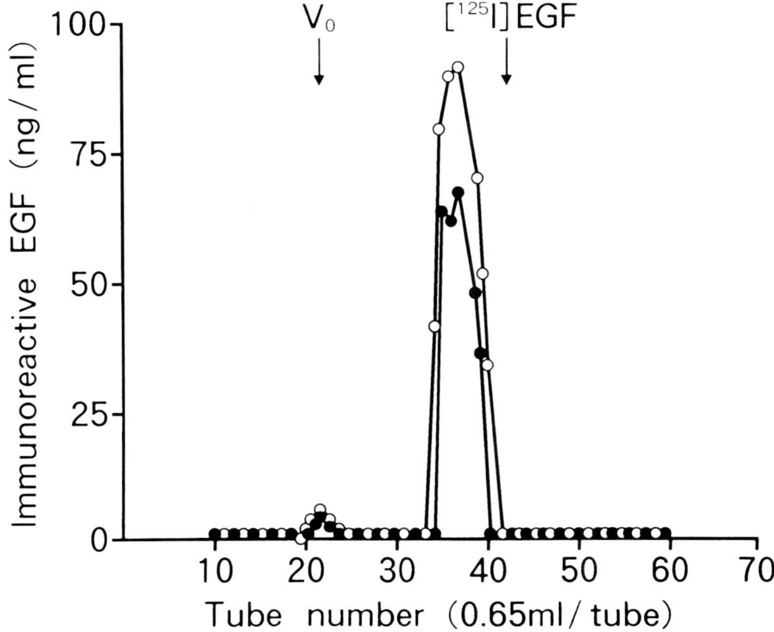

Figure 5. Elution profiles of human EGF immunoreactivity by Sephadex G-75 chromatography of serum-free media conditioned by early placental explants cultured in the absence (●-●) or presence (○-○) of 10^{-8} mol/l T3 for 96 hours. Fractions were assayed for human EGF immunoreactivity. The vertical arrow indicates the elution position of $[^{125}I]$human EGF. Vo, void volume.

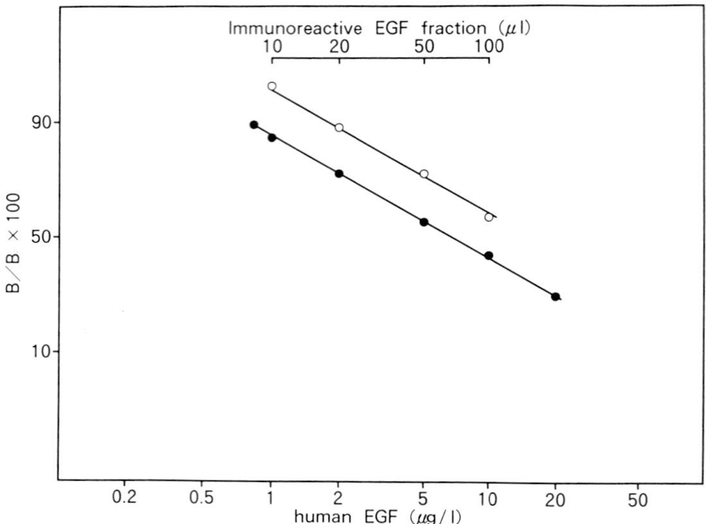

Figure 6. Displacement curves for Human EGF reference reparation and immunoreactive EGF fraction obtained through Sephadex G-75 chromatography of serum-free conditioned media in human EGF-RIA. Ordinate is a logit scale expressed as a percentage of B/Bo; abscissa is a log scale of human EGF reference preparation (●-●, micrograms per liter) or immunoreactive EGF fraction (○-○, microliters) obtained through Sephadex G-75 chromatography of serum-free conditioned media.

In contrast, Figure 3 represents the results of Northern blot analysis of total RNA preparations from 6-week, 20-week, and 37-week placental cells using c-erb B-1 cDNA probe. In case of the RNA preparations from 6-week and 20-week placental cells, four bands with the estimated size of 10.0-kb, 5.6-kb, 4.6-kb and 2.9-kb were detected, whereas RNA preparation from 37-week placental cells exhibited only the two bands with the estimated size of 10.0-kb and 5.6-kb. Densitometric analysis of the autoradiographic films revealed that the amounts of c-erb B mRNA varied during the course of gestation, being most abundant in 6-week placenta, more abundant in 20-week placenta, and least abundant in 37-week placenta.

To localize c-erb B-1 transcripts in the human placenta, *in situ* hybridization of a 7-week placental tissue visualized the abundant signals almost exclusively in the syncytiotrophoblast (Figure 4). Unlike the first-trimester placental tissue sections, the silver grain in the autoradiography of *in situ* hybridization of the third-trimester placental tissue sections appeared at the background level.

Local Production of an EGF-like Substance by Human Early Placenta in Synergy With Thyroid Hormone

In order to examine the local production of an EGF-like substance by human early placental tissues, serum-free media obtained following a 5-day culture of early placental tissues were pooled, dialyzed, acidified and chromatographed over a Sephadex G-75 column equilibrated with 1 mol/l acetic acid. Fractionation of the serum-free conditioned media resulted in the elution of immunoreactive EGF with an apparent molecular weight of 9,000 which is larger than [^{125}I]human EGF. Addition of T3 (10^{-8} mol/l) resulted in increased secretion of immunoreactive EGF by placental explants (Figure 5). By contrast, addition of cycloheximide (5×10^{-5} mol/l) drastically reduced the secretion of immunoreactive EGF. In human EGF-RIA, the combined immunoreactive EGF fraction gave inhibition curve parallel to authentic human EGF preparation (Figure 6), revealing that the material reacting in human EGF-RIA is an EGF-like substance.

EGF as a Local Regulator of Trophoblast Proliferation and Differentiated Function

The biological effects of EGF on trophoblast proliferation and function were investigated *in vitro* using an organ culture system of human placental tissues with a serum-free medium. Figure 7 represents the PCNA staining features of frozen sections of very early (4-5 week) placental explants cultured in the presence or absence of 100 mg/l EGF for 12 hours. In the explants cultured with EGF, a greater number of PCNA positive cytotrophoblast nuclei were observed relative to the control explants cultured without EGF. By contrast, in cultures of early (8-9 week) placental explants there was no apparent difference in the number of PCNA positive nuclei between EGF-treated and untreated explants.

In cultures of very early 4- to 5-week placental explants, there were no significant differences in hCG and hPL secretion into the medium throughout the whole culture period between EGF-treated and untreated cultures. By contrast, in cultures of early 6- to 11-week placental explants, EGF treatment significantly augmented hCG and hPL secretion compared to EGF-untreated cultures (Figure 8).

DISCUSSION

The simultaneous expression of EGF and EGF receptor in the cytotrophoblast of 4- to 5-week placentae and in the syncytiotrophoblast of 6- to 12-week placentae implies that EGF may act in an autocrine manner in first-trimester placentae. The dynamic change in cytologic localization of EGF and EGF receptor in developing human placentae may reflect the change in the possible role of EGF in the course of fetoplacental development (Ladines-Llave et al., 1991; Maruo et al., 1992).

EGF was found to enhance the proliferative activity of cytotrophoblast cells in culture of very early placental tissues obtained at 4-5 weeks gestation, without affecting the ability to secrete hCG and hPL. By contrast, in culture of 6- to 12-week placental tissues, EGF did not affect the proliferative activity of cytotrophoblast cells. Instead, EGF stimulated the secretion of hCG and hPL by cultured placental tissues. Our current findings conform with the result of Genbacev et al.(1994) who reported that EGF did not affect hCG production by villous explants of 5-6 weeks gestation, while it did stimulate production of hCG by explants of 8-10 weeks gestation. These findings suggest that EGF exerts gestational age-dependent dual actions on trophoblast in the first trimester placenta: one is the action to stimulate the proliferative potential of cytotrophoblast cells in very early placenta before 6 weeks gestation and the other is the action to stimulate hCG and hPL secretion by early placenta between 6 to 12 weeks gestation. Thus, EGF appears to exert two different actions independently in a stepwise manner on the trophoblast proliferation and differentiated trophoblast function during the course of first trimester gestation.

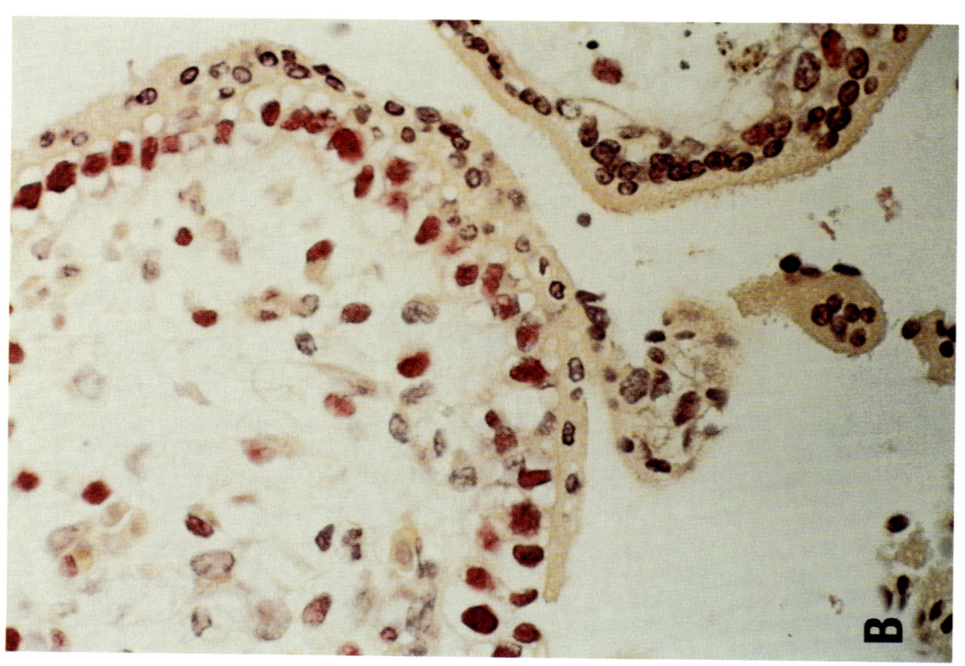
B

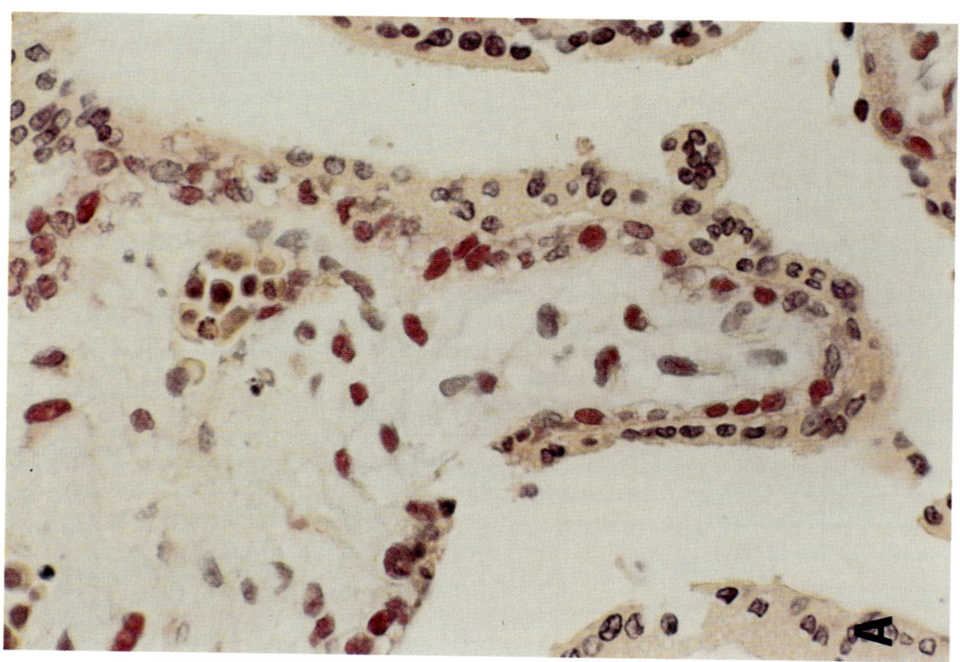
A

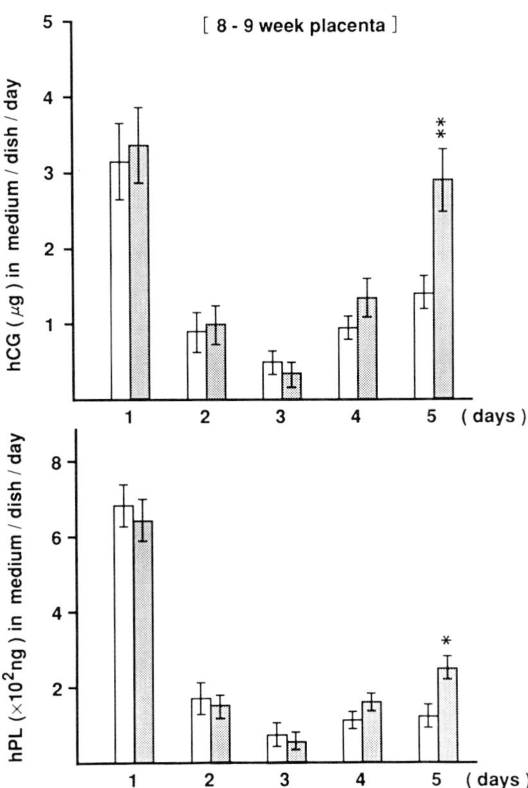

Figure 8. hCG and hPL secretion into the medium from early placental explants cultured in the absence or presence of EGF. Placental explants were cultured in the absence (open bars) or presence (solid bars) of 100 μg/l EGF during the first 48 hours of culture, and the cultures were continued for subsequent 72 hours in the absence of EGF. The results represent the mean ± SD of triplicate cultures. *, $p<0.05$; **, $p<0.01$

Figure 7. (opposite page) Immunohistochemical staining for PCNA of very early (4-5 week) placental explants cultured in the absence or presence of EGF (100 μg/l) for 24 hours. Frozen sections of placental explants cultured in the absence (A) or presence (B) of EGF for 24 hours were stained with a monoclonal antibody against PCNA using the avidin/biotin immunoperoxidase method. The lighter staining of nuclei is the hematoxylin counterstain. Original magnification x 200

The first step in the biological actions of EGF is represented by its binding to specific receptors present in the target cells. The binding process is thought to lead to receptor phosphorylation, internalization of the ligand-receptor complex, and generation of signals that mediate the actions of EGF. Consistent with the observation that EGF exerts the dual action in a stepwise manner on cytotrophoblast proliferation in very early placenta and on hCG and hPL secretion by early placenta, we have shown that EGF receptors in very early (4-5 week) placenta are primarily localized to cytotrophoblast, whereas EGF receptors in early (6-12 week) placenta are almost exclusively localized to syncytiotrophoblast. Thus, the change in cytologic localization of EGF receptors from cytotrophoblast to syncytiotrophoblast observed around 6 weeks of gestation is of great importance for the stepwise expression of the gestational age-dependent dual action of EGF on placental trophoblast cells during the course of first-trimester gestation.

It is now evident that a close similarity exists between EGF receptor and erb B oncoprotein. c-erb B oncogene family consists of c-erb B-1 and c-erb B-2 and it was demonstrated that c-erb B-1 codes EGF receptor protein while c-erb B-2 codes a protein which is closely related, but distinct (Semba et al., 1985). Consistent with the immunohistochemical findings, Northern blot analysis of placental RNA with c-erb B-1 cDNA probe demonstrated that placental RNA contains c-erb B-1 transcripts of 10.0-kb and 5.6-kb and that the amounts of the c-erb B-1 transcripts are most abundant in early placenta, more abundant in midterm placenta, and least abundant in term placenta. Although four bands with the estimated size of 10.0-, 5.6-, 4.6-, and 2.9-kb, were detected in Northern blot analysis of 6-week and 20-week placental RNA using a c-erb B-1 cDNA probe, it is assumed that the bands at 4.6-kb and 2.9-kb may be attributable to a cross-hybridization with c-erb B-related gene transcripts, 4.6-kb possibly with c-erb B-2 transcript (Yamamoto et al., 1986) and 2.9-kb possibly with truncated transcript of extracellular domain of c-erb B-1 (Ullrich et al., 1984). Thus, it is likely that the elevation of cellular levels of EGF receptor in early placental trophoblast cells results from the increased expression of c-erb B mRNA in early placental cells. Furthermore, *in situ* hybridization of placental tissue sections with ^{35}S-labeled c-erb B cDNA probe revealed that c-erb B mRNA is almost exclusively expressed in the syncytiotrophoblast of human placentae after six weeks of gestation.

Thus, the notions obtained in the immunohistochemical studies on the cytologic localization and cellular levels of EGF receptor in developing human placenta are consistent with the results obtained in the molecular studies which demonstrated that c-erb B mRNA expression in the human placenta after 6 weeks of gestation was localized to the syncytiotrophoblast and that c-erb B mRNA abundance in the human placental cells was highest in the first trimester and declined toward term.

Furthermore, results obtained in the present studies indicate that human early placental trophoblast is capable of producing an EGF-like substance and that thyroid hormone enhances the local production of the EGF-like substance (Matsuo et al., 1993). This suggests that human early placental trophoblast cells produce a EGF-like substance in synergy with thyroid hormone. Therefore, it can be concluded that an autocrine control system, wherein EGF serves as the signal in regulating placental growth and function in synergy with thyroid hormone, exists in the human early placenta. Much attention should be paid to the role of the interaction between endocrine factors and intraplacental growth factors in the regulation of feto-placental growth.

SUMMARY

In order to elucidate the role of EGF in human placental development, effects of EGF on the proliferation and endocrine function of trophoblast cells were investigated. Explants of trophoblastic tissues obtained from 4-5 week or 6-12 week placentae were, respectively, cultured with or without EGF, in the presence or absence of triiodo-L-thyronine (T3) in a serum-free condition. The proliferative activity was examined by immunocytochemical staining with an antibody to the proliferating cell nuclear antigen (PCNA), while the endocrine function was assessed by the ability to secrete hCG and hPL. In 4-5 week placentae, EGF and EGF receptor were localized in cytotrophoblast (C-cell) and EGF augmented the proliferation of C-cell without affecting the ability to secrete hCG and hPL. By contrast, in 6-12 week placentae, EGF and EGF receptor were localized in syncytiotrophoblast (S-cell) and EGF stimulated the secretion of hCG and hPL without affecting the proliferation of C-cell. *In situ* hybridization with c-erb B probe revealed that c-erb B mRNA is expressed in the S-cell after 6 weeks gestation. Column chromatography of the serum-free media obtained by 5-day culture of early placental tissues resulted in the elution of immunoreactive EGF. The addition of T3 (10^{-8} mol/l) resulted in increased secretion of immunoreactive EGF by placental explants. These findings suggest that EGF acts as an autocrine factor in regulating early placental growth and function in synergy with thyroid hormone.

ACKNOWLEDGEMENTS

This work was supported in part by Grants in Aid for Scientific Research 026704 and 05454451 from the Japanese Ministry of Education, Science and Culture, by the Ogyaa-Donation Foundation of the Japan Association of Maternal Welfare and by the International Committee of the Population Council, New York.

REFERENCES

Ashitaka, Y., Maruo, M,, Takeuchi, Y., Nakayama, H. and Mochizuki, M. (1988) 3,5,3'-triiodo-L-thyronine binding sites in nuclei of human trophoblastic cells. *Endocrinol. Jpn.* 35, 197-206.

Chirgwin, J.M., Przybyla, R., MacDonald, J. and Rutter, W.J. (1979) Isolation of biologically active ribonucleic acid from sources enriched in ribonuclease. *Biochemistry* 18, 5294-5299.

Downward, J., Yarden, Y., Mayes, E., Scrace, G., Totty, N., Ullrich, A., Schlessinger, J. and Waterfield, M.D. (1984) Close similarity of epidermal growth factor receptor and v-erb-B oncogene protein sequence. *Nature* 307, 521-527.

Genbacev, O., Powlin, S.S. and Miller, R.K. (1994) Regulation of human extravillus trophoblast (EVT) cell differentiation and proliferation in vitro - role of epidermal growth factor (EGF). *Trophoblast Research* 8, 427-442.

Kitagawa, M., Otani, T. and Mochizuki, M. (1992) Molecular biological studies on expression of placental corticotropin releasing factor (CRF) gene. *Acta Obst. Gynaecol. Jpn.* 44, 820-829.

Ladines-Llave, C.A., Maruo, T., Manalo, A.S. and Mochizuki, M. (1991) Cytologic localization of epidermal growth factor and its receptor in developing human placenta varies over the course of pregnancy. *Am J Obstet Gynecol.* 165, 1377-1382.

Maruo, T., Segal, S.J. and Koide, S.S. (1979) Studies on the apparent human chorionic gonadotropin-like factor in the crab ovalipes ocellatus. *Endocrinology* 104, 932-939.

Maruo, T., Matsuo, H., Ohtani, T., Hoshina, M. and Mochizuki, M. (1986) Differential modulation of chorionic gonadotropin subunit mRNA levels and chorionic gonadotropin secretion by progesterone in normal placenta and choriocarcinoma cultured in vitro. *Endocrinology* 119, 855-864.

Maruo, T and Mochizuki, M. (1987) Immunohistochemical localization of epidermal growth factor receptor and myc oncogene product in human placenta: Implication for trophoblast proliferation and differentiation. *Am. J. Obstet. Gynecol.* 156, 721-727.

Maruo, T., Matsuo, H., Oishi, T., Hayashi, M., Nishino, R. and Mochizuki, M. (1987) Induction of differentiated trophoblast function by epidermal growth factor: Relation of immunohistochemically detected cellular epidermal growth factor receptor levels. *J. Clin. Endocrinol. Metab.* 64, 744-750.

Maruo, T., Matsuo, H. and Mochizuki, M. (1991) Thyroid hormone as a biological amplifier of differentiated trophoblast function in early pregnancy. *Acta Endocrinol.* (Copenh) 125, 58-66.

Maruo, T., Matsuo, H., Murata, K. and Mochizuki, M. (1992) Gestational age-dependent dual action of epidermal growth factor on human placenta early in gestation. *J. Clin. Endocrinol. Metab.* 75, 1362-1367.

Matsuo, H., Maruo, T., Murata, K. and Mochizuki, M. (1993) Human early placental trophoblasts produce an EGF-like substance in synergy with thyroid hormone. *Acta Endocrinol.* (Copenh) 128, 225-229.

Mochizuki, M. (1988) Biology of trophoblast and placental protein hormones. In: *Placental Protein Hormones*, (eds.) M. Mochizuki and R.O. Hussa, Excerpta Medica, Amsterdam, pp. 3-18.

Nishii, H., Ashitaka, Y., Maruo, M. and Mochizuki, M. (1989) Studies on the nuclear 3,5,3'-triiodo-L-thyronine binding sites in cytotrophoblast. *Endocrinol Jpn* 36, 891-898.

Otani, T., Otani, F., Krych, M., Chaplin, D.D. and Boime, I. (1988) Identification of a promoter region in the CGb gene cluster. *J. Biol. Chem.* 263, 7322-7329.

Sap, J., Munoz, A., Damm, K., Goldberg, Y., Ghysdael, J. and Leutz, A. (1986) The c-erb A protein is a high-affinity receptor for thyroid hormone. *Nature* 324, 635-640.

Semba, K., Kamata, N., Toyoshima, K and Yamamoto, T. (1985) A v-erbB-related protooncogene, c-erbB-2 is distinct from the c-erbB-1/epidermal growth factor-receptor gene and is amplified in a human salivary gland adenocarcinoma. *Proc Natl Acad Sci USA* 82, 6497-6501.

Ullrich, A., Coussens, L., Hayflick, J.S., Dull, T.J., Gray, A., Tam, A.W., Lee, J., Yarden, Y., Liberman, T.A., Schlessinger, J., Downward, J., Mayers, E.L.V., Whittle, N., Waterfield, M.D. and Seeburg, P.H. (1984) Human epidermal growth factor receptor cDNA sequence and aberrant expression of the amplified gene in A431 epidermoid carcinoma cells. *Nature* 309, 418-425.

Yamamoto, T., Ikawa, S., Akiyama, T., Semba, K., Nomura, N., Miyajima, N., Saito, T. and Toyoshima, K. (1986) Similarity of protein encoded by the human c-erb-B-2 gene to epidermal growth factor receptor. *Nature* 319, 230-234.

Trophoblast Research 9:41-52, 1997

EXPRESSION OF EPIDERMAL GROWTH FACTOR (EGF) FAMILY AND EXPRESSION OF EPIDERMAL GROWTH FACTOR RECEPTOR(EGFR) IN HUMAN CHORIONIC VILLI

Tadashi Watanabe[1], Takao Fukaya[1], Akira Yajima[1] and Hironobu Sasano[2,3]

[1]Departments of Obstetrics and Gynecology and
[2]Pathology
Tohoku University School of Medicine
2-1, Seiryo- machi, Aoba-ku
Sendai, 980 Japan

INTRODUCTION

Epidermal growth factor (EGF), a peptide which consists of 53 amino acids, is a potent mitogen in many tissues (Carpenter, 1979). EGF has also been implicated in the production and secretion of human chorionic gonadotropin and human placental lactogen in *in vitro* trophoblast culture (Lai, 1984; Morrish et al., 1987). In addition, EGF deficiency during pregnancy in mice was reported to increase the incidence of abortion (Tsutsumi et al., 1987), which suggests a possible involvement of EGF in the maintenance of pregnancy. Transforming growth factor-α (TGF-α), a single polypeptide of 50 amino acids, was first purified from retrovirus-transformed cells in culture (Delarco, 1980) and shares a 30 percent sequence homology with EGF (Marquard et al., 1984). TGF-α is associated with cell proliferation and/or other biological activities of many tissues. The biological actions of EGF and TGF-α are considered to be exerted by binding of these growth factors to the epidermal growth factor receptor (EGFR) with equal affinity (Delarco, 1980, Pike et al., 1982, Derynck, 1986). Recently, amphiregulin (AR) and cripto (CR) were also identified as new members of EGF family. AR was originally purified from the conditioned medium of human breast carcinoma cell line, MCF-7, after treatment with phorbol 12-myristate 13-acetate (Shoyab et al., 1988). AR consists of a monomer of either 78 or 84 amino acids with 38% homology with EGF and partially competes for binding of EGF to the EGFR (Plowman et al., 1990). CR was originally isolated from human embryonal carcinoma cells (Ciccodicola et al., 1989). CR mRNA codes for 188 amino acids, which contains a 37-amino acid region that also shares structural homology with other members of the EGF/TGF-α family. CR exhibited transforming activity in transfected mouse NIH3T3 fibroblasts (Ciccodicola et al., 1989) and in mouse NOG-8 mammary epithelial cells (Ciardiello et al., 1991), as does for EGF and TGF-α, but interaction of CR with EGFR has not been well-established.

In this report, we examined immunolocalization of EGF family growth factors including TGF-α, AR and CR and EGFR in various stages of human chorionic villi from different gestational periods in order to elucidate the possible involvement of these growth factors during gestation.

[3]To Whom Correspondence Should Be Addressed.

MATERIALS AND METHODS

Subjects

First-trimester chorionic villi (6 to 12 weeks of gestation) were obtained by uterine curettage in 20 women undergoing an induced therapeutic abortion at the Nagaike Maternity Clinic, Sendai, Japan, and Sakata City Hospital, Sakata, Japan. Six cases from second trimester and eight cases from 32 to 40 weeks of gestation were obtained following normal spontaneous vaginal delivery at Tohoku University Hospital, Sendai, Japan and the Nagaike Maternity Clinic. Informed consent was obtained from each patient when the purpose of the study was fully explained by an obstetrician. Samples were cleared of blood clots immediately after curettage, fixed for 18 hours at 4°C in 4% paraformaldehyde buffered to pH 7.4, and then embedded in paraffin.

Primary Antibodies

Antibody against TGF-α (Ab-2, Oncogene Science Inc., Manhasset, NY, USA) was a mouse monoclonal IgG2a that reacts with denatured and native TGF-α of human and rat origin but shows no cross-reactivity with human or mouse EGF (Sorvillo et al., 1990). Antibody against AR was AR-Ab1, a rabbit polyclonal antibody kindly provided by Dr. Gibbes Johnson, Division of Cytokine Biology, National Institute of Health. Bethesda, MD, USA (Johnson et al., 1992). Antibody against CR was a rabbit polyclonal antibody C2-1, which was kindly provided by Dr. Wataru Yasui, First Department of Pathology, Hiroshima University School of Medicine. Hiroshima, Japan. Anti-CR was raised against a 12-amino acid synthetic peptide (GSVPHDTWLLPKK) which corresponds to amino-acid residues 116 to 127 in the human CR protein (Yasui et al., 1993). This peptide dose not exhibit any sequence homology with either EGF and TGF-α. The antibody was purified by affinity chromatography and its specificity was confirmed by immunoblotting and competition assays (Saeki et al., 1992). The antibody against EGFR was a mouse monoclonal antibody, 31G7 (Triton Diagnostics, Alameda, CA USA). It recognized the peptide backbone of the extracellular domain of the EGFR molecule.

Immunohistochemistry

Following deparaffinization, the sections were submerged in a bath of 0.3% hydrogen peroxide in 100% methanol for 30 minutes to block endogenous peroxidase activity, and then washed in three changes of 0.01 M PBS every 5 minutes. Sections for EGFR immunostaining were further incubated in 0.1% trypsin (Sigma Co., Ltd., St. Louis, MO) for 10 minutes for antigen retrieval (Sasano et al., 1994). They were treated with 1% normal rabbit serum (for TGF-α and EGFR) or 1% normal goat serum (for AR and CR) for 30 minutes at room temperature. All immunohistochemical procedures were performed by the streptoavidin-biotin-peroxidase method, using a Histofine kit (Nichirei Co., Ltd., Tokyo Japan). Deparaffinized sections were incubated with the primary antibodies at a concentration of 1:100 (TGF-α and EGFR), 1:25 (AR) and 1:1000 (CR) for 18 hours at 4°C in a humidified chamber. After rinsing in PBS, the sections were incubated with biotinylated rabbit anti-mouse (for TGF-α and EGFR) or goat anti-rabbit immunoglobulin (for AR and CR) and peroxidase-conjugated streptoavidin for 30 minutes each at room temperature in a humidified chamber. After an additional rinse, the sections were immersed for 5 minutes in a solution containing 0.05% Tris-HCl, pH 7.6, 0.06 mM 3,3'-diaminobenzidine and 2 mM hydrogen peroxidase at room temperature. The immunoreacted sections were counterstained with 1% methyl green and mounted in a glycerol-gelatin water-soluble medium. For negative controls, 0.01 M PBS and normal mouse or rabbit IgG were used instead of the primary antibody and neither of the control sections exhibited the immunostaining patterns observed when the primary antibodies were employed. No specific immunoreactivity was detected in these tissue sections. Human stomach cancer specimens were used for positive controls of immunostaining of TGF-α, AR, CR and EGFR.

RESULTS

TGF-α

In first trimester pregnancy, immunoreactivity of TGF-α was present predominantly in the cytoplasm of the cytotrophoblast (CT) (Figure 1A) and scattered weak immunoreactivity was also observed in the syncytiotrophoblast (ST). Among CTs, TGF-α immunoreactivity in intravillous CT was much more intense than that in cell islands or column (Figure 1B). Extravillous trophoblast cells which infiltrate the decidua or spiral arteries demonstrated no or relatively weak TGF-α immunoreactivity. In second trimester, immunoreactivity of TGF-α was weakly observed only in the CT (data not shown). In third trimester, immunoreactivity of TGF-α was not detected at all in either the CT and the ST.

Amphiregulin

AR immunoreactivity was sporadically observed predominantly in the nuclei and partly in the cytoplasm of the CT and ST of first trimester pregnancy (Figure 2A). Among CT, AR immunoreactivity was present in intravillous CT and CT in cell islands. AR immunoreactivity was also detected in extravillous trophoblast cells including intermediate trophoblast cells infiltrating the decidua or spiral arteries (Figure 2B). In second and third trimester, immunoreactivity of AR was not detected at all in either the CT or ST (data not shown).

Cripto

CR immunoreactivity was mainly observed in the cell membrane and/or the cytoplasm of the ST (Figure 3A), but the pattern of its expression was markedly heterogenous among individual chorionic villi (Figure 3A). Frequency of the immunoreactivity of CR in the ST of chorionic villi tended to increase as pregnancy progressed (Figure 3B).

EGFR

EGFR immunoreactivity was observed predominantly as membrane staining. In first trimester, EGFR immunostaining was intense in the ST, especially in the microvilli, and weak in the CT (Figure 4). The frequency of EGFR immunoreactivity in the ST and the CT in individual chorionic villi decreased as pregnancy progressed (data not shown).

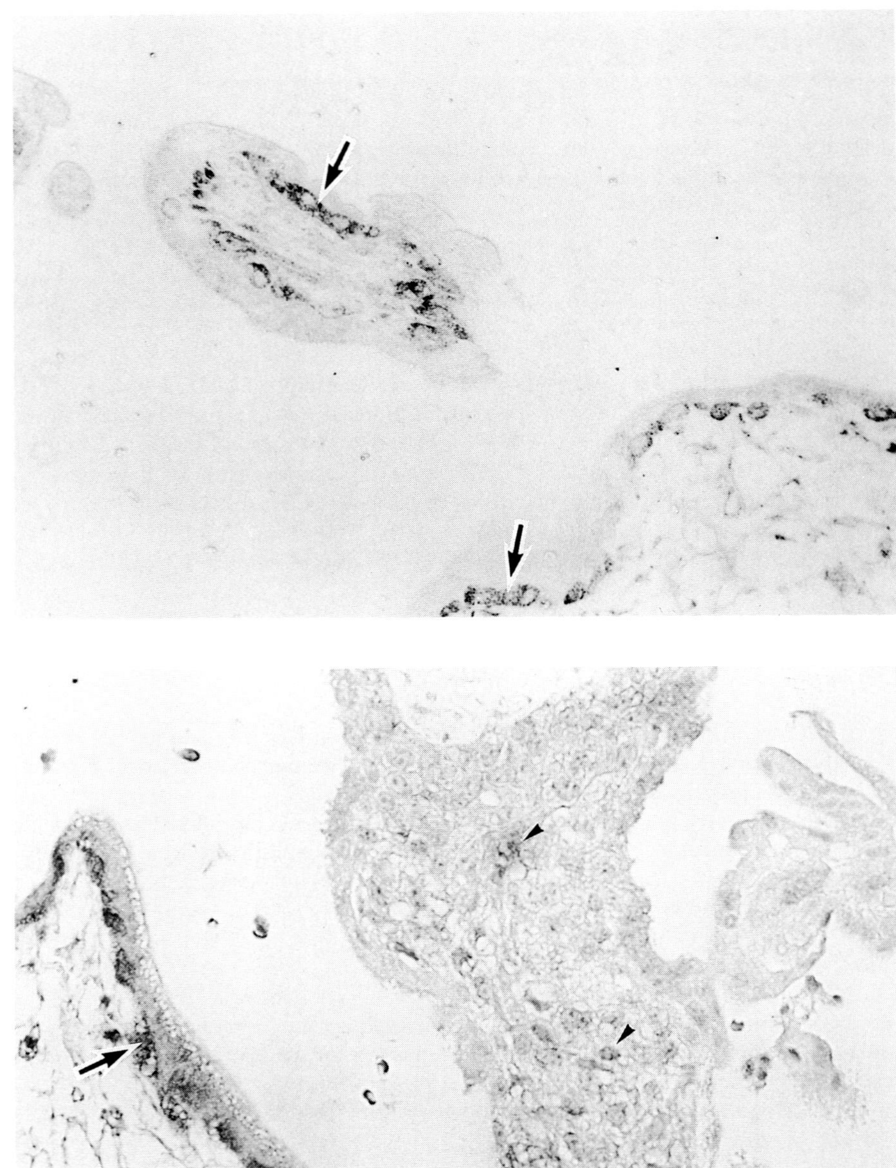

Figure 1. (A) (top) Immunostaining of TGF-α in chorionic villi of first trimester pregnancy. Immunoreactivity was observed predominantly in the cytoplasm of the CT as designated by arrows (X350). (B) (bottom) Immunostaining of TGF-α in chorionic villi of first trimester pregnancy. TGF-α immunoreactivity was much more intense in intravillous CT (arrow) than that in CT forming islands or columns (arrow heads) (X350).

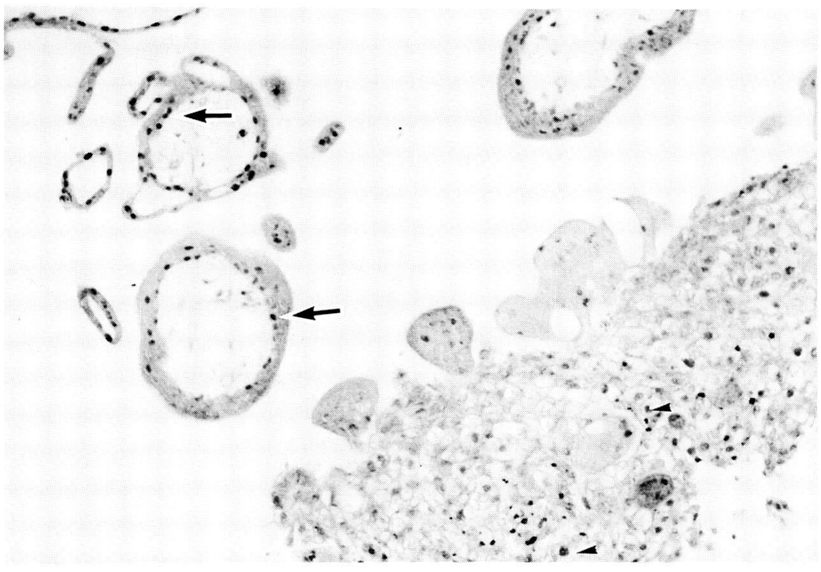

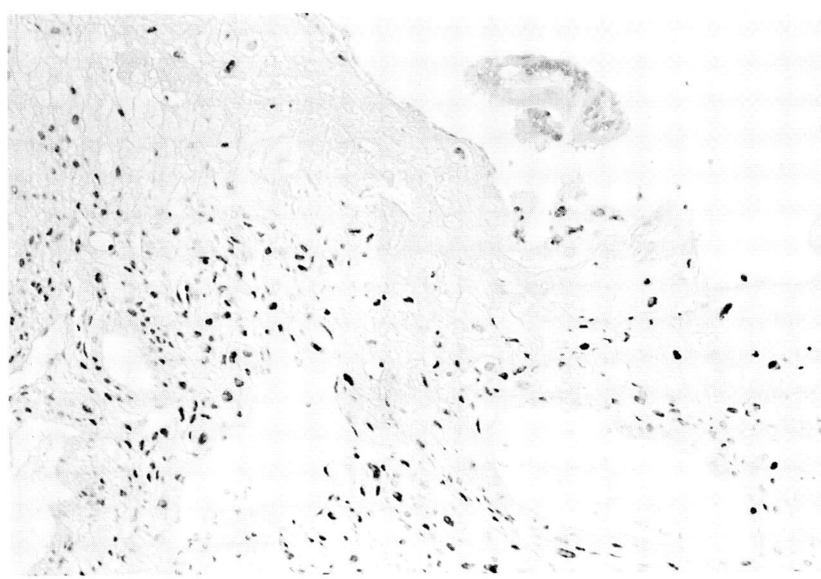

Figure 2. (A) (top) Immunostaining of AR in chorionic villi of first trimester pregnancy. Immunoreactivity was observed predominantly in the nucleus of the CT and some ST in the villi (arrow). CT forming cell islands also demonstrated AR immunoreactivity (arrow heads) (X250). (B) (bottom) Immunostaining of AR in decidua of first trimester pregnancy. AR nuclear immunoreactivity was detected in infiltrating extravillous trophoblast cells (X250).

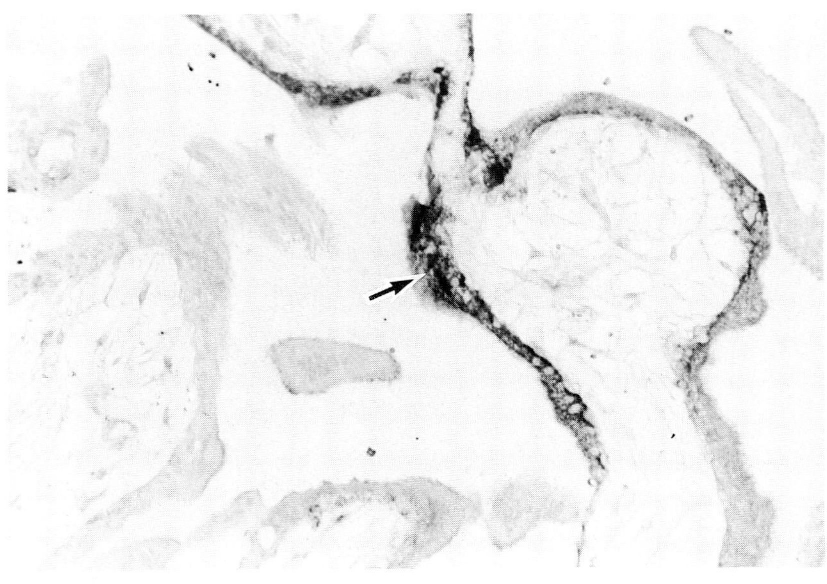

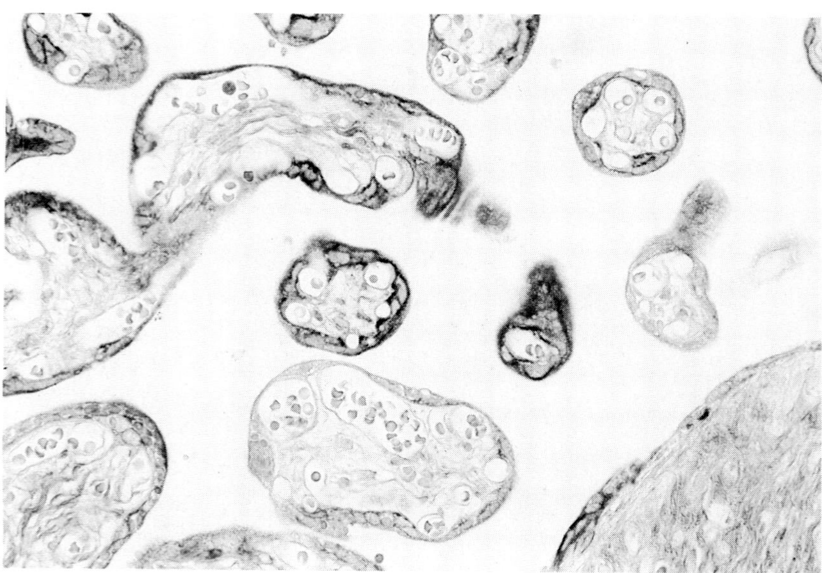

Figure 3. (A) (top) Immunostaining of CR in chorionic villi of first trimester pregnancy. CR immunoreactivity was observed in the ST (arrow) with heterogeneity (X300). (B) (bottom) Immunostaining of CR in chorionic villi of third trimester pregnancy. CR immunoreactivity was observed in the ST with heterogeneity (X300).

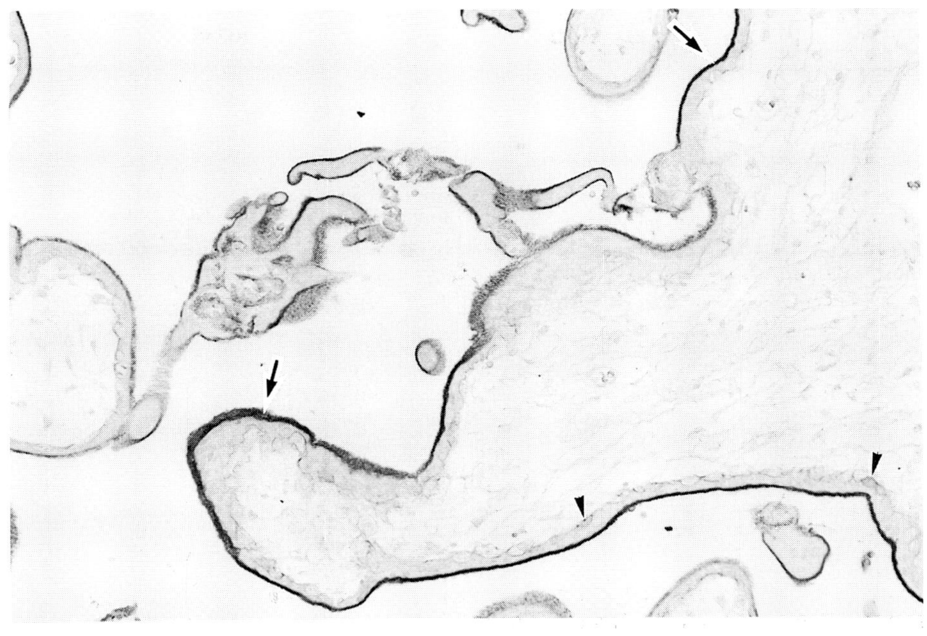

Figure 4. Immunostaining of EGFR in chorionic villi of first trimester pregnancy. Immunoreactivity was observed in the cell membrane of the ST especially microvilli (arrow). EGFR immunoreactivity was weak in the CT (arrow heads) (X200).

DISCUSSION

The expression of protein and mRNA of EGFR have been demonstrated to occur during pregnancy (Maruo et al., 1987; Chen et al., 1988; Bulmer et al., 1989, Kawagoe et al., 1989; Tavare, 1989; Hofmann et al., 1992). Function of EGF, which is one of the ligands of EGFR, has been reported to be associated with differentiation of trophoblast cells (Lai, 1984; Morrish et al., 1987). It is interesting to study how other ligands of EGFR or EGF-related growth factors are involved in trophoblast function during gestation.

In the present study, TGF-α immunoreactivity was detected in the CT of the chorionic villi of first trimester pregnancy. TGF-α immunoreactivity in the ST was weak and focal. Filla et al. (1993) reported TGF-α immunoreactivity was predominantly observed in the CT on frozen sections and cultured CT (Filla et al., 1993) which is consistent with the present study. In contrast, Horowitz et al. (1993) reported that TGF-α immunoreactivity ranged from moderate to intense in the ST, and from light to moderate in the CT in routinely processed tissues (Horowitz et al., 1993). This disparity in immunolocalization of TGF-α may derive from differences in tissue preparations or other variables. Our results suggest that the CT produce TGF-α, at least in first and second trimester. This hypothesis is supported by the result of Northern blot analysis by

Bissonnette et al. (1992), who detected TGF-α mRNA in human chorionic villi in first and second trimester pregnancy (Bissonnette et al.,1992). However, in third trimester pregnancy, TGF-α immunoreactivity was not detected in the current study, but TGF-α mRNA was detected in chorionic villi in third trimester pregnancy by Northern blot analysis (Bissonnette et al., 1992). Further investigations are required to clarify this difference.

AR immunoreactivity was restricted in first trimester pregnancy. An inverse correlation was reported between AR and TGF-α expression in human breast cancer cell lines (Todaro et al., 1990). Interestingly, the current results indicated that the distribution pattern of AR was similar to that of TGF-α. Furthermore, it is noteworthy that AR was detected predominantly in the nuclei. Recently, Kitadai et al. (1993) reported that AR immunoreactive protein was localized in the cytoplasm and/or nucleus in human gastric carcinoma (Kitadai et al., 1993). AR precursor contains both a signal peptide and two putative nuclear targeting sequences positioned at residues 26-29 and 40-43 (Shoyab et al., 1989). Therefore, nuclear immunoreactive AR may represent internalization (Waterfield et al., 1982; Brachmann et al., 1989) of AR/EGFR complexes. For AR during gestation, Plowman et al. (1988) reported that AR mRNA was detected in human placenta (Plowman et al., 1988), but they did not refer to a specific gestational period for AR mRNA expression.

EGFR immunoreactivity was observed in both the ST and the CT, but that in the ST was generally much more intense. In the ST, EGFR immunostaining was more intense on the microvilli than on the basal and lateral sides of the cell, which appears to be consistent with the fact that EGF/TGF-α derived from the circulation act mainly on the ST through binding to EGFR. Many studies have reported immunohistochemical localization of EGFR in placenta (Maruo et al., 1987; Bulmer et al., 1989; Kawagoe et al., 1989; Tavare, 1989; Hofmann et al., 1992). The present results were consistent with the previous reports. When the localization of both TGF-α and EGFR immunoreactivity are considered together, locally synthesized TGF-α, which is produced predominantly in the CT, may act mainly in a paracrine manner, i.e., from CT to ST, in first and second trimester human placenta. However, further investigations are required to clarify the biological significance of locally synthesized TGF-α.

The current results also revealed that CR was detected mainly in the ST, especially during third trimester pregnancy but expression patterns of CR were different among individual chorionic villi. *In vitro* studies suggest that CR expression is associated with a more undifferentiated state (Ciccodicola et al., 1989). CR mRNA was not detected in human placenta (Ciccodicola et al., 1989). However, CR immunoreactive protein was observed in the ST of chorionic villi in the current study. In human chorionic villi, the CT is generally considered to differentiate into the ST, and in third trimester, the great majority of the trophoblast cells are the ST. This finding is not consistent with the previous concept of association of CR with de-differentiation. Further investigations are required but immunoreactive CR in chorionic villi may be derived from circulation.

In summary, TGF-α mainly act possibly in a paracrine manner in first trimester chorionic villi. AR is considered to play roles in function of trophoblast cells in first trimester pregnancy. On the other hand, CR may play some roles in the ST, especially in third trimester pregnancy.

SUMMARY

Among epidermal growth factor (EGF) family growth factors, EGF, transforming growth factor-α (TGF-α) and amphiregulin (AR) are considered to exert their effects through binding to epidermal growth factor receptor(EGFR). Cripto (CR) is homologous to EGF in its structure, but it is postulated that it might not bind with EGFR. We examined localization of TGF-α, AR, CR and EGFR in human chorionic villi of various stages. Chorionic villi were collected from thirty-four cases. Immunohistochemistry was performed on 4% paraformaldehyde fixed and paraffin embedded specimens using an avidin biotin complex-peroxidase technique. TGF-α immunoreactivity was intense in the cytoplasm of the cytotrophoblast(CT), especially in first trimester. AR immunoreactivity was observed predominantly in the nuclei of the CT and some syncytiotrophoblast (ST) only in first trimester. EGFR immunoreactivity was intense in the ST, and weak in the CT and its frequency appeared to decrease as gestational period progressed. CR immunoreactivity was mainly observed in the ST, but marked heterogeneity was observed in its expression and its frequency appeared to increase as pregnancy progressed. Based on these findings, TGF-α is considered to act mainly in a paracrine manner in first trimester chorionic villi. AR is produced to modify the biological function of the trophoblast cells during the first trimester. In contrast, CR may modify ST function, especially during the third trimester.

ACKNOWLEDGMENTS

We would like to thank Dr. F. Nagaike, Nagaike Maternity Clinic and Dr. Y. Mandai, Department of Obstetrics and Gynecology, Sakata City Hospital for providing samples and appreciate Mr. Katsuhiko Ono and Ms. Fumiko Date, Department of Pathology, Tohoku University School of Medicine for their excellent technical assistance.

REFERENCES

Bissonnette, F., Cook, C., Geoghegan, T., Steffen, M., Henry, J., Yussman, M.A. and Schultz, G. (1992) Transforming growth factor-α and epidermal growth factor messenger ribonucleic acid and protein levels in early, mild, and late gestation human placenta. *Am. J. Obstet. Gynecol.* 166, 192-199.

Bulmer, J.N., Thrower, S. and Wells, M. (1989) Expression of epidermal growth factor receptor and transferrin receptor by human trophoblast populations. *Am. J. Reprod. Immunol.* 21, 87-93.

Brachmann, R., Lindquist, P.B., Nagahima, M., Kohr, W., Lipari, T., Napier, M. and Derynck, R. (1989) Transmembrane TGF-α precursors activates EGF/TGF-α receptors. *Cell* 56, 691-700.

Carpenter, G. and Cohen, S. (1979) Epidermal growth factor. *Ann. Rev. Biochem.* 48, 193-216.

Chen C.F., Kurachi, H., Fujita, Y., Terakawa, N., Miyake, A. and Tanizawa, O. (1988) Changes in epidermal growth factor receptor and its messenger ribonucleic acid levels in human placenta and isolated trophoblast cells during pregnancy. *J. Clin. Endocrinol. Metab.* 67, 1171-1177.

Ciardiello, F., Dono, R., Kim, N., Persico, M.G. and Salomon, D.S. (1991) Expression of *cripto*, a novel gene of the epidermal growth factor gene family, leads to *in vitro* transformation of a normal mouse mammary epithelial cell line. *Canc. Res.* 51, 1051-1054.

Ciccodicola, A., Dono, R., Obisi, S., Simeone, A., Zoll, M. and Persico, M.G. (1989) Molecular characterization of a gene of the 'EGF family' expressed in undifferentiated human NTERA2 teratocarcinoma cells. *EMBO J.* 8, 1987-1991.

Delarco, J.E. and Todaro, G.J. (1980) Sarcoma growth factor:specific binding to epidermal growth factor(EGF) membrane receptors. *J. Cell. Biol.* 102, 267-277.

Derynck, R. (1986) Transforming growth factor-alpha:structure and biological activities. *J. Cell. Biochem.* 32, 293-304.

Filla, M.S., Zhang, C.X. and Kaul, K.L. (1993) A potential transforming growth factor α/epidermal growth factor receptor autocrine circuit in placental cytotrophoblasts. *Cell Growth Different.* 4, 387-393.

Hofmann, G.E., Drews, M.R., Scott, Jr. R.T., Navot, D., Heller, D. and Deligdisch, L. (1992) Epidermal growth factor and its receptor in human implantation trophoblast: Immunohistochemical evidence for autocrine/paracrine function. *J. Clin. Endocrinol. Metab.* 74, 981-988.

Horowitz, G.M., Scott, Jr. R.T., Drews, M.R., Navot, D. and Hofmann, G.E. (1993) Immunohistochemical localization of transforming growth factor-α in human endometrium, decidua, and trophoblast. *J. Clin. Endocrinol. Metab.* 76, 786-792.

Johnson, G.R., Saeki, T., Gordon, A.W., Shoyab, M., Salomon, D.S. and Stromberg, K. (1992) Autocrine action of amphiregulin in a colon carcinoma cell line and immunocytochemical localization of amphiregulin in human colon. *J. Cell. Biol.* 118, 741-751.

Kawagoe, K., Akiyama, J., Kawamoto, T., Morishita, Y., and Mori, S. (1990) Immunohistochemical demonstration of epidermal growth factor (EGF) receptors in normal human placental villi. *Placenta* 11, 7-15.

Lai, W. H. and Guyda, H. J. (1984) Characterization and regulation of epidermal growth factor receptors in human placental cell cultures. *J. Clin. Endocrinol. Metab.* 58, 344-348.

Maruo, T., Matsuo, H., Oishi, T., Hayashi, M., Nishino, R. and Mochizuki, M. (1987) Induction of differentiated trophoblast function by epidermal growth factor: Relation of immunohistochemically detected cellular epidermal growth factor receptor levels. *J. Clin. Endocrinol. Metab.* 64, 744-750.

Morrish, D.W., Bhardwaj, D., Dabbagh, L. K., Marusyk, H. and Siy, O. (1987) Epidermal growth factor induces differentiation and secretion of human chorionic gonadotropin and placental in normal human placenta. *J. Clin. Endocrinol. Metab.* 65, 1282-1290.

Pike, L.J., Marquardt, H., Todaro, G.J., Gallis, B., Casnellie, J.E., Bornstein, P., and Krebs, E.G. (1982) Transforming growth factor and epidermal growth factor stimulate the phosphorylation of a synthetic, tyrosine-containing peptide in a similar manner. *J. Biol. Chem.* 257, 14628-14631.

Plowman, G.D., Green, J.M., McDonald, V.L., Neubauer, M.G., Disteche, C.M., Todaro, G. J. and Shoyab, M. (1990) The amphiregulin gene encodes a novel epidermal growth factor-related protein with tumor-inhibitory activity. *Mol. Cell. Biol.* 10, 1969-1981.

Saeki, T., Stromberg, K., Qi, C.F., Güllick, W.J., Tahara, E., Normanno, N., Ciardiello, F., Kenney, N., Johnson, G.R. and Salomon, D.S. (1992) Differential immunohistochemical detection of amphiregulin and cripto in human normal colon and colorectal tumors. *Canc. Res.* 52, 3467-3473.

Sasano, H., Suzuki, T., Shizawa, S., Kato, K. and Nagura, H. (1994) Transforming growth factor alpha, epidermal growth factor, and epidermal growth factor receptor in normal and diseased human adrenal cortex by immunohistochemistry and in situ hybridization. *Modern Pathol.* 7, 741-746.

Shoyab, M., McDonald, V.L., Bradley, J.G. and Todaro, G.J. (1988) Amphiregulin: A bifunctional growth-modulating glycoprotein produced by the phorbol 12-myristate 13-acetate-treated human breast adenocarcinoma cell line MCF-7. *Proc. Natl. Acad. Sci. USA* 85, 6528-6532.

Shoyab, M., Plowman, G.D., McDonald, V.L., Bradley, J.G. and Todaro, G.J. (1989) Structure and function of human amphiregulin: A member of the epidermal growth factor family. *Science* 243, 1074-1076.

Sorvillo, J.M., McCormack, E.S., Yanez, L., Valenzuela, D., and Reynolds, Jr., F.H. (1990) Preparation and characterization of monoclonal antibodies specific for human transforming growth factor-α. *Oncogene* 5, 377-386.

Tavare, J.M. and Holmes, C.H. (1989) Differential expression of the receptors for epidermal growth factor and insulin in the developing human placenta. *Cell Signall.* 1, 55-64.

Todaro, G.J., Rose, T.M., Spooner, C.E. and Shoyab, M. (1990) Cellular and viral ligands that interact with the EGF receptor. *Semin. Canc. Biol.* 1, 257-263.

Tsutsumi, O. and Oka, T. (1987) Epidermal growth factor deficiency during pregnancy causes abortion in mice. *Am. J. Obstet. Gynecol.* 156, 241-244.

Waterfield, M.D., Mayes, E.L.V., Stroobant, P., Bennet, P.L.P., Young, S., Goodfellow, P. N., Banting, G.S. and Ozanne, B. (1982) A monoclonal antibody to the human epidermal growth factor receptor. *J. Cell. Biochem.* 20, 149-161.

Yasui, W., Maki, J., Kuniyasu, H., Kitadai, Y., Yokozaki, H. and Tahara, E. (1993) Immunohistochemical detection of cripto protein in human cancers. *Proc. Am. Assoc. Canc. Res.* 34, 516.

Trophoblast Research 9:53-62, 1997

REGULATION OF DECIDUAL IGF-BINDING PROTEINS AND PROTEASE ACTIVITY BY PLACENTAL HORMONES

Mitsutoshi Iwashita[1,3], Yoshiki Kudo[2], Keiji Sakai[2], and Yoshihiko Takeda[1,2]

[1]Maternal and Perinatal Center
[2]Department of Obstetrics and Gynecology
Tokyo Women's Medical College
Tokyo 162, Japan

INTRODUCTION

Insulin-like growth factor-I (IGF-I) is a potent mitogen for a variety of *cells in vitro* and presumed to mediate growth hormone action *in vivo*. Several lines of evidence indicate important roles for maternal IGF-I in fetal growth. Maternal circulating IGF-I increases during pregnancy (Wilson et al., 1992) and correlates with birth weight (Han and Hill, 1992). Since IGF-I stimulates placental uptake of amino acid *in vitro* and enhances maternal amino acid transfer to the fetus *in vivo* (Takeda and Iwashita, 1993), it has been postulated that maternal IGF-I stimulates placental and fetal growth by activating transport system in the placenta.

Both IGFs (IGF-I and IGF-II) are associated with binding proteins (IGFBPs) in the circulation and these binding proteins are known to modify the actions of IGFs. At present, six distinct IGFBPs have been characterized based on their complete primary structure obtained by molecular cloning (Shimasaki and Ling, 1992). IGFBP-1 is most extensively studied IGFBP. Maternal IGFBP-1 levels have been found to be elevated during pregnancy (Than et al., 1983) and were inversely correlated with birth weights (Howell et al., 1985). Since IGFBP-1 has been shown to inhibit actions of IGF-I on trophoblast cells (Rutanen et al., 1988), it has been postulated that maternal IGFBP-1 prevents fetal growth by inhibiting IGF-I action in the placenta. Four IGFBPs including IGFBP-1, -2, -3 and -4 are detected in the circulation by Western ligand blot (Hossenlopp et al., 1986). However, binding activities of these IGFBPs except IGFBP-1 were markedly reduced during pregnancy due to increased specific protease activity for IGFBPs in the circulation (Hossenlopp et al., 1990).

Human decidual cells have been reported to produce several IGFBPs including IGFBP-1, -2 and -4 (Clemmons et al., 1990). The production of endometrial IGFBP-1 is increased during decidualization and positively regulated by progesterone (Bell et al, 1991). A large amount of IGFBP-1 produced in decidua during pregnancy may be responsible for increased IGFBP-1 in the maternal circulation. During pregnancy binding activities of IGFBPs in the circulation reduced markedly due to increased protease activity in the maternal circulation (Giudice et al., 1990; Hossenlopp et al., 1990). Recently decidua has been found to produce protease for IGFBPs (Deal and Lamson, 1991) that might be responsible for protease activity in the maternal circulation. Thus, decidual IGFBPs and protease may play an important role to modify IGF-I action on

[3]To Whom Correspondence Should Be Addressed.

placenta thereby influence fetal growth. The aim of this study is to elucidate the role of placental hormones including estradiol, progesterone and IGF-I in the regulation of IGFBPs produced by decidua.

MATERIALS AND METHODS

Hormones and Reagents

^{125}I-IGF-I and ECL western blotting kit were purchased from Amersham Japan (Tokyo, Japan). Collagenase and hyaluronidase were purchased from Worthington Biochem. Co. (Freehold, NJ, USA). Des-(1-3) IGF-I was from GroPep Ltd. (Adelaide, Australia). Horseradish peroxidase conjugated monoclonal antibody to IGFBP-1 was kindly provided from Medix Biochemica (Kauniainen, Finland). Polyclonal antibodies to IGFBP-2, -3, and -4 raised against rabbits were purchased from Upstate Biotechnology Inc. (Lake Placid, NY, USA). Estradiol and progesterone were from Sigma Chemical Co. (St. Louis, MO, USA). Nonglycosylated IGFBP-3 derived from E. coli labeled with ^{125}I was obtained from Diagnostic Systems Laboratories Inc. (Webster, TX, USA).

Cell Culture

Decidual cells were prepared from decidual tissue obtained from five term deliveries as previously described (Clemmons et al., 1990). Cells were plated onto Vitrogen 100 (Collagen Co., Palo Alto, CA, USA) in 12-well tissue culture plates (Costar, Cambridge, MA, USA) at concentrations of 1.0 million cells per well in RPMI-1640 medium (Gibco, Grand Island, NY, USA) supplemented with 10% fetal calf serum, 100U/ml penicillin and 100 μg/ml streptomycin. After 2 days culture, cells were washed in serum free medium and then cultured for 5 days in serum free medium in the absence or presence of estradiol (10 nmol/l), progesterone (1 μmol/l), or des (1-3) IGF-I (100 nmol/l). Each hormone was incubated with cells in quadruplicate cultures. The media were changed every 24 hours and IGFBPs in the pooled medium were analyzed by immunoblot and ligand blot.

Trophoblast cells obtained from term deliveries were prepared as previously reported (Iwashita et al., 1989). Enzymatically dispersed cells were cultured for 3 days in medium 1640 containing 10% fetal calf serum and then cultured for 2 days in serum free medium. Pooled medium was subjected to protease assay.

Western Immunoblot and Ligand Blot

For Western immunoblot, pooled medium from decidual cell culture was concentrated 5- to 10-fold in a Centricon-10 microconcentrator (Amicon, Danvers, MA, USA) and samples were subjected to Western immunoblot using chemiluminescence technique, as described previously (Hossenlopp et al., 1986). Briefly, 30 μl of concentrated medium was electrophoresed on 10% sodium dodecyl sulfate (SDS)-polyacrylamide gel under nonreducing conditions (Laemmli, 1970) except protein Mr standards (Rainbow Marker, Amersham Japan, Tokyo, Japan). After electrophoresis, proteins were transferred onto nitrocellulose membranes in a transblot cell (Bio-Rad Laboratories, Richmond, CA, USA) overnight at 70V in Towbin buffer (25 mmol/l Tris, 192 mmol/l glycine and 20% methanol). The nitrocellulose was blocked with 1% BSA (wt/vol) for 18 hours and washed three times with Tris buffer (0.1 mol/l Tris, 0.15 mol/l

NaCl, 0.1% Tween 20, pH 7.4). The nitrocellulose was incubated with a 1:5000 dilution of horseradish peroxidase conjugated monoclonal antibody to IGFBP-1 or with a 1:2500 dilution of polyclonal antibody to IGFBP-2, -3 or -4 for 2 hours at room temperature in above buffer, respectively. The nitrocellulose preincubated with antibody to IGFBP-2, -3, or -4 was then washed three times followed by incubation with a 1:5000 dilution of horseradish peroxidase conjugated Protein A (Amersham Japan, Tokyo, Japan) for 1 hour at room temperature and exposed to the ECL reagents for 1 minute and exposed to x-ray film (Kodak X-Omat AR, Eastman Kodak Co., Rochester, NY, USA) with Cronex Hi-Plus Intensifying Screens (Du Pont, Wilmington, DE, USA) for 1 to 5 minutes. The membrane preincubated with horseradish peroxidase conjugated monoclonal antibody to IGFBP-1 was directly exposed to the ECL reagents. For Western ligand blot, the nitrocellulose was blocked successively with 3% Nonidet P-40, 1% BSA and 0.1% Tween-20 in Tris-saline (pH 7.4) and incubated with a mixture of iodinated I-IGF-I and IGF-II (600,000 cpm) for 24 hours at 4°C. The membranes were washed and exposed to x-ray film with Cronex Hi-Plus Intensifying Screens for one week.

Protease Assay

Protease activity in the medium was assessed by the proteolysis of nonglycosylated iodinated IGFBP-3 according to the method described previously (Lamson et al., 1991). Pooled sera from term pregnancy (3 μl) and concentrated media conditioned by decidual and trophoblast cells (10 μl) were incubated with iodinated IGFBP-3 (50,000 cmp) in phosphate buffered saline containing 0.5 mmol/l $CaCl_2$ at 37°C for 5 hours. The mixture was subjected to SDS polyacrylamide gel electrophoresis and the dried gel was subjected to autoradiography. The intensity of each band was scanned on a densitometer and the amount of proteolysis was determined as the percentage of the proteolytic cleavage products over the total density in each lane.

Statistics

Western immunoblot, ligand blot and protease assay were repeated at least three times using decidua and trophoblast cells obtained from three different pregnant women and a representative picture from each experiment was shown. The effect of hormones on proteolysis of IGFBP-3 was repeated five times with decidual cells obtained from three different pregnant women and the results were expressed as the mean ± SE of five separate experiments. A student's t test was used to determine statistically significant differences as indicated in the table.

RESULTS

IGFBPs secreted by cultured decidual cells were assessed by Western ligand blot analysis. Figure 1 shows a representative immunoblot and ligand blot demonstrating the profiles of IGFBPs in conditioned media from decidual cells. Three IGFBP bands with Mr of 34,000, 30,000, and 24,000 were detected in control conditioned medium. The antibody to IGFBP-1 reacted only with 30,000 Mr form. Similarly, antibodies to IGFBP-2 and -4 reacted with 34,000 and 24,000 Mr band, respectively. Thus, decidual cells produce IGFBP-1, -2 and -4. Neither band reacted with antibody to IGFBP-3, suggesting that decidual cells did not release IGFBP-3 under these culture conditions. Figure 2 shows effect of des (1-3) IGF-I (100 nmol/l), progesterone (1 μmol/l) and estradiol (10 nmol/l) on IGFBPs profiles in medium. IGFBP-1 and -4 were diminished by des (1-3) IGF-I and

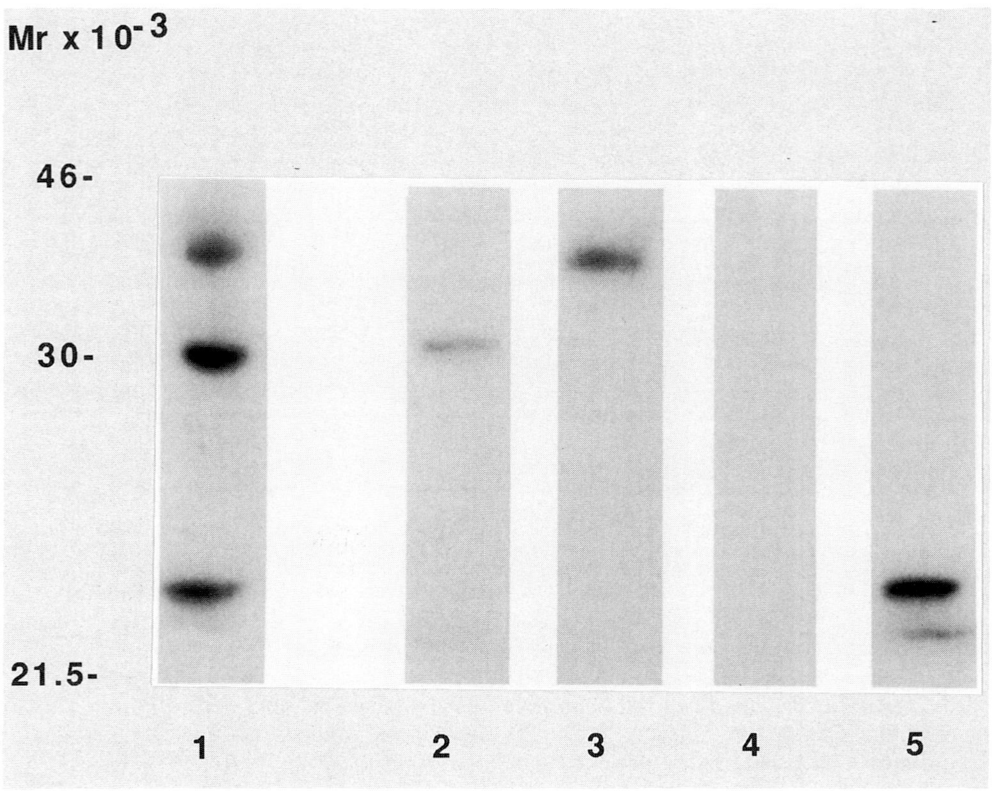

Figure 1. Analysis of IGFBPs in medium from decidual cell culture by Western ligand blot and immunoblot. Decidual cells were cultured for five days and conditioned media were pooled, concentrated and subjected to Western ligand blot and immunoblot. Details are given in Materials and Methods. Lane 1 shows ligand blot and lane 2 to 5 demonstrate immunoblot with anti IGFBP-1 to IGFBP-4 antibody, respectively.

all three IGFBPs were increased by progesterone while estradiol did not affect basal IGFBPs release into medium.

Subsequently, decidual cells conditioned media were assessed for proteolytic activity. The media were subjected to an IGFBP protease assay using iodinated IGFBP-3 as substrate. As shown in Figure 3, iodinated IGFBP-3 remained intact when incubated with medium from trophoblast cells. However, when IGFBP-3 was incubated with a pool of term pregnant sera and medium from decidual cell culture, intact IGFBP-3 was cleaved to smaller fragments with similar proteolytic pattern. Incubation of iodinated IGFBP-3 with medium from decidual cells showed approximately 56% degradation of IGFBP-3 (Figure 4 and Table 1). Des (1-3) IGF-I slightly increased protease activity and estradiol showed no effect on protease activity. In contrast, progesterone suppressed protease activity in which degradation of IGFBP-3 was completely inhibited.

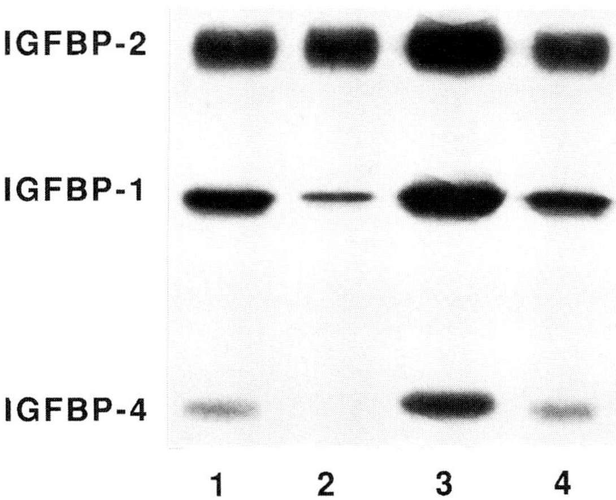

Figure 2. Analysis of medium from decidual cell culture by ligand blot. Decidual cells were incubated in the absence (control, lane 1) or presence of de(1-3) IGF-I (100 nmol/l, lane 2), progesterone (1 μmol/l, lane 3) , and estradiol (10 nmol/l, lane 4) for 5 days. Pooled media were concentrated and electrophoresed on SDS-polyacrylamide gel and proteins were transferred onto a nitrocellulose membrane. The membrane was incubated with 600,000 cpm of iodinated IGF-I and IGF-II for 24 hours and exposed to X-ray film for one week. Details are given in Materials and Methods.

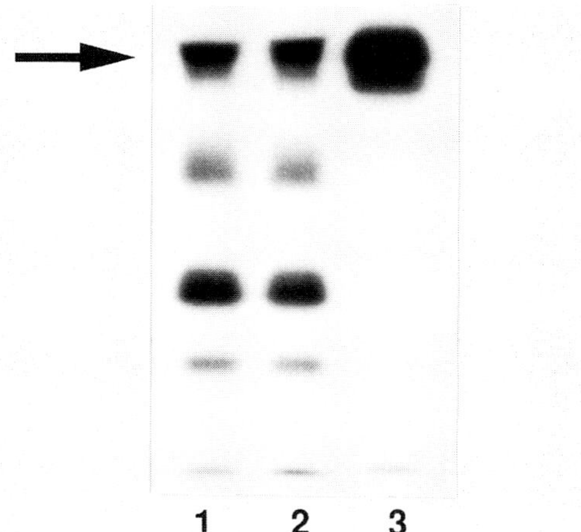

Figure 3. Proteolysis of iodinated IGFBP-3. Pooled sera from term pregnancy (3 μl, lane 1) and concentrated media conditioned by decidual cells (10 μl, lane 2) and trophoblast cells (10 μl, lane 3) were incubated with iodinated IGFBP-3 (50,000 cmp) at 37°C for 5 hours. The mixture was subjected to SDS-polyacrylamide gel electrophoresis and the dried gel was subjected to autoradiography. Details are given in Materials and Methods.

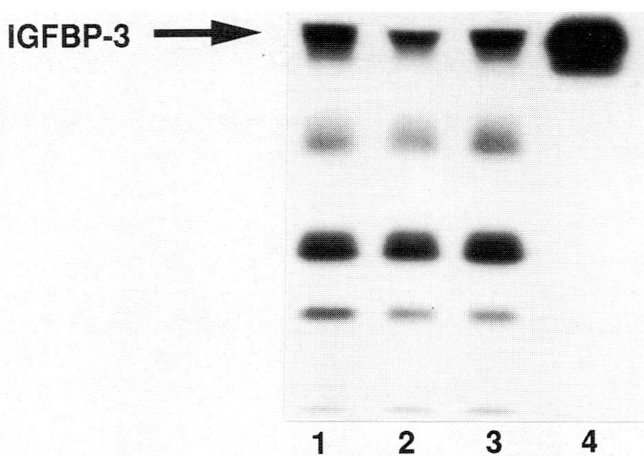

Figure 4. Proteolysis of iodinated IGFBP-3 by medium from decidual cell culture. Decidual cells were incubated in the absence (control, lane 1) or presence of des (1-3) IGF-I (100 nmol/l, lane 2) , estradiol (10 nmol/l, lane 3) , and progesterone (1 μmol/l, lane 4) for 5 days. Pooled media were concentrated and incubated with iodinated IGFBP-3 (50,000 cmp) at 37°C for 5 hours. The mixture was subjected to SDS-polyacrylamide gel electrophoresis and the dried gel was subjected to autoradiography. Details are given in Materials and Methods.

Table 1

Percent Proteolysis of Iodinated IGFBP-3 by Medium From Decidual Cell Culture

	Control	Des (1-3) IGF-I	Estradiol	Progesterone
proteolysis (%)	56 ± 4	75 ± 5[a]	60 ± 5	3 ± 1[b]

The intensity of each band on immunoblot was scanned on a densitometer and the amount of proteolysis was determined as the percentage of the proteolytic cleavage products over the total density in each lane. Data were expresses as mean ± SE of analysis from five separate experiments. [a]p<0.05 vs. control, [b]p<0.0001 vs. control.

DISCUSSION

Human decidual cells are known to produce several IGFBPs. IGFBP-1 is major secretory protein in luteal phase endometrium and decidua (Rutanen et al., 1985, Rutanen et al., 1986). Immunohistochemical study demonstrated that IGFBP-1 was localized to decidualized stromal cells (Waites et al, 1989). It has been presumed that decidual IGFBP-1 was responsible for high levels of IGFBP-1 in amniotic fluid and increased levels in maternal circulation during the first half of pregnancy (Rutanen et al., 1982). As demonstrated in this study and by other investigators (Clemmons et al, 1990), decidual cells produce IGFBP-2. The 24,000 Mr form of IGFBP secreted by decidual cells was identified as IGFBP-4 by immunoblot in our study and others (Myers et al., 1993). While we and other investigators (Myers et al., 1993; Thrailkill et al., 1990) did not detect IGFBP-3 in medium from decidual cell culture by immunoblot and ligand blot, Giudice et al. (1992) reported IGFBP-3 in medium from decidualized endometrium. The reason for this controversial result is not evident, however, one possible explanation is that *in vitro* decidualized endometrial cells and decidual cells obtained from term pregnancy may have different function.

Present study demonstrated that des(1-3)IGF-I suppressed IGFBP-1 and -4 levels in medium. Similar inhibitory effect of IGF-I on IGFBP-1 and -4 release by decidual cells has been reported previously (Thrailkill et al., 1990). Des (1-3) IGF-I which lacks the three amino acid residues at the N-terminus has high affinity for IGF-I receptor equivalent to those of IGF-I and has 50-500 fold less affinity for IGFBPs than IGF-I (Oh et al., 1993). Therefore, suppressive effect of this peptide is presumed to be mediated by IGF-I receptor without being influenced by IGFBPs in medium. Progesterone has been reported to induce IGFBP-1 in the endometrial tissue (Rutanen et al., 1986) and in decidualized endometrial cells (Bell et al., 1991). Progesterone also induced IGFBP-1 production in the endometrium in vivo (Pekonen et al., 1992). These stimulatory effects of progesterone on IGFBP-1 production are confirmed by our result. However, Giudice et al. (Giudice et al., 1992) failed to demonstrate IGFBP-1 increase in the endometrium treated with progesterone. These controversial results may be due to different experimental protocol such as duration of progesterone exposure and dose of progesterone used. Endometrial production of IGFBP-2 and -4 was found to be increased in the presence of progesterone and epidermal growth factor (Giudice et al., 1991, 1992).

We also observed increase in IGFBP-2 and -4 levels in the medium from decidual cell culture under the presence of progesterone. Julkunen et al. (1988) demonstrated expression of IGFBP-1 mRNA in secretory but not in proliferative endometrium and other investigators have shown that IGFBP-2 mRNA was induced in endometrial stromal cells by progesterone (Giudice et al., 1991).

Although these results indicate that progesterone may regulate these IGFBPs at transcriptional level, recent evidence suggests that protease are also involved in the regulation of these IGFBPs in decidua. As demonstrated in this study, decidual cell conditioned medium contains proteolytic activity. The proteolytic activity was first found in sera from pregnancy (Hossenlopp et al., 1990; Giudice et al., 1990). Serum IGFBPs except IGFBP-1 reduced their binding activities during pregnancy when analyzed by Western ligand blot and incubation of IGFBP-3 with sera from pregnant women degraded IGFBP-3 into smaller fragments as shown in Figure 3. IGFBP-3 was cleaved to smaller fragments with similar proteolytic pattern when incubated with medium from decidual cell culture suggesting that both protease in pregnancy serum and in medium may be similar or identical. Myers et al. (Myers et al., 1993) demonstrated that IGF-I decreased IGFBP-1 at transcriptional level while decreased IGFBP-4 by enhancing IGFBP-4 proteolysis. As demonstrated, des (1-3) IGF-I slightly enhanced proteolytic activity in medium and progesterone completely inhibited when analyzed by proteolysis of IGFBP-3. Therefore, IGF-I and progesterone may regulate decidual IGFBPs via protease activity. It is not sufficient to determine whether protease that cleaved iodinated IGFBP-3 in this study also cleaves other IGFBPs. Purification and characterization of protease from decidua are required to elucidate this point.

SUMMARY

Regulation of insulin-like growth factor-binding proteins (IGFBPs) produced by cultured decidual cells were evaluated. Ligand blot and immunoblot detected three distinct IGFBPs in medium conditioned by decidual cell culture including IGFBP-1, -2, and -4. Des (1-3) IGF-I suppressed IGFBP-1 and -4 while progesterone stimulated all three IGFBPs when medium was analyzed by ligand blot. Estradiol had no effect on these IGFBPs release by decidual cells. Proteolytic activity in medium conditioned by decidual cells was also evaluated by the proteolysis of nonglycosylated iodinated IGFBP-3. The medium from trophoblast cells obtained from term pregnancy did not degrade intact IGFBP-3. In contrast, the medium from decidual cells and pooled sera from term pregnancy displayed proteolysis of IGFBP-3 with similar proteolytic pattern. Densitometric analysis revealed medium from decidual cell culture degraded IGFBP-3 by 56%. While incubation of decidual cells with estradiol did not alter proteolytic activity in medium, des (1-3) IGF-I slightly stimulated and progesterone completely inhibited proteolysis of IGFBP-3.

Since IGF-I and gonadal steroids are produced by placenta, the present study indicates that these placental hormones differentially regulate decidual IGFBPs. Furthermore, these hormones may regulate IGFBPs activities in decidual microenvironment by both synthesis and degradation of these IGFBPs.

REFERENCES

Bell, S.C., Jackson, J.A., Ashmore, J., Zhu, H.H. and Tseng, L. (1991) Regulation of insulin-like growth factor-binding protein-1 synthesis and secretion by progestin and relaxin in long term cultures of human endometrial stroma cells. *J. Clin. Endocrinol. Metab.* 72, 1014-1024.

Clemmons, D.R., Thrailkill, K.M., Handwerger, S. and Busby, W.H. (1990) Three distinct forms of insulin-like growth factor binding proteins are released by decidual cells in culture. *Endocrinology* 127, 643-650.

Deal, N.C. and Lamson, G. (1991) Possible decidual origin of human IGFBP-3 protease activity during pregnancy. 73rd Annual Meeting, The Endocrine Society, Abstract 63.

Giudice, L.C., Farrell, E.M., Pham, H., Lamson, G. and Rosenfeld, R.G. (1990) Insulin-like growth factor binding proteins in maternal serum throughout gestation and in the puerperium; effects of a pregnancy-associated serum protease activity. *J. Clin. Endocrinol. Metab.* 71, 806-816.

Giudice, L.C., Milkowski, D.A., Lamson, G., Rosenfeld, R.G. and Irwin, J.C. (1991) Insulin-like growth factor binding proteins in human endometrium:steroid dependent messenger ribonucleic acid expression and protein synthesis. *J. Clin. Endocrinol. Metab.* 72, 779-787.

Giudice, L.C., Dsupin, B.A. and Irwin, J.C. (1992) Steroid and peptide regulation of insulin-like growth factor-binding proteins secreted by human endometrial stromal cells is dependent on stromal differentiation. *J. Clin. Endocrinol. Metab.* 75, 1235-1241.

Han, V.K.M. and Hill, D.J. (1992) *The Insulin-Like Growth Factors: Structure And Biological Functions*, Oxford University Press, pp. 178-220.

Hossenlopp, P., Seurin, D., Segovia-Quinson, B., Hardouin, S. and Binoux, M. (1986) Analysis of serum insulin-like growth factor binding proteins using Western blotting:use of method for titration of the binding proteins and competitive binding studies. *Anal. Biochem.* 154, 138-143.

Hossenlopp, P., Segovia, B., Lassarre, C., Roghani, M., Bredon, M. and Binoux, M. (1990) Evidence of enzymatic degradation of insulin-like growth factor-binding proteins in the 150K complex during pregnancy. *J. Clin. Endocrinol. Metab.* 71:797-805.

Howell, R.J.S., Perry, L.A., Choglay, N.S., Bohn, H. and Chard, T. (1985) Placental protein 12 (PP12): A new test for the prediction of the small-for-gestational-age infant. *Brit. J. Obstet. Gynaecol.* 92, 1141-1144.

Iwashita, M., Watanabe, M., Adachi, T., Ohira, A., Shinozaki, Y., Takeda, Y. and Sakamoto, S. (1989):Effect of gonadal steroids on gonadotropin-releasing hormones stimulated human chorionic gonadotropin release by trophoblast cells. *Placenta* 10, 103-112.

Julkunen, M., Koistinen, R., Aalto-Setälä, K., Seppälä, M. and Jänne, O.A. (1988) Primary structure of human insulin-like growth factor binding protein/placental protein 12 and tissue specific expression of its mRNA. *FEBS Lett.* 236, 295-302.

Laemmli, U.K. (1970) Cleavage of structure proteins during assembly of the head of bacteriophage T4. *Nature* 227, 680-685.

Lamson, G., Giudice, L.C. and Rosenfeld, R.G. (1991) A simple assay for proteolysis of IGFBP-3. *J. Clin. Endocrinol. Metab.* 72, 1391-1393.

Myers, S.E, Cheung P.T., Handwerger, S. and Chernausek, S.D. (1993) Insulin-like growth factor-I (IGF-I) enhanced proteolysis of IGF-binding protein-4 in conditioned medium from primary cultures of human decidua:independence from IGF receptor binding. *Endocrinology* 133, 1525-1531.

Oh, Y., Müller, H.L., Lee, D-Y., Fielder, P.J. and Rosenfeld, R.G. (1993) Characterization of the affinities of insulin-like growth factor (IGF)-binding proteins 1-4 for IGF-I, IGF-II, IGF/insulin hybrid, and IGF-I analogs. *Endocrinology* 132, 1337-1344.

Pekonen, F., Nyman, T., Lähteenmäki, P., Haukkamaa, M. and Rutanen, E-M. (1992) Intrauterine progestin induces continuous insulin-like growth factor-binding protein-1 production in the human endometrium. *J. Clin. Endocrinol. Metab.* 75, 660-664.

Rutanen, E-M., Bohn, H. and Seppälä, M. (1982) Radioimmunoassay of placental protein 12: Levels in amniotic fluid, cord blood and serum of healthy adults, pregnant women and patients with trophoblastic disease. *Am. J. Obstet. Gynecol.* 144, 460-463.

Rutanen, E-M., Koistinen, R., Wahlström, T., Bohn, H., Ranta, T. and Seppälä, M.(1985) Synthesis of placental protein 12 by human decidua. *Endocrinology* 116, 1304-1309.

Rutanen, E-M., Koistinen, R., Sjöberg, J., Julkunen, M., Wahlström, T., Bohn, H. and Seppälä, M. (1986) Synthesis of placental protein 12 by human endometrium. *Endocrinology* 118, 1067-1071.

Rutanen, E-M., Pekonen, F. and Makinen. T. (1988) Soluble 34K binding protein inhibits the binding of insulin-like growth factor to its cell receptors in human secretory phase endometrium-evidence for autocrine/paracrine regulation of growth factor action. *J. Clin. Endocrinol. Metab.* 66;173-180.

Shimasaki, S. and Ling, N. (1992) Identification and molecular characterization of insulin-like growth factor binding proteins (IGFBP-1, -2, -3, -4, -5 and -6). *Prog. Growth Res.* 3, 243-266.

Takeda, Y. and Iwashita, M. (1993) Role of growth factors on fetal growth and maturation. *Annals Acad. Med.* 22, 134-141.

Than, G.N., Csaba, I.F., Szabo, D.G., Bognar, Z.J., Arany, A. and Bohn, H. (1983) Levels of placenta-specific tissue protein 12 (PP12) in serum during normal pregnancy and in patients with trophoblastic tumor. *Arch. Gynecol.* 234, 39-46.

Thrailkill, K.M., Clemmons, D.R., Busby, W.H. and Handwerger, S. (1990) Differential regulation of insulin-like growth factor binding protein secretion from human decidual cells by IGF-I, insulin, and relaxin. *J. Clin. Invest.* 86, 878-883.

Waites, G.T., James, R.F.L. and Bell, S.C. (1989) Human pregnancy-associated endometrial α1-globulin, an insulin-like growth factor-binding protein: Immunohistological localization in the decidua and placenta during pregnancy employing monoclonal antibodies. *J. Endocrinol.* 120, 351-357.

Wilson, D.M., Bennett, A., Adamson, G.D., Nagashima, R.J., Liu, F., DeNatale, M.L., Hintz, R.L. and Rosenfeld, R.G. (1992) Somatomedins in pregnancy: a cross-sectional study of insulin-like growth factors I and II and somatomedin peptide content in normal human pregnancies. *J .Clin. Endocrinol. Metab.* 55, 858-891.

Trophoblast Research 9:63-74, 1997

Bcl-2 EXPRESSION AND APOPTOSIS IN HUMAN TROPHOBLAST

Noriaki Sakuragi, Min-Lian Luo, Itsuko Furuta, Hidemichi Watari, Norihiko Tsumura, Masashi Nishiya, Koji Hirahatake, Naoki Takeda, Toshihiro Ohkouchi, Hiroshi Ishikura[1], and Seiichiro Fujimoto

Department of Obstetrics and Gynecology
[1]Department of Pathology I
Hokkaido University School of Medicine
Sapporo, Japan

INTRODUCTION

The placenta is often referred to as 'pseudomalignant' because of its invasiveness, rapid cell proliferation and immune privilege (Ohlsson et al., 1993). Cytotrophoblast cells show marked proliferative activity early in pregnancy. They eventually differentiate and fuse to form quiescent syncytiotrophoblast cells which are terminally differentiated and have a major role in synthesis of many steroid and peptide hormones and in metabolic functions of the placenta.

Cell numbers of a tissue are regulated by a balance between proliferation, growth arrest and cell death. Apoptosis is the process of programmed cell death which is responsible for morphogenesis in vertebrate development, for the cell loss that accompanies atrophy of adult tissues, and for the death of cells in the course of certain tissue turnover (Wyllie, 1993). Bcl-2 is a cellular proto-oncogene which encodes a 26-kD protein, localized to mitochondrial membranes, endoplasmic reticulum, perinuclear membranes and the nucleus, that is believed to block apoptotic cell death and promote cell survival (Hockenbery et al., 1990; Garcia et al., 1992; Bissonnette et al., 1992; Lu et al., 1994). There are few studies on the expression of Bcl-2 in human trophoblast cells, and its relation to the proliferation and differentiation of human trophoblasts is unknown. We reported in a previous paper that Bcl-2 mRNA and protein are expressed in normal human trophoblast cells (Sakuragi et al., 1994). The aim of this study was to explore whether apoptotic cell death occurs in trophoblast cells and whether it is related to Bcl-2 expression.

MATERIALS AND METHODS

Tissue Preparations

Normal term placental tissues (four different placentae) were collected immediately after uncomplicated deliveries. First-trimester placental tissues (four different specimens), a second-trimester placental specimen and a complete hydatidiform mole (15-week of gestation) tissue were obtained after therapeutic pregnancy termination. Placental tissues and complete hydatidiform mole tissue were either fixed in 10% formalin or snap-frozen in liquid nitrogen and stored at -80°C.

Immunohistochemical Localization of Bcl-2 Protein and Proliferating Cell Nuclear Antigen (PCNA)

Frozen tissues were used for the immunohistochemical localization of Bcl-2 protein. Five µm cryotome sections were prepared from frozen placental tissue and hydatidiform mole embedded in OCT compound (Miles Laboratory). Sections were air-dried for 120 minutes and then fixed with acetone at -20°C for 10 minutes. After 10 minutes of air-drying, sections were immersed in methanol with 0.6% H_2O_2 (-20°C, 10 minutes) to block endogenous peroxidase activity. The sections were then washed with phosphate-buffered saline (PBS, 10 mM, pH 7.2) at room temperature for 5 minutes. After incubation with normal rabbit serum (37°C, 10 minutes) to block nonspecific binding, the sections were incubated with mouse anti-human Bcl-2 monoclonal antibody (clone 124, DAKO Corp.) diluted with PBS containing 1% bovine serum albumin to 1:20 for 20 minutes at 37°C. Following the incubation, sections were washed with PBS and incubated with biotin-bound anti-mouse IgG rabbit antibody (37°C, 10 minutes). After washing with PBS, sections were incubated with horseradish peroxidase-conjugated streptavidin (37°C, 5 minutes) followed by washing with PBS. The sections were then incubated with 3-amino-9-ethylcarbazole (AEC) at room temperature for 10 minutes. After washing with water for 5 minutes, sections were counterstained with hematoxylin, washed with water and mounted with coverslips.

PCNA is a 36-kD nuclear protein which appears mainly in the late G1 and S phase. It has been employed as a marker of cell proliferation (Mathews et al., 1984; Hall et al., 1991). Two of four first trimester placentae, two of four term placentae and a complete hydatidiform mole tissues were used for immunohistochemical detection of PCNA. Four µm sections were prepared from paraffin-embedded placenta and complete hydatidiform mole tissues. Sections were deparaffinized and were immersed in methanol with 0.6% H_2O_2 (-20°C, 10 minutes) to block endogenous peroxidase activity. The sections were then washed with phosphate-buffered saline (PBS, 10 mM, pH 7.2) at room temperature for 5 minutes. After incubation with normal rabbit serum (37°C, 10 minutes) to block nonspecific binding, the sections were incubated with mouse anti-PCNA monoclonal antibody (clone PC 10, DAKO Corp.) diluted with PBS containing 1% bovine serum albumin to 1:50 for 20 minutes at 37°C. Following the incubation, sections were washed with PBS and incubated with biotin-bound anti-mouse IgG rabbit antibody (37°C, 10 minutes). After washing with PBS, sections were incubated with horseradish peroxidase-conjugated streptavidin (37°C, 5 minutes) followed by washing with PBS. The sections were then incubated with 3-amino-9-ethylcarbazole (AEC) at room temperature for 10 minutes. After washing with water for 5 minutes, sections were counterstained with hematoxylin, washed with water and mounted with coverslips. More than one thousand nuclei of trophoblast cells (positive nuclei were observed in cytotrophoblast cells) were counted and the number of positive nuclei per 1000 trophoblastic nuclei was calculated (PCNA index).

Negative control study was carried out using Mouse IgG_1 (DAKO Corp.) instead of Bcl-2 antibody. Mouse IgG_1 was adjusted to the same concentration as Bcl-2 antibody. For PCNA staining, negative control slide was processed with primary antibody.

3'-End Labeling of Fragmented DNA

Apoptosis in trophoblast cells were examined in two of four first trimester placentae, two of four term placentae and hydatidiform mole. The fragmented DNA in apoptotic cells was detected using the DNA 3'-end labeling technique (Gavrieli et al., 1992) Four μm sections of sample tissues were prepared from paraffin-embedded specimens. After deparaffinization, the sections were incubated with 20 mg/ml proteinase K for 20 minutes at room temperature. The slides were washed in double distilled water. Endogenous peroxidase activity was blocked by 3% H_2O_2 for 5 minutes at room temperature. The sections were rinsed with distilled water, and immersed in TdT (terminal deoxynucleotidyl transferase) buffer (30 mM Trizma base, pH 7.2, 140 mM sodium cacodylate, 1 mM cobalt chloride). TdT (0.3 e.u./ml) and biotinylated dUTP in TdT buffer was then added to cover the sections, which were incubated at 37°C for 30 minutes. The reactions were terminated with TB buffer (300 mM sodium chloride, 30 mM sodium citrate) for 30 minutes at room temperature. The sections were rinsed with distilled water, covered with 2% BSA for 10 minutes at room temperature, rinsed in distilled water, and immersed in PBS for 5 minutes. The sections were reacted with peroxidase-conjugated streptavidin for 30 minutes at room temperature, washed with distilled water, immersed in PBS, and stained with diaminobenzidine (DAB). Cells with condensed chromatin and pyknotic nuclei which were strongly stained with 3'-end labeling were regarded as being in the process of apoptosis. More than 5000 nuclei of villous trophoblast cells and trophoblast cluster attaching to villi, which was observed in normal placentae of 5 and 7 week of gestation and complete hydatidiform mole, were counted with X40 objective lens (X400 magnification). Observation field were selected so that the marginal area of stained specimen was not included (more than half of the tissue was incorporated by the procedure). The number of the apoptotic nuclei per 1000 trophoblastic nuclei (Apoptosis index) was calculated.

A serial section was stained with hematoxylin-eosin and was served for identification of trophoblast cells. Cytotrophoblast cells were discriminated from syncytiotrophoblast cells with its frequent positivity for PCNA.

RESULTS

Detection of Bcl-2 Expression in Complete Hydatidiform Mole and Normal Placenta

We previously reported that there was a differentiation-dependent expression of Bcl-2 in the human placenta. In this study we repeatedly examined Bcl-2 expression in normal placentae of various gestational stages because we wanted to compare Bcl-2 expression and apoptosis index on the same tissue specimen. Bcl-2 protein was localized by immunoperoxidase staining using a monoclonal antibody. Weak to modest immunostaining was observed in a part of villi of hydatidiform mole (Figure 1A) and the first trimester placentae (Figure 1B). Bcl-2 seemed to localize to syncytiotrophoblast. More uniform, stronger staining was observed in the syncytiotrophoblast cytoplasm of the second trimester (Figure 1C) and the term placentae (Figure 1D). A focally prominent reaction was seen in the syncytiotrophoblast of the term placenta (Figure 1D). Placenta of 42-week of gestation showed reduced staining for Bcl-2 in syncytiotrophoblast although there was a focal strong reaction for Bcl-2. We could not see apparently positive staining

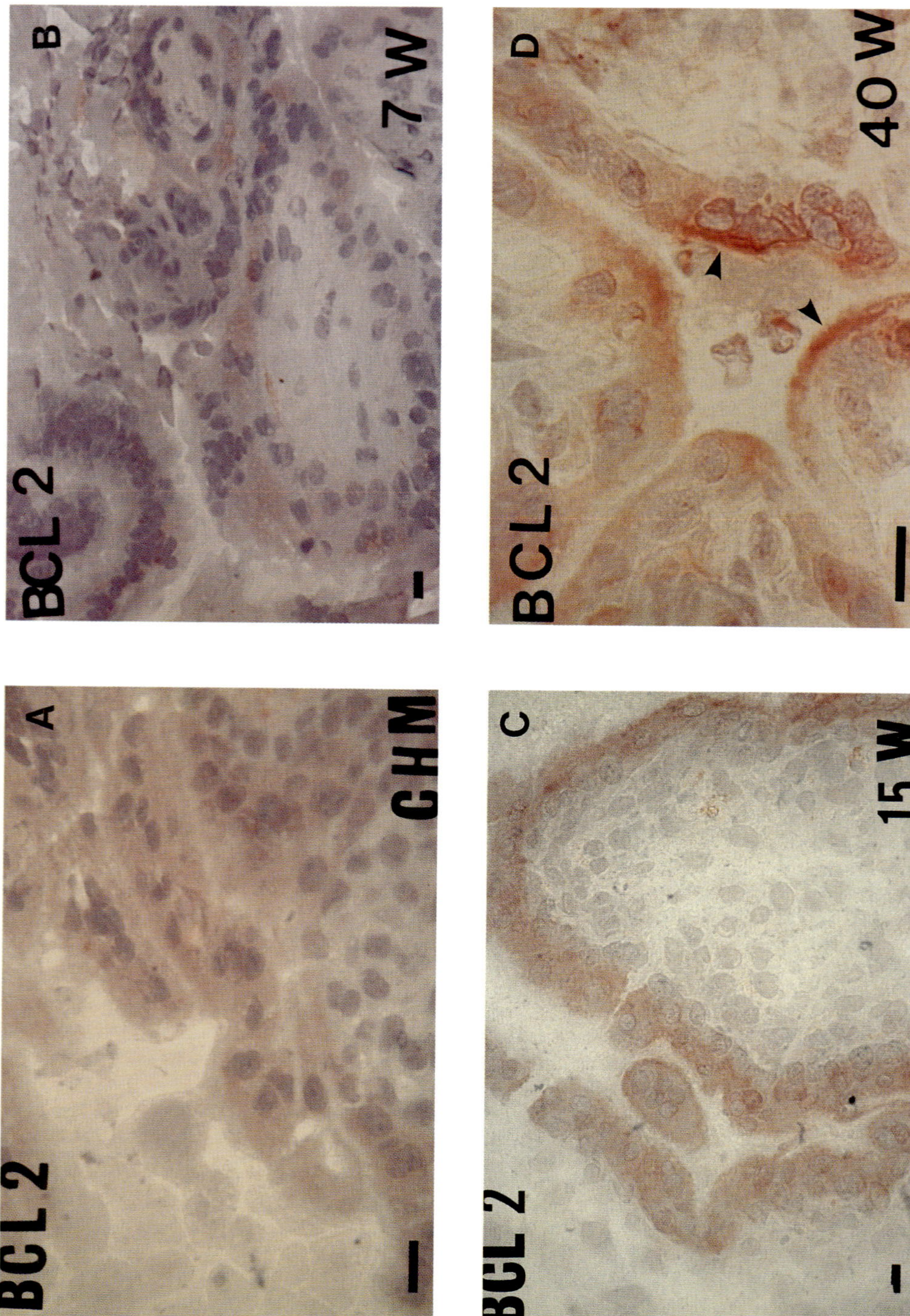
BCL 2
A
CHM
BCL 2
B
7 W
BCL 2
C
15 W
BCL 2
D
40 W

for Bcl-2 in the cytotrophoblast cells at any gestational stage. These results confirmed our previous observation on the localization of Bcl-2 protein in the human placenta which was done using paraffin-embedded specimens (Sakuragi et al., 1994).

Apoptotic Trophoblast Cells and Proliferation of Trophoblasts in Complete Hydatidiform Mole

Serial sections from complete hydatidiform mole tissue were stained with hematoxylin-eosin (Figure 2A) and 3'-end labeling (Figure 2B). The apoptotic nuclei seemed to be in the syncytiotrophoblast. Some apoptotic nuclei formed clusters (Figure 2B; arrow) which looked like so-called "syncytial sprout". This is consistent with the notion that some syncytial sprout is in the degenerative process (Benirschke and Kaufmann, 1990). 3'-end labeling-positive nuclei showed chromatin condensation and nuclear fragmentation (Figure 2C). We could not observe the obvious apoptotic nuclei in cytotrophoblasts.

Some cytotrophoblastic cells were positively stained for PCNA (Figure 2D). Both apoptosis index and PCNA index in complete hydatidiform mole were higher than those in the normal placenta (Table 1).

Apoptotic Trophoblast Cells and Proliferation of Trophoblasts in Normal Placenta

The apoptotic nuclei were observed mainly in the syncytiotrophoblasts of normal placenta (Figure 3A and B) and they appeared mostly in clusters (Figure 3B). Apoptosis was rarely detected in the cytotrophoblast cells of normal placenta. The incidence of apoptosis in the first trimester placentae tended to be higher than in the term placentae. The proliferative activity of trophoblasts determined by the PCNA index decreased as gestational age increased.

The results of apoptosis index and PCNA index were summarized in Table 1. Apoptosis index seemed to be higher in the trophoblast cells which have higher PCNA index.

Figure 1. (facing pages) Immunoperoxidase localization of Bcl-2 protein in the sections of complete hydatidiform mole (A), the first (B; 7-week), the second (C; 15-week) and the third (D; 40-week) trimester placenta. Cytoplasm of the syncytiotrophoblast reacted positively. A focal strong staining was seen in the syncytiotrophoblast of term placenta (D; arrowhead). Bar = 10 μm.

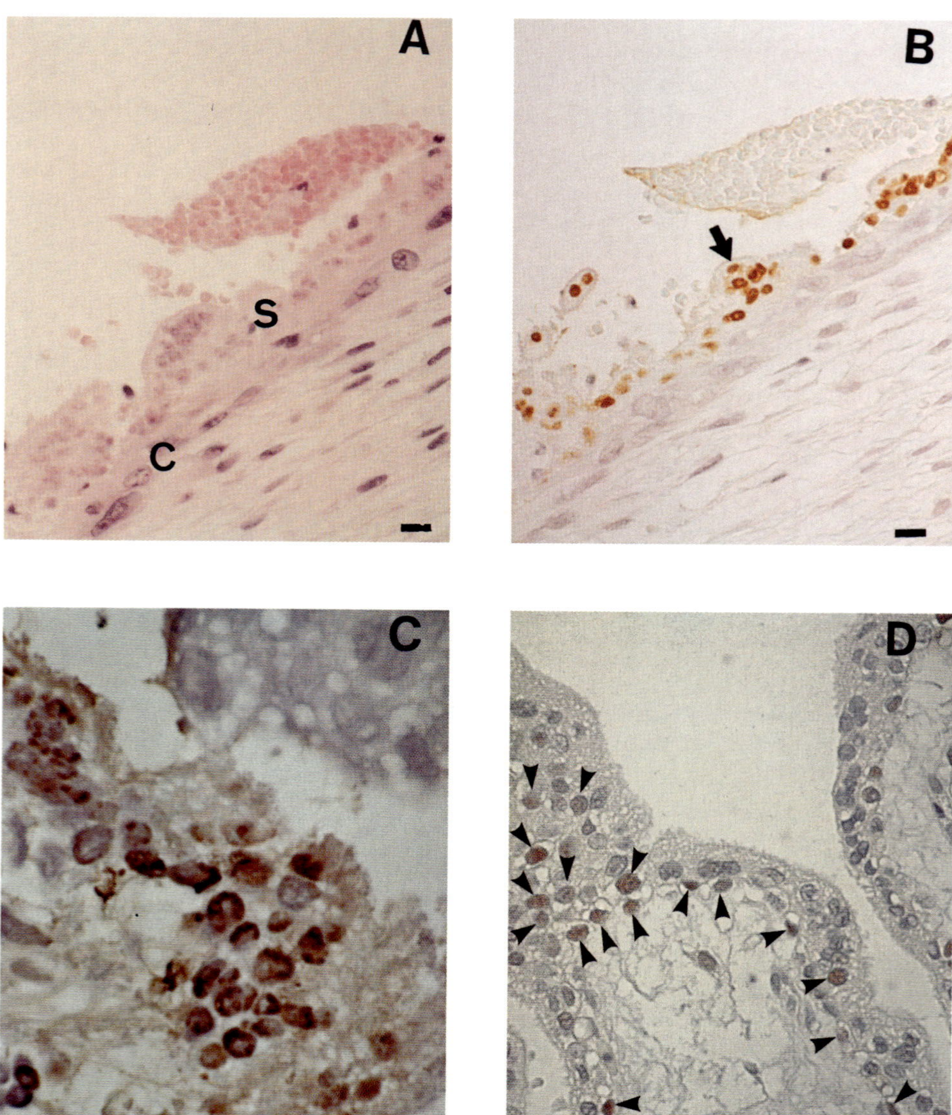
A
S
C
B
C
D

Table 1

Trophoblastic tissue specimens and the result of Bcl-2 immunostaining. Some of the specimens were used for determination of apoptosis index and PCNA index. apoptosis index: number of positive nuclei per 1000 nuclei, PCNA index: number of positive nuclei per 1000 nuclei, ±: weak to modest staining in a part of the trophoblast cells, +: positive staining in a large part of the trophoblasts, ++: strong staining

Specimen	Bcl-2 staining	Apoptosis index	PCNA index
Complete Hydatidiform Mole			
15 week	±	34.6	329.6
Normal Placenta			
5 week	±	10.5	150.1
7 week	±	4.6	168.7
9 week	±	N.D.*	N.D.
10 week	±	N.D.	N.D.
15 week	+	N.D.	N.D.
38 week	+ (focally ++)	N.D.	N.D.
40 week	+ (focally ++)	N.D.	N.D.
40 week	+ (focally ++)	3.6	82.0
42 week	± (focally ++)	1.8	82.4

*: not determined

Figure 2. Complete hydatidiform mole of 15-week gestation. Serial sections were stained with hematoxylin-eosin (A) and 3'-end labeling (B). Chromatin condensation and nuclear fragmentation were observed in the section stained with 3'-end labeling (C). Tissue section was also stained for PCNA (D). C: cytotrophoblastic cells, S: syncytiotrophoblasts, arrow in B: cluster of 3'-end-labeling positive nuclei, arrowhead in C: PCNA positive cells. Bar = 10 μm.

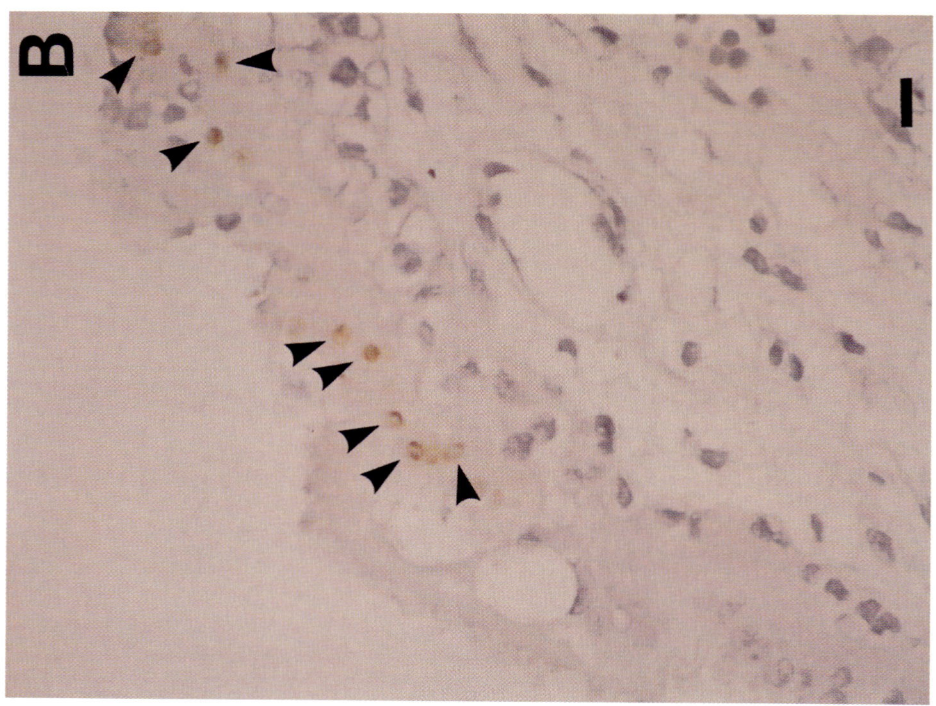
B

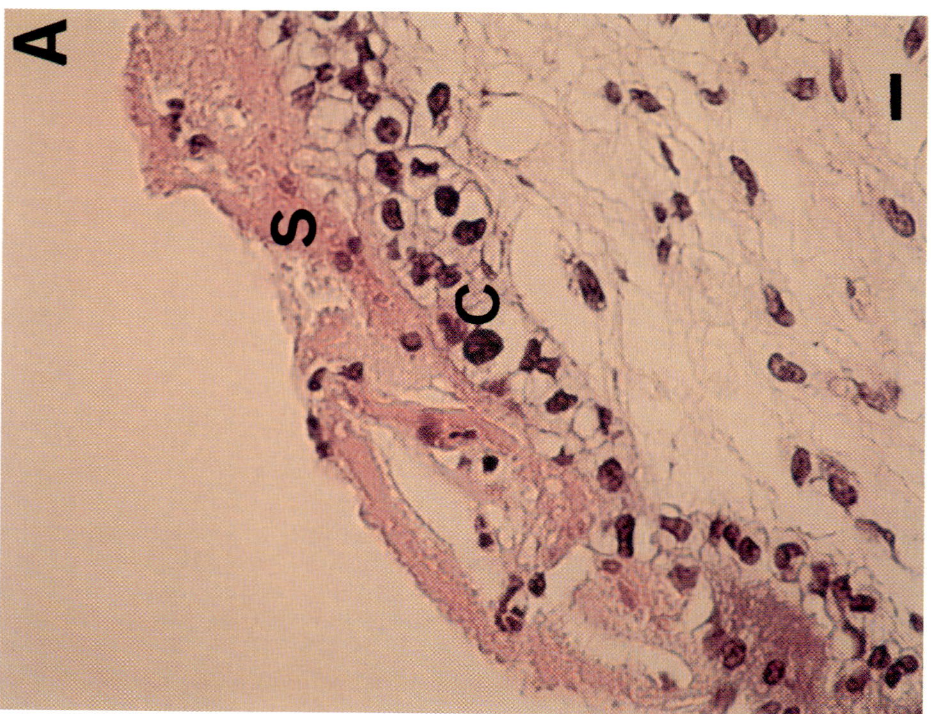
A
S
C

DISCUSSION

In a previous study, we reported that: 1) There is a differentiation-dependent pattern of Bcl-2 expression in the placenta, with the protein being most abundant in terminally differentiated trophoblast cells; 2) there appears to be an inverse relationship between Bcl-2 and p53 expression in trophoblast cells; and 3) cAMP regulates Bcl-2 protein in trophoblast cells (Sakuragi et al., 1994). In the present study, we observed that the proliferative activity of trophoblast cells which was shown as the ratio of proliferative cytotrophoblast to villous trophoblast declines in the late pregnancy when syncytiotrophoblast cells prevail in the trophoblast population in placenta, which is consistent with the findings of Kaufmann (1990). Apoptotic cell death of syncytiotrophoblast cells was found less frequently in the late pregnancy period. From these findings, we speculate that the high levels of Bcl-2 protein in syncytiotrophoblast may prevent or slow the death of these terminally differentiated cells. This would permit maintenance of trophoblastic cell mass (i.e., placental mass) as gestation progresses and the proliferative activity of cytotrophoblast decreases.

From the data of the PCNA index and the apoptosis index shown here, it seemed that the incidence of apoptosis was balanced with the proliferative activity of the trophoblasts. Apoptotic cell death occurs in the course of certain tissue turnover, some of which seems to be related to cellular senescence (Appleby and Modak, 1977). Apoptosis is also triggered by p53 in specific cells (Yonish-Rouach et al., 1991; Shaw et al., 1992; Lane, 1992). p53 plays an important role in apoptosis-inducing mechanism related with DNA damage (Lowe et al., 1993; Clarke, et al., 1993). Western blot analysis revealed the expression of p53 protein and Northern blot analysis showed p53 mRNA in placenta in the first trimester (Sakuragi et al., 1994). The incidence of DNA mutations is proportional to the proliferation rate (Parodi, 1992). Although we do not have direct evidence, p53 may function during the time of a rapid placental growth in the first trimester. Bcl-2 protein blocks p53-induced apoptosis (Hoffman and Liebermann, 1994). Thus, if we suppose that p53 functions in the apoptosis-inducing mechanism in the trophoblast cells, the low Bcl-2 protein expression in the first trimester would permit p53 to perform the cell death-promoting function.

Recent studies show that the Bcl-2 gene is under the control of wild-type p53 (Miyashita et al., 1994; Hoffman and Liebermann, 1994); P53 directly downregulates Bcl-2 expression. It has been shown that Bcl-2 is only a member of a large family of homologous proteins. A subgroup of Bcl-2 related gene, comprised of Bax (Oltvai et al., 1993), Bad (Yang et al., 1995), Bak (Chittenden et al., 1995; Farrow et al., 1995; Kiefer et al., 1995), antagonizes the apoptosis-suppressing effect of Bcl-2. Bax, which has been shown to be a p53-immediate early response gene, also downregulates the expression of Bcl-2. The present study does not elucidate whether Bcl-2 related proteins other than Bcl-2 are expressed in human trophoblast cells.

Figure 3. Normal placenta of 7-week gestation. Serial sections were stained with hematoxylin-eosin (A) and 3'-end labeling (B). C: cytotrophoblastic cells, S: syncytiotrophoblast cells, arrowhead in B: 3'-end labeling-positive nuclei in syncytiotrophoblast. Bar = 10 μm

Although the number of samples studied was small, and we have not examined the expression of p53 in complete hydatidiform mole, our previous results and this study suggest that there is a differentiation-dependent Bcl-2 and p53 expression in human trophoblast, which seems to be related to the apoptotic death of the trophoblast cells.

SUMMARY

The function of Bcl-2 in the differentiation and proliferation of human trophoblast cells is not yet elucidated. The expression of Bcl-2 in human placenta was examined immunohistochemically using a monoclonal antibody to Bcl-2 and apoptosis of trophoblast cells was investigated by the 3'-end labeling method to detect fragmented DNA. Bcl-2 protein was localized in syncytiotrophoblast and its amount seemed to increase as pregnancy advanced. The proliferation of trophoblast cells decreased by term, and apoptotic cell death was found less frequently in the placenta at term. This suggests that the higher expression of Bcl-2 protein in late pregnancy may prolong the life of trophoblast cells so that the placental mass is maintained although proliferactive activity of trophoblast cells has decreased. Combined with our previous result that there is an inverse relation between p53 protein expression and Bcl-2 protein expression, apoptotic death of trophoblast cells may be under the control of p53 and Bcl-2 proteins.

ACKNOWLEDGEMENTS

The authors wish to express gratitude to Dr. Jerome F. Strauss, III, University of Pennsylvania for his review of the manuscript.

REFERENCES

Appleby, D.W. and Modak, S.P. (1977) DNA degradation in terminally differentiating lens fiber cells from chick embryos. *Proc. Natl. Acad. Sci.* 74, 5579-5583.

Benirschke, K. and Kaufmann, P. (1990) *Pathology of the Human Placenta.* New York, Springer-Verlag.

Bissonnette, R.P., Echeverri, F., Mahboubi, A. and Green, D.R. (1992) Apoptotic cell death induced by c-myc is inhibited by Bcl-2. *Nature* 359, 552-554.

Chittenden, T., Harrington, E.A., O'Connor, R., Flemington, C., Lutz, R.J., Evan, G.I. and Guild, B.C. (1995) Induction of apoptosis by the Bcl-2 homologue Bak. *Nature* 374, 733-736.

Clarke, A.R., Purdie, G., Harrison, D.J., Morris, R.G., Bird, C.C., Hooper, M.L. and Wyllie, A.H. (1993) Thymocyte apoptosis induced by p53-dependent and independent pathways. *Nature* 362, 849-852.

Farrow, S.N., White, J.H.M., Martinou, I., Raven, T., Pun, K.T., Grinham, C.J., Martinou, J.C. and Brown, R. (1995) Cloning of a Bcl-2 homologue by interaction with adenovirus E1B 19K. *Nature* 374, 731-733.

Garcia, I., Martinou, I., Tsujimoto, Y. and Martinou, J-C. (1992) Prevention of programmed cell death of sympathetic neurons by the Bcl-2 proto-oncogene. *Science* 258, 302-304.

Gavrieli, Y., Sherman, Y. and Ben-Sasson, S.A. (1992) Identification of programmed cell death in situ via specific labeling of nuclear DNA fragmentation. *J. Cell Biol.* 119, 493-501.

Hall, P.A., Levison, D.A., Woods, A.L., Yu, C.C-W., Kellock, D.B., Watkins, J.A., Barnes, D.M., Gillett, C.E., Camplejohn, R., Dover, R., Waseem, N.H. and Lane, D.P. (1991) Proliferating cell nuclear antigen (PCNA) immunolocalization in paraffin sections. An index of cell proliferation with evidence of deregulated expression in some neoplasms. *J. Pathol.* 165, 356-357.

Hockenbery, D., Nunez, G., Milliman, C., Schreiber, R.D. and Korsmeyer, S.J. (1990) Bcl-2 is an inner mitochondrial membrane protein that blocks programmed cell death. *Nature* 348, 334-336.

Hoffman, B. and Liebermann, D.A. (1994) Molecular controls of apoptosis: Differentiation/growth arrest primary response genes, proto-oncogenes, and tumor suppressor genes as positive and negative modulators. *Oncogene* 9, 1807-1812.

Kaufmann, P. (1972) Untersuchungen uber die Langhanszellen in der menschlichen Placenta. *Z. Zellforsch* 128, 283-302.

Kiefer, M.C., Brauer, M.J., Powers, V.C., Wu, J.J., Umansky, S.R., Tomei, L.D. and Barr, P.J. (1995) Modulation of apoptosis by the widely distributed Bcl-2 homologue Bak. *Nature* 374, 736-739.

Lane, D.P. (1992) p53, guardian of the genome. *Nature* 358, 15-16.

Lowe, S.W., Schmitt, E.W., Smith, S.W., Osborne, B.A. and Jacks, T. (1993) p53 is required for radiation-induced apoptosis in mouse thymocytes. *Nature* 362, 847-849.

Lu, Q-L., Hanby, A.M., Nasser Hajibagheri, M.A., Gschmeissner, S.E., Lu, P-J., Taylor-Papadimitriou, J., Krajewski, S., Reed, J.C. and Wright, N.A. (1994) Bcl-2 protein localizes to the chromosomes of mitotic nuclei and is correlated with the cell cycle in cultured epithelial cell lines. *J. Cell Sci.* 107, 363-371.

Mathews, M.B., Bernstein, R.M., Franza Jr., B.R. and Garrels, J.I. (1984) Identity of proliferating cell nuclear antigen and cyclin. *Nature* 309, 374-376.

Miyashita, T., Krajewski, S., Krajewska, M., Wang, H.G., Lin, H.K., Liebermann, D.A., Hoffman, B. and Reed, J.C. (1994) Tumor suppressor p53 is a regulator of *Bcl-2* and *Bax* gene expression in vitro and in vivo. *Oncogene* 9, 1799-1805.

Ohlsson, R., Glaser, A., Holmgren, L., and Franklin, G. (1993) The molecular biology of placental development. In: The Human Placenta: Human Placental Development: A 'Pseudomalignant' Process, (eds.) C.W.G. Redman, I.L. Sargent, and P.M. Starkey, London. Blackwell Scientific Publications, pp. 34-39.

Oltvai, Z.N., Milliman, C.L. and Korsmeyer, S.J. (1993) Bcl-2 heterodimerizes in vivo with a conserved homolog, Bax, that accelerates programmed cell death. *Cell* 74, 609-619.

Parodi, S. (1992) AACR/EACR first joint conference: Concepts and molecular mechanisms of multistage carcinogenesis. *Ann. Oncol.* 3, 357-359.

Sakuragi, N., Matsuo, H., Coukos, G., Furth, E.E., Bronner, M.P., VanArsdale, C.M., Krajewski, S., Reed, J.C. and Strauss, III, J.F. (1994) Differentiation-dependent expression of the BCL-2 proto-oncogene in the human trophoblast lineage. *J. Soc. Gynecol. Invest.* 1, 164-172.

Shaw, P., Bovey, R., Tardy, S., Sahli, R., Sordat, B. and Costa, J. (1992) Induction of apoptosis by wild-type p53 in a human colon tumor-derived cell line. *Proc. Natl. Acad. Sci. USA* 89, 4495-4499.

Wyllie, A.H. (1993) Apoptosis (The 1992 Frank Rose Memorial Lecture). *Br. J. Cancer* 67, 205-208.

Yang, E., Zha, J., Jockel, J., Boise, L.H., Thompson, C.B. and Korsmeyer, S.J. (1995) Bad, a heterodimeric partner for Bcl-XL and Bcl-2, displaces Bax and promotes cell death. *Cell* 80, 285-291.

Yonish-Rouach, E., Resnitzky, D., Lotem, J., Sachs, K., Kim-Chi, A. and Oren, M. (1991) Wild-type *p53* induces apoptosis of myeloid leukaemic cells that is inhibited by interleukin-6. *Nature* 352, 345-347.

Trophoblast Research 9:75-85, 1997

SOME GESTATIONAL CHANGES IN PLACENTAL TRANSFER OF IONS
- A Review -

Robert Boyd, Jocelyn Glazier, Susan Greenwood and Colin Sibley

Department of Child Health
University of Manchester
St. Mary's Hospital
Hathersage Road
Manchester M13 0JH, United Kingdom

INTRODUCTION

It has been known that there is an anatomical barrier between the maternal and fetal circulation at least since the time of William Hunter (Boyd and Hamilton, 1970), a conclusion which depended on injection of the uterine circulation using the, then, high technology method of wax injection. It has also been known for nearly a century that there are different concentrations of some important nutrient solutes between maternal and fetal blood. The differences were reviewed by Needham (1931). He recorded, for example, that calcium had a higher concentration in fetal than in maternal plasma in woman, rabbit, dog, cow and pig by contrast with glucose which, in most species, was of lower concentration in the fetal circulation. He also emphasized the striking increase in rate of daily calcium accretion which takes place as gestation proceeds; two hundred-fold between the third and the ninth month in the human (Table 1). Flexner, in a series of classical studies using radioisotopes demonstrated dramatic gestational increases in the rate of sodium transfer to the fetus per gram of placenta in several species (summarized in a widely quoted review, Flexner and Gellhorn, 1942). Later work in women showed similar trends (Flexner et al., 1948). The purpose of this brief review is to revisit some gestational issues in placental transfer of ions in the light of recent data. In doing so, the well known morphological differences in the human between early gestation when there is a complete cytotrophoblastic layer between villous capillary and syncytiotrophoblast and term when fetal capillary and intervillous space may be in very close proximity and separated only by syncytiotrophoblast, basement membrane and capillary endothelium should be remembered. There is also the replacement of chorion frondosum by the definitive placenta (Boyd and Hamilton, 1970) and, near the end of the first trimester, the onset of maternal intervillous space circulation (Jauniaux et al., 1995) or, at least, of more vigorous circulation within the intervillous space (Moll, 1995).

Driving Forces and Mechanisms for Placental Transfer

Transport across the exchange barrier depends on the driving forces and on the routes or mechanisms available for molecules to move between the two circulations (Sibley and Boyd 1988). The most important forces that can drive a solute across the barrier appear to be, firstly, any differences between its chemical concentration (or more precisely activity) in the plasma water in the intervillous space and in the fetal capillary. The concentration in plasma water depends on the total plasma concentration and also on whether the nutrient is bound in some way in the circulation and on the kinetics of release from this binding. For molecules which cross the placenta rapidly it also depends on the rate and geometry of blood flows past the exchange area. Such dependence raises the important issue of flow limitation in placental transfer, a topic not considered here but well reviewed in Faber and Thornburg (1983). Secondly, electrical forces are important for charged solutes (e.g., sodium, chloride, potassium or calcium) or for molecules co-transported in electrogenic fashion (e.g., alanine with sodium using the system A amino acid transporter).

The mechanisms or routes which are available for transfer across the term syncytiotrophoblast are (Morriss et al., 1994) (a) *lipophilic diffusion*, particularly relevant to anesthetic gases and oxygen and carbon dioxide whose transfer is often flow limited. (b) *Receptor mediated endocytosis*: particularly important for immunoglobulin G and for iron. (c) *Facilitated diffusion*: in which permeability through the rate limiting lipid bilayer membrane of syncytiotrophoblast is enhanced by, for example, a glucose transporter. (d) *paracellular diffusion*, and (e) *active transport*.

In this review we focus on mechanisms (d) and (e), the key routes for ion transfer and, particularly, on the relationship of gestational age to the electrical forces present across the syncytiotrophoblast and their possible effects. Data is partly drawn from human studies but reference is also made to other species including the rat whose placental transport has recently been more fully reviewed by Sibley (1994). The rat placenta, like the human, is hemochorial, but its trophoblast is trilayered and fetal growth is of course faster proportionately than human fetal growth during its short 22-day gestation period.

Although some pieces in the puzzle which makes up placental ion transport are clear, how the puzzle is put together to make a coherent picture is, as yet, beyond us. So are the clinical implications of most placental transport studies. These may be long term, as well as short term, as has become clear through the work of Barker and colleagues (e.g., Barker, 1993).

Table I

Fetal Calcium OxideAccretion

Month	Daily Increment (mg)
3	1.5
4	10.0
5	4.1
6	10.0
7	192.7
8	207.0
9	399.0

From Needham 1931 p. 1285, after Schmitz, 1923

Relation Of Electrical Forces To Gestation

It has been known since 1958 that electrodes inserted into maternal and fetal tissues or vasculature and connected to a voltmeter give rise to a measurable deflection; the 'materno-fetal potential difference (pd)'. However, there are differences in both the sign and magnitude of this pd between species (Table II). The materno-fetal pd may not be generated in the placenta (McNaughton and Power, 1991). Its value also bears no simple relation to the transplacental steady state ion distribution and thus, presumably, to transplacental fluxes; this has led to the general belief that the pd is not expressed at the placental exchange area across which it would otherwise influence ion fluxes (Binder et al., 1978; Thornburg et al., 1979). However, in the sheep, the *in vivo* feto-maternal flux of, at least, iodide is a linear function of materno-fetal pd (Canning et al., 1986) so there does appear to be some relationship.

It is not widely remembered that there are reports of substantial gestational falls in the magnitude of the materno-fetal pd in some species by some investigators (Table III). Neither the mechanism nor any implication for ion transport of this fall is clear.

Because of the uncertainty as to whether materno fetal pd relates to transplacental pd across the materno-fetal exchange area, and thus to ion transport, direct measurement of transtrophoblast pd would be helpful. We (Sibley et al., 1986) and others have used pig placenta mounted in Ussing chambers as one approach to this. Such studies suggest that active sodium transport generates a local transplacental pd. While catecholamines stimulate this transplacental pd *in vitro* and also the materno-fetal pd *in vivo* (Boyd et al 1989) the magnitude and indeed the polarity of the pd may be very different between these two states. This supports the general thesis of Binder et al (1978). In the pig there are some gestational changes in its electrical behavior *in vitro* (Rice et al., 1991) but no *in vivo* data are available.

Table II

Materno-fetal pd
(mean ± SD; mV; mother as zero)

goat	-79 ± 9	Mellor	1970
sheep	-25 ± 20	Weedon	1978
guinea pig	-21 ± 6	Stulc	1972
rat (pre-term)	+17 ± 3	Mellor	1969
rabbit	NS	Mellor	1969
Human; pre-term	3 ± 1	Mellor	1969
term	NS	Stulc	1978

From Sibley and Boyd, 1988

Table III

Materno Fetal pd: decline with gestation (mother taken ref. zero)

Rat	15-20 days, + 15 mV; day 22, 0 mV (Mellor, 1969)
Guinea Pig	decline with fetal weight (Binder et al., 1978) probable decline: (Stulc et al., 1972, Figure 1) No change: (-18 mV) (Mellor, 1969)
Goat	79 days, - 133 mV; 131 days - 25 mV (Meschia et al., 1958) no change (Mellor, 1970)
Sheep	Pd = -164 ± gest. age (days) $p<0.05$ (Weedon et al., 1980) no change (Mellor, 1970)
Human	105 - 154 days, -2.7 + 0.4 mV (Stulc et al., 1978) term -0.3 ± 1.1 mV (Mellor et al., 1969)

In the human, materno-fetal pd has been measured (Table III). It is low in mid gestation and insignificant at term. Pd across the syncytiotrophoblast of isolated human villi has also been measured, more directly than has been done in any other species, using standard microelectrode techniques. Its value is not normally distributed but its mean of -3mV, fetal side negative, is significantly different from zero (Greenwood et al., 1993). Interestingly, this is in the range of the *in vivo* values for woman reported in mid gestation in Table III. To date, such measurements have only been made across syncytiotrophoblast of mature intermediate villi which site may or may not be the location of the materno-fetal exchange of ions.

The transmembrane potential of the microvillous membrane can also, and more easily, be recorded (Bara et al., 1988; Greenwood et al., 1993). There are gestational changes, with its value falling as gestation proceeds (Greenwood et al., 1991a). This may be because of greater sodium potassium ATPase electrogenic pumping in first trimester villi as these have a higher oxygen consumption than that of term villi. Furthermore, this oxygen consumption is partly inhibited by ouabain, an inhibitor of sodium potassium ATPase, in earlier gestation but not at term (Greenwood et al., 1991b). However, our recent observations suggest that a difference in membrane K^+ permeability is more likely to explain the different pd's at the two gestations (Birdsey et al., 1995). We do not as yet have any substantial results on the transyncytial pd in early gestation or of its value in sites other than in the intermediate villi.

Ion Channels

Ion channels are likely to be important in the transfer of electrolytes across syncytiotrophoblast "cell" membranes as well as in syncytiotrophoblast homeostasis. They have been studied by patch clamping. In intact villi the best delineated are the maxi chloride channels in the microvillous membrane which have been identified at term (Brown et al., 1993) and which may be commoner in early gestation villi (unpublished data). They and a number of other channels have been identified in the cell membrane of the precursors of syncytiotrophoblast, the cytotrophoblast cells (Bissonnette et al., 1995, Cronier et al., 1995, Greenwood et al., 1993; Byrne et al., 1993). The latter can be isolated from term placenta (Kliman et al., 1986) and syncytialization of the cells *in vitro*, as part of the process of differentiation is associated with a fall in membrane pd (Greenwood et al., 1996) and the expression of an inwardly rectifying potassium current (Clarson et al., 1995). There are thus, in terms of electrical forces and their impact, important developmental and gestational issues for transport of any charged solute or of any uncharged solute transported in an electrogenic mode but data are so far very preliminary.

The Paracellular Route

It has been apparent for some 25 years (e.g., Stulc et al., 1969) that unmetabolized hydrophilic uncharged molecules cross the placenta by a diffusional process without evidence of carrier mediation or active transport and which may therefore not be transcellular. The rates of transfer of the smaller of these molecules when normalized for their concentration difference across placenta, is a function of their diffusion coefficient in water. Sometimes, with larger molecules, there is restriction by what are thought to be steric factors between the molecule and a presumed paracellular pathway across the rate limiting barrier of the placenta (e.g., Boyd, et al., 1976). Gestational features of this transport route are well illustrated by permeability measurements across the intact rat placenta, made using the method of Flexner, in which inulin, mannitol and Cr EDTA were injected into the intact mother and their subsequent content within the fetus measured as a function of the elapsed time and of the maternal plasma concentrations (Robinson et al., 1988). Very similar values were obtained when the measurements were repeated using perfused placenta. Using such methods (Atkinson et al., 1991) or a similar approach in the guinea pig (King-Adams et al., 1988) it is apparent that over the latter part of gestation, two changes occur. Firstly, the overall permeability of the paracellular route increases but, secondly, it does so less for larger molecules than for smaller ones. Various explanations are possible (Stulc, 1989a) including that paracellular paths become greater in number but more restrictive in dimensions with increasing gestation or that the relationships between different layers within the barrier alter. Interestingly, as pointed out previously (Sibley, 1994) the paracellular permeability per gram of placenta, at least near term, is rather similar for the intact human at term and for the rat at a similar period.

The anatomical nature of the paracellular pathway has been a matter of considerable speculation (Stulc, 1989b). One possibility, suggested by Brownbill et al. (1995) is that areas of syncytiotrophoblast denudation in the human placenta, first described by Fox in the 1960s (Fox, 1968) might provide a locus for the pathway. It is not yet known if there are gestational changes in the density or dimensions of these denudations.

Calcium Transport

As already mentioned, calcium concentrations, both total and ionized, are higher in fetal than in maternal plasma and net calcium transport increases dramatically as gestation proceeds. The mechanism of this change is not known. It has been more thoroughly studied in the rat in which there is a great increase in both unidirectional materno-fetal calcium transfer and in net fetal calcium accretion over the last third of gestation (Glazier et al., 1992). During this increase the expression of mRNA for one potential rate limiting component of calcium transport, calcium ATPase, does not change substantially (Birdsey et al., 1994). It therefore looks unlikely to be a controlling feature in the rise. By contrast the expression of mRNA for calbindinD9_k rises strikingly, in parallel to the rise in calcium flux (Glazier et al., 1992). However, in the human while there is also a similar absence of any change in calcium ATPase mRNA there is no consistent expression of calbindinD9_k, unlike the rat, nor any change in its expression over gestation (Glazier et al., 1994). Perhaps the mechanism of calcium movement is different or merely a different calcium binding protein is involved.

Being charged, calcium flux will also be influenced by pds across the microvillous membrane and, if any, across the exchange barrier as a whole. The place of such changes in controlling calcium movement is unknown. So is the influence of intracellular calcium concentration in the syncytiotrophoblast on the movement of other ions through its ability to increase the open probability of ion channels such as chloride and potassium both of which are expressed in, at least, cytotrophoblast cell membranes (Kibble et al., 1996; Greenwood et al., 1993).

Amino Acid Transport

As is well known, the microvillous membrane of the syncytiotrophoblast can be sheared off by homogenization and isolated selectively. Enrichment can be measured by determining alkaline phosphatase activity, an enzyme localized to microvillous membrane. The basal plasma membrane can also be isolated although with some greater difficulty and both membranes can be reconstituted into vesicles. The uptake of various solutes into these vesicles can then be measured as a means of studying the transporters and channels in these two lipid bilayers which any solute crossing the placenta by a transcellular route must traverse. There has been a great deal of work in this area and current work on their amino acid transporters is summarized in Morriss et al. (1994).

The system A transporter which carries neutral amino acids with short side chains such as alanine and glycine is of interest in the context of ion transport because it is, as mentioned, electrogenic. Na^+ moves with each amino acid molecule transported and in a sense the movement can be considered to be that of an ion. The rate of alanine transport will not only be sensitive to sodium concentration (and vice versa as has been shown for the stimulation of sodium transport across the perfused rat placenta by alanine (Stulc et al., 1993; Sibley 1993), but also by the pd across the membrane in which the transporter lies. A clear gestational increase is apparent in the System A transporter in the microvillous membrane when its activity is compared in vesicles prepared from term and first trimester in the presence of the same transmembrane ion gradient (Iioka et al., 1992; Mahendran et al., 1994). However, as the syncytiotrophoblast transmembrane potential, at least *in vitro*, is greater in early gestation, the effect on transport may be less than this suggests. It is also interesting to note that the activity of the Na^+/H^+ exchanger appears to be greater in term than in first trimester microvillous membrane vesicles (Mahendran et al., 1994). But, as the activity of this transporter is measured in the presence of a H^+ gradient, it might be that the faster dissipation of this gradient by first trimester vesicles explains the difference in Na^+/H^+ activity (Mahendran et al., 1994).

The difficulty of drawing secure conclusions about *in vivo* transport rates from vesicle studies in the context of our present incomplete knowledge of electrical forces is clear. Nevertheless, it is exciting to note that the early work of Dicke and Henderson (1988) describing a relationship of amino acid transporter activity to pregnancy complication has now been considerably extended. Thus the system A transporter activity per mg of vesicle protein is not only lower in microvillous vesicles of placentae from SGA (Mahendran et al., 1993) and LGA babies (Kuruvilla et al., 1994) but in SGA fetuses there is a relationship between its activity and earlier measurements *in vivo* of fetal cord pH (Grey et al., 1995). It remains to be seen if this is a pathophysiological mechanism of etiological importance or whether it is a secondary effect of fetal disease. Further study of ion transport routes and of associated electrical forces *in vivo* and *in vitro* in health and disease and especially of how they change in balance over gestation should yield a rich harvest to investigators over the next few years.

ACKNOWLEDGEMENTS

We are most grateful to our collaborators in Manchester and elsewhere - Atkinson, Birdsey, Brownbill, Byrne, Clarson, Doughty, D'Souza, Grey, Kuruvilla, Mahendran, Robinson and Ward and to Cetin, Marconi, Nelson, Pardi, Stulc and Thornburg. This review is based on a lecture given at the second meeting of the Japan Placenta Group, Ohmiya, 10 November 1994.

REFERENCES

Atkinson, D.E., Robinson, N.R. and Sibley, C.P. (1991) Development of passive permeability characteristics of rat placenta during the last third of gestation. *Am. J. Physiol.* 261, R1461-R1464.

Bara, M., Challier, J.C. and Guiet-Bara, A. (1988) Membrane potential and input resistance in syncytiotrophoblast of human term placenta in vitro. *Placenta* 9, 139-146.

Barker, D.J.P. (Ed.) (1993) Fetal and Infant Origins of Adult Disease, London, British Medical Journal Publications.

Binder, N.D., Faber, J.J. and Thornburg K.L. (1978) The transplacental potential difference as distinguished in maternal fetal difference of the guinea pig. *J. Physiol.* 282, 561-570.

Birdsey, T.J., Clarson, L.H., Husain, S.M., Glazier, J.D., Garland, H.O. and Sibley, C.P. (1994) Ca ATPase and CalbindinD$_{9K}$ mRNA expression in placenta of diabetic rats. *Proc. Soc. Gynaecol. Invest.* 41, P188.

Birdsey, T.J., Greenwood, S.L., Boyd, R.D.H. and Sibley, C.P. (1995) Microvillous membrane potential (E_m) in isolated placental villi from the first trimester. *Placenta*, 16, A6.

Bissonnette, J., Maylie, B., Fiorillo, C. and Maylie, J. (1995) Volume regulated chloride channels in human cytotrophoblast. *Placenta* 16, A7.

Boyd, J.D. and Hamilton, W.J. (1970) The Human Placenta. Heffer, Cambridge.

Boyd, R.D.H., Haworth, C., Stacey, T.E. and Ward, R.H.T. (1976) Permeability of the sheep placenta to unmetabolised polar non-electrolytes. *J. Physiol.* 256, 617-634.

Boyd, R.D.H., Glazier, J.D., Sibley, C.P. and Ward, B.S. (1989) Materno-fetal potential in pig. *Am. J. Physiol.* 257, R37-R43.

Brown, P.D., Greenwood, S.L., Robinson, J. and Boyd, R.D.H. (1993) Chloride channels of high conductance in the microvillous membrane in the term human placenta. *Placenta* 14, 103-115.

Brownbill, P., Edwards, D., Jones, C., Mahendran, D., Owen, D., Sibley, C., Johnson, R., Swanson, P. and Nelson, M. (1995) Mechanisms of alphafetoprotein transfer in the perfused human placental cotyledon from uncomplicated pregnancy. *J. Clin. Invest.* 96, 2220-2226.

Byrne, S., Greenwood, S.L. and Sibley, C.P. (1993) Chloride channels in trophoblast cells cultured from human placentas. In: *Clinical Ecology of Cystic Fibrosis* (Eds H. Escobar, F. Baquero and L. Suarez) Elsevier Science Publishers B.V. pp. 3-8.

Canning, J.F., Stacey, T.E., Ward, R.T.H. and Boyd, R.D.H. (1986) Radioiodide transfer across sheep placenta. *Am. J. Physiol.* 250, R112-R119.

Clarson, L.H., Greenwood, S.L., Sides, M.K. and Sibley C.P. (1995) Emergence of an inwardly rectifying potassium current during differentiation of human cytotrophoblast cells. *Placenta* 16, A12.

Cronier, L., Bois, P., Herve, J.C. and Malassiné, A. (1995) Effect of human chorionic gonadotrophin on chloride current in human syncytiotrophoblasts in culture. *Placenta* 16, 611-622.

Dicke, J.M. and Henderson, G.I. (1988) Placental amino acid uptake in normal and complicated pregnancies. *Am. J. Med. Sci.* 295, 223-227.

Faber, J.J. and Thornburg, K.L. (1983) Placental Physiology. Raven Press, New York.

Flexner, L.B., Cowie, D.B., Hellman, L.M., Wilde, W.S. and Vosburgh, G.J. (1948) The permeability of the human placenta to sodium in normal and abnormal pregnancies and the supply of sodium to the human fetus as determined with radioactive sodium. Am. *J. Obstet. Gynecol.* 55, 469-480.

Flexner, L.B. and Gellhorn, A. (1942) The comparative physiology of placental transfer. *Am J. Obstet. Gynecol.* 43, 965-974.

Fox, H. (1968) Fibrinoid necrosis of placental villi. *J. Obstet. Gynec. Brit. Cwlth.* 75, 448-452.

Glazier, J.D., Atkinson, D.E., Thornburg, K.L., Sharpe, P.T., Edwards, D., Boyd, R.D.H., and Sibley, C.P. (1992) Gestational changes in Ca^{2+} transport across rat placenta and mRNA for calbinding$_{9k}$ and Ca^{2+} ATPase. *Am. J. Physiol.* 263, R930-935.

Glazier, J.D., Boyd, R.D.H. and Sibley, C.P. (1994) Calbindin$_{D9K}$ mRNA expression in first trimester and term human placenta. *Placenta* 15, A21.

Greenwood, S.L., Boyd, R.D.H. and Sibley, C.P. (1991a) Electrical potential differences (PD's) in isolated human placental villi from the first and third trimester. *J. Physiol.* 438, 267P.

Greenwood, S.L., Boyd, R.D.H. and Sibley, C.P. (1991b) Oxygen consumption by isolated placental villi from the first and third trimester. *J. Physiol.* 435, 89P.

Greenwood, S.L., Boyd, R.D.H. and Sibley, C.P. (1993) Transtrophoblast and microvillus membrane potential difference in mature intermediate human placental villi. *Amer. J. Physiol.* 265, (*Cell Physiol.* 34), C460-C466.

Greenwood, S.L., Brown, P.D., Edwards, D. and Sibley, C.P. (1993) Patch clamp studies of human placental cytotrophoblast cells in culture. *Trophoblast Research* 7, 53-68.

Greenwood, S.L., Clarson, L.H., Sides, M.K. and Sibley, C.P. (1996) Membrane potential difference and intracellular cation concentration in human placental trophoblast cells in culture. *J. Physiol.* 492, 629-640.

Grey, A.M., Glazier, J.D., Kuruvilla, A., Mahendran, D., D'Souza, S.W., Sibley, C.P., Perugino, G., Baggiani, A., Cetin, I., Marconi, A.M. and Pardi, G. (1995) A correlation between system A amino acid transporter activity and umbilical vein blood pH in placentas from growth retarded babies but not in placentas of normally grown babies. *Placenta* 16, A23.

Iioka, H., Hisanaga, H., Moriyama, I.S., Akada, S., Shimamoto, T., Yamada, Y. and Ichijo, M. (1992) Characterization of human placental activity for transport of l-alanine, using brush border (microvillous) membrane vesicles. *Placenta* 13, 179-190.

Jauniaux E., Jurkovic, D. and Campbell, S. (1995). In vivo investigation of the placental circulations by Doppler Echography. *Placenta* 16, 323-331.

Kibble, J.D., Greenwood, S.L., Clarson, L.H. and Sibley, C.P. (1996) Identification of a Ca^{2+} -activated whole cell Cl^- conductance in human placental cytotrophoblast cells. *J. Memb. Biol.* 151, 131-138.

King-Adams, A., Reid, D.L., Thornburg, K.L. and Faber, J.J. (1988) In vivo placental permeability to hydrophilic solutes as a function of fetal weight in the guinea pig. *Placenta* 9, 409-416.

Kliman, H.J., Nestler, J.E., Senmasi, E., Sanger, J.M. and Strauss III, J.F. (1986) Purification, characterisation and in vitro differentiation of cytotrophoblasts from human term placentae. *Endocrinology* 118, 1567-1582.

Kuruvilla, A.G., D'Souza, S.W., Glazier, J.D., Mahendran, D., Maresh, M.J. and Sibley, C.P. (1994) Altered activity of the system A amino acid transporter in microvillous membrane vesicles from placentas of macrosomic babies born to diabetic women. *J. Clin. Invest.* 94, 689-695.

Mahendran, D., Byrne, S., Donnai, P., D'Souza, S.W., Glazier, J.D., Jones, C and Sibley, C.P. (1994) Na^+ transport, H^+ concentration gradient dissipation, and system A amino acid transport activity in purified microvillous plasma membrane isolated from first trimester human placenta: comparison with the term microvillous membrane. *Am. J. Obstet. Gynecol.*, 171, 1534-1540.

Mahendran, D., Donnai, P., Glazier, J.D., D'Souza, S.W., Boyd, R.D.H. and Sibley, C.P. (1993) Amino acid (System A) transporter activity in microvillous membrane vesicles from the placentas of appropriate and small for gestational age babies. *Ped. Res.* 34, 661-665.

McNaughton, T. and Power, G.G. (1991) The maternal-fetal electrical potential difference: new findings and a perspective. *Placenta* 12, 185-197.

Mellor, D.J. (1969) A potential difference between mother & fetus at different gestational ages in the rat, rabbit and guinea pig. *J. Physiol.* 204, 395-405.

Mellor, D.J. (1970) Distribution of ions and electrical potential difference between mother & fetus at different gestational ages in goats and sheep. *J. Physiol.* 207, 133-150.

Mellor, D.J., Cockburn, F. and Lees, M.M. (1969) Distribution of ions and electrical potential differences between mother and fetus in the human at term. *J. Obstet. Gynaec. Brit. Commonwealth.* 76, 993-998.

Meschia, G., Wolkoff, A.S. and Baron, D.H. (1958) Difference in an electric potential across the placenta of goats. *Proc. Nat. Acad. Sci.* 44, 483-485.

Moll, W. (1995) Absence of intervillous blood flow in the first trimester of human pregnancy. *Placenta* 16, 333-334.

Morriss, F.M., Boyd, R.D.H. and Mahendran, D. (1994) Placental transport. In: The Physiology of Reproduction, 2nd Edn. Vol. II, (Eds.) E. Knobil and J.D. Neill, Raven Press, New York, pp. 814-861.

Needham, J. (1931) Chemical Embryology. Cambridge, Cambridge University Press (reprinted 1963 New York Hafner Publishing Co.)

Rice, G.E., Dantzer, V., Madsen, M.T. and Skadhauge, E. (1991) Gestational changes in electrolyte transport, electrical activity and permeability of the porcine placenta. *J. Comp. Physiol. B.* 171, 189-198.

Robinson, N.R., Atkinson, D.E., Jones, C.J.P. and Sibley, C.P. (1988) Permeability of the near term rat placenta to hydrophilic solutes. *Placenta* 9, 361-372.

Sibley, C.P., Ward, B.S., Glazier, J.D., Moore, W.M.O. and Boyd, R.D.H. (1986) Electrical activity and sodium transfer across in vitro pig placenta. *Am. J. Physiol.* 250. R474-484.

Sibley, C.P. and Boyd, R.D.H. (1988) Control of transfer across the mature placenta. *Oxford Revs. Reprod. Biol.* 10, 382-435.

Sibley, C.P. (1994) Mechanisms of ion transfer by the rat. Placenta: A model for the human placenta? *Placenta* 15, 675-692.

Stulc, J., Friedrich, R. and Jiricka, Z. (1969) Estimation of the equivalent poor dimensions in the rabbit placenta. *Life Sciences* 8, 167-180.

Stulc, J.J., Rietveld, W.J., Soteman, D.W. and Versprille, A. (1972) The transplacental potential difference in guinea pigs. *Biol. Neonate* 21, 130-147.

Stulc, J.J., Svihovec, J., Drabkova, J., Stribrny, J., Kobilkova, J., Vido, I. and Dolezal, A. (1978) Electrical potential difference across the mid-term human placenta. *Acta. Obstet. Gynaecol. Scan.* 57, 125-126.

Stulc, J. (1989a) Letter to Editor. *Placenta* 10, 427-428.

Stulc, J. (1989b) Extracellular transport pathways in the haemochorial placenta. *Placenta* 10, 113-119.

Stulc, J., Stulcova, B. and Sibley, C.P. (1993) Evidence for active maternal-fetal transport of Na+ across the placenta of the anaesthetised rat. *J. Physiol.* 470, 637-649.

Thornburg, K.L., Binder, N.D. and Faber, J.J. (1979) Distribution of ionic sulphate, lithium and bromide across the sheep placenta. *Am. J. Physiol.* 236, C58-C65.

Weedon, A.P., Stacey, T.E., Canning, Jane F., Ward, R.H.T. and Boyd, R.D.H. (1980) Maternal fetal electrical potential difference in conscious sheep: effect of fetal acidosis. *Am. J. Obstet. Gynecol.* 138, 422-428.

Trophoblast Research 9:87-98, 1997

MECHANISMS OF TRANSEPITHELIAL TRANSPORT OF AMINO ACIDS IN HUMAN PLACENTAL SYNCYTIOTROPHOBLAST

Yoshiki Kudo[1,2,3] and C. A. R. Boyd[1]

[1]Department of Human Anatomy
University of Oxford
Oxford, OX1 3QX, United Kingdom

[2]Department of Obstetrics and Gynecology
Tokyo Women's Medical College
8-1 Kawada, Shinjuku
Tokyo 162, Japan

INTRODUCTION

Placental transfer of amino acids from mother to fetus occurs across the syncytiotrophoblast which forms a complete cellular epithelium separating maternal from fetal compartments. The maternal-facing surface (brush border membrane) and the fetal-facing surface (basal membrane) of this epithelium possess the marked structural and functional polarity (Yudilevich and Boyd, 1987). The concentrations of amino acid in fetal plasma are roughly twice that found in maternal plasma (Yudilevich and Sweiry, 1985). The contrasting properties of the transport systems present in the two surfaces of the trophoblast underlie the ability of the placenta to produce overall net active transport from mother to fetus. Brush border membrane vesicles isolated from human full-term placental syncytiotrophoblast have been used extensively in the last decade to study the transport mechanisms of amino acids (Boyd and Lund, 1981; Asai et al., 1982; Ganapathy et al., 1986; Kudo et al., 1987; Karl et al., 1989) and other solutes (Bissonnette et al., 1981; Shennan and Boyd, 1987) into the epithelium. More recently a methodology for the isolation of basal plasma membrane vesicles from this epithelium has been developed (Kelley et al. 1983).

α-(Methylamino)isobutyrate and L-tyrosine were employed as representative substrates for Na^+-dependent transport system and Na^+-independent transport system at brush border surface, respectively (Yudilevich and Sweiry, 1985), and using the plasma membrane vesicles isolated respectively from the maternal-facing brush border and fetal-facing basal surface of the human full-term placental syncytiotrophoblast, transport studies were conducted to explore the mechanisms responsible for the transepithelial transfer of amino acids from mother to fetus. The results indicate marked differences in the distribution of amino acid transport systems between the brush border and basal surface of the human syncytiotrophoblast. This asymmetry may explain net transplacental transfer of amino acids to the fetus.

[3] To Whom Correspondence Should Be Addressed.

MATERIALS AND METHODS

Preparation of Brush Border and Basal Membrane Vesicles

Brush border and basal membrane vesicles were prepared from freshly obtained human full-term placenta according to previous reports (Smith et al., 1974; Kelly et al., 1983) with some modifications (Kudo et al., 1987; Kudo and Boyd, 1990). The orientation of brush border membrane vesicles is overwhelmingly right-side-out, however for basal side the vesicle preparation is mixed in orientation (Hoeltzli and Smith, 1989). Membrane vesicles were finally suspended in 2 mM Tris-Hepes buffer (pH 7.5) containing 0.1 mM $MgSO_4$ and 300 mM D-mannitol to give a final protein concentration of approximately 4-6 mg ml^{-1}. This brush border membrane vesicle preparation showed a degree of alkaline phosphatase (EC 3.1.3.1) enrichment 33 times greater than that of the starting homogenate, while acid phosphatase (EC 3.1.3.2) showed a one-fifth decrease in its specific activity. The purity of this membrane vesicle preparation was higher than that of those described previously (Ruzycki et al., 1978). The basal membrane vesicle preparation showed a degree of dihydroalprenolol binding activity enrichment 27 times greater than that of the villous tissue homogenate, while alkaline phosphatase showed no increase in its specific activity. The purity of basal membrane vesicle preparation is similar to that described in the original report (Kelley et al., 1983). Cross contamination appears not to be an appreciable problem since there is negative enrichment of the marker enzymes for the contralateral membrane. There was an inverse correlation between L-[^{3}H]tyrosine uptake at equilibrium and external osmolarity. Thus, it appears that amino acid uptake represents transport into the intravesicular space, rather than non-specific binding to the membrane.

Assay of Transport Activity

Amino acid uptake was measured according to the procedure described previously (Kudo et al., 1987). Uptake was initiated by adding 50 µl of the membrane suspension (approximately 200-300 µg of membrane protein) to 60 µl of an incubation medium composed of 36.7 µM α-[^{14}C](methylamino)isobutyrate or 3.67 µM L-[^{3}H]tyrosine, 0.1 mM $MgSO_4$, 50 mM D-mannitol, 20 mM Tris-Hepes (pH 7.5) and either 220 mM NaCl or 220 mM KCl. Other additions are described in the figure legends. Both the membrane suspension and the incubation medium were pre-incubated independently before mixing, followed by further incubation at the same temperature. The uptake of substrate was terminated by diluting the sample with a 40-fold excess of an ice-cold buffer composed of 150 mM NaCl, 50 mM $MgSO_4$, 30 mM D-mannitol and 10 mM Tris-Hepes buffer (pH 7.5). The diluted sample was immediately filtered through a Millipore cellulose filter (0.65 µm) and washed once with 3 ml of the same ice-cold buffer. The radioactivity of labeled substrate retained on the filter was counted by liquid scintillation spectroscopy.

Unlabeled L-Tyrosine Loaded Vesicles

Membrane vesicles were pre-loaded with unlabeled L-tyrosine by washing three times in 25 mM L-tyrosine, 0.1 mM $MgSO_4$, 100 mM D-mannitol and 2 mM Tris-Hepes (pH 7.5), followed by pre-incubation in the same medium for 30 minutes at 25°C.

Protein Estimation

The protein concentration of the vesicle preparation was determined by the method of Lowry et al. (1951) using bovine serum albumin as a standard.

Chemicals

α-[1-14C](Methylamino)isobutyrate (48 mCi/mmol) was purchased from New England Nuclear (Boston, MA) and L-[3,5-^{3}H]tyrosine (49 Ci/mmol) was from Amersham International (Amersham, U.K.). All other chemicals were of the highest purity commercially available.

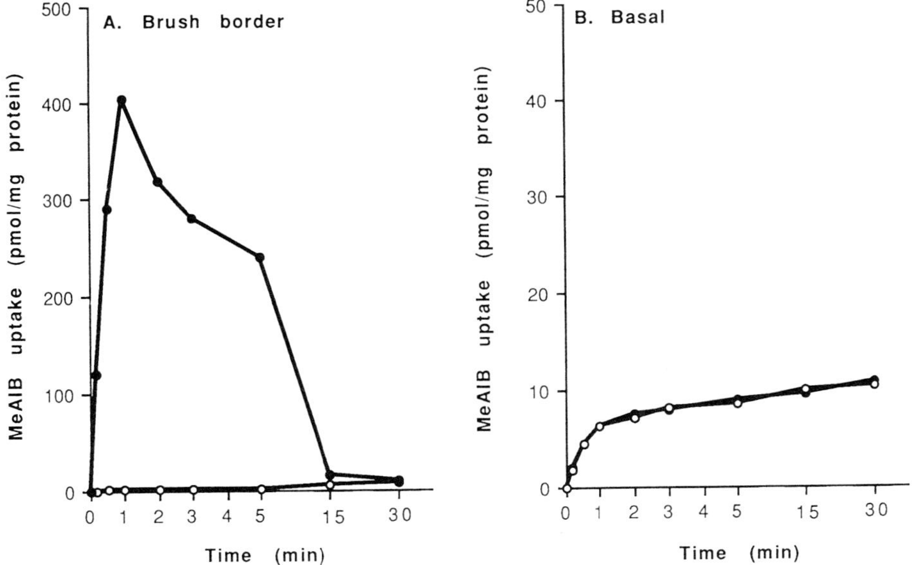

Figure 1. Time course of α-(methylamino)isobutyrate uptake by brush border (A) and basal (B) membrane vesicles. Membrane vesicles were suspended in 2 mM Tris-Hepes buffer (pH 7.5) containing 0.1 mM $MgSO_4$ and 300 mM D-mannitol to give a final protein concentration of approximately 4-6 mg ml^{-1}. α-(Methylamino)isobutyrate (MeAIB) uptake was initiated by adding 50 μl of the membrane suspension (approximately 200-300 μg of membrane protein) to 60 μl of an incubation medium composed of 36.7 μM α-[^{14}C](methylamino)isobutyrate, 0.1 mM $MgSO_4$, 50 mM D-mannitol, 20 mM Tris-Hepes (pH 7.5) and either 220 mM NaCl or 220 mM KCl. Both the membrane suspension and the incubation medium were pre-incubated independently at 37°C for 5 minutes before mixing, followed by further incubation at 37°C. NaCl (●), KCl (○). Each point represents the mean for three experiments.

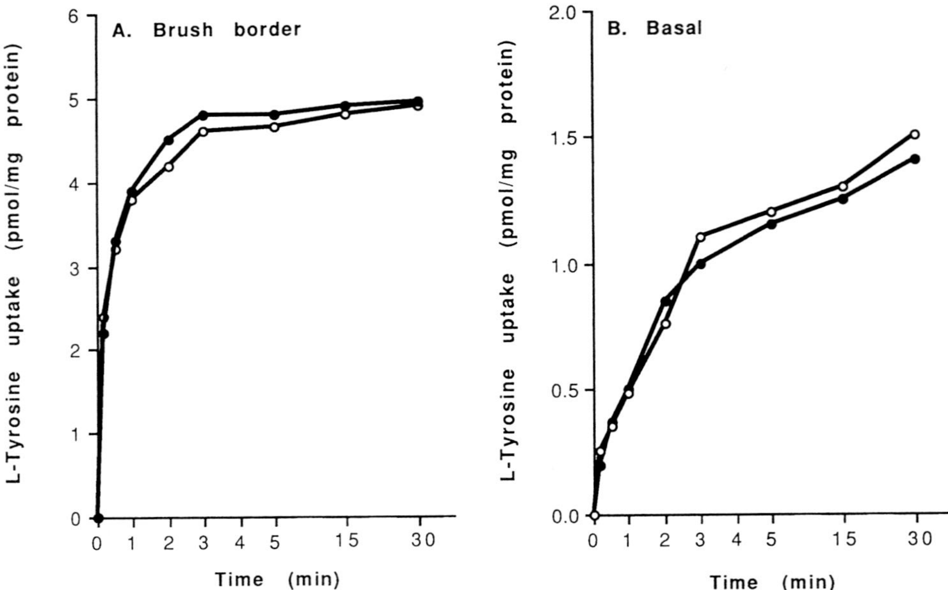

Figure 2. Time course of L-tyrosine uptake by brush border (A) and basal (B) membrane vesicles. Membrane vesicles were suspended in 2 mM Tris-Hepes buffer (pH 7.5) containing 0.1 mM $MgSO_4$ and 300 mM D-mannitol to give a final protein concentration of approximately 4-6 mg ml^{-1}. L-Tyrosine uptake was initiated by adding 50 μl of the membrane suspension (approximately 200-300 μg of membrane protein) to 60 μl of an incubation medium composed of 3.67 μM L-[^{3}H]tyrosine, 0.1 mM $MgSO_4$, 50 mM D-mannitol, 20 mM Tris-Hepes (pH 7.5) and either 220 mM NaCl or 220 mM KCl. Both the membrane suspension and the incubation medium were pre-incubated independently at 10°C for 5 minutes before mixing, followed by further incubation at 10°C. NaCl (●), KCl (○). Each point represents the mean for three experiments.

RESULTS

Properties of α-(Methylamino)isobutyrate Transport

The uptake of α-(methylamino)isobutyrate by brush border and basal membrane vesicles as a function of incubation time is shown in Figure 1. In brush border membrane vesicles, the presence of a Na^+ gradient toward the inside from the outside of the vesicles stimulated α-(methylamino)isobutyrate uptake. The uptake reached a maximum level at 60 s after the initiation of incubation, then decreased with time due to efflux to the steady-state level, a phenomenon known as the 'overshoot', indicating the existence of Na^+ coupled secondary active transport of α-(methylamino)isobutyrate in these membrane vesicles. In the absence of a Na^+ gradient, α-(methylamino)isobutyrate uptake was below 2 pmol (mg protein)$^{-1}$ at any period of incubation. There was no significant difference between α-(methylamino)isobutyrate uptake in the presence of an inwardly directed Na^+ or K^+ gradient in basal membrane vesicles.

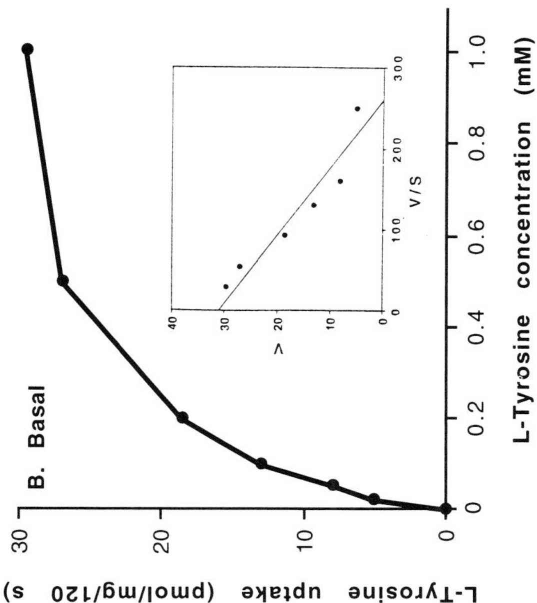
B. Basal
L-Tyrosine uptake (pmol/mg/120 s)
L-Tyrosine concentration (mM)
0
10
20
30
0
0.2
0.4
0.6
0.8
1.0
V
V/S
0
10
20
30
40
0
100
200
300

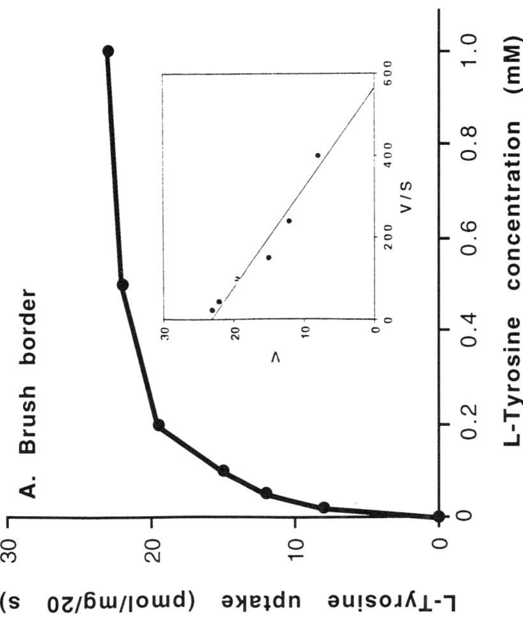
A. Brush border
L-Tyrosine uptake (pmol/mg/20 s)
L-Tyrosine concentration (mM)
0
10
20
30
0
0.2
0.4
0.6
0.8
1.0
V
V/S
0
10
20
30
0
200
400
600

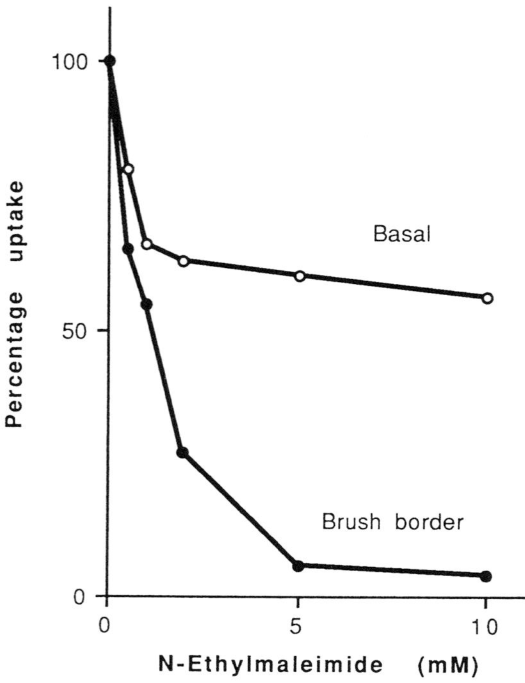

Figure 4. Effect of N-ethylmaleimide on L-tyrosine uptake by brush border and basal membrane vesicles. Membrane vesicles were pre-incubated with the indicated concentrations of N-ethylmaleimide for 10 minutes at 25°C prior to starting the assay. The uptake over a 20 s and 120 s period for brush border and basal membrane vesicles respectively was measured at 10°C in a medium composed of 2 μM L-[^{3}H]tyrosine, 0.1 mM $MgSO_4$, 300 mM D-mannitol and 10 mM Tris-Hepes (pH 7.5) in the presence of the concentration of N-ethylmaleimide indicated. Brush border membrane vesicles (●), basal membrane vesicles (○). Each point represents the mean for three experiments.

Properties of L-Tyrosine Transport

All of the L-tyrosine transport experiments were performed at 10°C since the rate of this amino acid transport was too fast to be determined accurately at 37°C. Figure 2 shows the time course of L-tyrosine transport in brush border and basal membrane vesicles. In neither membrane was inwardly directed Na^+ gradient able to alter the rate of L-tyrosine uptake. The uptake of L-tyrosine measured under these conditions was a linear function of time at least the first 20 s of incubation for brush border and 120 s for basal membrane vesicles. Therefore uptake measured after these incubation times was used to estimate the initial rate of L-tyrosine uptake in each preparation. The initial rate of uptake, corrected for the non-saturable component, as a function of L-tyrosine concentration showed saturable hyperbolic curves that obeyed Michaelis-Menten kinetics in both membranes (Figure 3).

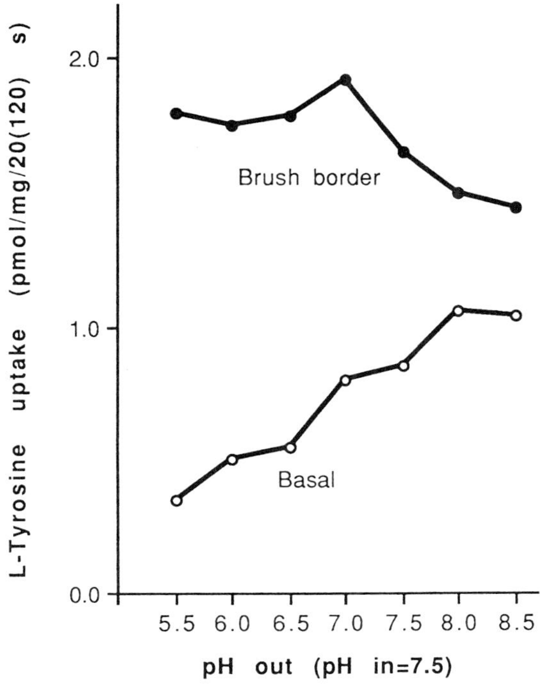

Figure 5. Effect of pH on the initial rate of L-tyrosine uptake by brush border and basal membrane vesicles. Membrane vesicles were suspended in 2 mM Tris-Hepes buffer (pH 7.5) containing 0.1 mM $MgSO_4$ and 300 mM D-mannitol. L-Tyrosine uptake was initiated by adding 20 μl of the membrane suspension (approximately 200-300 μg of membrane protein) to 90 μl of an incubation medium composed of 2.45 μM L-[^{3}H]tyrosine, 0.1 mM $MgSO_4$, 280 mM D-mannitol and either 20 mM Mes-Tris (pH 5.5-7.0) or 20 mM Tris-Hepes (pH 7.0-8.5). Brush border membrane vesicles (●), basal membrane vesicles (○). Each point represents the mean for three experiments. Mes, 2-(N-morpholino)ethanesulfonic acid; Hepes, 4-(2-hydroxyethyl)-1-piperazine ethanesulfonic acid.

The calculated values of the Michaelis constant, Km, and maximum velocity, Vmax, for L-tyrosine uptake in brush border membrane were 54.2 ± 6.0 μM and 1.28 ± 0.03 pmol (mg protein)$^{-1}$ s^{-1}, and in basal membrane were 168.9 ± 17.0 μM and 0.31 ± 0.01 pmol (mg protein)$^{-1}$ s^{-1}, respectively. The effect of the sulfhydryl modifying reagent N-ethylmaleimide on the initial rate of L-tyrosine uptake was determined. As shown in Figure 4, N-ethylmaleimide inactivated L-tyrosine uptake in a concentration-dependent manner in both membranes. In brush border membrane, the half-maximal inhibition of uptake was observed at about 1.1 mM and approximately 95% inhibition was observed at 5 mM N-ethylmaleimide. However, as high as 10 mM N-ethylmaleimide was needed to produce 45% inhibition of L-tyrosine uptake in basal membrane vesicles. L-Tyrosine uptake by brush border membrane vesicles was insensitive to external pH over the range of 5.5 - 8.5. In basal membrane vesicles, the uptake was stimulated by increasing pH in the extravesicular medium (Figure 5). In order to further characterize the transport of L-tyrosine, trans-stimulation of L-tyrosine influx was determined in these membrane preparations (Figure 6). L-Tyrosine uptake at 1 minute after the initiation of incubation by brush border membrane vesicles was enhanced 1.6-fold when the vesicles were pre-loaded with unlabeled L-tyrosine, which is known as the 'trans-stimulation' phenomenon. The final equilibrium levels obtained 30 minutes after incubation were not different in vesicles with or without pre-loading of unlabelled L-tyrosine. While in the basal membrane preparation trans-stimulation of L-tyrosine influx was not found. These results indicate that the increase of L-tyrosine uptake observed in the brush border membrane vesicles pre-loaded with unlabeled L-tyrosine is due to a specific increase in the activity of L-tyrosine transport system and not to an increase in the membrane permeability or to a difference in the intravesicular space accessible to L-tyrosine.

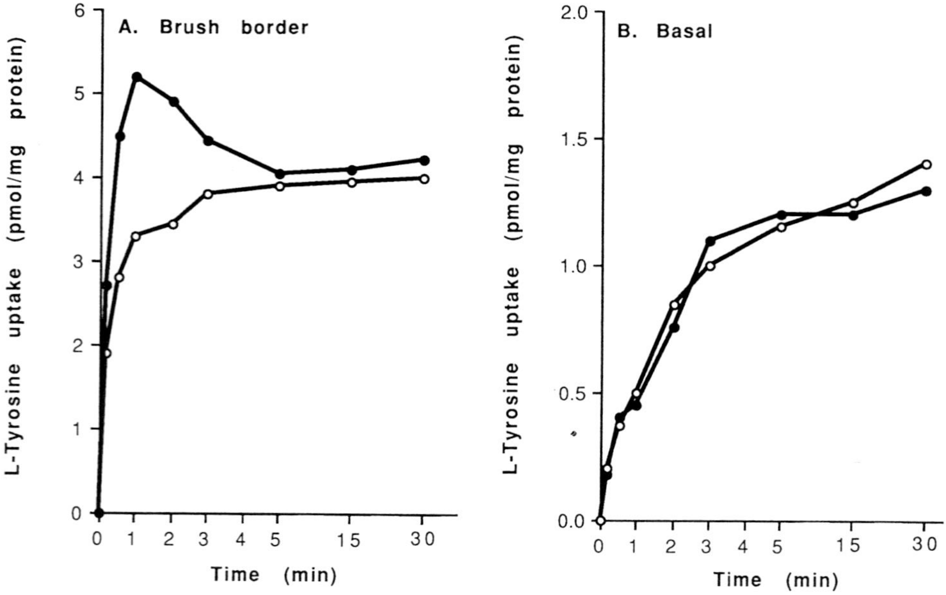

Figure 6. Trans-stimulation of L-tyrosine uptake by brush border (A) and basal (B) membrane vesicles. Membrane vesicles were pre-loaded with unlabeled L-tyrosine as described in the text. The uptake at the time indicated was measured at 10°C in a medium composed of 2 μM L-[^{3}H]tyrosine, 0.1 mM $MgSO_4$, 300 mM D-mannitol and 10 mM Tris-Hepes (pH 7.5). Unlabeled L-tyrosine loaded vesicles (●), control (○). Each point represents the mean for three experiments.

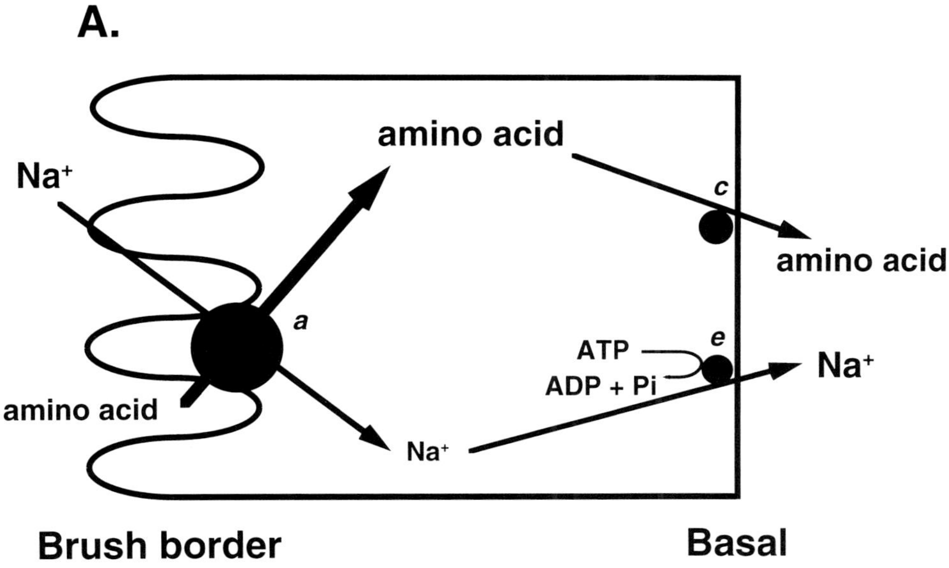

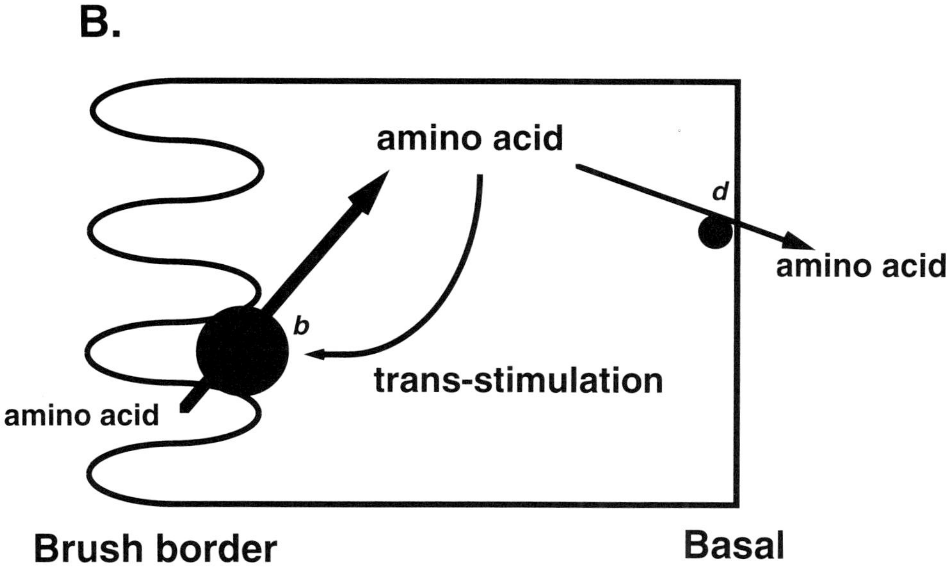

Figure 7. A schematic representation of the transepithelial amino acids transport from mother to fetus. a, Na^+-dependent transport system; b, Na^+-independent transport system with trans-stimulation; c, d, Na^+-independent transport system; e, Na^+, K^+-ATPase.

DISCUSSION

The placental transfer of amino acids from mother to fetus involves uptake of amino acids from maternal circulation across the brush border membrane of the syncytiotrophoblast, diffusion through the cytoplasm and exit into the fetal circulation across the basal membrane. Since the amino acid concentration of fetal plasma is higher than that of maternal plasma, the placental syncytiotrophoblast must be stimulating the transfer of amino acids at the input or output membrane of this epithelium. The mechanisms responsible for α-(methylamino)isobutyrate and L-tyrosine transport have now been studied at both the maternal-facing and fetal-facing surfaces of the human placenta using isolated brush border and basal membrane vesicles under conditions where a direct comparison of the transport properties of the two membranes can be made. The two types of membrane preparation used in these studies show a high degree of purity with respect to the activities of marker enzymes and both contain multiple Na^+-dependent and Na^+-independent transport systems for various amino acids (Kudo et al., 1987; Kudo and Boyd, 1990).

The results obtained show that there are carrier-mediated transport systems for each amino acid at both the maternal- and at the fetal-facing surfaces of the human placental syncytiotrophoblast. However, these systems are not identical on each side of the epithelium. The uptake of α-(methylamino)isobutyrate by brush border membrane vesicles in the presence of a Na^+ gradient from the outside to the inside of the vesicles showed a typical 'overshoot' phenomenon after reaching a maximal uptake level 1 minute after the start of incubation. This 'overshoot' of the uptake resulted from an electrochemical gradient formed by a Na^+ gradient, and its subsequent loss with time due to Na^+-substrate co-transport (Christensen, 1975). However, at the basal side of the syncytiotrophoblast the presence of an inwardly directed gradient of Na^+ did not influence α-(methylamino)isobutyrate transport. In contrast to the marked stimulatory effect of Na^+ gradient on α-(methylamino)isobutyrate uptake at brush border side, in the case of L-tyrosine at neither side of this epithelium was Na^+ able to enhance the rate of uptake. In both membranes the kinetics of L-tyrosine transport showed saturation. The amino acid transport systems for L-tyrosine are not identical in brush border and basal membranes since the kinetics of L-tyrosine transport are different. Other differences between the two membrane preparations with regard to L-tyrosine transport relate to the sensitivity of the sulfhydryl modifying reagent N-ethylmaleimide and the dependency of pH in the extra vesicular medium. Moreover, L-tyrosine uptake by brush border membrane vesicles exhibited trans-stimulation so that influx of labeled L-tyrosine was markedly stimulated when the vesicles were pre-loaded with unlabeled L-tyrosine. While in the basal membrane preparation trans-stimulation of L-tyrosine influx was not found.

This asymmetry of the distribution of amino acid transport systems in the brush border and the basal membrane of placental syncytiotrophoblast may explain net transplacental transfer of amino acids from mother to fetus. Thus, for α-(methylamino)isobutyrate (Figure 7A), the amino acid transporter energized by secondary active transport coupled to the electrochemical gradient of Na^+ between the external and internal environment of the syncytiotrophoblast is responsible for the active transport of this amino acid into the cell against the concentration gradient at brush border side. Na^+, K^+-ATPase located at basal membrane serves to maintain the electrochemical gradient of Na^+ (Shennan and Boyd, 1987). The other energy supply available to Na^+-independent transporter depends upon trans-stimulation of amino acid (e.g. for L-tyrosine) (Figure 7B). These active transport systems at the brush border surface, generate up hill transfer of amino acids from mother into the syncytiotrophoblast and facilitated or passive diffusion driven by the amino acid concentration gradient is responsible for the basal membrane exit step. Such mechanisms may underlie the observed gradients of amino acid concentration across the placenta, fetal concentrations being consistently higher than maternal.

SUMMARY

Brush border and basal plasma membrane vesicles prepared from human full-term placental syncytiotrophoblast have been used to study the mechanisms of transepithelial transfer of amino acids from mother to fetus. Such studies indicated marked differences in the distribution of amino acid transport systems between the brush border and the basal surface of the human syncytiotrophoblast. Transport of α-(methylamino)isobutyrate by brush border membrane vesicles was stimulated in the presence of an Na^+ gradient from the outside to the inside of the vesicles, while by basal membrane vesicles α-(methylamino)isobutyrate was transported only Na+-independently. L-Tyrosine transport was not enhanced by an inwardly directed gradient of Na^+ in either brush border or basal membrane vesicles. The initial rate of L-tyrosine transport as a function of concentration showed saturation and obeyed Michaelis-Menten kinetics in both membrane vesicles. The calculated values of Km and Vmax for the L-tyrosine transport in brush border membranes were 54.2 μM and 1.28 pmol (mg protein)$^{-1}$ s^{-1}, and in basal membranes were 168.9 μM and 0.31 pmol (mg protein)$^{-1}$ s^{-1}, respectively. N-Ethylmaleimide inactivated L-tyrosine transport in a concentration-dependent manner in both membranes, however the transport systems at the brush border were more sensitive to this reagent than those at the basal side. L-Tyrosine transport by basal membrane vesicles was stimulated at more alkaline pH, while at brush border side transport was insensitive to changes in external pH over the range of 5.5 to 8.5. Trans-stimulation of L-tyrosine transport was observed in brush border membrane vesicles but not in basal membrane vesicles.

REFERENCES

Asai, M., Keino, H. and Kashiwamata, S. (1982) L-Alanine uptake by microvillous brush border membrane vesicles prepared from human placenta. *Biochem. Int.* 4, 377-384.

Bissonnette, J.M., Black, J.A., Wickham, W.K. and Acott, K.M. (1981) Glucose uptake into plasma membrane vesicles from the maternal surface of human placenta. *J. Membr. Biol.* 58, 75-80.

Boyd, C.A.R. and Lund, E.K. (1981) L-Proline transport by brush border membrane vesicles prepared from human placenta. *J. Physiol.* 315, 9-19.

Christensen, H.N. (1975) *Biological Transport*, 2nd edition, W.A. Benjamin, M.A.

Ganapathy, E.M., Leibach, F.H., Mahesh, V.B., Howard, J.C., Devoe, L.D. and Ganapathy, V. (1986) Characterization of tryptophan transport in human placental brush-border membrane vesicles. *Biochem. J.* 238, 201-208.

Hoeltzli, S.D. and Smith, C.H. (1989) Alanine transport systems in isolated basal plasma membrane of human placenta. *Am. J. Physiol.* 256, C630-C637.

Karl, P.I., Tkaczevski, H. and Fisher S.E. (1989) Characteristics of histidine uptake by human placental microvillous membrane vesicles. *Pediat. Res.* 25, 19-26.

Kelley, L.K., Smith, C.H. and King, B.F. (1983) Isolation and partial characterization of the basal cell membrane of human placental trophoblast. *Biochim. Biophys. Acta* 734, 91-98.

Kudo, Y., Yamada, K., Fujiwara, A. and Kawasaki, T. (1987) Characterization of amino acid transport systems in human placental brush border membrane vesicles. *Biochim. Biophys. Acta* 904, 309-318.

Kudo, Y. and Boyd, C.A.R. (1990) Characterization of amino acid transport systems in human placental basal membrane vesicles. *Biochim. Biophys. Acta* 1021, 169-174.

Lowry, O.H., Rosebrough, N.J., Farr, A.L. and Randall, R.J. (1951) Protein measurements with the Folin phenol reagent. *J. Biol. Chem.* 193, 265-275.

Ruzycki, S.M., Kelley, L.K. and Smith, C.H. (1978) Placental amino acid uptake IV. Transport by microvillous membrane vesicles. *Am. J. Physiol.* 234, C27-C35.

Shennan, D.B. and Boyd, C.A.R. (1987) Ion transport by human placenta: A review of membrane transport system. *Biochim. Biophys. Acta* 906, 437-457.

Smith, N.C., Brush, M.G. and Luckett, S. (1974) Preparation of human placental villous surface membrane. *Nature* 252, 302-303.

Yudilevich, D.L. and Boyd, C.A.R. (1987) *Amino Acid Transport in Animal Cells.* Manchester University Press, Manchester.

Yudilevich, D.L. and Sweiry, J.H. (1985) Transport of amino acids in placenta. *Biochim. Biophys. Acta* 822, 169-201.

PRINT ON SEPARATE PAGE - SIDEWAY TO ACCOMMODATE FIGURE 3

Figure 3. Effect of L-tyrosine concentration on the initial rate of L-tyrosine uptake by brush border (A) and basal (B) membrane vesicles. The uptake over a 20 s and 120 s period for brush border and basal membrane vesicles respectively was measured at 10°C in a medium containing L-tyrosine at the concentrations indicated, 0.1 mM $MgSO_4$, 300 mM D-mannitol and 10 mM Tris-Hepes (pH 7.5) (final concentrations). Initial rate of uptake was determined by subtracting the non-saturable component from total uptake at each concentration. The non-saturable contribution to uptake was determined by employing the straight-line equation at higher L-tyrosine concentrations. Inset is an Eadie-Hofstee plot of the data in which V is the observed velocity at substrate concentration S; the line was determined by least-squares linear regression analysis. Each point represents the mean for three experiments.

Trophoblast Research 9:99-107, 1997

PLATELET-ACTIVATING FACTOR IN HUMAN DECIDUA: ITS ROLE IN PARTURITION AND PRETERM LABOR
- A Review -

Hisashi Narahara[1], Isao Miyakawa[1] and John M. Johnston[2]

[1]Department of Obstetrics and Gynecology
Oita Medical University
1-1 Hasama, Oita-gun
Oita 879-55, Japan

[2]Department of Biochemistry
The University of Texas
Southwestern Medical Center at Dallas
5323 Harry Hines Boulevard
Dallas, Texas 75235-9051 USA

INTRODUCTION

Platelet-activating factor (1-O-alkyl-2-acetyl-*sn*-glycero-3-phosphocholine, PAF) is one of the most potent lipid mediators described and has been shown to be involved in a wide variety of diseases including asthma, endotoxic shock, inflammation, diabetes, acute allergic reactions, thrombosis, ischemic bowel necrosis, etc. (Snyder, 1990; Hanahan, 1986). PAF has also been implicated in a number of reproductive processes, including ovulation, sperm motility, implantation, fetal lung maturation, and the initiation and maintenance of parturition (Narahara et al., 1994). The first suggestion that PAF may be involved in parturition was based on the observation that PAF appeared in increased amounts in the amniotic fluid obtained from women in labor (Billah and Johnston, 1983). Nishihira et al. (Nishihira et al., 1984) confirmed these observations. The origin of PAF in amniotic fluid is thought to be fetal lung and kidney (Nishihira et al., 1984; Hoffman et al., 1986). It has been reported that PAF stimulates prostaglandin E_2 production in fetal membranes (Billah et al., 1985; Morris et al., 1992). It has also been reported that PAF is one of the most potent stimuli of myometrial contraction (Nishihira et al., 1984; Tetta et al., 1986). We have demonstrated the presence of PAF receptors in human myometrium and found that PAF, at concentrations between 10^{-12} to 10^{-10} mol/L, stimulates Ca^{2+} uptake and myosin light chain phosphorylation in isolated human myometrial cells (Zhu et al., 1992).

PAF metabolism in fetal and maternal compartments may also be important throughout gestation. It has recently been suggested PAF is metabolized to ethanolamine plasmalogens, a rich source for arachidonate release, in amnion tissue before labor (Frenkel and Johnston, 1992). At term in labor, the arachidonate released from this lipid fraction by a phospholipase A_2 could be utilized in the formation of eicosanoids. The resulting lyso-plasmalogens might stimulate PAF synthesis via the transacylation reaction of the remodeling pathway (Uemura et al., 1991; Nieto et al., 1991; Toyoshima et al., 1994). Inactivation of PAF by the enzyme PAF-acetylhydrolase (PAF-AH) provides

[3]To Whom Correspondence Should Be Addressed.

an additional mechanism for the regulation of PAF metabolism. We have previously reported that during the latter stages of pregnancy PAF-AH activity is significantly decreased in maternal plasma (Maki et al., 1988). It has been suggested that the decrease in PAF-AH in maternal plasma occurs at a time when certain fetal tissues, *e.g.*, the fetal lung and kidney, have an increased capacity for PAF biosynthesis (Hoffman et al., 1986). The source of the plasma PAF-AH is thought to be the liver, the macrophages or both. Cultured rat hepatocytes (Prescott et al., 1990), HepG2 cells (Satoh et al., 1991), human peripheral blood derived macrophages (Prescott et al., 1990), rat alveolar macrophages (Yasuda et al., 1993), and phorbol ester-stimulated HL-60 cells (Narahara et al 1993a) secrete PAF-AH activity of the plasma type.

Gram-negative organisms elicit an inflammatory reaction which is largely induced by one of their cell wall constituents, endotoxin (lipopolysaccaride, LPS). It has been reported that LPS potentially activates cells in the monocyte/macrophage system, resulting in the production of inflammatory mediators such as tumor necrosis factor (TNF), interleukin-1 (IL-1), and eicosanoids (Hinshaw et al, 1990). It has also been shown that PAF is a key mediator of the inflammatory reaction caused by LPS (Braquet et al., 1987). Evidence to date suggests that LPS, cytokines, eicosanoids, and PAF are involved in preterm delivery or premature rupture of membranes associated with bacterial infections (Hoffman et al., 1990; Romero et al., 1991; Gibbs et al., 1992).

In order to clarify further the role of PAF in parturition and in preterm labor, we have investigated PAF metabolism in the decidua and its modulation by endotoxin or cytokines. The topic will be emphasized in this mini review.

THE ROLE OF MACROPHAGES IN PAF METABOLISM IN HUMAN DECIDUA

The inactivation of PAF produced in the fetal and maternal compartments prior to its contact with the myometrium would be of considerable importance, since PAF has been shown to be a potent stimulator of myometrial contraction (Nishihira et al., 1984; Tetta et al., 1986; Zhu et al., 1992). We have considered that decidua may be the tissue site of the PAF-inactivation. PAF produced in the fetal and maternal compartments would be inactivated by maternal plasma PAF-AH due to its abundant blood supply, thus preventing PAF from reaching the myometrium.

It has recently been demonstrated that cells isolated from human decidua by enzymatic digestion and Ficoll-Paque density centrifugation secrete PAF-AH into the culture medium (Narahara et al., 1993b). Human decidua was obtained from patients at term and not in labor following cesarean section. The cells were isolated by enzymatic digestion followed by Ficoll-Paque centrifugation or purified further by discontinuous Percoll density gradient centrifugation and cultured. The supernatant activity of PAF-AH) was assayed according to the method of Miwa et al. (Miwa et al., 1988) with minor modifications. Cell populations were analyzed by flow cytometry after labeling with macrophage-specific antibodies. The PAF-AH secreted by the decidual cell populations was characterized and found to be the plasma type; its synthesis and secretion were inhibited by Actinomycin D or cycloheximide. The presence of cells that secrete PAF-AH activity in the human decidua suggests that local regulatory mechanism may also contribute to the regulation of PAF accumulation. Therefore, in addition to the inactivation of PAF by plasma PAF-AH in the blood supply to this tissue, the PAF-AH secreted by decidual cells may play a role in the local metabolism of PAF in this tissue.

Table 1

PAF-AH Secretion Following The Isolation Of Decidual Cells

	CD-14(+) (% of total)	PAF-AH activity (units/106 cells)
Digestation and Ficoll-Paque density centrifugation	27 ± 11	193 ± 26
Discontinuous Percoll density centrifugation	52 ± 9	312 ± 20
Flow cytometrical isolation	96 ± 2	468 ± 35

Table 2

PAF-AH Secretion By Decidual Cells Pre-Treated With Monoclonal Antibodies And Complement

	PAF-AH Activity (% of control)		
Complement	Antibody (-)	Anti-CD-14	Anti-UPC10
(-)	100 ± 6	81 ± 10	79 ± 7
(+)	64 ± 6	8 ± 5	62 ± 8

Decidual cells were incubated with or without monoclonal antibodies (anti-CD14 or anti-UPC10; both are IgG2a) for 60 minutes at 4°C, followed by incubation with or without complement for 60 minutes at 37°C. After the treatment, cells were culture for six days. The accumulated PAF-AH activity in the medium was assayed. Values (mean ± SD) are shown as % of control (no antibody, no complement).

It is of particular importance to determine the cellular component that is responsible for the PAF-AH secretion in the human decidua, since the tissue consists of a number of cell populations with a diverse range of functions. It has been reported that cultured rat hepatocytes (Prescott et al., 1990), HepG2 cells (Satoh, et al., 1991), human macrophages (Prescott et al, 1990), and phorbol ester-stimulated HL-60 cells (Narahara et al., 1993a) secrete PAF-AH activity of the plasma type. In view of these observations, we have postulated that the macrophage population of the decidua may also function in the secretion of PAF-AH. In the isolation of cells from term decidual tissue three steps were employed: 1) enzymatic digestion and subsequent Ficoll-Paque density centrifugation, followed by either 2) discontinuous Percoll gradient centrifugation, or by 3) flow cytometric sorting using FITC-conjugated anti-CD14 monoclonal antibody. It was observed that the amounts of PAF-AH activity secreted into the culture medium correlated positively with the percentage of macrophages in the population (Table 1). In contrast, the medium obtained from CD14-negative cells did not secrete detectable amounts of PAF-AH activity. Moreover, pre-treatment of the cells with CD14 monoclonal antibody and complement specifically decreased the PAF-AH secretion by these cells (Table 2). Based on these observations, it is concluded that decidual macrophages are the major cell type that produces and secretes PAF-AH. Furthermore, it is clear that decidual macrophages secrete PAF-AH which has all the character of the plasma type.

PAF and Preterm Labor Caused by Chorioamnionitis

Bacterial endotoxins induce the release of a vast array of host mediators, including TNF-α, interleukins, interferons, colony stimulating factors, arachidonate metabolites (eicosanoids), and PAF (Braquet et al., 1987; Hinshaw et al, 1990). It has been proposed that the host mediators released from activated decidual monocytes/macrophages by endotoxins may play an important role in the onset of labor associated with infection (Romero et al., 1991; MacDonald et al., 1991; Gibbs et al., 1992). We have previously observed that PAF is present in the amniotic fluid of women at term and in labor but only present in trace quantities in women at term and not in labor (Billah et al., 1983). We also reported that PAF is present in the amniotic fluid of women with preterm labor and premature rapture of membranes (Hoffman et al., 1986). In view of these observations, we have postulated that PAF may be elevated at the fetal-maternal interface in response to bacterial infections which lead to preterm labor or premature rupture of membranes.

We have recently demonstrated that various LPS's inhibited the secretion of PAF-AH by decidual macrophages (Narahara and Johnston, 1993) (Figure 1). The inhibitory potency was affected by the stimulation index/μg LPS, a marker for the capacity of LPS to induce inflammatory reactions. The decrease in PAF-AH production by decidual macrophages may directly relate to the degree of the inflammatory responses caused by various LPS's. As previously discussed, LPS stimulates the monocyte/macrophage system to induce the release of cytokines such as TNF-α, IL-1α and IL-1β (Hinshaw et al, 1990). We tested the possibility that these cytokines might mediate the inhibitory effect of LPS on the PAF-AH secretion by decidual macrophages. The LPS-induced inhibition was partially reversed by IL-1 receptor antagonist or by neutralizing antibodies against IL-1α, IL-1β or TNF-α, which would be consistent with such a mechanism. In addition, the failure of treatment with excess concentrations of these blocking agents, singly or in combination, to obtain a complete reversibility of the LPS-induced inhibition suggests the presence of other LPS-induced mediators. These mediators might include other cytokines, eicosanoids, and PAF, since they also act as LPS-induced inflammatory mediators (Braquet et al., 1987; Hinshaw et al, 1990).

The findings that TNF-α, IL-1α, and IL-1β might participate in the LPS-induced inhibition of PAF-AH secretion directed us to examine the role of these cytokines in the PAF-AH secretion by decidual macrophages. It was demonstrated that TNF-α, IL-1α, and IL-1β decreased the PAF-AH secretion by decidual macrophages (Narahara and

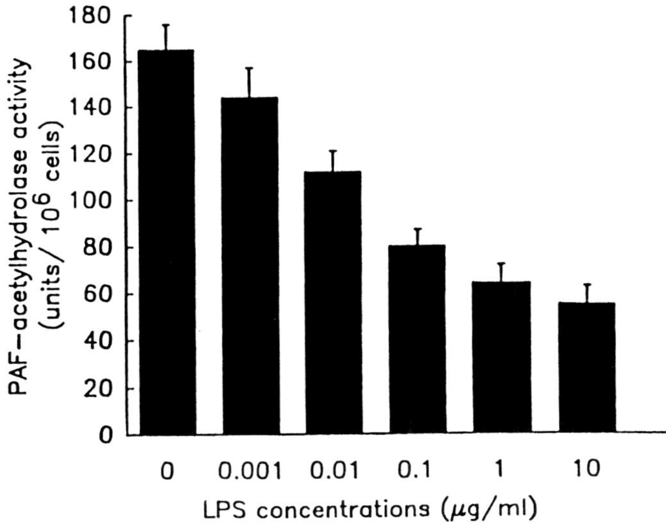

Figure 1. Effect of *E coli* LPS on the PAF-acetylhydrolase secretion by decidual macrophages. Cells were treated with various concentrations of *E coli* LPS on day 0 and cultured for six days. PAF-acetylhydrolase activity in the medium was assayed. One unit is equal to the release of 1 nmol of acetate per 1 hour at 37°C.

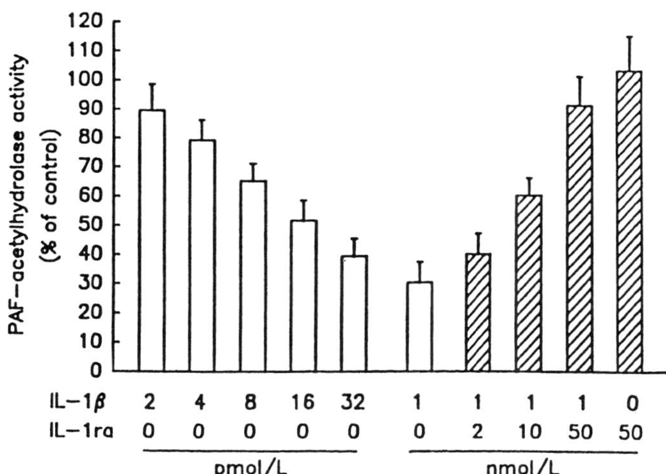

Figure 2. Effects of IL-1β and/or IL-1 receptor antagonist (IL-1ra) on the PAF-acetylhydrolase secretion by decidual macrophages. Cells were treated with various concentrations of IL-1β and/or IL-1ra on day 0 and cultured for six days. PAF-acetylhydrolase activity in the medium was assayed. Results are expressed as percentage of the enzyme activity of a non-treated control.

Table 3

Effect Of Neutralizing Antibodies Against TNF-α Or IL-1β On The Cytokine-Induced Inhibition Of PAF-AH Secretion By Decidual Macrophages

Treatment	PAF-acetylhydrolase activity (% of control)
Control	100 ± 8
TNF - α	45 ± 8*
+ anti-TNF-α	116 ± 12
+ anti-IL-1β	53 ± 7*
IL-1β	32 ± 5*
+ anti-TNF-α	41 ± 6*
+ anti-IL-1β	95 ± 8

Cells were treated with TNF-α (10 nmol/L), IL-1b (1 nmol/L), and/or neutralizing antibodies against TNF-α or IL-1β (1 μg/ml) on day 0 and cultured for six days. PAF-acetylhydrolase activity in the medium was assayed. Results are expressed as percentage of the enzyme activity of a non-treated control. * $p<0.02$ vs. control.

Johnston, 1993). Antibodies against TNF-α or IL-1β specifically and completely neutralized the inhibitory effects of the corresponding cytokines on the PAF-acetylhydrolase secretion (Table 3), supporting further the specific action of these cytokines. The inhibitory effect of IL-1α or IL-1β on the secretion was completely reversed by IL-1ra (Figure 2). It has been reported that IL-1ra, the naturally occurring one, is physiologically present in human amniotic fluid (Romero et al., 1992). Therefore, the complete reversibility also suggests the possible modulation of PAF metabolism by the IL-1ra in the intrauterine tissue: the antagonist might decrease PAF concentration by antagonizing the action of IL-1, a cytokine that not only stimulates PAF production (Braquet et al., 1987) but also inhibits PAF-AH secretion. Although PAF level in amniotic fluid would be relatively low compared with PAF-AH activity secreted by decidual macrophages, the local concentrations of PAF in the decidua might be much higher during parturition. The higher level of PAF in the decidua may not fully be inactivated by the PAF-AH activity as decreased by LPS and/or cytokines such as IL-1 and TNF-α. Based on these observations, it is suggested that PAF is involved in the pathogenesis of preterm labor or premature rupture of membranes caused by endotoxins and the subsequent activation of cytokine network.

SUMMARY

We have demonstrated that decidual macrophages produce and secrete PAF-AH of the plasma type. It is suggested that in addition to the PAF inactivation by plasma PAF-AH in the decidua, a localized PAF-AH secretion by decidual macrophages may exist for the autocrine or paracrine regulation of PAF concentration at the maternal-fetal interface. We have also demonstrated that LPS, TNF-α, IL-1α, and IL-1β inhibited the

PAF-AH secretion by decidual macrophages and suggested a synergy among these bioactive molecules in the regulation of PAF metabolism in the decidua. These observations lend additional support to the concept that PAF is pathophysiologically involved in human parturition and provide further insights into the mechanism underlying the pathogenesis in preterm labor or premature rupture of membranes associated with bacterial infections.

REFERENCES

Billah, M.M. and Johnston, J.M. (1983) Identification of phospholipid platelet-activating factor (1-*O*-alkyl-2-acetyl-*sn*-glycero-3-phosphocholine) in human amniotic fluid and urine. *Biochem. Biophys. Res. Commun.* 113, 51-8.

Billah, M.M., DiRenzo, G.C., Ban, C., Truong, C.T., Hoffman, D.R., Anceschi, M.M., Bleasdale, J.E. and Johnston, J.M. (1985) Platelet-activating factor metabolism in human amnion and the responses of this tissue to extracellular platelet-activating factor. *Prostaglandins* 30, 841-50.

Braquet, P., Touqui, L., Shen, T.Y., and Vargaftig, B.B. (1987) Perspectives in platelet-activating factor research. *Pharmacol. Rev.* 39, 97-146.

Frenkel, R.A. and Johnston, J.M. (1992) Metabolic conversion of platelet-activating factor into ethanolamine plasmalogen in an amnion-derived cell line. *J. Biol. Chem.* 267, 19186-19191.

Gibbs, R.S., Romero, R., Hillier, S.L., Eschenbach, D.A. and Sweet, R.L. (1992) A review of premature birth and subclinical infection. *Am. J. Obstet. Gynecol.* 166, 1515-1528.

Hanahan, D.J. (1986) Platelet-activating factor: A biologically active phosphoglyceride. *Ann. Rev. Biochem.* 55, 483-509.

Hinshaw, L.B. (1990) Pathophysiology of endotoxin action: An overview. In: Cellular and molecular aspects of endotoxin reactions, (eds.) A. Nowotny, J.J. Spitzer and E.J. Ziegler, New York: Elsevier Science, pp. 419-426.

Hoffman, D.R., Truong, T.C. and Johnston, J.M. (1986) Metabolism and function of platelet-activating factor in fetal rabbit lung development. *Biochim. Biophys. Acta* 879, 88-96.

Hoffman, D.R., Romero, R., and Johnston, J.M. (1990) Detection of platelet-activating factor in amniotic fluid of complicated pregnancies. *Am. J. Obstet. Gynecol.* 162, 525-528.

MacDonald, P.C., Koga, S. and Casey, M.L. (1991) Decidual activation in parturition: Examination of amniotic fluid for mediators of the inflammatory response. *Ann. NY Acad. Sci.* 622, 315-330.

Maki, N., Hoffman, D.R. and Johnston, J.M. (1988) PAF (platelet-activating factor) acetylhydrolase activities in maternal, fetal, and newborn rabbit plasma during pregnancy and lactation. *Proc. Natl. Acad. Sci. USA* 85, 728-732.

Miwa, M., Miyake, T., Yamanaka, T., Sugatani, J., Suzuki, Y., Sakata, S., Araki, Y., and Matsumoto, M. (1988) Characterization of serum platelet-activating factor (PAF) acetylhydrolase. Correlation between deficiency of serum PAF acetylhydrolase and respiratory symptoms in asthmatic children. *J. Clin. Invest.* 82, 1983-1991.

Morris, C., Khan, H., Sullivan, M.H.F. and Elder, M.G. (1992) Effects of platelet-activating factor on prostaglandin E_2 production by intact fetal membranes. *Am. J. Obstet. Gynecol.* 166, 1228-1231.

Narahara, H., Frenkel, R. and Johnston, J.M. (1993a) Secretion of platelet-activating factor acetylhydrolase (PAF-AH) following phobol ester-stimulated differentiation of HL-60 cells. *Arch. Biochem. Biophys.* 301, 275-281.

Narahara, H., Nishioka, Y. and Johnston, J.M. (1993b) Secretion of platelet-activating factor acetylhydrolase by human decidual macrophages. *J. Clin. Endocrinol. Metab.* 77, 1258-1262.

Narahara, H. and Johnston, J.M. (1993) Effects of enotoxins and cytokines on the secretion of platelet-activating factor-acetylhydrolase by human decidual macrophages. *Am. J. Obstet. Gynecol.* 169, 531-537.

Narahara, H., Frenkel, R.A. and Johnston, J.M. (1996) The role of PAF in reproductive biology. In: Advances in Molecular Biology, Vol. 1B, (ed.), R. Gross, JAI Press: Greenwich, CT, pp. 241-271.

Nieto, M.L., Venable, M.E., Bauldry, S.A., Greene, D.G., Kennedy, M., Bass, D.A. and Wykle, R.L. (1991) Evidence that hydrolysis of ethanolamine plasmalogens triggers synthesis of platelet-activating factor via a transacylation reaction. *J. Biol. Chem.* 266, 18699-18706.

Nishihira, J., Ishibashi, T., Mai, Y. and Muramatsu, T. (1984) Mass spectrometric evidence for the presence of platelet-activating factor (1-*O*-alkyl-2-acetyl-*sn*-glycero-3-phosphocholine) in human amniotic fluid during labor. *Lipids* 19, 907-910.

Prescott, S.M., Zimmerman, G.A. and McIntyre, T.M. (1990) Platelet-activating factor. *J. Biol. Chem.* 265, 17381-17384.

Romero, R., Avila, C., Brekus, C.A. and Morotti, R. (1991) The role of systemic and intrauterine infection in preterm parturition. *Ann. NY. Acad. Sci.* 622, 355-375.

Romero, R., Sepulveda, W., Mazor, M., Brandt, F., Cotton, D.B., Dinarello, C.A. and Mitchell, M.D. (1992) The natural interleukin-1 receptor antagonist in term and preterm parturition. *Am. J. Obstet. Gynecol.* 167, 863-872.

Satoh, K., Imaizumi, T., Kawamura, Y., Yoshida, H., Hiramoto, M., Takamatsu, S. and Takamatsu, M. (1991) Platelet-activating factor (PAF) stimulates the production of PAF acetylhydrolase by the human hepatoma cell line, HepG2. *J. Clin. Invest.* 87, 476-481.

Snyder, F. (1990) Platelet-activating factor and related acetylated lipids as potent biologically active cellular mechanisms. *Am. J. Physiol.* 259, C697-C708.

Tetta, G., Montruccho, G., Alloatti, G., Roffinello, C., Emanuelli, G., Benendetto, C., Camussi, G. and Massobrio, M. (1986) Platelet-activating factor contracts human myometrium *in vitro*. *Proc. Soc. Exp. Biol. Med.* 183, 376-381.

Toyoshima, K., Narahara, H., Frenkel, R.A. and Johnston, J.M. (1994) Coenzyme A-independent transacylation in amnion-derived (WISH) cells. *Arch. Biochem. Biophys.* 314, 224-228.

Uemura, Y., Lee, T-C., and Snyder, F. (1991) A coenzyme A-independent transacylase is linked to the formation of platelet-activating factor (PAF) by generating the lyso-PAF intermediate in the remodeling pathway. *J. Biol. Chem.* 266, 8268-8272.

Yasuda, K., Eguchi, H., Narahara, H., and Johnston, J.M. (1993) Platelet-activating factor: Its regulation in parturition. In: Eicosanoids and Other Bioactive Lipids in Cancer, Inflammation and Radiation Injury, (eds.) S. Nigam, K.V. Honn, L.J. Marnett, and T.L. Walden, Boston: Kluwer Academic, pp. 727-730.

Zhu, Y-P., Word, R.A. and Johnston, J.M. (1992) The presence of platelet-activating factor binding sites in human myometrium and their role in uterine contraction. *Am. J. Obstet. Gynecol.* 166, 1222-1228.

Trophoblast Research 9:109-120, 1997

THE ROLE OF PLATELET-ACTIVATING FACTOR AND PLATELET-ACTIVATING FACTOR ACETYLHYDROLASE IN FETAL AND PLACENTAL GROWTH IN RATS

Katsuhiko Yasuda, Takashi Matsubara, Tokuro Nakajima, Isamu Sawaragi, and Hideharu Kanzaki

Department of Obstetrics and Gynecology
Kansai Medical University
10-15 Fumizono-cho
Moriguchi, Osaka, 570, Japan

INTRODUCTION

Platelet-activating factor (PAF) was discovered as a chemical mediator released from sensitized basophils (Benveniste, 1972). Its structure was elucidated as 1-O-alkyl-2-acetyl-sn-glycero-3-phosphocholine (AGEPC) by Demopoulos et al. (1979), and Benveniste et al. (1979). Subsequent studies have demonstrated that PAF is produced upon appropriate stimulation by a variety of human cells, including neutrophils, eosinophils, monocytes and endothelial cells, and that PAF is related to allergy and inflammation (Hanahan, 1986; Braquet et al., 1987). PAF is also found in the normal brain (Tokumura et al., 1987), stomach (Sugatani et al., 1989), kidney (Camussi et al., 1989), and uterus (Yasuda et al., 1986), in which it may play a physiological role.

Recently, it has been suggested that PAF is associated with a number of clinical disorders, including cardiovascular disease, asthma, endotoxin shock, gastrointestinal ulceration, renal disease, and ischemic disorders (Snyder, 1989).

Farr and associates (1980) reported that an enzyme capable of inactivating PAF, PAF-acetylhydrolase (PAF-AH, EC3.1.1.48) was present in the low density lipoprotein (LDL) fraction of human plasma. Blank and associates (1981) further characterized PAF-AH using rat tissues.

In the present study, we investigated the adverse effects of PAF administration on fetal and placental growth and plasma PAF-AH activity during pregnancy.

MATERIALS AND METHODS

Experimental Animals

Pregnant Wistar rats were obtained from Oriental Bioservice Co., Kyoto, Japan. Gestation was timed to within 12 hours of mating. Pregnant rats were housed under controlled conditions (12 hours light and 12 hours dark) and were provided with water and rat chow *ad libitum*. All rats were euthanatized under ether anesthesia and fetal body weight and placental weight were measured. Blood samples from the tail vein of pregnant rats were collected in heparinized capillary tubes. The plasma was separated by centrifugation and stored at -20°C until assayed for PAF-AH activity.

Chemicals

1-O-Hexadecyl-2-[^{3}H]acetyl-sn-glycero-3-phosphocholine ([^{3}H]PAF, 10 Ci/mmol, 1 Ci=37 GBq) was purchased from New England Nuclear, Boston, MA, USA. Nonradiolabeled PAF was obtained from Bachem Feinchemikalien AG, Bubendorf, Switzerland and was purified by thin layer chromatography (solvent system; chloroform:methanol:acetic acid:water = 50:30:8:6)before use. Bovine serum albumin (BSA; fatty acid free) was purchased from Sigma Chemical, St. Louis, MO, USA.

Treatment of Pregnant Rats With PAF

PAF was dissolved in 0.1% BSA and made up to the desired concentrations (2 nmol/ml, 5 nmol/ml, and 10 nmol/ml) according to a previous report (Takashima et al., 1994). We employed less than 10 nmol/kg dose of PAF, because lethal dose of PAF was 10 nmol/kg in adult non-pregnant rats (Takashima et al., 1994; Yasuda et al., 1995). Pregnant rats were daily injected in the tail vein with either vehicle (0.1% BSA) or PAF (1 nmol/kg, 2.5 nmol/kg, and 5 nmol/kg) from 15 day to 20 day of pregnancy. All rats were euthanatized under ether anesthesia on 21 day of pregnancy (before the delivery), and fetal body weight and placental weight were measured to evaluate the effect of PAF on fetal and placental growth. Placental tissues were microscopically examined following hematoxylin and eosin staining.

Assay for Plasma PAF-AH Activity

The activity of plasma PAF-AH was assayed according to the of Yasuda and Johnston (1992). Briefly, the assay mixture contained 0.3 ml of Tris-HCl (50 mM, pH 7.4) containing BSA (2.0 mg/ml), 0.1 ml of the diluted plasma, and radiolabeled PAF substrate (0.05 mM); the final volume was 0.5 ml. [^{3}H]Acetyl-PAF and nonradiolabeled PAF were suspended in 0.1% BSA. Rat plasma was diluted 20-fold with 0.25 M sucrose prior to assay. The assay mixture was incubated for 20 minutes at 37°C. The reaction was terminated by the addition of 0.5 ml of trichloroacetic acid (14%) and the mixture was centrifuged for 10 minutes at 4°C (600 Xg). One-tenth of a milliliter of the supernatant was removed and mixed with 5 ml of scintillation fluid (New England Nuclear, Boston, MA, USA). The water-soluble [^{3}H]acetate released from [^{3}H]acetyl-PAF was assayed by liquid scintillation spectroscopy.

Statistical Analysis

Results are presented as means ± SD. Statistical analysis was performed using Wilcoxon's test (paired and nonpaired) and Student's t test. A level of $p < 0.05$ was considered statistically significant.

RESULTS

Fetal Body Weight and Placental Weight During Pregnancy

Fetal body weight and placental weight in non-treated pregnant rats were measured on 15, 16, 17, 18, 19, 20, and 21 day of pregnancy, respectively. As shown in

Figure 1 and 2, fetal body weight increased dramatically from 15 day to 21 day of pregnancy (0.241 ± 0.038 g to 5.570 ± 0.288 g), while placental weight increased gradually on the corresponding day (0.267 ± 0.051 g to 0.630 ± 0.071 g).

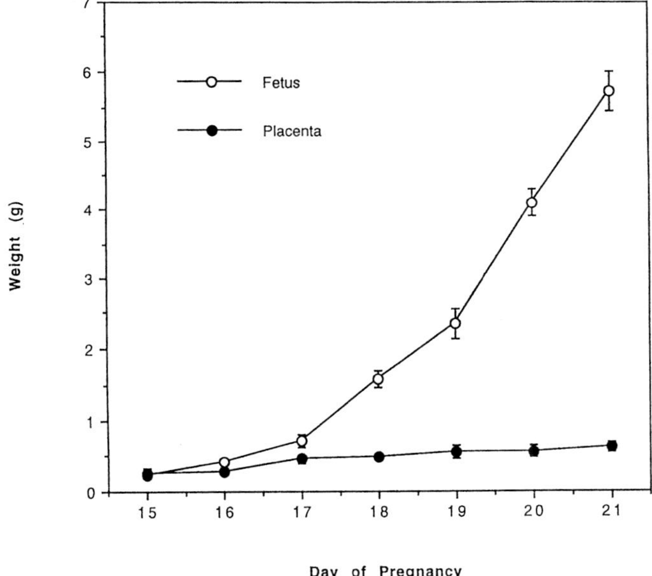

Figure 1. Growth curve of fetuses and placentae during pregnancy. Pregnant rats were euthanatized on 15, 16, 17, 18, 19, 20, 21 day of pregnancy and the weight of the fetus and the placenta was measured. Number of the fetus and the placenta was 16, 14, 16, 20, 19, 16, 17, respectively. The values are represented as means ± SD.

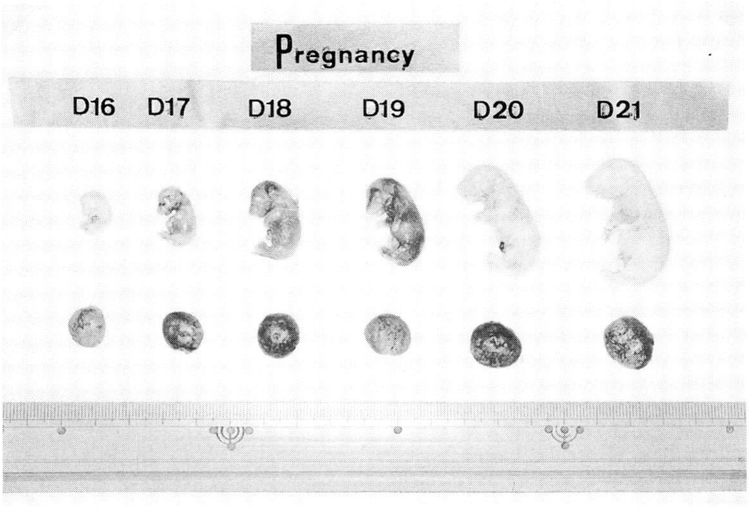

Figure 2. Comparison between fetal and placental size during pregnancy. The fetal and placental size on 15, 16, 17, 18, 19, 20, and 21 day of pregnancy was compared, respectively. The fetus and the placenta were same as in Figure 1.

Plasma PAF-AH Activity in Rats During Pregnancy

Plasma PAF-AH activity in non-treated pregnant rats on 13, 15, 16, 17, 18, 19, 20, and 21 day of pregnancy was assayed (Figure 3). Plasma PAF-AH activity decreased gradually during pregnancy from 58.8 ± 8.0 nmol/min/ml on 13 day of pregnancy to 36.5 ± 6.6 nmol/min/ml on 19 day of pregnancy ($p < 0.05$), and was still low on 20 and 21 day of pregnancy (39.1 ± 7.6 and 37.6 ± 5.0 nmol/min/ml, respectively; $p < 0.05$ vs. 13 day).

Effect of PAF on Fetal and Placental Growth

Pregnant rats treated with either 0.1% BSA (control group) or 1 nmol/kg, 2.5 nmol/kg, and 5 nmol/kg dose of PAF (PAF-treated group) from 16 day to 20 day of pregnancy were euthanatized on 21 day of pregnancy, and the weight of the fetus and the placenta was measured (Figures 4 and 5). In this experiment, none of the pregnant rats delivered before euthanasia on 21 day of pregnancy. Fetal body weight in the control group was 5.857 ± 0.437g (n=93). Fetal body weight in three PAF-treated groups (1 nmol/kg, 2.5 nmol/kg, and 5 nmol/kg dose of PAF) was 5.349 ± 0.466 g (n=51), 5.463 ± 0.540 g (n=64), and 5.444 ± 0.496 g (n=79), respectively and a significant difference was found between the control and each PAF-treated group ($p < 0.01$, respectively). However, no significant differences were found among PAF-treated groups. Intrauterine fetal death (IUFD) was found in pregnant rats treated with 5 nmol/kg dose of PAF (11.2%); however, no IUFD in pregnant rats treated with either 0.1% BSA or 1 nmol/kg, and 2.5 nmol/kg dose of PAF was found. Placental weight in pregnant rats treated with 5 nmol/kg dose of PAF was significantly lower than that in the control group (0.543 ± 0.093 g, n=93 vs. 0.590 ± 0.094g, n=79; $p < 0.01$). However, no significant difference in placental weight was found between the control group and the 1 nmol/kg or 2.5 nmol/kg dose of PAF-treated groups.

Plasma PAF-AH activity in pregnant rats treated with 0.1% BSA, 1 nmol/kg, 2.5 nmol/kg, and 5 nmol/kg dose of PAF was 36.5 ± 6.6, 37.8 ± 5.9, 34.1 ± 3.1, and 33.8 ± 4.2 nmol/min/ml on 20 day of pregnancy, respectively. No significant difference in the enzyme activity was found between each group.

Microscopic Findings of Placenta

The rat placenta is discoid and microscopically belong to the hemochorical type like the human. The area that is in immediate contact with the decidua is called the giant cell layer. The next is the spongiotrophoblast layer and is continuous with the labyrinth (Figures 6a and b, control group).

Histological differences between the control and the PAF treated group were observed. In the PAF-treated group, few erythrocytes were seen in the spongiotrophoblast layer and the labyrinth (Figure 6c). It resembled ischemic changes. Many thrombi existed in the vessels of the spongiotrophoblast layer and the labyrinth (Figures 6d and e). Bleeding and necrosis were partially seen in the labyrinth, where the cells were strongly degenerated (Figure 6f).

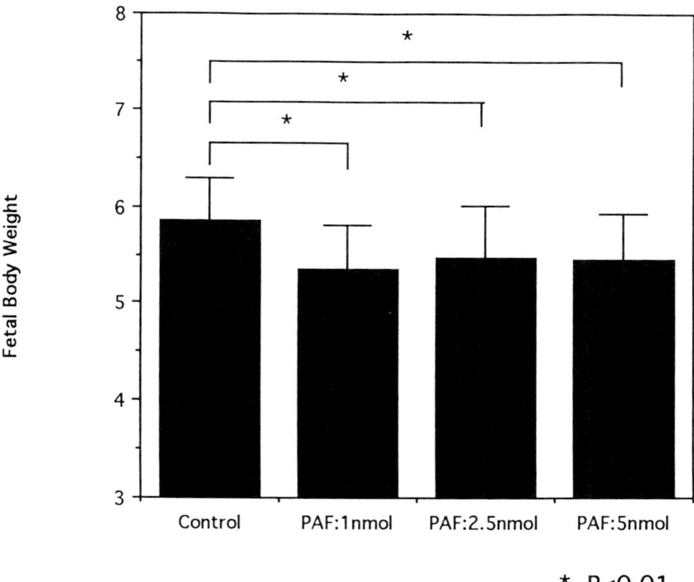

Figure 3. Change of plasma PAF-AH activity in pregnant rats. Blood samples were collected from the tail vein on 13, 15, 16, 17, 18, 19, 20, and 21 day of pregnancy. The values are means ± SD from 6 rats.

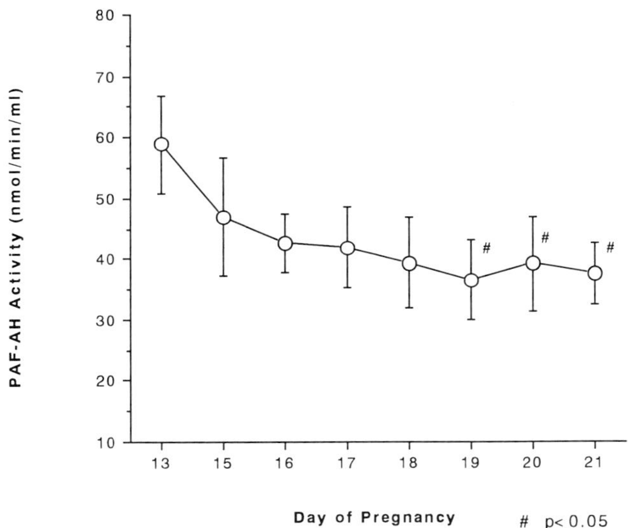

Figure 4. Effect of PAF on the fetal growth in rats. Pregnant rats received 0.1% BSA as vehicle (N=6), 1 nmol/kg of PAF (N=4), 2.5 nmol/kg of PAF (N=4), or 5 nmol/kg of PAF (N=6), by daily injection (16 to 20 day of pregnancy), and euthanatized after ether anesthesia on 21 day of pregnancy. The number of the fetus not included IUFD was 93, 51, 64, and 79, respectively. Ten IUFD was found in pregnant rats treated with 5 nmol/kg of PAF.

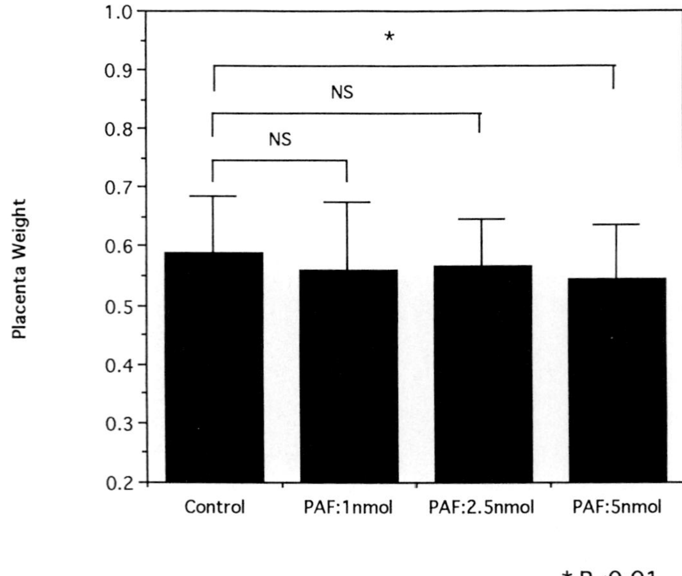

* P<0.01

Figure 5. Effect of PAF on the placental growth in rats. Pregnant rats were same as in Figure 4. The weight of the placenta not including fetal membrane and umbilical cord was measured.

DISCUSSION

PAF was discovered as a chemical mediator released from sensitized basophils. Subsequent studies have demonstrated that PAF is produced by a variety of human cells, including neutrophils, eosinophils, monocytes, and that PAF cause platelet aggregation, activation of blood cells (neutrophils, macrophages, monocytes), contraction of smooth muscle, increased vascular permeability, hypotension, reduction of cardiac output, glycogenolysis. PAF is also found in some organs, including the brain, stomach, kidney, and uterus. Recently, we have found PAF in rat placenta and elucidated its structure as AGEPC by using gas chromatography-mass spectrometer (unpublished data).

It is apparent that the level of PAF in plasma and tissue is regulated not only by its biosynthesis but also by its degradation. The content of PAF in the normal rat uterus is significantly reduced by bilateral oophorectomy and is restored by estrogen administration to the level seen in an untreated group (Nakayama et al., 1991). This report suggests that the PAF level in plasma or tissues is regulated in part by hormones that regulate the synthesis and/or the degradation of PAF. In the degradation of PAF, biologically active PAF is converted to the inactive lysoPAF by PAF-AH. Plasma PAF-AH activity decreases in maternal plasma during the later stages of pregnancy in rabbits (Maki et al., 1988) and humans (Johnston, 1989). We also found a decrease of plasma PAF-AH activity in pregnant rats. The enzyme activity decreased from 58.8 ± 8.0 nmol/min/ml on 13 day of pregnancy (almost same activity as that in non-pregnant rats) to 36.5 ± 6.6 nmol/min/ml on 19 day of pregnancy ($p < 0.05$), and was still low on 20 and 21 day of pregnancy (39.1 ± 7.6 and 37.6 ± 5.0 nmol/min/ml, respectively; $p < 0.05$ vs. 13 day). Thus, the biological activities of PAF intend to appear in later stage of pregnancy because of low PAF-AH activity.

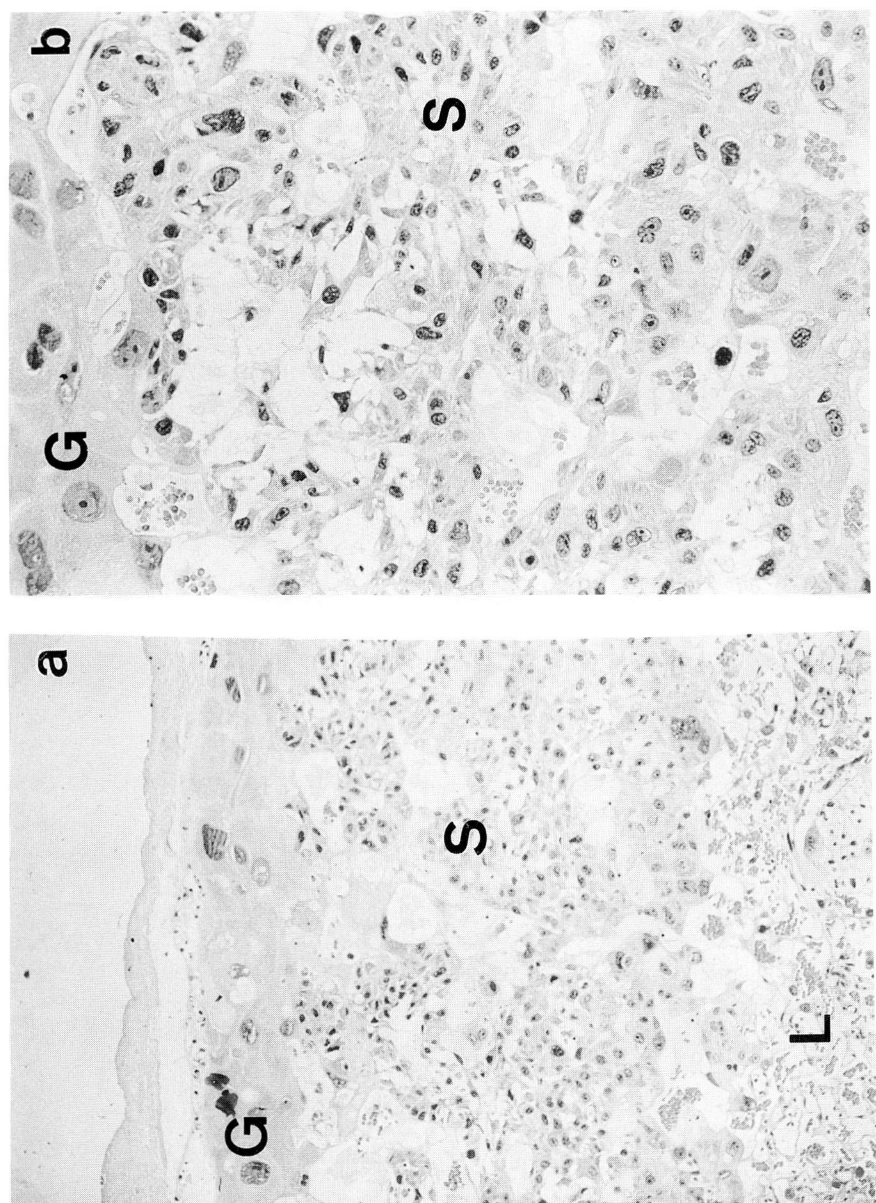
b
G
S
a
G
S
L

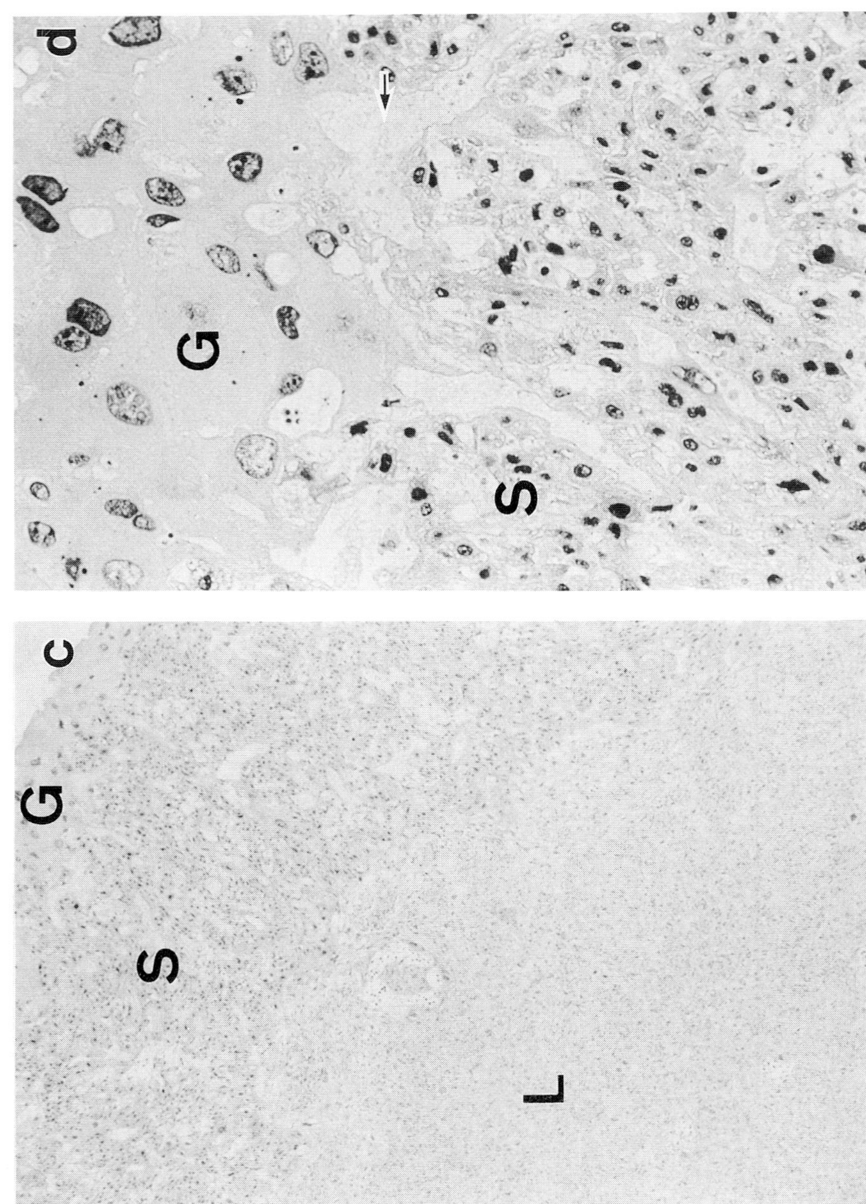
d
G
S
c
G
S
L

Figure 6e and f here - full page - sideways

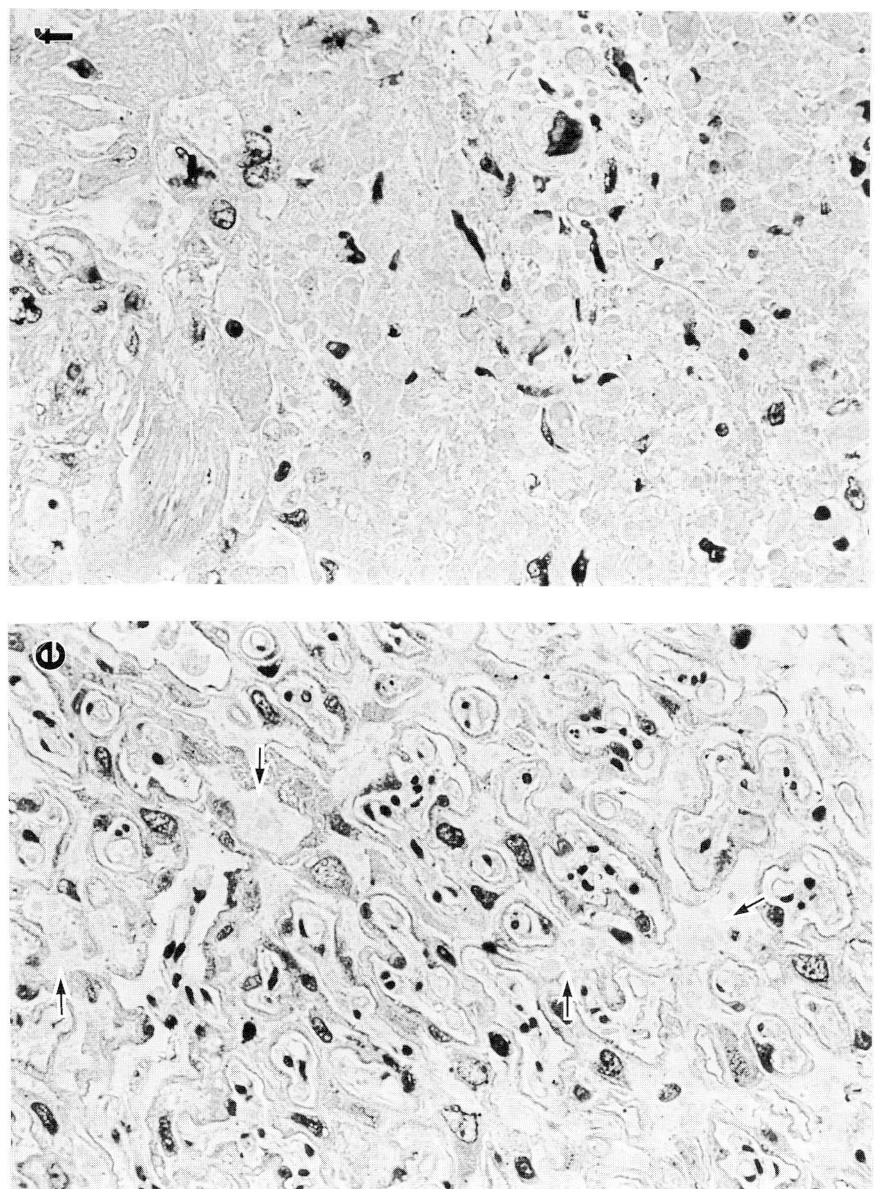

Figure 6. (previous three pages). Microscopic findings in the placentas of pregnant rats treated with PAF. Pregnant rats were same as in Figure 4. Rats in control and PAF-treated group were treated with 0.1% BSA as vehicle and 5 nmol/kg of PAF, respectively. The structure of the placenta in control group was shown on **a** (x100) and **b** (x200), and that in PAF-treated group was shown **c** (x40), **d** (x200), **e** (x200), and **f** (x200). Symbol of arrow shows thrombus. G, Giant cell layer; S, Spongiotrophoblast layer; L, Labyrinth;

Takashima et al. (1994) and Yasuda et al. (1995) have reported that the lethal dose of PAF in adult non-pregnant rat was 10 nmol/kg and demonstrated thrombi and ischemic changes in some organs, including the lung and heart in the PAF-treated rats. It is unknown whether the phenomenon in the PAF-treated rats was caused by PAF, itself or metabolites of PAF, because PAF-AH is present in excessive amounts in plasma and can immediately degrade PAF to lysoPAF. However, LysoPAF, a major metabolite of PAF, has few biological activities. Therefore, it appears that some of the PAF, escaped being inactivated by PAF-AH, and was related to thrombi and ischemic changes observed in PAF-treated rats.

In the present study, PAF (1 nmol/kg, 2.5 nmol/kg, and 5 nmol/kg) was administered in pregnant rats to evaluate the effect of PAF on fetal and placental growth. Fetal body weight in three PAF-treated groups was significantly lower than that in the control group, and IUFD was found in the group treated with 5 nmol/kg dose of PAF. This experimental result suggests that excessive PAF affects the fetal growth. Placental weight in pregnant rats treated with 5 nmol/kg dose of PAF was also significantly lower than that in the control group. However, a significant decrease that was seen in fetal body weight was not found in the placental weights of the group treated with 1 nmol/kg or 2.5 nmol/kg dose of PAF. This discrepancy may depend on the difference between the growth curve of the fetus and the placenta. The growth curve of the fetus greater than that of the placenta (Figure 1). Thus, significant differences in fetal body weight are easily found, and that for placental weights, differences are small. Microscopic findings demonstrated thrombi and ischemic changes including bleeding, necrosis, and degeneration in the placenta of the PAF-treated groups.

These results suggest that PAF, which is produced excessively in the uterus and placenta associated with inflammation, vascular diseases, and severe stress caused by mechanical or chemical stimulation, escape being inactivated by PAF-AH because the enzyme activity is low in the later stage of pregnancy. The escaped PAF may affect uteroplacental circulation or normal growth of placenta and then may cause IUGR, and IUFD.

SUMMARY

The adverse effects of PAF on the growth of rat fetuses and placentae were investigated. Plasma PAF-AH activity in pregnant rats was also investigated during the later stages of pregnancy.

The enzyme activity in pregnant rats decreased from 58.8 ± 8.0 nmol/min/ml on 13 day of pregnancy to 36.5 ± 6.6 nmol/min/ml on 19 day of pregnancy ($p < 0.05$), and was still low on 20 and 21 day of pregnancy.

When pregnant rats were treated with either 0.1% BSA (control group) or a 1 nmol/kg, 2.5 nmol/kg, and 5 nmol/kg dose of PAF (PAF-treated group) from 16 day to 20 day of pregnancy, fetal body weight in three PAF-treated groups was significantly lower than that in control group ($p < 0.01$). Intrauterine fetal death (IUFD) was found in pregnant rats treated with 5 nmol/kg dose of PAF, however no IUFD in pregnant rats treated with either 0.1% BSA or 1 nmol/kg and 2.5 nmol/kg dose of PAF was found. Placental weight in pregnant rats treated with 5 nmol/kg dose of PAF were also significantly lower than that in control group ($p < 0.01$). However, significant differences were not found between control group and 1 nmol/kg or 2.5 nmol/kg dose of PAF-treated groups. Microscopic findings demonstrated thrombi and ischemic changes including bleeding, necrosis, and degeneration in placentae in PAF-treated groups.

These results suggest that some of the PAF in tissues and plasma escape from PAF-AH and PAF levels increase during the later stages of pregnancy. The increased PAF may affect uteroplacental circulation or normal growth of placenta, and then may cause IUGR, and IUFD.

ACKNOWLEDGEMENTS

The authors appreciate the editorial assistance of Ms. Noriko Sugie and Ms. Itsuko Nakagawa.

REFERENCES

Benveniste, J., Henson, P.M. and Cochrane, C.G. (1972) Leukocyte-dependent histamine release from rabbit platelets: The role of IgE, basophils and platelet-activating factor. *J. Exp. Med.* 136, 1356-1377

Benveniste, J., Tence, M., Varenne, P., Bidault, J., Boullet, C. and Polonsky, J. (1979) Semi-synthese et structure purposee du facteur activant les plaquettes (PAF): PAF-acether, un alkyl ether analogue de la lysophosphatidylcholine. *C.R. Acad. Sci.* [D]. 289, 1037-1040.

Blank, M.L., Lee, T-C., Fitzgerald, V. and Snyder, F. (1981) A specific acetylhydrolase for 1-alkyl-2-acetyl-sn-glycero-3-phosphocholine (a hypotensive and platelet-activating lipid). *J. Biol. Chem.* 256, 175-178.

Braquet, P., Touqui, L., Shen, T.Y. and Vargaftig, B.B. (1987) Perspectives in platelet-activating factor research. *Pharmacol. Rev.* 39, 97-146.

Camussi, G., Bussolino, F. and Tetta, C. (1989) Platelet-activating factor and renal disease. In: *Platelet Activating Factor and Disease*, (eds.) K. Saito and D.J. Hanahan, Tokyo: International Medical Publishers, 85, pp. 113-128.

Demopoulos, C.A., Pinckard, R.N. and Hanahan, D.J. (1979) Platelet-activating factor: evidence for 1-O-alkyl-2-acetyl-sn-glyceryl-3-phosphorylcholine as the active component (a new class of lipid chemical mediators). *J. Biol. Chem.* 254, 9355-9358.

Farr, R.S., Cox, C.P., Wadlow, M.L. and Jorgensen, R. (1980) Preliminary studies of an acid-labile factor ALF in human sera that inactivates platelet-activating factor PAF. *Clin. Immunopathol.* 15, 318-330.

Hanahan, D.J. (1986) Platelet-activating factor: A biologically active phosphoglyceride. *Annu. Rev. Biochem.* 55, 483-509.

Johnston, J.M. (1989) The function of PAF in the communication between the fetal and maternal compartments during parturition. In: *Platelet-Activating Factor and Diseases* (eds.) K. Saito and D.J. Hanahan, Tokyo: International Medical Publishers, 85, pp. 129-138.

Maki, N., Hoffman, D.R. and Johnston, J.M. (1988) Platelet-activating factor acetylhydrolase activity in maternal, fetal, and newborn rabbit plasma during pregnancy and lactation. *Proc. Natl. Acad. Sci. USA* 85, 728-732.

Nakayama, R., Yasuda, K., Okumura, T. and Saito, K. (1991) Effect of 17β-estradiol on PAF and prostaglandin levels in oophorectomized rat uterus. *Biochim. Biophys. Acta* 1085, 235-240.

Sugatani, J., Fujimura, K., Miwa, M., Mizuno, T., Sameshima, Y. and Saito, K. (1989) Occurrence of platelet-activating factor (PAF) in normal rat stomach and alteration of PAF level by water immersion stress. *FASEB. J.* 3, 65-70.

Takashima, M., Yasuda, K., Umezaki, K., Nakajima, T. and Sawaragi, I. (1994) Effects of estrogen and progestin on platelet-activating factor acetylhydrolase. *Acta. Obstet. Gynaecol. Jpn.* 46, 481-488.

Tokumura, A., Kamiyasu, K., Takauchi, K. and Tsukatani, H. (1987) Evidence for existence of various homologues and analogues of platelet activating factor in a lipid extract of bovine brain. *Biochem. Biophys. Res. Commun.* 145, 415-425.

Yasuda, K., Satouchi, K. and Saito, K. (1986) Platelet-activating factor in normal rat uterus. *Biochem. Biophys. Res. Commun.* 138, 1231-1236.

Yasuda, K. and Johnston, J.M. (1992) The hormonal regulation of platelet-activating factor-acetylhydrolase in the rat. *Endocrinology* 130, 708-716.

Yasuda, K., Takashima, M. and Sawaragi, I. (1995) Influence of a cigarette smoke extract on the hormonal regulation of platelet-activating factor acetylhydrolase in rats. *Biol. Reprod.* 53, 244-252.

Trophoblast Research 9:121-129, 1997

ULTRASTRUCTURAL LOCALIZATION AND CYTOCHEMICAL CHARACTERISTICS OF HUMAN PLACENTAL ADP-DEGRADING ACTIVITY IN NORMAL AND PREECLAMPTIC PREGNANCY

Shigeki Matsubara[1,2,3], Ikuo Sato[1], and Takuma Saito[2]

[1]Department of Obstetrics and Gynecology
[2]Department of Anatomy
Jichi Medical School
Tochigi 329-04, Japan

INTRODUCTION

The placental intervillous space is a large blood-containing space. Although coagulation would be expected to occur in this space, significant placental thrombosis is relatively rare, suggesting that a potent anti-aggregatory mechanism for platelets exists in the human placenta.

Adenosine diphosphate (ADP), which is a powerful platelet aggregator, is hydrolyzed enzymatically to adenosine monophosphate (AMP), which is a platelet anti-aggregator. Degradation of ADP occurs not only in the vascular endothelium (Lieberman et al., 1977; Pearson et al., 1980) but also in the human placenta (Hutton et al., 1980a; O'Brien et al., 1987). The enzyme responsible for ADP degradation reportedly inhibits platelet aggregation, thus maintaining *in vitro* blood fluidity in the vascular endothelium and the placenta (Lieberman et al., 1977; Pearson et al., 1980; Hutton et al., 1980a; O'Brien et al., 1987).

In vivo platelet aggregation is believed to be significantly enhanced in the placenta during preeclampsia (Ahmed et al., 1991; Hutt et al., 1994). Increased platelet aggregation may induce microthrombus formation in the intervillous space, thus interfering with maternal-fetal transport of substances and leading to placental insufficiency. In a previous study, ultracytochemical analysis demonstrated that ADP-degrading activity in the normal human placenta was confined to the external surface of the microvillous membrane of the syncytiotrophoblast (Matsubara et al., 1987a). The details of the localization for this enzyme and the cytochemical characteristics of the ADP-degrading activity in normal and preeclamptic placentae have not yet been clarified.

In the present study, we investigated ADP-degrading activity in normal and preeclamptic placentae using an enzyme cytochemical method.

[3]To Whom Correspondence Should Be Addressed: Department of Obstetrics and Gynecology, Jichi Medical School, Minamikawachi-machi, Kawachi-gun, Tochigi, 329-04, Japan

MATERIALS AND METHODS

Placentae were obtained within 3 minutes of term delivery from 10 healthy, normotensive women and from 5 women with a clinical diagnosis of severe preeclampsia. Severe preeclampsia was defined according to the criteria of the Japanese Society of Obstetrics and Gynecology, as a maternal blood pressure (systolic/diastolic) > 160/110 mmHg on two separate occasions, a urinary protein level >200 mg/dl, or the presence of generalized edema. There were no significant differences between the normal and preeclamptic subjects in either gestational time or maternal age, but all infants born to mothers with preeclampsia were small for their gestational age. Subjects with preeclampsia had not received anti-platelet aggregation therapy.

Several pieces of tissue with a normal appearance and without visible areas of infarction were obtained from the central part of the maternal placental surface and fixed in 2% glutaraldehyde in a cacodylate buffer (0.1 M, pH 7.4) for 30 minutes at 4°C. After being washed in a cacodylate buffer (0.1 M, pH 7.4) for 12 hours, 40 μm sections were cut with a Vibratome (Oxford, California, USA) or a freezing microtome (MB 201, Komatsu Electric Inc., Japan). Sections were then incubated in a reaction medium consisting of 2.0 mM adenosine 5'-diphosphate (ADP) used as a substrate, 80 mM tris-maleate buffer (pH 7.4), 3.6 mM lead nitrate used as a capturing agent, and 5 mM manganese chloride used as an enzyme activator for 30 minutes at 37°C according to the method of Novikoff and Goldfischer (1961). To detect specific ADP-degrading activity uncontaminated by human placental alkaline phosphatase (HPAP), l-p-bromotetramisole oxalate (final concentration: 1.0 mM) and l-phenylalanine (final concentration: 30 mM) were added to the reaction medium (Borgers et al., 1973; Matsubara et al., 1987b). To further eliminate contamination by HPAP activity, sections were preincubated for 30 minutes at room temperature in a cacodylate buffer (0.1M, pH 7.4) containing 1.0 mM l-p-bromotetramisole and 30 mM l-phenylalanine (Matsubara et al., 1987b). Sections were then postfixed in Caulfield's solution (buffered 1 % osmium) for 60 minutes at 4°C, dehydrated in a series of graded alcohol, and embedded in epoxy resin, Quetol 812 (Nissin EM. Co., Japan). Ultrathin sections were obtained with an LKB Ultratome 3 (Stockholm, Sweden), lightly stained with uranyl acetate and lead citrate, and observed under a Hitachi H-7000 transmission electron microscope. For light microscopic histochemistry, the sections were rinsed in distilled water, treated with 1% ammonium sulfide for 2 minutes and then mounted on glass slides with glycerin jelly.

The following experiments were performed to confirm the cytochemical specificity and to further characterize the cytochemical features of ADP-degrading activity. a) To detect total ADP-degrading activity, tissue sections were incubated in medium with ADP as a substrate in the absence of HPAP inhibitors. b) To detect the heat-stable element of total ADP-degrading activity, sections were preheated at 65°C for 30 minutes, and then incubated in medium containing ADP as a substrate without HPAP inhibitors. c) To detect specific ADP-degrading activity uncontaminated by HPAP activity, sections were incubated for 30 minutes at 37°C in medium containing HPAP inhibitors, as mentioned in the preceding section. d) To detect HPAP at a neutral PH, sections were incubated with β-glycerophosphate, a substrate for HPAP, instead of ADP. e)Sections were also incubated in a substrate free medium to serve as substrate-free controls. f) To confirm that the inhibitors completely inhibited HPAP activity, sections were incubated with β-glycerophosphate as a substrate and HPAP inhibitors were added to the medium.

RESULTS

Placentae From Normal Pregnancies

Light microscopy showed marked staining with reaction products on the trophoblastic cell layers when ADP was used as a substrate in the absence of HPAP inhibitors (Figure 1a). No reaction products were observed in the fetal vessels or the villous stroma. The activity was homogeneously distributed within villi. There was no difference in the intensity or the pattern of staining between stem villi and terminal villi. Sections heated at 65°C continued to show positive staining, although heating reduced ADP-degrading activities (Figure 1b). When the HPAP inhibitors, l-p-bromotetramisole and l-phenylalanine, were added to the medium, specific reaction products were observed on trophoblast cells (Figure 1c). The activity present on the trophoblast cells, using β-glycerophosphate as a substrate was noted in Figure 1d. Substrate-free control gave no reaction (Figure 1e). When β-glycerophosphate was used as a substrate and HPAP inhibitors were added, reaction products disappeared completely (Figure 1f). These series of cytochemical experiments indicated that total ADP-degrading activity existed on the trophoblastic cell layers demonstrated in Figure 1a, which consisted of at least two enzymes; one as a heat stable element (demonstrated in Figure 1b), and the other as specific ADP-degrading activity (Figure 1c).

At the electron microscopic level, when sections were incubated in the medium containing ADP without HPAP inhibitors, marked electron dense reaction products indicating total ADP-degrading activity were observed on the external surface of the microvillous membrane of the syncytiotrophoblast (Figure 3). No reaction products were observed on the basal plasma membrane or the fetal capillary endothelium. When sections were incubated in medium containing ADP and HPAP inhibitors, activity was detected on the external surfaces of syncytial microvilli, indicating the presence of specific ADP-degrading activity uncontaminated by HPAP activity (Figure 4).

Placentae From Preeclamptic Pregnancies

Reaction products indicating total ADP-degrading activity were significantly decreased in placentae complicate by preeclampsia compared with normal placentae (Figure 2a). Electron microscopy demonstrated the presence of total ADP-degrading activity on the external surface of the membrane, but precipitates were decreased compared with normal placentae (Figure 5). Specific ADP-degrading activity was barely detectable in the preeclamptic placenta (Figures 2c and 6).

DISCUSSION

ADP-degrading activity and prostacyclin are the two main platelet anti-aggregatory substances. Recent reports suggest that ADP-degrading activity is a more important anti-aggregatory mechanism in the human placenta than prostacyclin (Hutton et al., 1980a; O'Brien et al., 1987). Hutton et al. demonstrated that an inhibitor of ADP-induced platelet aggregation was present in placental extracts and that this inhibition was produced by ADP-degrading activity and not by prostacyclin (Hutton et al., 1980a). Using biochemical method, Iioka et al. demonstrated that the human placental brush border membrane vesicle, which consisted of the microvillous membrane of the

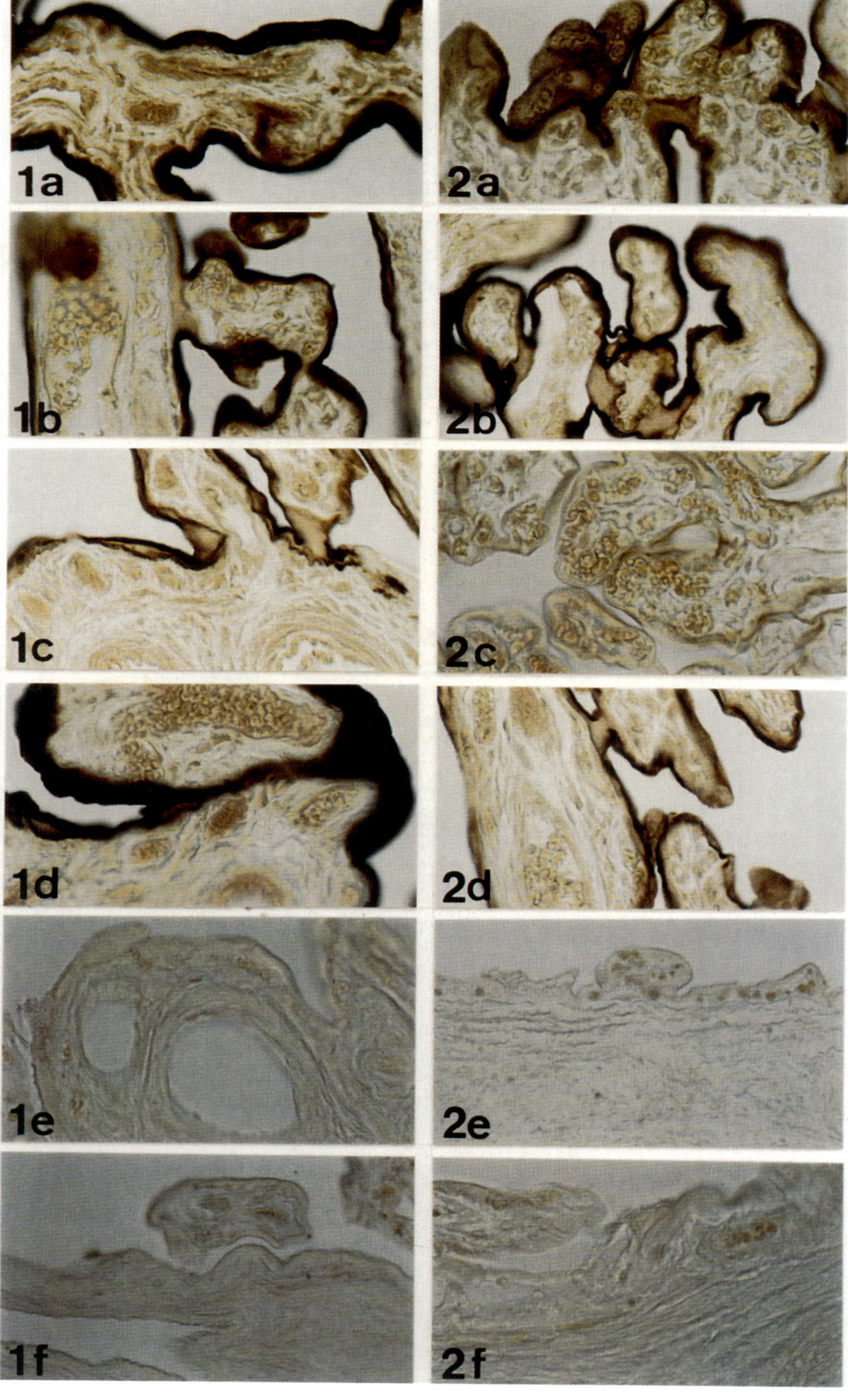
1a
2a
1b
2b
1c
2c
1d
2d
1e
2e
1f
2f

syncytiotrophoblast, possessed both a potent platelet anti-aggregatory ability and a high level of ADP-degrading activity (Iioka et al. 1993a). Iioka et al. have suggested that the ADP-degrading enzyme may be present as an ectoenzyme within the plasma membrane, with the catalytic unit located outside the membrane (Iioka et al., 1993b). In the present study, ADP-degrading activity was exclusively confined to the microvillous membrane of the syncytiotrophoblast of the human placenta. Electron-dense reaction products were observed outside the membrane, indicating that this enzyme was an ectoenzyme. These cytochemical observations are consistent with the biochemical data obtained by Iioka et al (1993a; 1993b).

Human placenta contains a very high level of HPAP, which is located mainly on the microvillous membrane of the syncytium (Matsubara et al., 1987b). Conflicting data have been obtained concerning whether HPAP and ADP-degrading activity represent different enzymes (O'Brien et al., 1987; Iioka et al., 1993b) or the expression of a single enzyme (Hutton et al., 1980b). In the present study, we demonstrated cytochemically that HPAP also degraded ADP at a neutral pH. Although part of the ADP-degrading activity might be due to this HPAP, a specific ADP-degrading enzyme was also present. The current results suggest that total ADP-degrading activity in the human placenta consists of at least two enzymes, HPAP and a specific ADP-degrading activity. The physiological role of HPAP has not been elucidated, but the current findings suggest that one of its functions is ADP degradation.

Studies have shown that platelet function is altered in preeclampsia (Whigham et al., 1978; O'Brien et al., 1986;) and anti-platelet therapy has been found to be beneficial in women at risk of developing preeclampsia (Elder et al., 1988), suggesting that alterations in platelet function may contribute to the development of preeclampsia. Preeclampsia is believed to be associated with *in vivo* activation of platelets, leading to clumping and aggregation (O'Brien et al., 1986). In the present study, both total and specific ADP-degrading activity was significantly decreased in placentae from women with preeclampsia.

Figures 1a-f. Light microscopy of normal term human placentae. a. When ADP was used as a substrate, reaction products indicating total ADP-degrading activity were observed on trophoblastic cell layers. b. Sections preheated at 65°C for 30 minutes continued to demonstrate positive reaction products. c. When sections were incubated with HPAP inhibitors (1.0 mM bromotetramisole and 30 mM l-phenylalanine) specific ADP-degrading activity was observed on trophoblastic cell layers. d.-f. Control experiments. See text for details.

Figures 2a-f. Light microscopy of the term human placentae of patients with preeclampsia. a. When ADP was used as a substrate, reaction products indicating total ADP-degrading activity was observed on trophoblastic cell layers, but its activity was significantly decreased compared with normal placentae. b. Sections preheated continued to demonstrate positive reaction products. c. When sections were incubated with HPAP inhibitors, specific ADP degrading activity was barely detectable in the preeclamptic placenta. d.-f. Control experiments. See text for details.

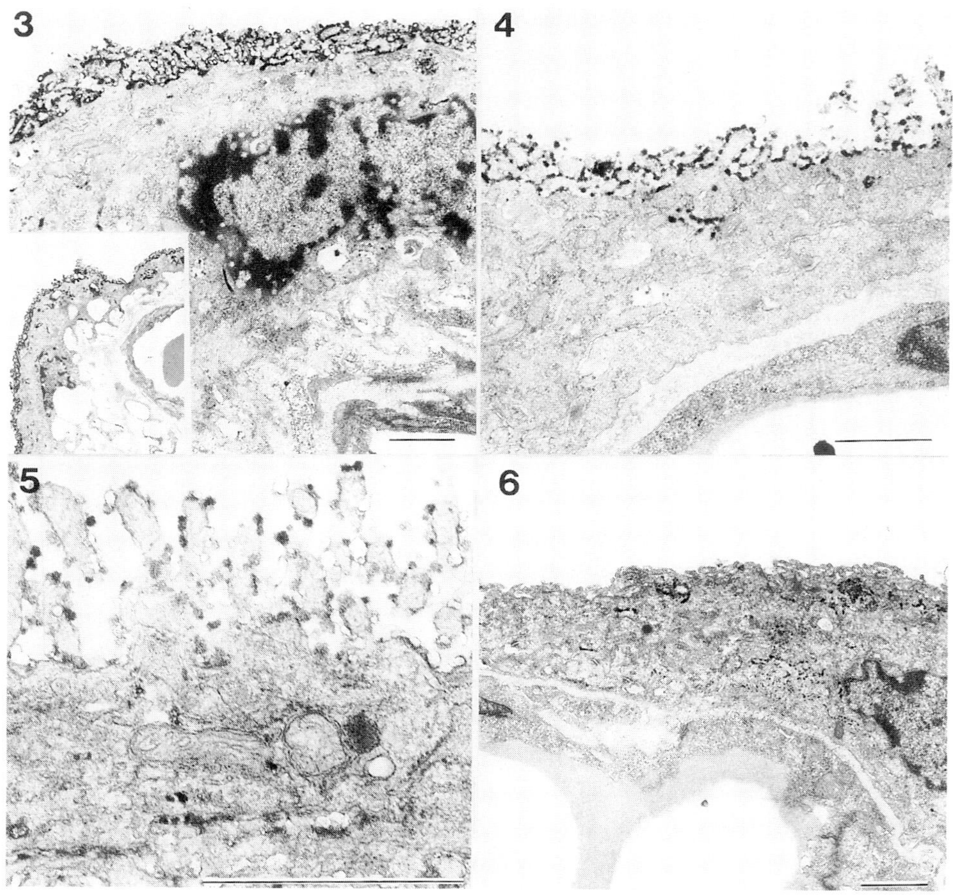

Figures 3-6. Electron microscopy of ADP-degrading activity in normal (Figures 3 and 4) and preeclamptic (Figures 5 and 6) term human placentae. (Bar = 1 μm)

3. Electron microscopy showed total ADP-degrading activity on the microvillous membrane of the syncytiotrophoblast in the normal placenta.

4. Electron-dense precipitates indicating specific ADP-degrading activity were detected on the microvillous membrane of the syncytium in the normal placenta.

5. Electron microscopy demonstrated decreased precipitates on the external surface of the microvillous membrane of the syncytium of the preeclamptic placenta.

6. Specific ADP-degrading activity was barely detectable in the preeclamptic placenta.

Jones and Fox (1980) demonstrated that HPAP activity was decreased in placentae from women with preeclampsia using cytochemical methods. Whigham et al. (1978) found that women with severe preeclampsia had increased plasma levels of adenine nucleotide. These previous observations support the present finding that ADP-degrading activity was decreased in preeclampsia. If ADP-degrading activity, which was a potent inhibitor of platelet aggregation, was decreased in the syncytiotrophoblastic microvillous membrane adjacent to the maternal blood, platelets would be expected to aggregate and become partially exhausted. Such aggregation would lead to chronic disseminated intravascular coagulation (DIC). In a study by Hirano et al. (1993), electron microscopy revealed microthrombus formation in the intervillous space near the syncytium in preeclamptic placenta. The microthrombi were trapped an phagocytosed by adjacent syncytiotrophoblast cells. In preeclampsia, however, platelets may be involved only passively in the process of DIC or may be trapped passively in the microcirculation of the placenta (Mckay, 1981). Further, it is possible that some mechanism other than an ADP-degrading enzyme may contribute to the increased platelet aggregation associated with preeclampsia. For example, Walsh (1985) observed an increase in the production of the thromboxane and a decrease in the production of prostacyclin in the placenta of preeclampsia, which could enhance platelet aggregation. Although the exact mechanism of the increase in platelet aggregation in preeclampsia has not been fully elucidated, the current findings suggest that platelets and the ADP-degrading enzyme are involved in the pathogenesis of preeclampsia.

In conclusion, ADP-degrading activity was exclusively confined to the external surface of the syncytiotrophoblast in human term placentae. The activity of this enzyme was significantly decreased in placentae complicated by preeclampsia. Decreased enzyme activity may play an important role in alteration in hemostasis associated with preeclampsia.

SUMMARY

The cellular and subcellular localization of adenosine diphosphate (ADP)-degrading activity, a potent platelet anti-aggregator, was investigated. Its cytochemical characteristics in normal and preeclamptic placentae were determined. Ultrastructural enzyme cytochemistry demonstrated that ADP-degrading activity was confined exclusively to the external surface of the microvillous membrane of the syncytiotrophoblast in normal placentae. ADP-degrading activity was significantly reduced in preeclamptic placentae.

REFERENCES

Ahmed, Y., Sullivan, M.H.F. and Elder, M.G. (1991) Detection of platelet desensitization in pregnancy-induced hypertension is dependent on the agonist used. *Thromb. Haemostas.* 65, 474-477.

Borgers, M. (1973) The cytochemical application of new potent inhibitors of alkaline phosphatase. *J. Histochem. Cytochem.* 21, 812-824.

Elder, M.G., de Swiet, M., Robertson, A., Elder, M.A., Floyd, E. and Hawkins, D.F. (1988) Low-dose aspirin in pregnancy. *Lancet* (i), 410-412.

Hirano, M. (1993) Pathological changes of chorionic villi in cases of toxemic pregnancy. *Abst. of 1st Meeting of the Japan Placenta Group*, p. 32.

Hutt, R., Ogunniyi, S.O., Sullivan, M.H.F. and Elder, M.G (1994) Increased platelet volume and aggregation precede the onset of preeclampsia. *Obstet. Gynecol.* 83, 146-149.

Hutton, R. A., Chow, F.P.R., Craft, I.L. and Dandona, P. (1980a) Inhibitors of platelet aggregation in the fetoplacental unit and myometrium with particular reference to the ADP-degrading property of placenta. *Placenta* 1, 125-130.

Hutton, R.A., Dandona, P., Chow, F.P.R. and Craft, I.L. (1980b) Inhibition of platelet aggregation by placental extracts. *Thromb. Res.* 17, 465-471.

Iioka, H., Akada, S., Shimamoto, T., Yamada, Y., Sakamoto, Y., Moriyama, S.I. and Ichijo, M. (1993a) Platelet aggregation inhibiting activity of human placental chorioepithelial brush border membrane vesicles. *Placenta* 14, 75-83.

Iioka, H., Akada, S., Sakamoto, Y., Shimamoto, T., Yamada, Y., Moriyama, I.S. and Ichijo, M. (1993b) Properties of ADP-degrading activity of human placental syncytiotrophoblast brush border membrane vesicles. *Placenta* 14, 333-339.

Jones, C.P.J. and Fox, H. (1980) An ultrastructural and ultrahistochemical study of the human placenta in maternal pre-eclampsia. *Placenta* 1, 61-76.

Lieberman, G.E., Lewis, G.P. and Peters, T.J. (1977) A membrane-bound enzyme in rabbit aorta capable of inhibiting adenosine-diphosphate-induced platelet aggregation. *Lancet* 2, 330-332.

Matsubara, S., Tamada, T. and Saito, T. (1987a) Ultracytochemical localization of ADP-degrading enzyme activity in the human term placenta -direct cytochemical evidence-. *Acta Obst. Gynaecol. Jpn.* 39, 2195-2196.

Matsubara, S., Tamada, T. and Saito, T. (1987b) Ultracytochemical localization of alkaline phosphatase and acid phosphatase activities in the human term placenta. *Acta Histochem. Cytochem.* 20, 283-294.

Mckay, D.G. (1981) Chronic intravascular coagulation in normal pregnancy and preeclampsia. *Contrib. Nephrol.* 25,108.

Novikoff, A.B. and Goldfischer, S. (1961) Nucleotide diphosphatase activity in the Golgi apparatus and its usefulness for cytochemical studies. *Proc. Natl. Acad. Sci. USA* 47, 802-810.

O'Brien, W.F., Saba, H.I., Knuppel, R.A., Scerbo, J.C. and Cohen, G.R. (1986) Alterations in platelet concentration and aggregation in normal pregnancy and preeclampsia. *Am. J. Obstet. Gynecol.* 155, 486-490.

O'Brien, W.F., Knuppel, R.A., Saba, H.I., Angel, J.L., Benoit, R. and Bruce, A. (1987) Platelet inhibitory activity in placentas from normal and abnormal pregnancies. *Obstet. Gynecol.* 70, 597-600.

Pearson, J.D., Carleton, J.S. and Gordon, J.L. (1980) Metabolism of adenine nucleotides by ectoenzymes of vascular endothelial and smooth-muscle cells in culture. *Biochem. J.* 190, 421-429.

Walsh, S.W. (1985) Preeclampsia: An imbalance in placental prostacyclin and thromboxane production. *Am. J. Obstet. Gynecol.* 152, 335-340.

Whigham, K.A.E., Howie, P.W., Drummond, A.H. and Prentice, C.R.M. (1978) Abnormal platelet function in pre-eclampsia. *Br. J. Obstet. Gynaecol.* 85, 28-32.

Trophoblast Research 9:131-140, 1997

THE ROLE OF ACTIVATED PROTEIN C (APC) IN FIBRINOLYSIS OF HUMAN PLACENTA

Nobuhiko Moniwa, Hiroshi Kobayashi, Ming You She, Takako Kobayashi and Toshihiko Terao

Department of Obstetrics and Gynecology
Hamamatsu University School of Medicine
Hamamatsu, Shizuoka, 431-31, Japan

INTRODUCTION

During pregnancy, a state of hypercoagulation and suppressed fibrinolytic activity are present in the placenta (Beller et al., 1962). This may occur due to the abundance of plasminogen activator inhibitors (PAIs) which exist in the placenta (Kawano et al., 1970). Suppression of fibrinolytic activity plays an important role in the prevention of hemorrhage during pregnancy and labor (Kruithof et al., 1987). There should be a protection system which prevents thrombus formation including factors produced by placenta thrombomodulin, blocking factors of platelet aggregation, coagulation blocking factors, extrinsic pathway inhibitors, heparin sulfate (Paschke et al., 1979), plasminogen and plasminogen activator (Bonnar et al., 1990).

Activated protein C (APC) is a serine protease derived from the vitamin K-dependent zymogen, protein C, via, activation by a complex of thrombin and thrombomodulin (Esmon et al., 1982). Plasminogen activator inhibitor type-1 (PAI-1) can be neutralized by plasminogen activators or APC, which is the physiological inhibitor of Factors Va and VIIIa (Krishnamurti et al., 1991). We previously reported that APC dissociated the urokinase type plasminogen activator/plasminogen activator inhibitors type-1 (uPA-PAI) complex in vitro (Kobayashi et al., 1988). As there is an abundance of APC in intervillous spaces of the placenta, APC is thought to be involved in fibrinolysis of placenta.

The expression of uPA, PAI-1 and APC by human placenta *in vivo* was examined. Therefore, we used specific monoclonal antibodies (MAbs) to stain villous trophoblast within placentae of early gestational ages. Immunohistochemistry was used to characterize the expression of these antigens in villous trophoblast. These antigens were abundantly expressed during the first trimesters of gestation by syncytiotrophoblast. The expression of uPA, PAI-1 and PC/APC by villous trophoblast during the first trimester, which are actively invasive *in vivo*, may serve to facilitate the generation of plasmin at the interface of these cells with maternal plasma, thereby limiting the deposition of fibrin within the placental intervillous spaces. Increased uPA expression by villous trophoblast at term may represent a physiological adaptation to diminish local fibrinolysis and limit hemorrhage at parturition (Nakashima et al., 1996).

In this study, the abundance of immunoreactive APC was determined in normal placenta with the associated role of human APC in fibrinolysis by dissociating uPA-PAI-1 complex in the intervillous space.

MATERIALS AND METHODS

Reagents

All chemicals are the best analytical grade commercially available. A highly purified preparation of human APC (M.W. 53 kDa) was kindly supplied by Teijin Pharmaceutical Co. Ltd., Tokyo, Japan. Mouse MAb against human APC (JTC-1) was also kindly supplied by Teijin. Mouse MAbs to PAI-1 (MAb 3783) and uPA (MAb 377, which reacts with A chain of uPA) were purchased from American Diagnostics (New York, USA). HMW-uPA [urokinase], an active form of two-chain high-molecular-weight form of uPA, was obtained from The Green Cross (Osaka, Japan). PAI-1 was purified from conditioned media collected from HUVEC as described previously (Wun et al., 1989).

Cells And Cell Culture

Human umbilical vein endothelial cells (HUVEC) were prepared as described previously (Jaffe et al., 1973). Umbilical cords were obtained after either cesarean section or vaginal delivery from pregnant women without any complications. HUVEC were cultured on a thin cover glass in a CO_2 incubator. Serum-free RPMI 1640 medium was used for HUVEC. HUVEC were washed with PBS, and incubated in 1 ml of serum-free medium for 30 minutes in a CO_2 incubator before experiments.

Immunohistochemical Staining Of uPA, PAI-1 And APC In Human Placenta

First trimester placentae were routinely submitted for pathologic inspection after pregnancy termination. Serial sections of 10% formalin-fixed and paraffin-embedded tissue were placed on glass slides, dried at temperatures not greater than 60°C, and stored at room temperature until used. The sections were dewaxed and rehydrated in a graded series of alcohol solutions. Labeled Streptavidin Biotin staining (DAKO LSAB kit, DAKO, Tokyo) was performed as described previously (Guesdon et al., 1979).

Analysis Of APC-Dependent Fibrinolytic Activity On uPA-PAI-1 Complexes

Analysis of APC-dependent fibrinolytic activity on uPA-PAI-1 complexes was performed as described previously with slight modification (Kobayashi, 1994). Ninety-six well microtiter plate wells were coated with uPA by incubation of 100 µl aliquots of protein solution (10 µg/ml protein in 0.1 M sodium carbonate, pH 9.5; 4°C, overnight). After washing the wells, nonspecific binding sites were blocked with 200 µ1 aliquots of PBS, 2% BSA, pH 7,4. Wells were washed extensively with Tris-buffered saline with Tween 20 (TBST) and then incubated with TBS, 2% BSA for 1 hour at 23°C to reduce background labeling. After that the wells were incubated with PAI-1 (25 µg/100 µl per well) in TBS, 2 % BSA for 1 hour at 23°C. Unbound PAI-1 was removed by washing the wells extensively with TBST. The wells were incubated with the different concentration of APC (0 - 10 µg/100 µl per well) for 30 minutes at 23°C, to study the APC-dependent fibrinolytic activity on uPA-PAI-1 complexes. After the reaction was over, the unbound proteins were removed.

uPA activity was assayed with the chromogenic substrate S-2444 (Chromogenix, Molndal, Sweden). The background determined with BSA-coated wells was subtracted from each value. Absorbance was read at 405 nm with an automated 96-well spectrophotometer (Bio-Rad, Richmond, CA). In a parallel examination Anti-PAI-1 antibodies (MAb 3783) diluted in 50 μ1 PBS, 2% BSA was allowed to bind for 2 hours at 23°C. Then the plates were washed three times with PBS, and bound antibodies were incubated with biotinylated second antibody (0.5 μg/ml for 1 hour at 23°C; Dako), followed by incubation with avidin-peroxidase (0.5 μg/ml for 1 hour at 23°C; Dako). After washing, wells were incubated with enzyme substrate (3,3'-5,5'-tetramethyl benzidine). Absorbance was read at 450 nm with the spectrophotometer. All samples were tested in triplicate, and the results shown are the means obtained in a typical experiment from a series. The data is expressed in terms of the mean ± SD.

Effect of APC on HUVEC-associated uPA-PAI-1 Complexes

To investigate the mechanism of the APC-dependent fibrinolysis on HUVEC, we cultured HUVEC in the absence or presence of APC (100 μg/ml) and assayed the culture media for PAI-1 and APC proteins.

Electrophoresis and Western Blot

SDS-PAGE was performed in slab gels with 12 % polyacrylamide gel under nonreducing conditions. After electrophoresis, the gels were electroblotted onto a polyvinylidine difluoride (PVDF) membrane (Millipore), using a semi dry electroblotting apparatus (Marysol, Japan; 40 mA/gel, 90 minutes, 23°C), according to the procedure described previously (Kobayashi et al., 1991).

Statistical Analysis

Differences between group means were statistically tested for significance using Wilcoxon's U-test, with separate variance estimates if the standard deviations were significantly different. Differences were considered significant if $P<0.05$.

RESULTS

Immunohistochemical Staining of uPA, PAI-1 and APC in Human Placenta

We initially assessed the pattern of expression of uPA (Figure 1a), PAI-1 (Figure 1b), and APC (Figure 1c) by villous trophoblast during early gestation. Examination of these specimens revealed that uPA was diffusely expressed in the syncytiotrophoblast and cytotrophoblast, while PAI-1 and APC were expressed predominantly on the syncytiotrophoblast, although diffuse, less prominent staining was observed in the remainder of the cells. These results demonstrate that uPA, PAI-1 and APC are expressed by villous trophoblast, with significantly greater expression occurring at early pregnancy.

APC-Dependent Fibrinolytic Activity on uPA-PAI-1 Complexes

We tested whether the complex between PAI-1 and immobilized uPA, once formed, are resistant to dissociation by APC. PAI-1 reacted with 96-well microtiter plates coated-uPA to yield stable complexes. Binding of PAI-1 to microtiter plates coated-uPA

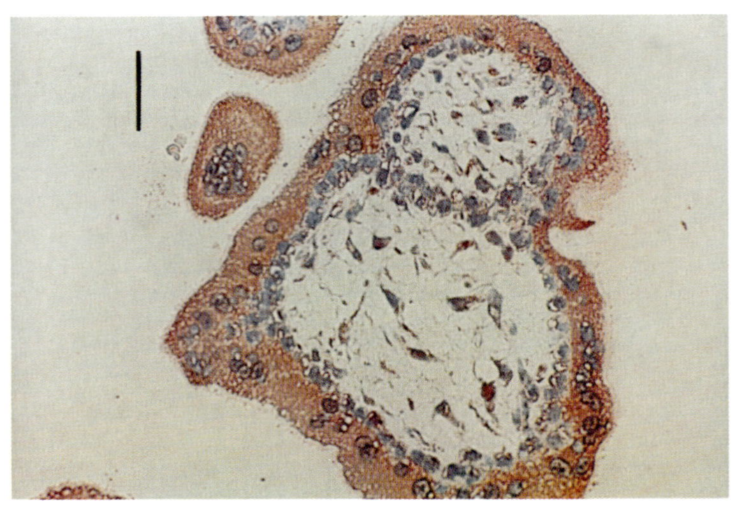

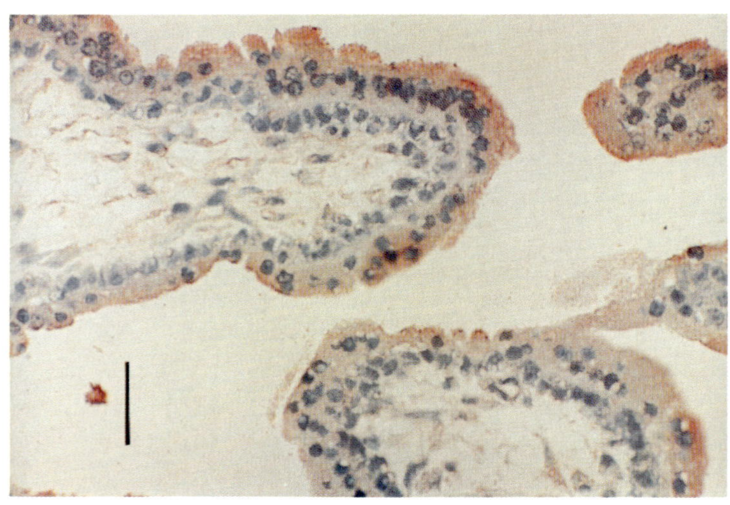

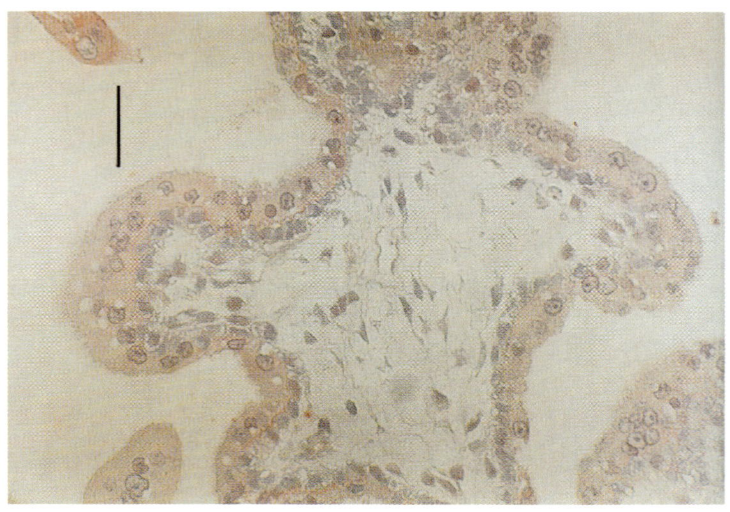

occurred in a dose-dependent and saturable manner (data not shown). The specific activity of microtiter plate well-coated uPA was assayed after treated with increasing concentrations of APC. The uPA activity was determined using chromogenic substrate S-2444. uPA activity was recovered in the presence of higher concentrations of APC (Figure 2a). PAI-1 bound to the fixed uPA was dissociated in the presence of higher concentrations of APC (Figure 2b). We have previously reported that APC could be unable to compete with equal amounts of uPA for complex formation with PAI-1 (Kobayashi, 1994). Under the experimental conditions, in which the concentration of APC was more than 2 μg/ml, PAI-1 was dissociated from uPA. In the absence of PAI-1, however, APC did not stimulate uPA activity (data not shown).

Effect of APC on HUVEC-Associated uPA-PAI-1 Complex

HUVEC have been shown to express functional uPA receptor (Barnathan et al., 1990). We examined whether complexes between uPA and PAI-1 on the cell surface, once formed, are resistant to dissociation by exogenously applied APC. To investigate the mechanism of the APC-dependent fibrinolysis on HUVEC, the cells were incubated with APC (100 μg/ml) and assayed the culture media for PAI-1 and APC proteins. The amount of APC-PAI-1 complexes formed in conditioned medium was evaluated by immunoblotting (Figure 3). Under the experimental conditions, in which the concentration of APC in the conditioned medium was more than 1 μg/ml, PAI-1 was dissociated from receptor-bound uPA. APC-PAI-1 complexes were observed in culture medium. HUVEC incubated with APC (10 μg/ml) had significantly more PAI-1 released into the conditioned medium compared with the cells incubated with 0.1 ug/ml of APC (data not shown).

DISCUSSION

Thrombotic lesions in placenta may occur due to the hypofibrinolytic and hypercoagulation changes. PAI-1 is an abundant protein in the placenta that inhibits free and cell-associated plasminogen activators, uPA and tPA (Philips et al., 1984). Thus, the formation of a thrombus on the trophoblastic cell membrane is easy, leading to a reduction in placental function or worse. Consequently, preeclampsia and intrauterine fetal growth restriction (IUGR) can sometimes occur (DeBoer et al., 1988). For the treatment of such diseases, human APC was recently used in a clinical trial for disseminated intravascular coagulation (DIC) or severe toxemia of pregnancy. APC has been suggested to be useful for treating the patients with thrombosis (Krishnamurti et al., 1991).

Figure 1. Immunohistochemical staining for uPA, PAI-1 and APC in the first trimester placenta (x 100). Representative sections of 10 % formalin-fixed and paraffin-embedded tissues were stained using anti-uPA (MAb 377), anti PAI-1 (MAb 3783) and anti-APC (MAb JTC-1) antibody. (a): localization of uPA at week 6 of pregnancy, (b): PAI-1 at week 6 of pregnancy, (c): APC at week 6 of pregnancy. Examination of these specimens under high power magnification revealed that uPA was diffusely expressed in the syncytiotrophoblast and cytotrophoblast, while PAI-1 and APC were expressed predominantly on the syncytiotrophoblast, although diffuse, less prominent staining was observed over the remainder of the cells. Bar: 50 μm.

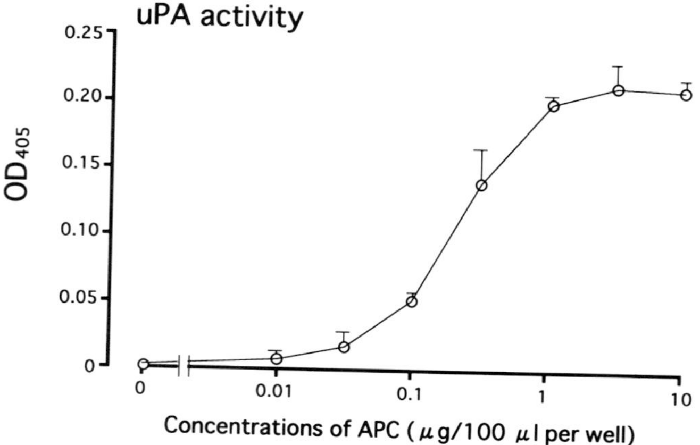

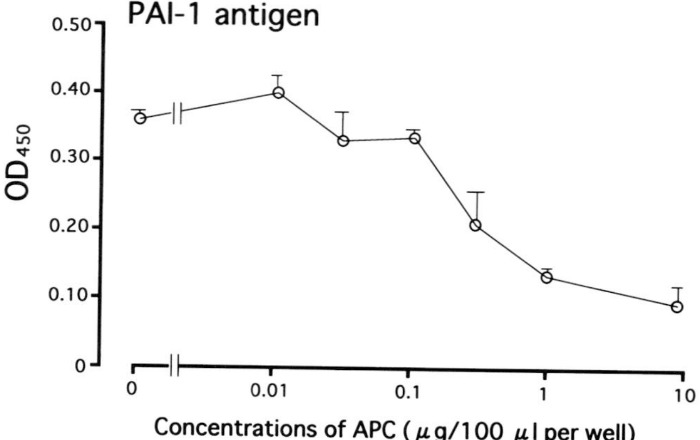

Figure 2. APC-dependent fibrinolytic activity on uPA-PAI-1 complexes. Binding of PAI-1 to microtiter plates coated-uPA was incubated with APC at varying concentrations (0, 0.01, 0.1, 1.0 and 10 μg/100 μl per well). The uPA activity was determined using chromogenic substrate S-2444. Absorbance was read at 405 nm with an automated spectrophotometer. The PAI-1 antigen was determined by ELISA. Absorbance was read at 450 nm with the spectrophotometer. All samples were tested in triplicate. The data are expressed as mean ± SD. The specific activity of microtiter plate well-coated uPA was assayed after treatment with increasing concentrations of APC. PAI-1 bound to the fixed uPA was dissociated in the presence of high concentrations of APC.

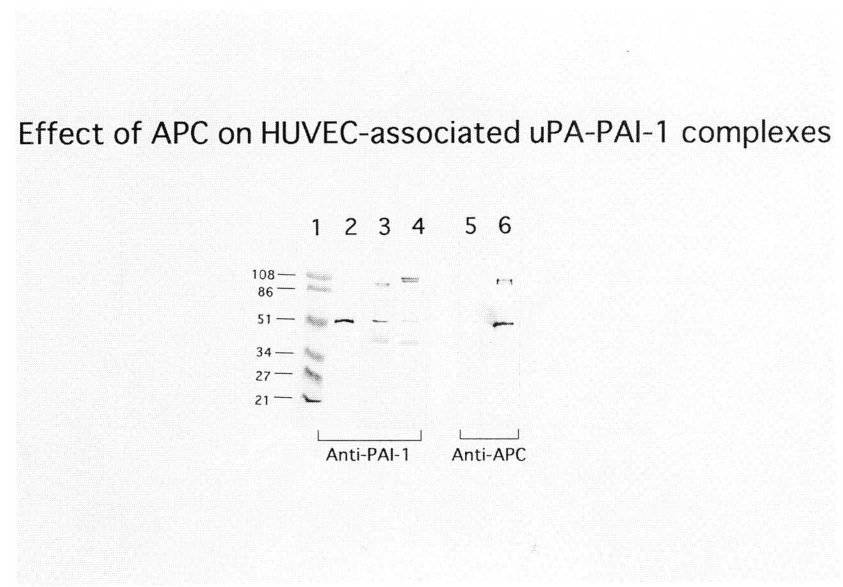

Figure 3. Effect of APC on HUVEC-associated uPA-PAI-1 complexes Samples from the culture media of HUVEC in the absence or presence of APC (100 μg/ml) were separated on 12 % polyacrylamide gels and stained with anti PAI-1 antibody (MAb 3783) and anti-APC antibody (MAb JTC-1). Lane 1, molecular weight marker; lane 2: purified PAI-1, lanes 3 and 5, the culture media in the absence of APC: and lanes 4 and 6: culture media in the presence of APC (100 μg/ml). APC-PAI-1 complex (100 kDa) could be detected in the culture media of HUVEC in the presence of APC (100 μg/ml).

In this study, we report that APC is also present the trophoblast cells in the first trimester placentae. APC is localized on syncytiotrophoblast as revealed by immunohistochemical analysis. APC is abundantly present in normal placenta and may have an important homeostatic function in thrombosis and fibrinolysis in placenta. Moreover, plasminogen activator could be blocked by the formation of PA-PAIs complexes which may occur thrombosis or infarction in the normal placenta.

In addition, we have found that APC could dissociate the uPA-PAI-1 complexes resulting in the reappearance of uPA activity. When uPA-PAI-1 complexes were incubated with an excess of APC, PAI-1 was dissociated from the complex and uPA activity was regained. Until now, APC was known to as a regulator of coagulation by the inhibition of Factor Va and VIIIa. Fay and Owen reported that bovine APC could neutralize PAI-1 activity, but human APC may not be a sensitive inhibitor of PAI-1 (Fay and Owen 1989). The current results demonstrated that excess APC could neutralize uPA-PAI complexes *in vitro*. We demonstrated the presence of uPA and PAI-1 in the syncytiotrophoblast by immunohistochemical methods. Immunoreactive APC was also stained in the syncytiotrophoblast. These results support our in vitro data.

We also measured the concentration of tissue type plasminogen activator (tPA), uPA, PAI-1 and PAI-2 in placental extracts via ELISA. The specimens showed a high concentration of PAI-1 compared with uPA. Therefore, we suggest that fibrinolytic activity is maintained by APC in intervillous space of placenta in spite of an abundance of PAIs. We have demonstrated that APC is present in abundance in the syncytiotrophoblast and appears to play a critical role in the regulation of fibrinolysis in the placenta.

In conclusion, APC was localized on the syncytiotrophoblast at the first trimester placenta by immunohistochemistry. APC-PAI-1 complexes were demonstrated in serum-free culture medium when HUVEC were incubated with APC, indicating that PAI-1 was dissociated from receptor-bound uPA on cell surface of HUVEC. This was also confirmed by ELISA. Taken together, these results suggest that APC could possibly play a significant role in fibrinolytic activity during early stages of gestation.

SUMMARY

Suppression of the fibrinolytic activity plays an important role in the prevention of hemorrhage during pregnancy and labor. A hypofibrinolytic and hypercoagulable state may be established in the placenta during pregnancy. However, little infarction is present in the normal placenta. This evidence demonstrates that the placenta maintains a fibrinolytic activity in spite of a hypercoagulation state. We studied the role of APC (activated protein C) on fibrinolysis in the placenta. (1) APC-PAI-1 (plasminogen activator inhibitor type-1) complexes were demonstrated in serum-free culture medium when human umbilical vein endothelial cells (HUVEC) were incubated with APC, indicating that PAI-1 was dissociated from receptor-bound uPA (urokinase type plasminogen activator) on the cell surface of HUVEC. This was also confirmed by enzyme-linked immunosorbent assay (ELISA), (2) APC antigen is located on the syncytiotrophoblast in first trimester placenta by immunohistochemistry. Taken together, these results suggest that APC could possibly play a significant role in fibrinolytic activity during early stages of gestation.

REFERENCES

Beller, F.K., Goessner, W., and Herrschlein, H.J. (1962) Tissue activator of the fibrinolytic system in placental tissue. *Obstet. Gynecol.* 20, 117-122.

Bonnar, J., Daly, L, and Sheppard, B.L. (1990) Changes in the fibrinolytic system during pregnancy. *Semin. Thromb. Hemost.* 16, 221-229.

Barnathan, E.S., Kuo, A., Rosenfeld, L., Kariko, K., Leski, M., Robbiati, F., Nolli, M.L. Henkin, J. and Cines, D.B. (1990) Interaction of single-chain urokinase-type plasminogen activator with human endothelial cells. *J. Biol. Chem.* 265, 2865-2872.

DeBoer, K., Lecander, I., ten Cate, J.W., Borm, J.J.J. and Treffers, P.E. (1988) Placental-type plasminogen activator inhibitor in preeclampsia. *Am. J. Obstet. Gynecol.* 158, 158-518.

Erickson, L.A., Lawrence, D.A. and Loskutoff, D.J. (1984) Reberse fibrin autography: A method to detected and partially characterize protease inhibitors after sodium dodecyl sulfate-polyacrylamide gel electrophoresis. *Anal Biochem.* 137, 454-463.

Esman, N.L., Owen, W.G. and Esmon, C.T. (1982) Isolation of a membrane-bound cofactor for thrombin-catalyzed activation of protein C. *J. Biol. Chem.* 257, 859-864.

Fay, W.P. and Owen, W.G. Platelet plasminogen activator inhibitor: purification and characterization with plasminogen activators and activated protein C. *Biochemistry* 28, 5773-5778.

Feinberg, R.F., Kao, L.C., Haimowitz, J.E., Queenan, J.T. Jr., Wun, T-C., Strauss III, J.F. and Kliman, H.J. (1989) Plasminogen activator inhibitor types 1 and 2 in human trophoblasts. *Lab. Invest.* 61, 20-26.

Guesdon, J-T., Ternynck, T. and Avrameas, S. (1979) The use of avidin-biotin interaction in immunoenzymatic techniques. *J. Histochem. Cytochem.* 27, 1131-1139.

Jaffe, E.A., Nachman, R.L., Becker, C.G. and Minick, C.R. (1973) Culture of human endothelial cells derived from umbilical veins. Identification by morphologic and immunologic criteria. *J. Clin. Invest.* 52, 2745-2756

Kawano, T., Morimoto, K. and Uemura, Y. (1970) Partial purification and properties of urokinase inhibitor from human placenta. *J. Biochem.* 67, 333-342.

Kobayashi, H., Schmitt, M., Goretzki, L., Chucholowski, N., Calvete, J., Karmer, M., Günzler, W.A., Jänike, F. and Graeff, H. (1991) Cathepsin B efficiently activates the soluble and the tumor cell receptor-bound form of the proenzyme urokinase-type plasminogen activator (pro-uPA). *J. Biol. Chem.* 266, 5147-5152.

Kobayashi, H., Moniwa, N., Gotoh, J., Sugimura, M. and Terao T. (1994) Role of activated protein C in facilitating basement membrane invasion by tumor cells. *Canc. Res.* 54, 261-267.

Kobayashi, T. and Terao, T. (1988) A study of placental plasminogen activator - the effect of activated protein C on placental plasminogen activator inhibitor complex in vitro. *Fibrinolysis: Current Prospects* 101-106.

Krishnamurti, C., Young, G.D., Barr, C.F., Colleton, C.A. and Alving, B.M. (1991) Enhancement of tissue plasminogen activator-induced fibrinolysis by activated protein C in endotoxin-treated rabbits. *J. Lab. Clin. Med.* 118, 523-530.

Kruithof, E.K.O., Tran-Thang, C., Gudinchet, A., Hauert, J., Nicoloso, G., Genton, C., Welti, H. and Bachmann, F. (1987) Fibrinolysis in pregnancy: A study of plasminogen activator inhibitors. *Blood* 69, 460-466

Nakashima, A., Kobayashi, T. and Terao, T. (1996) Fibrinolysis during normal pregnancy and severe preeclampsia relationships and plasma levels of plasminogen activators and inhibitors. *Gynecol. Obstet. Invest.* 42, 95-101.

Paschke, E. And Kresse, H. (1979) Multiple forms of 2-deoxy-D-glucoside-2-sulphamate sulphohydrolase from human placenta. *Biochem. J.* 181, 677-684.

Philips, M., Juul, A.G. and Thorsen, S. (1984) Human endothelial cells produce a plasminogen activator inhibitor and a tissue-type plasminogen activator-inhibitor complex. *Biochim Biophys Acta* 802, 99-110

Wun, T-C., Palmier, M.O., Siegel, N.R. and Smith, C.E. (1989) Affinity purification of active plasminogen activator inhibitor 1 (PAI-1) using immobilized anhydrourokinase. *J. Biol. Chem.* 264, 7862-7868.

Trophoblast Research 9:141-153, 1997

NITRIC OXIDE SYNTHASE FROM HUMAN PLACENTA

Sachio Iida[1], Hiroyuki Ohsawa[1], Hiroaki Soma[1], Toshio Hata[1], Yukiko Kurashima[2], and Hiroyasu Esumi[2]

[1]Department of Obstetrics and Gynecology
Saitama Medical School
38 Morohongo, Moroyama, Iruma-gun
Saitama, Japan

[2]Biochemistry Division
National Cancer Center Research Institute
5-1-1, Tsukiji, Chuo-ku
Tokyo, Japan

INTRODUCTION

Nitric oxide (NO) has been regarded as one of the most intriguing molecules because of its variety of biological functions. Among the biological roles for NO are the activation of guanylate cyclase and the biological properties that accounts for endothelium-derived relaxing factor (EDRF) (Palmer et al., 1987), the inhibition of platelet coagulation and adhesion (Radomski et al., 1987), as a role of neurotransmitter in nervous system (Knowles et al., 1989). Furthermore, NO plays a role as a cytotoxic molecule in macrophage (Hibbs et al., 1988) and granulocyte. NO is endogenously formed by NO synthase (NOS) from l-arginine. Some types of NOS have been purified from numerous tissues (Bredt and Snyder, 1990; Pollock et al., 1991; Stuher et al., 1991; Hevel et al., 1991) and cloned its cDNA (Bredt et al., 1991; Jassens et al., 1992; Lamas et al., 1992; Lyons et al., 1992; Xie et al., 1992; Lowenstein et al., 1992; Adachi et al., 1993). Those NOSs has been classified as three isoforms: brain type (neuronal; nNOS) and endothelial type (eNOS) of NOS those expressed constitutively, and macrophage type (inducible; iNOS) that is induced by endotoxin and cytokines. All NOSs have the binding sites for FMN, FAD, NADPH, and calmodulin (CaM). Reports of purification for NOS from human tissues are limited because of the difficulty in obtaining sufficient quantities of tissue. Schmitt and Moncada (1992) reported the purification from human brain tissue. On the other hand, cDNA cloning of NOS from human tissue has been performed from the library of brain, endothelium, cultured hepatocyte (Marsden et al., 1992). Myatt et al. (1993) has already identified the endothelial type of NOS from human placenta vascular tree. Its distribution was recognized in fetus endothelium and syncytiotrophoblast (Buttery et al., 1993). Indeed, NO synthase from human placenta has been purified as a single protein band with a molecular weight of 135 kDa that was identified by calmodulin binding (Garvey et al. 1994). Morris et al. (1995) showed lower NOS activities in placental villi in pregnancies complicated by preeclampsia and growth retardation compared to villi from normal placentae. Thus, NO that is endogenously produced by NOS seems to be very important to maintain pregnancy and to be deeply associated with the pathogenesis of various obstetrical disease. Unfortunately, it is not fully understood how NOS function in human placenta. The endothelial type of NOS was completely purified from both soluble (cytosolic) and insoluble (membrane) fractions of human placenta and examined the localization of NOS on chorionic villi in early and late pregnancy.

MATERIALS AND METHODS

Human Placenta and Reagents

Five human term placentae were obtained after spontaneous vaginal delivery from women with normal pregnancies. Placentae were delivered and immediately separated into 10 gram pieces, ice-packed to store at -80°C until use for protein purification, and a part of them were used for immunohistochemistry. Two villi of the first trimester of gestation were also obtained after dilatation and curettage and were used for immunohistochemical analysis. L-[2,3-^{3}H] arginine (spec. act. 55 Ci/mmol; 1 Ci = 37 GBq) was obtained from NEN (DuPont de Nemours, Wilmington, DE). N^G-monomethyl-1-arginine and (6R)-5,6,7,8-tetrahydro-L-biopterin ((6R)-BH4) were from Calbiochem (La Jolla, CA) and Dr. B. Schircks (Jona, Switzerland), respectively. Other reagents were of analytical grade and were purchased from Sigma (St. Louis, MO) or Wako (Osaka, Japan).

Antisera Against NOSs

Two type of antisera, those are namely anti-common NOS and anti-eNOS respectively, were raised in rabbits to the synthetic peptides. The first one was based on deduced amino acids sequences of the common region that exist in the upstream of putative CaM binding site of three isoforms of NOS. The peptide sequence was RSAITVFPQRSDGKHDFR-C, correspondingly to amino acids (aa) 461-478 in case of human nNOS, to aa 225-242 of human eNOS and to aa 238-255 of rat liver iNOS, respectively. This sequence is coserved in three isoforms of NOSs. The other peptide was based on deduced amino acids of cDNA encoding human endothelial NOS (Lames et al., 1992; Marsden et al., 1992). The sequence was RGAVPWAFDPPGSDTNP-C, corresponding to amino acids 1186-1203 of the deduced amino acids sequence of human eNOS. It was the typical sequence of eNOS and having no homology with those of human or rat nNOS and iNOS. Those peptides were conjugated with a keyhole limpet hemocyanin (KLH) at the C-terminal to assist coupling to the carrier. Anti-nNOS and anti-iNOS antisera were as previously reported (Ohshima et al., 1992)

Measurement of NOS Activities

The activities of NOS were measured by determining either the conversion of L-[2,3-^{3}H] arginine to L-[2,3-^{3}H] citrulline essentially based on a method of Bredt and Snyder (1990). Shortly, 25 μl of enzyme extract and 25 μl of 100 nM[^{3}H] arginine to 100 μl of buffer containing 50 mM Hepes (pH 7.4), 1 mM NADPH, 1 mM EDTA, 1.25 mM CaCl2, 1 mM dithiothreitol (DTT), and 10 μg of calmodulin per ml. After incubation for five minutes at 22°C, assay were terminated with 2 ml of 20 mM Hepes, pH 5.5/2 mM EDTA, and were applied to 1 ml of Dowex AG50WX-8 (Na+ form), which were eluted with 2 ml of water. [^{3}H] citrulline was quantified by liquid scintillation spectroscopy of the 4-ml flow-through. To confirm whether the conversion activity is by NOS or not, the enzyme activities were simultaneously measured under the existence of 60 mM of N^G-monomethyl-l-arginine (L-NMMA), competitive inhibitor.

Purification of NOS From Human Placenta

Fifty grams of human term placentae were homogenized in a blender in 250 ml of ice cold buffer 1 [50 mM Tris-HCl pH 7.4, containing 0.5 mM EDTA and EGTA, 1 μM leupeptin, 0.1 mM phenylmethylsulfonyl fluoride (PMSF) and 1 mM DTT]. All subsequent procedures were performed at 4°C . The homogenate was centrifuged at 900 *g* for 30 minutes to remove the rough nuclear fractions and other connective tissues of placenta. The supernatant was centrifuged at 105,000 *g* for 60 minutes and separated to the soluble (hPl-sF) and insoluble fraction (hPl-iF) of human placenta. The 100 μl of 2',5'-ADP agarose (Sigma) per 100 μl of the hPl-sF were added and incubated for 30 minutes with gentle agitation. The 2',5'-ADP agarose was collected and packed into column having 5 ml of bed volume by centrifugation and washed with a 50 column volume of buffer 1 containing 0.5 M NaCl, and then with 30 column volume of buffer 1 only. Partially purified NO synthase was then eluted with 10 mM NADPH in 3 column volume of buffer 1. This eluate (sF-ADP pool) was diluted in 3 times volume of buffer 2 (buffer 1 containing 1 mM CaCl instead of 0.5 mM EDTA and EGTA) and added 30 μl of calmodulin agarose(Sigma) per 1 ml of the diluted sF-ADP pool. The mixture was incubated for 30 minutes with gentle agitation and washed with 30 volumes of buffer 2 containing 0.3 M NaCl , and then with 10 volumes of buffer 2 only. Finally, purified NOS was eluted with 3 column volume of buffer 1 containing 5 mM EGTA. On the other hand, insoluble fraction of human placenta (hPl-iF) was reconstituted in 5 ml of buffer 1 containing 1 M KCl per gram of the starting tissue and centrifuged at 105,000 g and repeated 2 times. KCl-washed insoluble fraction was then solubilized with detergent 3-[(3-cholamide propyl) dimethyl ammonio]-1-propanosulfonate (CHAPS; Sigma, 20 mM) in buffer 1 for 30 minute with gentle rotation of sample and centrifuged at 105,000 *g* for 60 minutes. All subsequent procedures was conducted in same manner with soluble fraction using buffer 1 containing 10 mM CHAPS. Above methods were modified and combined for brain NOS and endothelial NOS purification those were previously reported (Pollock et al., 1991; Schmitt et al., 1991). ADP pool of rat liver as inducible type of NOS and rat cerebellum as brain type of NOS were prepared as previously reported(Iida et al., 1992; Ohshima et al., 1992).

SDS/PAGE and Western Blot

The purity of each fraction was monitored with SDS/PAGE using 6% polyacrylamide gel (Laemmli, 1970). Gels were stained with sliver nitrate (Morrissey, 1981). After SDS/PAGE, partial purified NOS were also transferred to a nitrocellulose membrane. Subsequently, the membrane was incubated with antibodies against various type of NOS, those were anti-nNOS, anti-iNOS, anti-eNOS and anti-common NOS at 1:200, 1:1000, 1:4000 and 1:100 dilution, respectively after blocking by 10% BSA. After washing by PBS(-), it was incubated with anti-rabbit F(ab2)' conjugated peroxidase at 1:400 dilution and the immunoreactive bands were detected with 3, 3'-diaminobenzidine as a substrate for peroxidase.

Immunohistochemistry of Human Placenta for NOS

The tissues were fixed in periodate-lysine-paraform aldehyde solution for six hours at 4°C , rinsed with PBS(-) containing 10%, 15%, and 20% sucrose each for six hours at 4°C , placed in OCT compound freezing medium, and frozen in dry ice-acetone. The frozen sections, 4 μ thick, were stained using an indirect peroxidase method. After blocking of endogenous peroxidase activity with 0.3% hydrogen peroxide in 50% methanol for 15 minutes and non-specific reactive site with 1:10 normal goat serum each for 30 minutes, the sections were incubated with the primary rabbit anti-eNOS at 1:50 dilution for one hour followed by the 1:100 dilution of horseradish peroxidase-labeled second antibodies against rabbit immunoglobulins (Amersham International, Buckinghamshire, England) for 30 minutes at room temperature. The sections were thoroughly washed in PBS(-) after each incubation step. The peroxidase coloring reaction was mediated by 0.02% 3,3'-diaminobenzidine (DAB) solution, pH 7.4, containing 0.005% hydrogen peroxide. The sections were counterstained with hematoxylin, dehydrated, and mounted. For electron microscopy, after DAB reaction, the sections were postfixed in 1% osmium tetroxide, dehydrated through increasing ethanol concentration and embedded in epon. Control tissue sections were stained using normal rabbit serum at a 1:10 dilution.

RESULTS

Purification of NOS From Human Placenta

The proteins with the molecular weight of 135, 105 and 80 kDa, respectively were recovered from hPl-sF by 2',5'-ADP agarose. With further purification, using calmodulin agarose columns, only 135 kDa protein was recovered and eluted as a single protein band. The proteins of 135 kDa and 80 kDa was recovered by 2',5'-ADP agarose from hPl-iF solubilized by 20 mM CHAPS buffer. Only 135 kDa proteins was purified by CaM agarose from ADP pool in the same manner as hPl-sF (Figure 1).

Measurement of NOS Activities

NOS activities for completely purified 135 kDa protein after calmodulin-agarose was relatively low because the activities of NOSs from both fractions may be lost rapidly. ADP pools from both fractions had high conversion activities from L-[2,3-^{3}H] arginine to L-[2,3-^{3}H] citrulline and that was inhibited by 60 mM L-NMMA (Table 1).

Table 1

Conversion Activities From L-{2,3-3H]arginine to L-{2,3-3H]citrulline Of Each Fractions From The Human Placenta

Fractions	Activity [cpm]	+60 μM L-NMMA [cpm]
sF-ADP pool	43320	3250
iF-ADP pool	41580	3120
CaM-agarose eluate (soluble)	5480	-----
CaM-agarose eluate (insoluble)	3820	-----

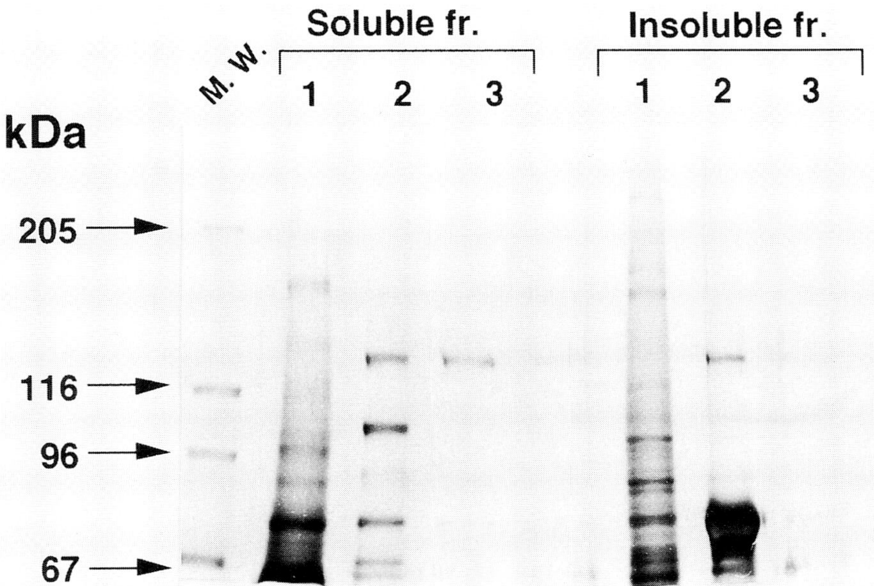

Figure 1. SDS-PAGE of the soluble (cytosolic) and insoluble (membrane) fractions of human placenta NOS partially purified by 2', 5'-ADP agarose affinity chromatography. Six percent SDS-polyacrylamide gel was stained with silver. Soluble fraction (fr.) and insoluble fr. of Lane 1, 2 and 3 are crude extract, ADP pool and CaM agarose column eluate, respectively.

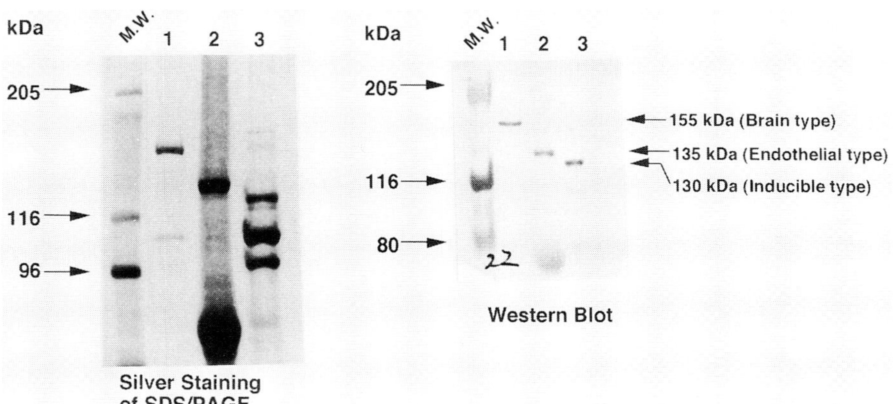

Figure 2A. (left) SDS-PAGE of ADP pool from various organs. Six percent of SDS-polyacrylamide gel was stained with silver. Lane 1 and 2 are the ADP pool of rat brain and insoluble fraction from human placenta(hPl-iF). Lane 3 is an ADP pool of rats liver treated with *Propionibacterium acnes(Pa)* and *Escherichia coli* lipopolysaccharide (LPS). B (right). Immunoblotting analysis of ADP pool from various organs using by anti-common NOS antibody. Lanes are as in Figure 2A (left).

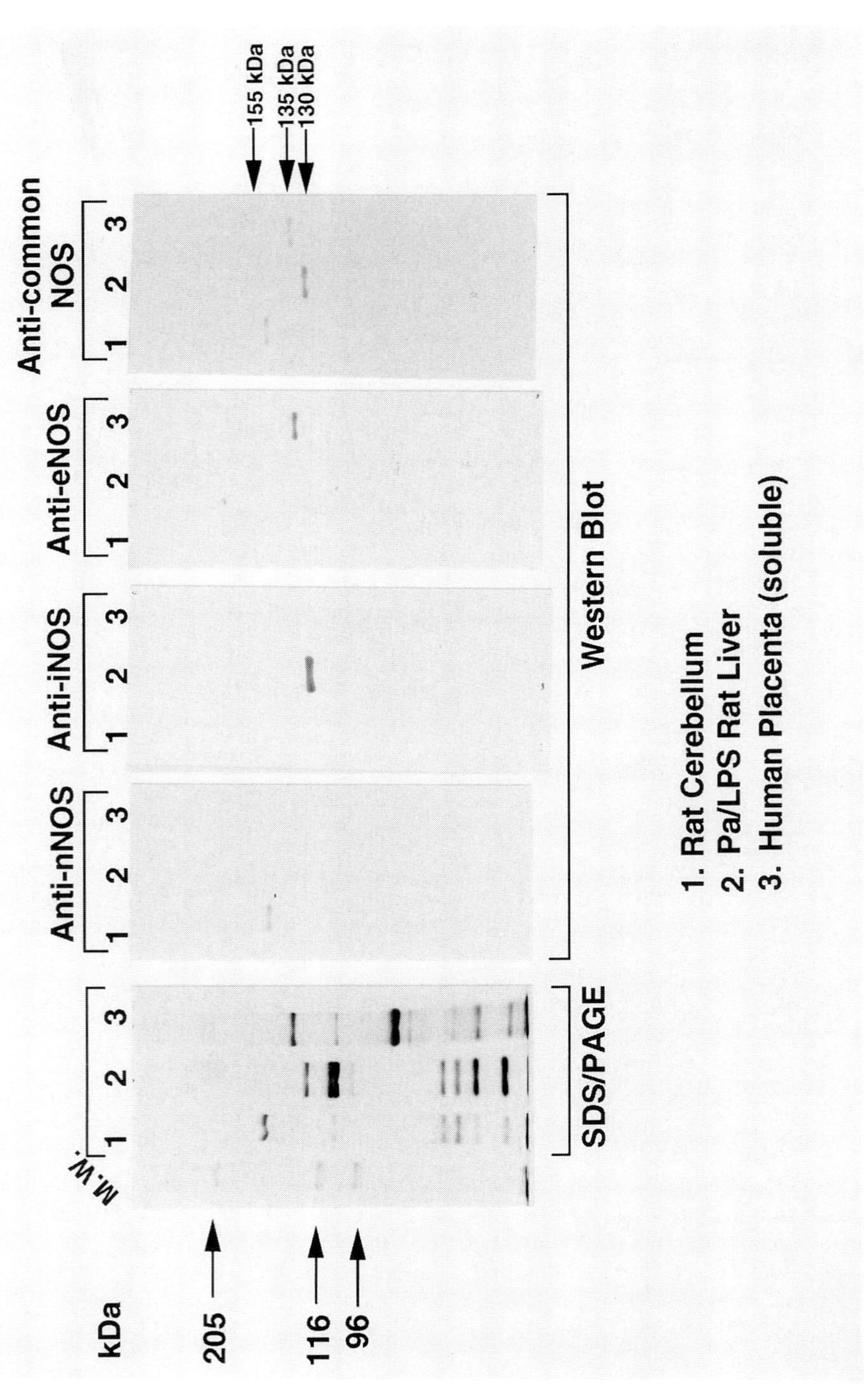
kDa
205
116
96
M.W.
1 2 3
SDS/PAGE
Anti-nNOS
1 2 3
Anti-iNOS
1 2 3
Anti-eNOS
1 2 3
Anti-common NOS
1 2 3
Western Blot
155 kDa
135 kDa
130 kDa
1. Rat Cerebellum
2. Pa/LPS Rat Liver
3. Human Placenta (soluble)

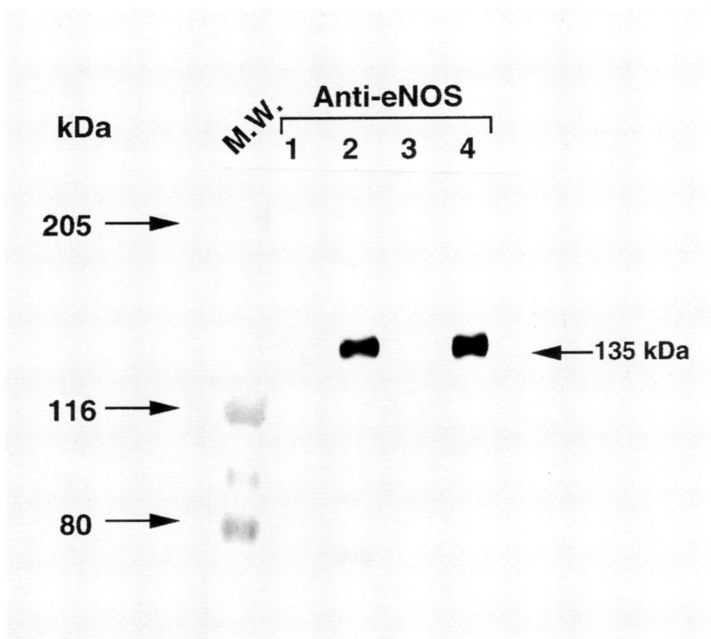

Figure 3A. (opposite page) Immunoblotting analysis of ADP pool from various organ using antisera to isoforms of NOSs. Lane 1, 2, and 3 are rat cerebellum, hPl-sF and Pa/LPS rat liver, respectively. 3B. (this page) Immunoblotting analysis of crude extracts and ADP pools from human placenta using anti-eNOS. Lane 1, and 3 are 4 mg of crude extracts from hPl-iF and hPl-sF, lane 2, and 4 are 400 ng of ADP pools from hPl-iF and hPl-sF of human placenta, respectively.

Western Blot Analysis

Anti-common NOS antibody that recognized the highly conserved sequence of three types of NOSs equally reacted with the 155 kDa protein bands of ADP pool from rats cerebellum, 135 kDa from human placenta (hPl-iF), and 130 kDa from rats liver treated with *Propionibacterium acnes* (P. a.) and *Escherichia coli* lipopolysaccharide (LPS) (Figure 2). For the 80 kDa protein was recovered too much, anti-common NOS was slightly reacted with the protein for non-specific or with degradation products of NOS. Furthermore, Figure 3A show the results of immunoblotting of NO synthase from various organs by using polyclonal antibodies against nNOS, iNOS, eNOS and anti-common NOS. The immunoreactive bands were observed at 135 kDa in both of ADP pools from hPl-sF by using polyclonal antibodies against eNOS and this antibody did not recognize rat nNOS and iNOS. Both anti-nNOS antibody and anti-iNOS did not recognize the 135 kDa protein bands of ADP pools from hPl-sF of human placenta. Furthermore, anti-eNOS did not react with crude extracts from hPl-sF and hPl-iF from human placenta, and equally recognized with 135 kDa proteins of ADP pool from hPl-sF and hPl-iF, respectively (Figure 3B).

fig. 4a

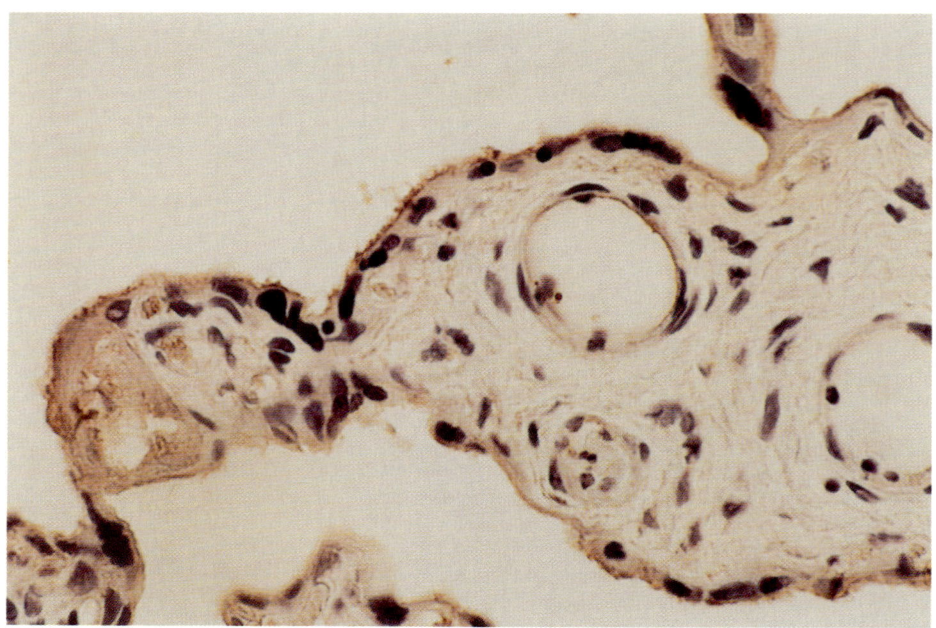

fig. 4b

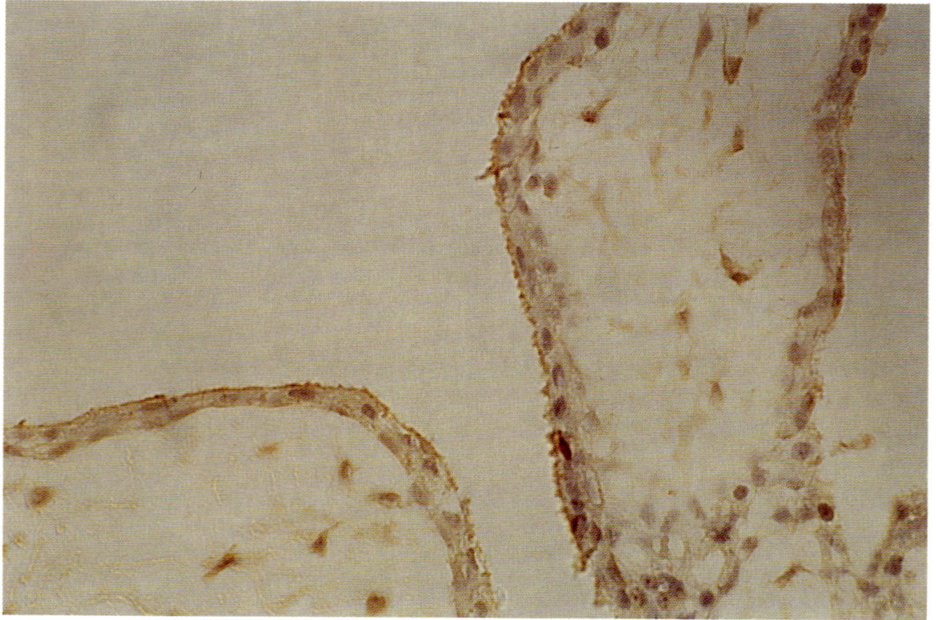

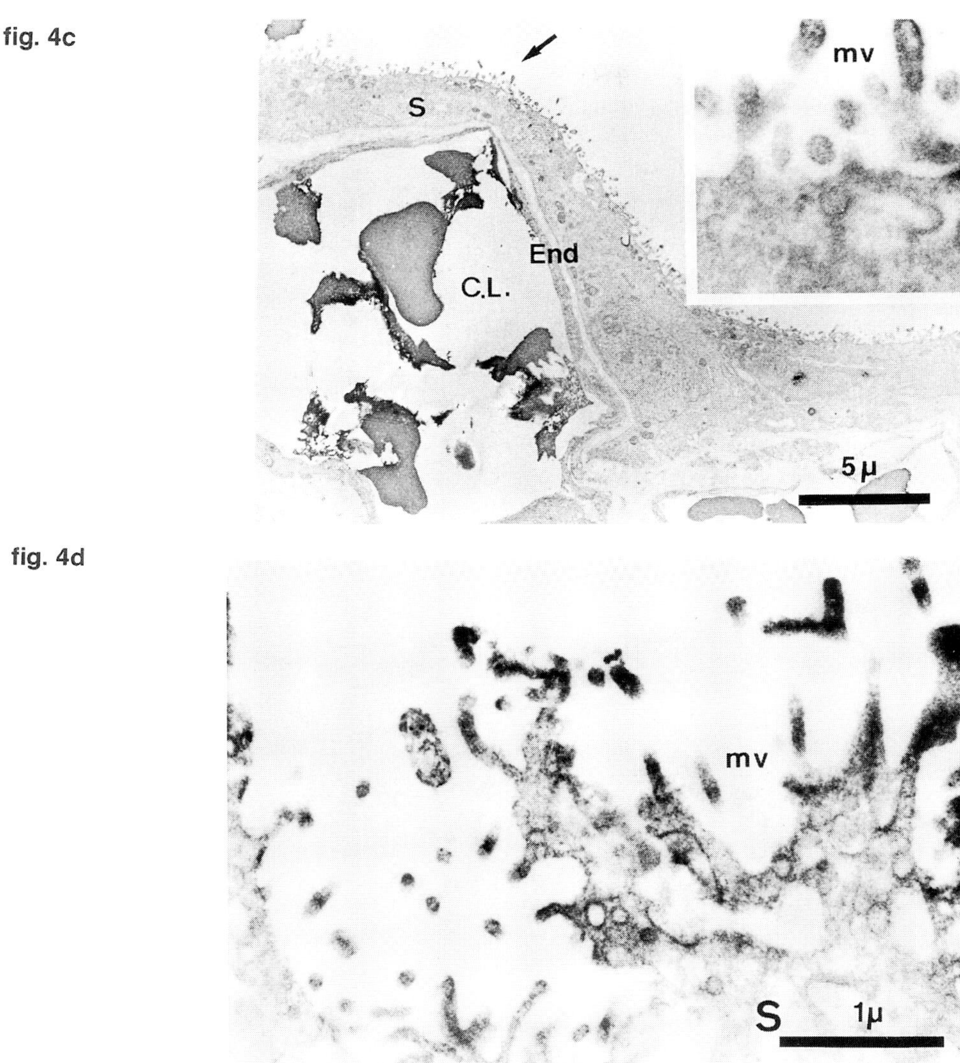

Figure 4A. (opposite page) Immunohistochemical localization of an eNOS in human placenta at 37 gestational weeks normal pregnancy. The presence of NOS are recognized in the endothelial cells and the surface of syncytium. B. (opposite page) Immunohistochemical localization of an eNOS in villi at 6 gestational weeks normal pregnancy. The presence of NOS are recognized in the surface of syncytiotrophoblast cells. C. (this page, top) Pre-embedding immunoelectron microscopy for eNOS using a frozen section of human placenta at 37 gestational weeks of normal pregnancy. Electron-dense reaction products are located in microvilli at syncytial cell. Upper right inset shows the magnification of microvilli on the trophoblast cell. Electron-dense reaction products are located in the surface of microvilli. D. (this page, bottom) Pre-embedding immunoelectron microscopy for the eNOS using a frozen section of villi at six gestational weeks normal pregnancy. Electron-dense reaction products are located in the surface of villi.

Immunohistochemistry of Human Placenta

Immunohistochemically, in the third trimester placenta, the localization of endothelial type of NOS was obtained both on the surface of fetal endothelium and in microvilli of syncytium, on the side of the maternal intervillous space (Figure 4A). Furthermore, in the villous of the first trimester of gestation, the products of DAB reaction were deposited on the surface of microvilli of syncytium using anti-eNOS serum (Figure 4B). The fibroblast nuclei in the villous stroma was also positive but the reason is not clear. Ultrastructually, in regions of vascular syncytial membrane where the junction of syncytium and fetal endothelium were thin, the deposits of DAB were found on the surface of syncytial cell (Figure 4C) and strongly observed in the cell membrane and in the membrane structure of vesicle associated with the microvilli (Figure 4C inset) in the third trimester placenta. In the same manner, DAB products were found in microvilli and on the surface of cell membrane from villi of six weeks gestation, as well as the membrane structure of vesicle (Figure 4D). The endothelial types NOS was obtained in the microvilli of syncytial surface from first trimester to third trimester placenta.

DISCUSSION

In the current study: 1) A 135 kDa single protein band from soluble and insoluble fractions of human placenta was purified using by CaM-agarose after 2',5'-ADP agarose affinity purification; the ADP pool from both fractions had high conversion activities from L-[2,3-^{3}H] arginine to L-[2,3-^{3}H] citrulline, which were completely inhibited by L-arginine analog for the specific inhibitor of NOS; these 135 kDa proteins reacted with polyclonal antibodies against c-terminal peptide of endothelial type of NOS using western blot; localization of eNOS on human placenta was on the surface of syncytiotrophoblast and in the cytosol of fetal endothelium.

These facts clearly demonstrate that 135 kDa proteins from both the soluble and insoluble fractions of human term placenta are the endothelial type of NOS. Myatt and associates have previously published the identification of NOS from vascular tree of human placenta and its molecular weight was 135 kDa and reacted with monoclonal antibody against bovine endothelial type of NOS on western blotting. The same NOS was completely purified in the current study and the proteins from both fractions were identified as the endothelial type of NOS using polyclonal antibody against c-terminal peptide of endothelial NOS which has no cross reactivity against inducible or brain NOS. Those proteins from ADP pool of both fractions were equally reacted with anti-eNOS. Buttery et al. (1994) also reported the distribution of eNOS in human placenta using antisera to constitutive NOS. In current study, the localization of eNOS was identified using specific eNOS antisera both in gestational villi and term placentae. Furthermore, we excised both proteins from gel and immunized rabbits were isolated. Separate polyclonal antibodies were made against them. Both antibodies were cross-reacted with 135 kDa protein from soluble and insoluble fraction each other (data is not shown). These facts demonstrate that 135 kDa proteins from both fractions must be same molecule protein. NO has been an essential molecule for micro blood flow in placenta circulation. In the current study, endothelial type of NOS was expressed in microvilli of syncytial trophoblast cell using by immunohistochemical methods. Thus, NO may be continuously released to blood flow of intervillous space and may prevent the coagulation of platelet.

SUMMARY

Nitric Oxide Synthase (NOS) from both soluble and insoluble fractions of human placenta was purified completely to a single protein band having molecular weight of 135 kDa by using calmodulin affinity chromatography after 2',5'-ADP agarose affinity chromatography. The Western Blot analyses using the three types of antibodies against NOSs clearly demonstrated that NOSs from human placenta were endothelial type of NOS, and not brain or macrophage type. Using immunohistochemistry at the light and electron microscopy level, NOS was localized in syncytiotrophoblast of human chorionic villi in both early and third trimester of pregnancy. NOS was localized in microvilli of syncytiotrophoblast. In placenta from third trimester of pregnancy, the localization of NOS was at the microvilli of syncytiotrophoblast from the maternal side and the cytoplasm of endothelial cell from the fetal side. These results demonstrated that NO is produced during pregnancy mainly by endothelial type of NOS in human placenta and may play an important role in maintaining pregnancy.

REFERENCES

Adachi, H., Iida., S., Oguchi, S., Ohshima, H., Suzuki, H, Nagasaki, K., Kawasaki, H., Sugimura, T. and Esumi, H. (1993) Molecular cloning of cDNA encoding an inducible calmodulin-dependent nitric oxide synthase from rat liver and its expression in COS-1 cells. *Eur. J. Biochem.* 217, 37-43.

Bredt, D.S., Hwang, P.M., Glatt, C.E., Lowenstein, C., Reed, R.R. and Snyder S. H. (1991) Cloned and expressed nitric oxide synthase structurally resembles cytochrome P-450 reductase. *Nature* 351, 714-718.

Bredt, D.S. and Snyder S.H. (1990) Isolation of nitric oxide synthase, a calmodulin-requiring enzyme. *Proc. Natl. Acad. Sci. USA* 87, 682-685.

Busconi, L. and Michel, T. (1993) Endothelial nitric oxide synthase. *J. Biol. Chem.* 268, 8410-8413.

Buttery, L.D.K., McCarthy, A., Springall, D.R., Sullivan, M.H.F., Elder, M.G., Michel, I. and Polak, J.M. (1994) Endothelial nitric oxide synthase in the human placenta: Regional distribution and proposed regulatory role at the fetomaternal interface. *Placenta* 15, 257-265.

Conrad, K.P., Joffe, G.M., Kruszyna, R., Rochelle, L.G., Smith, R.P., Chavez, J.E. and Mosher, M.D. (1993) Identification of increased nitric oxide biosynthesis during pregnancy in rats. *FASEB* 7, 566-571.

Garvey, E.P., Tuttle, J.V., Covington, K., Merrill, B.M., Wood, E.R., Baylis, S.A. and Charles, I.G. (1994) Purification and characterization of the constitutive nitric oxide synthase from human placenta. *Arch. Biochem. Biophys.* 311, 235- 241.

Hevel, J.M., White, K.A., and Marletta, M.A. (1991) Purification of the inducible murine macrophage nitric oxide synthase. *J. Biol. Chem.* 266, 22789-22798.

Hibbs, J.B., Jr., Tainor, R.R., Vavrin, Z. and Rachlin, E.M. (1988) Nitric oxide: A cytotoxic activated macrophage effector molecule. *Biochem. Biophys. Res. Commun.* 157, 87-94.

Iida, S., Ohshima, H., Oguchi, S., Hata, T., Suzuki, H., Kawasaki, H., and Esumi, H. (1992) Identification of inducible calmodulin-dependent nitric oxide synthase in the liver of rats. *J. Biol. Chem.* 267, 25385-25388.

Jassens, S.P., Shimouchi, A., Quertermous, T.Q., Bloch, D.B., and Bloch, K.D. (1992) Cloning and expression of a cDNA encoding human endothelium-derived relaxing factor/nitric oxide synthase. *J. Biol. Chem.* 267, 14519-14522.

Knowles, R.G., Palacios, M., Palmer, R.M.J. and Moncada, S. (1989) Formation of nitric oxide from L-arginine in central nervous system.: a transduction mechanism for stimulation of the soluble guanylate cyclase. *Proc. Natl. Acad. Sci. USA* 86, 5159-5162.

Laemmeli, U.K. (1970) Cleavage of the structural proteins during the assembly of the head of bacteriophage T4. *Nature* 227, 680-685

Lamas, S., Marsden, P.A., Li, G. K., Tempst, P. and Michel, T. (1992) Endothelial nitric oxide synthase: Molecular cloning and characterization of a distincter constitutive enzyme isoform. *Proc. Natl. Acad. Sci. USA* 89, 6348-6352.

Lowenstein, C., Glatt, C.S., Bredt, D.S. and Snyder S.H. (1992) Cloned and expressed macrophage nitric oxide synthase contrasts with the brain enzyme. *Proc. Natl. Acad. Sci.* US. 89, 6711-6715.

Lyons, C.R., Orloff, G.J. and Cunningham, J.M. (1992) Molecular cloning and functional expression of an inducible nitric oxide synthase from a murine macrophage cell line. *J. Biol. Chem.* 267, 6370-6374.

Marsden, P.A., Schappert, K.T., Chen, H.S., Flowers, M., Sundell, C.L., Wilcox, J.N., Lamas, S. and Michel, T. (1992) Molecular cloning and characterization of human endothelial nitric oxide synthase. *FEBS Lett.* 307, 287-293.

Mayer, B., John, M. and Bohme, E. (1990) Purification of a Ca2+/ calmodulin-dependent nitric oxide synthase from porcine cerebellum. *FEBS Lett.* 277, 215-219.

Myatt, L., Brockman, D.E., Langdon, G. and Pollock J.S. (1993) Constitutive calcium-dependent isoform nitric oxide synthase in the human placenta villous vascular tree. *Placenta* 14, 373-383.

Moncada, S., Palmer, R.M.J., and Higgs, E.A. (1991) Nitric oxide: Physiology, pathology, and pharmacology. *Pharmacol. Rev.* 43, 109-142.

Morris, N.H., Sooranna, S.R., Learmount, J.G., Poston, L., Ramsey, B., Pearson, J.D. and Steer, P.J. (1995) Nitric oxide synthase activities in placental tissue from normotensive, preeclamptic and growth retarded pregnancies. *Br. J. Obstet. Gynecol.* 102, 711-714.

Morrissey, J.H. (1981) Silver stain for proteins in poyacrlamide gels: A modified procedure with enhanced uniform sensitivity. *Analy. Biochem.* 117, 307-310.

Ohshima, H., Oguchi, S., Adachi, H., Iida, S., Suzuki, S., Sugimura, T. and Esumi, H. (1992) Purification of nitric oxide synthase from bovine brain: Immunological characterization and tissue distribution. *Biochem. Biophys. Res. Comm.* 183, 283-244.

Palmer, R.M.J., Ferrige, A G. and Moncada, S. (1987) Nitric oxide release accounts for the biological activity of endothelium-derived relaxing factor. *Nature* 327, 524-526.

Pollock, J.S., Fostermann, U., Mitchel, J.A., Warner, T.D., Schmidt, H.H.H.W., Nakane, M. and Murad, F. (1991) Purification and characterization of particulate endothelium-derived relaxing factor synthase from cultured and native bovine aortic endothelial cells. *Proc. Natl. Acad. Sci. USA* 88, 10480-10484.

Radomski, M.W., Palmer, R.M.J. and Moncada, S. (1987) Comparative pharmacology of endothelium-derived relaxing factor, nitric oxide and prostacyclin in platelets. *Br. J. Pharmacol.* 92, 181-187.

Schmidt, H.H.H.W. and Murad, F. (1991) Purification and characterization of a human NO synthase. *Biochem. Biophys. Res. Comm.* 181, 1372-1377.

Schmidt, H.H.H.W., Pollock, J.S., Nakane, M., Gorsky, L., Fostermann, U. and Murad, F. (1991) Purification of a soluble isoform of guanylyl cyclase-activating factor synthase. *Proc. Natl. Acad. Sci. USA* 88, 356-369.

Sessa, W.C., Harrison, J.K., Barber, C.H., Zeng, D., Durieux, M.E., D'Angelo, D.D., Lynch, K.R. and Peach, M.J. (1992) Molecular cloning and expression of a cDNA encoding endothelial cell nitric oxide synthase. *J. Biol. Chem.* 267, 15274-15276.

Stuehr, D.J., Cho, H.J., Kwon, N.S, Weise, M.F. and Nathan, C.F. (1991) Purification and characterization of the cytokine-induced macrophage nitric oxide synthase: An FAD- and FMN-containing flavoprotein. *Proc. Natl. Acad. Sci.* USA 88, 7773-7777.

Xie, Q., Cho, H., Calacay, J., Mumford, R.A., Swiderek, K.M., Lee, T.D., Ding, A., Troso, T. and Nathan, C. (1992) Cloning and characterization of inducible nitric oxide synthase from mouse macrophage. *Science* 256, 225-228.

Trophoblast Research 9:155-164, 1997

STRUCTURAL AND FUNCTIONAL CHANGE OF BLOOD VESSEL LABYRINTH IN MATURING PLACENTA OF MICE

Kazuyo Suzuki[1], Miya Kobayashi[2], Kunihiko Kobayashi[1], Yosuke Shiraishi[2], Setsuko Goto[1] and Takeshi Hoshino[2]

[1]Nagoya University
College of Medical Technology
Nagoya 461, Japan

[2]Department of Anatomy
Nagoya University School of Medicine
Nagoya 466, Japan

INTRODUCTION

Placentation means bringing two vascular systems, maternal and fetal, into close proximity. Since there are several similarities in early placental development in the rodents and men (Pijnenborg et al., 1981), basic biological principles governing placental development can be established only by comparative studies of some species.

In rodents, the placenta is discoidal and is known to be constructed by fetal blood vessels and maternal blood spaces forming a labyrinth (Jollie, 1964a, 1964b; Carpenter, 1975). The placental vascular architecture of the labyrinth is important as a basis for functional interpretation. Since one pregnant mouse has ten or more fetuses, each associated with a placenta, it would be interesting to know how she can nourish so many fetuses at one time in the uterus. Thus, knowledge of the structure and function of the placenta would give us a valuable clues to understand this phenomenon in detail. The special arrangement of maternal blood space and fetal blood vessels in the placental exchange is one important criterion for the effectiveness of transplacental transport in mice. To investigate the efficiency of the labyrinth, we histologically examined the formation of placental labyrinth as pregnancy advanced.

MATERIALS AND METHODS

Mice

Mice of ddY strain were used. After mating, the day when a vaginal plug first appeared was considered as day 0 of gestation. Care of the animals in this investigation conformed to the Guide for Animal Research, Nagoya University School of Medicine. The pregnant mice were anesthetized with ether, and fetuses as well as placentae were dissected on each fetal day (10 to 18), and placentae just after delivery were also taken. Three independent experiments using three mothers in each period were performed.

Measurement of Size of Placentae and Fetuses

Diameter and thickness of the placentae as well as fetal crown-rump-length (CRL) were measured just after dissection. The weight of each placenta and fetus was also estimated. Subplacental tissues including cord were removed carefully under the dissection microscope. Mean value and standard deviation were calculated from ten or more placentae and fetuses in each experiments.

Light Microscopy

Fresh placentae were cut into blocks and were fixed in 4% paraformaldehyde (PFA) in 0.1 M cacodylate buffer (pH 7.4) at 4° C for 2 days, then dehydrated and embedded in paraffin. Four micron-thick paraffin sections were deparaffinized and stained with hematoxylin and eosin, before being examined by light microscopy.

Light Microscopic Observation of Mitotic Trophoblast Cells

One µm-thick resin sections stained with toluidine blue were observed by light microscopy and the mitotic trophoblast cells were counted per unit area of the placental labyrinth (number of cells/mm^2).

Electron Microscopy

Fresh placentae were cut into blocks and fixed with Karnovsky's fixative (pH 7.4) for 4 hours at room temperature and then postfixed with 1% osmium tetroxide in 0.1 M cacodylate buffer (pH 7.4) for 90 minutes at room temperature. After dehydration with graded ethanol concentrations, they were embedded in Quetol 812 (Nissin EM, Tokyo). Ultrathin sections were cut on a Porter Blum MT-1 ultramicrotome using a diamond knife. Sections were stained with uranyl acetate and lead citrate and examined by transmission electron microscope (Hitachi H-7100).

RESULTS

Size Development in Mouse Fetuses and Placentae

Profiles of placentae and fetuses from fetal day 12 to just after birth are demonstrated in Figure 1. In contrast to the rapid growth of fetuses, placentae did not show much growth, but were maintained rather constant in size and weight (Figs. 2A and 2B, respectively). The weight ratio of fetus/placenta reached to 30 : 1 at birth (Table 1).

Process of Complex Labyrinth Formation

Figures 3A, 3B and 3C show the light microscopic profiles of fetal blood vessel and maternal blood space at 10, 14 and 18 days, respectively. The fetal blood vessels as well as the maternal blood space, branched finely close to one another as pregnancy advanced. The width of both the fetal blood vessel and the maternal blood space were gradually decreased and formed a labyrinth. The change is clearly demonstrated in Figures 3D, 3E and 3F, which are schematic profiles of Figures 3A, 3B and 3C, respectively.

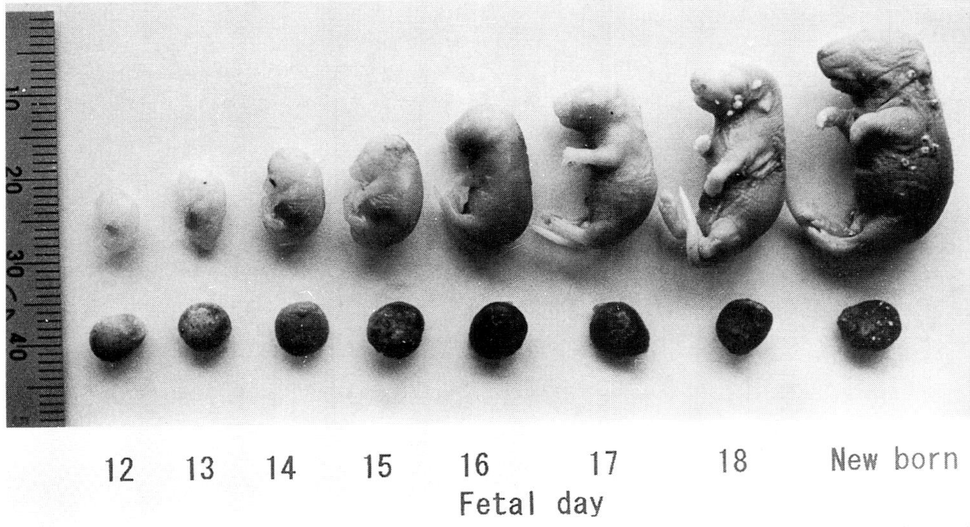

Figure 1. Size development of mouse placentae (bottom) and fetuses (top) from fetal day 12 to just after birth.

Table 1

Fetal day	Fetal weight (g)	Placental weight (g)	Weight ratios fetus/placenta
11	0.04 ±* 0.01 (15)**	0.04 ± 0.01 (15)	1
12	0.1 ± 0.01 (10)	0.06 ± 0.02 (10)	2
13	0.14 ± 0.02 (12)	0.07 ± 0.01 (12)	2
14	0.29 ± 0.03 (12)	0.09 ± 0.01 (12)	3
15	0.4 ± 0.04 (13)	0.1 ± 0.01 (13)	4
16	0.72 ± 0.04 (11)	0.11 ± 0.01 (11)	7
17	0.89 ± 0.07 (14)	0.09 ± 0.01 (14)	10
18	1.4 ± 0.11 (16)	0.09 ± 0.01 (16)	16
Newborn	2.09	0.07 ± 0.01 (6)	30

The weight ratios of fetus/placenta on each fetal day. The weight of each placenta as well as fetus was estimated and mean value and standard deviation were calculated. The values are representatives of three independent experiments. * Mean ± SD **Numbers of fetus or placenta examined are appeared in the parentheses.

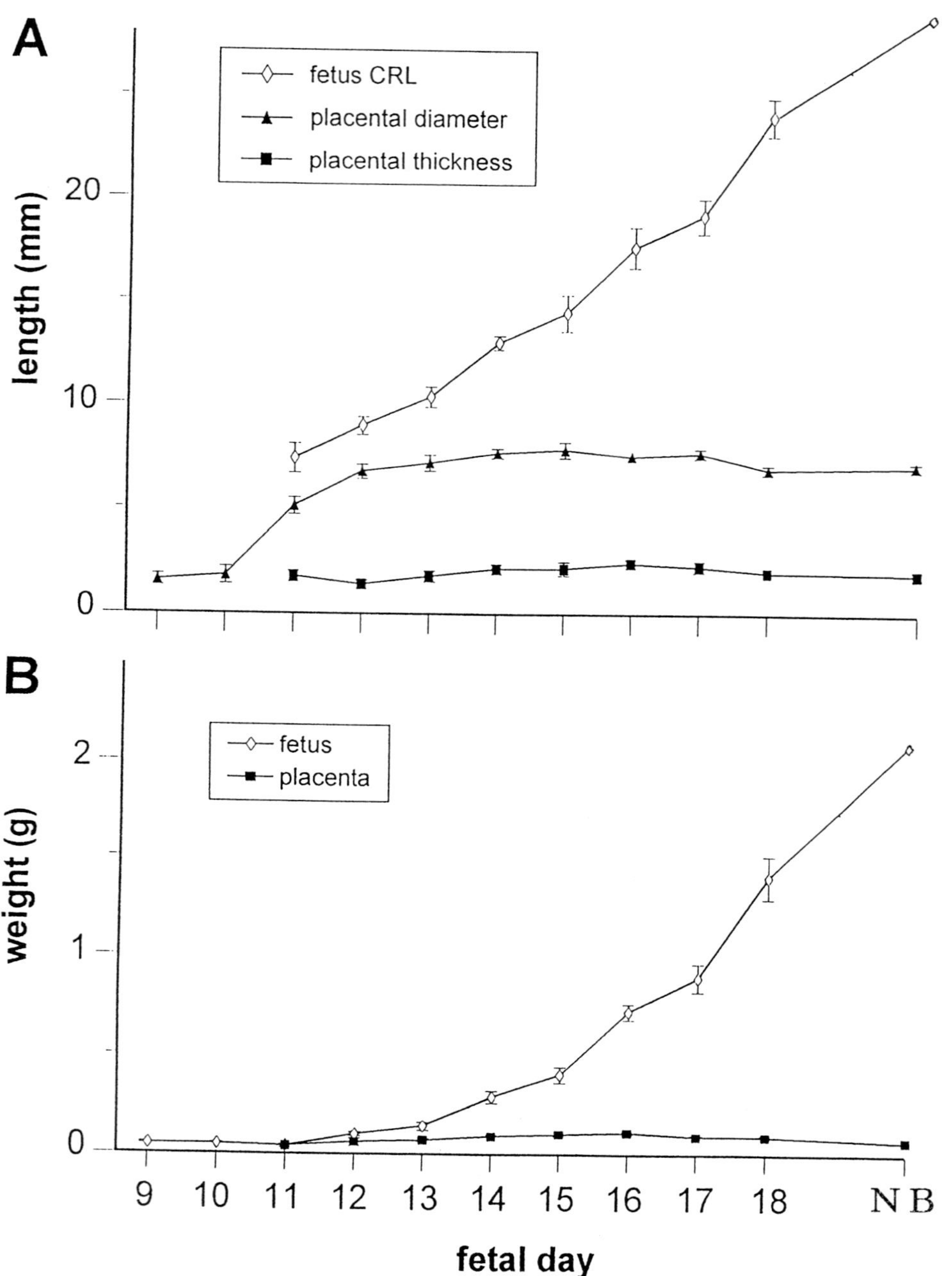

Figure 2. Growth of mouse placenta and fetus. A: length of fetal CRL as well as diameter and thickness of placenta. B: weight of fetus and placenta on each fetal day. Each point shows the mean and S.D. The values are representative of three independent experiments.

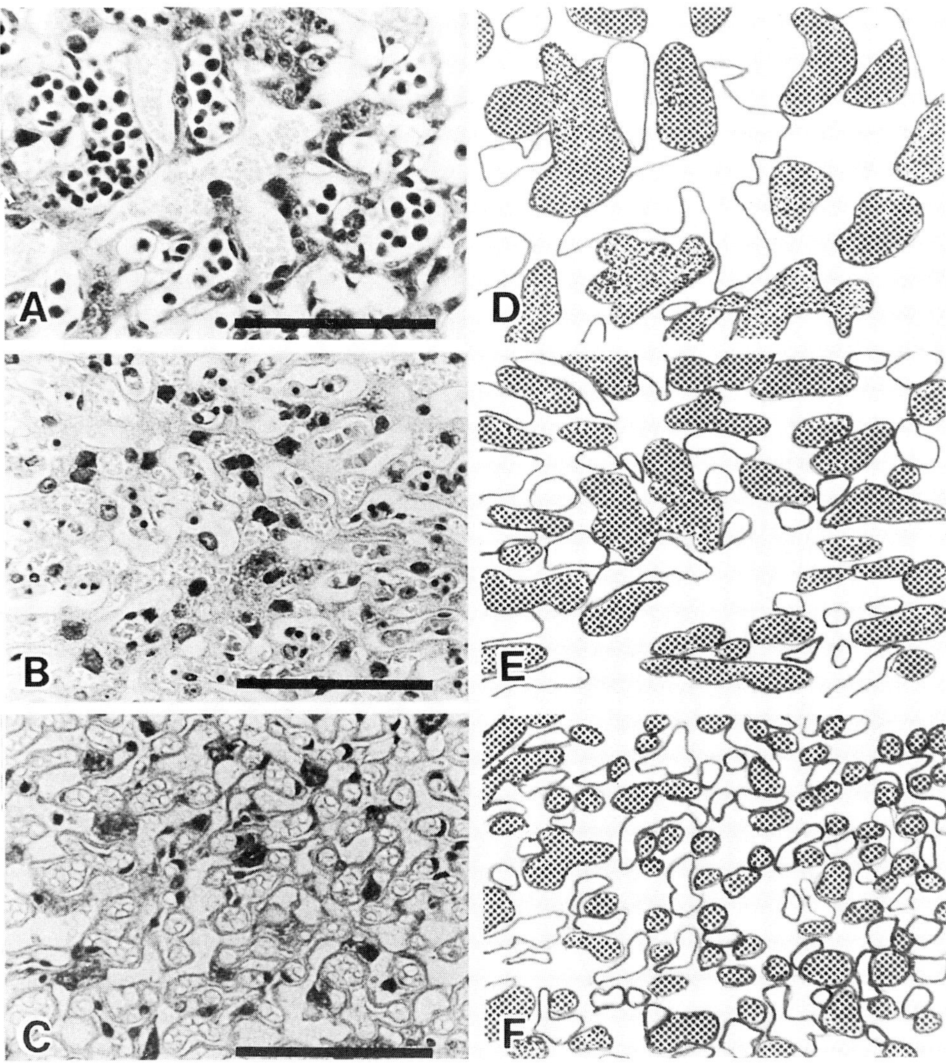

Figure 3. Light microscopy and corresponding schematic representation of growing labyrinth of fetal days 10 (A, D), 14 (B, E) and 18 (C, F). Hematoxylin and eosin staining. Bar indicates 100μ. In the scheme, dotted area indicates fetal vascular space, and undotted enclosed area, maternal blood space.

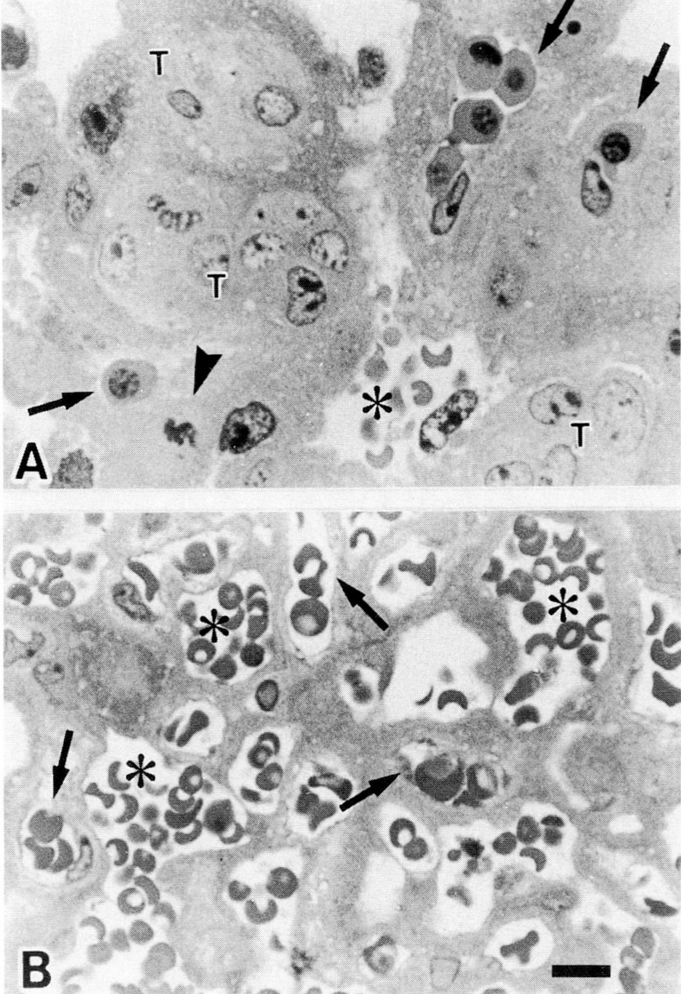

Figure 4. Light microscopy of semithin resin sections at fetal days 12 (A) and 18 (B). Toluidine blue staining. (A) Fetal blood vessels are recognizable by nucleated erythrocytes and endothelial cells (arrows). Trophoblast cells (T) show large and clear nuclei, sometimes at mitotic stage (arrowhead). (B) Fetal blood vessels are recognizable only by the endothelial cells (arrows). Trophoblasts are difficult to recognize because they are very thin. *; maternal blood space. Bar indicates 10μ.

Mitotic Activity of Trophoblast Cells in Developing Placental Labyrinth

Light microscopic observations of semi-thin resin sections indicated that mitosis in trophoblast cells were frequently observed on fetal day 12 (Figure 4A), but they were rarely observed at fetal day 18 (Figure 4B). Table 2 summarizes the occurrence of mitotic cells in developing placental labyrinth.

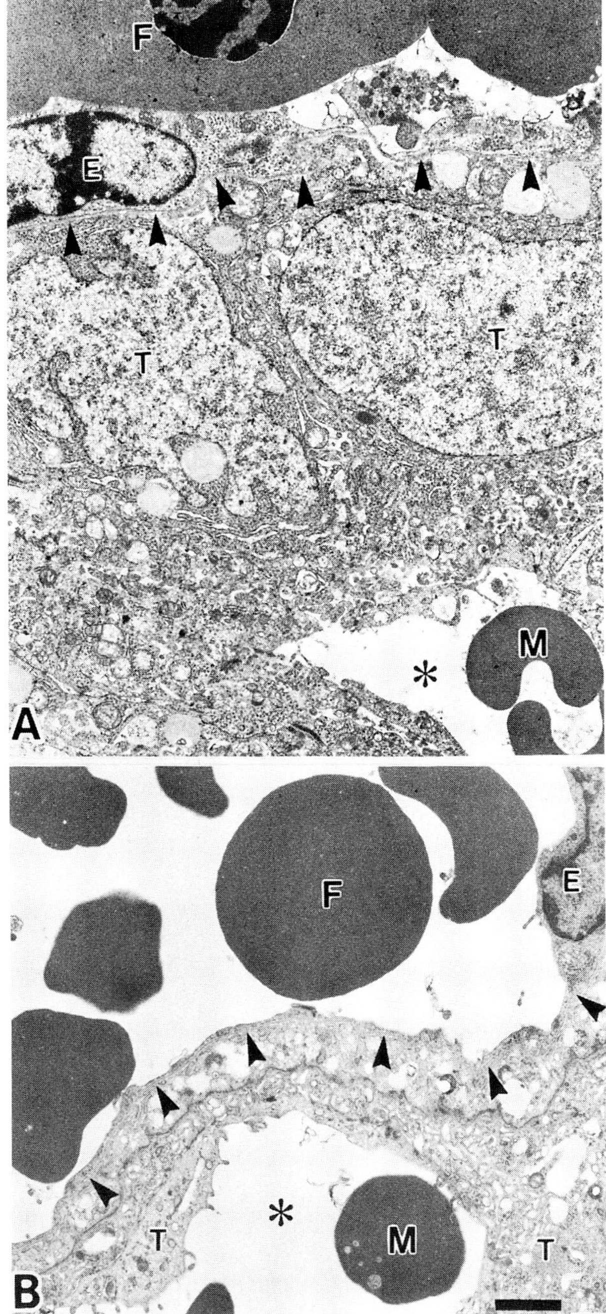

Figure 5. Ultrastructure of trophoblast cells at fetal days 12 (A) and 18 (B). (A) Cytoplasmic wall of the trophoblast (T) on fetal day 12 is apparently thick. Nucleated fetal erythrocytes (F) are seen. Arrowheads line basal lamina of endothelial cells (E). (B) On fetal day 18, the cytoplasmic wall of trophoblast cells is very thin. M; maternal erythrocyte; *; maternal blood space. Bar indicates 2μ.

Table 2

Occurrence Of Mitotic Trophoblast Cells Per Unit Area Of Maturing Placental Labyrinth Of Mice

Fetal Day	Mitotic cells/mm^2
12	6.9 ± 1.8
13	2.0 ± 0.8
14	0.1 ± 0 2
15	0
16	0
17	0
18	0

Ultrastructure of Fetal Vascularization and Maternal Blood Space

Electron microscopy demonstrated, round, large trophoblast cells which formed a thick cytoplasmic wall in the early stage (fetal day 12) (Figure 5A), whereas a thin cytoplasmic wall of the trophoblast cells was evident at the later stage (fetal day 18) (Figure 5B).

DISCUSSION

The present study demonstrated that maturing architecture of the mouse placenta is accompanied by the formation of a more complex labyrinth rather than on any increase in size. Whereas many chorionic villi are bathed in maternal blood in the intervillous space of the human placenta, the nutritional efficiency of a small mouse placenta may be attributed to the construction of the labyrinth. This may enable ten or more fetuses to co-exist in the uterus. The fetus/placenta weight ratio of the mouse reached 30 : 1 at birth, higher than that of man which is 6 : 1 (Dantzer et al., 1988). The higher ratio would indicate that a smaller mouse placenta can more efficiently nourish the fetus.

The complex construction of the placental labyrinth in mice was formed as the mouse fetus matured. Maternal blood spaces and fetal blood vessels, both branched off in close proximity and may contribute to the efficient exchange of nutrients and wastes between the fetus and mother. To clearly demonstrate the fine vascularization, three-dimensional observations using corrosion casts (Dantzer and Leiser, 1994) have been performed. In addition, computer-assisted three-dimensional reconstruction from serial sectional profiles is currently being conducted in our laboratory.

The exchange of nutrients and wastes also depend upon the transport activity of trophoblast cells (Bersu et al., 1989; Asai et al., 1993). The trophoblast cells in the later stages demonstrated very thin cytoplasmic membrane which may well facilitate the exchange of materials. In the early stages, however, a major activity of the placenta appears to proliferate in order to construct a complex labyrinth. This was confirmed in the present study by the frequent occurrence of mitotic trophoblast cells in the early stages (Table 2). It is reported that mice lacking hepatocyte growth factor fail to complete development and die *in utero* (Schmidt et al., 1995). The mutation affects the embryonic liver, which is reduced in size and shows extensive loss of parenchymal cells. In addition, development of the placenta, particularly the trophoblast cells, are impaired.

SUMMARY

The mouse placenta consists of fetal blood vessels, interpolating trophoblast cells and maternal blood spaces forming a labyrinth. It was observed that in contrast to the rapid growth of fetuses, the placentae maintained a constant size through pregnancy. The weight ratio of fetus/placenta was 30 : 1 at birth, about 5 times that in humans. To investigate the efficiency of the labyrinth in materno-fetal exchange in maturing placenta, we histologically examined the formation of the labyrinth as pregnancy advanced. In the late stage, fetal blood vessels and maternal blood spaces had fine, close-knit branches that formed a complex labyrinth, which may explain the efficient exchange of nutrients and wastes between fetus and mother. The maturation of mouse placenta is accompanied by the formation of a more complex labyrinth rather than on any increase in size.

REFERENCES

Asai, M., Faber, W., Neth-Jessee, L., di Sant'Agnese, P.A., Nakanishi, M. and Miller, R.K. (1993) Human placental transport and metabolism of all-trans retinoic acid in vitro. *Trophoblast Res.* 7, 25-33.

Bersu, E.T., Mossman, H.W. and Kornguth, S.E. (1989) Altered placental morphology associated with murine trisomy 19. *Teratology* 40, 513-523.

Carpenter, S.J. (1975) Ultrastructural observations on the maturation of the placental labyrinth of the golden hamster (days 10 to 16 of gestation). *Am. J. Anat.* 143, 315-348.

Dantzer, V., Leiser, R., Kaufmann, P. and Luckhardt, M. (1988) Comparative morphological aspects of placental vascularization. *Trophoblast Res.* 3, 235-260.

Dantzer, V. and Leiser, R. (1994) Initial vascularization in the pig placenta: I. Demonstration of nonglandular areas by histology and corrosion casts. *Anat. Rec.* 238, 177-190.

Jollie, W.P. (1964a) Radioautographic observations on variations in desoxyribonucleic acid synthesis in rat placenta with increasing gestational age. *Am J. Anat.* 114, 161-171.

Jollie, W.P. (1964b) Fine structural changes in placental labyrinth of the rat with increasing gestational age. *J. Ultrastruct. Res.* 10, 27-47.

Pijnenborg, R., Robertson, W.B., Brosens, I. and Dixon, G. (1981) Review article: Trophoblasts invasion and the establishment of haemochorial placentation in man and laboratory animals. *Placenta* 2, 71-92.

Schmidt, C., Bladt, F., Goedecke, S., Brinkmann, V., Zschiesche, W., Sharpe, M., Gherardi, E. and Birchmeier, C. (1995) Scatter factor/hepatocyte growth factor is essential for liver development. *Nature* 373, 699-702.

LIST OF CONTRIBUTORS

Robert Boyd
Department of Child Health
University of Manchester
St. Mary's Hospital
Hathersage Road
Manchester M13 OJH, United Kingdom

C.A.R. Boyd
Department of Human Anatomy
University of Oxford
Oxford OX1 3QX, United Kingdom

Hiroyasu Esumi
Biochemistry Division
National Cancer Center Research Institute
5-1-1, Tsukiji, Chuo-ku
Tokyo, Japan

Seiichiro Fujimoto
Department of Obstetrics and Gynecology
Hokkaido University
School of Medicine
Sapporo, Japan

Takao Fukaya
Department of Obstetrics and Gynecology and Pathology
Tohoku University School of Medicine
Sendai 980, Japan

Itsuko Furuta
Department of Obstetrics and Gynecology
Hokkaido University
School of Medicine
Sapporo, Japan

Jocelyn Glazier
Department of Child Health
University of Manchester
St. Mary's Hospital
Hathersage Road
Manchester M13 OJH, United Kingdom

Setsuko Goto
Nagoya University College of Medical Technology
Nagoya 461, Japan

Susan Greenwood
Department of Child Health
University of Manchester
St. Mary's Hospital
Hathersage Road
Manchester M13 OJH, United Kingdom

J. Gedis Grudzinskas
Department of Obstetrics and Gynecology
London Hospital Medical College
London, United Kingdom

Toshio Hata
Department of Obstetrics and Gynecology
Saitama Medical School
38 Morohongo, Moroyama, Iruma-gun
Saitama, Japan

Naka Hattori
Laboratory of Cellular Biochemistry
Animal Resource Sciences/Veterinary Medical Sciences
The University of Tokyo
1-1-1 Yayoi, Bunkyo-ku
Tokyo 113, Japan

Koji Hirahatake
Department of Obstetrics and Gynecology
Hokkaido University
School of Medicine
Sapporo, Japan

Mitsuko Hirosawa
Laboratory of Cellular Biochemistry
Animal Resource Sciences/Veterinary Medical Sciences
The University of Tokyo
1-1-1 Yayoi, Bunkyo-ku
Tokyo 113, Japan

Takeshi Hoshino
Department of Anatomy
Nagoya University
School of Medicine
Nagoya 466, Japan

Sachio Iida
Department of Obstetrics and Gynecology
Saitama Medical School
38 Morohongo, Moroyama, Iruma-gun
Saitama, Japan

Keiichi Isaka
Department of Obstetrics and Gynecology
Tokyo Medical College
Tokyo, Japan

Hiroshi Ishikura
Department of Pathology I
Hokkaido University
School of Medicine
Sapporo, Japan

Mitsutoshi Iwashita
Maternal and Perinatal Center
Tokyo Women's Medical College
Tokyo 162, Japan

John M. Johnston
Department of Biochemistry
The University of Texas Southwestern
Medical Center at Dallas
5323 Harry Hines Boulevard
Dallas, Texas 75235-9051 USA

Hideharu Kanzai
Department of Obstetrics and Gynecology
Kansai Medical University
10-15 Fumizono-cho, Moriguchi
Osaka 570, Japan

Hiroshi Kobayashi
Department of Obstetrics and Gynecology
Hamamatsu University School
of Medicine
Hamamatsu, Shizuoka 431-31, Japan

Kunihiko Kobayashi
Nagoya University College of
Medical Technology
Nagoya 461, Japan

Miya Kobayashi
Department of Anatomy
Nagoya University
School of Medicine
Nagoya 466, Japan

Takako Kobayashi
Department of Obstetrics and Gynecology
Hamamatsu University School
of Medicine
Hamamatsu, Shizuoka 431-31, Japan

Yoshiki Kudo
Department of Obstetrics and Gynecology
Tokyo Women's Medical College
8-1 Kawada, Shinjuku
Tokyo 162, Japan

Yukiko Kurashima
Biochemistry Division
National Cancer Center Research
Institute
5-1-1, Tsukiji, Chuo-ku
Tokyo, Japan

Min-Lian Luo
Department of Obstetrics and Gynecology
Hokkaido University
School of Medicine
Sapporo, Japan

Takeshi Maruo
Department of Obstetrics and Gynecology
Kobe University School of Medicine
Kobe 650, Japan

Shigeki Matsubara
Department of Obstetrics and Gynecology
and Anatomy
Jichi Medical School
Tochigi 329-04, Japan

Takashi Matsubara
Department of Obstetrics and Gynecology
Kansai Medical University
10-15 Fumizono-cho, Moriguchi
Osaka 570, Japan

Hiroya Matsuo
Department of Obstetrics and Gynecology
Kobe University School of Medicine
Kobe 650, Japan

Kwan-Sik Min
Laboratory of Cellular Biochemistry
Animal Resource Sciences/
Veterinary Medical Sciences
The University of Tokyo
1-1-1 Yayoi, Bunkyo-ku
Tokyo 113, Japan

Ryuichi Miura
Laboratory of Cellular Biochemistry
Animal Resource Sciences/
Veterinary Medical Sciences
The University of Tokyo
1-1-1 Yayoi, Bunkyo-ku
Tokyo 113, Japan

Isao Miyakawa
Department of Obstetrics and Gynecology
Oita Medical University
1-1 Hasama, Oita-gun
Oita 879-55, Japan

Matsuto Mochizuki
Department of Obstetrics and Gynecology
Kobe University School of Medicine
Kobe 650, Japan

Nobuhiko Moniwa
Department of Obstetrics and Gynecology
Hamamatsu University School
of Medicine
Hamamatsu, Shizuoka 431-31, Japan

Tokuro Nakajima
Department of Obstetrics and Gynecology
Kansai Medical University
10-15 Fumizono-cho, Moriguchi
Osaka 570, Japan

Hisashi Narahara
Department of Obstetrics and Gynecology
Oita Medical University
1-1 Hasama, Oita-gun
Oita 879-55, Japan

Masashi Nishiya
Department of Obstetrics and Gynecology
Hokkaido University
School of Medicine
Sapporo, Japan

Ken Noda
Laboratory of Cellular Biochemistry
Animal Resource Sciences/
Veterinary Medical Sciences
The University of Tokyo
1-1-1 Yayoi, Bunkyo-ku
Tokyo 113, Japan

Tomoya Ogawa
Laboratory of Cellular Biochemistry
Animal Resource Sciences/
Veterinary Medical Sciences
The University of Tokyo
1-1-1 Yayoi, Bunkyo-ku
Tokyo 113, Japan

Toshihiro Ohkouchi
Department of Obstetrics and Gynecology
Hokkaido University
School of Medicine
Sapporo, Japan

Hiroyuki Ohsawa
Department of Obstetrics and Gynecology
Saitama Medical School
38 Morohongo, Moroyama, Iruma-gun
Saitama, Japan

Tetsuo Otani
Department of Obstetrics and Gynecology
Kobe University School of Medicine
Kobe 650, Japan

Takuma Saito
Department of Obstetrics and Gynecology
Jichi Medical School
Tochigi 329-04, Japan

Keiji Sakai
Department of Obstetrics and Gynecology
Tokyo Women's Medical College
Tokyo 162, Japan

Noriaki Sakuragi
Department of Obstetrics and Gynecology
Hokkaido University
School of Medicine
Sapporo, Japan

Hironobu Sasano
Department of Obstetrics and Gynecology
and Pathology
Tohoku University School of Medicine
Sendai 980, Japan

Ikuo Sato
Department of Anatomy
Jichi Medical School
Tochigi 329-04, Japan

Isamu Sawaragi
Department of Obstetrics and Gynecology
Kansai Medical University
10-15 Fumizono-cho, Moriguchi
Osaka 570, Japan

Ming You She
Department of Obstetrics and Gynecology
Hamamatsu University School
of Medicine
Hamamatsu, Shizuoka 431-31, Japan

Kunio Shiota
Laboratory of Cellular Biochemistry
Animal Resource Sciences/
Veterinary Medical Sciences
The University of Tokyo
1-1-1 Yayoi, Bunkyo-ku
Tokyo 113, Japan

Yosuke Shiraishi
Department of Anatomy
Nagoya University
School of Medicine
Nagoya 466, Japan

Colin Sibley
Department of Child Health
University of Manchester
St. Mary's Hospital
Hathersage Road
Manchester M13 OJH, United Kingdom

Hiroaki Soma
Department of Obstetrics and Gynecology
Saitama Medical School
38 Morohongo, Moroyama, Iruma-gun
Saitama, Japan

Kazuyo Suzuki
Nagoya University College of
Medical Technology
Higashi-ku, Daiko Minami 1-1-20
Nagoya 461, Japan

Yoshichika Suzuki
Department of Obstetrics and Gynecology
Tokyo Medical College
Tokyo, Japan

Junko Takada
Department of Obstetrics and Gynecology
Tokyo Medical College
Tokyo, Japan

Masaomi Takayama
Department of Obstetrics and Gynecology
Tokyo Medical College
Tokyo, Japan

Naoki Takeda
Department of Obstetrics and Gynecology
Hokkaido University
School of Medicine
Sapporo, Japan

Yoshihiko Takeda
Maternal and Perinatal Center
Department of Obstetrics and Gynecology
Tokyo Women's Medical College
Tokyo 162, Japan

Toshihiko Terao
Department of Obstetrics and Gynecology
Hamamatsu University School
of Medicine
Hamamatsu, Shizuoka 431-31, Japan

Norihiko Tsumura
Department of Obstetrics and Gynecology
Hokkaido University
School of Medicine
Sapporo, Japan

Tadashi Watanabe
Department of Obstetrics and Gynecology
Tohoku University School of Medicine
1-1, Seiryo-machi, Aoba-ku
Sendai 980, Japan

Hidemichi Watari
Department of Obstetrics and Gynecology
Hokkaido University
School of Medicine
Sapporo, Japan

Akira Yajima
Department of Obstetrics and Gynecology
and Pathology
Tohoku University School of Medicine
Sendai 980, Japan

Katsuhiko Yasuda
Department of Obstetrics and Gynecology
Kansai Medical University
10-15 Fumizono-cho, Moriguchi
Osaka 570, Japan

INDEX

All page numbers listed below represent the first page of each chapter where subject is listed.

Trophoblast Research
Volume 10

EARLY PREGNANCY

Trophoblast Research Volume 10

EARLY PREGNANCY

Edited by

Jean-Michel Foidart
University of Liege
Liege, Belgium

John Aplin
University of Manchester
Manchester, United Kingdom

Peter Kaufmann
Technical University of Aachen
Aachen, Germany

Jean-Pierre Schaaps
University of Liege
Liege, Belgium

UNIVERSITY OF ROCHESTER PRESS

First published 1997

University of Rochester Press
668 Mount Hope Avenue
Rochester, New York 14620 USA
and at PO Box 9, Woodbridge, Suffolk IP12 3DF, UK

ISBN 1-58046-017-8

Library of Congress Cataloging-in-Publication Data

Early pregnancy / edited by Jean-Michel Foidart ... [et al.]
p. cm. -- (Trophoblast research ; v.10)
Includes index
ISBN 1-58046-017-8 (alk. paper)
1. Pregnancy--Trimester, First, 2. Placenta. 3. Trophoblast.
I. Foidart, Jean-Michel, 1949- II. Series.
RG558.E27 1997 97-28673
812.6'3--dc21 CIP

British Library Cataloguing-in-Publication Data

A catalogue record for this book is available
from the British Library

This publication is printed on acid-free paper

Printed in the United States of America

Trophoblast Research

Series Editor

Richard K. Miller and Henry A. Thiede
University of Rochester Medical Center
Rochester, New York

Volume 1	FETAL NUTRITION, METABOLISM AND IMMUNOLOGY The Role of the Placenta Edited by Richard K. Miller and Henry A. Thiede Published: 1983
Volume 2	CELLULAR BIOLOGY AND PHARMACOLOGY OF THE PLACENTA Techniques and Applications Edited by Richard K. Miller and Henry A. Thiede Published: 1987
Volume 3	PLACENTAL VASCULARIZATION AND BLOOD FLOW Basic Research and Clinical Applications Edited by Peter Kaufmann and Richard K. Miller Published: 1988
Volume 4	TROPHOBLAST INVASION AND ENDOMETRIAL RECEPTIVITY - Novel Aspects of the Cell Biology of Embryo Implantation Edited by Hans-Werner Denker and John D. Aplin Published: 1990
Volume 5	THE MOLECULAR BIOLOGY AND CELL REGULATION OF THE PLACENTA Edited by Richard K. Miller and Henry A. Thiede Published: 1991
Volume 6	PLACENTAL SIGNALS Endocrine and Paracrine Control of Pregnancy Edited by Lise Cedard and J. Anthony Firth Published: 1992
Volume 7	FETAL GROWTH AND THE PLACENTA - From Implantation to Delivery Edited by Henning Schneider, Paul Bischof, and Rudolf Leiser Published: 1993

Volume 8	HIV, PERINATAL INFECTIONS AND THERAPY The Role of the Placenta Edited by Richard K. Miller and Henry A. Thiede Published: 1994
Volume 9	PLACENTAL MOLECULES IN HEMODYNAMICS, TRANSPORT AND CELLULAR REGULATION Edited by Toshio Hata, Masaomi Takayama and Ichiro Taki Published: 1997
Volume 10	EARLY PREGNANCY Edited by Jean-Michel Foidart, John Aplin Peter Kaufmann and Jean-Pierre Schaaps Published: 1997

TROPHOBLAST RESEARCH

Trophoblast Research publishes contributions concerning the placenta and the extraembryonic membranes as they relate to embryonic and fetal development and to trophoblastic neoplasia. Original articles, reviews, and reports are published in single bound volumes. All articles are peer-reviewed.

The Editorial Office for
Department of Obstetrics and Gynecology
University of Rochester School of Medicine and Dentistry
601 Elmwood Avenue, Rochester, New York 14642-8668 USA
716-275-3638 or trophres@obgyn.rochester.edu

Derived from the
Vith Meeting of the
European Placenta Group
and
The Rochester Trophoblast Conference
held September 20-23, 1995
in Spa, Belgium

PREFACE

This volume represents the report from the VIth Meeting of the European Placenta Group, Joint Meeting with the Rochester Trophoblast Conference which was held in Spa, Belgium from September 20-23, 1995. The meeting was devoted to the study of the placenta during early pregnancy. Colleagues from many different disciplines contributed. The Editors of this volume have selected twenty-five papers which reflect either scientifically exciting novel research or present important state of the art reviews.

Naturally, trophoblast invasion and related events are a central theme during early pregnancy. Accordingly, this subject dominated the meeting in Spa, and the majority of contributions to this volume are devoted to implantation. The section on trophoblast invasion is introduced by a summary of the basics including a definition of terms, structural aspects of implantation and trophoblast invasion. There follows a series of detailed contributions dealing with cell biological aspects. They include the action of matrix metalloproteinases, their inhibitors, the influence of regulating cytokines, the importance of cell adhesion molecules, and interactions with maternal immune cells.

Other aspects of the placenta in early pregnancy include angiogenesis as well as endocrine and paracrine signaling. In a series of six articles, basic and clinical aspects of human placental capillary development are described in normal and pathological pregnancies with challenging new concepts. A further six papers deal with various endocrine and paracrine aspects of the placenta as a secretory and target organ.

Availability of human early pregnancy material for research depends on ethical and legal considerations. The more sophisticated the methods used to work with such tissue *in vitro* and to manipulate its constituent cells biologically and genetically, the greater the demands for moral and legal definitions. During the meeting, Claude Sureau dealt with these problems from the viewpoint of a practicing obstetrician/gynecologist. We feel his contribution is a worthy introduction to this volume.

The Editors are grateful to Richard K. Miller and Henry A. Thiede as Series Editors and to the Editorial Board of *Trophoblast Research* for including this volume in the series, to the reviewers for their helpful comments, and to Jackie White and Elaine Welliquet, for taking up the immense burden of the secretarial work. We also acknowledge the financial support provided by the Belgian National Research Foundation which helped us to organize the meeting which was the basis for this volume.

It is hoped that these reviews and research reports provide a useful reference for the participants of the meeting in Spa and the basis for all readers for his to extend their own future investigations in this field.

Jean-Michel Foidart, Liege
John Aplin, Manchester
Peter Kaufmann, Aachen
Jean-Pierre Schaaps, Liege

CONTENTS

TROPHOBLAST INVASION

Trophoblast Research 10:1-12, 1997

STATUS OF EMBRYO AND FETUS, A MATTER OF CONTRACT OR OF CIVIL RIGHT ?

Claude Sureau

Theramex Institute
Bioethics, Women's Health, and Society
38-40 Avenue de New-York
75016 Paris, France

INTRODUCTION

Two preliminary points have to be made: I am not a philosopher and nor do I feel competent in presenting you with the solution of this long standing philosophical discussion. I am also not a lawyer and, although I may occasionally make allusion to specific legal issues, I will not attempt the impossible synthesis of the legal status of embryo and fetus in different countries of the world.

I am going to talk as a practicing obstetrician/gynecologist who has been confronted with the dilemmas arising from the conflicts between clinical situations associated with the corresponding human distresses, and moral as well as legal rules or political decisions.

Factual Clinical Situations

Let us start with the description of some clinical situations, involving the conceptual status of the embryo. Of course such list is by no means exhaustive.

Voluntary abortion (by instrumental or biochemical methods): claims have been made in favor of respect for the woman's freedom of choice, or for the life of the embryo considered already as a "human being". This problem concerns not only overt pregnancy (after delayed menstruation) but also canceled pregnancy, before any delay, the interruption of which, is sometimes called contragestion.

Medically indicated termination of pregnancy (sometimes called "therapeutic"):

The ethical conflicts or choices are the followings:

• life of the mother versus life of the baby. This situation is rare nowadays and usually easy to manage

• termination of pregnancy may be supposedly protecting the interests of the child to be

because its death is unavoidable (ex: anencephalus). An ethical problem arises here, linked to the possible use of its organs, including the heart, which militates to the postponement of termination on the one hand, and on the other the imperative to obtain these organs, before the "complete" death of the baby, the definition of which is difficult to ascertain (Beller, 1994).

because its life will be considered as an unbearable burden. A perfect example is given by the bullous epidermolysis. However the difficulty resides in determining the limits of such a definition, particularly when considering the evolution of knowledge over time. This is particularly so when considering late occurrence of diseases (e.g., Huntington disease).

because its life will be too much of a burden to its family or to society. This is a frightening evolution leading to the slippery slope, not of some kind of biological eugenism, but of actual economic eugenism if economic considerations lead to a pressure towards the acceptance of interruption.

In all these situations practitioners may be tempted to withdraw themselves from the decision making process and to refer the case to courts or ethical committees. Such an attitude can lead the medical profession into abnegating responsibility, furthermore it serves to illustrate that in certain circumstances the embryo or fetus is not considered as an actual human being.

Another aspect of this local differences in approach is demonstrated by the fact that in some countries there is no time limit for termination of pregnancy. It can be done up to the last moment before delivery, as in case of craniotomy on hydrocephalus, showing clearly that the status of the fetus is fundamentally not the same before and after birth.

Fetal medicine and surgery: despite the remarks above, a concept has evolved over time leading to consider the fetus as a "patient". This means that this patient may be submitted to investigations or treatment aimed exclusively for its benefit.

However it is clear that this "patient" keeps a specificity, due to its dependence on the well being of its "envelop", the mother. This can obviously lead to possible conflicts of interest.

It follows that it may be difficult to determine what level of risks has to be accepted by the mother in order to potentially improve the chances of survival in good condition of the baby to be. The point of extraction by cesarean section will be considered below. The acceptance of treatments or investigation which may bear more or less heavy risks is not always obvious and needs to be carefully considered. This is particularly true for surgery *in utero*.

The refusal of cesarean section is a case of particular difficulty where the aim of protecting fetal life is in opposition with the will of the competent mother. The legal solutions to this dilemma vary by country. Again they deal with the consideration to be given to the protection of the life of a not yet born, i.e. the status of this "individual", and the respect for freedom of the mother, supposed to be a competent person.

IVF (In Vitro Fertilization) has promoted more than any other clinical situation the consciousness of the "rights" of the embryo. This is particularly so with the fate of the "spare" embryos and their availability for donation, research, or death. Are these eggs only by-products of infertility treatments, or already human beings? Under what regulations can they be donated? Can they be used for research, what kind of research, and if so up to what stage of development, and more controversially can they be made specifically for research purposes? (Baird, 1995).

The same kind of questions and reflections applies to use of embryonic or fetal tissues for transplantation.

At the frontier between "natural" childbirth and "artificial" procreation remains the problem of surrogacy. Since it deals with the parental relationship between the child and the parents (both biological, or genetic, and commissioning) it raises the point of the "ownership" of this child, and thus of its status. Again different legal approaches have been looked at, which clearly demonstrates the lack of a precise link between ethical and legal aspects of human dilemmas.

All these aspects, namely natural childbirth and artificial procreation, AID (Artificial Insemination with Donor), IVF, and PID (Pre Implantatory Diagnosis), prenatal diagnosis as well, are involved within the problem of sex selection, when done, not for medical reasons, but following personal choices, stimulated sometimes by cultural habits.

Behind all these medical or sociological situations, the main ethical interrogation which requires the attention of the individuals concerned, including those belonging to the medical profession, is the one concerning the definition of the something/someone present in the uterus, whose life may or may not need to be protected, whose filiation may or may not need to be taken into consideration.

In the face of this situation different answers have been brought, related to religious influences, cultural habits, structures of society, legal attitudes, even sometimes political concerns.

The above differences clearly illustrate that it would appear impossible to arrive at a universal answer acceptable to all.

In fact there are three main sets of answers: two of them are highly dogmatic, the third one more pragmatic.

"Duty-Based" Ethics

The first set of answers corresponds to a "duty-based ethics" and is founded on a fundamental dogma: the absolute respect for human life. Although this dogma, mainly of religious origin, is not always and is still not perfectly adhered to everywhere (wars, including religious wars, death sentences, etc.), this concept is widely accepted, when it means the respect for individual human life, and body, i.e. the avoidance of harmful attitudes which could provoke total or partial destruction of the human body. This applies to body of competent adults as well as "vulnerable" persons like infants. The fundamental point under discussion is effectively to determine if the someone/something *in utero* is a human being, and if yes, from what point. Within the above concept the answer is clearly yes and more precisely since "conception".

Frequently associated with this concept is the one of genetic filiation: strictly linked to religious beliefs it condemns and forbids any kind of introduction in the filiation process of genetic material not belonging to the parents.

The practical consequences of this attitude are quite clear and coherent.

Almost all modern advances of reproductive medicine are rejected when adhered to their extreme limits: no termination of pregnancy, even for medical reasons (at a time, even the surgery for ectopic pregnancy was questioned), no medically assisted procreation, even by IVF, of course neither surrogacy nor AID, consequently absolutely no research on embryos, no PID, and even no PND (Prenatal Diagnosis), no fetal tissue transplant, and possibly forced cesarean section.

Such an attitude based on authentically held beliefs deserves respect and to some extent could be admired. However some points still require clarification. The first is the definition of the precise, supposedly clear cut, point of inception. "When did I begin?", as the priest N.M. Ford has entitled his book (Ford, 1988). To this difficult question many different answers can be given, as shown at the symposium on the beginning of life (Iowa City, 1990; Beller and Weir, 1994). If we follow the strictest attitude, i.e. to say at fertilization, we are facing a difficulty. When can we say that the fertilization has occurred? In the human species this process lasts more than 30 hours, between the first contact spermatozoa-oocyte to the division of the two cells egg. This definition is important if we are to follow this philosophical way of thinking since any kind of medical interference made in order to prevent pregnancy would be contraception, before it, and abortion later on. A difference which can be considered important unless we recognize that abortion and contraception are only two aspects of the same evil attitude. Moreover at this stage and even later some pregnancies cannot really be accepted as bearing human individuals, like ectopic pregnancies, already alluded to, molar pregnancies, blight ova, and still twins, who are able to divide till around the 14th day.

In fact much consideration is given in such attitude to the concept of ensoulment, which is a philosophical concept. The difficulty here is to find a link between the philosophical concept and biological facts. It must be added that even the most strict believers do not claim that the embryo *in utero is* a human being. They only maintain that the embryo deserves respect and attention *like* a human being. It must also be added that if we consider for example the position of the Vatican, this position is a matter of moral rule, and not a matter of dogma.

Finally it appears that this very restrictive position meets some difficulties and limitations, associated with the difficulty of harmonizing biological facts and philosophical concepts, with the recognition of an evolution of nature (anatomical, biochemical, genetic, etc.) of the embryo with time. It must be added that it may be considered realistic (with the restriction of the conflicts between pragmatic reality and idealistic, or transcendent rules) that contemporary societies are more and more pluralistic. It may be questionable to impose on all the constituent members of a given society the respect for principles espoused by one particular section even if this section forms a majority.

It must be realized that such an authoritative attitude leads unavoidably to escape, either in secret (e.g., clandestine abortion with its consequences) or by leaving to more liberal countries, a decision which can be taken only by the "happy few" financially able to afford it. A consequence which questions seriously the principle of justice.

It must also be pointed out that even the more coercive moral attitudes may change with time, as cultural habits do. It is difficult at the same time to claim that a cultural habit, shared by the majority of the population, like genital mutilations, must change, and to refuse in the name of "truth", exceptions to a principle whose grounds are essentially philosophical. I feel personally unhappy to think that in the name of such principle a French woman was sentenced to death and executed for abortion in 1942.

Liberty Oriented Approach

The second set of answers is also linked to principles. This principle, perfectly respectable, is the one of respect for persons, which includes mainly the respect for the freedom of choice of the competent individual, and also, at least for some, the right for someone to know its origins.

The respect for freedom of choice, when involving more than one individual, means than between those individuals any kind of activity, decision, behavior, must be regulated by clear contracts. In other words there is no "moral", or "transcendent" rules; there are only contracts between free individuals; in this sense the responsibility of the authorities particularly governmental and legal is only to check the respect of the contract by parties and not to express a judgment on their legal validity or moral foundation.

Consequently several solutions evoked above can be reached providing the parties involved express their mutual agreement. In some cases they are decisions which in the end lead to the birth of a child, and the advocacy for the interests of this child is supposed to be best made by the parents to be: surrogacy, egg donation, fetal surgery. Sometimes a conflict occurs (late refusal of surrogacy by the pregnant woman, death of the father of an embryo, irresponsible behavior of the mother, refusal of cesarean section, etc.) and the society, represented by the legal authority, is supposed to give the right solution, unfortunately often versatile. In some other cases these decisions lead to destruction of the embryo or fetus: voluntary abortion, medical termination of pregnancy, fetal tissue donation, research on embryos, etc.). The principle of respect for freedom appears to be in opposition with the principle of protection of the vulnerable, and the role of advocacy may be considered as imperfectly fulfilled by the parents.

Of course it may be contended that the embryo or fetus in utero cannot be considered as a "vulnerable person" since it is not yet born. However it becomes difficult to reconcile this view with the possibility of prosecution of a pregnant woman for irresponsible behavior which may affect this "thing" *in utero* if it is not considered as a vulnerable individual.

The situation is still worse when considering the use of diagnostic tools, nowadays available at the request of parents, in order to determine the sex of the fetus and to decide obvious sexual discrimination.

A contract usually deals with a service and the compensation for this service. There is always some reluctance to analyze clearly this aspect of the agreements. Even the word "compensation" bears frequently a somehow ambiguous connotation. It covers the fees for services by the technicians involved (ex.: *in vitro* fertilization, insemination, etc.) and the true "compensation", for traveling, various costs, lost hours of work, etc., for the individuals involved in the contract (commissioned mother, for instance). These considerations do not raise too difficult ethical questions. More difficult is the determination of the line which must not be transgressed, if one considers mandatory to avoid this activity to become a "commercial" one: such commercial activity has too obvious consequences : the involvement of commercial companies, specifically forbidden for example in the UK ; payment of surrogates or of donors, which is accepted in several countries where surrogacy and gametes/eggs donation is authorized.

Within the discussion about this matter the argument of necessity is frequently used: if there were no compensation nor financial incentive, there would be no surrogacy or donation. This argument is questionable as demonstrated by the French experience with sperm donation in the CECOS (Centre d'étude et de Conservation du Sperme).

At any rate it leads to an important aspect, which is linked itself to the organization of the system of health care provision.

In a system of "liberty oriented" health care provision, the financial burden is supported by the individuals directly or (with some difficulties) indirectly through insurance services. It is clear that such organization put in question the ethical principle of justice.

The situation is worse in a system of "welfare oriented" health care provision, where this medical activity is taken in charge by collective social security organizations. Because this organization means that the whole community is involved in activities which can be morally condemned by some of its members. A perfect example of such situation is shown by the covering of expenses linked to voluntary abortion by the general social security systems, as in France, or by the expenditure of Federal funds for embryo research, as requested in the USA.

Taking the argument to its logical extreme, Engelhardt (Engelhardt, 1995) has proposed to spread the financial participation to health care provision, according to the moral principles involved. Some members of the community would follow strictly the Vatican rules and be exempted to pay for abortion, surrogacy, egg donation or research, some others would be ready to accept to pay and at the same time to be allowed to use these services.

Of course this is an extreme position which may appear somehow difficult to put into practice.

It remains that the absolute respect for freedom bears the risk of leading to a form of society ever more commercially orientated, with its unavoidable and already present deleterious consequences: exploitation and exclusion of human beings.

Another aspect of freedom is the right for an individual to know his genetic origin. It is interesting to note that this right is considered as of paramount importance within the context of religious considerations and leads in some faiths to the forbidding of any kind of sperm donation from donor or of egg donation or of surrogacy.

At the same time sperm donation considered as a civil right is linked with the possibility of identification of donors as enacted in the Swedish law of artificial insemination of March 18, 1985, n° 1140/1984. The ethical aspect of such decision is interesting to consider : this legislation was introduced, as Daniels and Lalos pointed out (1995), because "the Swedish parliament considered that secrecy and lack of access to donor information was not in the best interests of the offspring". Although this affirmation remains to be demonstrated, it is interesting to note that an important decision concerning the "interests" of somebody not only not yet born, but even not yet conceived have been taken into consideration, and viewed as of more "value" than the possible wish for secrecy of the parents and shortage of donors. This attitude is linked to the concept of moral obligation toward next generation or generations.

Pragmatic Attitude

The third set of attitudes is without any doubt the most difficult one to bring into action since it tries to reconcile contradictions in order to allow life in a community of people who do not share the same moral values.

Two levels of reflection have to be considered, the individual level ("micro ethics") and the general level ("macro" and meso" ethics).

The individual level is frequently more restrictive than the general one, allowing those who wish to follow more strict rules to do so. A problem arises however when those individuals want to interfere with the behavior of some others who do not share their convictions, sometimes violently. Such attitudes clearly require legal and sometimes criminal prosecution.

In contrast, the individuals may be faced with an excessively restrictive general approach. The individual's duty in this context is to attempt to change or influence the general policy using available democratic means.

One of the points which deserves a very careful consideration is the possibility that a given country establish mandatory a medical attitude. For instance, one can imagine an obligation of prenatal diagnosis of a given congenital abnormality, either hereditary or accidental, in order to avoid the medical and economic consequences of a handicapped baby. The pressure in such instance could be direct, or perhaps more subtly indirect, through the intervention or non intervention of social security help.

A fundamental point needs to be stressed here: such actions aiming at destruction of the unborn baby must not in any circumstances, be imposed on a pregnant woman. In a similar way neither can pressure (social, etc.) be imposed on the medical profession when a particular form of treatment becomes a matter of individual conscience (in French, this is called the respect for the "clause de conscience").

Concerning its behavior it must be emphasized that within the management of such delicate matters, the medical profession has still to remain personally involved. Without interfering in the patient's own decision, without adopting an "imperialistic" behavior, respecting as much as possible the principle of honest information, and of consent, despite all the legal, social, economic, administrative obstacles, it remains the moral duty of medical profession to continue to share with the patients the psychological burden of the difficult choices.

This means, in particular, that the reference to courts, ethical committees, any kind of "outside" bodies, is to be avoided as much as possible. The medical decisions ought to be made within the medical framework, i.e. within the "colloque singulier", between patient and practitioner. If not, the way is opened wide to the abnegation of responsibility, on the part of medical profession, leaving the patient alone, submitted to the anonymous, sometimes arrogant, and aggressive power of society.

As far as the general regulations (macro-ethical decisions) are concerned, this "third way" tries to harmonize the needs associated with the personal choices of a pluralistic society, allowing life within a community, avoiding tensions, particularly those of religious origin, with their deleterious consequences, the necessary organization of public health and protection of the vulnerable and the obligations toward next generations.

Taking all these constraints together leads unavoidably to conflicts of interests. It has to be realized that in such circumstances the only possible solution is founded in humanistic pragmatism, rather than on supposedly transcendent dogmatic rules.

Making possible the life in community, in agreement with the principles of beneficence and justice, means taking measures to avoid exploitation and exclusion, and reinforcing solidarity. The vulnerable is primarily the infant whose interests have to be properly appreciated. It may be also the woman whose situation within society is so frequently undermined and this means also according due consideration to the structure of the family in agreement with cultural habits. It is also the embryo and the fetus who is increasingly considered not as a thing or an actual human being, but rather as a "potential human person" as described by the French ethical committee or a "potentiality of human person", warranting particular moral consideration, which evolves over time. The duties toward next generations correspond also to a subtle concept. It is frequently considered as the obligation to take the necessary precautions in order to avoid adverse effects, of any kind, of medical progress on the next generations. This is perfectly true, although it is also clear that the nil risk can never be obtained, and that avoiding progress has its own inherent risks (for instance, the risk of multiple pregnancy, if hyperstimulation can not be avoided). The next generations have also the right to have the benefits of further progress as the present one did with the progresses made before.

With these principles in mind the practical situations quoted earlier may find a fair, although always temporary solution.

Voluntary abortion: is it a conflict between the life of the embryo (or fetus) and the woman's freedom of choice? No. Such arguments lead unavoidably to conflicts which cannot be resolved. The important factor is to ensure the protection of the women's well being; forbidding abortion leads to illegal methods which have ill-consequences on health and life. The macro-ethical decision must be factual. On the other hand, the micro-ethical restrictive attitude of some as well as the conscience clause for the practitioners must be acknowledged.

Medical termination of pregnancy in order to protect maternal life does not raise too many ethical dilemmas. The same applies to embryonic reduction. In order to avoid the so called "wrongful life", the decision is much more difficult to take. The risk of a slippery slope is obvious, when consideration is given to less important risks on the one hand, or to psychological, social, or economic aspects on the other hand. Once more within the general rule tolerance must prevail, leaving individual decision-making in the hands of the parents and the practitioner, with a possible reference to the conscience clause.

The same applies obviously to fetal medicine and surgery.

The refusal of cesarean section is a perfect example of such difficulties, since individual decisions may lead to litigation. We may consider the example of the current situation which exists in France. No prosecution has as yet been initiated either for having carried out a forced cesarean section, or for having respected the mother's refusal of a cesarean section and putting the babies life at risk. Similarly the authorization or court order has never been requested. In other words, by decision of the legal authorities, the final medical decision is a matter of reflection between the parents and practitioners, removed from the media and public opinion. The same applies to the still more sensitive issue of the care given to a pregnant women in a state of coma.

It is no longer necessary to discuss the legitimacy of IVF. Amongst the ethical points still under discussion and dealing with the status of the embryo, two will be considered here: one very common concern is the research on pre-embryos, the other, rare, concerning the status and rights of the embryo after the death of its father.

Research on pre-embryos is still a matter of controversy despite, or because of all the consideration given to it by the authorities of most countries, and particularly the UK, Germany, Spain, France and the European Community, as well as Canada and USA.

More and more the legitimacy of research on embryos, up to 14th day, is recognized (Baird, 1995). Of course it is emphasized that these researches have to be accepted by the progenitors, scientifically grounded, preceded by experimental research if possible, and carefully checked by medical, scientific and administrative bodies. The needs for such researches are obvious for numerous medical reasons. They have to be conducted either on embryos which stand no chance of survival, or on "spare" embryos.

The problem which remains is whether it is acceptable to "create" such embryos for the sole purpose of research. On this specific point the ethical committee of FIGO was unable to reach an unanimous agreement (FIGO recommendations 1994). The embryo research panel of the US National Institute of Health, has recently established a distinction between acceptable research, research warranting additional review and unacceptable research (for federal funding), and what is more important, has accepted the principle of fertilization *for research alone* in special circumstances, for special purposes and with specific control (Fletcher, 1995). This attitude is much more permissive than the one followed by the French Law on Bioethics (July 94) which practically forbids any kind of research on embryos.

The same law has adopted a questionable position concerning the fate of frozen embryo after the death of the father: it becomes mandatory in such a case either to destroy the embryo or to donate it to another couple.

This case illustrates the difficulties which occur when trying to find a "medium" way between restrictive and permissive attitudes, and particularly when non medical, namely political instances, try to go deeply into scientific reflection which goes far beyond their field of competence. In this particular case such a decision is obviously in strong opposition with several ethical principles: justice, respect for the right of filiation, respect for the natural, moral, highly respectable wish of the mother, and finally and simply right of this embryo to live.

As far as surrogacy is concerned several attitudes are adopted: from complete freedom, including a "compensation" system very close to incentive, to complete proscription. The main point is the desire to avoid maternal exploitation, and to ascertain the "belonging" of the baby to be, to the commissioned or commissioning couple.

The use of embryonic or fetal tissue concerns the status of the embryo and fetus if it leads to abortion, and to pregnancy, in order to obtain these tissues. In this case, the embryo or fetus, clearly becomes a thing, analogous to a biochemical kind of treatment.

The matter of sex selection (post conceptional) is also related to the status of the embryo. If the decision of a voluntary abortion may be left totally at the discretion of the woman concerned, whatever the reasons, which have not to be revealed, it is certainly legitimate to make every effort to avoid discrimination against the female sex, by abortion after prenatal diagnosis as well as by infanticide. In this respect there is no difference in status between the baby and the baby to be. However the problem here, as for genital mutilation, is mainly a cultural problem. Once more it must be emphasized that cultures may change and must change, and that there is no reason to respect them and to help them to perpetuate such acts.

In this respect, the medical profession has great responsibilities. It has to help this evolution, instead of using the cultural taboos as a mean of improving its social and sometimes financial status.

CONCLUSION

The medical profession faces serious challenges. On the one hand, it has to meet the ever increasing demand of individuals for a better, faster, more efficient care, for the satisfaction of needs or desires, clearly linked to life style choices. This demand is frequently very selfish ; it does not pay always attention to the nature, the "status" of the embryo and fetus, nor to the true interests of the child to be.

In contrast, the medical profession is confronted with restrictions arising from transcendent religious rules, philosophical fears, cultural habits, economic constraints, legal and administrative regulations which either neglect the status of the embryo or fetus, or over emphasize it. Facing these opposite trends the medical profession is led to recognize :

The difficulty to accept the strict obedience to transcendent rules, or to cultural habits, which may change with time.

The lack of strict link between Law and Ethics. The main goal of Law is to render possible the life in community, to protect the vulnerable, and to be concerned by the interest of the next generations. As it has been said, ethics is about thinking, choosing, and being able to explain the choices (Fagot-Largeault, 1995).

Finally the embryo and the fetus can be recognized as a specific entity, certainly not a thing, but neither a "full" human person, a potential for human being, highly vulnerable and warranted specific protection ; the attention deserved by this individual does not start at a clear-cut point, during the biological process, but follows an evolution with time, each evolutive step requiring specific ethical attitudes. If I may end by a final and quite personal remark I am tempted to say that there is no convergence nor coherence between philosophical concepts (including those of religious origin), public regulations, biologic phenomenons and personal ethical principles. As far as the medical profession is concerned, I take the risk of looking "old fashion", but I am still convinced of the persisting value of its mission of care and help, with modesty and compassion.

SUMMARY

The medical profession faces serious challenges. On the one hand, it has to meet the ever increasing demand of individuals for a better, faster, more efficient care, for the satisfaction of needs or desires, clearly linked to life style choices. This demand is frequently very selfish ; it does not pay always attention to the nature, the "status" of the embryo and fetus, nor to the true interests of the child to be.

In contrast, the medical profession is confronted with restrictions arising from transcendent religious rules, philosophical fears, cultural habits, economic constraints, legal and administrative regulations which either neglect the status of the embryo or fetus, or over emphasize it.

Facing these opposite trends the medical profession is led to recognize:

The difficulty to accept the strict obedience to transcendent rules, or to cultural habits, which may change with time.

The lack of strict link between Law and Ethics. The main goal of Law is to render possible the life in community, to protect the vulnerable, and to be concerned by the interest of the next generations. As it has been said, ethics is about thinking, choosing, and being able to explain the choices (Fagot-Largeault, 1995).

Finally the embryo and the fetus can be recognized as a specific entity, certainly not a thing, but neither a "full" human person, a potential for human being, highly vulnerable and warranted specific protection. The attention deserved by this individual does not start at a clear-cut point, during the biological process, but follows an evolution with time, each evolutive step requiring specific ethical attitudes.

REFERENCES

Baird, P. (1995) Research on pre-embryos (zygotes). In: *Ethical Aspects of Human Reproduction.* (eds.) C. Sureau and F. Shenfield, J. Libbey Eurotext Publ.: Paris pp. 327-340.

Beller, F. and Weir, R. (1994) *The Beginning Of Human Life.* Kluwer Acad. Publ.

Daniels, K. and Lalos, O. (1995) The Swedish insemination act and the availability of donors. *Hum. Reprod.* 10, 1871-1874.

Engelhardt Jr., H.T. (1995) Human Reproduction: Conflicts at the roots of bioethics and health care policy In: *Ethical Aspects of Human Reproduction,* (eds.) C. Sureau and F. Shenfield, Edit. J. Libbey Eurotext Publ. Paris, pp. 49-60.

Fagot-Largeault, A, (1995) Procréation responsable. In: *Aspects éthiques de la Reproduction Humaine*, (eds.) C. Sureau and F. Shenfield, Edit. J. Libbey Eurotext Publ., Paris, pp. 3-18.

FIGO (1994) Recommendations on ethical issues in obstetrics and gynecology. *The FIGO Committee For The Study Of Ethical Aspects Of Human Reproduction*, FIGO, London.

Fletcher, J.C. (1995) US public policy on embryo research *Human Reprod.* 10, 1875-1878.

Ford, N.M. (1988) When did I begin? Cambridge Univ. Press. Cambridge.

Trophoblast Research 10:13-18, 1997

CYTOGENETICS OF GAMETES, ZYGOTES AND EARLY PREGNANCY

Joep P.M. Geraedts

Department of Molecular Cell Biology and Genetics
University of Limburg
P.O. Box 616
6200 MD Maastricht, The Netherlands

INTRODUCTION

Of all human conceptions only 25%-30% progress successfully to delivery (Leridon, 1977). Early loss of pregnancy may occur between fertilization and implantation, during the time of implantation, or after implantation at various stages of pregnancy. Chromosome abnormalities constitute the major cause of embryonic loss. In chromosome studies of spontaneous abortions abnormalities are encountered in more than 50% (Boué et al., 1975; Eiben et al., 1990). Both abortion rates and chromosome abnormalities are maternal age dependent as a direct result of chromosomal nondisjunction during meiosis (Hook, 1981). Until 1978 the events between ovulation and pregnancy were hidden in a 'black box'. The development of *in vitro* fertilization (IVF) provided the opportunity to determine the frequency of chromosome abnormalities in human gametes and in early stages of embryo development. Because of ethical and legal restrictions in most countries human embryos will only originate in the framework of infertility treatment and therefore will hardly ever become available for investigation . In most studies, including ours, only zygotes and embryos are used which are not suitable for transfer to the uterus. The results of these studies will be presented. Recently it has also become possible to address some questions in the framework of pre-implantation diagnosis.

MATERIALS AND METHODS

Cytogenetic studies were carried out on metaphase chromosomes as well as interphase cells. In the first case classical karyotypes were made following mitotic arrest, while non-dividing cells were studied by *in situ* hybridization.

For chromosome studies of oocytes and zygotes, Tarkowsky (1966) introduced a method which is used by many authors. For the fixation of oocytes to the microscope slide excellent results were obtained using the technique of Mikamo and Kamigushi (1983).

Protocols used for *in situ* hybridization, i.e. target cell preparation and fixation, probe labeling, hybridization, washing procedures and signal visualization are very diverse.

For fluorescent *in situ* hybridization (FISH) the DNA from chromosome-specific probes was chemically modified by nick-translation (e.g., thymine was replaced by deoxyuridine triphosphate labeled with biotin or digoxigenin) or by a transamination reaction resulting in a sulfone group on the cytosine moieties. Although some probes were directly conjugated with fluorescent molecules, the most widespread approach was the labeling of probes with reporter molecules that, after hybridization, bind fluorescent affinity reagents. *In situ* hybridization was achieved by the simultaneous denaturation of both the DNA of the target cell and the labeled probe. Incubation of the probe and target DNA together at a temperature below the melting point allowed the chemically modified probe to anneal to complementary sequences in the target. Following their hybridization with target DNA sequences, these probes were visualized immunocytochemically using fluorochromes such as fluorescein isothiocyanate (yellow-green fluorescence) or tetramethyl rhodamine isothiocyanate (red fluorescence) as reporter molecules. Counterstaining of cells was performed with DNA-specific dyes such as 4'-6-diamidino-2-phenylindole (DAPI) or propidium iodide. Binding of several was detected simultaneously by properly employing different probe labels (e.g., biotin and digoxygenin).

The use of FISH on sperm cells is more complex due to the inaccessibility of sperm DNA and the interpretation of the results. Our group has introduced a new method, consisting of a standard, routinely applicable protocol for sperm decondensation, restoring morphology. The detection method is based on brightfield microscopy, allowing counterstaining after ISH. This allows simultaneous nondisjunction and morphology studies (Martini et al., 1995).

RESULTS

Spermatozoa

Aneuploidy rates have been shown to vary from 0.03% for the X-chromosome to about 1% for chromosome No. 1 (Pieters and Geraedts, 1994). A wide range was observed in the percentages of disomy for the different chromosomes. The median value for all chromosomes reported is in the range of about 0.3%.

In a recent study it was shown by Pang et al. (1995) that the percentage of numerical abnormalities was more than ten-fold increased in a group of patients showing oligoasthenoteratozoospermia (OAT).

Oocytes

Since the first report on the study of meiosis in human oocytes in 1968, a large number of publications deal with the cytogenetic analysis of inseminated oocytes that failed to fertilize in the framework of an IVF program. Data on 2434 oocytes, reported by 12 groups have shown that the total incidence of chromosome anomalies ranges from 8.1 to 54.2% with an average of 26.5%: 13.3% hypohaploidy, 8.1% hyperhaploidy, 1.6 % structural abnormalities and 3.5% diploidy (Plachot, 1995). These results are different from those obtained by Angell (1994). She studied 400 oocytes from 125 women showing no sign of fertilization or cleavage. Of these 124 were unanalyzable and 79 were normal haploid. The remaining were all abnormal. However, not a single case of hyperhaploidy was noted.

Zygotes

About 20 hours after insemination during IVF the zygotes are studied for the presence of pronuclei. The two main categories of abnormalities are parthenogenetic activation (1 pronucleus and triploidy (3 pronuclei). Tripronulear zygotes are eliminated from culture and frequently used for studies. It was shown that the vast majority results from dispermy (Plachot et al., 1989).

Preimplantation Embryos

The first cytogenetic studies of early embryos were carried out using classical techniques. To obtain sufficient metaphases cleaving embryos were required. Only a minority of the embryos studied this way could be analyzed. Furthermore, the proportion of cases in which results were obtained in all cells was even much smaller (Jamieson et al., 1994; Pellistor et al., 1994). The results of these studies are different with respect to the percentages of abnormalities, which were 22.5% and 90%, respectively.

From the study of preimplantation embryos arising from tripronuclear zygotes it has become clear that they display a variety of chromosomal abnormalities (Pieters et al., 1992). They include 1) complete triploidy in all cells after regular division, 2) gross abnormalities in all cells due to chaotic chromosome movement after multipolar spindle division, 3) cell subpopulations with either a haploid or a diploid chromosomal content because of extrusion of a haploid nucleus during the first cleavage division and 4) cell subpopulations with a diploid or a triploid chromosomal content as a result of extrusion of a haploid nucleus during the first cleavage division and subsequent incorporation in one of the two nuclei. FISH studies performed on nuclei of embryos resulting from abnormal fertilization revealed mostly mosaic chromosome complements (Coonen et al., 1994; Harper et al., 1994).

The development of FISH technology resulted in the use of arrested human embryos as study material. Since interphase cells could be studied with high efficiency it was possible to draw conclusions about the presence of normal and abnormal chromosome copy numbers. Besides these, in quite a proportion of embryos it was shown that more than one cell type was present. These so called mosaicisms are generated in almost all cases at the second or later divisions when they are monospermic diploid. However, when they are polyploid or haploid mosaic they usually originate at the first divisions (Munné et al., 1994).

DISCUSSION

Cytogenetic analysis of human gametes, zygotes and early embryos is an important tool for the investigation of formation, transmission and etiology of chromosomal investigations. However, for cytogenetic analysis, metaphase-arrested nuclei are essential. In contrast, when using appropriate DNA probes, *in situ* hybridization methodologies can rapidly and efficiently be applied for the identification of numerical and even structural abnormalities in interphase nuclei. ISH and FISH play an increasingly important role in a variety of research areas, including cytogenetics, prenatal diagnosis, tumor biology, gene amplification and gene mapping.

In situ hybridization of sperm nuclei proved to be efficient and promising with respect to the detection of a broad variety of studies, including the effects of environmental, occupational and therapeutic agents.

It can also be applied to study the sex chromosome constitution of spermatozoa, for example to control the results obtained with sperm separation methods.

The frequencies of disomic sperm cells show large variations between studies. This can be attributed to the following technical factors: decondensation and denaturation of the DNA in the specimen, distortion of the morphology after pretreatment leading to difficulties in the interpretation of the distinction between disomic and diploid sperm. This problem can be circumvented or solved by the combined use of an ISH method and a morphological staining method or double hybridization (Martini et al., 1995).

If the observed 0.3% is truly representative for the disomy rate among sperm, an overall percentage of 6.9 would be found using the FISH technique. This figure is in the range of the findings using the zona free hamster technique (Martin et al., 1990; Pellestor, 1991).

Besides technical factors explaining a proportion of the variation observed, it should not be forgotten that also differences might be observed between patients. The finding of the frequency of abnormalities in the patients showing OAT has to be considered as an important complication for ICSI, which nowadays is the method of choice for the treatment of the infertility problem of these patients.

For ethical reasons only polypronuclear embryos can be studied. As expected the majority of these is chromosomally abnormal. However, about 18% of the cells might be normal (Pellestor, 1995).

With respect to the abnormality rate of embryos a discrepancy is clearly present when one compares the results obtained by Jamieson et al. (1994) and Pellestor et al. (1994). This seems to result from one important factor being the morphology. The percentage of abnormalities was more than two times higher in poor quality embryos than in good quality embryos (Pellestor, 1995).

Furthermore it was shown that aneuploidy increases significantly ($p<0.001$) with maternal age. This corroborates the hypothesis that oocytes of older women are more prone to non-disjunction caused by meiotic errors at the gamete level (Munné et al., 1995).

The observation that mosaicism is present in a large proportion of embryos has consequences for preimplantation diagnosis. It is recommended to use two blastomeres if available. In case results from both cells are conflicting, the embryo should not be considered suitable for transfer. However, it should be kept in mind that if no mosaicism has been detected, chromosomal mosaicism in the remaining cells can never be totally excluded. Furthermore, the presence of mosaicism also depends on the developmental capacity.

SUMMARY

By the introduction of *in vitro* fertilization and other artificial reproductive technology methods it has become possible to study, gametes, zygotes and preimplantation embryos. It has been estimated that the contribution of aneuploid oocytes and spermatozoa is about 26% and 7%, respectively. Furthermore, between 5 and 10% of all *in vitro* fertilizations result in a zygote showing three in stead of two pronuclei after 20 hours. In most cases the origin of triploidy is dispermy. In total about 40% of all zygotes is chromosomally abnormal. Also after the first cleavage division abnormalities might occur. These will result in an early embryo showing mosaicism.

REFERENCES

Angell, R.R. (1994) Cytogenetic analysis of unfertilized human oocytes. In: *Gamete and Embryo Quality* (ed.) L. Mastroianni Jr., H.J.T. Coeling Bennink, S. Suzuki, and H.M. Vemer, London. Parthenon, pp. 47-62.

Boué, J., Boué, A., and Lazar, P. (1975) Retrospective and prospective epidemiological studies of 1500 karyotyped abortions. *Teratology* 12, 11-26.

Coonen, E., Harper, J.C., Delhanty, J.D.A., Ramaekers, F.C.S., Geraedts, J.P.M., Hopman, A.H.N. and Handyside, A.H. (1994) Presence of chromosomal mosaicism in abnormal preimplantation embryos detected by fluorescence in situ hybridization. *Hum. Genet.* 94, 609-615.

Eiben, B., Bartels, I., Bähr-Porsch, S., Borgmann, S., Gatz, G., Gellert, G., Goebel, R., Hammans, W., Hentemann, M., Osmers, R., Rauskolb, R., and Hansmann, I. (1990) Cytogenetic analysis of 750 spontaneous abortions with the direct preparation method of chorionic villi and its implications for studying genetic causes of pregnancy wastage. *Am. J. Hum. Genet.* 47, 656-663.

Harper, J.C. Coonen, E., Handyside, A.H., Winston, R.L.M., Hopman, A.H.N. and Delhanty J.D.A. (1995) Mosaicism of autosomes and sex chromosomes in morphologically normal, monospermic, preimplantation human embryos. *Pren. Diagn.* 15, 41-50.

Hook, E.B. (1981) Rates of chromosome abnormalities at different maternal ages. *Obstet. Gynaecol.* 58, 282-285.

Jamieson, M.E., Couts, J.R.T. and Connor, J.M. (1994) The chromosome constitution of human preimplantation embryos fertilized in vitro. *Hum. Reprod.* 9, 709-715.

Leridon H. (1977) *Human Fertility*, Chicago. University of Chicago Press.

Martin, R.H. and Rademaker, A. (1990) The frequency of aneuploidy among individual chromosomes in 6821 human sperm chromosome complements. *Cytogenet. Cell Genet.* 53, 103-107.

Martini, E., Speel, E.J.M., Geraedts, J.P.M., Ramaekers, F.C.S., and Hopman, A.H.N. (1995) Application of different in-situ hybridization detection methods for human sperm analysis. *Hum. Reprod.* 10, 855-861.

Mikamo, K. and Kamiguchi, Y. (1983) A new assessment system for chromosomal mutagenicity using oocytes and early zygotes of the Chinese hamster. In: *Radiation-induced Damage in Man,* (eds.) T. Ishihara, T. and M. Sasaki, New York: Alan R. Liss, pp. 411-432.

Munné, S., Grifo, J., Cohen, J. and Weier, H.-U.G. (1994) Chromosome abnormalities in human arrested preimplantation embryos: A multiple-probe FISH study. *Am. J. Hum. Genet* 55, 150-159.

Pang, M.G., Zackowski, J.L., Hoegerman, S.F., Moon, S.Y., Cutticchia, A.J., Acosta, A.A. and Kearns, W.G. (1995) Detection of fluorescence in situ hybridization of chromosome 7, 11, 12, 18, X and Y abnormalities in sperm from oligoasthenoteratozoospermic patients of in vitro fertilization program. *J. Asso. Reprod. Genet.* 12, 53S.

Pellestor, F. (1991) Differential distribution of aneuploidy in human gametes according to their sex. *Hum. Reprod.* 6, 1252-1258.

Pellestor, F., Dufor, M-C., Arnal, F. and Humeau, C. (1994) Direct assessment of the rate of chromosome abnormalities in grade IV human embryos produced by in vitro fertilization procedure. *Hum. Reprod.* 9, 293-302.

Pellestor, F., (1995) The cytogenetic analysis of human zygotes and preimplantation embryos. *Hum. Reprod. Update* 1, 581-585.

Pieters, M.H.E.C., Dumoulin, J.C.M., Ignoul-Vanvuchelen, R.C.M., Bras, M., Evers, J.L.H. and Geraedts, J.P.M. (1992) Triploidy after in vitro fertilization: Cytogenetic analysis of human zygotes and embryos *J. Assist. Reprod. Fertil.* 9, 68-76.

Pieters, M.H.E.C. and Geraedts, J.P.M. (1994) In situ hybridization studies of human gametes, zygotes and pre-embryos. In: *Gamete and Embryo Quality,* (eds.) L. Mastroianni, Jr., H.J.T. Coeling Bennink, S. Suzuki, and H.M. Vemer, London. Parthenon, pp. 35-47.

Plachot, M., Mandelbaum, J., Junca, A.M., de Grouchy, J., Salat-Baroux, J. and Cohen, J. (1989) Cytogenetic analysis and developmental capacity of normal and abnormal embryos after IVF. *Hum. Reprod.* 4 (Suppl.) 99-103.

Plachot, M. (1995) Oocyte. Genetic aspects. In: *Gametes, The Oocyte,* (eds.) J.G. Grudzinskas and J.L. Yovich, Cambridge Reviews in Human Reproduction, pp. 95-107.

Robbins, W.A., Segraves, R., Pinkel, D. and Wyrobek, A.J. (1993) Detection of aneuploid sperm by fluorescence in situ hybridization: Evidence for a donor difference in frequency of sperm disomic for chromosomes 1 and Y. *Am. J. Human Genet.* 52, 799-807.

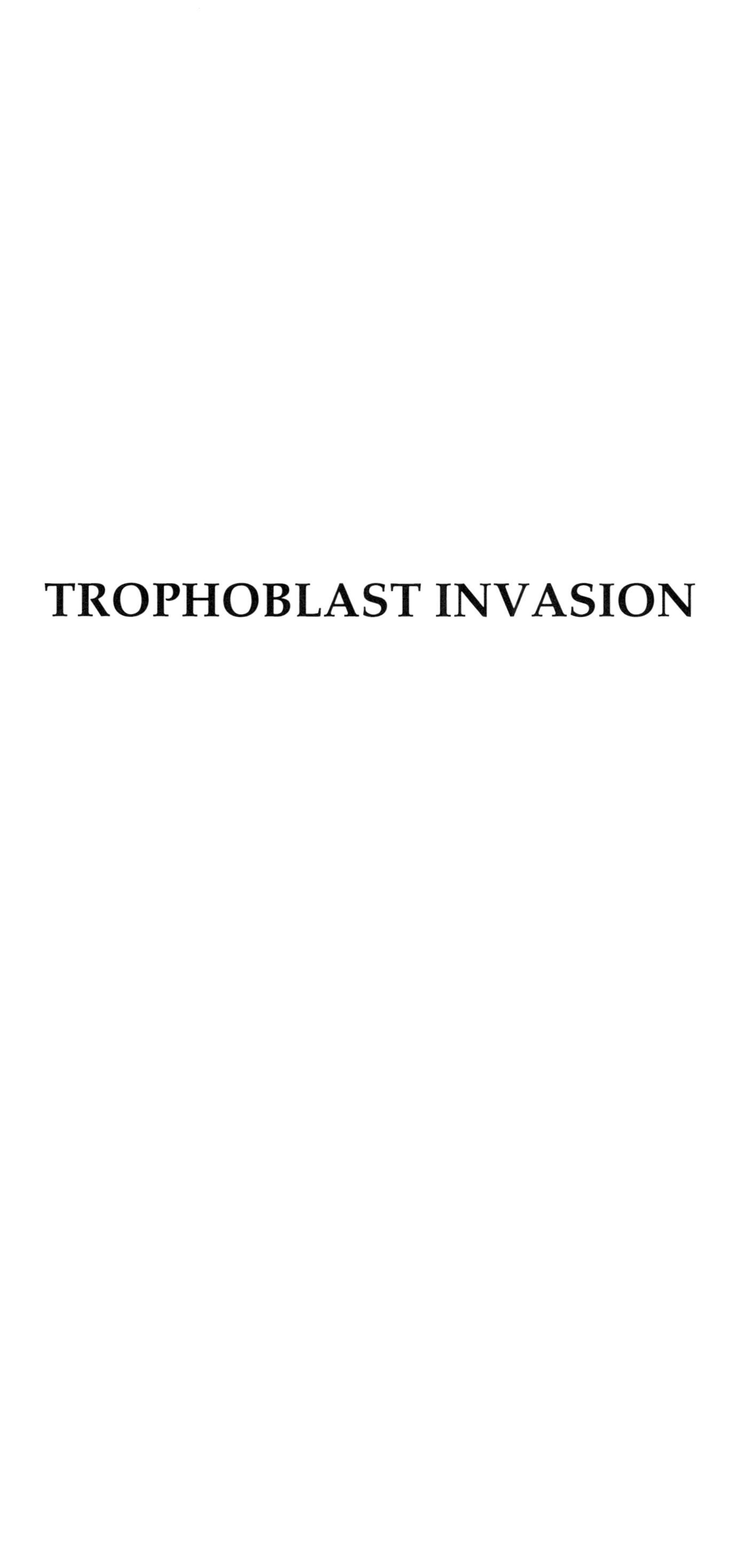

TROPHOBLAST INVASION

Trophoblast Research 10:21-65, 1997

EXTRAVILLOUS TROPHOBLAST IN THE HUMAN PLACENTA
- A Review -

Peter Kaufmann[1,3] and Mario Castellucci[2]

[1]Department of Anatomy
Technical University of Aachen
Wendlingweg 2
D-52057 Aachen, Germany

[2]Dipartimento di Scienze Mediche
Anatomia Umana
Via Solaroli, 17
I-28100 Novara, Italy

INTRODUCTION

By definition, extravillous trophoblast comprises all those trophoblastic elements which can be found outside the villi (Figure 1). This term is more or less synonymous with the classical term "X-cell" (Scipiades and Burg, 1930) which implies that the origin of these cells for decades was under dispute. Currently, there is no doubt as to the trophoblastic, fetal origin of these cells which was proven by Y-specific fluorescence of the cells in placentae from male infants (Faller and Ferenci, 1973; Khudr et al. 1973; Maidman et al. 1973). Also the widely used term "intermediate trophoblast" (Kurman et al., 1984a,b) largely overlaps with the term extravillous trophoblast. Since the former is only poorly defined, we suggest the exclusive use of the descriptive and well-defined name extravillous trophoblast.

Extravillous trophoblast is largely composed of uninuclear cells. Syncytial or multinucleated elements are present, too, mainly in the deeper parts of the junctional zone. These elements were only rarely studied. It is a matter of dispute whether these multinucleated cells are remainders of early invasive syncytiotrophoblast or the result of syncytially fusing deeply invasive extravillous trophoblast cells. Due to paucity of respective knowledge, we will not further deal with these multinucleated elements.

Extravillous trophoblast cells can be found in various locations: in the chorionic plate, the smooth chorion, the cell islands, the cell columns, the basal plate, the placental septa, and in walls and lumina of uteroplacental vessels (Figure 2). The cells of all these locations seem to be quite similar to each other, however, with one exception. Whereas extravillous trophoblast cells in most locations are opposed to the maternal decidua which they are invading, those of the chorionic plate and the cell islands usually are not facing decidua but rather fibrin-type fibrinoid as maternal blood clot product (Frank et al., 1994; Lang et al., 1994). Despite this difference, it is interesting to note that extravillous trophoblast in all locations seems to behave similarly and to express similar patterns of extracellular matrix molecules, integrins, receptors, growth factors, proteases, etc.

[3]To Whom Correspondence Should Be Addressed.

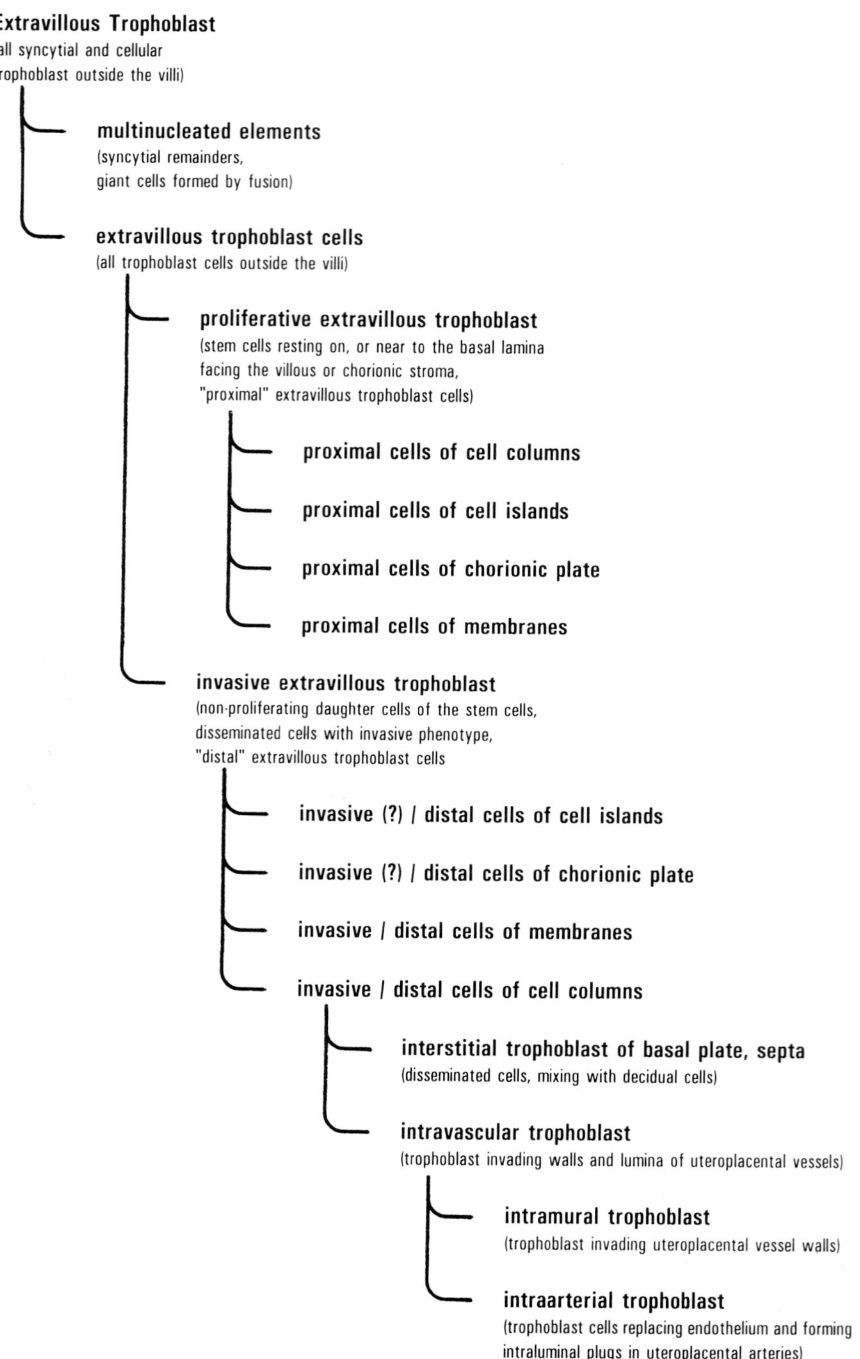

Figure 1. Classification and definition of the various stages and local subtypes of extravillous trophoblast. The question marks behind "invasive" point to the still open question whether the distal extravillous trophoblast cells of cell islands and chorionic plate, which do not infiltrate the endometrium, have a truly invasive phenotype.

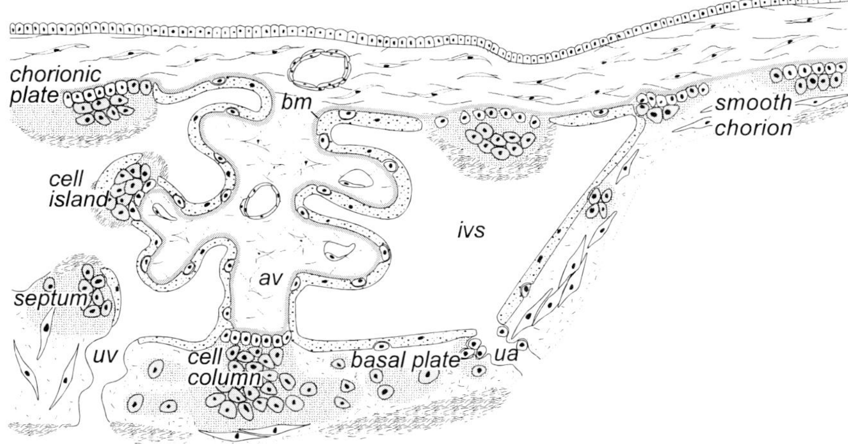

Figure 2. Scheme of the marginal region of the human placenta showing the distribution of extravillous trophoblast cells embedded in their self-produced extracellular matrix (matrix-type fibrinoid, point-shading). The proliferating stem cells (slightly lighter shading) of the extravillous trophoblast face the basal lamina (bm) which separates fetal stroma from trophoblast. Extravillous trophoblast and its extracellular matrix, the matrix-type fibrinoid are delimited from the intervillous space (ivs) by fibrin-type fibrinoid (line-shading). av: anchoring villous; ua: uterine artery with intravascular trophoblast, uv: uterine vein.

Since most detailed studies on extravillous trophoblast have been performed on cell columns together with the neighboring interstitial trophoblast cells in the basal plate and with the trophoblast cells in uteroplacental arteries, we will focus for a detailed description of extravillous trophoblast on this invasive pathway. The few respective studies on cell islands, chorionic plate, and membranes will be dealt with later.

THE INVASIVE PATHWAY FROM CELL COLUMNS VIA BASAL PLATE TO MYOMETRIUM OR UTEROPLACENTAL VESSELS

Proliferation Patterns

When extravillous trophoblast of first or second trimester placentae is stained with proliferation markers such as Ki 67, MIB-1, antibodies against PCNA or ^{3}H-thymidine, only those subsets are marked which are close to the basal lamina of the neighboring fetal stroma of the anchoring villous (Bulmer et al., 1988; Kohnen et al., 1993; Mühlhauser et al., 1993; Blankenship and King, 1994a; Kosanke, 1994) (Figure 3).

It is important to note that the various proliferation markers show highly characteristic differences in their staining patterns which may lead to misinterpretations (Kosanke, 1994): *^{3}H-thymidine* is only incorporated during the S-phase of the cell cycle. Because of this it is the most reliable proliferation marker. ^{3}H-thymidine is usually found only in those trophoblast cells directly resting on the basal lamina (Figure 3) and sometimes also within the next line of cells. General applicability of ^{3}H-thymidine is, however, restricted by the fact that in human material this method can only be applied in postpartal incubation experiments which impair the structural preservation. Moreover, incomplete diffusion of the incubation medium may lead to false negative results, in particular in the compact cell columns.

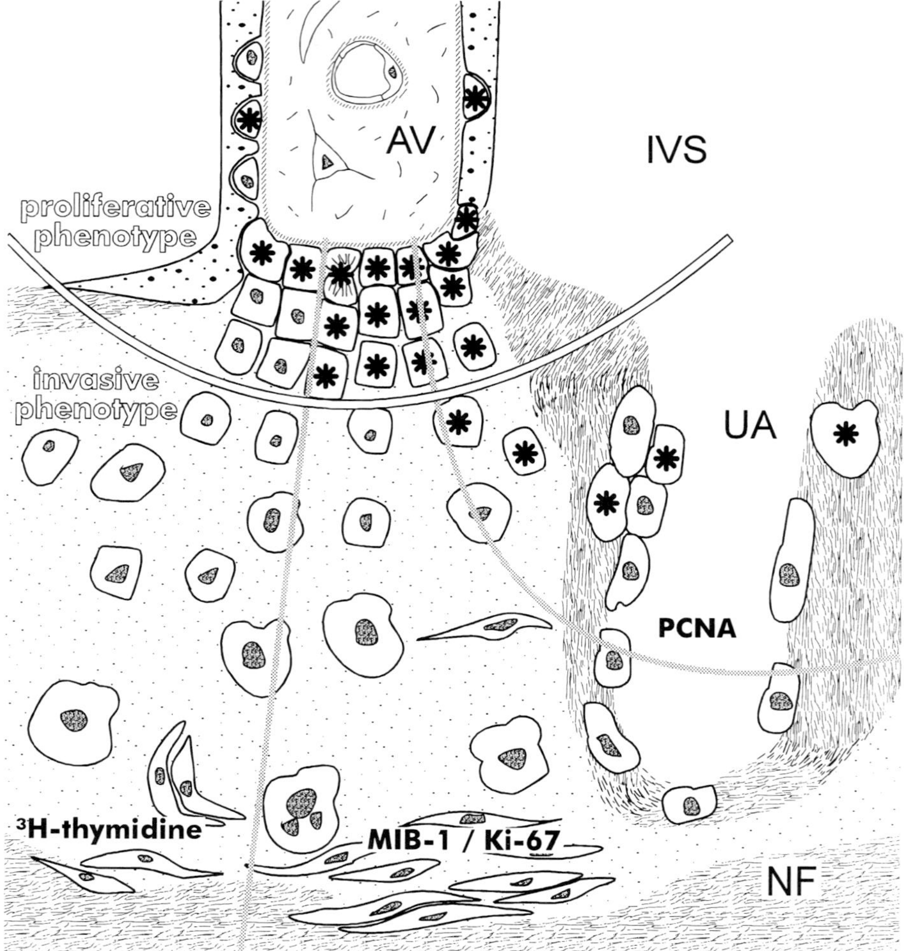

Figure 3. Scheme of a cell column of the first or second trimester, connecting an anchoring villous (AV) to the basal plate (lower half of the figure), the latter structure containing a uterine artery (UA). Extravillous trophoblast cells (polygonal) derived from the cell column, mix with decidual cells (spindle-shaped) in the depth of the basal plate. The extravillous trophoblast cells are embedded in matrix-type fibrinoid (light point shading). The latter is delimited from the intervillous space (IVS) and from the arterial lumen (UA) by fibrin-type fibrinoid (line-shading). Also Nitabuch fibrinoid (NF), covering the basal placental surface, belongs to the fibrin-type of fibrinoids. The asterisks indicate those trophoblast cells which are marked by the proliferation markers ^{3}H-thymidine, MIB-1/Ki-67, and anti-proliferating cell nuclear antigen (PCNA). Note that depending on the biological half-life of the respective antigen and depending on the stage of cell cycle which is identified by the marker, stained cells may even be found deeply invading the basal plate. In spite of the PCNA staining pattern, cells of the proliferative phenotype can only be found close to the anchoring villous (AV).

Ki-67 and MIB-1 are both directed against the same nuclear antigen, which is expressed in nearly all phases of the cell cycle except G_0 and first G_1 phase following a G_0 stage. As a consequence, this marker shows higher staining indices since it does not only recognize actually dividing cells, but rather all cells which did not leave the cell cycle. Immunoreactivity for this antigen in most cases can be found in the two to six proximal layers of cell columns (Figure 3). Deeply invasive and intravascular trophoblast cells are always negative for 3H-thymidine and Ki-67/MIB-1.

The various antibodies (clones PC10, 19F4, and 19A2) against the proliferating cell nuclear antigen (*PCNA*) behave quite differently from the above listed proliferation markers. This antigen, a DNA-polymerase δ related protein, is not only expressed throughout the entire cell cycle. Due to its long biological half-life of about 20 hours, it may even be immunoreactive in deeper invasive trophoblast cells which left the cell cycle already some days before (cf., Kosanke, 1994; Blankenship and King, 1994a). For example, PCNA-expression has been found in intravascular trophoblast (King and Blankenship, 1993) (Figure 3) where it has led to the erroneous assumption that this group of trophoblast cells represents a self-replicating population (Blankenship and King, 1993; Blankenship et al., 1993 a,b) rather than a final stage of columnar trophoblastic invasion. Later the same authors repeated their studies with Ki-67 and concluded that endovascular trophoblastic immunoreactivity of PCNA is a consequence of the antigen's long half-life rather than an indication of proliferation (Blankenship and King, 1994a).

It is important to note that the same proliferation patterns are also valid for all other locations of extravillous trophoblast. Proliferation can only be found in those extravillous trophoblast cells which are adjacent to a basal lamina separating the trophoblast from fetal villous or chorionic stroma (Figures 3, 10). These proliferating trophoblast cells are either in immediate contact with the basal lamina or separated from it by other proliferating trophoblast cells. This is valid for cell islands, chorionic plate and chorion laeve of the first two trimesters of pregnancy. For placental septa, containing many extravillous trophoblast cells, attached cell columns of anchoring villi serve as a growth zone. In respective structures of term pregnancies, extravillous trophoblast shows considerably reduced proliferative activities and locally it may be impossible to find any of the proximal cells still being in the cell cycle (cf., Figure 13).

We conclude from these findings that cells close to the villous basal lamina represent the stem cells whereas all the other extravillous trophoblast cells are more differentiated stages, derived from the nearest proliferative center by migration or invasion. The study of growth factor receptors and their ligands, integrins, extracellular matrix secretion, etc underlines this view.

Growths Factor Receptors And Their Ligands

We focus here on oncogene products, growth factor receptors and growth factors which show obvious relations to the various stages of the invasive pathway from proliferation via differentiation to invasion.

Epidermal growth factor receptor, *EGF-R,* encoded by the proto-oncogene c-erbB-1 is only expressed in those few proximal layers of trophoblast columns which also express proliferation markers (Mühlhauser et al., 1993; Duello et al., 1994; Johki et al., 1994) (Figure 4). The latter group has described some additional immunoreactivity for

EGF-R in the deeply invasive multinucleated trophoblast cells. One of its at least five ligands, EGF, is a potent epithelial mitogen. Moreover, its stimulatory influence on trophoblast invasion has been stressed (Bass et al., 1994). EGF is produced by numerous tissues. Its secretion by extravillous trophoblast has been described, suggesting autocrine/paracrine regulatory loops (Hofmann et al., 1992).

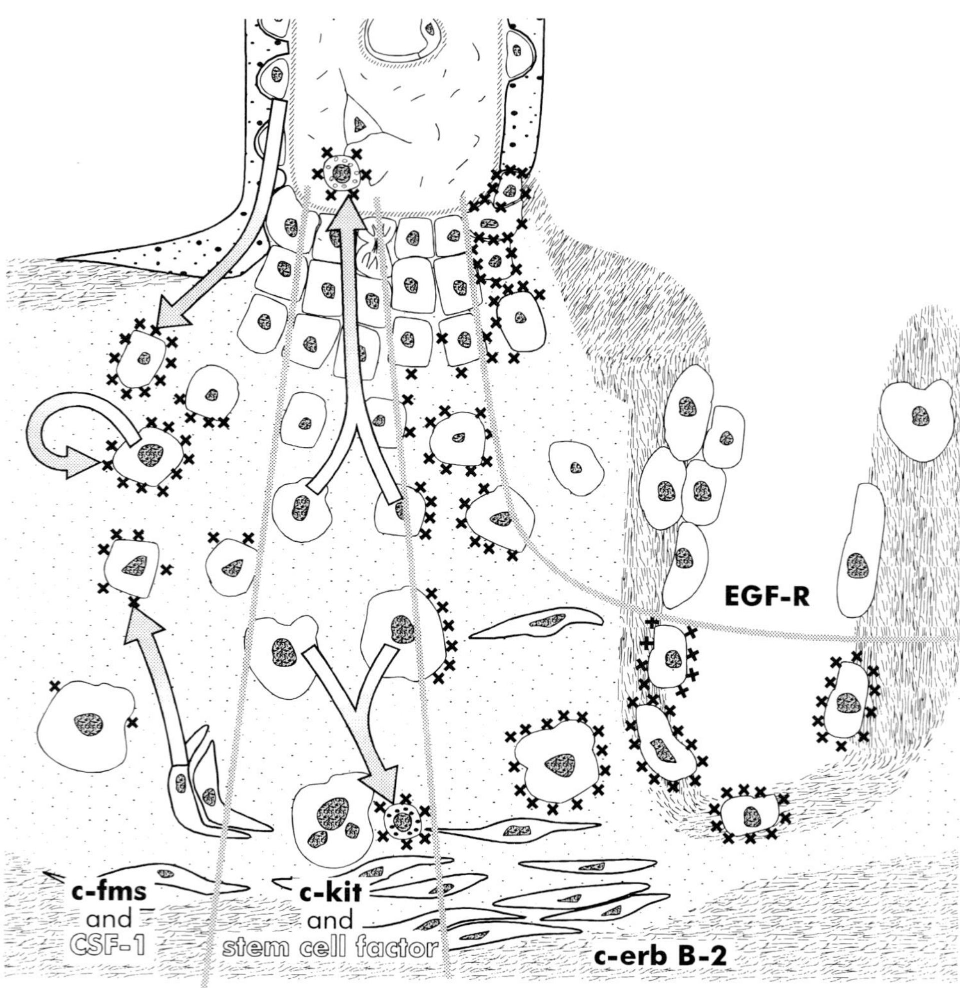

Figure 4. The expression patterns of some growth factors and their receptors underline the differentiation gradient from proliferating stem cells (above) to invasive trophoblast cells (below) in a cell column from the first or second trimester of pregnancy. Expression of the receptors is symbolized by crosses. For c-fms protein and c-kit protein, the sources of the corresponding ligands, colony stimulating factor-1 and stem cell factor, are indicated by arrows. These arrows suggest that the invasive pathway of extravillous trophoblast is controlled by autocrine and paracrine regulatory loops.

The receptor encoded by the proto-oncogene *c-erbB-2* is thought to cooperate with other receptors of the erbB family in neuregulin/heregulin-mediated signaling (Holmes et al., 1992; Peles et al., 1992; Carraway and Cantley, 1994). This receptor shows a reciprocal expression pattern as compared to EGF-R (Figure 4). It is expressed at the surfaces of all those extravillous trophoblast cells which are negative for EGF-receptor (Mühlhauser et al., 1993; Jokhi et al., 1994). It thus characterizes all differentiated and invasive stages.

Also the expression of the gene product of the oncogene *c-fms*, the receptor for colony stimulating factor-1 (CSF-1) supports the concept of a differentiation gradient. According to Johki et al. (1993) it is strongly expressed in early invasive stages but weaker during deeper invasion including the intravascular trophoblast (Figure 4). The corresponding growth factor, CSF-1 may be derived from three different sources, from villous trophoblast (Shorter et al., 1992a), from endometrial cells (Shorter et al., 1992b), and from invasive extravillous trophoblast cells themselves (Hamilton et al., 1995). CSF-1 is not only a potent mitogen but also its effects on upregulation of gelatinase B and TIMP-1 is under discussion (Hamilton et al., 1995). This means that neighboring endometrial and villous tissue as well as extravillous trophoblast cells themselves may influence the proliferative and invasive behavior of these cells.

C-kit ligand, also known as *stem cell factor*, another blood cell mitogen, is expressed by invasive extravillous trophoblast cells (Sharkey et al., 1994) (Figure 4); the receptor for this growth factor, c-kit protein, is expressed by villous macrophages and by the large granular lymphocytes (NK-cells) of the junctional zone.

The expression of both latter growth factors and their receptors indicates that there are several paracrine and autocrine loops, which may be responsible for the regulation of the proliferative and invasive process.

Extracellular Matrix Receptor Integrins

Integrins are integral membrane proteins of the plasmalemmas of all cells. They act as receptors for extracellular matrix molecules or other cell adhesion molecules (cf., Humphries, 1990). They are heterodimeric proteins consisting of an α and a β subunit. For both subunits a large variety of distinct gene products is known. The combination of α and β subunit forming the dimer, defines the specificity of the receptor for its ligand, which is either an extracellular matrix molecule or another cell adhesion molecule. The functional importance of those integrins binding to extracellular matrix molecules has been shown *in vitro* by blocking them specifically by antibodies. In the placenta, this results in loss of trophoblast cell adhesion to the extracellular matrix (Librach et al., 1991a; Burrows et al., 1995; Irving and Lala, 1995).

The following integrins seem to be important extracellular matrix receptors in extravillous trophoblast. $\alpha6\beta4$ integrin represents the usual basal lamina receptor. It has been localized to hemidesmosomes in a few specialized epithelial cells. In contrast, many epithelia and endothelia that express this integrin, lack hemidesmosomes. It binds to laminin which is then linked to collagen IV (Ruoslahti, 1991; Flug and Köpf-Maier, 1995). $\alpha3\beta1$ integrin is a receptor for laminin that also localizes to focal contacts in cells cultured on other substrates; it seems to have a role in matrix deposition (DiPersio et al., 1995; Wu et al., 1995). $\alpha5\beta1$ and $\alpha1\beta1$ integrins are usually found as receptors for interstitial matrix molecules, α5β1 binding to fibronectins and α1β1 to interstitial collagens besides laminin and collagen IV. Moreover, extracellular matrix receptors of the extravillous trophoblast comprise the integrins αvβ3 and αvβ5, both of which have a repertoire of ligands that includes vitronectin but also fibronectin, osteopontin and in the former case several others.

α3β1 is only weakly expressed in villous trophoblast and in few of the most proximal columnar trophoblast cells resting on the basal lamina of anchoring villi (Damsky et al., 1992) (Figure 5). The dominating integrin of the proliferating subset of trophoblast is *α6β4* (Damsky et al., 1992; Aplin, 1993; Burrows et al., 1993a). Its expression is, interestingly, not restricted to the basal trophoblastic surface of the most proximal cells, but rather can be found all around the trophoblast cells of the proximal half of the cell columns of first and second trimester pregnancies (Figure 5).

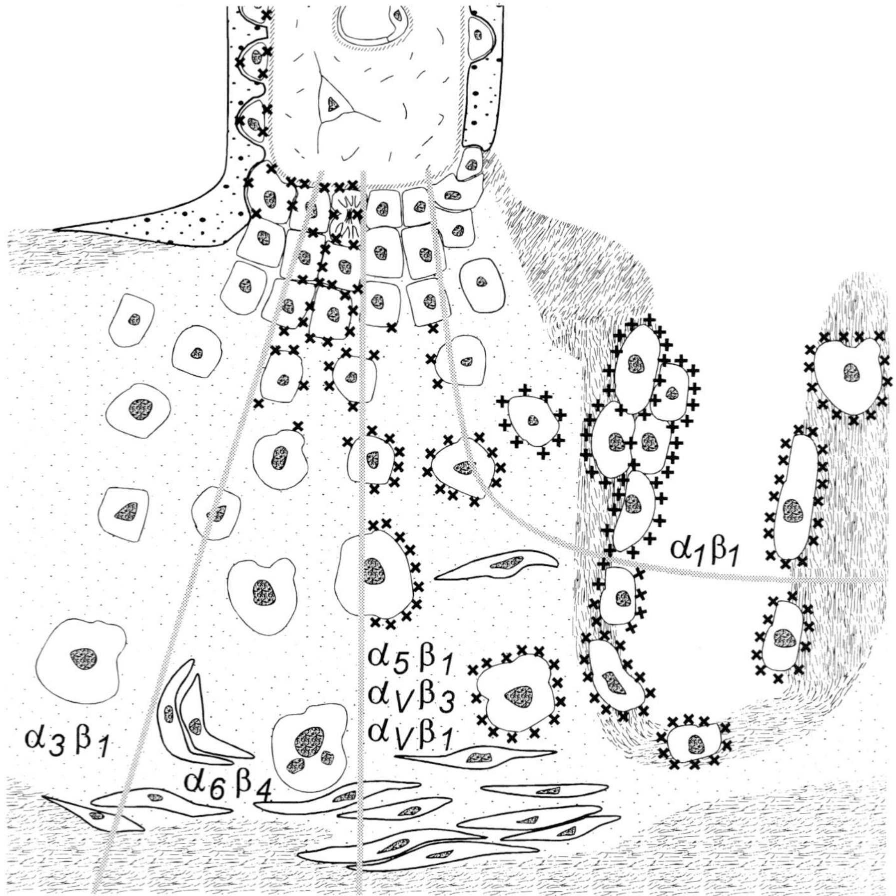

Figure 5. Normal pregnancy: Integrin expression (crosses) along the invasive pathway. Whereas α3β1 integrin is restricted to the most proximal stem cells, α6β4 expression extends slightly deeper into the early invasive stages. There it is gradually replaced by α5β1, αvβ1 and αvβ3 integrins and finally by α1β1 integrins, the latter indicating fully invasive stages.

In term pregnancies with reduced proliferative activities of extravillous trophoblast, also the α6β4 expression is more restrained (Figure 13). α6β4 is not restricted to the proliferative phenotype; rather its expression extends under continuous downregulation slightly deeper into the junctional zone. At the same time, expression of integrin α5β1, a fibronectin receptor, is upregulated and kept constant until the deepest stages of invasion (Figure 5). Slightly later, expression of integrin α1β1, a receptor for laminin and collagens I and IV is turned on (Figure 5) (Aplin, 1991; Fisher et al., 1991; Damsky et al., 1992, 1994; Lim et al., 1993). According to *in vitro* findings by Irving et al. (1995), also αvβ3 and αvβ5 are expressed by the invasive phenotype (Figure 5).

The gradual switch from the basal lamina receptor α6β4 to interstitial receptors such as α5β1, α1β1, and αv integrins, marks the transition from the proliferative to the invasive phenotype of extravillous trophoblast and has been called the *integrin switch*. It has to be admitted that this switch is difficult to identify in mature placental tissues; e.g. in term chorion laeve both steps of differentiation may be so close to each other that the integrin patterns of proliferative and infiltrative trophoblast may overlap (Malak and Bell, 1994). In our experience the basal trophoblast cells of mature membranes are mostly non-proliferating and thus share at least some similarities with the invasive phenotype of earlier stages of pregnancy. Recently, the integrin switch raised increasing interest since it seems to be involved in the regulation of trophoblast invasion in the placental junctional zone (Zhou et al., 1993; Damsky et al., 1993; Lim et al., 1995).

Bischof et al. (1995) found α6 expression by extravillous trophoblast to be positively correlated with gelatinase secretion whereas the secretion ceased as soon as α5 expression was turned on. However, cause and effect are not clear. Irving and Lala (1995) described that blocking of α5β1 by specific antibodies inhibited trophoblast migration *in vitro*. It is likely, however, that the results of such *in vitro* studies strongly depend on unknown experimental parameters; e.g. Librach et al. (1991a) came to opposite experimental findings. In their *in vitro* experiments, antibody blocking of the α1β1 integrin impaired trophoblast invasion whereas antibody blocking of α5β1 even stimulated invasion.

Studies on extravillous trophoblast invasion in preeclamptic pregnancies underline the importance of the integrin shift. In *preeclampsia* trophoblast invasion is generally reduced and limited to the superficial decidua; moreover, the impaired invasion is linked to an abnormal integrin switch (Zhou et al., 1993; Damsky et al., 1993; Lim et al., 1995). In preeclampsia as compared to normal pregnancies, the extravillous trophoblast cells continue expressing integrin α6β4 along the limited invasive pathway rather than downregulating it (Figure 6). As in normal pregnancy, they upregulate α5β1 after leaving the mitotic cycle, but they largely or completely fail to express integrin α1β1. In contrast to these findings, Divers et al. (1995) did not find any such differences between normal and preeclamptic pregnancies. Recently, Genbacev et al. (1996) have shown that hypoxia *in vitro* partly inhibits the integrin switch. The respective trophoblast cells failed to invade ECM substrates and thus mimic preeclampsia changes.

If the findings of Damsky and coworkers are correct, the *balance of α5β1 and α1β1* expression is critical for the regulation of trophoblast invasion. This concept is supported by the above mentioned *in vitro* studies of the same group. According to Librach et al. (1991a) α5β1 binding to fibronectins restrained trophoblast invasion whereas blocking of α5β1 by antibodies stimulated trophoblast invasion, as addition of exogenous fibronectin did. In contrast, α1β1 promoted invasion *in vitro* whereas blocking of this integrin or its ligands (laminin, collagens I and IV) by antibodies resulted in impaired invasion. Also insulin-like growth factor binding protein-1, IGFBP-1, may be involved in these mechanisms. Irving and Lala (1994, 1995) have shown that decidua-derived IGFBP-1 binding to α5β1 by means of its RGD-sequence (and thus blocking the integrin for other ligands?) promoted invasion, an effect that was inhibited by the presence of α5β1 blocking antibodies. Even though partly contradictory, all these data suggest that the balance between α5β1 integrin including its ligands and α1β1 including its ligands may be important for regulation of trophoblast invasion. Imbalance may result in insufficient invasion and seems to be a basic pathogenetic event in preeclampsia (Damsky et al., 1993).

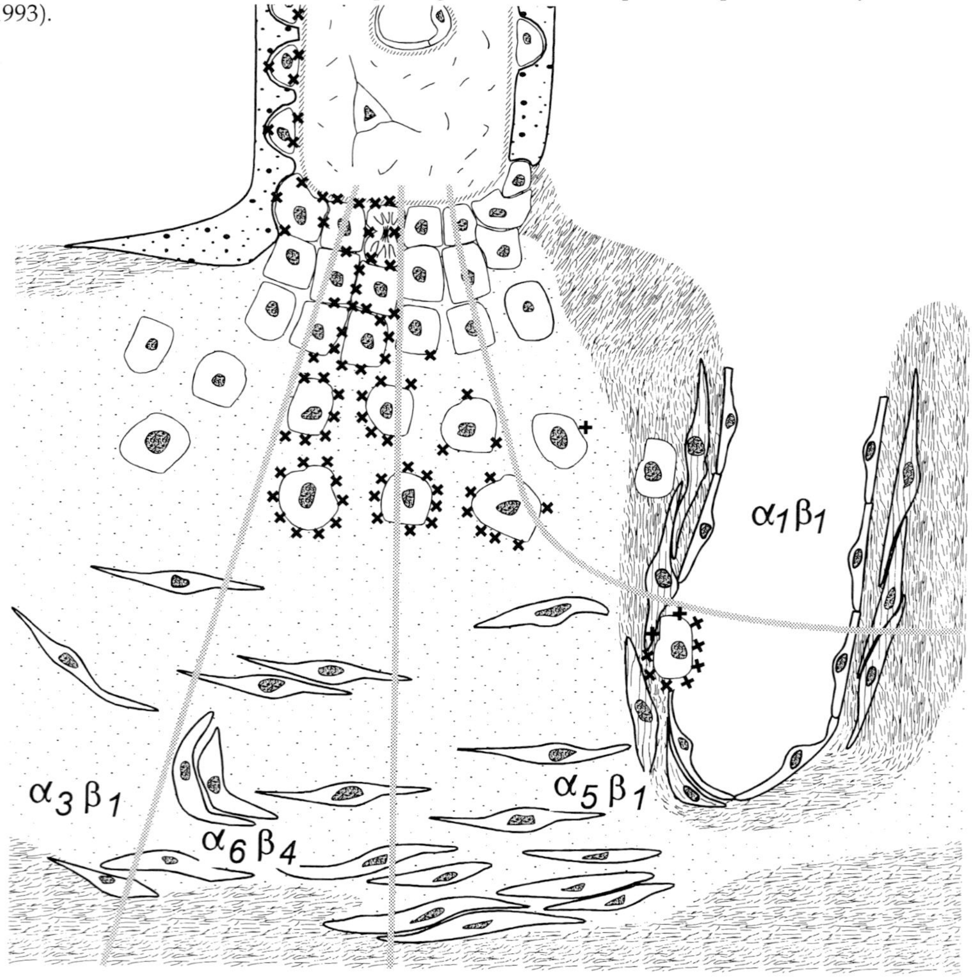

Figure 6. Preeclampsia: Integrin expression (crosses) along the invasive pathway. In this condition, the trophoblast invasion is only superficial, the uteroplacental vessels remain largely intact and mostly void of trophoblast invasion. Whereas integrins α3β1, α6β4 and α5β1 are expressed as in normal pregnancy, expression of integrin α1β1 is considerably reduced if present at all.

There is general agreement that integrin expression is compatible with both adhesion and invasion (cf.. Fisher and Damsky, 1993). This is in accordance with the essentially dynamic nature of integrin-ligand interactions during cell adhesion and migration. Cells plated onto matrix substrates attach, spread, polarize and migrate by sequentially forming, breaking and reforming interactions with substrate at the leading and trailing edges. As discussed above, the balance of α5β1 and α1β1 expression may be critical for the regulation of trophoblast invasion. It still needs to be established whether this phenomenon is linked to peculiarities of the trophoblastic pericellular matrix; e.g. α1β1-binding to laminin and to various collagens can result in effective adhesion only if the respective ligands are peripherally linked to other interstitial macromolecules. As will be discussed below, the invasive extravillous trophoblast cells are not surrounded by a conventional basal lamina (cf., Flug and Köpf-Maier, 1995; Miosge et al., 1995). Rather, collagen IV and laminin are either absent from the trophoblast pericellular matrix (and thus the ligands for α1β1 are missing) or they are present forming huge accumulations of basal lamina material in which other extracellular matrix molecules could not be found yet (Frank et al., 1995; Kertschanska et al., 1995). Similar relations may be valid for integrin α5β1 which usually binds to fibronectins, the latter cross-linking among others to collagens. Within the extracellular matrix of the extravillous trophoblast, co-distributed with the other fibronectins, we found fibronectin-like "i"-glycosylated molecules which contained the cell-binding domain, but which showed reduced collagen binding affinity *in vitro* (Frank et al., 1995; Huppertz et al., 1995). These fibronectin-like molecules may block integrin α5β1 for other fibronectins and thus promote invasion.

The *regulation of integrin expression* and of the integrin switch in many aspects is still a mystery. It is interesting to note that tumor necrosis factor-α, TNFα, a decidual secretory product upregulates the expression of integrin α1β1 (Defilippi et al., 1992). According to Irving and Lala (1995), transforming growth factor-β, TGF-β, upregulates expression of integrin α5β1 and reduces the migratory effects of invasive trophoblast. Finally, according to the latter group, insulin-like growth factor II, IGF-II, has no obvious effect on integrin expression, but stimulates migration. Furthermore, *in vitro* data provided by Lim et al. (1995) raise the interesting possibility that also heparin or heparan sulfate, present in the extracellular matrix surrounding the invasive extravillous trophoblast (cf., below), play a role in trophoblast differentiation and invasion by enhancing the effects of heparin-binding growth factors such as basic fibroblast growth factor, bFGF. Recent *in vitro* findings by Genbacev et al. (1996) suggest that oxygen partial pressure is a potent regulator of the integrin switch (cf. above).

Other Cell Adhesion Molecules and Gap Junction Molecules

E-cadherin is an adhesion molecule that establishes cell-cell contacts. Proliferative cells of the proximal cell columns express this adhesion molecule, whereas it is turned off with increasing invasiveness (MacCalman et al., 1995; Winterhager et al., 1996) (Figure 7). In intravascular cytotrophoblast expression is again upregulated. Whether this is true also for the deeply invasive aggregations of trophoblast cells outside the uteroplacental vessels, remains open. This finding is supported and extended by *in vitro* findings by Babawale et al. (1995) who found E-cadherin not only expressed in cytotrophoblast lining the stromal basal lamina but still to some extent in invasive cells. In tumor pathology, reduction of E-cadherin expression is regarded an indicator of premalignant changes of epithelial tumors (Pizarro et al., 1995).

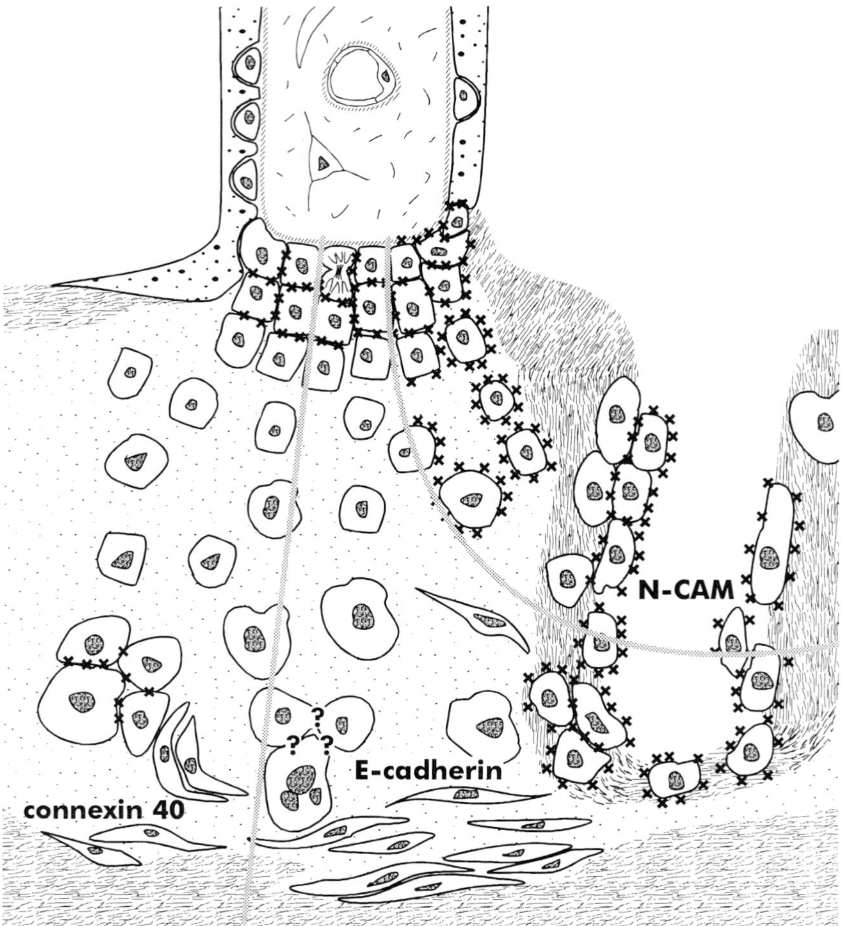

Figure 7. Expression patterns of connexin 40, E-cadherin and neural cell adhesion molecule (N-CAM) along the invasive pathway. Question marks indicate unclear findings.

The neural cell adhesion molecule *N-CAM* is not only responsible for cell-cell contacts but also for matrix adhesion. In early pregnancy N-CAM-positive extravillous trophoblast cells are abundant in all parts of the basal plate including intraluminal trophoblast cells in the uteroplacental arteries (Figure 7); only the intramural trophoblast cells of the arteries stained weaker if at all (King and Blankenship, 1995, findings in the rhesus monkey). With advancing pregnancy N-CAM expression is restricted to the more proximal, proliferative phenotype. Burrows et al. (1994) confirmed these findings for the human extravillous trophoblast. Moreover, according to their findings, it is the unique polysialylated form of N-CAM which is responsible for cell-cell adhesion within the intraluminal trophoblastic plugs in uteroplacental arteries and within trophoblastic aggregates in the neighborhood of uteroplacental arteries. The authors speculate that maternal serum factors percolating through the occluding trophoblastic plugs, are responsible for induction of N-CAM expression and for its regulation.

Adhesion of the trophoblast cells to the maternal endothelium which originally lined these arteries, may be established by trophoblastic expression of the cell surface carbohydrate sialyl-Lewis^x (sialyl-Le^x) (King and Loke, 1988). The respective receptors are E- and P-selectins. Both of the latter are lectins which are expressed by the maternal endothelial cells only in the implantation site (Burrows et al., 1994). Usually, both selectins are expressed by endothelium during inflammatory reactions. They enable leukocytes expressing sialyl-Le^x to attach and to leave the vessel wall. It may be discussed that at the implantation site, trophoblastic invasion of the arterial walls initiates the respective "inflammatory reaction" resulting in selectin expression and intraluminal trophoblast adhesion. Interestingly, the sialyl-Le^x/E-selectin adhesive interaction is also involved in the adhesion of human cancer cells to human umbilical vein endothelial cells *in vitro* (Takada et al., 1993).

Burrows et al. (1994) found intercellular cell adhesion molecule, *ICAM-1*, not only expressed by all maternal endothelial cells, large granular lymphocytes, macrophages, but also to a lesser degree by some subpopulations of extravillous trophoblast, including interstitial trophoblast encircling the uteroplacental vessels and intravascular trophoblast. The ligand for ICAM-1 is the β2 integrin heterodimer CD11a/18 (Springer, 1990), which is expressed by large granular lymphocytes (Burrows et al., 1993b), but not by trophoblast cells. The same authors did not find expression of vascular cell adhesion molecule (VCAM) in extravillous trophoblast cells.

It can be concluded that several adhesion molecules are involved in the mechanisms of vascular trophoblast invasion. In general, these mechanisms seem to correspond to those found in transvascular migration of leukocytes in inflammatory reaction and in tumor metastases in distant organs (cf., Burrows et al., 1994). To our best knowledge, a convincing hypothesis regarding the sequence of events leading to the trophoblast-endothelial encounter, is still missing. Moreover, it is still uncertain whether all vascular penetration occurs as "intravasation" via the interstitial route. Data obtained from baboon and macaques (Ramsey et al., 1976; Blankenship et al., 1993b) clearly indicate that at least in these species spreading of the trophoblast within the vessels takes place by intraluminal migration once early penetration has occurred.

The distribution of *connexins* shares some similarities with that of E-cadherin. Connexins are gap junction molecules which support cell-cell communication and seem to be important for cell proliferation and differentiation. Among the connexins studied so far, only connexin 40 (cx 40) has been found in extravillous trophoblast. Its expression is most prominent in the proximal, proliferative parts of the cell columns (Figure 7) and cell islands where the cells are still in contact with each other (Hellmann et al., 1995; von Ostau et al., 1995; Winterhager et al., 1996). Interestingly, also some villous trophoblast cells may express connexin 40 at cell-cell contacts, and thus behave similarly to the columnar stem cells. With transition to the invasive phenotype, cx40 is downregulated. Deeply invasive groups of aggregated trophoblast cells again express cx40 (Figure 7), whereas intravascular trophoblast cells probably do not (Winterhager et al., 1996). In agreement with the localization of cx40 in the proliferative subset of trophoblast, transfection of JEG3 choriocarcinoma cells *in vitro*, which normally do not express cx40, with various connexin genes, showed that cx40 did not reduce the high proliferative activity of these cells whereas transfection of JEG3 with cx43 drastically reduced proliferation (Hellmann et al., 1995). Cronier et al. (1994) described a stimulating effect of human chorionic gonadotropin on connexin expression in the human placenta.

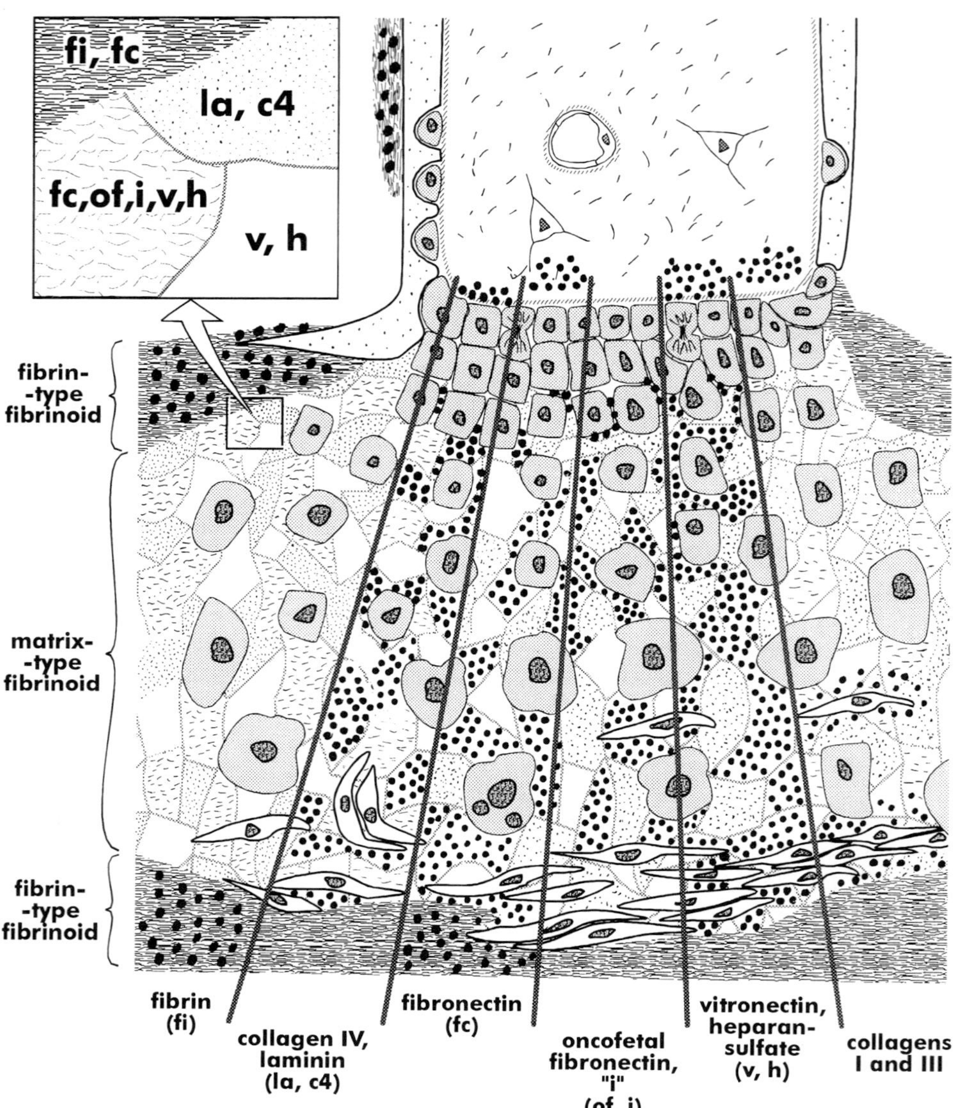

Figure 8. Distribution of extracellular matrix molecules and types of fibrinoid along the invasive pathway of cell columns. The topographical distribution of the respective immunoreactivities is indicated by fat black dots. For coexpression patterns compare the inset. Fibrin-type fibrinoid (dark line-hatching), mostly a blood clot product, is immunoreactive for fibrin (fi) (left column) and non-oncofetal fibronectins (fc) (third column). Matrix-type fibrinoid, embedding the extravillous trophoblast cells is composed of a patchwork of three different subtypes of matrix, symbolized by different shading (left column). These subtypes are composed of (1) laminin (la) and collagen IV (c4), or (2) vitronectin (v) and heparan sulfate (h), or (3) non-oncofetal fibronectins (fc), oncofetal fibronectins (of), "i"-glycosylated fibronectin-like molecules (i), vitronectin and heparan sulfate (compare inset). Collagens I and III are only found within the villous stroma and close to decidual cells.

Secretion Of Extracellular Matrix

The human placenta in all stages of pregnancy contains large amounts of a special condensed glossy extracellular matrix within which the extravillous trophoblast cells are embedded. In the classical literature, this matrix has been subsumed under the historical term "fibrinoid" (cf., Benirschke and Kaufmann, 1995). Even routine histological studies easily reveal that parts of the fibrinoid are coarsely fibrillar in structure and do not encase extravillous trophoblast cells (Figure 8). This subtype of fibrinoid is immunoreactive for fibrin and for some fibronectin epitopes (Frank et al., 1994). It is likely to be primarily a blood clot product. We call it the *fibrin-type fibrinoid*. In routine histological studies this subtype can be easily identified by orange staining in PAF-Halmi stained sections (a modified paraldehyde-fuchsin stain; Lang et al., 1994).

The remaining fibrinoid is histologically more glossy or granular in nature and stains violet to green with the PAF-Halmi stain. Moreover, it is specifically stained by the lectin from lycopersicon esculentum (LEA) (Lang et al., 1994). Different from fibrin-type fibrinoid, this material encapsulates non-proliferating, invasive extravillous trophoblast cells which have not yet reached the uteroplacental arteries (Figure 8). This stage of extravillous trophoblast differentiation has also been named interstitial trophoblast (Pijnenborg et al., 1980). Immunohistochemically, this latter type of fibrinoid is void of fibrin and rather shows immunoreactivities for basal lamina molecules (Frank et al., 1994). Because of this composition and its ultrastructural similarities with basement membranes, it has been named pseudo-basement membrane (Aplin and Campbell, 1985) or basement membrane-like layer (King and Blankenship, 1994b). We prefer the term *matrix-type fibrinoid* since it differs in several important aspects from basement membranes.

Matrix-type fibrinoid is the typical extracellular matrix (ECM) of the invasive stages of extravillous trophoblast. Semithin sections and ultrastructural evaluation of this ECM reveal that it is not homogeneous but rather consists of a patchy mosaic of three different subtypes. Each of those can be identified by typical ultrastructural features (Frank et al., 1994) and by a highly characteristic composition of extracellular matrix molecules (Frank et al., 1995; Kertschanska et al., 1995). We did not find any differences regarding the composition between the matrix-type fibrinoid from various placental structures and locations (Frank et al., 1994). The description given below for the extracellular matrix of the cell columns, is valid also for cell islands, membranes, chorionic plate, etc.

Amount, composition and orientation of extracellular matrix deposed by trophoblast of the cell columns, depends on the stage of trophoblastic differentiation. The most proximal layer of *proliferating trophoblast* cells which rest on the basal lamina facing the stroma of the anchoring villous in first and second trimester pregnancies, are clearly polarized cells which in concert with the stromal cells secrete extracellular matrix molecules only along their basal plasmalemma (Damsky et al., 1992; King and Blankenship, 1994b) (Figure 8). This basal lamina is composed of collagen IV, laminin and heparan sulfate (Damsky et al., 1992; Rukosuev, 1992; Castellucci et al., 1993; Frank et al., 1994) and thus does not differ from usual trophoblastic basal laminas (Duance and Bailey, 1983; Nanaev et al., 1991; Flug and Köpf-Maier, 1995). Fibronectins, tenascin and vitronectin are found within the closely neighboring stroma of the anchoring villous (Castellucci et al., 1991; Frank et al., 1995).

The *later proliferative generations*, which have lost contact to the basal lamina, do not show light and electron microscopical signs of extracellular matrix deposition (Okudaira et al., 1991; Damsky et al., 1992; Frank et al., 1994, 1995) (Figure 8). This does not necessarily rule out the matrix secretion in these layers. Rather, the presence of matrix metalloproteinases in these cells (see below; Polette et al., 1994; Huppertz et al., 1997) suggests that also increased degradation of matrix proteins may be the reason for the absence of matrix accumulations. *In situ* hybridization studies could be useful to further investigate this matter.

Deeper in the junctional zone, immunoreactivities for ECM molecules appear around the extravillous trophoblast cells which gradually switch to the *invasive phenotype* (Damsky et al., 1992; Castellucci et al., 1993; Frank et al., 1994, 1995). In mature placentae showing low proliferative activities of extravillous trophoblast, this may already be apparent around cells of the first trophoblastic layer. In earlier placentae in which trophoblast proliferation extends deeper into the junctional zone, the first 2 to 4 layers of extravillous trophoblast form compact epithelial aggregates with narrow intercellular spaces without light microscopically visible ECM deposition. The re-appearance of extracellular matrix molecules is accompanied by an increasing degree of dilatation of the extracellular spaces which finally lend to complete isolation of the invasive cells (Figure 8). Different from the basal layer of stem cells, ECM secretion is no longer a polar phenomenon restricted to the basal plasmalemma but rather appears all over the surface of the extravillous trophoblast cells. Moreover, the matrix is no longer homogeneous, but rather shows the patchy mosaic pattern characteristic of matrix-type fibrinoid. Immunohistochemically, not only basal lamina molecules such as collagen IV, laminin, heparan sulfate, but also fibronectins and vitronectin are detectable, but no collagens I, III and VI nor fibrin (Frank et al., 1994). Also in this respect our definition is comparable to that given for basement membrane-like structures (Aplin and Campbell, 1985; Damsky et al., 1992; King and Blankenship, 1994b).

An exception of the above description can be found in all those locations of extravillous trophoblast in which proliferation ceased (e.g., cell columns and membranes of the mature placenta). In these cases secretion of pericellular matrix started already in the most proximal layer of cells resting on the basal lamina (Aplin and Campbell, 1985). According to our own observations, most of the respective non-proliferative "proximal" extravillous trophoblast cells express integrin subunits $\alpha1$, $\alpha5$ and $\beta1$ and c-erbB-2 protein rather than integrin $\alpha6\beta4$ and EGF-R (Figure 13), thus indicating that they have left the proliferative phenotype and behave similarly to the invasive trophoblast cells as described above.

Based on the ultrastructure and the molecular composition, three different types of patches have been described (Huppertz et al., 1996):

Basal lamina-like patches show ultrastructurally an electron dense, granular appearance (Frank et al., 1994). Immunocytochemically, only laminin and collagen IV could be detected in these areas, but no fibronectins, heparan sulfate nor vitronectin (Figure 8) (Kertschanska et al., 1995; Huppertz et al., 1996). These patches seem to be irregularly distributed, sometimes in direct contact with the surface of extravillous trophoblast cells, sometimes forming bridges between the cells. Some of the patches are seemingly isolated.

Patches of glossy, *homogeneous ground substance* form the second subtype. Ultrastructurally, this matrix does not show any fibrillar or granular structures. Immunohisto- and -cytochemically, we found evidence for the presence of heparan sulfate (Castellucci et al., 1993; Kertschanska et al., 1995) and vitronectin (Kertschanska et al., 1995; Huppertz et al., 1996) (Figure 8). The low density of immunogold labeling for both molecules may be due to technical problems (e.g., we used an antibody against the largely hidden core protein of heparan sulfate); it also may be a hint that other yet unidentified molecules contribute to the composition of this ground substance. However, we were unable to identify e.g., hyaluronan or chondroitin sulfate in this location.

Patches of *fine fibrillar networks* embedded in the same glossy, homogeneous ground substance described above, make up the third subtype of matrix-type fibrinoid. Electron-microscopically, the fibrils measure 10 to 14 nm in diameter and show an electron-translucent core resulting in a seemingly tubular appearance. These fibrils are very similar to those derived from the co-assembly of cellular and plasma fibronectins as described by Pesciotta-Peters et al. (1990). Because of this, it was not surprising that all anti-fibronectin antibodies tested were found to be associated with these fibrils rather than with any other component of the matrix-type fibrinoid (Frank et al., 1995; Kertschanska et al., 1995; Huppertz et al., 1996) (Figure 8).

The antibody IST-4, directed against the III-5 domain of fibronectins, is the most general marker for all types of fibronectins (Borsi et al., 1987). It stains all of the above fibrils with high density. The antibody IST-9 directed against the ED-A domain of cellular fibronectins (Carnemolla et al., 1987) shows a comparable staining pattern and thus indicates that most, if not all of these fibrils represent cellular fibronectin. Also the antibodies BC-1 (against the ED-B domain, Carnemolla et al., 1989) and FDC-6 (against the O-glycosylated III-CS region, Matsuura and Hakomori, 1985), both of which recognize epitopes expressed only in oncofetal fibronectins, bind exclusively to the same fibrils. The high density of FDC-6 labeling which was comparable to that of IST-4 and IST-9 may point to a high proportion of glycosylated fibronectins within the matrix-type fibrinoid. On the other hand, the antibody IST-6 (against the type III-7 domain) only recognizes fibronectin isoforms lacking the ED-B domain of oncofetal fibronectins and thus is indicative of the presence of non-oncofetal fibronectins (Carnemolla et al., 1992). This antibody only occasionally labeled a few fibrils. These immunocytochemical findings, obtained at the electron microscopical level (Huppertz et al., 1996), are in good agreement with earlier immunohistochemical results describing the presence of cellular and oncofetal fibronectins around extravillous trophoblast cells (Aplin and Campbell, 1985; Virtanen et al., 1988; Feinberg et al. 1991; Blankenship et al., 1992, 1993b; Damsky et al., 1992; Blankenship and King, 1993; Castellucci et al., 1993; Feinberg and Kliman 1993a,b; Nanaev et al., 1993; Frank et al. 1994, 1995; King and Blankenship, 1994 a,b). Since of all these studies were performed on cryostat and paraffin sections, they could not differentiate the various subtypes of extracellular matrix within the matrix-type fibrinoid.

As already stated above, fibronectin immunoreactivities are always restricted to the 10 to 14 nm fibrils. The glossy, homogeneous matrix embedding the fibrils seems to be identical to the second subtype described above; so far we have detected only heparan sulfate and vitronectin in this ground substance.

The size and the number of patches increases with increasing depth of invasion thus further separating the invasive cells by the accumulated masses of extracellular matrix. Since our ultrastructural studies have focused on the distal parts of the cell columns, we cannot exclude that the composition of this matrix is different in more proximal parts of the cell columns or deep in the junctional zone where fetal/trophoblastic and decidual extracellular matrices mix. We found some preliminary evidence that in the latter zone among others, increasing amounts of collagens I and III can be found. Also the extracellular matrix of the intramural and intravascular trophoblast cells awaits further evaluation.

The functional relevance of this peculiar extracellular matrix and of its compartmentalization is still open. Feinberg et al. (1991) have speculated about the glue-character of oncofetal fibronectins, anchoring the placenta at the implantation site. Implications of these fibronectins in preterm labor (Feinberg and Kliman, 1992) and in preeclampsia are under discussion (Friedman et al., 1992; Kupferminc et al., 1995). Moreover, the cellular receptors (integrins) for the various extracellular matrix molecules making up this ECM, are known to mediate both invasiveness and anti-adhesiveness, depending on the type of integrin and possibly also on the kind of surrounding environment (Librach et al., 1991a; Damsky et al., 1993; Irving and Lala, 1995). It needs to be tested whether this heterogeneity of the extracellular matrix alone or in concert with the integrin shift is involved in regulation of trophoblast invasiveness (cf., chapter on integrins).

Blood Group Antigens

Carbohydrate epitopes on cell surfaces and on extracellular matrix molecules represent an important subgroup of oncofetal antigens that often have close molecular relations to normal adult blood group antigens. Their metabolic relationship is either based on blocking of the normal synthetic pathway or on unusual pathways like chain extension or atypical sialylation and fucosylation (cf., Frank et al., 1995). Such carbohydrate chains may act as immunologically relevant cell surface markers or can represent ligands for cell adhesion molecules (Lewis^x/ selectin-interaction, King and Loke, 1988; Burrows et al., 1994), or, if expressed on extracellular matrix molecules, may decrease collagen binding and increase resistance to proteolytic cleavage (Zhu et al., 1984; Zhu and Laine, 1985).

Blood group antigen "i" is the most primitive precursor of the AB0 system of blood groups. It is not only expressed by immature red blood cells, but also by all stages of invasive extravillous trophoblast rather than by the proliferating stages (Frank et al., 1995). It is to be discussed whether the cell surface expression of such an immature blood group antigen which every human organism has expressed throughout embryonic and fetal life, prevents immune recognition of extravillous trophoblast cells by the maternal immune system. Interestingly, the identical carbohydrate chain "i" was also found extracellularly in the surrounding of the invasive extravillous trophoblast cells (Figure 8), bound to fibronectin-like molecules (Frank et al., 1995; Huppertz et al., 1996).

The slightly more mature *blood group antigen "I"* could be detected in human trophoblast only after sialidase pretreatment, thus indicating that not "I" but rather sialyl-"I" is expressed, an unusual type of blood group antigen maturation (Frank et al., 1995). Sialyl-"I" immunoreactivity was found on villous syncytiotrophoblast and only in few cases on the surfaces of invasive extravillous trophoblast cells, where it was coexpressed with "I".

The cell surface carbohydrate *sialyl-Le*[x] belonging to the Lewis family of blood group antigens, is expressed by the intraarterial trophoblast cells which form huge trophoblastic plugs (King and Loke, 1988). This molecule acts as ligand for E- and P-selectins which are expressed by the endothelium of the uteroplacental vessels, thus mediating trophoblastic adhesion to the vessel walls (Burrows et al., 1994) (see section on adhesion molecules).

Proteinases, Activators and Inhibitors

Trophoblast invasion requires not only breaching of endometrial structures such as epithelium, glands, vessels and stroma but also degradation of endometrial extracellular matrix (cf., Graham and Lala, 1991; Bischof and Martelli, 1992). These invasive activities of trophoblast have been demonstrated by *in vitro* studies (Fisher et al., 1985). Numerous immunohistochemical, biochemical and experimental studies suggest that a variety of proteinases, their activators and inhibitors are involved. However, only few of these studies have related the expression patterns to the various stages of the invasive pathway.

Among the best studied proteinases are the matrix metalloproteinases (MMPs). It has been proven *in vitro* that they are secreted by trophoblast cells (Bischof et al., 1991). Respective immunoreactivities can be found not only intracellularly, but with increasing invasive depth also in the surrounding extracellular matrix. There they partly can be found directly bound to their respective ECM substrates (Huppertz et al., 1997).

MMP-1 (interstitial collagenase) degrades various interstitial collagens, among others collagens I and III which are most abundant in the endometrium. Because of this, it is not surprising that invasive extravillous trophoblast expresses interstitial collagenase (Moll and Lane, 1990). Its expression has also been proven in trophoblast cell culture (Emonard et al., 1990). However, detailed information concerning the expression patterns of this enzyme is missing. In preliminary studies, we found very weak immunoreactivity within the most proximal, proliferating layers. Early invasive cells were not reactive, but with increasing invasive depth, immunoreactivity increased again and was also most prominent in intravascular trophoblast and in decidual cells (Figure 9).

MMP-2 (72 kD-type IV collagenase, gelatinase A) degrades mainly collagen IV. Its expression in extravillous trophoblast has been demonstrated by immunohistochemistry and *in situ* hybridization (Autio-Harmainen et al., 1992; Fernandez et al., 1992; Blankenship and King, 1994b; Polette et al., 1994) as well as by *in vitro* approaches (Emonard et al., 1990). According to the *in situ* hybridization studies by Polette et al. (1994), this proteinase is not expressed in proliferating trophoblast, but only in the invasive phenotype throughout all stages of pregnancy (Figure 9).

MMP-3 (stromelysin-1) degrades fibronectins, laminin, various collagens and core proteins of proteoglycans. According to our own unpublished studies, it is weakly expressed within the proximal proliferating layers; no immunoreactivity is found in early invasive stages. In later invasion, trophoblast cells and extracellular matrix are again immunoreactive, most of the extracellular immunoreactivities linked to the fibronectin fibrils in the ECM (Figure 9).

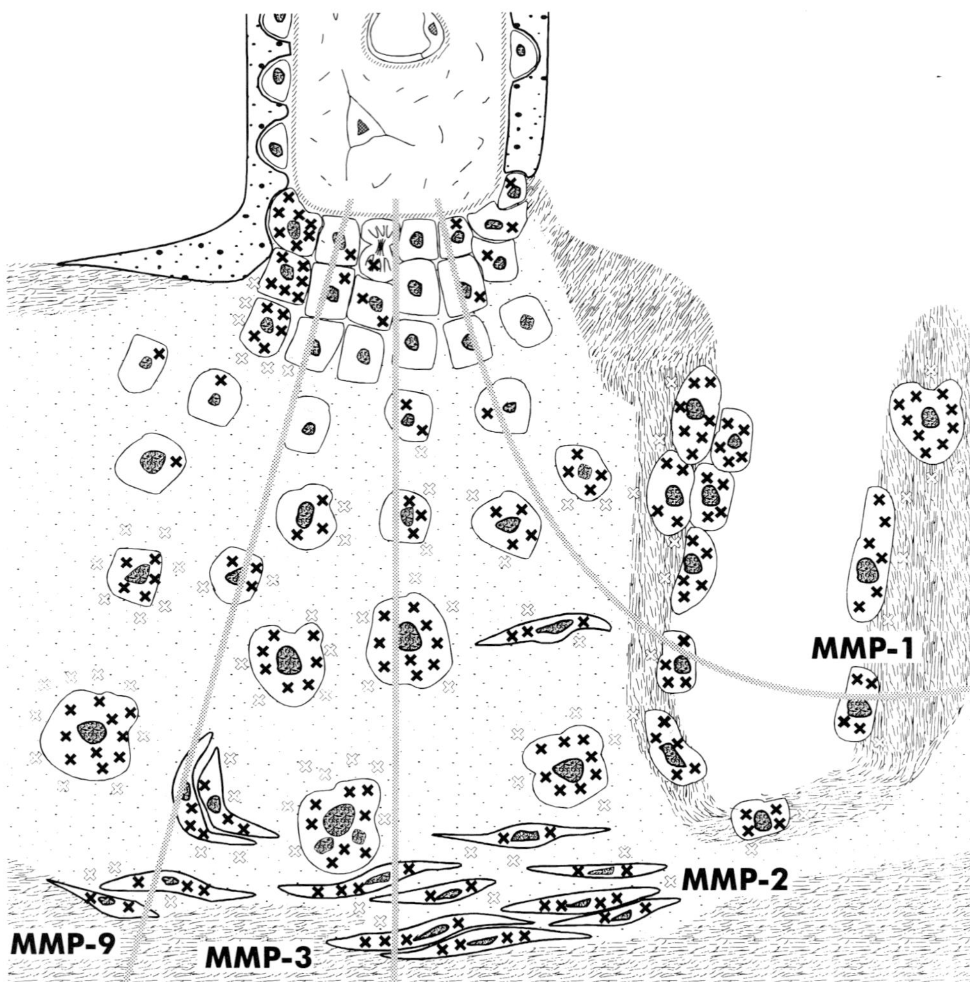

Figure 9. Matrix metalloproteinases 1, 2, 3 and 9 are immunoreactive in extravillous trophoblast cells and decidual cells (dark crosses) and in the surrounding extracellular matrix (light crosses). The distribution patterns depend on the type of metalloproteinase.

MMP-7 (matrilysin) has a broad range of ECM substrates which include fibronectins, collagens, laminin and proteoglycans. According to Vettraino et al. (1996) it is the most active MMP during trophoblast invasion and is over expressed in preeclampsia.

MMP-9 (92 kD-type IV collagenase, gelatinase B) mainly degrades collagen IV and in this respect is similar to MMP-2. However, its expression pattern differs. It is present at both, mRNA and protein levels in the proximal, proliferating layers of extravillous trophoblast (Polette et al., 1994). Expression is down-regulated in the early invasive stages (unpublished results) and then upregulated again in the deeper stages of invasion (Figure 9). An antibody against MMP-9 completely inhibited trophoblast invasion *in vitro* (Librach et al., 1991b). Expression decreases towards term (Librach et al., 1991b; Fisher et al., 1989; Polette et al., 1994).

MMP-11 (stromelysin-3) is a laminin-, collagen IV-, and proteoglycan-degrading proteinase. According to immunohistochemical and in-situ hybridization findings by Maquoi et al. (1995) and Polette et al. (1994), it is expressed only by the invasive phenotype of extravillous trophoblast. Its expression decreases with advancing gestation.

MT-MMP (membrane-type matrix metalloproteinase) activates MMP-2 and is expressed by proliferating and invasive trophoblast cells in first and third trimester (Nawrocki et al., 1996).

It is interesting to compare the expression of these matrix metalloproteinases with the distribution of extracellular matrix molecules along the invasive pathway. In the proximal, proliferative zone where basal lamina molecules are secreted, MMP-9 prevails, whereas MMP-2 is absent and MMP-1 and MMP-3 are only weakly expressed. The activities of the latter MMPs increase with increasing invasive depth where besides basal lamina molecules also fibronectins and proteoglycans, and later interstitial collagens need to be degraded (cf., section on extracellular matrix molecules). The dependence of MMP secretion on the stage of differentiation was also demonstrated *in vitro* by measuring gelatinolytic activity as related to integrin expression of cultured cells (Bischof et al., 1995): integrin $\alpha6$-positive trophoblast cells show much higher gelatinolysis and less oncofetal fibronectin expression as compared to integrin $\alpha5$-positive cells. Similar relations have been described for human melanoma cell invasion where MMP-2 expression is modulated by the expression of integrins $\alpha v\beta3$ and $\alpha5\beta1$ (Seftor et al., 1993). As described above, both integrins are also expressed by invasive extravillous trophoblast.

Some of the publications concerning MMP expression have focused on extravillous trophoblast in spiral arteries. MMP-2 was found only in a subset of intramural and intraluminal trophoblast cells of the rhesus monkey throughout pregnancy (Blankenship and King, 1994b). These findings in the rhesus monkey support those obtained already before in the human (Fernandez et al., 1992; Autio-Harmainen et al., 1992). Also MMP-1 expression has been described for the intravascular trophoblast (Moll and Lane, 1990) whereas data on all other matrix metalloproteinases are missing.

Expression of matrix metalloproteinases is regulated by their tissue inhibitors (*TIMPs*). In the human placenta so far TIMP1, an inhibitor of all known MMPs, and TIMP2 which preferentially interacts with MMP-2, have been described. Both are expressed by decidual cells (Polette et al., 1994), and to a certain degree by extravillous trophoblast cells (Ruck et al., 1995, 1996; Huppertz et al., 1997). In the decidua, TIMP2 is constantly expressed throughout pregnancy, whereas TIMP1 showed increasing activities towards term (Polette et al., 1994). These results suggest that decidual cells limit trophoblast invasion in the deeper parts of the uterine wall. The importance of TIMPs for the regulation of trophoblast invasion is underlined by the experimental findings by Librach: TIMP-1 and TIMP-2 completely inhibited human cytotrophoblast invasion *in vitro* (Librach et al., 1991b).

Matrix metalloproteinases probably are not only involved in trophoblast invasion but are also related to the separation of the placenta from the uterine wall. During spontaneous labor, but before delivery, MMP-3 and MMP-9 mRNA levels are significantly increased whereas after spontaneous delivery tPA (see below) and TIMP-1 mRNA levels were increased (Bryant-Greenwood and Yamamoto, 1995).

Plasminogen activators have only limited direct matrix degrading activity. However, they convert plasminogen to plasmin which activates other proteinases. Two plasminogen activators are known, urokinase-type PA (uPA) and the tissue-type PA (tPA). In malignant cells uPA is secreted in higher amounts than in their normal counterparts.

The expression of uPA by extravillous trophoblast has been shown by Hofmann et al. (1994). However, at present these data are questioned (Pierleoni, unpublished data). In contrast, *in vitro* uPA m-RNA has been found in cytotrophoblast (Eldar Geva et al., 1993) and, moreover, also secretion of uPA into the culture medium (cf. Bischof and Martelli, 1992). Moreover, uPA-receptor is expressed in extravillous trophoblast (Multhaupt et al., 1994; Pierleoni, unpublished data). There are no hints as to the expression of tissue-type plasminogen activator (tPA) by extravillous trophoblast.

Also the inhibitors of uPA and tPA i.e. PAI-1 and PAI-2, have been found in human trophoblast (Astedt et al., 1986; Feinberg et al., 1989). Corresponding to the uPA expression, only PAI-1 is expressed by invasive cytotrophoblast of cell islands and cell columns (Feinberg et al., 1989). On the other hand tPA and PAI-2 were found only in villous syncytiotrophoblast where they are thought to regulate fibrinolysis (Feinberg et al., 1989).

Although several aspects of the functional importance of uPA and PAI-1 remain to be clarified, *in vitro* studies have shown that the addition of PAI-1 or uPA-antibody to trophoblast cell culture inhibited trophoblast invasion only partially (Librach et al., 1991b).

Few data have been published concerning *other serine protease inhibitors*, such as α1-anti-chymotrypsin, α1-antitrypsin, and inter-α-trypsin inhibitor. None of them was immunoreactive within the extravillous trophoblast, however, the surrounding matrix-type fibrinoid in cell columns and basal plate showed ample reaction product (Castellucci et al., 1994). The fibrinoid of cell islands was less reactive. The authors suggest that the protein is so rapidly exported into the ECM that no intracellular immunoreactivity is detectable. These data are largely consistent with those by Earl et al. (1989) who could detect neither α1-anti-chymotrypsin nor α1-anti-trypsin in invasive trophoblast except in uteroplacental arteries.

Nitric Oxide Synthase

In lower concentrations, as produced by the endothelial isoform of the enzyme nitric oxide synthase (eNOS), NO causes vasodilatation. Higher NO levels as produced by the macrophage isoform (mNOS) are said to be cytotoxic.

Nitric oxide synthase immunoreactivities in extravillous trophoblast were described for the first time by Morris et al. (1993). This finding raised little attention. Recently, we have repeated these studies and found evidence not only for mNOS but also for eNOS expression by invasive extravillous trophoblast cells (unpublished). The potential importance of these results is stressed by recent experimental findings in rats and guinea pigs: Yallampalli and Garfield (1993), Chwalisz and Garfield (1994) and Garfield et al. (1994) have shown that preeclampsia-like symptoms including hypertension, proteinuria and fetal growth retardation, can be induced in rats and guinea pigs by blocking nitric oxide synthase with L-NAME. The experiments were successfully repeated by Osawa et al. (1995) in rats.

These findings led us to study trophoblast invasion in relation to NOS expression by means of immune- and enzyme-histochemistry in guinea pigs (Nanaev et al., 1995). The results indicate that uteroplacental vessel dilation in this species is related to eNOS expression in periarterial trophoblast cells, the latter corresponding to the interstitial subpopulation of extravillous trophoblast in the human. Invasion and destruction of the vessel walls as well as formation of intraarterial trophoblastic plugs in the guinea pig, seem to be secondary effects. Similar to the extravillous trophoblast cells in the human, the invasive trophoblast cells in the guinea pig also expressed mNOS, thus raising the question whether the high NO levels provided by this inducible isoform of nitric oxide synthase, are responsible for cytotoxic effects and thus contribute to the invasive phenotype of the extravillous trophoblast.

HLA-G

Villous cyto- and syncytiotrophoblast separating maternal blood and villous stroma express neither class I nor class II antigens of the major histocompatibility complex (MHC) at their surfaces (cf., Loke, 1989). The proliferating extravillous trophoblast cells of the cell columns behave like villous cytotrophoblast and are MHC I-negative; in contrast, invasive extravillous trophoblast cells express a non-classic MHC I variant, namely HLA-G (Schmidt and Orr, 1993; Chumbley et al., 1994a; Colbern et al., 1994; McMaster et al., 1995).

From the morphological point of view, the situation becomes even more interesting when one compares the immunohistochemical distribution of MHC I with the distribution of the class I MHC mRNA (Hunt et al., 1990, 1991). Hunt and coworkers were able to divide placental trophoblast into three subpopulations:

- villous cytotrophoblast (Langhans' cells) that contains class I mRNA but does not express the MHC I antigen;
- villous syncytiotrophoblast that neither contains class I mRNA nor expresses MHC I antigen;
- invasive extravillous cytotrophoblast that contains the mRNA message and expresses the variant MHC I antigen (HLA-G).

These results are in agreement with later findings by Lata et al. (1992), concluding that MHC-I expression is regulated by posttranscriptional events. Moreover, these findings support the view, that the trophoblast cells resting on villous and extravillous basal laminae represent a uniform population of proliferating stem cells for both, villous syncytiotrophoblast and extravillous cytotrophoblast (Benirschke and Kaufmann, 1995).

The descriptions concerning the topography of HLA-G expression are not fully consistent. Whereas Chumbley et al. (1994a) pointed out that the molecule is expressed in all populations of extravillous trophoblast, McMaster et al. (1995) found it only in the invasive phenotype. This is in agreement with earlier results by Butterworth et al. (1985) and Shorter et al. (1993) who also did not reveal respective immunoreactivities within the proliferating subset. Moreover, McMaster et al. (1995) described that extravillous trophoblast cells with reduced invasive capacity in term pregnancy showed also reduced HLA-G expression. Also in preeclampsia HLA-G expression levels in the materno-fetal junctional zone are decreased; this reduction appears to be related to reduced numbers of HLA-G-positive invasive trophoblast cells (Colbern et al., 1994).

The question how extravillous trophoblast cells that bear fetal MHC class I antigens can survive in a maternal environment, has caused much speculation (cf., Loke, 1989; Loke and King, 1991). One possible explanation is that the trophoblast expresses this unusual form of MHC class I molecule, the HLA-G, which obviously is only expressed by extravillous trophoblast and which is considered to be unlikely to provoke immune responses even at the maternal-fetal interface. Moreover, it has been shown that the extravillous trophoblast cells do not express surface structures that can be recognized by natural killer cells from the endometrial population of large granular lymphocytes (King et al., 1989, 1990); possibly HLA-G expression makes these cells resistant to lysis by the surrounding decidual natural killer cells (Chumbley et al., 1994b).

From the morphological point of view, it is interesting to note that the cell surface antigen HLA-G seems to be the only antigen detected so far, that specifically distinguishes extravillous trophoblast cells from any other type of cell. Because of this, a specific HLA-G antibody such as that one recently raised by the Loke group (Chumbley et al., 1994a) is of particular importance for the identification and isolation of these cells.

CELL ISLANDS

Cell islands are roundish or irregularly shaped accumulations of extravillous trophoblast cells, at least partly embedded in fibrinoid, attached to free floating tips of larger villi. Their basic histology and ultrastructure has been only rarely studied (Krönicher, 1975; Steininger, 1978). There are even less reliable data concerning the cell biology of trophoblastic cell islands. This is mainly due to the fact that cell islands are difficult to distinguish from cell columns in first trimester specimens derived from interruption of pregnancy. Both cell islands and cell columns represent similar accumulations of extravillous trophoblast embedded in matrix-type fibrinoid and both are peripherally attached to larger villi. The major difference is a topographical one: cell islands are freely floating structures whereas cell columns are anchored to the basal plate or to septa, their differentiated trophoblast cells invading the latter structures (for further details cf. Benirschke and Kaufmann, 1995). However, this latter connection is usually disrupted during termination of pregnancy, the basal plate remaining *in utero* until curettage.

A further difference between cell islands and cell columns may be derived from the fact that the extravillous trophoblast of cell columns is closely apposed to endometrial tissues. It has been shown in *in vitro* experiments by Vicovac et al. (1995a) that this contact may be essential for trophoblast cells to enter the invasive pathway. In contrast, cells that accumulate in the centers of the islands degenerate under liquefaction and formation of cysts. These latter extravillous trophoblast cells never come into contact with maternal tissues but rather with maternal blood, if at all.

Due to the above mentioned problems of identification, we suspect that many data seemingly obtained on cell columns, in fact were pooled from cell columns and cell islands. Few studies carefully tried to handle both structures separately; most of these could not identify relevant differences concerning proliferation patterns, oncogene product expression (Mühlhauser et al., 1993), distribution patterns of adhesion molecules and connexins (von Ostau et al., 1995; Winterhager et al., 1996) and extracellular matrix secretion (Castellucci et al., 1993; Nanaev et al., 1993; Frank et al., 1994). When staining cell islands with the proliferation marker MIB-1, we found proliferating trophoblast cells only close to the basal lamina of the adjacent villous stroma (Figure 10), as also for cell columns.

Explants of cell islands from first trimester placentae have been cultivated in order to study their biology *in vitro* (DeMesy Jensen et al., 1993). The authors concluded that the extravillous trophoblast from cell islands in culture behaves like that derived from cultivated cell columns (Genbacev et al., 1993).

CHORIONIC PLATE

The chorionic plate is another rich source of extravillous trophoblast. Similar to cell islands, its extravillous trophoblast has been studied in detail only by means of histology, enzyme histochemistry and electron microscopy (Weser and Kaufmann, 1978; Wiese, 1975; Bourne, 1962; cf., Benirschke and Kaufmann, 1995), but immunohistochemical studies are largely missing. The only exception is a recent study concerning the expression patterns of type IV collagen, laminin and fibronectins in the macaque chorionic plate (King and Blankenship, 1994a). It revealed that the distribution of ECM molecules embedding the extravillous trophoblast, does not differ from that described for cell columns (see above). Own unpublished data showed that also in the chorionic plate only those trophoblast cells proliferate which rest on the basal lamina facing the chorionic mesoderm (Figure 10). In term placentae, it is difficult to find any proliferating trophoblast cells in the chorionic plate. In conclusion, there are no obvious differences concerning the extravillous trophoblast of this location as compared to that of cell columns.

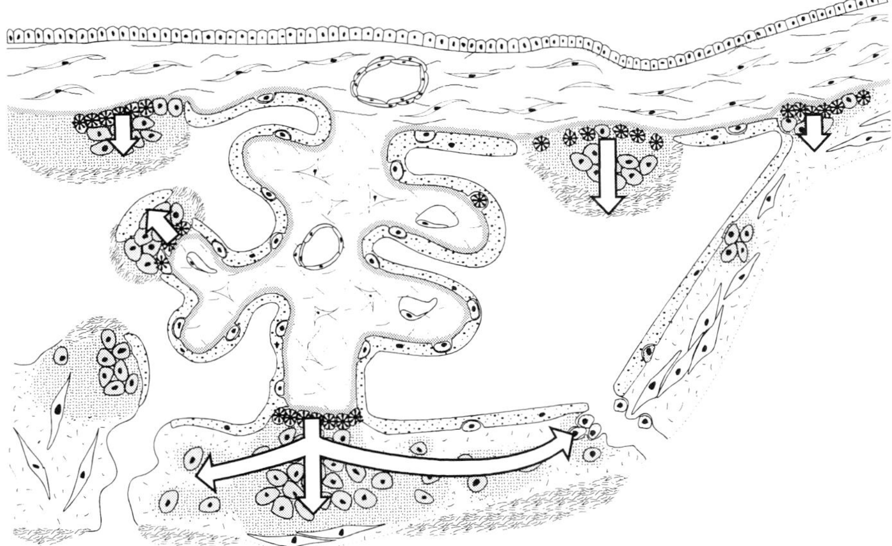

Figure 10. Scheme of the placental margin showing the routes of differentiation and invasion of the extravillous trophoblast cells. The proliferating stem cells are marked by asterisks. The arrows indicated the differentiation and invasion pathway. Note that the columnar extravillous trophoblast may even reach placental septs (left) and uteroplacental vessels.

SMOOTH CHORION

The mesodermal structures of the smooth chorion have been studied in detail since they were thought to be involved in pathogenesis of rupture of membranes. On the other hand, the extravillous trophoblast of this location has raised mostly marginal attention (Bourne, 1962; Bou Resli et al., 1981; Lister, 1968; Wang and Schneider, 1987; Bartels and Wang, 1983; Sakbun et al., 1990 a, b). Our own unpublished data on trophoblast proliferation show that similar to all other locations of extravillous trophoblast, only those cells resting on the basal lamina proliferate (Figure 10). The extracellular matrix surrounding the extravillous trophoblast cells, has been studied in more detail (Aplin and Campbell, 1985; Malak et al., 1993). The data suggest again, that there are no relevant immunohistochemical differences as compared to other locations of extravillous trophoblast, provided that one compares samples of comparable stages of pregnancy. However, different from the extravillous trophoblast cells of the placental site, those of the smooth chorion do not infiltrate much into the adjacent decidua. According to our own experience, the amnion and large parts of the decidua can be stripped off easily. Because of this, the smooth chorion is an easily accessible and rich source of extravillous trophoblast cells and their matrix for purposes of cell culture and biochemistry (Lewis et al., 1996; Gaus et al., 1997).

ONE STEM CELL ORIGIN FOR VILLOUS SYNCYTIOTROPHOBLAST AND THE EXTRAVILLOUS TROPHOBLAST

When studying the above receptors, integrins, matrix molecules, etc. in various extravillous locations, we did not find any relevant differences between the stem cells of the extravillous trophoblast and the Langhans' cells underlying the villous syncytial trophoblast. This is in agreement with the findings of many other groups (for literature see foregoing chapters). Because of this, it is very likely that both kinds of stem cells represent a uniform stem cell population. Their future fate, whether they fuse to form syncytiotrophoblast or whether they acquire an invasive phenotype, probably only depends on the surrounding environment (blood or extracellular matrix molecules) (Figure 11).

This hypothesis would explain why peripheral villi transform into cell islands or cell columns, as soon as they are embedded into extracellular matrix, e.g. maternal blood clot attached to their surfaces. This is not only a common pathohistologic pattern (Benirschke and Kaufmann, 1995). Rather the same behavior of trophoblast cells was also shown when cultivating villous explants in extracellular matrices such as fibrin, collagens and matrigel (Castellucci et al., 1990; Genbacev et al., 1991, 1992; Vicovac et al., 1993, 1995a). Interestingly, Vicovac et al. (1995b) found, that after embedding peripheral villi in agar, the trophoblast maintained its villous phenotype. Differentiation of extravillous trophoblast out of villous explants cultivated on extracellular matrices, has only been observed when cultivating first trimester villi; second trimester villi in culture did not give rise to extravillous trophoblast cells (Genbacev et al., 1992). It is still uncertain whether this is due to reduced proliferative activity of second trimester trophoblast or whether villous cytotrophoblast of this stage of pregnancy no longer comprises a uniform stem cell population.

When one accepts this hypothesis of a uniform trophoblastic stem cell population, it is to be expected that the proliferating cells of the cell columns do not only contribute to the extravillous population. Rather by lateral migration some of the cells may come into a villous position and may be transformed into syncytiotrophoblast. Thus the highly proliferative cell columns can also contribute to longitudinal growth of stem villi. Even though merely speculative, this possibility would explain why in quickly growing structures such as stem villi, only very few proliferating trophoblast cells can be found beneath the syncytial trophoblast (Kosanke, 1994).

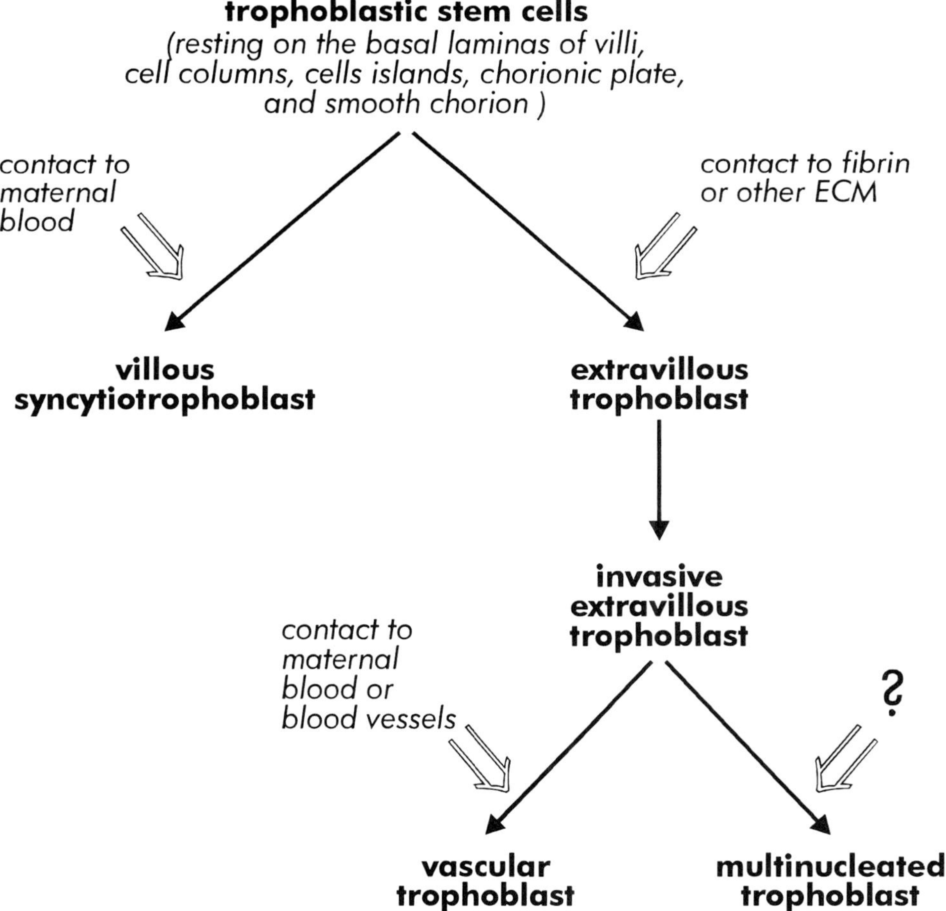

Figure 11. The routes of trophoblast differentiation from stem cells to villous or extravillous trophoblast are controlled by exogeneous factors, the exact nature of which is still largely a mystery. Environmental conditions such as contact to maternal blood, contact to extracellular matrix molecules or contact to maternal blood vessels seem to be decisive.

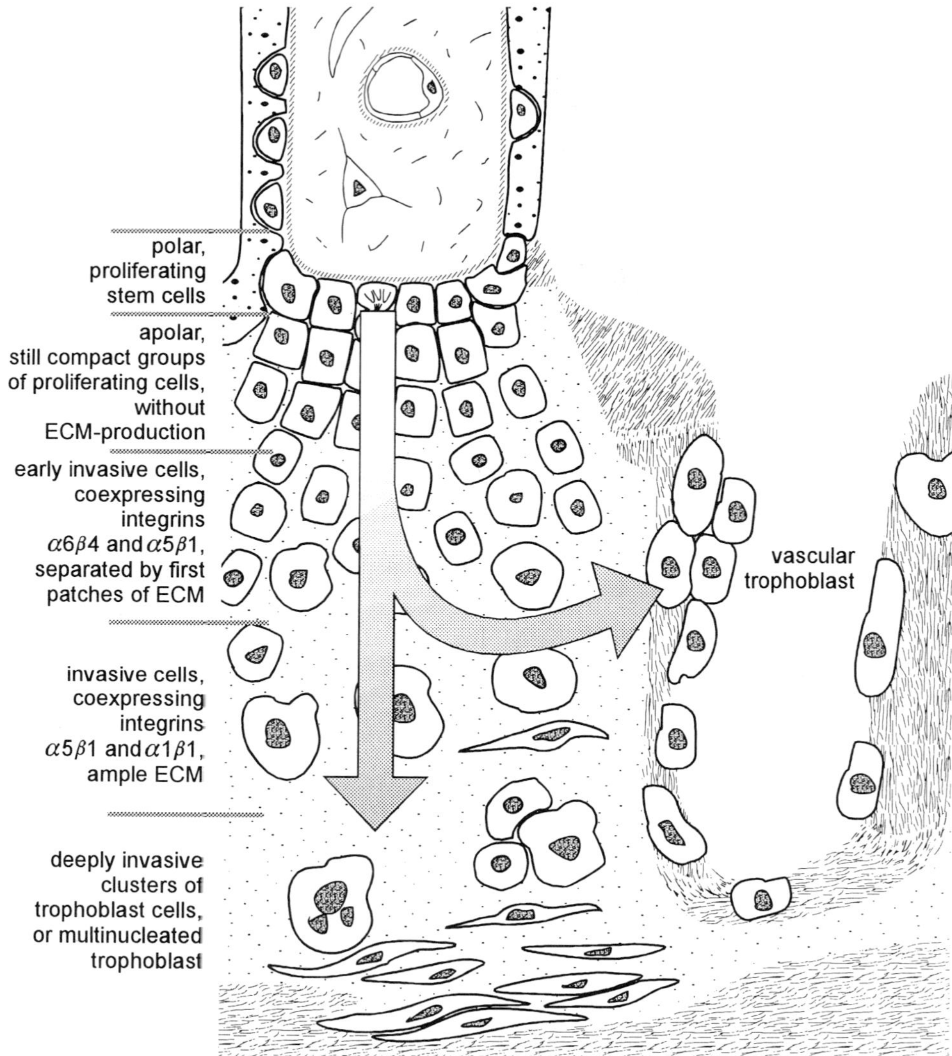

Figure 12. The typical topography, morphology and antigen expression patterns of extravillous trophoblast cells derived from a cell column in the first and second trimester of pregnancy, help to define the stages of the invasive pathway.

CONCLUSIONS

Considering the phenotypic changes along the invasive pathway, it is no longer justified to attribute results simply to the "extravillous trophoblast" in general, without any definition of the stage and of the line of differentiation, as this unfortunately still nowadays quite often is done. Where possible, one should clearly define location of the cells and stages of differentiation. For the cell columns and for the neighboring parts of the basal plate, these stages comprise (Figure 12):

(a) The polar, proliferating stem cells, which express α6β4 integrins and connexin 40, and secrete a basal lamina on which they are resting.

(b) The successive apolar, proliferating generations of cells, which due to absence of obvious ECM secretion, still form a compact aggregate, still expressing α6β4 integrins and being coupled by connexin 40-positive gap junctions.

(c) The postproliferative, early invasive phenotype of cells, usually coexpressing α6β4 and α5β1 integrins and expressing the first patches of apolar extracellular matrix, which separates the cells.

(d) Later stages of invasion, which coexpress α5β1 and α1β1 integrins and are embedded in ample ECM.

(e) Final stages of invasion, which continue coexpressing α5β1 and α1β1 integrins and form clusters of cells coupled by connexin 40-positive gap junctions, or multinucleated elements which are positive for EGF- receptor.

(f) Those late invasive stages, which as a side road of invasion, enter the walls and lumina of uteroplacental vessels and migrate within the arterial lumina.

In spite of our assumption that all extravillous trophoblast of the placenta belongs to one population, we cannot exclude local phenotypic differences which are due to environmental factors. Because of this the following topographically defined subpopulations need to be kept in mind, each of which represents a complex system of cells belonging to the same pathway of differentiation and invasion. These subpopulations comprise (Figure 10):

(a) Proliferating trophoblast of cell columns together with interstitial trophoblast of the basal plate and final stages of invasion either into decidua and myometrium or into uteroplacental vessels.

(b) Extravillous trophoblast of placental septa (as folded parts of the basal plate) together with attached cell columns and enclosed uteroplacental vessels; this subpopulation is directly comparable to that described as group (a).

(c) Cell islands with proliferative trophoblast lining the basal laminas beneath the villous stroma, and with migrating stages within the centers of the islands, but without final invasive stages within maternal decidua and uteroplacental vessels.

(d) Chorionic plate with proliferative trophoblast lining the basal lamina of the chorionic mesenchyme, and migrating stages within the Langhans fibrinoid stria, but without final invasive stages within maternal decidua and uteroplacental vessels.

(e) Smooth chorion with proliferative trophoblast lining the basal lamina of the chorionic mesenchyme, and invasive stages within fibrinoid and capsular decidua.

The proliferating cells of all these extravillous subpopulations together with the villous Langhans cells, are very likely to represent a uniform group of stem cells, the future postproliferative fate of which, whether villous syncytiotrophoblast or extravillous trophoblast, depends on the local environment.

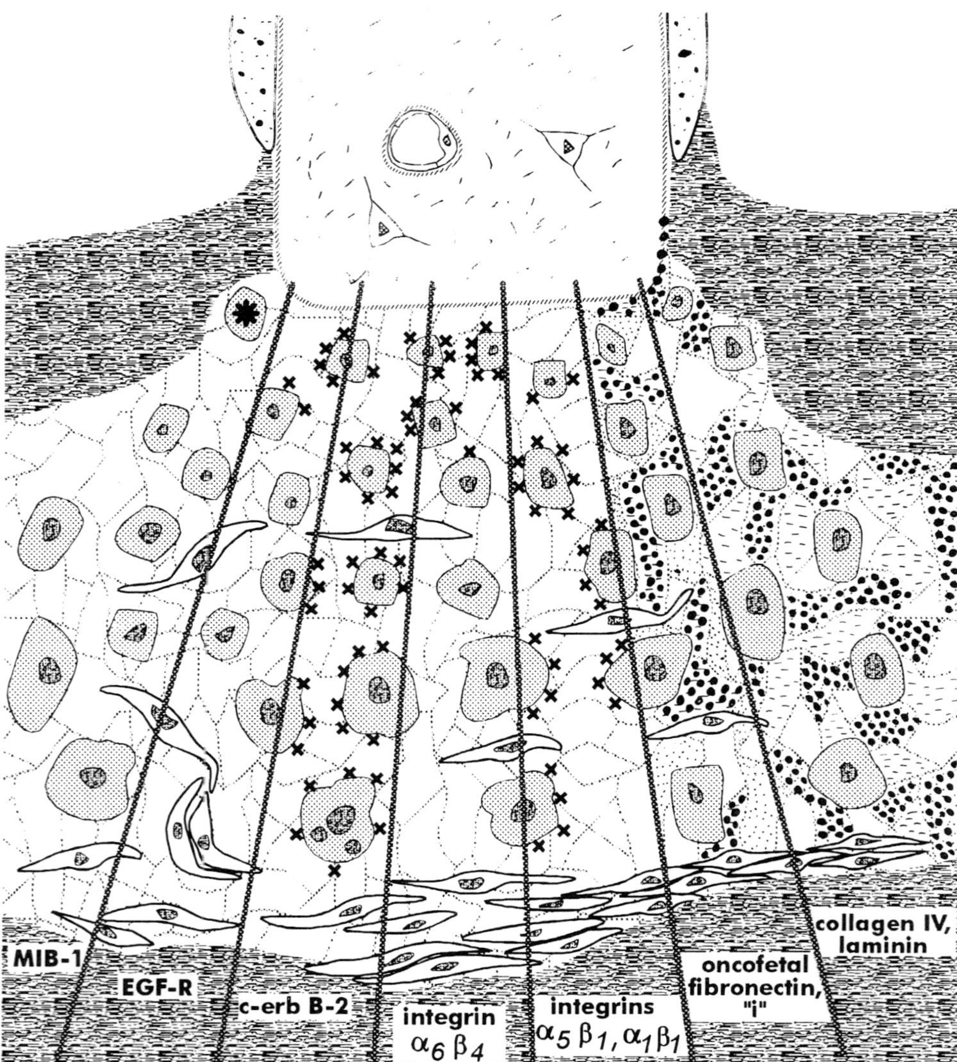

Figure 13. Scheme of a cell column of the human placenta at term. The structure and the expression patterns are different from earlier stages of pregnancy since the proliferation of the stem cells has largely ceased. Because of this, only few cells if any, still express proliferation markers (MIB-1), epidermal growth factor receptor (EGF-R), and integrin α6β4. Rather, most cells have already passed the integrin switch and belong to the invasive phenotype, expressing c-erbB-2 gene product and integrins α5β1 and α1β1. Note that even most of the proximal cells localized close to basal lamina, have already lost the polar pattern of matrix secretion and are surrounded by a thin pericellular sheet of matrix-type fibrinoid. The structural situation and the expression patterns in chorion laeve, chorionic plate and cell islands of term placentae are comparable.

SUMMARY

Extravillous trophoblast is the general term for all those trophoblastic elements residing outside the villi. It can be found in the smooth chorion, chorionic plate, cell islands, septa, cell columns, and basal plate. Focusing on the cell columns, the various stages of the "invasive pathway" from proliferating, polarized stem cells to non-proliferative, unpolarized invasive cells, is analyzed regarding the expression patterns of several antigenic determinants. These comprise growth factors and their receptors, cell adhesion molecules, connexins, extracellular matrix molecules, blood group antigens, proteinases, their activators and inhibitors, nitric oxide synthase and HLA-G. When comparing these expression patterns for various locations of extravillous trophoblast, no relevant differences between the proliferating stem cells of the extravillous trophoblast and the villous Langhans' cells were found. Therefore it is suggested that both kinds of stem cells represent a uniform stem cell population, the future fate of which, whether syncytial fusion or extravillous invasion, depends on the surrounding environment. Differences and similarities of the expression patterns in various locations of extravillous trophoblast are described, as far as data were available. They suggest that extravillous trophoblast in all locations generally behaves the same, but that due to special local factors, such as presence or absence of endometrial tissues, phenotypic differences between various local subpopulations may exist.

ACKNOWLEDGEMENTS

We gratefully acknowledge the artistic help of Wolfgang Graulich. The unpublished findings presented in this review are based on studies performed by Hans-Georg Frank, Berthold Huppertz and Sonya Kertschanska, Aachen, as well as by Caterina Crescimanno, Novara and Claudia Pierleoni, Ancona. It is a pleasure to thank these coworkers for contributing their findings, stimulating discussion and critical reading of the manuscript. Research was supported by grants No. Ka 360-7/2 and 3 from the Deutsche Forschungsgemeinschaft and No. 322-vigoni-dr from the VIGONI programme.

REFERENCES

Aplin, J.D. (1991) Loss of integrin alpha-6-beta-4 from extravillous trophoblast. *Placenta* 12, 366.

Aplin, J.D. (1993) Expression of integrin alpha 6 beta 4 in human trophoblast and its loss from extravillous cells. *Placenta* 14, 203-215.

Aplin, J. D. and Campbell, S. (1985) An immunofluorescence study of extracellular matrix associated with cytotrophoblast of the chorion laeve. *Placenta* 6, 469-479.

Astedt, B., Hagerstrand, I. and Lecander, I. (1986) Cellular localisation in placenta of placental type plasminogen activator inhibitor. *Thromb. Haemost.* 56, 63-65.

Autio-Harmainen, H., Hurskainen, T., Niskasaari, K., Hoyhtya, M. and Tryggvason, K. (1992) Simultaneous expression of 70 kilodalton type IV collagenase and type IV collagen alpha 1 (IV) chain genes by cells of early human placenta and gestational endometrium. *Lab. Invest.* 67, 191-200.

Babawale, M.O., Van Noorden, S., Stamp, G.W.H., Pignatelli, M., Elder, M. G. and Sullivan, M.H.F. (1995) In vitro invasion of first trimester human trophoblast: Localisation of E-cadherin and MMP-9. *Placenta* 16, A5.

Bartels, H. and Wang, T. (1983) Intercellular junctions in the human fetal membranes. A freeze-fracture study. *Anat. Embryol.* 166, 103-120.

Bass, K. E., Morrish, D. W., Roth, I., Bhardwaj, D., Taylor, R., Zhou, Y. and Fisher, S. J. (1994) Human cytotrophoblast invasion is up-regulated by epidermal growth factor: Evidence that paracrine factors modify this process. *Dev. Biol.* 164, 550-561.

Benirschke, K. and Kaufmann, P. (1995) *Pathology Of The Human Placenta.* New York: Springer. Ed. 3

Bischof, P., Friedli, E., Martelli, M. and Campana, A. (1991) Expression of extracellular matrix-degrading metalloproteinases by cultured human cytotrophoblast cells: effects of cell adhesion and immunopurification. *Am. J. Obstet. Gynecol.* 165, 1791-1801.

Bischof, P., Haenggeli, L. and Campana, A. (1995) Gelatinase and oncofetal fibronectin secretion is dependent on integrin expression on human cytotrophoblasts. *Hum. Reprod.* 10, 734-742.

Bischof, P. and Martelli, M. (1992) Proteolysis in the penetration phase of the implantation process. *Placenta* 13, 17-24.

Blankenship, T.N., Enders, A. C. and King, B.F. (1992) Distribution of laminin, type IV collagen, and fibronectin in the cell columns and trophoblastic shell of early macaque placentas. *Cell Tissue Res.* 270, 241-248.

Blankenship, T.N., Enders, A.C. and King, B.F. (1993a) Trophoblastic invasion and modification of uterine veins during placental development in macaques. *Cell Tissue Res.* 274, 135-144.

Blankenship, T.N., Enders, A.C. and King, B.F. (1993b) Trophoblastic invasion and the development of uteroplacental arteries in the macaque: Immunohistochemical localization of cytokeratins, desmin, type IV collagen, laminin, and fibronectin. *Cell Tissue Res.* 272, 227-236.

Blankenship, T.N. and King, B.F. (1993) Developmental changes in the cell columns and trophoblastic shell of the macaque placenta: an immunohistochemical study localizing type IV collagen, laminin, fibronectin and cytokeratins. *Cell Tissue Res.* 274, 457-466.

Blankenship, T.N. and King, B.F. (1994a) Developmental expression of Ki-67 antigen and proliferating cell nuclear antigen in macaque placentas. *Develop. Dynam.* 201, 324-333.

Blankenship, T.N. and King, B.F. (1994b) Identification of 72-Kilodalton Type-IV Collagenase at Sites of Trophoblastic Invasion of Macaque Spiral Arteries. *Placenta* 15, 177-187.

Borsi, L., Carnemolla, B., Castellani, P., Rosellini, C., Vecchio, D., Allemanni, G., Chang, S. E., Taylor-Papadimitriou, J., Pande, H. and Zardi, L. (1987) Monoclonal antibodies in the analysis of fibronectin isoforms generated by alternative splicing of mRNA precursors in normal and transformed human cells. *J. Cell Biol.* 104, 595-600.

Bou Resli, M.N., Al Zaid, N.S. and Ibrahim, M.E.A. (1981) Full-term and prematurely ruptured fetal membranes. An ultrastructural study. *Cell Tissue Res.* 220, 263-278.

Bourne, G.L. (1962) *The Human Amnion And Chorion.* London: Lloyd-Luke.

Bryant-Greenwood, G.D. and Yamamoto, S.Y. (1995) Control of peripartal collagenolysis in the human chorion-decidua. *Am. J. Obstet. Gynecol.* 172, 63-70.

Bulmer, J.N., Morrison, L. and Johnson, P.M. (1988) Expression of the proliferation markers Ki67 and transferrin receptor by human trophoblast populations. *J. Reprod. Immunol.* 14, 291-302

Burrows, T.D., King, A. and Loke, Y.W. (1993a) Expression of integrins by human trophoblast and differential adhesion to laminin or fibronectin. *Hum. Reprod.* 8, 475-484.

Burrows, T.D., King, A. and Loke, Y.W. (1993b) Expression of adhesion molecules by human decidual large granular lymphocytes. *Cell Immunol.* 147, 81-94.

Burrows, T.D., King, A. and Loke, Y. W. (1994) Expression of adhesion molecules by endovascular trophoblast and decidual endothelial cells - Implications for vascular invasion during implantation. *Placenta* 15, 21-33.

Burrows, T.D., King, A. and Loke, Y.W. (1995) Trophoblast - matrix interactions and integrin-mediated signal transduction in vitro. *Placenta* 16, A9.

Butterworth, B.H., Khong, T.Y., Loke, Y.W. and Robertson, W.B. (1985) Human cytotrophoblast populations studied by monoclonal antibodies using single and double biotin-avidin-peroxidase immunocytochemistry. *J. Histochem. Cytochem.* 33, 977-983.

Carnemolla, B., Balza, E., Siri, A., Zardi, L., Nicotra, M. R., Bigotti, A. and Natali, P. G. (1989) A tumor-associated fibronectin isoform generated by alternative splicing of messenger RNA precursors. *J. Cell Biol.* 108, 1139-1148.

Carnemolla, B., Borsi, L., Zardi, L., Owens, R. J. and Baralle, F.E. (1987) Localization of the cellular-fibronectin-specific epitope recognized by the monoclonal antibody IST-9 using fusion proteins expressed in E. coli. *FEBS Lett.* 215, 269-273.

Carnemolla, B., Leprini, A., Allemanni, G., Saginati, M. and Zardi, L. (1992) The inclusion of the type-III repeat ED-B in the fibronectin molecule generates conformational modifications that unmask a cryptic sequence. *J. Biol. Chem.* 267, 24689-24692.

Carraway, K.L. and Cantley, L.C. (1994) A new acquaintance for erbB3 and erbB4: A role for receptor heterodimerization in growth signaling. *Cell* 78, 5-8.

Castellucci, M., Classen-Linke, I., Mühlhauser, J., Kaufmann, P., Zardi, L. and Chiquet Ehrismann, R. (1991) The human placenta: a model for tenascin expression. *Histochemistry* 95, 449-458.

Castellucci, M., Crescimanno, C., Schröter, C. A., Kaufmann, P. and Mühlhauser, J. (1993) Extravillcus trophoblast: immunohistochemical localization of extracellular matrix molecules. In: *Frontiers in Gynecologic and Obstetric Investigation,* (eds.) A.R. Genazzani, F. Petraglia, and A.D. Genazzani, New York: Parthenon pp. 19-25.

Castellucci, M., Kaufmann, P. and Bischof, P. (1990) Extracellular matrix influences hormone and protein production by human chorionic villi. *Cell Tissue Res.* 262, 135-142.

Castellucci, M., Theelen, T., Pompili, E., Fumagalli, L., Derenzis, G. and Mühlhauser, J. (1994) Immunohistochemical localization of serine-protease inhibitors in the human placenta. *Cell Tissue Res.* 278, 283-289.

Chumbley, G., King, A., Gardner, L., Howlett, S., Holmes, N. and Loke, Y.W. (1994a) Generation of an antibody to HLA-G in transgenic mice and demonstration of the tissue reactivity of this antibody. *J. Reprod. Immunol.* 27, 173-186.

Chumbley, G., King, A., Robertson, K., Holmes, N. and Loke, Y. W. (1994b) Resistance of HLA-G and HLA-A2 transfectants to lysis by decidual NK cells. *Cell Immunol.* 155, 312-322.

Chwalisz, K. and Garfield, R. . (1994) Role of progesterone during pregnancy: models of parturition and preeclampsia. *Geburtsh. Perinatol.* 198, 170-180.

Colbern, G.T., Chiang, M.H. and Main, E.K. (1994) Expression of the nonclassic histocompatibility antigen HLA-G by preeclamptic placenta. *Am. J. Obstet. Gynecol.* 170, 1244-1250.

Cronier, L., Bastide, B., Traub, O., Herve, J.C., Deleze, J. and Malassine, A. (1994) Stimulation of connexin expression and gap junctional communication by hCG during human trophoblast differentiation. *Placenta* 15, A10.

Damsky, C.H., Fitzgerald, M.L. and Fisher, S.J. (1992) Distribution patterns of extracellular matrix components and adhesion receptors are intricately modulated during first trimester cytotrophoblast differentiation along the invasive pathway, in vivo. *J. Clin. Invest.* 89, 210-222.

Damsky, C.H., Sutherland, A. and Fisher, S. (1993) Extracelllar matrix 5: Adhesive interactions in early mammalian embryogenesis, implantation, and placentation. *FASEB J.* 7, 1320-1329.

Damsky, C.H., Librach, C., Lim, K.H., Fitzgerald, M.L., McMaster, M.T., Janatpour, M., Zhou, Y., Logan, S.K. and Fisher, S.J. (1994) Integrin switching regulates normal trophoblast invasion. *Development* 120, 3657-3666.

Defilippi, P., Silengo, L. and Tarone, G. (1992) Alpha 6 beta 1 integrin (laminin receptor) is down regulated by tumor necrosis factor alpha and interleukin 1 beta in human endothelial cells. *J. Biol. Chem.* 267, 18303-10307.

DeMesy Jensen, K., Genbacev, O., Penney, D., Maltby, K. and Miller, R.K. (1993) Human cytotrophoblast cell islands - morphological characteristics in vitro. *Placenta* 14, A14.

DiPersio, C.M., Shah, S. and Hynes, R.O. (1995) α3β1 integrin localizes to focal contacts in response to diverse extracellular matrix proteins. *J. Cell Sci.* 108, 2321-2336.

Divers, M.J., Bulmer, J.N., Miller, D. and Lilford, R.J. (1995) Beta 1 integrins in third trimester human placentae: No differential expression in pathological pregnancy *Placenta* 16, 245-260.

Duance, V.C. and Bailey, A.J. (1983) Structure of the trophoblast basement membrane. In *Biology of Trophoblast*, (ed.) Y.W. Loke and A. Whyte, Amsterdam: Elsevier, p. 597.

Duello, T.M., Bertics, P.J., Fulgham, D.L. and Vaness, P. . (1994) Localization of epidermal growth factor receptors in first- and third-trimester human placentas. *J. Histochem. Cytochem.* 42, 907-915.

Earl, U.M., Morrison, L., Gray, C. and Bulmer, J.N. (1989) Proteinase and proteinase inhibitor localization in the human placenta. *Int. J. Gynecol. Pathol.* 8, 114-124.

Eldar Geva, T., Rachmilewitz, J., De Groot, N. and Hochberg, A.A. (1993) Interaction between choriocarcinoma cell line (JAr) and human cytotrophoblasts in vitro. *Placenta* 14, 217-223.

Emonard, H.P., Christiane, Y., Smet, M., Grimaud, J.A. and Foidart, J.M. (1990) Type IV and interstitial collagenolytic activities in normal and malignant trophoblast cells are specifically regulated by the extracellular matrix. *Invasion Metastasis* 10, 170-177.

Faller, T.H. and Ferenci, P. (1973) Der Aufbau der Placenta-Septen. Untersuchungen mit Hilfe der Quinacrinfluorescenzfärbung des Y-Chromatins. *Z. Anat. Entwickl. Gesch.* 142, 207-217.

Feinberg, R.F., Kao, L.C., Haimowitz, J.E., Queenan, J.T.,Jr., Wun, T.C., Strauss, J.F. and Kliman, H.J. (1989) Plasminogen activator inhibitor types 1 and 2 in human trophoblasts. PAI-1 is an immunocytochemical marker of invading trophoblasts. *Lab. Invest.* 61, 20-26.

Feinberg, R.F. and Kliman, H.J. (1992) Fetal fibronectin and preterm labor [letter; comment]. *N. Engl. J. Med.* 326, 708.

Feinberg, R.F. and Kliman, H.J. (1993a) Human trophoblasts and tropho-uteronectin (TUN): a model for studying early implantation events. *REP* 3, 19-25.

Feinberg, R.F. and Kliman, H.J. (1993b) Tropho-uteronectin (TUN): A unique oncofetal fibronectin deposited in the extracellular matrix of the tropho-uterine junction and regulated in vitro by cultured human trophoblast cells. *Troph. Res.* 7, 167-181.

Feinberg, R.F., Kliman, H.J. and Lockwood, C.J. (1991) Is oncofetal fibronectin a trophoblast glue for human implantation? *Am. J. Pathol.* 138, 537-543.

Fernandez, P.L., Merino, M.J., Nogales, F.F., Charonis, A.S., Stetler Stevenson, W.G. and Liotta, L. (1992) Immunohistochemical Profile of Basement Membrane Proteins and 72 Kilodalton Type-IV Collagenase in the Implantation Placental Site - An Integrated View. *Lab. Invest.* 66, 572-579.

Fisher, S.J., Cui, T.Y., Zhang, L., Hartman, L., Grahl, K., Zhang, G.Y., Tarpey, J. and Damsky, C.H. (1989) Adhesive and degradative properties of human placental cytotrophoblast cells in vitro. *J. Cell Biol.* 109, 891-902.

Fisher, S.J. and Damsky, C.H. (1993) Human cytotrophoblast invasion. *Semin. Cell Biol.* 4, 183-188.

Fisher, S.J., Leitch, M.S., Kantor, M.S., Basbaum, C.B. and Kramer, R.H. (1985) Degradation of extracellular matrix by the trophoblastic cells of first-trimester human placentas. *J. Cell Biochem.* 27, 31-41.

Fisher, S.J., Librach, C.L., Fitzgerald, M.L. and Damsky, C.H. (1991) Human cytotrophoblast invasion is mediated by metalloproteinases and $\alpha 1$ integrins and is accompanied by changes in the expression of adhesion molecules and HLA-G. *Placenta* 12, 387-388.

Flug, M. and Köpf-Maier, P. (1995) The basement membrane and its involvement in carcinoma cell invasion. *Acta Anat.* 152, 69-84.

Frank, H.-G., Huppertz, B., Kertschanska, S., Blanchard, D., Roelcke, D. and Kaufmann, P. (1995) Anti-adhesive glycosylation of fibronectin-like molecules in human placental matrix-type fibrinoid. *Histochem. Cell Biol.* 104, 317-329.

Frank, H.G., Malekzadeh, F., Kertschanska, S., Crescimanno, C., Castellucci, M., Lang, I., Desoye, G. and Kaufmann, P. (1994) Immunohistochemistry of two different types of placental fibrinoid. *Acta Anat.* 150, 55-68.

Friedman, S.A., Degroot, C.J.M., Taylor, R.N. and Roberts, J.M. (1992) Circulating concentrations of fetal fibronectin do not reflect reduced trophoblastic invasion in preeclamptic pregnancies. *Am. J. Obstet. Gynecol.* 167, 496-497.

Garfield, R.E., Yallampalli, C., Buhimschi, I. and Chwalisz, K. (1994) Reversal of pre-eclampsia symptoms induced in rats by nitric oxide inhibition with L-arginine, steroid hormones and an endothelin antagonist. *Soc. Gynecol. Invest.*, Chicago (III) Abstract No. 384.

Gaus, G., Funayama, H., Huppertz, B., Kaufmann, P. and Frank, H.G. (1997) Parent cells for trophoblast hybridization I: Isolation of extravillous trophoblast cells from human term chorion laeve. *Tropho. Res.* 10, 181-190, this volume.

Genbacev, O., Joslin, R., Damsky, C.H., Polliotti, B.M. and Fisher, S.J. (1996) Hypoxia alters early gestation human trophoblast differentiation/invasion in vitro and models the placental defects that occur in preeclampsia. *J. Clin. Invest.* 97, 540-550.

Genbacev, O., DeMesy Jensen, K., Schubach-Powlin, S. and Miller, R.K. (1993) In vitro differentiation and ultrastructure of human extravillous trophoblast (EVT) cells. *Placenta* 14, 463-475.

Genbacev, O., Schubach, S.A. and Miller, R.K. (1992) Villous culture of first trimester human placenta--model to study extravillous trophoblast (EVT) differentiation. *Placenta* 13, 439-461.

Genbacev, O., Papic, N., Cuperlovic, M., Vicovac, L.J., Vuckovic, M. and Miller, R.K (1991) First trimester chorionic villous explants in culture as a model to study the origin and characteristics of extravillous cytotrophoblast. *Placenta* 12, 389.

Graham, C.H. and Lala, P.K. (1991) Mechanism of control of trophoblast invasion in situ. *J. Cell Physiol.* 148, 228-234.

Hamilton, G.S., Lysiak, J.J., Watson, A.J. and Lala, P.K. (1995) Colony-stimulating factor-1 provides an autocrine signal for first trimester extravillous trophoblast cell proliferation. *Placenta* 16, A24.

Hellmann, P., von Ostau, C., Grümmer, R. and Winterhager, E. (1995) Connexin40 expression in the human trophoblast: Implicator for proliferation and invasion properties. *Placenta* 16, A26.

Hofmann, G.E., Drews, M.R., Scott, R.T., Jr., Navot, D., Heller, D.S. and Deligdisch, L. (1992) Epidermal growth factor and its receptor in human implantation trophoblast: immunohistochemical evidence for autocrine/paracrine function. *J. Clin. Endocrinol. Metab.* 74, 981-988.

Hofmann, G.E., Glatstein, I., Schatz, F., Heller, D. and Deligdisch, L. (1994) Immunohistochemical localization of urokinase-type plasminogen activator and the plasminogen activator inhibitors 1 and 2 in early human implantation sites. *Am. J. Obstet. Gynecol.* 170, 671-676.

Holmes, W.E., Sliwkowski, M.X., Akita, R.W., Henzel, W.J., Lee, J., Park, J.W., Yansura, D., Abadi, N., Raab, H., Lewis, G.D., Shepard, H.M., Kuang, W.-J., Wood, W I., Goeddel, D.V. and Vandlen, R.L. (1992) Identification of heregulin, a specific activator of $p185^{erbB2}$. *Science* 256, 1205-1210.

Humphries, M.J (1990) The molecular basis and specificity of integrin-ligand interactions. *J. Cell Science* 97, 585-592.

Hunt, J.S., Fishback, J.L., Chumbley, G. and Loke, Y.W. (1990) Identification of class I MHC mRNA in human first trimester trophoblast cells by in situ hybridization. *J. Immuncl.* 144, 4420-4425.

Hunt, J.S., Hsi, B.L., King, C.R. and Fishback, J.L. (1991) Detection of class I MHC mRNA in subpopulations of first trimester cytotrophoblast cells by in situ hybridization. *J. Reprod. Immunol.* 19, 315-323.

Huppertz, B., Kertschanka, S., Demire, A., Frank, H.G. and Kaufmann, P. (1997) Immunohistochemistry of matrix metalloproteinases (MMP), their substrates and their inhibitors (TIMP) during trophoblast invasion in the human placenta. *Cell Tiss. Res.*, in press.

Huppertz, B., Kertschanska, S., Frank. H.G., Gaus, G., Funayama, H. and Kaufmann, P. (1996) Extracellular matrix components of placental extravillous trophoblast: Immunocytochemistry and ultrastructural distribution. *Histochem. Cell Biol.* 106, 291-301.

Huppertz, B., Kertschanska, S., Gaus, G., Frank, H.-G. and Kaufmann, P. (1995) Fibronectin-like immunoreactivities in the matrix-type fibrinoid of the human placenta. *Placenta* 16, A30

Irving, J.A. and Lala, P.K. (1994) Decidua-derived IGFBP-1 stimulates human intermediate trophoblast migration by binding to the alpha5/beta1 integrin subunits. *Placenta* 15, A31.

Irving, J.A. and Lala, P.K. (1995) Functional role of cell surface integrins on human trophoblast cell migration: Regulation by TGF-beta, IGF-II, and IGFBP-1. *Exp. Cell Res.* 217, 419-427.

Irving, J.A., Lysiak, J.J., Graham, C.H., Hearn, S., Han, V.K.M. and Lala, P.K. (1995) Characteristics of trophoblast cells migrating from first trimester chorionic villous explants and propagated in culture. *Placenta* 16, 413-433.

Jokhi, P.P., Chumbley, G., King, A., Gardner, L. and Loke, Y.W. (1993) Expression of the colony stimulating factor-1 receptor (c-fms product) by cells at the human uteroplacental interface. *Lab. Invest.* 68, 308-320.

Jokhi, P.P., King, A. and Loke, Y.W. (1994) Reciprocal expression of epidermal growth factor receptor (EGF-R) and c-erbB2 by non-invasive and invasive human trophoblast populations. *Cytokine.* 6, 433-442.

Kertschanska, S., Frank, H.-G., Huppertz, B., Funayama, H. and Kaufmann, P. (1995) Immunocytochemistry of matrix-type fibrinoid in the human placenta. *Placenta* 16, A34.

Khudr, G., Soma, H. and Benirschke, K. (1973) Trophoblastic origin of the x cell and the placental site giant cell. *Am. J. Obstet. Gynecol.* 115, 530-533.

King, A., Kalra, P. and Loke, Y.W. (1989) Human trophoblast resistance to decidual NK lysis is due to lack of NK cell target. *Placenta* 10, 492.

King, A., Kalra, P. and Loke, Y.W. (1990) Human trophoblast cell resistance to decidual NK lysis is due to lack of NK target structure. *Cell Immunol.* 127, 230-237.

King, A. and Loke, Y.W. (1988) Differential expression of blood-group-related carbohydrate antigens by trophoblast subpopulations. *Placenta* 9, 513-521.

King, B.F. and Blankenship, T.N. (1993) Expression of proliferating cell nuclear antigen (PCNA) in developing macaque placentas. *Placenta* 14, A36.

King, B.F. and Blankenship, T.N. (1994a) Differentiation of the chorionic plate of the placenta: Cellular and extracellular matrix changes during development in the macaque. *Anat. Rec.* 240, 267-276.

King, B.F. and Blankenship, T.N. (1994b) Ultrastructure and development of a thick basement membrane-like layer in the anchoring villi of macaque placentas. *Anat. Rec.* 238, 498-506.

King, B.F. and Blankenship, T.N. (1995) Neural cell adhesion molecule is present on macaque intra-arterial cytotrophoblast. *Placenta* 16, A36.

Kohnen, G., Kosanke, G., Korr, H. and Kaufmann, P. (1993) Comparison of various proliferation markers applied to human placental tissue. *Placenta* 14, A38.

Kosanke, G. (1994) Proliferation, Wachstum und Differenzierung der Zottenbäume der menschlichen Placenta. Aachen, FRG: Verlag Shaker.

Krönicher, W.D. (1975) Ein Beitrag zur Genese der Inseln in der menschlichen Placenta. *Z. Mikrosk. Anat. Forsch.* 89, 777-803.

Kupferminc, M.J., Peaceman, A.M., Wigton, T.R., Rehnberg, K.A. and Socol, M.L. (1995) Fetal fibronectin levels are elevated in maternal plasma and amniotic fluid of patients with severe preeclampsia. *Am. J. Obstet. Gynecol.* 172, 649-653.

Kurman, R.J., Main, C.S. and Chen, H. C. (1984a) Intermediate trophoblast: A distinctive form of trophoblast with specific morphological, biochemical and functional features. *Placenta* 5, 349-370.

Kurman, R.J., Young, R.H., Norris, H.J., Main, C.S., Lawrence, W.D. and Scully, R.E. (1984b) Immunocytochemical localization of placental lactogen and chorionic gonadotropin in the normal placenta and trophoblastic tumors, with emphasis on intermediate trophoblast and the placental site trophoblastic tumor. *Int. J. Gynecol. Pathol.* 3, 101-121.

Lang, I., Hartmann, M., Blaschitz, A., Dohr, G. A., Kaufmann, P., Frank, H.-G., Hahn, T., Skofitsch, G. and Desoye, G. (1994) Differential lectin binding to the fibrinoid of human full- term placenta: Correlation with a fibrin antibody and the PAF-Halmi method. *Acta Anat.* 150, 170-177.

Lata, J.A., Tuan, R.S., Shepley, K.J., Mulligan, M.M., Jackson, L.G. and Smith, J.B. (1992) Localization of major histocompatibility complex class I and II mRNA in human first-trimester chorionic villi by in situ hybridization. *J. Exp. Med.* 175, 1027-1032.

Lewis, M.P., Clements, M., Takeda, S., Kirby, P.L., Seki, H., Lonsdale, L.B., Sullivan, M.H.F., Elder, M.G. and White, J.O. (1996) Partial characterization of an immortalized human trophoblast cell-line, TCL-1, which possesses a CSF-1 autocrine loop. *Placenta* 17, 137-146.

Librach, C L., Fisher, S.J., Fitzgerald, M.L. and Damsky, C.H. (1991a) Cytotrophoblast-fibronectin and cytotrophoblast-laminin interactions have distinct roles in cytotrophoblast invasion. *J. Cell Biol.* 115, 6a.

Librach, C.L., Werb, Z., Fitzgerald, M. L., Chiu, K., Corwin, N.M., Esteves, R.A., Grobelny, D., Galardy, R., Damsky, C.H. and Fisher, S.J. (1991b) 92-kD type IV collagenase mediates invasion of human cytotrophoblasts. *J. Cell Biol.* 113, 437-449.

Lim, K. H., Bass, K.E., Kosten, K., Damsky, C.H. and Fisher, S.J. (1993) Developmental delay of cytotrophoblast differentiation along the invasive pathway. *Am. J. Obstet. Gynecol.* S166.

Lim, K.H., Damsky, C.H. and Fisher, S.J. (1995) Basic fibroblast growth factor and heparin stimulate integrin-alpha-1 expression by cytotrophoblasts. *J. Soc. Gynecol. Invest.* 2, P287.

Lister, U.M. (1968) Ultrastructure of the human amnion, chorion and fetal skin. *J. Obstet. Gynaecol. Br. Cmwlth.* 75, 327-341.

Loke, Y.W. (1989) Trophoblast antigen expression. *Curr. Opin. Immunol.* 1, 1131-1134.

Loke, Y.W. and King, A. (1991) Recent developments in the human maternal-fetal immune interaction. *Curr. Opin. Immunol.* 3, 762-766.

MacCalman, C.D., Omigbodun, A., Bronner, M.P. and Strauss, J.F. (1995) Identification of the cadherins present in the human placenta. *J. Soc. Gynecol. Invest.* 2, 146.

Maidman, J.E., Thorpe, L.W., Harris, J.A. and Wynn, R.M. (1973) Fetal origin of x-cells in human placental septa and basal plate. *Obstet. Gynecol.* 41, 547-552.

Malak, T.M. and Bell, S.C. (1994) Differential expression of the integrin subunits in human fetal membranes. *J. Reprod. Fertil.* 102, 269-276.

Malak, T.M., Ockleford, C.D., Bell, S.C., Dalgleish, R., Bright, N.A. and MacVicar, J. (1993) Confocal immunofluorescence localization of collagen type- I, type-III, type-IV, type-V and type-VI and their ultrastructural organization in term human fetal membranes. *Placenta* 14, 385-406.

Maquoi, E., Polette, M., Nawrocki, B., Bischof, P. Noel, A., Pintiaux, A., Birembaut, P., Basset, P. and Foidart, J.M. (1995) Expression of stromelysin-3 in the human implantation placental site. *Placenta* 16, A 46.

Matsuura, H. and Hakomori, S. (1985) The oncofetal domain of fibronectin defined by monoclonal antibody FDC-6: its presence in fibronectins from fetal and tumor tissues and its absence in those from normal adult tissues and plasma. *Proc. Natl. Acad. Sci. USA* 82, 6517-6521.

McMaster, M.T., Librach, C.L., Zhou, Y., Lim, K.H., Janatpour, M.J., DeMars, R., Kovats, S., Damsky, C.H. and Fisher, S.J. (1995) Human placental HLA-G expression is restricted to differentiated cytotrophoblasts. *J. Immunol.* 154, 3771-3778.

Miosge, N., Günther, E., Heyder, E., Manshausen, B., Herken, R. (1995) Light and electron microscopic localization of α1-chain and the E1 and E8 domains of Laminin-1 in mouse kidney using monoclonal antibodies to establish the orientation of laminin-1 within basement membranes. *J. Histochem. Cytochem.* 43, 675-680.

Moll, U.M. and Lane, B.L. (1990) Proteolytic activity of first trimester human placenta: Localization of interstitial collagenase in villous and extravillous trophoblast. *Histochemistry* 94, 555-560.

Morris, N.H., Eaton, B.M., Sooranna, S.R. and Steer, P.J. (1993) NO synthase activity in placental bed and tissues from normotensive pregnant women. *Lancet* 342, 679-680.

Mosher, D.F., Sottile, J., Wu, C. and McDonald, J.A. (1992) Assembly of extracellular matrix. *Curr. Opin. Cell Biol.* 4, 810-818.

Mühlhauser, J., Crescimanno, C., Kaufmann, P., Höfler, H., Zaccheo, D. and Castellucci, M. (1993) Differentiation and proliferation patterns in human trophoblast revealed by c-erbB-2 oncogene product and EGF-R. *J. Histochem. Cytochem.* 41, 165-173.

Multhaupt, H.A.B., Mazar, A., Cines, D.B., Warhol, M.J. and McCrae, K.R. (1994) Expression of urokinase receptors by human trophoblast. A histochemical and ultrastructural analysis. *Lab. Invest.* 71, 392-400.

Nanaev, A.K., Milovanov, A.P. and Domogatsky, S.P. (1993) Immunohistochemical localization of extracellular matrix in perivillous fibrinoid of normal human term placenta. *Histochemistry* 100, 341-346.

Nanaev, A.K., Rukosuev, V.S., Shirinsky, V.P., Milovanov, A.P., Domogatsky, S.P., Duance, V.C., Bradbury, F.M., Yarrow, P., Gardiner, L., d'Lacey, C. and Ockleford, C.D. (1991) Confocal and conventional immunofluorescent and immunogold electron microscopic localization of collagen types III and IV in human placenta. *Placenta* 12, 573-595.

Nanaev, A., Chwalisz, K., Frank, H.-G., Kohnen, G., Hegele-Hartung, C. and Kaufmann, P. (1995) Physiological dilation of uteroplacental arteries in the guinea pig depends on nitric oxide synthase activity of extravillous trophoblast. *Cell Tissue Res.* 282, 407-421.

Nawrocki, B., Polette, M., Marchand, V., Maquoi, E., Beorchia, A., Tournier, J.M., Foidart, J.M. and Birenbaut, P. (1996) Membrane-type matrix metalloproteinase-1 expression at the site of human placentation. *Placenta* 17, 565-572.

Okudaira, Y., Matsui, Y. and Kanoh, H. (1991) Morphological variability of human trophoblasts in normal and neoplastic conditions. An ultrastructural reappraisal. In: *Placenta: Basic Research and Clinical Application,* (ed.) H. Soma, Basel: Karger, pp. 176-187.

Osawa, H., Iida, S., Tomioka, Y. and Hirano, M. (1995) The effects of NOS inhibitor on pregnancy. *Placenta* 16, A54.

Peles, E., Bacus, S.S., Koski, R.A., Lu, H.S., Wen, D., Ogden, S.G., Ben-Levy, R. and Yarden, Y (1992) Isolation of the neu/Her-2 stimulatory ligand: a 44 kd glycoprotein that induces differentiation of mammary tumor cells. *Cell* 69, 205-216.

Pesciotta-Peters, D.M., Portz, L. M., Fullenwider, J. and Mosher, D.F. (1990) Co-assembly of plasma and cellular fibronectins into fibrils in human fibroblast culture. *J. Cell Biol.* 111, 249-256.

Pijnenborg, R., Dixon, G., Robertson, W.B. and Brosens, I.A. (1980) Trophoblastic invasion of human decidua from 8 to 18 weeks of pregnancy. *Placenta* 1, 3-19.

Pizarro, A., Gamallo, C., Benito, N., Palacios, J., Quintanilla, M., Cano, A. and Contreras, F. (1995) Differential patterns of placental and epithelial cadherin expression in basal cell carcinoma and in the epidermis overlying tumours. *Br. J. Cancer* 72, 327-332.

Polette, M., Nawrocki, B., Pintiaux, A., Massenat, C., Maquoi, E., Volders, L., Schaaps, J. P., Birembaut, P. and Foidart, J.M. (1994) Expression of gelatinases A and B and their tissue inhibitors by cells of early and term human placenta and gestational endometrium. *Lab. Invest.* 71, 838-846.

Ramsey, E.M., Houston, M.L. and Harris, J.W.S. (1976) Interactions of the trophoblast and maternal tissues in three closely related primate species. *Am. J. Obstet. Gynecol.* 124, 647-652.

Ruck, P., Marzusch, K., Horny, H.P., Dietl, J. and Kaiserling, E. (1996) The distribution of tissue inhibitor of metalloproteinases-2 (TIMP-2) in the human placenta. *Placenta* 17, 263-266.

Ruck, P., Marzusch, K., Horny, H.-P., Dietl, J. and Kaiserling E (1995) Distribution of tissue inhibitor of metalloproteinases-2 (TIMP-2) in trophoblast of early human pregnancy. *Placenta* 16, A62.

Rukosuev, V.S. (1992) Immunofluorescent localization of collagen types I, III, IV, V, fibronectin, laminin, entactin, and heparan sulphate proteoglycan in human immature placenta. *Experientia* 48, 285-287.

Ruoslahti, E. (1991) Integrins as receptors for extracellular matrix. In: *Cell Biology of Extracellular Matrix*, (ed.) E.D. Hay, New York: Plenum Press, Ed 2, pp 343-363.

Sakbun, V., Ali, S.M., Greenwood, F.C. and Bryant Greenwood, G.D. (1990a) Human relaxin in the amnion, chorion, decidua parietalis, basal plate, and placental trophoblast by immunocytochemistry and northern analysis. *J. Clin. Endocrinol. Metab.* 70, 508-514.

Sakbun, V., Ali, S. M., Lee, Y.A., Jara, C.S. and Bryant Greenwood, G.D. (1990b) Immunocytochemical localization and messenger ribonucleic acid concentrations for human placental lactogen in amnion, chorion, decidua, and placenta. *Am. J. Obstet. Gynecol.* 162, 1310-1317.

Schmidt, C. M. and Orr, H.T. (1993) Maternal/Fetal Interactions - The Role of the MHC Class I Molecule HLA-G. *Crit. Rev. Immunol.* 13, 207-224.

Scipiades, E. and Burg, E. (1930) Über die Morphologie der menschlichen Plazenta mit besonderer Rücksicht auf unsere eigenen Studien. *Arch. Gynecol.* 141, 577-619.

Seftor, R.E.B., Seftor, E.A., Stetler Stevenson, W.G. and Hendrix, M.J C. (1993) The 72 kDa Type-IV collagenase is modulated via differential expression of alpha-v-beta-3-integrin and alpha-5-beta-1-integrin during human melanoma cell invasion. *Cancer Res.* 53, 3411-3415.

Sharkey, A.M., Jokhi, P.P., King, A., Loke, Y.W., Brown, K.D. and Smith, S.K. (1994) Expression of c-kit and kit ligand at the human maternofetal interface. *Cytokine* 6, 195-205.

Shorter, S.C., Clover, L.M. and Starkey, P.M. (1992a) Evidence for both an autocrine and paracrine role for the colony-stimulating factors in regulating placental growth and development. *Placenta* 13, A58.

Shorter, S.C., Starkey, P.M., Ferry, L.M., Clover, L.M., Sargent, I.L. and Redman, C.W.G. (1993) Antigenic heterogeneity of human cytotrophoblast and evidence for the transient expression of MHC class I antigens distinct from HLA-G. *Placenta* 14, 571-582.

Shorter, S.C., Vince, G.S. and Starkey, P.M. (1992b) Production of granulocyte colony-stimulating factor at the materno-foetal interface in human pregnancy. *Immunology* 75, 468-474.

Springer, T.A. (1990) Adhesion receptors of the immune system. *Nature* 346, 425-434.

Steininger, H. (1978) Über die Herkunft von Septen und Inseln der menschlichen Plazenta. *Arch. Gynecol.* 226, 261-275.

Takada, A., Ohmori, K., Yoneda, T., Tsuyuoka, K., Hasegawa, A., Kiso, M. and Kannagi, R. (1993) Contribution of carbohydrate antigens sialyl Lewis^a and sialyl Lewis^x to adhesion of human cancer cells to vascular endothelium. *Cancer Res.* 53, 354-361.

Vettraino, I.M., Roby, J., Tolley, T. and Parks, W.C. (1996) Collagenase-1, stromelysin-1, and matrilysin are expressed within the placenta during multiple stages of human pregnancy. *Placenta* 17, 557-563.

Vicovac, L.J., Jones, C.J.P. and Aplin, J.D. (1993) Morphogenesis of human placental anchoring villi in culture. *Placenta* 14, A80.

Vicovac, L.J., Jones, C.J.P. and Aplin, J.D. (1995a) Trophoblast differentiation during formation of anchoring villi in a model of the early human placenta in vitro. *Placenta* 16, 41-56.

Vicovac, L.J., Jones, C.J.P., Church, H.J. and Aplin, J.D. (1995b) Factors controlling trophoblast entry into the extravillous lineage. *Placenta* 16, A74.

Virtanen, I., Laitinen, L. and Vartio, T. (1988) Differential expression of the extra domain-containing form of cellular fibronectin in human placentas at different stages of maturation. *Histochemistry* 90, 25-30.

Von Ostau, C., Grümmer, R., Kohnen, G., Kaufmann, P. and Winterhager, E. (1995) Expression verschiedener Connexine während der entwicklung der humanen Plazenta. *Ann. Anatomy* 2, A102/103.

Wang, T. and Schneider, J. (1987) Cellular junctions on the free surface of human placental syncytium. *Arch. Gynecol.* 240, 211-216.

Weser, H. and Kaufmann, P. (1978) Lichtmikroskopische und histochemische Untersuchungen an der Chorionplatte der reifen menschlichen Placenta. *Arch. Gynecol.* 225, 15-30.

Wiese, K.H. (1975) Licht- und elektronenmikroskopische untersuchungen an der Chorionplatte der reifen menschlichen Plazenta. *Arch. Gynecol.* 218, 243-259.

Winterhager, E., von Ostau, C., Grümmer, R., Kaufmann, P. and Fisher, S.J. (1996) Connexin und E-cadherin Expression während der Differenzierung des humanen Trophoblasten. 91. Versammlung der Anat. Gesellschaft, Jena, März 1996

Wu, C., Chung, A.E. and McDonald, J.A. (1995) A novel role for α3β1 integrins in extracellular matrix assembly. *J. Cell Sci.* 108, 2511-2523.

Yallampalli, C. and Garfield, R.E. (1993) Inhibition of nitric oxide synthesis in rats during pregnancy produces signs similar to those of preeclampsia. *Am. J. Obstet. Gynecol.* 169, 1316-1320.

Zhou, Y., Damsky, C.H., Chiu, K., Roberts, J.M. and Fisher, S.J. (1993) Preeclampsia is associated with abnormal expression of adhesion molecules by invasive cytotrophoblasts. *J. Clin. Invest.* 91, 950-960.

Zhu, B.C.R., Fisher, S.F., Pande, H., Calaycay, J., Shively, J.E. and Laine, R.A. (1984) Human placental (fetal) fibronectin: increased glycosylation and higher protease resistance than plasma fibronectin. *J. Biol. Chem.* 259, 3962-3970.

Zhu, B.C.R. and Laine, R.A. (1985) Polylactosamine glycosylation on human fetal placental fibronectin weakens the binding affinity of fibronectin to gelatin. *J. Biol. Chem.* 260, 4041-4045.

Trophoblast Research 10:67-82, 1997

IN VITRO MODELS USED TO STUDY IMPLANTATION, TROPHOBLAST INVASION AND PLACENTATION

-A Review-

Paul Bischof

Department of Obstetrics and Gynaecology
Laboratoire d'Hormonologie Maternité
University of Geneva
1211 Genève 14, Switzerland

Implantation In The Human Is Unique

Despite the fact that some examples of viviparity exist in invertebrates, fish, amphibians and reptiles, implantation is a relatively new acquisition in evolution. The establishment of an intimate trophic connection between mother and embryo is a characteristic of mammals. Implantation is a new strategy in reproduction which allows the development of a small number of embryos in the protective maternal organism. This became possible with the establishment of a functional uterus. Although viviparity is an evolutionary advantage, it has one important limitation: the absolute necessity of a synchronization between embryonic and uterine development. The uterus of almost all mammals is ready for implantation only during a limited period of time (implantation window). Before or after that period the uterus is either indifferent or hostile to the embryo. The period during which an embryo is capable of implanting is usually much longer but varies considerably from species to species.

Implantation can be *superficial* or *interstitial* according to the extent of blastocyst penetration into the endometrium. Implantation is said to be *lateral, mesometrial,* or *antimesometrial* according to the orientation of the embryo in the uterine cavity. In humans, implantation is interstitial and antimesometrial: the embryo implants deeply in the body of the uterus, most frequently in the upper part of the posterior wall near the mid sagittal plane (Boyd and Hamilton, 1970). In all mammals, the implantation process can be divided in two phases: 1) an *attachment* phase and 2) a *penetration* phase.

The attachment phase consists of 2 steps: *apposition* and *adhesion*. During apposition, no visible connections are established between the blastocyst and the endometrium, and the blastocyst can be dislodged by simple washing of the uterine cavity. Adhesion is the step during which functional connections are established although the nature of these connections is still very speculative. The attachment phase is common to almost all mammals with noticeable differences: in rhesus monkeys attachment occurs simultaneously at two opposite poles of the blastocyst (Enders et al., 1983).

The penetration phase proceeds along three general lines with minor modifications. In *intrusive* penetration as in the ferret, the guinea pig, or the rhesus monkey penetration occurs by thin folds of trophoblast (called invadopodia) progressing between adjacent epithelial cells. In *displacement* penetration as in the rat and mouse, the trophoblast phagocytes dead epithelial cells and progresses to the basal lamina. Epithelial cells die by autolysis probably induced by embryonic factors (Weitlauf, 1988; Schlafke and Enders, 1975). In *fusion* penetration as in the rabbit, the abembryonic trophoblast fuses with the apical membrane of uterine epithelial cells. Cytoplasmic confluence converts the epithelial cells to a syncytium which proliferates and invades the basal lamina.

Although human blastocyst cells resulting from IVF programs have been examined by electron microscopy, neither the stage of adhesion to the uterine surface nor the penetration phase have been observed in the human. Consequently the nature of these crucial events are deduced from information gathered in other primates or from the very few observations *in vitro*. Implantation in the human seems to be intrusive (Enders and Schlafke, 1972). *In vitro*, trophoblast cells penetrate an endometrial epithelial cell monolayer by forming long slender ectoplasmic protrusions which insinuate themselves between the endometrial cells and disrupt their desmosomes (Lindenberg et al., 1992). Although it is clear from observations of early implantation sites in the human uterus (Hertig et al., 1956) that the syncytiotrophoblast is the invasive embryonic tissue once the embryo is embedded, it is still unclear if syncytium formation occurs as a consequence of the first contact between the trophectoderm and the uterine lining (as in rhesus monkeys, Enders, 1991) or if it is formed later when the trophectodermal cells invade the endometrial interstitium. This of course has important implications in the choice of a suitable model to mimic trophoblast invasion at implantation (see below).

Implantation in the human is unique (Lindner, 1972). Mice, rats, or rabbits are not suited to study human implantation (see above). Guinea pigs have an intrusive type of implantation but syncytiotrophoblast forms before hatching. Even rhesus monkeys are different (superficial implantation, limited decidualization, dual insertion at opposite poles, etc...). This uniqueness is perhaps best illustrated by the fact that extrauterine pregnancies are not uncommon in humans whereas they are almost unknown in other mammals. There is thus something very peculiar about the human trophoblast which drastically limits the use of animal models to study implantation. One is thus left with developing *in vitro* systems with human trophoblastic cells.

The Steps Of The Implantation Process

As said above, implantation can be divided in two steps: attachment and penetration. From a biological point of view, these two steps are completely different. Studying attachment implies the identification of molecules which will allow two different epithelial cells (the trophectodermal cells of the blastocyst and the endometrial epithelial cells) to make contact through their apical membranes. Studying penetration implies to understand the mechanisms which allow a cell (or a group of cells) to invade a neighboring tissue. Thus, in the first case one will have to study cell-cell adhesion molecules whereas in the second case one will study the expression of proteolytic enzymes which allow a cell to digest the extracellular matrix in which it is embedded.

Attachment of two epithelial cells through their apical membrane domain is a biological paradox (Denker, 1993). Epithelia are formed by polarized cells and the trophectoderm of the blastocyst as well as the uterine lining are no exceptions. Polarized cells have two membrane domains: *a baso-lateral domain* which anchors the cells to each other (through desmosomes and other specialized areas) and to the basement membrane (through integrins), and an *apical domain* which exhibits microvilli but which is usually devoid of adhesion molecules. Thus understanding attachment of trophectodermal cells to uterine epithelial cells through their respective apical domains is a challenging new field of investigations. It must be noticed that since blastocyst attachment is a common step in the implantation process of most mammals (Schlafke and Enders, 1975) this process can probably be safely studied with animal cells. The only limitation being the use of polarized cultures of uterine epithelial cells in contrast to monolayer cultures (Glasser et al., 1991). It has been known for some time that a decrease in the thickness of the uterine epithelial glycocalyx precedes the time of implantation in rodents (Enders and Schlafke, 1974; Chavez and Anderson, 1985). Lectin binding studies indicate that modulation of cell surface glycoconjugates of both uterine epithelium and trophectoderm accompanies these changes (Wu and Chang, 1978; Chavez and Enders, 1981; Chavez, 1986). The appearance of new glycoproteins on the apical surface of the uterine luminal epithelium during the period of receptivity has been demonstrated biochemically in the rabbit (Anderson et al., 1986). Lindenberg and co-workers (1988) tested seven oligosaccharides for their effect in an *in vitro* model of mouse blastocyst attachment and trophoblast outgrowth on endometrial epithelial monolayers. Only one compound, lacto-N-fucopentaose 1 (LNF-1) significantly inhibited attachment and outgrowth. It was later shown that mouse blastocyst carry LNF-1 receptors (Lindenberg et al., 1990; Yamagata and Yamazaki, 1991). LNF-1 is specifically expressed on the mouse uterine epithelial lining during the first four days of pregnancy and expression can be induced in castrated animals by estrogen treatment (Kimber and Lindenberg, 1990). LNF-1 is a pentasaccharide which contains the H-type-1 epitope. This epitope is a fucosylated blood group antigen in humans but a tissue antigen in rodents. Alpha (1-2)-fucosyltransferase is the enzyme that catalyzes the final step in the formation of H-type-1 epitope. Since estrogen stimulate and progesterone inhibit this enzyme in the uterine luminal epithelium it is believed that this fucosyltransferase is the rate limiting factor of the hormonal control of blastocyst attachment in mice (White and Kimber, 1994). There is probably more then only one glycoprotein involved in blastocyst attachment in mice (Fenderson et al., 1990) but clearly estrogen induced carbohydrates play a crucial role in this important step. In rodents, implantation is notoriously dependent on estrogen. This is however not the case in primates (Ghosh et al., 1994). Nevertheless, oligosaccharides and particularly glycosaminoglycans have been implicated in the attachment of the human blastocyst to the endometrium (Rohde and Carson., 1993). Perlecan is such a glycosaminoglycan appearing on the surface of mouse blastocyst at the time of attachment (Carson et al., 1993). Since perlecan can bind to the integrin $\alpha v\beta 3$ (Hayashi et al., 1992) and since this integrin is specifically expressed on human uterine epithelium only at the time of implantation (Lessey et al., 1992), it is tempting to speculate that these molecules are implicated in the attachment of the human blastocyst (Damsky et al., 1993). Recently, more direct evidences have implicated E-cadherin (an epithelial cell adhesion molecule) and integrin $\alpha 6\beta 1$ in human trophoblast-epithelial cell adhesiveness (Thie et al., 1995).

Penetration (or invasion) of trophoblastic cells into the endometrium is not due to passive growth pressure but to an active biochemical process. This has been the subject of several recent reviews (Bischof and Martelli., 1992; Fisher and Damsky, 1993; Cross et al., 1994; Tabibzadeh and Babaknia, 1995; Loke and King, 1995; Bischof and Campana, 1996) and will only be briefly summarized here. The invasive human trophoblastic cells secrete an array of proteolytic enzymes among which certain metalloproteinases (MMP). These enzymes are the only ones capable of digesting the different constituents (several collagen types, laminin, fibronectin) of the endometrial basement membrane and extracellular matrix and are thus considered as the rate limiting steps in trophoblastic invasion. MMP form a family of nine homologous enzymes classified according to their substrate specificity into *gelatinases, collagenases,* and *stomelysins.* These enzymes are secreted as inactive proenzymes which are secondarily activated in a cascade-like process. The activity of the activated enzymes is controlled by local inhibitors called tissue inhibitors of metalloproteinases (TIMP). Although the human trophoblast secretes at least four of these MMP, only gelatinase B (MMP-9) has been shown *in vitro* to be implicated in trophoblast invasion (Librach et al., 1991; Bischof et al., 1995a). The regulation of this gelatinolytic activity and hence of the invasive behavior of trophoblast is under study in different laboratories. Recent *in vitro* data suggest that it is under the control of endometrial paracrine factors such as epidermal growth factor (EGF, Bass et al., 1994), leukemia inhibitory factor (LIF, Bischof et al., 1995b) or transforming growth factor β (TGFβ, Graham and Lala, 1991) among others.

Which Type Of Cell Should Be Used?

Depending on the implantation step to be studied several types of *in vitro* systems can be used. They all require trophoblastic and/or endometrial cells.

Trophoblastic cells can be choriocarcinoma cell lines or can be primary cells obtained either by enzymatic disruption of placental tissue or by outgrowth from explant cultures of trophoblastic villi. Three different *choriocarcinoma cell lines* are routinely used in in-vitro studies: BeWo, JAr, and JEG 3. Both human malignant (BeWo) and normal trophoblast cells have an invasive potential. However while BeWo cells exhibit an unlimited invasive capacity, normal human trophoblastic cell invasion is both limited in space (endometrium and proximal myometrium) and in time (the first trimester of pregnancy). Curiously, the malignant trophoblastic cells have a very limited invasive potential as compared to normal trophoblastic cells when tested *in vitro* (Simon et al., 1992; Grummer et al., 1994; Bischof, unpublished observations). This seems to be due to their reduced capacity of secreting gelatinolytic enzymes (Fisher et al., 1989). Furthermore, the regulation of their gelatinase production is different from that of normal cytotrophoblast cells (Emonard et al., 1990). Although these trophoblast cell lines secrete human chorionic gonadotropin (hCG) they rarely fuse to form syncytiotrophoblast (Coutifaris et al., 1991) and in contrast to primary cytotrophoblast cells they actively divide when cultured *in vitro*. These malignant trophoblast cells seem thus inappropriate to study trophoblast invasion *in vitro*. *Primary trophoblast cells* can be obtained by several techniques from first, second, or third trimester placenta. While cytotrophoblastic cells (CTB) are present throughout gestation, it is essentially during the first trimester that they express their full invasive and degradative properties (Fisher et al., 1989). Third trimester CTB (which are more easily available) secrete lower amounts of gelatinases and have a reduced capacity to invade matrigel *in vitro* (Shimonovitz et al., 1994; Polette et al., 1994). For these reasons, first trimester CTB are certainly the cells to be chosen to study trophoblast invasion. It must be realized however that during the early steps of implantation in the human it is the syncytiotrophoblast (STB) which is the invasive cell and it is only once the primary placental villi are formed (about three weeks after ovulation) that the CTB become invasive. Thus by using primary cultures of first trimester CTB to study trophoblast invasion it will be very difficult to draw any conclusion on the mechanisms that govern syncytial invasion during implantation. Unfortunately, first trimester STB does not secrete gelatinases (Bischof et al., 1994) and obtention of invasive STB has never been described. Furthermore, not all CTB are invasive. Villous CTB are mononuclear epithelial cells attached to the villous basement membrane and separated from the intervillous blood by a layer of syncytium. These cells are considered as stem cells which differentiate along two pathways: by fusion they terminally differentiate into syncytium (Kliman et al 1986, Kao et al 1988) or else they differentiate into transiently invasive extravillous CTB (Yeh and Kurman, 1989) which can secondarily fuse to form syncytia. Thus extravillous CTB should be used to study trophoblast invasion *in vitro*. Several methods have been described to isolate these cells. CTB outgrowth was obtained from explant cultures (Yagel et al., 1989; Castellucci et al., 1990; Genbacev et al., 1993) and long term cultures established (Yagel et al., 1989). Since these cells proliferate *in vitro*, it is believed that they represent a particular population of extravillous CTB, those situated at the proximal end of the cell columns (Loke and King, 1995). Primary cultures of CTB after enzymatic digestion of chorionic villi have been first described by Thiede (1960) and used extensively ever since. This procedure releases a variety of cells among which are the extravillous CTB. A clean preparation of mononuclear cells can be obtained by submitting the cell preparation obtained after digestion of the placental tissue to a percoll gradient (Kliman et al., 1986). This yields a cell suspension which contains about 98% CTB (both villous and extravillous) when starting from term placenta. Since first trimester trophoblast is obtained from legal abortions performed mostly by suction-curettage, trophoblast is contaminated by blood cells and these are found in the final CTB preparation. This is why it is advisable to use a further purification step after the percoll gradient. We use an immunological separation procedure which allows to discard all contaminating leukocytes from the CTB suspension (Bischof et al., 1991). While this CTB preparation is fairly pure (95-98 %) it still contains a mixture of villous and extravillous CTB. Although this type of culture is extensively used to study trophoblast invasion and attachment (Fisher et al., 1989; Kliman and Feinberg, 1990) *in vitro* it has the disadvantage that the proportion of villous and extravillous CTB cannot be controlled precisely and this contributes to increasing the variability of the results. Methods to obtain pure extravillous CTB have been described (Loke et al., 1989; Bischof et al., 1995c) but are much more labor intensive.

Endometrial cells can be obtained from decidualized or non decidualized endometrium. Since trophoblast invasion occurs in a decidualized endometrium, culture of this tissue seems more relevant. The best source (because uncontaminated by trophoblast cells) of this tissue is curettage material obtained from a patient with an ectopic pregnancy. For attachment studies however, secretory phase endometria should be preferred. Endometrial *explants* have been used to study trophoblast attachment and invasion (Kliman and Feinberg, 1990). Although this type of culture preserves rather well the three dimensional structure of the tissue, it cannot provide information on fine paracrine effects between one type of endometrial cell and the invading trophoblast. Each endometrium is composed of a mixture of cell types and isolation of *purified cell* populations is necessary to identify which cells are involved in interaction with trophoblast. Separation of stromal and glandular (epithelial) cells is based on the observation that collagenase digestion of endometrial tissue, dissociates stromal cells leaving the glandular structures intact (Liu and Tseng, 1979). Glands can be separated by gravity, dissociated by trypsin digestion and cultured separately. Different types of leukocytes are hosted in decidualized and non decidualized endometria and contaminate stromal as well as glandular cell preparations. These can be removed by using magnetic particles coupled to an antibody that recognizes an epitope common to all leukocytes (CD45, Martelli et al., 1993). Conversely, a similar type of approach allowed Shi and co-workers (1995) to separate the different endometrial leukocyte subsets and to culture them separately. Purified endometrial cells can be cultured on (or in) different substrates like plastic, collagen, fibronectin, laminin, matrigel, or agarose. The cells attach to these substrates except to agarose (Martelli et al., 1993). Endometrial glandular cells are used as surrogates for the uterine surface epithelium in studies of blastocyst-endometrial interactions. Although these cells are phenotypically different (ciliated) successful separation of surface epithelial cells has not yet been achieved. To mimic the surface epithelium, the glandular cells are usually cultured on extracellular matrices as a polarized monolayer (Schatz et al., 1990).

The Role of Adhesion Molecules

Cell adhesion molecules (CAM) are represented by four subfamilies of molecules: Cadherins, Immunoglobulins, Selectins, and Integrins. Cadherins and selectins are membrane glycoproteins involved in Ca-dependent cell-cell binding, immunoglobulins allow cell-cell binding in a Ca-independent way whereas integrins regulate mainly (but not exclusively) Ca-dependent cell-substrate interactions.

Integrins are widely expressed cell-surface adhesion receptors. They are all α β heterodimers. So far 8 β and 14 α subunits have been cloned and sequenced but several alternatively spliced form have also been described (cf., Hynes, 1992). Depending on the type of α β combination, the integrin will bind to one or another matrix-glycoprotein i.e., $\alpha5\beta1$ to fibronectin, $\alpha2\beta1$ to laminin, etc. Both integrin subunits are transmembrane glycoproteins with a short cytoplasmic domain, a single transmembrane segment and a large extracellular domain. Integrins are considered as transducers signaling the nature of the extracellular environment to the interior of the cell. This signal is then translated into events which allow the cell to change shape, to migrate, to adhere to other matrices, and to release protease. This will in turn modify the micro environment of the cell to which it will readapt by changing its integrin repertoire. How the transduction mechanism works precisely is unclear to date but it involves the clustering of integrins at focal contacts on the cell membrane, the phosphorylation of the kinase FAK 125 (Focal Adhesion Kinase) and possibly the induction of gene transcription (Kornberg and Juliano, 1992). Integrins are causally related to invasion as demonstrated by transfection of Chinese hamster ovary cells with cDNA for the integrin $\alpha5\beta1$ (Giancotti and Ruoslahti, 1990). Increased expression of $\alpha5\beta1$ (a fibronectin receptor) on these cells inhibited their capacity to form tumors and to migrate. This suggests that enhancing adhesion of a cell reduces its invasive potential.

Trophoblast and particularly CTB also express integrins. Immunohistochemical studies (Korhonen et al., 1991; Damsky et al., 1992; Aplin, 1993; Bischof et al., 1993) and functional studies (Burrows et al., 1993) have shown a fascinating new property of CTB: they modulate their integrin repertoire during invasion of the endometrium. The villous CTB are immotile stem cells resting on the villous basement membrane, they express the integrin $\alpha6\beta4$ (a laminin receptor) in a clustered manner towards the basement membrane. When these cells leave the villous tree to form CTB cell columns they still express the integrin $\alpha6\beta4$ but in an unclustered way. Delocalization of $\alpha6\beta4$ probably allows the CTB to become motile and to start invasion of the endometrium. CTB located deeper in the decidualized endometrium have lost their capacity to express the integrin $\alpha6\beta4$ but express instead the integrin $\alpha5\beta1$, the major fibronectin receptor. CTB that have invaded the endometrial blood vessels express yet another integrin: the integrin $\alpha1\beta1$, a known collagen receptor. These observations show that during invasion CTB adapt to their successive environments, the placental basement membrane, the cell columns, the placental bed, and the endometrial blood vessels. This adaptation does not only change the integrin repertoire of the CTB but also their metabolism. We observed recently (Bischof et al., 1995a; c) that CTB isolated from first trimester trophoblast and expressing the $\alpha6$ integrin subunit secreted significantly higher concentrations of gelatinases and significantly lower concentrations of fibronectin than the CTB expressing the $\alpha5$ integrin subunit. In contrast both CTB subsets secreted similar amounts of hCG. We conclude from these *in vitro* studies that during trophoblast invasion, extravillous CTB expressing the $\alpha6\beta4$ integrin (in an unclustered way) represent the invasive population of CTB. Once the cells express the $\alpha5\beta1$ integrin their invasive behavior and their gelatinase (MMP-9) secretion has almost stopped and the cells become immotile and secrete fibronectin. This fibronectin secretion is deposited in the extracellular matrix and contributes at anchoring the CTB into the endometrium. It is tempting to speculate that delocalization of the $\alpha5\beta1$ integrin and acquisition of the $\alpha1\beta1$ integrin will secondarily allow the CTB to invade the endometrial spiral arteries.

Interpretation of Metalloproteinases and Invasion Assays

It is now accepted that CTB invade their environment by degrading the constituents of the extracellular matrix that surrounds them (Fisher et al., 1985, 1989; Librach et al., 1991; Bischof et al., 1991, 1995 a,c; Bischof and Martelli, 1992; McMaster et al., 1994). The *metalloproteinases* which are responsible for this degradative activity are generally measured by zymography (a semi-quantitative assay) and rarely by a proper quantitative assay. Zymography is an electrophoresis run on a polyacrylamide gel containing a substrate. Depending on the substrate, collagenases and stromelysins (if the substrate is casein) or gelatinases (if the substrate is gelatine) can be visualized as digestion areas on a blue background (the undigested substrate being stained with Coomassie Blue). The fact that electrophoresis is run in presence of SDS (sodium dodecyl sulphate) has important implications for the interpretation of the results. SDS dissociates non covalent complexes of molecules including those formed by MMP and TIMP. Under these conditions the size of the digestion band on the zymogram will reflect the total enzymatic activity (free activated and inhibitor complexed MMP). Thus in the culture medium of cells that do not degrade their environment and secrete inactive MMP (complexed to TIMP) a digestion zone will be seen on the zymogram. Furthermore, SDS partially activates (without loss of the propeptide) the proform of MMP so that proMMP (normally inactive) will digest the substrate on the zymogram and appear as a digestion band. These SDS induced artefacts need to be taken into account when interpreting zymography data. Quantitative assays have also their limitations. They are based on the measurement of the amount of substrate degraded by the MMP. Substrates are labeled radioactively and the degradation products estimated after size chromatography (Fisher et al., 1989) or after precipitation by trichloracetic acid (Bischof et al., 1995 a, b). The specificity of these assays must be questioned because it relies essentially on the substrate used. Cell supernatants do not only contain MMPs but also other enzymes which under certain circumstances might degrade the chosen substrate. In our hands (Bischof, unpublished observations) ^{3}H Gelatine is partially degraded by CTB even in the presence of gelatinase inhibitors such as EDTA or phenanthroline. This means that to measure "specifically" the activity of MMPs this background activity must be subtracted. Quantitative assays can be run in presence or absence of *in vitro* MMP activators such as APMA (p-aminophenyl mercuric acetate). In absence of activators the assay measures the contribution of *in vivo* activated enzymes whereas in presence of activators the total activity is measured. Again these details must be taken into account before interpreting the data.

Invasion assays measure the net effect of enzyme activation. These assays are based on the capacity of tumor cells (here of CTB) to invade natural and reconstituted basement membranes (Mignatti et al., 1986). The human amnion after removal of its epithelium provides a natural basement membrane (Yagel et al., 1989), other natural membranes are:

chicken chorioallantoic membrane, mouse urinary bladder, or eye lens capsules. These assays are difficult to standardize and reconstituted membranes are preferred. The most well known invasion assay uses matrigel as the reconstituted basement membrane (Kliman and Feinberg, 1990). This substrate is placed on a porous (8 μm) filter situated at the bottom of a plastic insert introduced in the culture dish. There are batch to batch variations in the composition of matrigel which can lead to uncontrolled variability. We have observed (Bischof, unpublished results) that the volume of matrigel is a very important factor since volumes which are too small (10 to 30 μl) do not cover uniformly the surface of the filter and the system is leaky. Quantification of the assay is done by counting the cells that invaded the matrigel either by using the microscope (Fisher et al., 1989), counting radioactively labeled cells (Yagel et al., 1989) or by using a colorimetric assay which detects mitochondrial dehydrogenase (Bischof and Kessi, unpublished results). The advantage of this last method lies in the fact that it measures only living cells that invaded the matrigel.

The Role of Endometrial Secretions

The importance of endometrial factors in the control of trophoblast invasiveness was demonstrated quite a number of years ago. By transplanting mice embryos to extrauterine sites, Kirby (1960) showed that in contrast to uterine sites, "implantation" outside the uterus could occur at any time during the cycle and that trophoblast invasion was unlimited. Thus, in mice uterine receptivity to an implanting blastocyst is clearly hormone (estrogen) dependent and the decidualized endometrium curtails the invasiveness of trophoblast. In humans, the hormonal requirements for uterine receptivity are not clearly defined but the duration of progesterone exposure seems to be the determinant factor (de Ziegler, 1995). *In vitro* observations indicate that in humans as in mice, decidua limits trophoblast invasion. Conditioned medium from human first trimester decidual cells suppressed invasion of CTB in the amnion invasion assay (Graham and Lala, 1991) and we observed that a similar conditioned medium inhibited the activity of trophoblastic gelatinases A and B (Bischof, 1995a). The identification of endometrial regulators of trophoblast invasiveness is the subject of intense research in several laboratories. Endometrial leukocytes and the other endometrial cells produce and secrete an array of cytokines (cf., Tabibzadeh, 1991; Smith, 1994), here we will consider only some endometrial cytokines which have been shown to affect trophoblast: interleukin-1 (IL-1), transforming growth factor-β (TGF-β), epidermal growth factor (EGF), and leukemia inhibitory factor (LIF).

IL-1 consists of 2 distinct but related molecules (IL-1α, IL-1β) which mediate similar effects. IL-1 is a known product of monocytes and macrophages but is also produced by endometrial cells and CTB (Simon et al., 1994). Human endometrial epithelial and stromal cells as well as trophoblast express IL-1 receptors (Tabibzadeh, 1991). IL-1 has been shown to increase the activity of MMP-1, MMP-3, and TIMP in human fibroblast cells (Unemori et al., 1994) and of gelatinase activity and invasive behavior in human cytotrophoblast cells (Librach et al., 1994). Endometrial and/or trophoblastic IL-1 seems to be a stimulator of trophoblast invasion. It is very interesting to note that while IL-1 inhibits decidualization of stromal cells *in vitro* (Kariya et al., 1991) it stimulates also the production of decidual MMP-1, MMP-3, and MMP-9 but not MMP-2 (Rawdanowicz et al., 1994). One would thus like to speculate that trophoblast derived IL-1 can stimulate the release of stromal cell MMP thereby promoting trophoblast invasion of the endometrium.

TGFβ is represented by five homodimeric polypeptides which share 70 to 80% structural homologies. TGFβ 1, 2, and 3 are produced by many mammalian cells including macrophages, fibroblasts, activated T cells, decidual cells, endometrial stromal and epithelial cells and cytotrophoblast cells (Graham et al., 1992). In human fibroblasts and in keratinocytes it was shown that TGFβ increases MMP-2 and MMP-9 activity while it decreases TIMP-2 (Salo et al., 1991; Overall et al., 1991). This however, is not the case for CTB because the inhibitory effect that decidual cell conditioned medium exerts on the invasive behavior of CTB seems to be due to TGFβ since antibodies to this cytokine inhibit its effect (Graham and Lala., 1991). TGFβ exerts this anti-invasive effect by stimulating the secretion of TIMP by CTB. Thus here again, TGFβ could be an endometrial signal which controls trophoblast invasion during implantation and placentation.

EGF is a fifty-three amino acid peptide produced by many cells including human endometrial cells but not CTB. This last cell however expresses the EGF receptor. It has been shown recently (Bass et al., 1994) that EGF dramatically increases invasiveness of first but not third trimester CTB. The authors proposed that decidual EGF plays an important role in up-regulating trophoblast invasiveness.

LIF is a pleiotropic cytokine which was initially identified as a factor promoting differentiation and inhibiting proliferation of the murine myeloid leukemia cell line M1 (Tomida et al., 1984). In mice LIF expression appears in the endometrium on the fourth day of pregnancy just before implantation of the blastocyst (Bhatt et al., 1991). This transient expression of LIF is essential for pregnancy since in transgenic female mice lacking the LIF gene implantation does not occur. Furthermore, when the blastocyst cells of these transgenic mice are transferred to wild type pseudopregnant recipients, the blastocyst implants and leads to a normal pregnancy (Stewart et al., 1992). The human endometrium also produces LIF (Kojima et al., 1994) and LIF mRNA is more abundant in a secretory endometrium as compared to a proliferative one (Kojima et al., 1994). Since LIF could also be a regulator of trophoblast invasion in the human, we examined the effects of LIF on the gelatinolytic activity of CTB (Bischof et al., 1995b). Recombinant human LIF (rhLIF) inhibits the gelatinolytic activity secreted by CTB. LIF exerts profound effects on cytotrophoblast differentiation since at the same time that this cytokine inhibits MMP it also inhibits hCG secretion (Bischof et al., 1995b). We would thus conclude that endometrial LIF inhibits differentiation of CTB into syncytium (decrease of hCG) inhibiting at the same time the acquisition of an invasive phenotype (inhibition of gelatinases). If this is true *in vivo*, then the net result of LIF would be to maintain the CTB in an undifferentiated state.

ACKNOWLEDGEMENTS

The author would like to thank the Swiss National Science Foundation for their continuous support over the last 6 years.

REFERENCES

Anderson, T.L., Olsen, G.E. and Hoffman, L.H. (1986) Stage specific alterations in the apical membrane glycoproteins of endometrial epithelial cells related to implantation in the rabbit. *Biol. Reprod.* 34, 701-720.

Aplin, J.D. (1993) Expression of integrin alpha6 beta4 in human trophoblast and its loss from extravillous cells. *Placenta* 14, 203-215.

Bass, K.E., Morrish, D., Roth, I., Bhardwaj, D., Taylor, R., Zhou, Y. and Fisher, S.J. (1994) Human cytotrophoblast invasion is up-regulated by epidermal growth factor. Evidence that paracrine factors modify this process. *Dev. Biol.* 164, 550-561.

Bhatt, H., Brunet, L.J. and Stewart, C.L. (1991) Uterine expression of leukaemia inhibitory factor coincides with the onset of blastocyst implantation. *Proc. Natl. Acad. Sci.* 88, 11402-11412.

Bischof, P., Friedli, E., Martelli, M. and Campana, A. (1991) Expression of extracellular matrix-degrading metalloproteases by cultured human cytotrophoblast cells. Effect of cell adhesion and immunopurification. *Am. J. Obstet. Gynecol.* 65, 1791-1801.

Bischof, P. and Martelli, M. (1992) Proteolysis in the penetration phase of the implantation process. *Placenta* 13, 17-24.

Bischof, P., Redard, M., Gindre, P., Vassilakos, P. and Campana, A. (1993) Localisation of alpha 2, alpha 5 and alpha 6 integrin subunits in human endometrium, decidua and trophoblast. *Eur. J. Obstet. Gynecol. Reprod. Biol.* 51, 217-226.

Bischof, P., Martelli, M. and Campana, A. (1994) La régulation des métalloprotéases endométriales et trophoblastique lors de l'implantation du blastocyste. *Contracep. Fertil. Sex.* 22, 48-52.

Bischof, P., Martelli, M., Campana, A., Itoh, Y., Ogata, Y. and Nagase, H. (1995a) Importance of metalloproteinases (MMP) in human trophoblast invasion. *Early Pregn. Biol. Med.* 1, 263-269.

Bischof, P., Haenggeli, L. and Campana, A. (1995b) Effect of leukaemia inhibitory factor on human cytotrophoblast differentiation along the invasive pathway. *Am. J. Reprod. Immunol.* 34, 225-230.

Bischof, P., Haenggeli, L. and Campana, A. (1995c) Gelatinase and oncofetal fibronectin secretion are dependent upon integrin expression on human cytotrophoblasts. *Human Reprod.* 10, 734-742.

Bischof, P. and Campana, A. (1996) A model for implantation of the human blastocyst and early placentation. *Human Reprod. Update* 2, 262-270.

Boyd, J.D. and Hamilton, W.J. (1970) The human placenta. In: Placenta, (eds.) J.D. Boyd and W.J. Hamilton, Heffer and Sons:Cambridge, pp 1.

Burrows, T.D., King, A. and Loke, Y.W. (1993) Expression of integrins by human trophoblast and differential adhesion to laminin and fibronectin. *Human Reprod.* 8, 475-484.

Carson, D.D., Tang, J.P. and Julian, J. (1993) Heparan sulphate proteoglycan (perlecan) expression by mouse embryo during acquisition of attachment competence. *Dev. Biol.* 155, 97-106.

Castellucci, M., Kaufmann, P. and Bischof, P. (1990) Extracellular matrix influences hormone and protein production by human chorionic villi. *Cell Tissue Res.* 262, 135-142.

Chavez, D.J. and Enders, A.C. (1981) Temporal changes in lectin binding of preimplantation mouse blastocyst. *Dev. Biol.* 87, 267-276.

Chavez, D.J. and Anderson, T.L. (1985) The glycocalyx of the mouse uterine luminal epithelium during oestrus, early pregnancy, the peri-implantation period, and delayed implantation. I Acquisition of ricinus communis binding sites during pregnancy. *Biol. Reprod.* 32, 267-276.

Chavez, D.J. (1986) Cell surface of the mouse blastocysts at the trophectoderm-uterine interface during the adhesive stage of implantation. *Am. J. Anat.* 176, 153-158.

Coutifaris, C., Kao, L.C., Sehdev, H.M., Chin, U., Babalola, G.O. Blaschuk, O.W. and Strauss III, J.F. (1991) E-cadherin expression during the differentiation of human trophoblasts. *Develop.* 113, 767-777.

Cross, J.C., Werb, Z. and Fisher, S.J. (1994) Implantation and the placenta : key pieces of the development puzzle. *Science* 266, 1508-1518.

Damsky, C.H., Fitzgerald, M. and Fisher, S.J. (1992) Distribution patterns of extracellular matrix components and adhesion receptors are intricately modulated during first trimester cytotrophoblast differentiation along the invasive pathway in vivo. *J. Clin. Invest.* 89, 210-222.

Damsky, C., Sutherland, A. and Fisher, S. (1993) Extracellular matrix 5: Adhesive interactions in early mammalian embryogenesis, implantation and placentation. *FASEB J.* 7, 1320-1329.

De Ziegler, D. (1995) Hormonal control of endometrial receptivity. *Human Reprod.* 10, 4-7.

Denker, H.W. (1993) Implantation: A cell biological paradox *J. Exp. Zool.* 266, 541-558.

Emonard, H., Christiane, Y., Smet, M., Grimaud, J.A. and Foidart, J.M. (1990) Type IV and interstitial collagenolytic activities in normal and malignant trophoblast cells are specifically regulated by the extracellular matrix. *Invasion Metast.* 10, 170-177.

Enders, A.C. and Schlafke, S. (1972) Implantation in the ferret: Epithelial penetration. *Am. J. Anat.* 133, 291-316.

Enders, A.C. and Schlafke, S. (1974) Surface coat of the mouse blastocyst and the uterus during the preimplantation period. *Anat. Rec.* 181, 31-46.

Enders, A.C., Henrickx, A.G. and Schlafke, S. (1983) Implantation in the rhesus monkey. *Am. J. Anat.* 176, 275-298.

Enders, A.C. (1991) Implantation. In: *Encyclopedia of Human Biology*, Vol. 4, Academic Press pp. 423-430.

Fenderson, B.A., Eddy, E.M, and Hakomori, S.I. (1990) Glycoconjugate expression during embryogenesis and its biological significance. *BioEassy* 12, 173-180.

Fisher, S.J., Leitch, M.S., Kantor, M.S., Basbaum, C.B. and Kramer, R.H. (1985) Degradation of extracellular matrix by the trophoblastic cells of first trimester placenta. *J. Cell. Biochem.* 27, 31-41.

Fisher, S.J., Cui, T., Zhang, L., Hartmann, L., Grahl, K., Guo-Yang, Z., Tarpey, J. and Damsky, C.H. (1989) Adhesive and degradative properties of human placental cytotrophoblast cells in vitro. *J. Cell. Biol.* 109, 891-902.

Fisher, S.J. and Damsky, C.H. (1993) Human cytotrophoblast invasion. *Seminar in Cell Biol.* 4, 183-188

Genbacev, O., De Mesy-Jensen, K., Schubach-Powlin, S. and Miller, R.K. (1993) In vitro differentiation and ultrastructure of human extravillous trophoblast (EVT) cells. *Placenta* 14, 463-475.

Ghosh, D., De, P. and Sengupta, J. (1994) Luteal phase ovarian oestrogen is not essential for implantation and maintenance of pregnancy from surrogate embryo transfer in rhesus monkey. *Human Reprod.* 9, 629-637.

Giancotti, F.G. and Ruoslahti, E. (1990) Elevated levels of the alpha5 beta1 fibronectin receptor suppress the transformed phenotype of Chinese hamster ovary cells. *Cell* 60, 849-859.

Glasser, S., Mani, S.K. and Mulholland, J. (1991) In vitro models of implantation. In: *Uterine and Embryonic Factors in Early Pregnancy*, (eds.), J.F. Strauss III, and C.R. Lyttle, Plenum Press:New York, pp. 33-50.

Graham, C.H. and Lala, P.K. (1991) Mechanism of control of trophoblast invasion in situ. *J. Cell. Physiol.* 148, 228-234.

Graham, C.H., Lysiak, J.J., Mc Crae, K.R. and Lala, P.K. (1992) Localisation of transforming growth factor at the human fetal-maternal interface. Role in trophoblast growth and differentiation. *Biol. Reprod.* 46, 561-572.

Grummer, R., Hohn, H.P., Mareel, M.M. and Denker, H.W. (1994) Adhesion and invasion of three human choriocarcinoma cell lines into human endometrium in a three dimensional organ culture system. *Placenta* 15, 411-429.

Hayashi, K., Madri, J. and Yurchenco, P. (1992) Endothelial cells interact with the core protein of basement membrane perlecan through alpha 1 and beta 3 integrins: an adhesion modulated by glycosaminoglycan. *J. Cell. Biol.* 119, 945-955.

Hertig, A.T., Rock, J. and Adams, E.C. (1956) A description of 34 human ova within the first 17 days of development *Am. J. Anat.* 98, 435-494.

Hynes, R.O. (1992) Integrins :Versatility, modulation and signalling in cell adhesion. *Cell* 69, 11-25.

Kao, L.C., Caltabiano, S., Wu, S., Strauss, III J.F. and Kliman, H. (1988) The human villous cytotrophoblast interaction with extracellular matrix proteins, endocrine function and cytoplasmic differentiation in the absence of syncytium formation. *Dev. Biol.* 130, 693-702.

Kariya, M., Kanzaki, H., Takakura, K., Imai, K., Okamoto, N., Emi, N. Kariya, Y. and Mori, T. (1991) Interleukin 1 inhibits in vitro decidualisation of human endometrial stromal cells. *J. Clin. Endocrinol. Metab.* 73, 1170-1174.

Kimber, S.J. and Lindenberg, S. (1990) Hormonal control of a carbohydrate epitope involved in implantation in mice. *J. Reprod. Fertil.* 89, 13-21.

Kirby, D.R.S. (1960) Development of mouse eggs beneath the kidney capsule. *Nature* 187, 707-708.

Kliman, H., Nestler, J.E., Sermasi, E., Sanger, J.M. and Strauss, III J.F. (1986) Purification, characterization and in vitro differentiation of cytotrophoblast from human term placenta. *Endocrinology* 118, 1567-1582.

Kliman, H.J. and Feinberg, R.F. (1990) Human trophoblast-extracellular matrix (ECM) interactions in vitro: ECM thickness modulates morphology and proteolytic activity. *Proc. Natl. Acad. Sci.* 87, 3057-3061.

Kojima, K., Kanzaki, H., Iwai, M., Hatayama, H., Fujimoto, M., Inoue, T., Horie, K., Nakayama, H., Fujita, J. and Mori, T. (1994) Expression of leukaemia inhibitory factor in human endometrium and placenta. *Biol. Reprod.* 50, 882-887.

Korhonen, M., Ylänne, J., Laitineen, L., Cooper, H.M., Quaranta, V. and Virtanen, I. (1991) Distribution of the alpha1-beta6 integrin subunits in human developing and term placenta. *Lab. Invest.* 65, 347-356.

Kornberg, L. and Juliano, R.L. (1992) Signal transduction from the extracellular matrix: The integrin-tyrosine kinase connection. *TIPS* 13, 93-95.

Lessey, B.A., Damjanovich, L., Coutifaris, C., Castelbaum, Albelda, S.M. and Buck, C.A. (1992) Integrin adhesion molecules in human endometrium. *J. Clin. Invest.* 90, 188-195.

Librach, C.L., Feigenbaum, S.L., Bass, K.E., Cui, T.Y., Verastas, N., Sadovsky, Y., Quigley, J.P., French, D.L. and Fisher, S.J. (1994) Interleukin-1 beta regulates human cytotrophoblast metalloproteinase activity and invasion in vitro. *J. Biol. Chem.* 269, 17125-17131.

Librach, C.L., Werb, Z., Fitzgerald, M.L., Chiu, K., Corwin, N.M., Esteves, R.A., Grobelny, D., Galardy, R. and Damsky, C.H. (1991) 92 kDa type IV collagenase mediates invasion of human cytotrophoblasts. *J. Cell Biol.* 113, 437-449.

Lindenberg, S., Sundberg, K., Kimber, S.J. and Lundblad, A. (1988) The milk oligosaccharide, lacto-N-fucopentaose I, inhibits attachment of mouse blastocysts on endometrial monolayers. *J. Reprod. Fertil.* 83, 149-158.

Lindenberg, S., Kimber, S.J. and Kallin, E. (1990) Carbohydrate binding properties of mouse embryos. *J. Reprod. Fertil.* 89, 431-439.

Lindenberg, S., Pedersen, B., Hamberger, L. and Kimber, S.J. (1992) Models for human implantation derived from implantation in vitro. *Reprod. Fertil. Dev.* 4, 653-670.

Lindner, H.R. (1972) Choice of an animal model for the study of ovum implantation. *Acta. Endocrinol.* 166, 93-99

Liu, H.C. and Tseng, L. (1979) Estradiol metabolism in isolated human endometrial epithelial glands and stromal cells. *Endocrinology* 104, 1674-1681.

Loke, Y.W., Gardner, L. and Grabowska, A. (1989) Isolation of human extravillous trophoblast cells by attachment to laminin-coated magnetic beads. *Placenta* 10, 407-415.

Loke, Y.W. and King, A. (1995) Human Implantation Cell Biology and Immunology, Cambridge University Press: Cambridge, New York, Melbourne, p. 1.

Martelli, M., Campana, A. and Bischof, P. (1993) Secretion of matrix metalloproteinase by human endometrial cells in vitro. *J. Reprod. Fertil.* 98, 67-76.

McMaster, M.T., Bass, K.E. and Fisher, S.J. (1994) Human trophoblast invasion autocrine control and paracrine modulation. *Ann. N.Y. Acad. Sci.* 734, 122-131.

Mignatti, P., Robbins, E. and Rifkin, D.B. (1986) Tumour invasion through the human amniotic membrane: Requirement for a proteinase cascade. *Cell* 47, 487-498.

Overall, C.M., Wrana, J.L. and Sodek, J. (1991) Transcriptional and post transcriptional regulation of 72 kDa gelatinase type IV collagenase by transforming growth factor ß1 in human fibroblasts. *J. Biol. Chem.* 266, 14064-14071.

Polette, M., Nawrocki, B., Pintiaux, A., Massenat, C., Maquoi, E., Volders, L., Schaaps, J.P., Birembaut, P. and Foidart, J.M. (1994) Expression of gelatinases A and B and their tissue inhibitors by cells of early and term human placenta and gestational endometrium. *Lab. Invest.* 71, 838-846.

Rawdanowicz, T.J., Hampton, A.L., Nagase, H., Wooley, D.E. and Salamonsen, L.A. (1994) Matrix metalloproteinase production by cultured human endometrial stromal cells: identification of interstitial collagenase, gelatinase-A, gelatinase-B, and stromelysin-1 and their differential regulation by interleukin-1 alpha and tumour necrosis factor-alpha. *J. Clin. Endocrinol. Metab.* 79, 530-536.

Rohde, L.H. and Carson, D.D. (1993) Heparin like glycosaminoglycans participate in binding of human trophoblastic cell line (Jar) to human uterine epithelial cell line (RL 95). *J. Cell Physiol.* 155, 185-196.

Salo, T., Lyons, J.G., Rahemtulla, F., Birkedal-Hasen, H. and Larjava, H. (1991) Transforming growth factor-beta up-regulates type IV collagenase expression in cultured human keratinocytes. *J. Biol. Chem.* 266, 11436-11441.

Schatz, F., Gordon, R.E., Laufer, N. and Gurpide, E. (1990) Culture of human endometrial cells under polarised conditions. *Differentiation* 42, 184-190.

Schlafke, S. and Enders, A. (1975) Cellular basis of interaction between trophoblast and uterus at implantation. *Biol. Reprod* 12, 41-65.

Shi, W.L., Mognetti, B., Campana, A. and Bischof, P. (1995) Metalloproteinase secretion by endometrial leukocyte subsets. *Am. J. Reprod. Immunol.* 34, 299-310.

Shimonovitz, S., Hurwitz, A., Dushnik, M., Anteby, E., Geva-Eldar, T. and Yagel, S. (1994) Developmental regulation of the expression of 72 and 92 kd type IV collagenases in human trophoblasts. A possible mechanism for control of trophoblast invasion. *Am. J. Obstet. Gynecol.* 171, 832-838.

Simon, N., Noël, A., and Foidart, J.M. (1992) Evaluation of in vitro reconstituted basement membrane assay to assess the invasion of tumour cells. *Invasion Metast.* 12, 156-167.

Simon, C., Frances, A., Piquette, G., Hendrickson, M. Milki, A. and Polan, M.L. (1994) Interleukin-1 system in the materno-trophoblast unit in human implantation: Immunohistochemical evidence for autocrine/paracrine function. *J. Clin. Endocrinol. Metab.* 78, 847-854

Smith, S.K. (1994) Growth factors in the human endometrium. *Human Reprod. Update* 9, 936-946.

Stewart, C.L., Kaspar, P., Brunet, L.J., Bhatt, H., Gadi, I., Köntgen, F. and Abbondanzo, S. (1992) Blastocyst implantation depends on maternal expression of leukaemia inhibitory factor. *Nature* 359, 76-79.

Tabibzadeh, S. (1991) Human endometrium: An active site of cytokine production and action. *Endocr. Reviews* 12, 272-290.

Tabibzadeh, S. and Babaknia, A. (1995) The signals and molecular pathways involved in implantation, a symbiotic interaction between blastocyst and endometrium involving adhesion and tissue invasion. *Mol. Hum. Reprod.* 10, 1579-1602.

Thie, M., Harrach-Ruprecht, B., Sauer, H., Fuchs, P., Albers, A. and Denker, H.W. (1995) Cell adhesion to the apical pole of epithelium: A function of cell polarity. *Eur. J. Cell. Biol.* 66, 180-191.

Thiede, H.A. (1960) Studies on human trophoblast tissue in culture. I Cultural methods and histological staining. *Am. J. Obstet. Gynecol.* 79, 636-647.

Tomida, M., Yamamoto-Yamaguchi, Y. and Hozumi, M. (1984) Purification of a factor inducing differentiation of mouse myeloid leukaemia M1 cells from conditioned mouse fibroblast L929 cells. *J. Biol. Chem.* 259, 10978-10982.

Unemori, E.U., Ehsani, N., Wang, M., Lee, S., Mc Guire, J. and Amento, E.P. (1994) Interleukin-1 and transforming growth factor alpha synergistic stimulation of metalloproteinases, PGE2 and proliferation in human fibroblasts. *Exp. Cell Res.* 210, 166-171.

Weitlauf, H.M. (1988) Biology of Implantation. In: *Physiology of Reproduction*, (eds.), E. Knobil and J.D. Neill, Raven Press: New York, pp. 231-262.

White, S. and Kimber, S.J. (1994) Changes in a(1-2)-fucosyltransferase activity in the murine endometrial epithelium during the oestrous cycle, early pregnancy, and after ovariectomy and hormone replacement. *Biol. Reprod.* 50, 73-81.

Wu, J.T. and Chang, M.C. (1978) Increase in concanavalin binding sites in mouse blastocyst during implantation. *J. Exp. Zool.* 105, 447-453.

Yagel, S., Casper, R.F., Powell, W., Parhar, R.S., and Lala, P.K. (1989) Characterisation of pure human first trimester cytotrophoblast cells in long term culture: Growth pattern, markers, and hormone production. *Am. J. Obstet. Gynecol.* 160, 938-945.

Yamagata, T. and Yamazaki, K. (1991) Implanting mouse embryo stain with a LNF-I bearing fluorescent probe at their mural trophectodermal side. *Biochem. Biophys. Res. Comm.* 181, 1004-1009.

Yeh, I.T. and Kurman, R.J. (1989) Functional and morphological expression of trophoblast. *Lab. Invest* 61, 1-4.

Trophoblast Research 10:83-95, 1997

CYTOTROPHOBLAST INVASION OF THE ENDOMETRIUM IN THE HUMAN AND MACAQUE EARLY VILLOUS STAGE OF IMPLANTATION

Allen C. Enders

Department of Cell Biology and Human Anatomy
University of California School of Medicine
Davis, California 95616 USA

INTRODUCTION

In common with early implantation stages in other primates such as the baboon and macaque, the human conceptus has a trophoblastic plate stage during which the mass of syncytial trophoblast and cytotrophoblast abutting the endometrium spreads along the plane of the uterine luminal epithelium before penetrating maternal vessels to initiate the lacunar stage (Enders, 1989, 1993). Subsequently cytotrophoblast proliferates to form primary villi. Before the primary villi extend beyond the lacunae into the endometrium, some of the villi are invested with a mesenchymal core, resulting in secondary villi. Only after these events does the cytotrophoblast expand at the fetal-maternal junction to form the trophoblastic shell. Consequently it is only at this transition period that there begins to be substantial extravillous cytotrophoblast.

A large number of studies in the last decade demonstrated the utility of immunological methods to aid in sorting out cell types in the basal plate of the human placenta (Wells and Bulmer, 1988). Identification of intermediate filaments proved particularly useful in identifying extravillous cytotrophoblast in the basal plate (Ockleford et al., 1981; Clark and Damjanov, 1985; Loke et al., 1986). This method is commonly used in both the rhesus monkey and human to separate epithelial cytotrophoblast from maternal endometrial stroma (Douglas and King, 1990; Aplin, 1993; Blankenship and King, 1993; King and Blankenship, 1993; Vicovac et al., 1995). The usefulness of a variety of antibodies to intermediate filaments in studying human placentation has also been demonstrated (Mirecka and Hanarz, 1990). More recently the differentiation of different cytokeratins during the course of gestation has been examined (Muhlhauser et al., 1995).

In this manuscript we used staining of extravillous cytotrophoblast cells with an antibody to cytokeratin (CAM 5.2) to compare the way in which such cells were distributed in the implantation site at the time of formation of the trophoblastic shell in archival human implantation specimens and freshly obtained material from the cynomolgus macaque.

MATERIALS AND METHODS

Four human early implantation stages were available from archival collections. A primary villous stage was estimated to be approximately day 13 after fertilization; the diameter of the conceptus was 1.7 mm, and was estimated to be Carnegie Stage 7. Two early secondary villous stages, approximately day 14-15, were about 5 mm wide and 2.5-3 mm deep (Carnegie stage 7-8). The conceptus of a larger early villous stage, (day 18-19), was 18 mm wide and 12 mm deep. Early implantation stages in the macaque, including previllous and early villous stages, have been described previously (Enders, 1995)

Tissues from the human implantation sites were fixed initially in 10% formalin, and processed for routine paraffin sectioning. Since the slides had been stained previously with H&E, the coverslips were removed and the sections were rehydrated, and destained prior to immunostaining. Sections to be stained for cytokeratin were preincubated in 0.1% pepsin to expose antigenic sites. Endogenous peroxidase activity was blocked with hydrogen peroxide/methanol. The slides were incubated overnight at 4°C in either anti-human cytokeratin (CAM 5.2; Becton Dickinson, San Jose, CA) or anti-human SP1 (pregnancy-specific Beta-1-glycoprotein; Zymed, San Francisco, CA). After a rinse, the sections were incubated in biotin-conjugated secondary antibody, then streptavidin peroxidase using routine immunohistochemical procedures. The antibodies were visualized using diaminobenzidine or aminoethyl carbazole, counterstained with hematoxylin, and examined. Implantation sites from the cynomolgus macaque were fixed in 4% formaldehyde for 4 hours, then embedded in paraffin. Immunostain procedures were the same as those used for the human implantation sites.

RESULTS

Invasive Pathway Of Cytotrophoblast At The Early Villous Stage In The Human

Primary Villous Stage

The primary villous stage was represented by a single specimen, estimated to be about thirteen days after fertilization (Figure 1a). Although cytotrophoblast was abundant in the septae between the lacunae, the lacunae were very close to the endometrium, with only a thin layer of syncytial trophoblast intervening over much of the circumference. Other areas showed clusters of cytotrophoblast and syncytium, but there was no continuous mass of cytotrophoblast constituting a trophoblastic shell. In addition decidualization of the endometrial stroma, as indicated by rounding and enlargement of stromal fibroblast cells, was just beginning.

Examination of the junctional zone after staining for cytokeratin revealed that, although much of the trophoblast stained only lightly, and the staining of the endometrial glands was more irregular than in some preparations, individual cytotrophoblast cells at the junction had moderate to heavy staining (Figures 1b, c). Cytokeratin positive cytotrophoblast cells could be seen at varying distances from the fetal-maternal junction into the endometrium. These cells were entirely within the stromal matrix, and none were found within the adjacent vessels.

Early Secondary Villous Stage

Three early secondary villous stage specimens were available (secondary villi formed but prior to fetal circulation). The younger two of these specimens (day 14-15) demonstrated the formation of a thick and apparently continuous trophoblastic shell surrounding simple secondary villi.

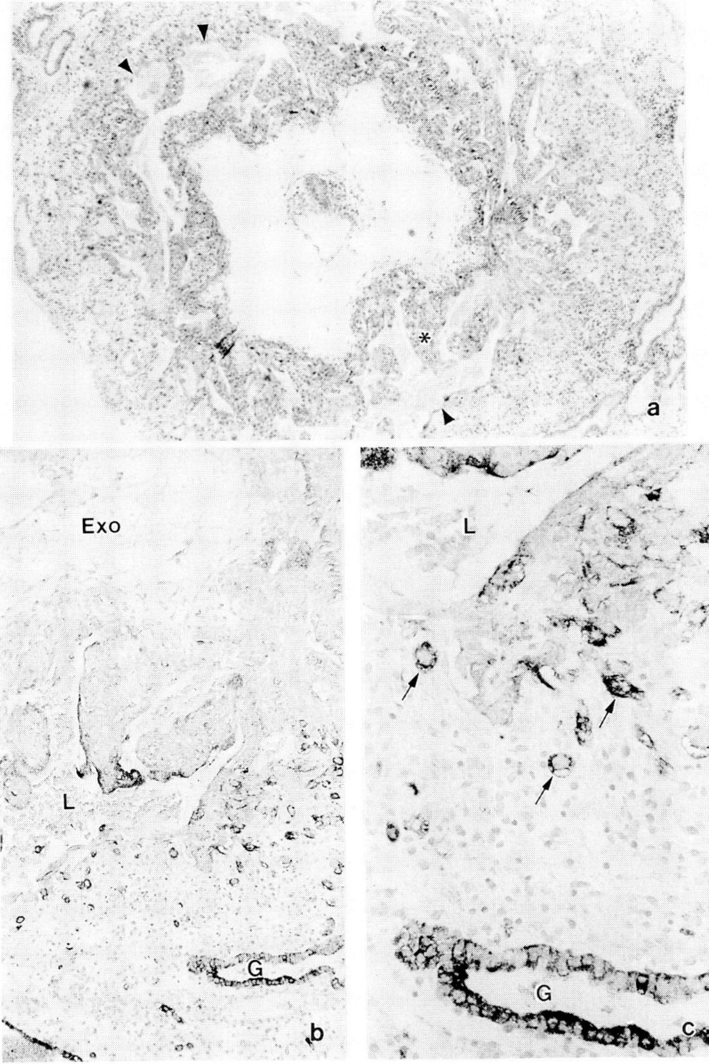

Figure 1. a) Micrograph of an H&E stained archival section of a human implantation site at the transition from the lacunar to the villous stage. The embryo and its yolk sac are seen in the middle of the exocelom. Note the relatively abrupt transition from lacunae to endometrium at a number of places (arrowheads) around the chorionic vesicle. The area marked with the asterisk is seen in the micrographs below. b) A region from the section above after destaining and reaction with antibody to cytokeratin. Reactivity of the cytotrophoblast and syncytium is light. However, the individual interstitial cytotrophoblast cells are clearly seen extending into the endometrium towards the glands (G). Exocelom (Exo); lacunae (L) c) At a higher magnification the large size of some of the interstitial cytotrophoblast cells (arrows) as compared to the size of the gland cells can be seen. a) X55, b) X100, c) X270

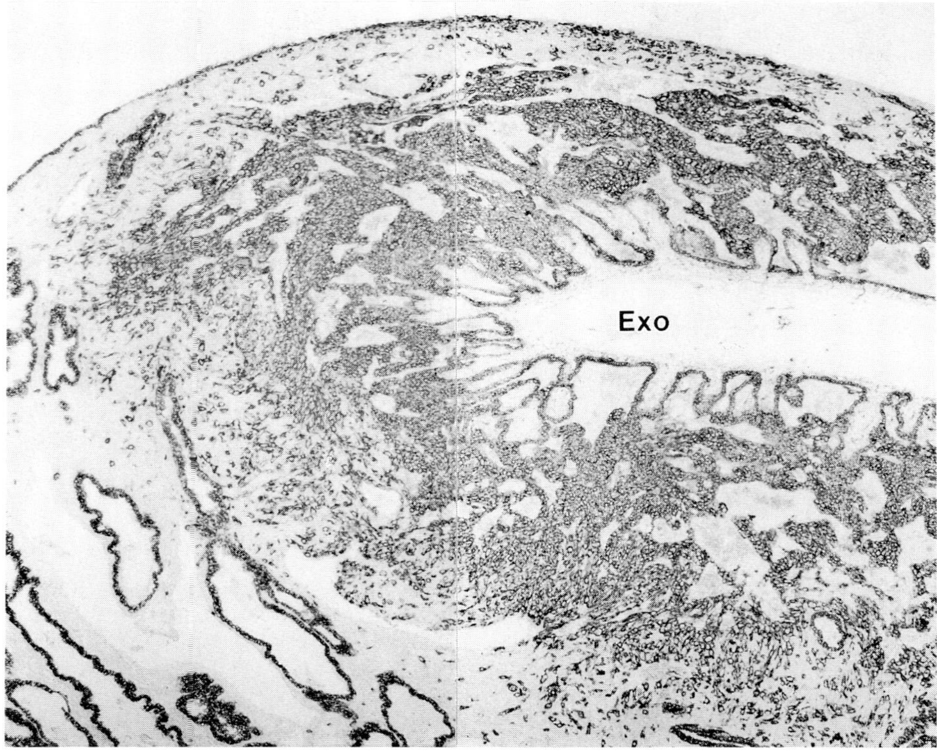

Figure 2. An archival human implantation site destained and reacted with cytokeratin from approximately day 14 of pregnancy. The secondary villi abut solid cords of cytotrophoblast constituting the trophoblastic shell. Note that in the decidua capsularis as well as basalis, individual cytotrophoblast cells are present in the endometrial stroma between the trophoblastic shell and the glands. Exocelom (Exo). X45

In a conceptus estimated to be about day 14, cytokeratin staining revealed a pronounced zonation of cytotrophoblast encircling the chorionic vesicle (Figure 2). The cytotrophoblast cells immediately adjacent to the secondary villi were in an arrangement of branching anastomotic cords of tightly packed, relatively uniform cells. This area constituted a thick trophoblastic shell. Radiating from this compact region were numerous individual cytotrophoblast cells which stained especially heavily for cytokeratin (Figure 3a). These cells extended all around the surface of the chorionic vesicle, and penetrated within the stroma to and beyond the immediate coils of the uterine glands. Although many of these cells were rounded in shape, others were elongated, and others were large and irregular (Figure 3b). The cells showed little tendency to enter the endometrial glands, and although a few instances were found where cytotrophoblast cells were adjacent to endometrial blood vessels, no cytokeratin positive cells were seen within any of the endometrial blood vessels at this stage.

Staining of sections from these conceptuses with antibody to SP1 demonstrated that this material was restricted to the syncytial trophoblast (Figure 3c). Although a few patches of SP1-positive trophoblast could be seen in the trophoblastic shell, clearly the vast majority of individual trophoblast cells radiating from the shell into the stroma were SPl-negative.

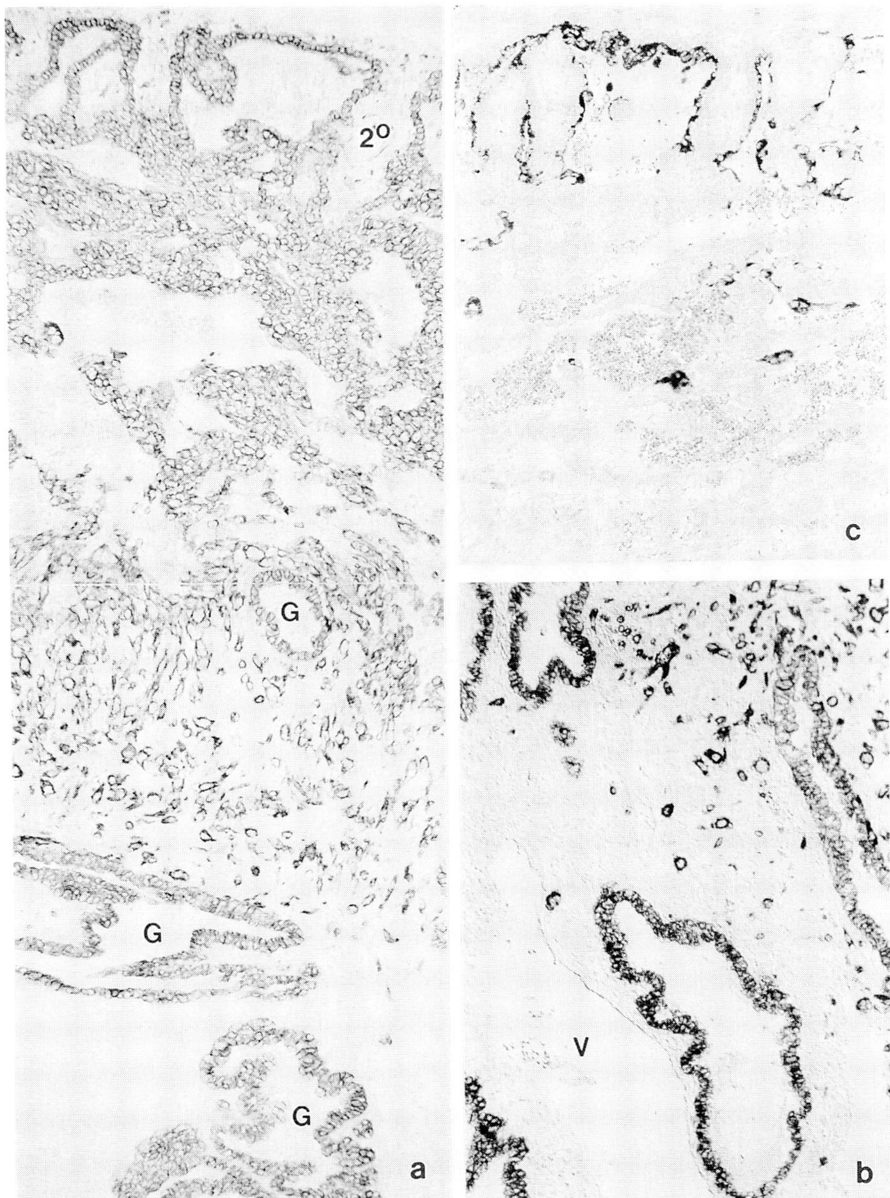

Figure 3. a) A montage from the cytokeratin-stained specimen shown in Figure 2, extending from the exocelom above into the endometrium. Note the richly cellular trophoblastic shell beneath the secondary villi (2°), and that the individual cytotrophoblast cells penetrate beyond the most superficial glands (G) in the center of the micrograph. b) At a higher magnification it can be seen that the individual cytotrophoblast cells are abundant within the stroma and do not appear to have entered either the glands or the dilated venule (V). c) A section of the same specimen in **a** but reacted with antibody to SP1. Note that the syncytial trophoblast lining lacunar spaces is the only tissue stained with this antibody. a) c) X100, b) X120

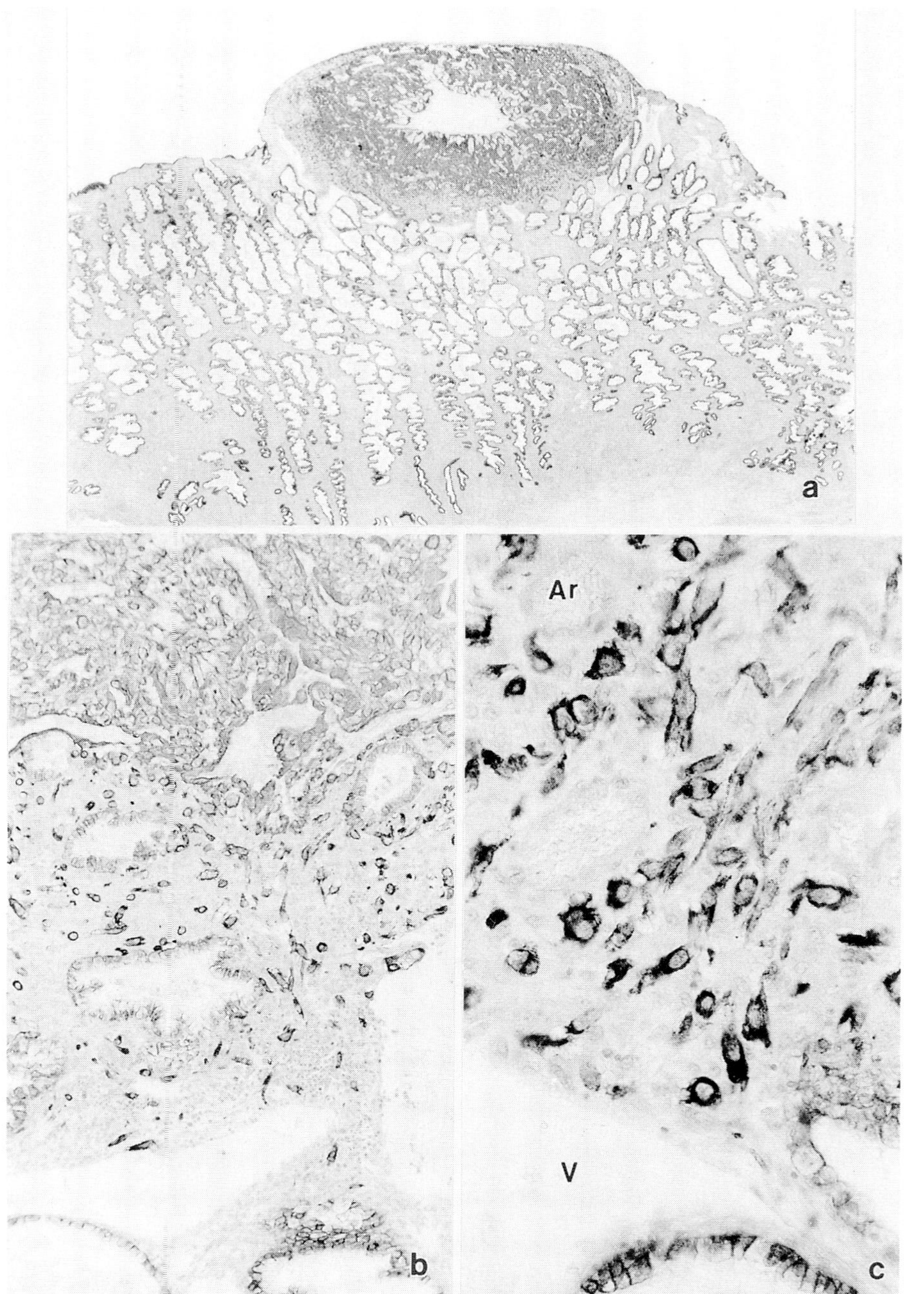

Figure 4. Archival specimen of a human implantation site approximately 15 days postfertilization. a) A dense trophoblastic shell completely surrounds the secondary villi. b) A higher magnification of this specimen illustrates the extensive infiltration of the endometrium by cytokeratin-positive cytotrophoblast cells. c) Micrograph of a region of endometrium adjacent to the trophoblastic shell which has been thoroughly infiltrated by large irregular cytotrophoblast cells. Note however that these cells show no tendency to enter the arterioles (Ar) or the venule (V). a) X10, b) X100, c) X270

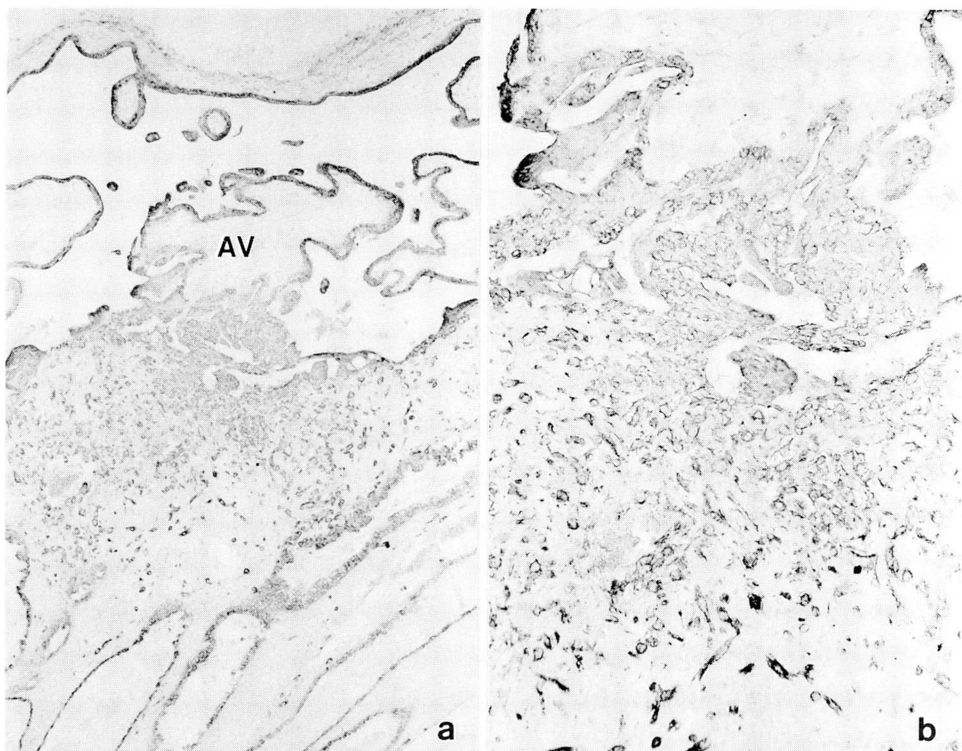

Figure 5. Archival specimen of a human implantation site approximately 18-19 days of gestation, stained for cytokeratin. a) The trophoblastic shell is more dispersed, and individual anchoring villi (AV) are readily discernible. b) Higher magnification of the anchoring villous. Note that the underlying trophoblastic shell is dispersed but that there is still a wealth of individual cytotrophoblast cells extending into the endometrium. a) X45 b) X100

A slightly older chorionic vesicle stage had an especially thick trophoblastic shell (Figure 4a). This specimen demonstrated clearly a wreath of interstitial cytotrophoblast cells radiating from the trophoblastic shell and penetrating beyond some of the more superficial coils of the glands (Figure 4b). It could be seen that these individual cytokeratin-positive cytotrophoblast cells showed no tendency to enter either small arterioles or dilated venules (Figure 4c).

In a chorionic vesicle from an estimated postfertilization stage of 18-19 days, the trophoblastic shell appeared to be more dispersed and irregular with associated broadly spaced individual anchoring villi (Figure 5a). The impression received was that the cytotrophoblast cells of the trophoblastic shell had dispersed more widely into the endometrium during growth of the conceptus. The individual cytotrophoblast cells were more concentrated toward the intervillous space, becoming more widely spaced further into the endometrium (Figure 5b).

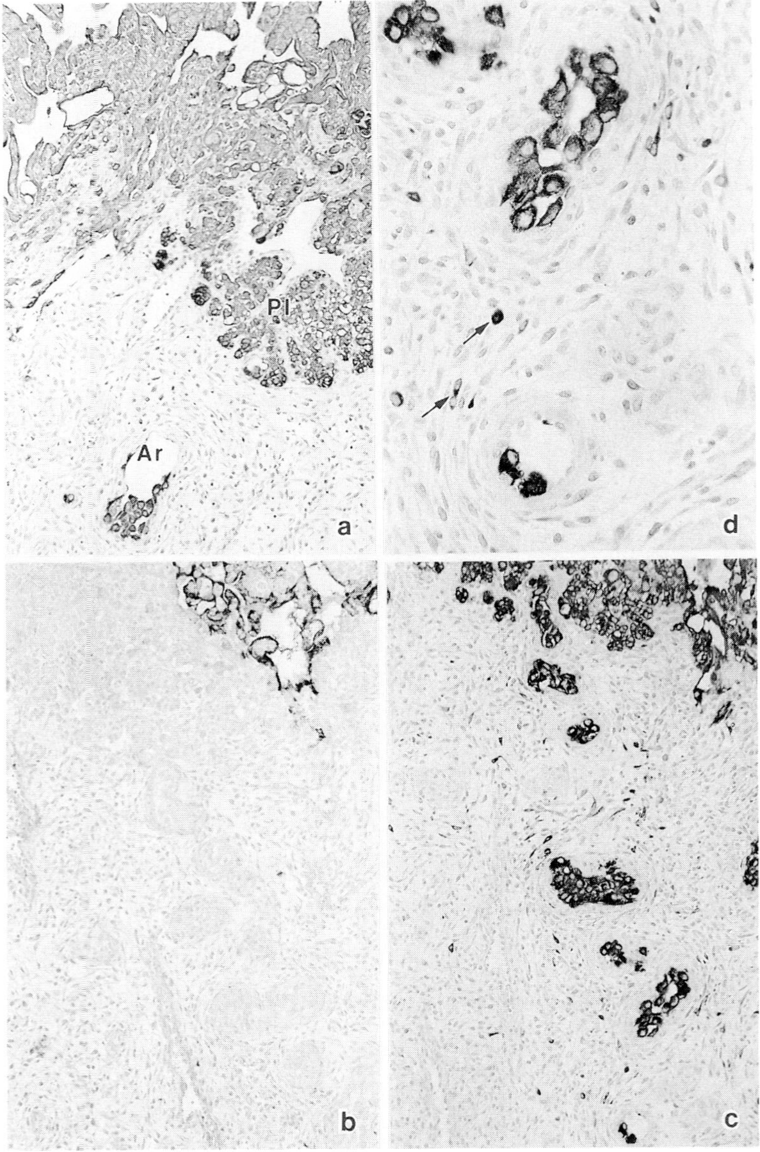

Figure 6. A 13 day implantation site of the cynomolgus monkey. a) In addition to trophoblast reacting to cytokeratin, a lobulated mass of epithelial plaque cells is seen (Pl). Note the intravascular cytotrophoblast cells in the arteriole (Ar). b, c) Illustrate an arteriole extending beneath the day 13 macaque site. The section in **b** is reacted with anti-SP1 demonstrating that syncytial trophoblast is limited to the lacunar region, and that the cells within the arteriole (Ar) are not reactive. In **c** however intravascular cytotrophoblast (arrows) is seen within the coiled arteriole. **d**. Higher magnification of a portion of the arteriole seen in c. Note that the upper two coils of the arteriole are largely occluded by cytotrophoblast cells, whereas the lower coils have few or no cells. Note also the rather small cytokeratin-positive interstitial cells (arrows). Day 13). a-c) X100, d) X270

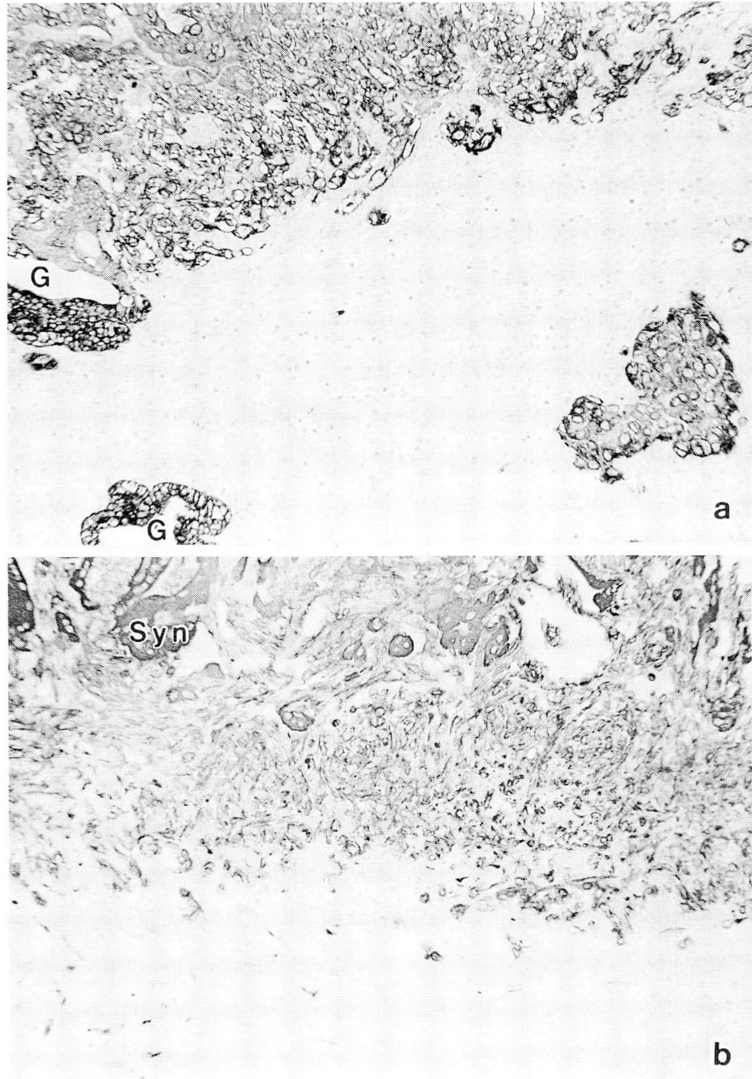

Figure 7. a) Trophoblastic shell of a day 15 macaque implantation site stained for cytokeratin. Note that there are few cytotrophoblast cells extending into the endometrium except for those filling the arteriole (Ar). Two coils of the gland are seen (G). b) Trophoblastic shell of day 18 macaque implantation site. A few cells of approximately the same size as those within the trophoblastic shell extend into the endometrium below syncytial trophoblast (Syn). X100

Early Invasion Of Cytotrophoblast In The Macaque

A description of the morphological changes occurring in the conceptus during trophoblastic shell formation in the macaque has appeared elsewhere (Enders, 1995). Similar to the human at the start of this stage, there is largely but not entirely syncytial trophoblast as a thin layer between the maternal blood-filled lacunae and the endometrium. Subsequently the cytotrophoblast proliferates and invades the endometrium. However there are two additional complications. Over most of the maternal surface of the implantation site and extending around the site there is a region of enlarged luminal epithelial cells and cells of the gland necks called the epithelial plaque. The central area of the implantation site is devoid of plaque cells, and when the cytotrophoblast penetrates the endometrium the plaque cells are largely by-passed by the invading clusters of cytotrophoblast cells. Subsequently a region of necrosis develops between much of the trophoblastic shell and the endometrium.

Cytokeratin staining of a comparable (day 13 post fertilization) primary villous stage of the macaque showed that there was a relatively abrupt transition between the trophoblast and the endometrium, with very few individual cells progressing any distance into the endometrial stroma except within blood vessels (Figure 6a). Even at this early stage trophoblast cells as a cluster had penetrated well into adjacent endometrial arterioles (Figures 6c, d). In the macaque as in the human, anti-SP1 antibody was confined to syncytial trophoblast. This antigen thus marked the extent of the intervillous spaces (Figure 6b).

With formation of the trophoblastic shell, large masses of cytotrophoblast cells penetrated the endometrium, especially within venules and arterioles. While still in secondary villous stages (prior to circulation of maternal blood), cytotrophoblast penetrates and at least partially occludes the lumen of arterioles as deep as 1 mm from the placental-endometrial interface (Figure 7a). In contrast to comparable human implantation sites, there is little penetration by individual cytotrophoblast cells into the endometrial stroma. A few individual cells that are small and fusiform react with antibody to cytokeratin Most of these cells are found near the invaded arterioles. Occasional regions also show a few cells near the border of the trophoblastic shell, but the widespread radiation of individual cytotrophoblast cells into the endometrial stroma so characteristic of human implantation sites was not found in the macaque (Figure 7b).

DISCUSSION

Early implantation stages of both the macaque and human progress from a lacunar stage to a villous stage and establish the trophoblastic shell with a very similar time sequence. However, there would appear to be some qualitative differences in that the wreath of cytotrophoblast cells entering the stroma, which is so pronounced in the human, appears to be minor if present at all in the macaque. On the other hand, the invasion of trophoblast into blood vessels, especially the arterioles, in the macaque is not found in these earliest stages of the human conceptus. Possibly the more interstitial position of the human conceptus makes it somewhat less susceptible to disruption by excess maternal blood flow into the intervillous spaces than is the macaque conceptus. Consequently establishment of partial regulation of maternal blood flow by fetal trophoblast cells can be delayed by the human conceptus but not the macaque conceptus (Enders and King, 1991). These observations also suggest that the peripheral cytotrophoblast of the human has more extensive ligands for stromal constituents whereas the macaque would have ligands allowing it to adhere to arteriolar endothelium at this early stage.

It is interesting that many of the interstitial trophoblast cells are large and some are multinucleate but these cells were not SP1-positive, indicating that they were not merely displaced syncytial trophoblast. This supports the contention of Pijnenborg and colleagues (Pijnenborg et al., 1981; Pijnenborg, 1994) that the placental bed giant cells are derived from interstitial infiltration of cytotrophoblast, not syncytium. The recent finding that cytotrophoblast from free villi, when explanted onto endometrium in culture, will also migrate into endometrial stroma (Vicovac et al., 1995) suggests that development of migratory phenotype with its accompanying matrix adhesion properties is retained for considerable time by these cells. The differences in migratory behavior between the macaque and human may also help explain the relatively uniform basal plate of the macaque and more irregular interrelationship in the human (Ramsey et al., 1976).

Finally it should be pointed out that although the invasion of the endometrium in macaque and human at this stage is overwhelmingly accomplished by cytotrophoblast, the original invasion of both the epithelium and the maternal vessels is accomplished in both species by syncytial trophoblast (Knoth and Larsen, 1972 ; Enders and King, 1991; Enders, 1989, 1995).

SUMMARY

Archival specimens of human implantation sites were destained and reacted with antibodies to cytokeratin and SP1. These specimens were compared to recently collected macaque implantation sites of similar age. It was found that, in the human, individual cytokeratin positive cytotrophoblast cells from the forming trophoblastic shell penetrated into the endometrial stroma and around the more superficial coils of the uterine glands. However, these interstitial cytotrophoblast cells showed no tendency to enter either the endometrial blood vessels or glands. In contrast, in the macaque, cytokeratin positive cells entered the arterioles even before formation of the trophoblastic shell, and showed a substantially lesser tendency to migrate into the stroma as individual interstitial cytotrophoblast cells. It is suggested that, despite similarities in trophoblastic shell formation, the population of migrating cytotrophoblast must have different adhesion capabilities in the two species.

ACKNOWLEDGEMENTS

These studies were supported by grant HD10342 from the National Institute of Child Health and Human Development, and NIH RR00169 to the California Regional Primate Research Center. The author thanks the staff of the primate center, especially David Robb and Dr. Alice Tarantal, for timing and arranging for macaque implantation specimens, and Katy Lantz for excellent assistance in tissue preparation and sectioning for light and electron microscopy.

REFERENCES

Aplin, J.D. (1993) Expression of integrin α6β4 in human trophoblast and its loss from extravillous cells. *Placenta* 14, 203-215.

Blankenship, T. and King, B.F. (1993) Developmental changes in the cell columns and trophoblastic shell of the macaque placenta: An immunohistochemical study localizing type IV collagen, laminin, fibronectin and cytokeratins. *Cell Tissue Res.* 274, 457-466.

Clark, R.K. and Damjanov, I. (1985) Intermediate filaments of human trophoblast and choriocarcinoma cell lines. *Virch. Archiv A: Pathol. Anat. Histol.* 407, 203-208.

Douglas, G.C. and King, B.F. (1990) Differentiation of human trophoblast cells in vitro as revealed by immunocytochemical staining of desmoplakin and nuclei. *J. Cell Sci.* 96, 131-141.

Enders, A.C. (1989) Trophoblast differentiation during the transition from trophoblastic plate to lacunar stage of implantation in the rhesus monkey and human. *Am. J. Anat.* 186, 85-98.

Enders, A.C. (1993) Overview of the morphology of implantation in primates. In: *In Vitro Fertilization and Embryo Transfer in Primates,* (eds.) D.P. Wolf, R.L. Stouffer, and R.M. Brenner, New York: Springer-Verlag, pp. 145-157.

Enders, A.C. (1995) Transition from lacunar to villous stage of implantation in the macaque, including establishment of the trophoblastic shell. *Acta Anat.* 152, 151-169.

Enders, A.C. and King, B.F. (1991) Early stages of trophoblastic invasion of the maternal vascular system during implantation in the macaque and baboon. *Am. J. Anat.* 192, 329-346.

King, B.F. and Blankenship, T.N. (1993) Development and organization of primate trophoblast cells. In: *Trophoblast Cells: Pathways for Maternal-Embryonic Communication,* (eds.) J.M. Soares, S. Handwerger, and F. Talamantes, New York: Springer-Verlag, pp. 13-30.

Knoth, M. and Larsen, J.F. (1972) Ultrastructure of a human implantation site. *Acta Obstet. Gynec. Scand.* 51, 385-393.

Loke, Y.W., Butterworth, B.H., Margetts, J.J. and Burland, K. (1986) Identification of cytotrophoblast colonies in cultures of human placental cells using monoclonal antibodies. *Placenta* 7, 221-231.

Mirecka, J. and Hanarz, M. (1990) Localization of intermediate filament proteins in various structures of the human placenta as revealed by immunoperoxidase technique. *Gegenb. Morphol. Jahrb.* 136, 695-708.

Muhlhauser, J., Crescimanno, C., Kasper, M., Zaccheo, D. and Castellucci, M. (1995) Differentiation of human trophoblast populations involves alterations in cytokeratin patterns. *J. Histochem. Cytochem.* 43, 579-589.

Ockleford, C.D., Wakely, J. and Badley, R.A. (1981) Intermediate filament proteins in human placenta. *Cell Biol. Int. Reports* 212, 305-316.

Pijnenborg, R., Bland, J.M., Robertson, W.B., Dixon, G. and Brosens, I. (1981) The pattern of interstitial trophoblastic invasion of the myometrium in early human pregnancy. *Placenta* 2, 303-316.

Pijnenborg, R. (1994) Trophoblast invasion. *Reprod. Med. Rev.* 3, 53-73.

Ramsey, E.M., Houston, M.L. and Harris, J.W.S. (1976) Interactions of the trophoblast and maternal tissues in three closely related primate species. *Am. J. Obstet. Gynecol.* 124, 647-652.

Vicovac, L., Jones, C.J.P. and Aplin, J.D. (1995) Trophoblast differentiation during formation of anchoring villi in a model of the early human placenta in vitro. *Placenta* 16, 41-56.

Wells, M. and Bulmer, J.N. (1988) The human placental bed: histology, immunohistochemistry and pathology. *Histopathology* 13, 483-498.

Trophoblast Research 10:97-113, 1997

EXPRESSION OF MATRIX METALLOPROTEINASES AND THEIR INHIBITORS DURING HUMAN PLACENTAL DEVELOPMENT

Béatrice Nawrocki[1,3], Myriam Polette[1], Erik Maquoi[2] and Philippe Birembaut[1]

[1]I.N.S.E.R.M. U.314
CHU Maison Blanche
45, rue Cognacq-Jay
51100 Reims, France

[2]Laboratoire de Biologie Générale
Tour de Pathologie
CHU Liège, Belgique

INTRODUCTION

Human placental implantation represents one of the most highly invasive physiological process and can be therefore related to the invasion of tumor cells. However, unlike tumor invasion, placentation is tightly regulated during gestation both spatially and temporally (Lala and Graham, 1990; Graham and Lala, 1991). In particular, trophoblast cells and especially extravillous trophoblast cells, which arise from the proliferative trophoblast cell columns, are responsible for the anchoring of the placenta. These cells express therefore the invasive phenotype necessary to penetrate deeply into the maternal decidualized endometrium (Kurman et al., 1984; Yeh and Kurman, 1989; Graham and Lala, 1992; Lewis and Bernirschke, 1992; Foidart et al., 1993).

The fact that during implantation invasive trophoblast cells have to break through the maternal extracellular matrix suggests that matrix degrading enzymes may play important roles in trophoblast invasiveness. Indeed, serine proteases like the plasmin system (Sherman et al., 1976; Strickland et al., 1976; Sappino et al., 1989; Castelluci et al., 1994; Hofman et al., 1994; Multhaup et al., 1994) and matrix metalloproteinases (MMPs) (Fisher et al., 1985; Brenner et al., 1989; Moll and Lane, 1990; Librach et al., 1991; Autio-Harmeinen et al., 1992; Fernandez et al., 1992; Werb et al., 1992) have been shown to be secreted by trophoblast cells *in vitro* and *in vivo*. Among these proteases, MMPs, which are able to degrade all components of the extracellular matrix, seem to have a particularly pivotal role during placentation.

MMPs form a family of enzymes commonly divided into three groups according to their substrate specificities: 1) interstitial collagenases including three enzymes, fibroblast collagenase (MMP-1), polymorphonuclear collagenase (MMP-8) and a recently isolated tumor-associated collagenase 3 (Freije et al., 1994) that digest the fibrillar collagens type I, II, III; 2) type IV collagenases including a 72 KDa form also called gelatinase A (MMP-2) and a 92 KDa form also called gelatinase B (MMP-9) (Tryggvason et al., 1993) which degrade denatured collagens (gelatins) and native type IV collagen (Harris et al., 1972; Murphy et al., 1985); 3) stromelysins including stromelysin-1 (MMP-

[3]To Whom Correspondence Should Be Addressed.

3), stromelysin-2 (MMP-10), stromelysin-3 first described in breast cancers (Basset et al., 1990), the matrilysin (PUMP-1 or MMP-7), and a metallo-elastase isolated from human lung alveolar macrophages (Shapiro et al., 1993) which have the broadest substrate range (laminin, fibronectin, elastin, proteoglycan core protein and collagens (Nagase et al., 1991) (cf., Kleiner and Stetler-Stevenson, 1993; Ray and Stetler-Stevenson, 1994; Cawston, 1995). A new subclass of MMPs has been recently identified, MT-MMPs (membrane-type MMP), which include actually MT1-MMP, MT2-MMP and MT3-MMP (Sato et al., 1994; Will and Hinzmann, 1995; Takino et al., 1995). MT1-MMP is a unique MMP that contains a transmembrane domain which localizes this enzyme at the cell surface. All MMPs are generally secreted as inactive proenzymes and require activation into their mature forms before they can degrade their substrates. Among them, gelatinase A activation appears as a membrane-associated mechanism which requires the presence of MT1-MMP (Atkinson et al., 1995; Cao et al., 1995).

MMP genes are selectively activated during implantation and MMPs are expressed by invasive trophoblast cells, particularly type I collagenase (Yagel et al., 1988; Moll and Lane, 1990) and gelatinases (Bishof et al., 1991; Autio-Harmeinen et al., 1992; Fernandez et al., 1992) but also stromelysin-1 (Brenner et al., 1989; Dieron and Bryant-Greenwood, 1991). It has been shown that first trimester trophoblast cells secrete gelatinases *in vitro* (Librach et al., 1991; Fisher et al., 1989; Bishof et al., 1991) and cultured first trimester trophoblast cells display an invasive phenotype associated with gelatinase B expression (Fisher et al., 1985; Emonard et al., 1990; Emonard et al., 1993). Moreover Librach et al. (1991) have demonstrated that an anti-gelatinase B antibody can block invasion of matrigel by the early trophoblast cells. Also in agreement with the finding that cultured third trimester are no longer invasive, Fisher et al. (1989) have shown that gelatinase B expression is strongly reduced later in gestation. Gelatinase A is also secreted by early trophoblast cells *in vivo* and *in vitro* (Emonard et al., 1993; Fernandez et al., 1992; Autio-Harmeinen et al., 1992). As MT1-MMP is a potential gelatinase A activator, it is implicated in the implantation process.

Even though stromelysin-3 has to date only been described in term placenta (Basset et al., 1990) it is involved in many other invasive processes such as invasive cancers, cutaneous wound healing, uterine involution (Basset et al., 1990; Muller et al., 1993; Polette et al., 1993b; Wagner et al., 1992; Wolf et al., 1992; Wolf et al., 1993). Taken together, these data suggest that all the above mentioned MMPs could play an essential role in the invasion of trophoblast.

The activity of MMPs is tightly controlled by a specific class of tissue inhibitors known as TIMPs (Tissue Inhibitors of Metalloproteinases). Three distinct TIMPs have been isolated. TIMP-1 inhibits all activated MMPs and both latent and active forms of progelatinase B (Goldberg et al., 1992). TIMP-2 binds either latent or activated forms of progelatinase A. It also has an inhibitory activity against other members of the MMP family (Stetler-Stevenson et al., 1989; De Clerck et al., 1989). TIMP-3 which has been recently discovered also has inhibitory activity against MMPs (Leco et al., 1994; Uria et al., 1994; Apte et al., 1994).

TIMPs (TIMP-1 and TIMP-2) have been shown to be expressed *in vitro* by maternal decidual cells (Graham and Lala, 1991) and *in vivo* in the murine decidua (Nomura et al., 1989; Werb et al., 1992). These data have suggested that implantation and consecutive placentation could depend on a balance between MMPs and TIMPs expression.

In order to identify the cells responsible for the production *in vivo* of MMPs and their inhibitors in implantation process but also to verify the hypothesis that a cooperation exists between maternal decidual cells and fetal trophoblast cells, we have undertaken immunohistochemical and *in situ* hybridization studies for gelatinases A/B, MT1-MMP, stromelysin-3 in parallel with their inhibitors TIMP-1, TIMP-2 and TIMP-3 in human placental bed and placental villi of first and third trimesters of gestation. We also quantified the level of mRNAs of MMPs and TIMPs to evaluate the possible changes in their levels of expression in first and third trimesters.

MATERIALS AND METHODS

Tissue Preparation

Samples of seven human placental villi were obtained from first trimester (10-12 weeks) voluntarily interrupted pregnancies. Five first trimester (10-12 weeks) placental bed biopsies were obtained *in situ* by echoguided chorionic villi sampling forceps in normal pregnancies. This procedure allows the precise anatomical localization of the biopsy and prevents tissue damage during placental aspiration or delivery. One part of each tissue sample was fixed in 10% formalin and embedded in paraffin for immunoperoxidase technique and *in situ* hybridization, while the other part was frozen in liquid nitrogen for immunofluorescence technique and Northern blotting.

Immunohistochemistry

MT1-MMP, Gelatinase A And Stromelysin-3 Localizations

Five µm tissue paraffin sections were rehydrated and treated with 0.3% hydrogen peroxide for five minutes to quench endogenous peroxidase activity. Nonspecific binding was blocked with serum (Dako, Carpinteria, CA, USA) for 20 minutes. Slides were incubated with anti-MT1-MMP 113-5B7 monoclonal antibody (2 µg/ml) (directed against the peptide CDGNFDTVAMLRGEM - residues 310-333 - located in the extracellular domain of the enzyme) (Fuji Chemical Industries, Ltd., Takaoka, Japan) or anti-gelatinase A monoclonal antibody (2 µg/ml) (Oncologix, Gaithersburg, MD, USA) or mouse monoclonal antibody raised against the hemopexin domain of stromelysin-3 (1/4000) (MAb 5ST-4A9; a gift from Pr. P. Basset, Strasbourg, France) overnight at 4°C. Negative controls were carried out by replacing the primary antibody with non-immune immunoglobulin G.

After three ten-minute washes in PBS, further steps were performed with the peroxidase LSAB kit (labeled streptavidin biotin method, Dako, Carpinteria, CA, USA). Peroxidase activity was revealed with AEC (3-amino-9-ethylcarbazole) chromogen providing a red-brown color product. All slides were counterstained with Mayer's hematoxylin, mounted and examined under a Zeiss Axiophot microscope.

Epithelial cells and trophoblast cells were identified with an anti-cytokeratin 8/18/19 monoclonal antibody (dilution 1/20) (Monosan/Sanbio, Netherlands) using the same detection methodology as above.

Gelatinase B, TIMP-1 And TIMP-2 Localizations

Nonspecific binding was blocked with 3% bovine serum albumin-phosphate buffered saline (PBS) for 30 minutes. Cryosections (5 µm) were treated with affinity purified antibodies against gelatinase B, TIMP-1 (gifts from Dr. G. Murphy, Cambridge, United Kingdom) and TIMP-2 (a gift from Y. De Clerck, Los Angeles, USA). Dilutions were 20 µg/ml for TIMP-2 and 30 µg/ml for TIMP-1 and gelatinase B. Negative controls consisted of slides incubated with either normal sheep or rabbit immune sera or with PBS. After a one hour incubation, tissue sections were washed three times for five minutes each in PBS and incubated with a biotinylated secondary antibody for one hour (1/50) (goat anti-rabbit antibody for TIMP-2; goat anti-sheep antibody for gelatinase B and TIMP-1 (Amersham, United Kingdom) and followed by streptavidin fluorescein complex (30 minutes, 1/50, Amersham, United Kingdom).

In situ Hybridization

Tissue paraffin sections (5 µm) were rehydrated and treated with 0.2 N HCl for 20 minutes at room temperature, followed by 15 minutes in proteinase K (2 µg/ml in Tris-EDTA-NaCl, 37°C, Sigma Chemical Company, St. Louis, MO, USA) to remove basic proteins. Sections were washed in 2xSSC (Standard Saline Citrate), acetylated in 0.25% acetic anhydride in 0.1M triethanolamine for 10 minutes and then hybridized overnight (50°C) with (^{35}S) labeled antisense RNA transcripts.

MT1-MMP cDNA (1760 bp), prepared by PCR amplification as described in the paper of Gilles et al (1996), gelatinase A cDNA (1500 bp) (a gift from G. Murphy Cambridge, UK), gelatinase B cDNA (1700 bp) (a gift from K. Tryggvason, Oulu, Finland), TIMP-1 cDNA (670 bp), TIMP-2 cDNA (1035 bp) (gifts from Y. De Clerck, Los Angeles, USA), stromelysin-3 cDNA (1600 bp) and TIMP-3 cDNA (1976 bp) (gifts from P. Basset, Strasbourg, France) inserts were subcloned into Bluescript plasmids and used to prepare (^{35}S) RNA probes. Hybridizations were followed by RNase treatment (20 µg/ml, 1 hour, 37°C) to remove unhybridized probes and two stringent washes (50% formamide/2xSSC, 2 hours, 60°C). Autoradiography using D19 emulsion (Kodak) lasted 15 days at -20°C. After exposure, sections were developed and counterstained with HPS (hematoxylin phloxin safranin), mounted and examined under a Zeiss Axiophot microscope. Controls were performed under the same conditions using (^{35}S) labeled sense RNA probes.

Northern Blot Analysis

Total RNA was extracted from tissues by RNAzol treatment (Biogenesis, Bournemouth, UK). Fifteen micrograms of each RNA was analyzed by electrophoresis on 1% Agarose gels containing 10% formaldehyde and transferred onto nylon membranes (Hybond™-N, Amersham, UK). Membranes were hybridized with the cDNA probes (as described above) labeled with ^{32}P using random-priming synthesis (5x10^{8} cpm/µg) (Dupont de Nemours, Brussels, Belgium). The filters were exposed for 10 days, at -80°C. The membrane was rehybridized to an oligonucleotide probe for human 28S rRNA (Cloneth, Palo Alto, CA, USA), which served as a control. Signal intensities were recorded using a Ultrascan XL laser-scanning densitometer.

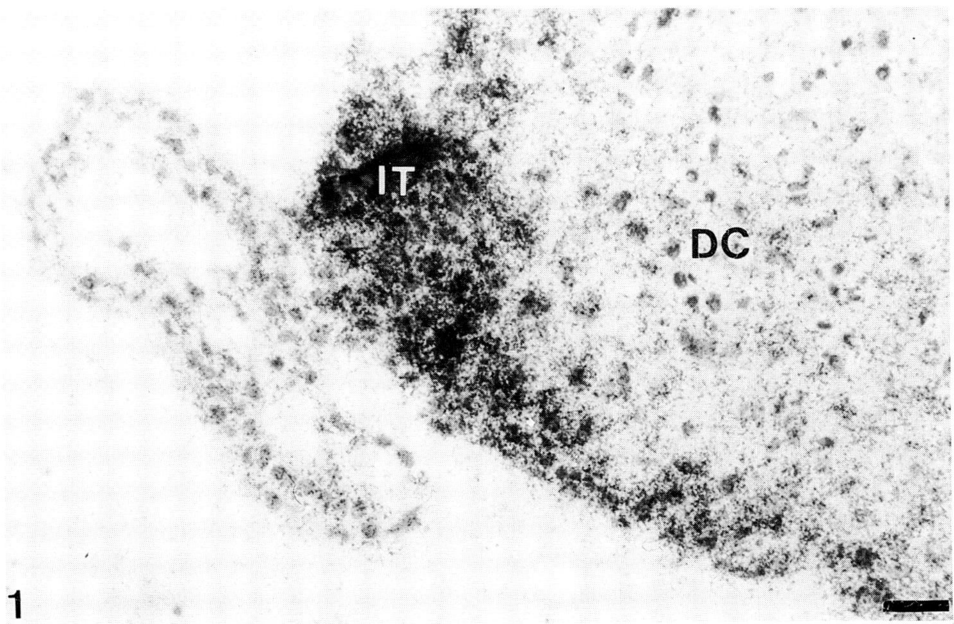

Figure 1. MT1-MMP mRNAs are detected in invasive trophoblast cells (IT) and in decidual cells (DC) in the first trimester placental bed (bar = 70 μm).

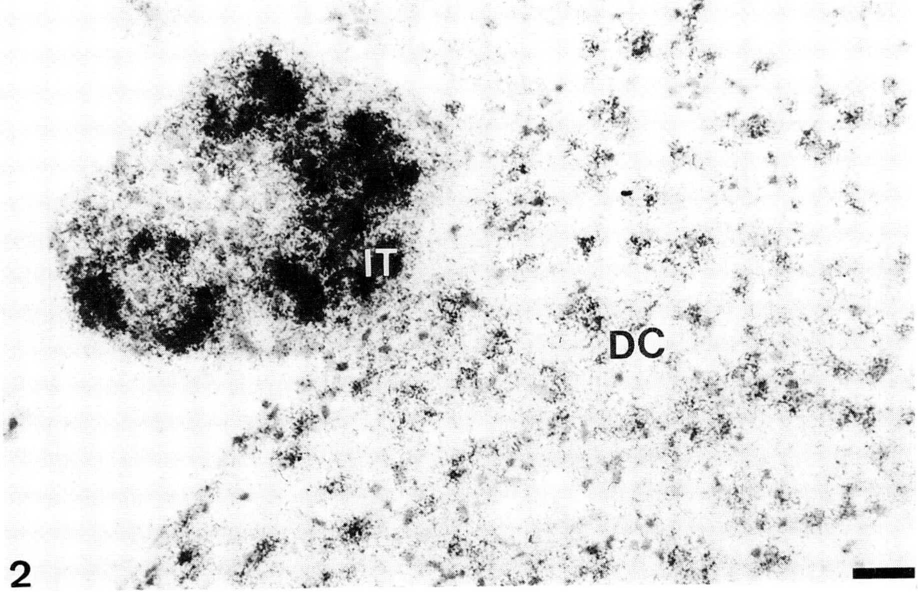

Figure 2. Gelatinase A mRNAs are also distributed in invasive trophoblast cells (IT) and in decidual cells (DC) in the first trimester placental bed (bar = 70 μm).

RESULTS

Placental Beds

In first trimester placental beds, MT1-MMP (Figure 1), gelatinase A (Figure 2) and gelatinase B (data not shown) were detected by *in situ* hybridization and immunohistochemistry in extravillous anchoring trophoblastic cell columns in close contact with the decidua, in the isolated invasive extravillous trophoblast cells (recognized with the anti-cytokeratin antibody) which are scattered deeply in the maternal compartment and also in maternal decidual cells. In addition, intravascular trophoblast cells and placental bed giant cells were also positive for mRNAs and protein of these three MMPs (data not shown). *In situ* hybridization and immunohistochemistry displayed a strong expression of stromelysin-3 mRNAs and protein only in extravillous trophoblast cells infiltrated among negative decidual cells in the maternal compartment and in intravascular trophoblast cells and placental bed giant cells (data not shown). In parallel, first trimester maternal decidual cells were highly positive for TIMP-1, TIMP-2 and TIMP-3 by immunohistochemical detection and *in situ* hybridization (data not shown).

At term, MT1-MMP and gelatinase A expression persisted in some invasive trophoblast cells in the decidua and also in stromal cells for gelatinase A and MT1-MMP and in trophoblast (syncytiotrophoblast and cytotrophoblast) which delimits the villous surface of anchoring villi for MT1-MMP (data not shown). Stromelysin-3 mRNAs and protein were still localized in most of the infiltrated extravillous trophoblast cells (data not shown). In contrast, gelatinase B was no longer detected, either by *in situ* hybridization or by immunohistochemistry in third trimester placental beds (data not shown).

Expression of TIMP-1, TIMP-2 and TIMP-3 was always present in maternal decidual cells and TIMP-1 especially showed a higher level of expression of mRNAs in decidual cells at term than during the first trimester (Figure 3) (Table 1).

Floating Villi

In first trimester floating villi, MT1-MMP mRNAs and protein were expressed in villous trophoblast cells (cytotrophoblast and syncytiotrophoblast). A faint labeling of stromal core was observed too. Gelatinase B and TIMP-2 were found by the both methods in trophoblast cells (cytotrophoblast and syncytiotrophoblast) which delimit the villous surface. For gelatinase A, protein was localized in both villous trophoblast and stromal cells of first trimester floating villi, but mRNAs were only distributed over the stromal core (including the fetal fibroblast cells, Hofbauer's cells and vascular endothelial cells). Expression of stromelysin-3 was only restricted in syncytiotrophoblast whereas TIMP-1 and TIMP-3 were positive in the stromal compartment of the floating villi (data not shown).

Third trimester floating villi showed the same pattern of expression as during the first trimester for MT1-MMP, gelatinase A, stromelysin-3, TIMP-1, TIMP-2 and TIMP-3 except for gelatinase B which was detected neither by *in situ* hybridization nor by immunohistochemistry (data not shown).

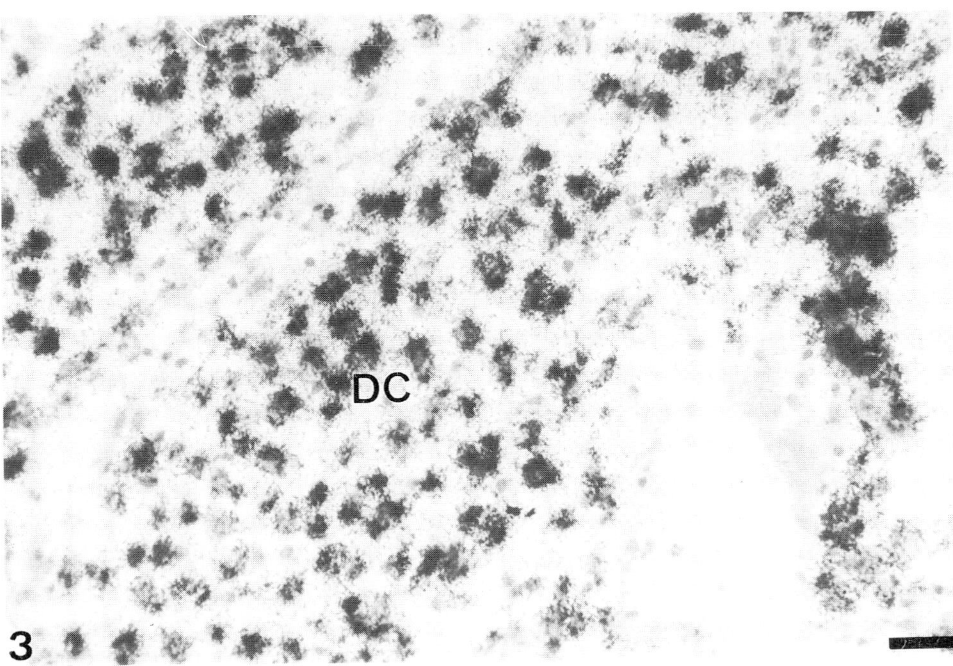

Figure 3. TIMP-1 transcripts are expressed in numerous maternal decidual cells (DC) in the third trimester placental bed (bar = 70 μm).

Northern Blot Analysis

Consistent results were obtained by Northern blot analysis.

Placental beds expressed 4 fold more MT1-MMP mRNAs than floating villi but no variation was observed between first and third trimesters of gestation.

There was no variation between first and third trimesters for gelatinase A and TIMP-2 mRNAs in placental beds or floating villi.

Gelatinase B mRNA was not detected by Northern blotting in third trimester placental beds or floating villi.

Third trimester floating villi expressed 4-fold more stromelysin-3 mRNA than during the first trimester, whereas first trimester placental beds expressed 4-fold more stromelysin-3 mRNA than placental beds of third trimester.

Northern blot analysis indicated a 3 fold increase in TIMP-1 and TIMP-3 mRNA levels from samples of placental bed biopsies at term pregnancy. No significant variations of TIMP-1 and TIMP-3 mRNA levels were detected between first and third trimester floating villi (data not shown).

Table 1

Principal localizations of MMPs and TIMPs in placental beds during the pregnancy (TCC: trophoblast column cells; IT: invasive extravillous trophoblast cells; DC: decidual cells; PBGC: placental bed giant cells; IVT: intravascular trophoblast cells). Arrows indicate the variations of mRNA expression levels between first and third trimesters of gestation for each MMP and TIMP.

	First Trimester Placental Beds		Third Trimester Placental Beds
MT1-MMP	TCC IT DC PBGC IVT	↔	IT
Gelatinase A	TCC IT DC PBGC IVT	↔	IT
Gelatinase B	TCC ITT DC	↘ ↘	
Stromelysin-3	IT PBGC IVT	↘	IT
TIMP-1	DC	↗	DC
TIMP-2	DC	↔	DC
TIMP-3	DC	↗	DC

DISCUSSION

Our observations indicate that MT1-MMP, gelatinase A, gelatinase B and stromelysin-3 are highly expressed in first trimester placental bed and especially in invasive extravillous trophoblast cells. Therefore, MMP expression seems to be associated with an invasive phenotype of trophoblast (Librach et al., 1991; Mignatti and Rifkin, 1993). This is in agreement with findings describing a key role of MMPs in other invasive processes such as wound healing (Wolf et al., 1992) and tumor invasion (Bernhard et al., 1990; Levy et al., 1991; Davies et al., 1993; Gilles et al., 1994). As suggested by our own studies, other authors have established the important role of gelatinases *in vivo* or *in vitro* in animal or human implantation processes (Blankenship and King, 1994; Shimonovitz et al., 1994). Indeed gelatinase A and gelatinase B have been found in extravillous anchoring trophoblastic cell columns in close contact with the decidua and in extravillous trophoblast cells deeply scattered in the decidua in the first trimester placental bed samples. Interestingly, gelatinase A mRNA is particularly strongly expressed in the tips of the trophoblastic columns, i.e. at the front of invasion (Polette et al., 1994, Nawrocki et al., 1996). Furthermore, we have observed that MT1-MMP pattern is similar to the one described for progelatinase A especially in trophoblastic cell columns and infiltrated extravillous trophoblast cells of first trimester of gestation (Polette et al., 1994; Nawrocki et al., 1996). Indeed, the same cells produce simultaneously the proteolytic enzyme progelatinase A and its cell surface activator MT1-MMP. This result pleads in favor of a coordinated transcriptional regulation of the genes encoding both MT1-MMP and gelatinase A in human implantation sites. Strongin et al. (1995) have suggested that MT1-MMP could be activated by the plasmin cascade leading to the subsequent activation of progelatinase A. This activation is mediated by a ternary complex consisting of activated MT1-MMP/TIMP-2/progelatinase A. The activated MT1-MMP serves as a receptor for TIMP-2 that can bind progelatinase A (Strongin et al., 1993, 1995). Our findings strongly support the idea that trophoblast cells producing MT1-MMP could then activate progelatinase A in order to degrade the basement membranes and extracellular matrix components of the endometrium.

Even though stromelysin-3 would have a weak activity as a MMP (Murphy et al., 1993), it seems to be involved in the invasive placentation process (Maquoi et al., 1997). Indeed, Lefebvre et al. (1995) have detected stromelysin-3 in developmental murine placenta. Moreover, the pattern of stromelysin-3 distribution is closely similar to that of u-PA (urokinase type- plasminogen activator) (Hofmann et al., 1994) as it has been described in tumors (Basset et al., 1993). These observations suggest that both enzymes could cooperate during implantation through a proteolytic cascade involving the plasmin in order to degrade directly extracellular matrix. Stromelysin-3 has recently been demonstrated to cleave the proteinase inhibitors α2-macroglobulin, α1-PI (Plasmin Inhibitor α1) and α2-AP (α2-antiplasmin) (Pei et al., 1995). Stromelysin-3 could be implicated in a cascade leading to the activation of other proteases involved in trophoblast implantation. As other members of stromelysin subclass, stromelysin-3 has previously been described to be exclusively expressed in the stromal compartment of invasive human carcinomas and in other human tissues (Basset et al., 1993). This study constitutes one of the first non stromal localization of stromelysin-3 so far reported (Maquoi et al., 1997). Taken together our results suggest that the expression of MMPs in trophoblast cells is temporally associated with an invasive phenotype. These findings are further supported by our data showing vimentin expression in some trophoblast cells (data not shown). Indeed, the unusual vimentin synthesis in epithelial cells has been associated with the acquisition of a migratory phenotype and the ability to synthesize MMPs implicated in other processes such as tumor invasion (Gilles and Thompson, 1996) or wound healing (Buisson et al., 1996).

The invasive and migratory phenotype of trophoblast cells could be tightly regulated by the MMP inhibitors. Indeed, first and third trimester maternal decidual cells have been shown to produce TIMP-1, TIMP-2 and TIMP-3 (Polette et al., 1994). In particular, the increased level of TIMP-1 mRNA during the third trimester can be correlated with the dramatic decrease of gelatinase B level (Polette et al., 1994). TIMPs, by inhibiting MMPs, could limit the trophoblast invasion to the inner third of the uterus (Graham and Lala. 1991) and stop it at term of gestation. Indeed, decidua is known to have a protective role against extensive invasion (Kirby et al., 1960, 1963a, b, 1965; Lala and Graham, 1990; Graham and Lala, 1991; Werb et al., 1992).

Paradoxically, we observed that first trimester maternal decidual cells also produce MT1-MMP, gelatinase A and gelatinase B mRNAs and proteins (Polette et al., 1994; Nawrocki et al., 1996). As already suggested by Rawdanowicz (1992), Martelli (1993) and Salamonsen (1994) studies, our data support the idea that, during the first trimester, maternal decidual cells could cooperate with fetal trophoblast cells to produce MMPs, leading to the migration of intermediate trophoblast cells.

MMPs were also distributed in stromal core and/or trophoblast cells which delimit the surface of floating villi during first (gelatinases A/B, MT1-MMP, stromelysin-3) and third (gelatinase A, MT1-MMP, stromelysin-3) trimesters of gestation (Polette et al., 1994; Nawrocki et al., 1996; Maquoi et al., 1997). They could be involved in the growth and the branching of villi. In parallel, their specific inhibitors, TIMP-1, TIMP-2 and TIMP-3, which were also present in floating villi (Polette et al., 1994), could limit the excessive activity of MMPs.

This study suggests that human trophoblast implantation results *in vivo* from a balance between MMPs and TIMPs secreted respectively by fetal trophoblast cells and maternal decidual cells. Like in tumor invasion, MMPs seem to be directly involved in the invasion process of placentation since they are overexpressed in invasive fetal cells at the maximal time of invasion. In contrast with tumor invasion, placentation is tightly both spatially and temporally regulated. Moreover, in this physiological invasive phenomenon, MMPs are produced by infiltrating cells themselves whereas in the most epithelial tumors, the tumor cells induce the synthesis of these enzymes in adjacent host stromal cells (Polette et al., 1993a, Polette et al., 1996).

In conclusion, MMPs may have a dramatic role *in vivo* in trophoblast implantation and consecutive placentation. In this particular physiological process, MMPs expression is associated with an invasive phenotype.

SUMMARY

MMPs (Matrix Metalloproteinases) are thought to play important roles in invasion processes (normal or pathological) by degrading basement membranes and extracellular matrix components which constitute a barrier for the migrating cells. Human trophoblast implantation is a physiological invasive process which is both spatially and temporally regulated. We have studied by immunohistochemistry, *in situ* hybridization and Northern blotting the expression of several MMPs (gelatinases A/B,

MT1-MMP and stromelysin-3) and their specific inhibitors (TIMP-1, TIMP-2, TIMP-3) in first and third trimester human placenta in order to specify their role *in vivo* in this type of invasion. All the MMPs studied were highly expressed especially in first trimester invasive trophoblast cells. In contrast, TIMPs were strongly expressed in decidual cells especially during the third trimester of pregnancy. These data suggest that trophoblast implantation depends on a spatio-temporal balance between MMPs expressed by fetal cells and TIMPs expressed by maternal cells. As in other invasion processes, MMP expression is associated with an invasive phenotype.

ACKNOWLEDGEMENTS

We gratefully thank Drs. G. Murphy (Strangeways Research Laboratory, Cambridge, UK), K. Iwata (Fuji Chemical Industries Ltd, Takaoka, Japan), P. Basset (Strasbourg, France), Y. De Clerck (Los Angeles, USA) and K. Tryggvason (Oulu, Finland) for their respective generous gifts of probes and antibodies. This work was supported by a grant from the A.R C. n° 1096, and the Lions Club of Soissons. Béatrice Nawrocki is the recipient of a grant from the A.R.C. (Association de Recherche contre le Cancer).

REFERENCES

Apte, S.S., Mattei, M.G. and Olsen, B.R. (1994) Cloning of the cDNA encoding human tissue inhibitor of metalloproteinases-3 (Timp-3) and mapping of the timp-3 gene to chromosome 22. *Genomics* 19, 56-90.

Atkinson, S., Crabbe, T., Cowell, S., Ward R. V., Butler, M. J., Sato, H., Seiki, M., Reynolds, J.J. and Murphy, G. (1995) Intermolecular autolytic cleavage can contribute to the activation of progelatinase A by cell membranes. *J. Biol. Chem.* 270, 30471-30485.

Autio-Harmeinen, H., Hurkainen, T., Niskasaari, K., Höyhtya, M. and Tryggvason, K. (1992) Simultaneous expression of 70 kilodalton type IV collagenase and type IV collagen alpha 1 (IV) chain genes by cells of early human placenta and gestational endometrium. *Lab. Invest.* 67, 143-153.

Basset, P., Bellocq, J.P., Wolf, C., Stoll, I., Hutin, P., Limacher, J.M., Podhajcer, O.L., Chenard, M.P., Rio, M.C.and Chambon, P. (1990) A novel metalloproteinase gene specifically expressed in stromal cells of breast carcinomas. *Nature* 348, 699-704.

Basset, P., Wolf, C. and Chambon, P. (1993) Expression of the stromelysin-3 gene in fibroblastic cells of invasive carcinomas of the breast and other human tissues: A review. *Br. Cancer Res. Treat.* 24, 185-193.

Bernhard, E.J., Muschel, R.J. and Hughes, E.N. (1990) Mr 92,000 gelatinase release correlates with the metastatic phenotype in transformed rat embryo cells. *Cancer Res.* 50, 3872-3877.

Bishof, P., Friedli, E. Martelli, M. and Campana, A. (1991) Expression of extracellular matrix-degrading metalloproteinases by cultured human cytotrophoblast cells: Effects of cell adhesion and immunopurification. *Am. J. Obstet. Gynaecol.* 165, 1791-1801.

Bischof, P. and Martelli, M. (1992) Proteolysis in the penetration phase of the implantation process. *Placenta* 13, 17-24.

Blankenship, T.N. and King, B.F. (1994) Identification of 72-kilodalton type IV collagenase at sites of macaque spiral arteries. *Placenta* 15, 177-187.

Brenner, C.A., Adler, R.R., Rappolee, D.A., Pedersen, R.A. and Werb, Z. (1989) Genes for extracellular matrix-degrading metalloproteinases and their inhibitor, TIMP, are expressed during early mammalian development. *Genes Dev.* 3, 849-859.

Buisson, A.C., Gilles, C., Polette, M., Zahm, J.M., Birembaut, P. and Tournier, J.M. (1996) Wound repair-induced expression of stromelysins is associated with the acquisition of a mesenchymal phenotype in human respiratory epithelial cells. *Lab. Invest.* 74, 1-12.

Cao, J., Sato, H., Takino, T. and Seiki, M. (1995) The C-terminal region of membrane type matrix metalloproteinase is a functional transmembrane domain required for pro-gelatinase A activation. *J. Biol. Chem.* 270, 801-805.

Castellucci, M., Theelen, T., Pompili, E., Fumagalli, L., De Renzis, G. and Mühlauser, J. (1994) Immunohistochemical localization of serine-protease inhibitors in the human placenta. *Cell Tissue Res.* 278, 283-289.

Cawston, T.E. (1995) Proteinases and inhibitors. *Brit. Med. Bull.* 51, 385-401.

Davies, B., Miles, D.W., Happerfield, L.C., Naylor, M.S., Bobrow, L.G., Rubens, R.D. and Balkwill, F.R. (1993) Activity of type IV collagenases in benign and malignant breast disease. *Br. J. Cancer* 67, 1126-1131.

De Clerck, Y., Yean, T., Ratzkin, B., Lu, H. and Langley, K. (1989) Purification and characterization of two related and distinct metalloproteinase inhibitors secreted by bovine aortic endothelial cells. *J. Biol. Chem.* 264, 17445-53.

Dieron, D. and Bryant-Greenwood, G.D. (1991) Collagens, collagenolytic enzymes, and inhibitors in the human fetal membranes and decidua. *Troph. Res.* 5, 205-216.

Emonard, H., Christiane, Y; Munaut, C. and Foidart, J.M. (1990) Reconstituted basement membrane matrix stimulates interstitial procollagenase synthesis by human fibroblasts in culture. *Matrix* 10, 373-377.

Emonard, H., Aghayan, M., Smet, M., Schaaps, J.P., Grimaud, J.A., Christiane Y. and Foidart, J.M. (1993) Role of extracellular matrix in regulation of type IV collagenases synthesis by human trophoblast cells and their malignant counterparts. *Tropho. Res.* 7, 201-210.

Fernandez, P.L., Merino, M.J., Nogales, F.F., Charonis, A.S., Stetler-Stevenson, W.G. and Liotta, L. (1992) Immunohistochemical profile of basement membrane proteins and 72 kilodalton type IV collagenase in the implantation placental site. *Lab. Invest.* 66, 572-579.

Fisher, S.J., Leitch, M.S., Kantor, M.S., Basbau, C.B. and Kramer, R.H. (1985) Degradation of extracellular matrix by the trophoblastic cells of first trimester human placentas. *J. Cell. Biochem.* 27, 31-40.

Fisher, S.J., Cui, T.Y., Zhang, L., Hartman, L., Grahl, K., Guo-Yang, Z., Tarpey, J. and Damsky, C.H. (1989) Adhesive and degradative properties of human placental cells in vitro. *J. Cell Biol.* 109, 891-902.

Foidart, J.M., Hustin, J., Dubois, M. and Schaaps, J.P. (1993) The human placenta becomes haemochorial at the 13th week of pregnancy. *Int. J. Dev. Biol*, 36, 451-453.

Freije, J.M.P., Diez-Itza, I., Balbin, M., Sanchez, L.M., Blasco, R., Tolivia, J. and Lopez-Otin, C. (1994) Molecular cloning and expression of collagenase-3, a novel human matrix metalloproteinase produced by breast carcinomas. *J. Biol. Chem.* 269, 16766-16773.

Gilles, C., Polette, M., Piette, J., Birembaut, P. and Foidart, J.M. (1994) Epithelial-to-mesenchymal transition in HPV-33-transfected cervical keratinocytes is associated with increased invasiveness and expression of gelatinase A. *Int. J. Cancer* 59, 661-666.

Gilles, C. and Thompson, E.W. (1996) The epithelial to mesenchymal transition and metastatic progression in carcinoma. *Breast J.* 2, 83-96.

Goldberg, G.I., Strongin, A., Collier, I.E., Genrich, L.T., Marmer, B.L. (1992) Interaction of 92 kDa type IV collagenase with the tissue inhibitor of metalloproteinases prevents dimerization, complex formation with interstitial collagenase, and activation of the proenzyme with stromelysin. *J. Biol. Chem.* 267, 4583-91.

Graham, C.H. and Lala, P.K. (1991) Mechanism of control of trophoblast invasion in situ. *J. Cell Physiol.* 148, 228-234.

Graham, C.H. and Lala, P.K. (1992) Mechanisms of placental invasion of the uterus and their control. *Biochem. Cell Biol.* 70, 867-874.

Harris, E.D. and Krane, S.M. (1992) An endopeptidase from rheumatoid synovial tissue culture. *Biochem. Biophys. Acta* 258, 566-576.

Hofmann, G.E., Glastein, I., Schatz, F., Heller, D. and Deligdish, L. (1994) Immunohistochemical localization of urokinase-type plasminogen activator inhibitors 1 and 2 in early human implantation sites. *Am. J. Obstet. Gynaecol.* 170, 671-676.

Kirby, D.R.S. (1960) The development of mouse eggs beneath the kidney capsule. *Nature* (London) 187, 707-708.

Kirby, D.R.S.(1963a) The development of the mouse blastocyst transplanted to the spleen. *J. Reprod. Fertil.* 5, 1-12.

Kirby, D.R.S. (1963b) The development of the mouse blastocysts transplanted to the cryptochird and scrotal testis. *J. Anat. Lond.* 97, 119-130.

Kirby, D.R.S. (1965) The invasiveness of the trophoblast. In: *The Early Conceptus Normal and Abnormal*, (ed.) W.W. Park, Edinburgh: University of St Andrews Press, pp. 68-74.

Kleiner Jr., D.E. and Stetler-Stevenson, W.G. (1993) Structural biochemistry and activation of matrix metalloproteases. *Curr. Opin. Cell. Biol.* 5, 891-897.

Kurman, R.J., Main, C.S. and Chen, H-C. (1984) Intermediate trophoblast: A distinctive form of trophoblast with specific morphological, biochemical and functional features. *Placenta* 5, 349-370.

Lala, P.K. and Graham, C.H. (1990) Mechanisms of trophoblast invasiveness and their control: The role of proteases and protease inhibitors. *Cancer Metastasis Rev.* 9, 369-379.

Lefebvre, O., Régnier, C., Chenard, M-P., Wendling, C., Chambon, P., Basset, P. and Rio, M-C. (1995) Developmental expression of mouse stromelysin-3 mRNA. *Development* 121, 947-955.

Leco, K.J., Khokha, R., Pavloff, N., Hawkes, S.P. and Edwards, D.R. (1994) Tissue inhibitor of metalloproteinases-3 (TIMP-3) is an extracellular matrix associated protein with a distinctive pattern of expression in mouse cells and tissues. *J. Biol. Chem.* 269, 9352-9360.

Levy, A.T., Cioce, V., Sobel, M.E., Garbisa, S., Grigioni, W.F., Liotta, L. A. and Stetler-Stevenson, W.G. (1991) Increased expression of the Mr 72,000 type IV collagenase in human colonic adenocarcinoma. *Cancer Res.* 51, 439-444.

Lewis, S.H. and Bernirschke, K. (1992) Placenta. In: *Histology for Pathologists*, (ed.) S.S. Sternberg, New York: Raven Press Ltd., pp. 855-863.

Librach, C.L., Werb, Z., Fitzgerald, M.L., Chiu, K., Corwin, N.M., Esteves, R., Grobelny, D., Galardy, R., Damsky, C.H. and Fisher, S.J. (1991) 92-KD type IV collagenase mediates invasion of human cytotrophoblasts. *J. Cell Biol.* 113, 437-449.

Maquoi, E., Polette, M., Nawrocki, B., Bishof, P., Noel, A., Pintiaux, A., Santavica, M., Schaaps, J.P., Pijnenborg, R., Birembaut, P. and Foidart, J.M. (1997) Expression of stromelysin-3 in the human placenta and placenta bed. *Placenta* 18, in press.

Martelli, M., Campana, A. and Bischof, P. (1993) Secretion of matrix metalloproteinases by human endometrial cells in vitro. *J. Reprod. Fertil.* 98, 67-76.

Mignatti, P. and Rifkin, D.B. (1993) Biology and biochemistry of proteases in tumour invasion. *Physiol. Rev.* 73, 161-195.

Moll, U.M. and Lane, B.L. (1990) Proteolytic activity of first trimester human placenta: Localization of interstitial collagenase in villus and extravillus trophoblast. *Histochemistry* 94, 555-560.

Muller, D., Wolf, C., Abecassis, J., Millon, R., Engelmann, A., Bronner, G., Rouyer, N., Rio, M.C., Eber, M., Methlin, G., Chambon, P. and Basset, P.(1993) Increased stromelysin 3 gene expression is associated with increased local invasiveness in head and neck squamous cell carcinomas. *Cancer Res.* 53, 165-169.

Multhaup, H.A.B., Mazar, A., Cines, D.B., Warhol, M.J. and McCrae, K.R. (1994) Expression of urokinase receptors by human trophoblast. A histochemical and ultrastructural analysis. *Lab. Invest.* 71, 392-400.

Murphy, G., Mc Alpine, C.G., Poll, C.T. and Reynolds, J.J. (1985) Purification and characterization of a bone metalloproteinase that degrade gelatin and types IV and V collagen. *Biochem Biophys Acta* 831, 49-58.

Murphy, G., Segain, J.P., O'Shea, M., Cockett, M., Ionnou, C., Lefebvre, O., Chambon, P. and Basset, P. (1993) The 28 kDa N-terminal domain of mouse stromelysin-3 has the general properties of a weak metalloproteinase. *J Biol Chem* 268, 15435-15441.

Nagase, H., Ogata, Y., Suzuki, K., Enghild, J.J. and Salvesen, G. (1991) Substrate specificities and activation mechanisms of matrix metalloproteinases. *Biochem Soc Trans.* 19, 715-718.

Nawrocki, B., Polette, M., Marchand, V., Maquoi, E., Beorchia, A., Tournier, J.M., Foidart, J.M. and Birembaut, P. (1996) Membrane-type matrix metallo-proteinase-1 expression at the site of human placental implantation. *Placenta* 17, 565-572.

Nomura, S., Hogan, L.M., Wills, A.J., Health, J.K. and Edward, D.R. (1989) Developmental expression of tissue inhibitor of metalloproteinase (TIMP) RNA. *Development* (Cambridge) 105, 575-584.

Pei, D. and Weiss, S.J. (1995) Furin-dependent intracellular activation of the human stromelysin-3 zymogen. *Nature* 375, 244-247.

Polette, M., Clavel, C., Cockett, M., Girod De Bentzmann, S., Murphy, G. and Birembaut, P. (1993a) Detection and localization of mRNAs encoding matrix metalloproteinases and their tissue inhibitor in human breast pathology. *Invasion Metastasis* 13, 31-37.

Polette, M., Clavel, C., De Clerck, A. and Birembaut, P. (1993b) Localization by in situ hybridization of mRNAs encoding stromelysin 3 and tissue inhibitors of metallo-proteinases TIMP-1 and TIMP-2 in human head and neck carcinomas. *Pathol. Res. Pract.* 189, 1052-1057.

Polette, M., Nawrocki, B., Pintiaux, A., Massenat, C., Maquoi, E., Volders, L., Schaaps, J.P., Birembaut, P. and Foidart, J.M. (1994) Expression of gelatinases A and B and their tissue inhibitors by cells of early and term human placenta and gestational endometrium. *Lab. Invest.* 71, 838-846.

Polette, M., Nawrocki, B., Gilles, C., Sato, H., Seiki, M., Tournier, J.M. and Birembaut, P. (1996) MT-MMP expression and localisation in human lung and breast cancers. *Vichows Arch.* 428, 29-35.

Rawdanowicz, T.J.. Hampton, A.L., Nagase, H., Woolley, D.E. and Salamonsen, L.A. (1992) Matrix metalloproteinase production by cultured human endometrial stromal cells: identification of interstitial collagenase, gelatinase A, gelatinase B and stromelysin 1 and their differential regulation by interleukin-1α and tumor necrosis factor α. *Proc. Aust. Soc. Med. Res.* 1-8.

Ray, J.M. and Stetler-Stevenson, W.G. (1994) The role of matrix metalloproteases and their inhibitors in tumor invasion, metastasis and angiogenesis. *Eur. Respir. J.* 7, 2062-2072.

Salamonsen, L.A. (1994) Matrix metalloproteinases and endometrial remodelling. *Cell Biol. Intern.* 18, 1139-1144.

Sappino, A.P., Huarte, J., Belin, D. and Vassali, J.D. (1989) Plasminogen activator in tissue remodeling and invasion: mRNA localization in mouse ovaries and implanting embryos. *J. Cell Biol.* 109, 2471-2479.

Sato, H., Takino, T , Okada, Y., Cao, J., Shinagawa, A., Yamamoto, E. and Seiki, M. (1994) A matrix metalloproteinase expressed on the surface of invasive tumour cells. *Nature* 370, 61-65.

Shapiro, S.D., Kobayashi, D.K. and Ley T.J. (1993) Cloning and characterization of a unique elastolytic metalloproteinase produced by human alveolar macrophages. *J. Biol. Chem.* 268, 23824-23829.

Sherman, M.I., Strickland, S. and Reich, E. (1976) Differentiation in early mouse embryonic and teratocarcinoma cells in vitro: plasminogen activator production. *Cancer Res.* 36, 4208-4216.

Shimonovitz, S., Hurwitz, A., Dushnik, M., Anteby, E., Geva-Eldar, T. and Yagel, S. (1994) Developmental regulation of the expression of 72 and 92Kd type IV collagenases in human trophoblasts: a possible mechanism for control of trophoblast invasion. *Am. J. Obstet. Gynecol.* 171, 832-838.

Stetler-Stevenson, W.G., Krutzsch, H.C .and Liotta, L.A. (1989) Tissue inhibitor of metalloprcteinase-2 (TIMP-2), a new member of metalloproteinase inhibitor family. *J. Biol. Chem.* 264, 17374-17378.

Strickland, S., Reich, E. and Sherman, M.I. (1976) Plasminogen activator in early embryogenesis: enzyme production by trophoblast and parietal endoderm. *Cell* 9, 231-240.

Strongin, A.Y., Marmer, B.L., Grant, G.A. and Goldberg, G.I. (1993) Plasma membrane dependent activation of the 72 KDa type IV collagenase is prevented by complex formation with TIMP2. *J. Biol. Chem.* 268, 14033-14039.

Strongin, A.Y., Collier, I., Bannikov, G., Marmer, B.L., Grant, G.A. and Goldberg, G.I. (1995) Mechanism of cell surface activation of 72-KDa type IV collagenase. *J. Biol. Chem.* 270, 5331-5338.

Takino, T., Sato, H., Shinagawa, A. and Seiki, M. (1995) Identification of the second membrane type matrix metalloprotinase (MT-MMP 2) gene from a human placenta cDNA library. *J. Biol. Chem.* 270, 23013-23020.

Tryggvason, K., Höyhtyä, M. and Pyke, C. (1993). Type IV collagenase in invasive tumors. *Breast Cancer Res. Treat.* 24, 209-218.

Uria, J.A., Ferrando, A.A., Velasco, G., Freije, J.M.P. and Lopez-Otin, C. (1994) Structure and expression in breast tumors of TIMP-3, a new member of the metalloproteinase inhibitor family. *Cancer Res.* 54, 2091-2094.

Wagner, S.N., Ruhri, C., Kunth, K., Holecek, B.U., Goos, M., Höfler, H. and Atkinson, M.J (1992) Expression of stromelysin 3 in the stromal elements of human basal cell carcinoma. *Diagn. Mol. Pathol.* 1, 200-205.

Werb, Z., Alexander, C.M. and Adler, R.R. (1992) Expression and function of matrix metalloproteinases in development. *Matrix (Suppl .)* 1, 337-343.

Will, H. and Hinzmann, B. (1995) cDNA sequence and mRNA tissue distribution of a novel human matrix metalloproteinase with a potential transmembrane segment. *Eur. J. Biochem.* 231, 602-608.

Wolf, C., Chenard, M.P., Durand De Grossouvre, P., Bellocq, J.P., Chambon, P. and Basset, P. (1992) Breast-cancer-associated stromelysin-3 gene is expressed in basal cell carcinoma and during cutaneous wound healing. *J. Invest. Dermatol.* 99, 870-872.

Wolf, C., Rouyer, N., Lutz, Y., Adida, C., Loriot, M., Bellocq, J.P., Chambon, P. and Basset, P. (1993) Stromelysin 3 belongs to a subgroup of proteinases expressed in breast carcinoma fibroblastic cells and possibly implicated in tumor progression. *Proc. Natl. Acad. Sci. USA* 90, 1843-1847.

Yagel, S., Parar, R.S., Jeffrey, J.J. and Lala, P.K. (1988) Normal non-metastatic trophoblast cells share in vitro invasive properties of malignant cells. *J. Cell Physiol.* 136, 455-464.

Yeh, I.T. and Kurman, R.J. (1989) Functional and morphologic expressions of trophoblast. *Lab. Invest.* 61, 1-4

Trophoblast Research 10:115-121, 1997

THE DISTRIBUTION OF TISSUE INHIBITOR OF METALLOPROTEINASES-2 (TIMP-2) IN DECIDUA AND TROPHOBLAST OF EARLY HUMAN PREGNANCY

Peter Ruck[1,3], Klaus Marzusch[2], Stefan Kröber[1], Hans-Peter Horny[1], Johannes Dietl[2] and Edwin Kaiserling[1]

[1]Institute of Pathology
[2]Department of Obstetrics and Gynaecology
University of Tubingen
Liebermeisterstr. 8
72076 Tubingen, Germany

INTRODUCTION

Placentation in human pregnancy involves invasion of the endometrial epithelium and its basement membrane, the decidual stroma, the myometrium and blood vessels by the trophoblast. Thus, trophoblast invasion resembles in some respects the invasive growth of malignant tumors (Lala and Graham, 1990). However, the invasive growth of the trophoblast, unlike that of malignant tumors, is strictly controlled and normally does not proceed beyond the inner third of the myometrium (Pijnenborg, 1990), although the mechanisms by which it is controlled have not been fully clarified.

Studies based on experimental models have shown that matrix metalloproteinases (MMPs) synthesized by cytotrophoblast cells are crucial for the invasive growth of the human trophoblast (Fisher et al., 1989; Librach et al., 1991). The same studies also revealed that tissue inhibitors of metalloproteinases (TIMPs), a group of proteins that inhibit the proteolytic activity of MMPs, inhibit the invasive growth of the trophoblast *in vitro,* and it has been suggested that TIMPs could play a role in regulating trophoblast invasion *in vivo* (Brenner et al., 1989; Librach et al., 1991). There are at least three structurally different TIMPs: TIMP-1, TIMP-2 and TIMP-3 (Statler-Stevenson et al., 1989; Boone et al., 1990; Apte et al., 1994). In mice, the various mRNAs for all these TIMPs have been found in placental tissue, and mRNAs for TIMP-1 and TIMP-2 have also been detected in the decidua (Waterhouse et al., 1993; Apte et al., 1994). In humans, expression of TIMP-2 mRNA has been found in cultured trophoblast cells (Graham et al., 1994) and in non-pregnant endometrium (Hampton and Salamonsen, 1994). Polette et al. (1994) demonstrated mRNA and immunoreactivity for TIMP-1 and TIMP-2 in human decidua and placenta. The aim of this immunohistochemical study was to investigate the distribution of TIMP-2 in the various different placental and decidual cell populations at the fetomaternal interface in early human pregnancy.

MATERIALS AND METHODS

The study was performed on six specimens of placental tissue and five of decidua. The tissue derived from healthy women undergoing therapeutic abortion of a normal pregnancy at 7 - 11 weeks' gestation and had been submitted for routine

[3]To Whom Correspondence Should Be Addressed.

histological investigation. Part of each specimen was shock-frozen in liquid nitrogen. Frozen sections were cut at 6 μm and air-dried overnight. They were then fixed in acetone for 10 minutes at room temperature and air-dried for 20 minutes. The sections were immunostained by the avidin-biotin-peroxidase complex method (Hsu et al., 1981) with a mouse monoclonal antibody (subclass IgGla) against TIMP-2 (Oncogene Sciences, Cambridge, MA, USA) diluted 1:50 in Tris-buffered saline (TBS; pH 7.6). They were exposed to the primary antibody for one hour at room temperature. 3,3'-diaminobenzidine was used as the chromogen. After immunostaining, the sections were counterstained for 20 seconds with hematoxylin. Immunostaining with the anti-keratin antibody KL1 (Immunotech, Marseille, France) diluted 1:100 in TBS was also carried out on corresponding sections to differentiate invading trophoblast from decidual stromal cells. A negative control was performed by replacing the primary antibody with mouse IgGla (Dako, Glostrup, Denmark) diluted 1:50 in Tris-buffered saline.

RESULTS

In the decidua, moderate to strong staining for TIMP-2 was found in nearly all the stromal cells in all the specimens (Figure 1). Immunoreactivity was also seen in the walls of blood vessels, there being strong staining of nearly all the arterioles and venules and weak to moderate staining of some capillary-type blood vessels. In two specimens, some of the endometrial glands were immunoreactive. Staining here was of weak or moderate intensity and was often confined to the apical parts of the epithelial cells. No unequivocal staining of leukocytes was found. KL1 stained only endometrial glands and invading trophoblast, thus allowing the latter to be differentiated from decidual stromal cells.

In the placenta, immunoreactivity for TIMP-2 was found in the cytotrophoblasts of villi and columns in all the cases investigated (Figure 2). The staining intensity of the cytotrophoblast cells of villi was usually moderate, with marked accentuation at the cell membrane, especially at the interfaces with the syncytiotrophoblast and the villus stroma. The cytotrophoblast cells in the columns exhibited strong cytoplasmic staining. There was only weak, focal staining of the cytoplasm and apical border of the syncytiotrophoblast. A few scattered cells in the villus stroma and some vascular endothelial cells were weakly stained, as were smooth muscle cells in the wall of a large blood vessel in a stem villus in one case.

When the primary antibody was replaced by mouse IgGla diluted 1:50 in Tris-buffered saline, there was weak staining of a few scattered cells in the villus stroma and some endothelial cells in the placenta, but no staining was seen in the decidua.

DISCUSSION

Cultured human trophoblast cells have been shown to produce both the 72-kD and the 92-kD type IV collagenase (Librach et al., 1991; Graham et al., 1994; Shimonovitz et al., 1994), MMPs that degrade type IV collagen present in basement membranes. These MMPs have been suggested to play a crucial role in trophoblast invasion and in the invasive growth of malignant tumors (Librach et al., 1991; Albini et al., 1991; Furcht et al., 1994; Shimonovitz et al., 1994). The proteolytic activity of both these MMPs can be

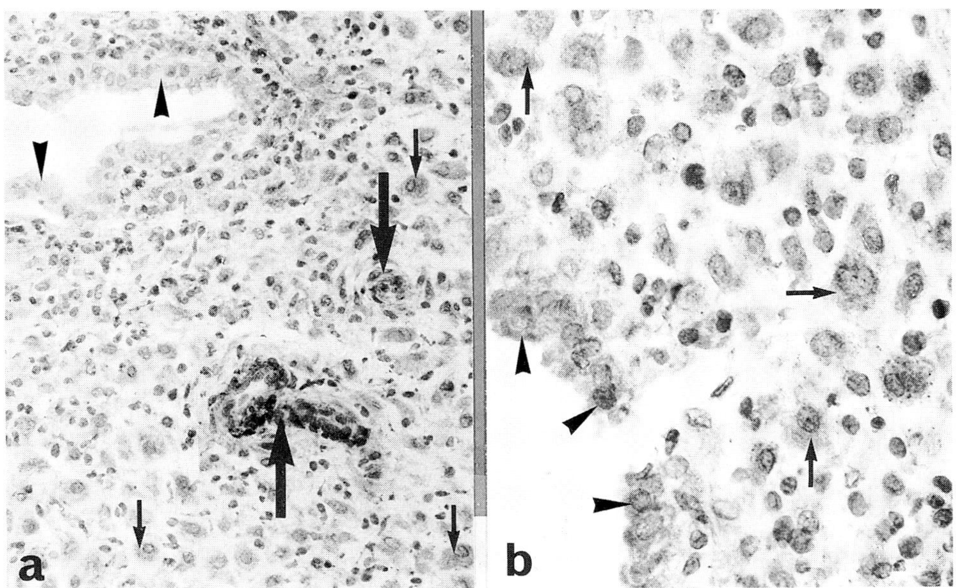

Figure 1. First trimester decidua. (a) Decidual stromal cells exhibit immunoreactivity for TIMP-2 (small arrows). There is also staining of the walls of blood vessels (large arrows), with particularly strong staining of an arteriole in the lower half of the illustration. The epithelium of an endometrial gland (arrowheads) is not stained. (b) In this case the epithelium of an endometrial gland (arrowheads) exhibits immunoreactivity for TIMP-2. As in (a) nearly all the stromal cells (small arrows) are stained. (a) x180; (b) x360.

inhibited by both TIMP-1 and TIMP-2, and the invasive growth of the cytotrophoblast in invasion assays can also be inhibited by the addition of TIMP-1 or TIMP-2 to the culture medium (Librach et al., 1991).

In this study, as in the study of Polette et al. (1994), immunoreactivity for TIMP-2 was found in both first trimester human decidua and placenta.

The distribution of TIMP-2 in the various different cell populations in these tissues was evaluated. In the decidua, TIMP-2-immunoreactivity was found in stromal cells and in the walls of blood vessels in all five specimens investigated. By contrast, endometrial epithelium stained in only two specimens, and no unequivocal staining of decidual leukocytes was found. Polette et al. (1994) demonstrated TIMP-2 mRNA in the decidua by both *in situ* hybridization and Northern blot analysis. The decidual stroma and maternal blood vessels are both invaded by the trophoblast during human placentation. TIMP-2 produced by decidual stromal cells and smooth muscle cells of blood vessels could thus inactivate the type IV collagenases secreted by the trophoblast and play a role in regulating trophoblast invasion. Decidual stromal cells and spiral arteries have also been found to express the 72-kD type IV collagenase (Autio-Harmainen et al., 1992), so that TIMP-2 in the decidua may not only regulate trophoblast invasion, but may also have an autoregulatory function.

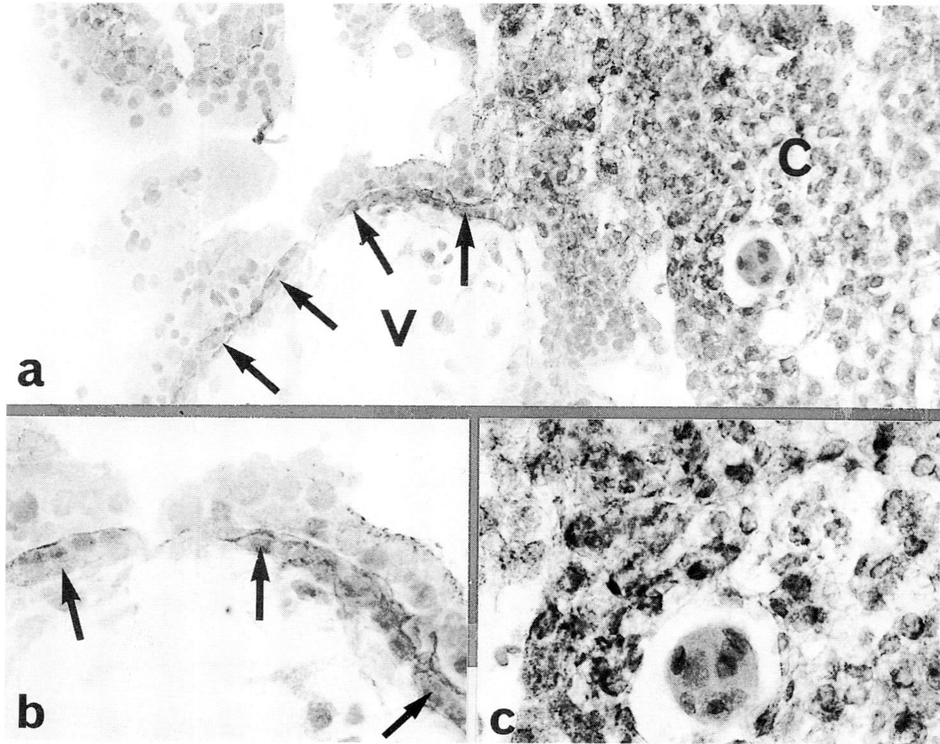

Figure 2. First trimester placenta. A villus (V) with a column of proliferating cytotrophoblast (C) is seen in (a). Staining for TIMP-2 of moderate intensity with accentuation at the cell membrane is seen in villus cytotrophoblasts (arrows), which is shown at higher magnification in (b). The cytotrophoblasts in the column exhibit strong cytoplasmic staining, which is seen at higher magnification in (c). (a) x180; (b) and (c) x360.

In the placenta, the strongest staining was observed in cytotrophoblast columns. The staining intensity of these first trimester cytotrophoblast columns was much stronger than that of the corresponding villus cytotrophoblast and syncytiotrophoblast, and than that of both the cytotrophoblast and syncytiotrophoblast of term placenta, which is described in detail elsewhere (Ruck et al., in press). As it is the first trimester cytotrophoblast columns that invade the placental bed, our findings demonstrate that the strongest immunoreactivity for TIMP-2 in the trophoblast is found in cells that are known to produce MMPs and exhibit an invasive growth pattern (Fisher et al., 1989; Librach et al., 1991). On the basis of these findings and the fact that Polette et al. (1994) demonstrated mRNA for TIMP-2 in the trophoblast, it seems possible that TIMP-2 produced by cytotrophoblast cells may be involved in autoregulation of the invasive growth of these cells. This notion is further supported by the fact that the pattern of distribution of TIMP-2-immunoreactivity in first trimester trophoblast corresponds largely to that of the 72-kD type IV collagenase: immunostaining for this MMP is strongest and mainly cytoplasmic in cytotrophoblast columns, and is membrane-associated in villus cytotrophoblast, with accentuation at the border with the syncytiotrophoblast (Autio-Harmainen et al., 1992).

In summary, our findings, in conjunction with published data, indicate that TIMP-2 produced by certain cell populations at the feto-maternal interface could play a role in regulation and autoregulation of the invasive growth of the trophoblast. However, the findings of some studies suggest that the invasive growth of the trophoblast is mediated mainly by the 92-kD type IV collagenase (Librach et al., 1991; Behrendtsen et al., 1992; Waterhouse et al., 1993), whereas TIMP-2 binds preferentially to the 72-kD type IV collagenase (Statler-Stevenson et al., 1989; Goldberg et al., 1989). It has also been shown that complexes between TIMP-2 and certain MMP may exhibit proteolytic activity (Kolkenbrock et al., 1994; Strongin et al., 1995). Thus, the extent to which TIMP-2 produced by cells in the decidua or cytotrophoblast cells plays a role in regulation or autoregulation of the invasive growth of the trophoblast cannot be determined with certainty on the basis of this study. Further investigation of the distribution and level of expression of all three types of TIMPs in the placenta and placental bed and of their inhibitory effect in invasion assays will be necessary to further elucidate the role of TIMP-2 in human placentation.

SUMMARY

Matrix metalloproteinases (MMPs) secreted by human cytotrophoblast cells are crucial for the invasion of the placental bed by these cells. The invasive growth of the trophoblast is similar to that of malignant tumors in many respects, but, unlike the latter, it is strictly controlled. Tissue inhibitors of metalloproteinases (TIMPs) have been postulated to play a role in the control of trophoblast invasion. In this immunohistochemical study, the distribution of TIMP-2 at the feto-maternal interface in first trimester human pregnancy was investigated. In the decidua, staining for TIMP-2 was found in stromal cells and the walls of blood vessels in all five specimens, but decidual leukocytes were nonreactive and staining of the endometrial epithelium was found in only two specimens. In the placenta, TIMP-2 immunoreactivity was found mainly in cytotrophoblasts, the strongest staining being observed in cytotrophoblast columns. The results suggest that TIMP-2 produced by certain cell populations in the decidua could play a role in regulating trophoblast invasion. As it is the first trimester cytotrophoblast columns that invade the placental bed, the strongest immunoreactivity for TIMP-2 in the placenta is found in cells that are known to produce MMPs and exhibit an invasive growth pattern. These findings indicate that TIMP-2 may also be involved in autoregulation of the invasive growth of the trophoblast.

ACKNOWLEDGEMENTS

This study was supported by a grant from the Else-Übelmesser-Stiftung, Tübingen, Germany.

REFERENCES

Albini, A., Melchiori, A., Santi, L., Liotta, L.A, Brown, P.D. and Stetler-Stevenson, W.G. (1991) Tumor cell invasion inhibited by TIMP-2. *J. Natl . Cancer Inst.* 83, 740-742.

Apte, S.S., Hayashi, K., Seldin, M.F., Mattei, M-G., Hayashi, M. and Olsen, B. R. (1994) Gene encoding a novel murine tissue inhibitor of metalloproteinases (TIMP), TIMP-3, is expressed in developing mouse epithelia, cartilage, and muscle, and is located on mouse chromosome 10. *Dev. Dyn* . 200, 177-197.

Autio-Harmainen, H., Hurskainen, T., Niskasaari, K., Hoyhtya, M. and Tryggvason, K. (1992) Simultaneous expression of 70 kilodalton type IV collagenase and type IV collagen alphal(IV) chain genes by cells of early human placenta and gestational endometrium. *Lab. Invest.* 67, 191-200.

Behrendtsen, O., Alexander, C. M. and Werb, Z. (1992) Metalloproteinases mediate extracellular matrix degradation by cells from mouse blastocyst outgrowths. *Development* 114, 447-456.

Boone, T.C., Johnson, M.J., De Clerck, Y.A. and Langley, K.E. (1990) cDNA cloning and expression of a metalloproteinase inhibitor related to tissue inhibitor of metalloproteinases. *Proc. Natl . Acad . Sci.* USA 87, 2800-2804.

Brenner, C.A., Adler, R.R., Rappolee, D.A., Pedersen, R.A., and Werb, Z. (1989) Genes for extracellular matrix-degrading metalloproteinases and their inhibitor, TIMP, are expressed during early mammalian development. *Genes Dev.* 3, 848-859.

Fisher, S.J., Cui, T.-Y., Zhang, L., Hartman, L., Grahl, K., Zhang, G.-Y., Tarpey, J. and Damsky, C.H. (1989) Adhesive and degradative properties of human placental cytotrophoblast cells in vitro. *J. Cell Biol.* 109, 891-902.

Furcht, L.T., Skubitz, A.P.N. and Fields, G.B. (1994) Tumor cell invasion, matrix metalloproteinases, and the dogma. *Lab. Invest. 70,* 891-902.

Goldberg, G.I., Marmer, B.L., Grant, G.A., Eisen, A.Z., Wilhelm, S. and He, C. (1989) Human 72-kilodalton type IV collagenase forms a complex with a tissue inhibitor of metalloproteases designated TIMP-2. *Proc. Natl. Acad. Sci.* USA 86, 8207-8211.

Graham, C.H., Conelly, I., MacDougall, J.R., Kerbel, R.S., Stetler-Stevenson, W.G. and Lala P.K. (1994) Resistance of malignant trophoblast cells to both the anti-proliferative and anti-invasive effects of transforming growth factor-beta. *Exp. Cell Res.* 214, 93-99.

Hampton, A. and Salamonsen, L. A. (1994) Expression of messenger ribonucleic acid encoding matrix metalloproteinases and their tissue inhibitors is related to menstruation. *J. Endocrinol.,* 141, 141: R1-R3.

Hsu, S.-M., Raine, L., and Fanger, H. (1981) Use of avidin-biotin peroxidase complex (ABC) in immunoperoxidase techniques: A comparison between ABC and unlabeled antibody (PAP) procedures. *J. Histochem. Cytochem.* 29, 577-580.

Kolkenbrock, H., Hecker-Kia, A., Orgel, D., Ruppitsch, W., and Ulbrich, N. (1994) Activity of ternary gelatinaseA-TIMP-2-matrix metallo-proteinase complexes. *Biol. Chem. Hoppe-Seyler* 375, 589-595.

Lala, P.K. and Graham, C.H. (1990) Mechanisms of trophoblast invasiveness and their control: .The role of proteases and protease inhibitors. *Cancer Metastasis Rev.* 9, 369-379.

Librach, C.L., Werb, Z., Fitzgerald, M.L. Chiu, K., Corwin, N.M., Esteves, R.A., Grobelny, D., Galardy, R., Damsky, C.H. and Fisher, S.J. (1991) 92-kD type IV collagenase mediates invasion of human cytotrophoblasts. *J. Cell Biol.* 113, 437-449.

Pijnenborg, R. (1990) Trophoblast invasion and placentation: Morphological aspects. *Troph. Res.* 4, 33-50.

Polette, M., Nawrocki, B., Pintiaux, A., Massenat, C., Maquoi, E., Volders, L., Schaaps, J.P., Birembaut, P. and Foidart, J.M. (1994) Expression of gelatinases A and B and their tissue inhibitors by cells of early and term human placenta and gestational endometrium. *Lab.* Invest. 71, 838-846.

Ruck, P., Marzusch, K., Horny, H.-P., Dietl, J. and Kaiserling, E. The distribution of tissue inhibitor of metalloproteinases-2 (TIMP-2) in the human placenta. *Placenta,* in press. (PLEASE UPDATE)

Shimonovitz, S., Hurwitz, A., Dushnik, M., Anteby, E., Geva-Eldar, T. and Yagel, S. (1994) Developmental regulation of the expression of 72 and 92 kd type IV collagenases in human trophoblasts: A possible mechanism for control of trophoblast invasion. *Am. J. Obstet. Gynecol.* 171, 832-838.

Stetler-Stevenson, W.G., Krutzsch, H.C.,- and Liotta, L.A. (1989) Tissue inhibitor of metalloproteinases-2 (TIMP-2): A new member of the metalloproteinase inhibitor family. *J. Biol. Chem.* 264, 17374-17378.

Strongin, A.Y., Collier, I., Bannikov, G., Marmer, B.L., Grant, G.A. and Goldberg, G.I. (1995) Mechanisms of cell surface activation of 72-kDa type IV collagenase. Isolation of the activated form of the membrane metalloprotease. *J. Biol. Chem.* 270, 5331-5338.

Waterhouse, P., Denhardt, D. T. and Khoka, R. (1993) Temporal expression of tissue inhibitors of metalloproteinases in mouse reproductive tissues during gestation. *Mol. Reprod. Dev.* 35, 219-226.

Trophoblast Research 10:123-142, 1997

MATRIX METALLOPROTEINASES IN CHORIOCARCINOMA CELL LINES: A POTENTIAL REGULATORY ROLE OF EXTRACELLULAR MATRIX COMPONENTS

Erik Maquoi[1], Agnès Noël, and Jean-Michel Foidart

Laboratory of Biology
University of Liège
Tower of Pathology (B23 Sart Tilman)
B-4000 Liège, Belgium

INTRODUCTION

Human placentation is a dynamic process which requires a series of complex, coordinated interactions between maternal and fetal cells. During the first four months of gestation, groups of differentiated trophoblastic cells, also known as extravillous or intermediate trophoblast cells (Kurman et al., 1984), sprout from the basal tip of anchoring villi, migrate through the decidualized endometrium, and invade the wall of the spiral arteries where they replace the endothelial lining to become the endovascular trophoblasts. This trophoblastic invasiveness is characterized by the breaching of multiple basement membranes (BM) and the degradation of the interstitial extracellular matrix (ECM) of the decidualized endometrium.

Many steps of trophoblast infiltration are similar to events that occur during tumor cell invasion. According to the classical model of Liotta and coworkers (1986), invasion by tumor cells involves three main steps: (1) binding of the cell surface receptors to matrix glycoproteins; (2) activation of proteinases and pericellular degradation of matrix components; (3) migration of the cell within the matrix. However, the analogy between tumor cell invasion and trophoblasts penetration of the decidualized endometrium is limited by the fact that during normal pregnancy, trophoblast infiltration is tightly regulated.

Proteins in the ECM, such as laminin, fibronectins, and collagens, have been shown to have dynamic functions in the regulation of cell adhesion, spreading, differentiation and migration (Hynes, 1992; Damsky et al., 1993). Interactions between cells and ECM proteins are mediated by adhesion receptors expressed at the surface of the cell. Among the different adhesion receptors, integrins, a family of heterodimeric transmembrane glycoproteins, are the most ubiquitous and versatile receptors (Hynes, 1992; Pignatelli and Stamp, 1995). In normal pregnancies, a progressive modification of integrin expression at the surface of trophoblast cells has been shown to occur during the differentiation of non-invasive villous cytotrophoblast cells into invasive extravillous trophoblasts (Damsky et al., 1994; Loke and King, 1995). Such a remodeling of the integrin profile enables cells to interact with different ECM components encountered during decidual infiltration.

[1]To Whom Correspondence Should Be Addressed.

This infiltrative process requires not only the capacity for cells to adhere to ECM components but also the ability to digest these components. Several *in vivo* and *in vitro* experiments have suggested that trophoblast invasiveness is mediated by a regulated synthesis of matrix degrading enzymes (Lala and Graham, 1990; Bischof and Martelli, 1992; Graham and Lala, 1992). Among these proteolytic enzymes, members of matrix metalloproteinases (MMPs), which share functional and structural characteristics, have been shown to play key roles during placentation (Yagel et al., 1988; Brenner et al., 1989; Fisher et al., 1989; Feinberg et al., 1989; Moll and Lane, 1990; Bischof et al., 1991; Librach et al., 1991; Autio-Harmainen et al., 1992; Fernandez et al., 1992; Blankenship and King, 1994; Hofmann et al., 1994; Polette et al., 1994; Shimonovitz et al., 1994; Maquoi et al., 1997, in press). The MMPs are zinc- and calcium-dependent endopeptidases which degrade components of the ECM at physiological pH, and are inhibited by specific tissue inhibitors (TIMPs). Most of them are produced as inactive proenzymes which are activated by proteolytic cleavage of an amino-terminal pro-domain. The net proteolytic activity will thus depend on the balance between active proteinases and the TIMPs. MMPs play important roles in normal processes such as tissue morphogenesis and wound healing, but are also associated with several pathological conditions involving tissue destruction, including rheumatoid arthritis, and cancer invasion and metastasis (cf., Ries and Petrides, 1995). To date, eleven members of the MMPs family have been identified and classified into three main groups based on substrate specificity (cf., Birkedal-Hansen et al., 1993; Stetler-Stevenson et al., 1993; Cawston, 1995). There are some clear differences in substrate specificity between family members, although in many cases there is significant overlap. The *collagenases* mainly digest fibrillar collagens; the *gelatinases* have a high affinity against gelatin (denaturated fibrillar collagens) and cleave triple-helical type IV collagen (the major constituent of basement membranes); and the *stromelysins* degrade a wide variety of ECM substrates such as proteoglycans, laminin or fibronectin. Recently, new MMPs were described. A macrophage metalloelastase (Belaaouaj et al., 1995), transmembrane-MMPs (MT-MMPs for membrane type MMPs, Sato et al., 1994; Takino et al., 1995; Will and Hinzmann, 1995) and stromelysin-3, a proteinase initially identified in the stroma of breast carcinoma (cf., Basset et al., 1993), have been identified.

The acquisition of an invasive phenotype by early trophoblast cells has been correlated with the ability of these cells to synthesize a large array of MMPs: interstitial collagenase (Yagel et al., 1988; Moll and Lane, 1990), gelatinases A and B (Bischof et al., 1991; Autio-Harmainen et al., 1992; Fernandez et al., 1992; Blankenship and King, 1994; Polette et al., 1994), stromelysins -1, -2, -3 (Brenner et al., 1989; Maquoi et al., 1997), and MT-MMP-1 (Nawrocki et al., 1996). Among these proteinases, cultured first trimester trophoblasts essentially secrete gelatinases, and their invasive potential appears to be mediated by the activity of these enzymes (Librach et al., 1991; Shimonovitz et al., 1994).

However, little is known about the influence of ECM encountered during trophoblast infiltration on the proteolytic activity of these cells. We have therefore analyzed the influence of purified ECM components upon the production of MMPs by extravillous trophoblast cells. However, due to the difficulty to isolate large amounts of pure first trimester extravillous trophoblast cells, we have used choriocarcinoma cell lines (malignant counterparts of invasive extravillous trophoblast cells) which express certain of the characteristics of normal trophoblast (Aplin et al., 1992). Cells were cultured on plastic or thin coats of laminin, fibronectin, collagen I, and collagen IV in order to evaluate the potential effect of these various substrates upon gelatinases secretion.

MATERIALS AND METHODS

Cell Lines

Human choriocarcinoma cell lines, BeWo and JAr were obtained from the American Type Culture Collection (Rockville, MD, USA). BeWo cells were routinely passaged in HAM's-F12 medium (TechGen) supplemented with 20% fetal calf serum (FCS, Gibco). JAr cells were maintained in RPMI 1640 medium (TechGen) containing 10% FCS.

Coating Of Culture Dishes With ECM Components

Laminin and fibronectin were purified from mouse Engelbreth-Holm-Swarm (EHS) tumors (Timpl et al., 1979) and from human plasma by gelatin affinity chromatography (Yamada, 1983), respectively. Purified type I collagen was prepared from fetal calf skin (Delvoye et al., 1983) and conserved in 0.1M acetic acid at -20°C. Type IV collagen was extracted from EHS tumors according to the procedure described by Timpl et al. (1979). Fibronectin, laminin and collagen types I (Col I) and IV (Col IV) were diluted at 30 μg/ml in Phosphate-Buffered Saline (PBS) and used to coat 24-well (NUNC Multidishes, Denmark) and 96-well (Becton-Dickinson) plates. Dishes were left to air-dry overnight in a laminar flow hood under UV light.

Culture Conditions

In order to monitor the influence of ECM components on gelatinases secretion, the two choriocarcinoma cell lines were plated on plastic, laminin, fibronectin, Col I and IV. Coated dishes were washed twice with sterile H_2O and incubated for 1 hour in serum-free medium. Cell suspensions ($2x10^5$ cells per well) in serum-containing medium were seeded into ECM-coated 24-well plates and incubated at 37°C in a humid atmosphere (5% CO_2 - 95% air). After 24 hours, medium was removed, cells were washed twice with 1 ml of serum-free medium and cultured in 500 μl of serum-free medium for 48 hours. Control wells were processed in the same way except that cells were omitted. Conditioned media were collected, centrifugated to eliminate cellular fragments, and stored at -80°C. In order to detect ECM-bound gelatinases, ECM coats were extracted in lysis buffer (Brij35 2g/l in Tris-HCl, pH 7.6) and dissolved in 10% SDS non reducing sample buffer.

Cell Adhesion Assay

Multiwell culture plates (96-wells, Becton Dickinson) were coated with 50 μl of PBS containing Bovine Serum Albumin (BSA, fraction V, Sigma), fibronectin, laminin, Col I or Col IV at a concentration of 30 μg/ml (see above). Coated wells were washed twice in sterile water and saturated for 1 hour with 1% heat denaturated (3 minutes at 80°C) BSA. Cell suspensions (10^4 cells per well) in serum-free media containing 1% BSA were seeded into protein-coated plates, and at specific times the unattached cells were removed by washing the wells with 100 μl of medium containing 1% BSA and by swirling at 350 rpm for 5 minutes. Attached cells were then incubated for 2 hours at 37°C in 100 μl of culture medium supplemented with 10% WST-1 (Boehringer Mannheim). WST-1 is a tetrazolium salt which is cleaved by metabolically active cells and released as a water-soluble formazan dye product. The extent of adhesion was determined by measuring the absorbance of cell-containing wells at 450 nm with a reference filter of 595 nm (Titertek Multiscan, Labsystem, Finland). Cell-free blank values were subtracted from each value. In preliminary studies, the relationship between formazan production and cell number was determined by plating out cells from 5-200 x 10^2 cells per well in 96-well microtiter plates, allowing cells to attach for two hours and then determining formazan production. Optical densities showed a linear relationship to cell numbers over the range of densities used (data not shown).

We also evaluated the spreading capacity of the cells on the different matrices. For this purpose, adherent cells (see above) were observed under a phase contrast inverted microscope (Olympus) and the percentage of spread adherent cells was visually determined (at least 100 cells were counted in each condition). Each assay point was derived from quadruplicate wells and the experiments were repeated twice. Statistical analysis of the results was performed using the *Student's T* test. Differences are considered as statistically significant for *p* values lower than 0.05.

Antibodies

Monoclonal antibodies specific for α2 (P1E6), α3 (P1B5), α4 (P4G9), α5 (P1D6) integrin subunits were purchased from Dako (Denmark); α1 subunit (TS2/7) from T Cell Diagnostics, Inc. (MA, USA); α6 subunit from Cappel (Organon Teknika, Belgium); and β1 subunit (mAb 13) from Becton Dickinson (California, USA). Controls included omission of primary antibody or its replacement with non-specific IgG of the same isotype. These negative controls provided fluorescence patterns indistinguishable from one another.

Flow Cytometry

Cell suspensions of the two choriocarcinoma cell lines (10^6 cells/50 μl) were incubated with anti-integrin antibodies diluted in serum-containing media for 30 minutes at 4°C. Cells were then washed, incubated in fluorescein-conjugated rabbit anti-mouse IgG (Dako, Denmark) for 30 minutes at 4°C, washed again, and fixed in 1% paraformaldehyde. Fluorescence signals from 10^4 cells were acquired on a FACStar^{PLUS} flow cytometer (Becton Dickinson) to quantify cell surface staining.

Gelatin Zymography And Reversed Zymography

Substrate polyacrylamide gel electrophoresis (zymography) was used to identified type IV collagenolytic enzymes secreted in conditioned media as described previously (Remacle et al., 1995). Five hundred μl of cell-free conditioned media were lyophilized, reconstituted in 50 μl of water and mixed with an equal volume of 10% SDS non reducing sample buffer. Samples were separated at 4°C on 10% SDS-polyacrylamide gels containing 1 mg/ml gelatin (Sigma). After removal of SDS from the gel by incubation in 2% (w/v) Triton X-100 (2 washes at room temperature), the gels were incubated at 37°C for 18 hours in 50 mM Tris-HCl buffer, pH 7.6, containing 10 mM $CaCl_2$ and 0.02 % NaN_3. Gels were stained for 30 minutes in 0.1% Coomassie Brillant Blue G-250 in 40% methanol, 10% glacial acetic acid, and destained in the same solution without dye. Enzyme activity was evidenced by clear zones of lysis against a blue background. Conditioned media of fibrosarcoma HT1080 cells treated with 1 mM *para*-aminophenyl-mercuric acetate (APMA) allowed us to identify most of the different proteolytic activities. HT1080 cells are known to secrete a 92 kDa band corresponding to latent pro-gelatinase B, as well as a 66 kDa band corresponding to latent pro-gelatinase A (Wilhelm et al., 1989). Upon APMA treatment, pro-gelatinases A and B are partially processed into lower molecular weight species of 84 kDa and 59 kDa, which correspond respectively to activated gelatinase B and completely activated gelatinase A.

Reverse zymography was performed according to Remacle et al. (1995) with some modifications. Gelatin substrate gels with incorporated recombinant gelatinase A were prepared and concentrated samples were separated and processed as described for gelatin zymography. Gelatinase inhibitors were detected as dark bands against a clear background, indicating the inhibition of lysis of the substrate by gelatinases included in the gel.

RESULTS

Choriocarcinoma Cell Adhesion To Extracellular Matrix Components

In order to study the interactions between invasive trophoblast cells and decidual ECM components, we first performed adhesion assays in the presence of purified matrix components which are known to be produced by first trimester decidual cells (Wewer et al., 1985; Aplin and Jones, 1989; Korhonen et al., 1991). Two human choriocarcinoma cell lines (BeWo and JAr) were seeded in the absence of serum onto plastic wells coated with matrix components including laminin, fibronectin, and Col I and IV. Attachment and spreading were monitored for up to four hours. The results obtained after two and four hours are summarized in Figure 1. Although the two cell lines attached to each of the matrix components, adhesion to fibronectin was the most efficient (Figures 1A and C). However, BeWo cells differed from JAr cells in their ability to interact with collagens. Indeed, Col I and IV promoted cell attachment more efficiently in BeWo than in JAr cells.

Microscopic examination of cells adhering to the different matrix components revealed that cell shape was clearly modulated by the nature of the substrate (Figure 1B and D). BeWo cells spread very extensively and rapidly on fibronectin and, to a lower extent, on Col I and IV (Figure 1B). More than 95 % of adherent cells spread within two hours on fibronectin. On collagen types I and IV, about 70-80% of cells have spread. In contrast, on laminin, the percentage of spread cells never exceeded 35%, with most cells remaining refractile even after 24 hours (data not shown). The ability of JAr cells to spread on the different ECM components is clearly distinct from that of BeWo cells. Although JAr cells adhered efficiently on all substrates (Figure 1 C), spreading was much more extensive on fibronectin (96%) than on laminin, Col I, and Col IV (15%, 7%, and 20% respectively, Figure 1D). It appears therefore that the two choriocarcinoma cell lines tested in this study have different capacities to interact with ECM components: BeWo cells have a high affinity for fibronectin and collagens, while JAr cells present an affinity restricted to fibronectin.

Integrins Expression On Choriocarcinoma Cells

As most of cell-ECM interactions are mediated by specific integrin receptors, we analyzed the expression of these receptors in our two cell lines. Integrins are heterodimeric transmembrane glycoproteins consisting in an α and a β chain. At least 14 different α chains and 8 different β subunits have been identified so that more than 20 different heterodimeric complexes can be formed (Sonnenberg, 1993). Each of these complexes, however, does not only bind to one substrate, but most of the integrins can bind to several different matrix components.

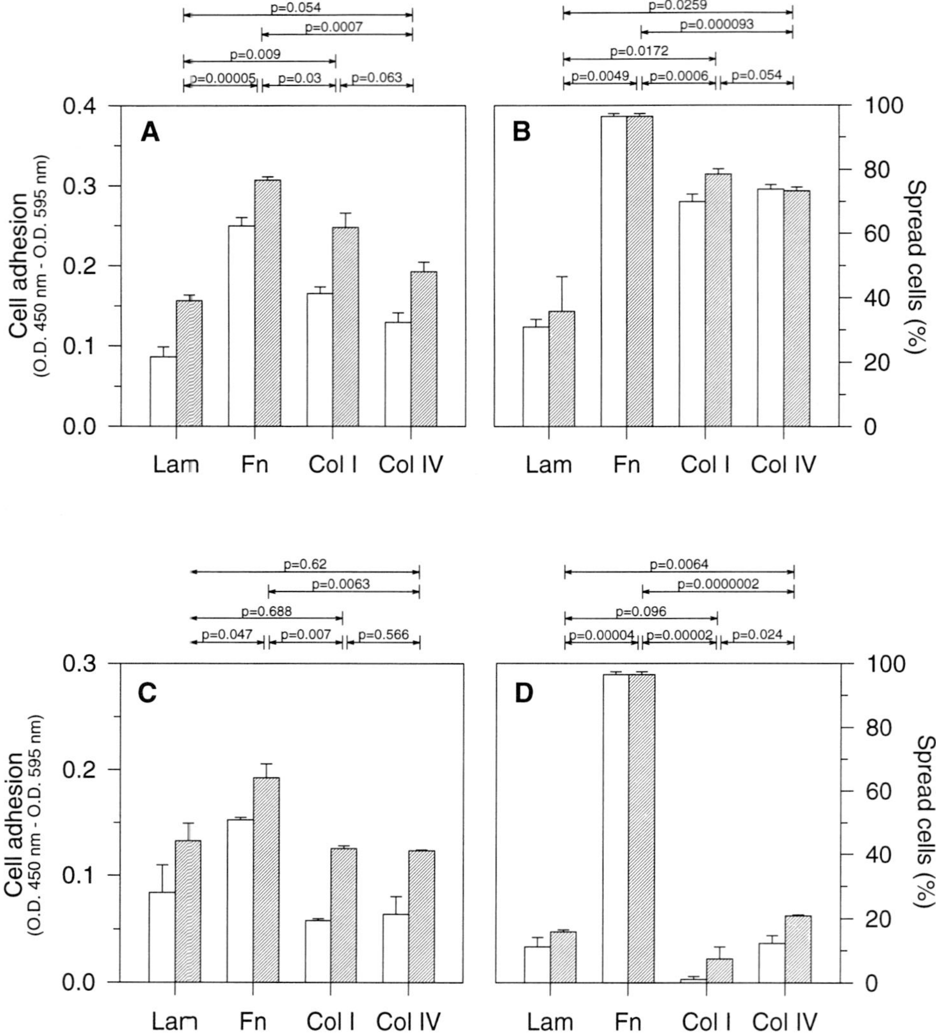

Figure 1. Adhesion and spreading of BeWo (A and B) and JAr (C and D) cells on Laminin (Lam), Fibronectin (Fn), type I collagen (Col I) and type IV collagen (Col IV). The extend of cell adhesion (A and C) was determined after 2 (open bars) and 4 (hatched bars) hours. Adhesion was evaluated by measuring the absorbance of the cell-dependent cleavage product of WST-1 at 450 nm with a reference filter of 595 nm. Spreading (B and D) was evaluated by counting the percentage of adherent cells which spread on the different substrates during the incubation periods used for the adhesion assay. Each assay point was derived from quadruplicate wells and experiments were repeated twice. Results are shown as mean ± standard errors. Statistical analysis of the results are shown for the 4 hours data.

By using flow cytometry, we have evaluated the integrin profile of the two choriocarcinoma cell lines. Indirect immunofluorescence staining with monoclonal antibodies, which recognize the $\alpha1$-$\alpha6$, and $\beta1$ integrin subunits, revealed that both cell lines express $\alpha3$, $\alpha5$, $\alpha6$, and $\beta1$ subunits. However, they differently express the $\alpha1$ and $\alpha2$ subunits. BeWo cells express the $\alpha2$ subunit but not the $\alpha1$ subunit while JAr cells display at their surface the $\alpha1$ but not the $\alpha2$ subunit (Figure 2).

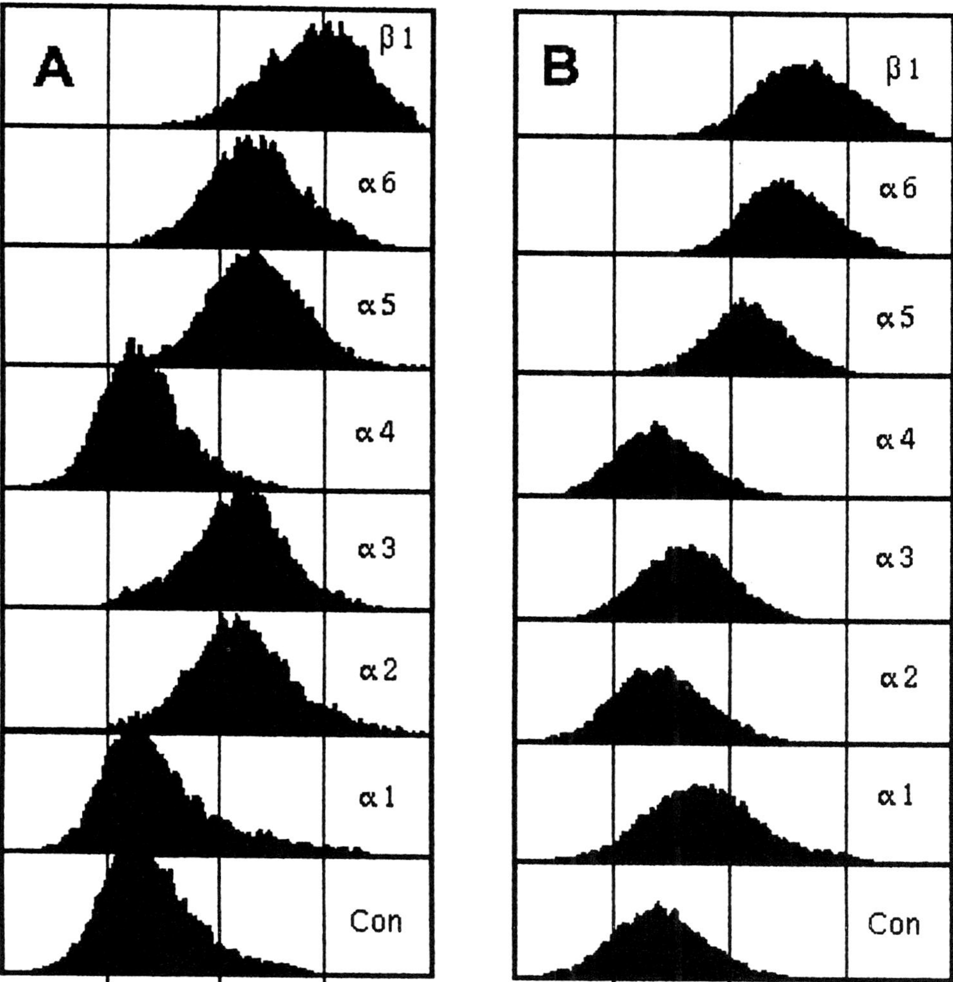

Figure 2. Flow cytometric analysis of integrin expression by BeWo (A) and JAr (B) cells. Horizontal axis represents the log fluorescence intensity, and vertical axis indicates the relative cell counts. Negative controls (Con) include omission of the primary antibody or use of an unrelevant isotype-matched IgG.

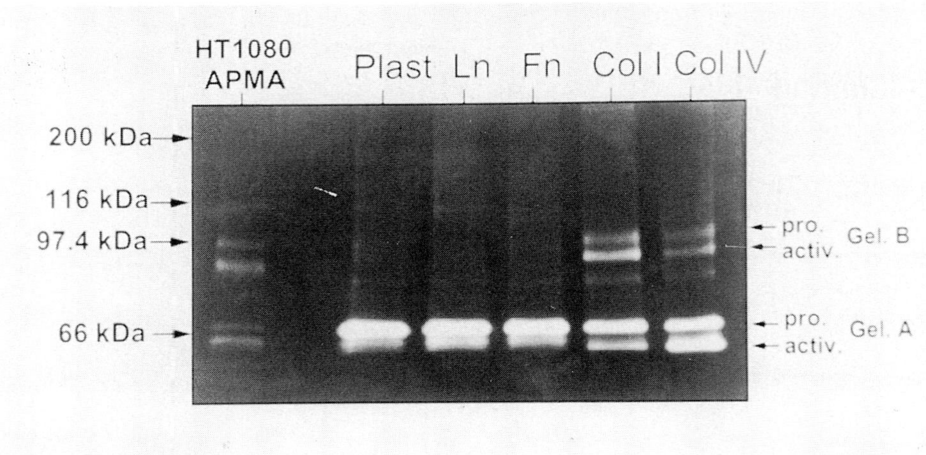

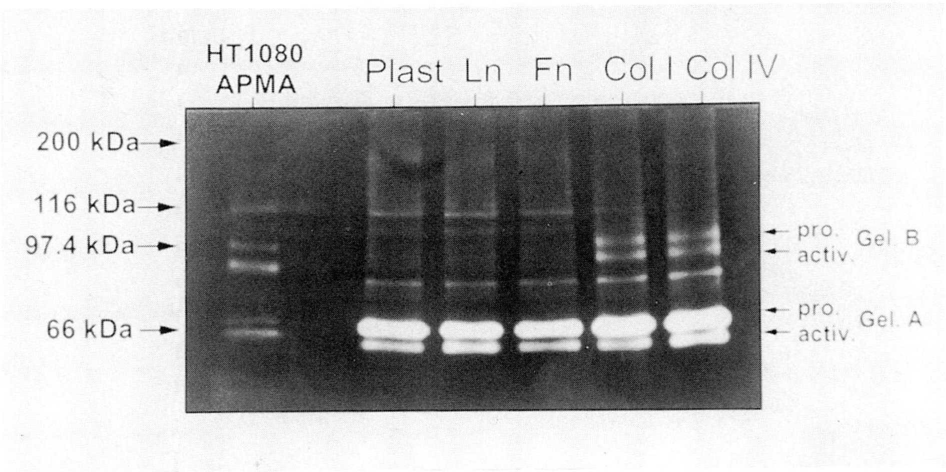

Figure 3. Zymographic analysis of conditioned media from BeWo (A) and JAr (B) cells cultured for two days on plastic (Plast), laminin (Lam), fibronectin (Fn), collagen I (Col I), and collagen IV (Col IV). Gelatinolytic activities were identified by comparing their migration patterns with APMA-activated conditioned media from HT1080 cells (HT1080 APMA). Five major bands were identified in conditioned media of choriocarcinoma cells cultured on collagens: pro-gelatinase B (pro. Gel B; 92 kDa), active gelatinase B (activ. Gel B; 84 kDa), pro-gelatinase A (pro. Gel A; 66 kDa), intermediate gelatinase A (62 kDa) and fully activated gelatinase A (activ. Gel A; 59 kDa). Two unidentified minor bands of 116 and 80 kDa were also detected, especially in conditioned media of JAr cells (B).

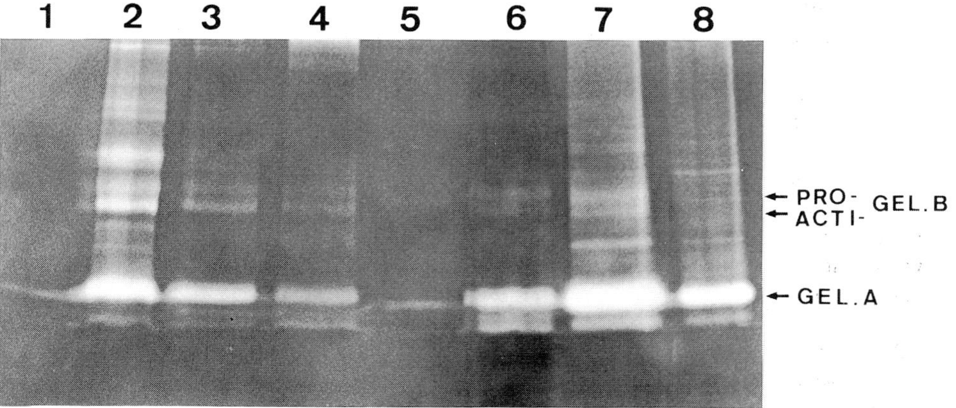

Figure 4. Zymographic analysis of conditioned media and Col IV coats extracts obtained in presence of serum-containing medium or JAr cells. Col IV coats were incubated with JAr cells in serum-containing medium or with serum-supplemented medium alone during 24 hours. Coats were washed twice in serum-free medium and then incubated during 48 hours in 500 µl of serum-free medium. Serum-containing medium, conditioned media obtained in presence or absence of JAr cells and the corresponding Col IV coats extracts were collected and analyzed by gelatin zymography. Serum-derived pro- and activated gelatinase B (pro- and activ- Gel B, respectively) as well as gelatinase A (Gel A), initially bound to Col IV coats, are observed in lanes 2, 3, 4, 6 and 7. **Lane 1**: pure Col IV; **lane 2**: RPMI medium containing 10% serum (RPMI-10% serum); **lane 3**: extract of Col IV coat incubated with RPMI-10% serum for 24 hours; **lane 4**: extract of Col IV coat incubated for 24 hours with cell-free RPMI-10% serum and for 48 hours in serum-free medium; **lane 5**: conditioned medium from Col IV coat incubated for 24 hours with cell-free RPMI-10% serum and for 48 hours in serum-free medium; **lane 6**: extract of Col IV coat incubated for 24 hours with JAr cells in RPMI-10% serum and for 48 hours in serum-free medium; **lane 7**: conditioned medium from Col IV coat incubated for 24 hours with JAr cells in RPMI-10% serum and for 48 hours in serum-free medium; **lane 8**: conditioned medium from JAr cells cultivated on plastic for 48 hours in serum-free medium.

Influence Of ECM Components On Metalloproteinases Secretion

Conditioned media of the two choriocarcinoma cell lines plated on plastic or different purified ECM components were assayed for gelatinolytic activities by gelatin zymography. Several proteinases varying in molecular weight were detected (Figure 3). Addition of 10 mM EDTA in the activation buffer completely inhibited gelatin degradation (data not shown), therefore indicating that all detected proteinases belong to the family of MMPs. On plastic, the conditioned medium of BeWo cells exhibited a strong gelatinolytic band of apparent 66 kDa, corresponding to latent pro-gelatinase A, as well as minor bands of 62 kDa, corresponding to an intermediate activated form of gelatinase A, and an unidentified band of 80 kDa (Figure 3A). In JAr cells conditioned media, a similar gelatinolytic pattern was observed, except that the 62 kDa band was replaced by a completely activated gelatinase A form of 59 kDa (Figure 3B).

In order to investigate the influence of ECM components on MMPs secretion, the two choriocarcinoma cell lines where plated on fibronectin-, laminin-, Col I-, or Col IV-coated wells (Figure 3). Cultures of these cells on either fibronectin or laminin did not modify the gelatinolytic pattern observed on plastic. When the choriocarcinoma cells were plated on Col I or Col IV, changes in gelatinolytic activities were observed. In BeWo cells, 66 kDa and/or 62 kDa bands were partially converted into the 59 kDa species. However, in conditioned medium of JAr cells, which constitutively processed pro-gelatinase A (66 kDa) into its fully activated form (59 kDa), no qualitative modification of pro-gelatinase A activation was detected. Conditioned media of the two cell lines also contained two additional bands of 92 kDa and 84 kDa, corresponding respectively to pro- and activated gelatinase B. However, these two forms of gelatinase B were not secreted by the cells. Indeed, when Col I and IV coats were incubated overnight with cell-free serum-containing media, in order to mimic the plating of cells, we observed that pro- and activated gelatinase B, as well as gelatinase A, were trapped from the serum (Figure 4, lanes 2-4). When cells were added to these matrices, bound gelatinases were released into cells conditioned media. The release of gelatinase B was cell-dependent since no spontaneous release was detected in the absence of cells (Figure 4, lanes 5, 7). Altogether, these results show that, in choriocarcinoma cells, the processing of latent gelatinase A into lower molecular weight active forms was influenced by interactions between cells and specific ECM components such as collagens I and IV.

Secretion Of Metalloproteinase Inhibitors

Reverse zymography was performed to evaluate the secretion of metalloproteinase inhibitors. One inhibitor band of 21 kDa (corresponding to TIMP-2) was detected in conditioned media of the two cell lines, however, at much higher concentration in JAr cells than in BeWo cells (data not shown). In contrast to gelatinase A activation, TIMP-2 production did not appeared to be influenced by the nature of the ECM.

DISCUSSION

The processes of implantation and placentation both depend on the penetration and remodeling of the uterine endometrium and vasculature by first trimester trophoblasts. During pregnancy, under the influence of the implanting embryo and the hormonal environment, uterine stromal cells initiate a series of characteristic physiological changes which condition the uterus for infiltration by trophoblasts. Among the numerous changes which take place during the decidualization process, extensive modifications of the composition and organization of endometrial ECMs have been well documented (Wewer et al., 1985, 1986; Kisalus et al., 1987; Aplin et al., 1988; Aplin and Jones, 1989). These modifications include the secretion by decidual cells of a distinct pericellular basement membrane (which contains laminin, type IV collagen, fibronectin, and heparan sulfate proteoglycans; Wewer et al., 1985; Kisalus et al., 1987; Korhonen et al., 1991; Divers et al., 1995) as well as the remodeling of the decidual interstitial matrix characterized by the disappearance of interstitial collagen VI (which is thought to cross-link fibrils of the major collagens), the reduction in fibril density, and the synthesis of new collagens (Aplin and Jones, 1989; Mylona et al., 1995). Thus, during the first trimester of pregnancy, trophoblast infiltration will take place into a sparsely fibrillar network of several matrix proteins (including laminin, fibronectin, and types-I, -III, -IV, -V collagens) which separates the villous cytotrophoblast cells from the maternal vasculature. Concomitant to the endometrial decidualization, another maturation process takes place at the basal tip of anchoring villi. At this precise location, immature proliferative cytotrophoblast anchored to the villous basement membrane progressively differentiate into motile and highly invasive extravillous trophoblasts. During this process, several changes in cell-matrix interactions occur. While cytotrophoblast cells migrate from the villi into the decidua, they modulate their integrin repertoire from being exclusively α6β4 positive to progressively becoming α6β4 negative and α1β1, α3β1, α5β1, and α6β1 positive (Aplin, 1993; Damsky et al., 1993, 1994; Korhonen et al., 1991; Burrows et al., 1993). *In vitro* studies have demonstrated that such an integrin switch was clearly linked to trophoblastic invasiveness. Indeed, interactions between α1β1 integrin and its ligands (type-IV collagen, laminin) promoted invasion, whereas interactions between α5β1 and fibronectin reduced trophoblast invasiveness (Damsky et al., 1994).

In this study, we have used two different choriocarcinoma cell lines (BeWo and JAr) as a model system for investigating trophoblast interactions with ECMs components. We first evaluated the capacity of these cells to adhere and spread on purified components of the decidual ECM. Our data demonstrate that BeWo cells adhere most efficiently to both fibronectin and type-I and -IV collagens, whereas adhesion of JAr cells was restricted to fibronectin. Since fibronectin is abundant in decidua in locations where trophoblast infiltration occurs (Wewer et al., 1986; Kisalus et al., 1987; Aplin and Jones, 1989; Damsky et al., 1992, 1993), and may thus provide a transient anchorage for migrating cells, the strong affinity of these choriocarcinoma cells for fibronectin may be considered as a requirement for an extravillous phenotype (Aplin et al., 1992). However, the differential affinity of these two cell lines for collagens suggests that they could present certain phenotypical similarities with two distinct subpopulations of extravillous trophoblast cells. This hypothesis is further illustrated by the analysis of integrin profiles of the two cell lines. Both cell lines express high levels of fibronectin receptor (α5β1 integrin) but differ by the expression of collagens/laminin receptors. JAr cells express the α1β1 integrin characteristic of invasive first trimester trophoblast cells (Damsky et al., 1993, 1994), whereas, in BeWo cells, α1β1 is replaced by α2β1, an integrin observed *in vivo* at the base of trophoblast columns where the differentiation process begins (Loke and King, 1995). Our data are very similar to the integrin patterns previously reported (Hall et al., 1990; Aplin et al., 1992), except that these studies failed to detect the α3 subunit. The combination of our results and data from the literature show that BeWo cells express the α2β1, α3β1, α5β1, and α6β4 integrins, a profile similar to that displayed *in vivo* by proliferating cytotrophoblast cells during their progression through the trophoblastic columns (Loke and King, 1995). In contrast, JAr cells mainly express the α1β1, α5β1, and α6β1 integrins, which are characteristic of interstitial trophoblast cells (Loke and King, 1995). Altogether, these results reveal that BeWo and JAr cells are clearly distinct in terms of both adhesive behavior and integrins expression, and might thus mimic two different trophoblastic subpopulations at distinct stages of invasion.

The cell-matrix interactions required for a successful infiltration of the decidua are not restricted to anchorage processes. Since trophoblast cells have to migrate through the decidua, they must degrade ECM components to which they attach. Such a degradation of decidual ECMs requires the action of different proteinases among which members of the MMPs family are known to play key roles. *In vivo*, invasive first trimester trophoblast cells are known to synthesize several MMPs including interstitial collagenase (Yagel et al., 1988; Moll and Lane, 1990), gelatinases A and B (Bischof et al.,1991; Autio-Harmainen et al., 1992; Fernandez et al., 1992; Blankenship and King, 1994; Polette et al., 1994), stromelysins -1, -2, and -3 (Brenner et al., 1989; Maquoi et al., 1997), and MT-MMP-1 (Nawrocki et al., 1996).

In vitro, several groups have reported the synthesis of gelatinases A and B by cultured first trimester cytotrophoblast cells (Fisher et al., 1989; Bischof et al., 1991; Librach et al., 1991; Shimonovitz et al., 1994). In our choriocarcinoma model, we also demonstrated the secretion of high levels of gelatinase A in both BeWo and JAr cells. In contrast, no gelatinase B was detected in standard culture conditions (e.g., in the absence of ECM components). Similar observations have been previously reported for BeWo cells by Kato and coworkers (1995). The ability of cells to produce gelatinase A has been often correlated with the invasive and migratory capacity of several normal as well as tumoral cell types, thus suggesting that this proteinase is important in the biology of cellular invasion (Stetler-Stevenson, 1990). The production of the proenzyme species appears to be necessary but not sufficient for invasion. Indeed, the aquisition of an invasive phenotype relies on the activation of the secreted latent proenzyme. This activation has been demonstrated to be a two-step proteolytic process leading to the appearance of a partially activated intermediate species, which is subsequently converted into a fully activated enzyme by an autocatalytic cleavage (Strongin et al., 1993; Atkinson et al., 1995).

Recent reports have demonstrated that this activation process is catalyzed by an integral plasma membrane protein, the membrane-type MMP (MT-MMP) to which pro-gelatinase A binds (Sato et al., 1994; Atkinson et al., 1995; Strongin et al., 1995; Stetler-Stevenson, 1995). Due to its location at the surface of the cell, MT-MMP can bind and activate secreted latent pro-gelatinase A and, thus localize the degradation of the ECM by concentrating active membrane associated gelatinase A at sites of cellular invasion (Monsky et al., 1993; Vassalli and Pepper, 1994). We have recently reported the expression of MT-MMP-1 by both first trimester trophoblast cells and decidual cells, suggesting that this new MMP could play a role during placentation (Nawrocki et al., 1996). The analysis of conditioned media of the two choriocarcinoma cell lines revealed that pro-gelatinase A and TIMP-2 secretion were independent of the composition of the ECM on which they were grown. In contrast, pro-gelatinase A activation was clearly dependent on both cell line and nature of the substrate, thus suggesting the existence of distinct activation processes. BeWo cells secreted large amounts of pro-gelatinase A which was constitutively processed into the intermediate species of 62 kDa. However, in the presence of collagen (types I and IV), the fully activated form of 59 kDa was detected, suggesting that collagen can either directly (via collagen-gelatinase A interactions) or indirectly (via the induction of a cellular activation process) induce the processing of the intermediate gelatinase A form. In the later case, the collagen-induced cellular responses have been shown to be largely mediated by the integrin family of collagen receptors such as $\alpha1\beta1$, $\alpha2\beta1$, and $\alpha3\beta1$ which can transduce information from the ECM to the nucleus of the cell (Thompson et al., 1994). Similar substrate-dependent inductions of pro-gelatinase A processing have already been documented in different cell types including rat testicular cells (Sang et al., 1991), as well as human fibroblasts (Azzam and Thompson, 1992; Seltzer et al., 1994). In contrast, JAr cells constitutively process the pro-gelatinase A into its fully activated form, independently of the nature of the ECM on which they were grown. Whether these differences in the ability of the two cell lines to process pro-gelatinase A results from their distinct integrin patterns is currently unknown. However, it is interesting to note that these two cell lines, which potentially mimic two distinct trophoblastic subpopulations, could exhibit distinct mechanisms in order to activate pro-gelatinase A. We are actually investigating the potential involvement of MT-MMP in these activation processes.

In the presence of collagen (I and IV), additional gelatinolytic activities corresponding to both pro- and activated gelatinase B were detected in the conditioned media of both cell lines. These proteinases were shown to be released, in the presence of cells, from serum-derived gelatinases which have been trapped in collagen coats. This observation is of great physiological relevance since decidual cells have been shown, both *in vivo* and *in vitro*, to secrete interstitial collagenase, gelatinases and stromelysin (Dieron and Bryant-Greenwood, 1991; Rawdanowicz et al., 1992; Martelli et al., 1993; Polette et al., 1994; Salamonsen, 1994). Once released by decidual cells as inactive proenzymes, these proteinases could be trapped in decidual ECMs. In these locations, they provide a potent, rapidly and targetedly activable store of proteolytic enzymes for infiltrating trophoblast cells, as well as for the uterine ECM remodeling observed during decidualization. Thus, it appears likely that, during the first trimester of pregnancy, the remodeling of the decidua results from the combined activity of both trophoblasts-, as well as decidual-derived proteinases. Moreover, both cell types are also known to secrete TIMPs (the tissue inhibitors of MMPs), which can bind to and inhibit MMPs activities (Lala and Graham, 1990; Polette et al., 1994).

Altogether, these data suggest that the infiltration of the maternal decidua by first trimester extravillous trophoblast cells relies on a precisely regulated balance between MMPs and TIMPs which are secreted by both the invading and invaded compartments.

In conclusion, the combination of our observations and data from the literature suggests that the extensive decidualization of the uterine stroma observed during the first trimester of pregnancy is not a simple protective mechanism to limit trophoblast infiltration (Kirby, 1960, 1965). On the contrary, as suggested by Aplin (1991), it seems likely that decidual tissue is designed for directing trophoblast infiltration rather than acting as a physical barrier to invasion. Indeed, decidualization results in the secretion by stromal cells of unusual ECMs characterized by a reduced collagen fibril density and the presence of several matrix components which can promote trophoblast adhesion, and constitute a reservoir for decidual derived products such as growth factors, proteinases and proteinase inhibitors. In parallel, trophoblast cells modify their expression of ECM receptors which enable them to interact with decidual ECMs components and modulate the trophoblast-derived proteolytic activities in order to penetrate the decidua.

SUMMARY

BeWo and JAr choriocarcinoma cell lines were used to study cell adhesion and proteolysis, properties thought to be relevant during invasion of maternal decidua by first trimester extravillous trophoblast cells. Both cell lines adhere and spread efficiently on fibronectin, a major component of decidual ECMs. In contrast, they differ by their capacity to interact with laminin and types-I and -IV collagens, which constitute a good substrate for BeWo cells only. These differences are mirrored by the integrin profiles of these cells. BeWo cells express $\alpha2\beta1$, $\alpha3\beta1$, $\alpha5\beta1$, and $\alpha6\beta4$ integrins, a profile displayed *in vivo* by proliferating cytotrophoblast cells during their progression through trophoblastic columns, whereas JAr cells mainly express α1β1, α5β1, and α6β1 integrins which are characteristic of interstitial trophoblast cells. Interactions between trophoblast cells and ECMs are not simply restricted to adhesion processes since they also involve the remodeling of ECMs by tightly regulated proteolytic activities. From this point of view, both cell lines have been shown to actively secrete pro-gelatinase A which is subsequently activated, depending on the nature of the substrate encountered by the cell (different substrate specificities are observed depending on the cell line). These cells are also able to induce the release of proteinases previously trapped by the collagenous component of the ECMs. Altogether, our observations suggest that the infiltration of the maternal decidua by first trimester trophoblast cells rely on extremely complex interactions among which ECMs-derived information appear to play pivotal roles.

ACKNOWLEDGEMENTS

We gratefully thank Dr. R Greimers for his expert knowledge in flow cytometric analysis. Erik Maquoi is the beneficiary of a grant from the Belgian National Fund for Scientific Research-Télévie Foundation (No. 7.4567.95). This work was supported by grants from the "Communauté Française de Belgique" (Actions de Recherche Concertées 93/98-171), The Commission of European Communities (Concerted European Action BIOMED 1 n° PL931346), the "Fonds de la Recherche Scientifique Médicale" (No. 3.9003.92), the "Association contre le Cancer", the "Association Sportive contre le Cancer", the "CGER-Assurances" and "asbl VIVA" 1993/1996, the industry (Boehringer Mannheim GmbH, Penzberg, Germany); the "Centre Anticancéreux près de l'Université de Liège" and the "Fondation Léon Frédéricq" - University of Liège, Belgium.

REFERENCES

Aplin, J.D., Charlton, A.K. and Ayad, S. (1988) An immunohistochemical study of human endometrial extracellular matrix during the menstrual cycle and first trimester of pregnancy. *Cell Tissue Res.* 253, 235-240.

Aplin, J.D. and Jones, C.J.P. (1989) Extracellular matrix in endometrium and decidua. In: *Placenta As A Model And A Source,* (ed.) O. Genbacev and A. Klopper, Plenum Publishing Corporation, pp. 115-128.

Aplin J.D. (1991) Implantation, trophoblast differentiation and haemochorial placentation: Mechanistic evidence *in vivo* and *in vitro*. *J Cell Science*, 99, 681-692.

Aplin, J.D., Sattar, A. and Mould, P. (1992) Variant choriocarcinoma (BeWo) cells that differ in adhesion and migration on fibronectin display conserved patterns of integrin expression. *J. Cell Science* 103, 435-444.

Aplin J.D. (1993) Expression of integrin α6β4 in human trophoblast and its loss from extravillous cells. *Placenta* 14, 203-216.

Atkinson, S.J., Crabbe, T., Cowell, S., Ward, R., Butler, M.J., Sato, H., Seiki, M., Reynolds, J.J. and Murphy, G. (1995) Intermolecular autolytic cleavage can contribute to the activation of progelatinase A by cell membranes. *J. Biol. Chem.* 270, 30479-30485.

Autio-Harmainen, H., Hurkainen, T., Niskasaari, K., Höyhtya, M. and Tryggvason, K. (1992) Simultaneous expression of 70 kilodalton type IV collagenase and type IV collagen alpha 1 (IV) chain genes by cells of early human placenta and gestational endometrium. *Lab. Invest.* 67, 191-200.

Azzam, H.S., and Thompson, E.W. (1992) Collagen-induced activation of the *Mr* 72,000 type IV collagenase in normal and malignant human fibroblastoid cells. *Cancer Res.* 52, 4540-4544.

Basset, P., Wolf, C. and Chambon, P. (1993) Expression of the stromelysin-3 gene in fibroblastic cells of invasive carcinomas of the breast and other human tissues: A review. *Br. Cancer Res. Treat.* 24, 185-193.

Belaaouaj A., Shipley, J.M., Kobayashi, D.K., Zimonjic, D.B., Popescu, N., Silverman, G.A. and Shapiro, S.D. (1995) Human macrophage metalloelastase. *J. Biol. Chem.* 270, 14568-14575.

Birkedal-Hansen, H., Moore, W.G.I., Bodden, M.K., Windsor, L.J., Birkedal-Hansen, B., DeCarlo, A. and Engler, J.A. (1993) Matrix metalloproteinases: A review. *Crit. Rew. Oral Biol. Med.*, 4, 197-250.

Bischof, P. and Martelli, M. (1992) Current topic: Proteolysis in the penetration phase of the implantation process. *Placenta* 13, 17-24.

Bischof, P., Friedli, E., Martelli, M. and Campana, A. (1991) Expression of extracellular matrix-degrading metalloproteinases by cultured human cytotrophoblast cells: Effects of cell adhesion and immunopurification. *Am. J. Obstet. Gynecol.* 165, 1791-1801.

Blankenship, T.N. and King, B.F. (1994) Identification of 72-kilodalton type IV collagenase at sites of trophoblastic invasion of macaque spiral arteries. *Placenta* 15, 177-187.

Brenner, C.A., Adler, R.R., Rappolee, D.A., Pedersen, R.A. and Werb, Z. (1989) Genes for extracellular matrix-degrading metalloproteinases and their inhibitor, TIMP, are expressed during early mammalian development. *Genes Dev.* 3, 848-859.

Burrows, T.D., King, A. and Loke, Y.W. (1993) Expression of integrins by human trophoblast and differential adhesion to laminin or fibronectin. *Hum. Reprod.* 8, 475-484.

Cawston, T.E. (1995) Proteinases and inhibitors. *Brit. Med. Bull.* 51, 385-401.

Damsky, C.H., Fitzgerald, M.L. and Fisher, S.J. (1992) Distribution patterns of extracellular matrix components and adhesion receptors are intricately modulated during first trimester cytotrophoblast differentiation along the invasive pathway, in vivo. *J. Clin. Invest.* 89, 210-222.

Damsky, C.H., Librach, C., Lim, K.H., Fitzgerald, M.L., McMaster, M.T., Janatpour, M., Zhou, Y., Logan, S.K. and Fisher, S.J. (1994) Integrin switching regulates normal trophoblast invasion. *Development* 120, 3657-3666.

Damsky, C.H., Sutherland, A. and Fisher, S.J. (1993) Extracellular matrix 5: Adhesive interactions in early mammalian embryogenesis, implantation, and placentation. *FASEB J.* 7, 1320-1329.

Delvoye, P., Nusgens, B. and Lapière, Ch.M. (1983) The capacity of retracting a collagen matrix is lost by dermatosparactic skin fibroblasts. *J. Invest. Dermatol.* 81, 267-270.

Dieron, D. and Bryant-Greenwood, G.D. (1991) Collagens, collagenolytic enzymes, and inhibitors in the human fetal membranes and decidua. *Troph. Res.* 5, 205-216.

Divers, M.J., Bulmer, J.N., Miller, D. and Lilford, R.J. (1995) Beta 1 integrins in third trimester human placentae: No differential expression in pathological pregnancy. *Placenta* 16,245-260.

Feinberg, R.F., Kao, L.-C., Haimowitz, J.E., Queenan, J.T., Wun, T.-C., Strauss III, J.F. and Kliman, H.J. (1989) Plasminogen activator inhibitor type 1 and 2 in human trophoblasts. *Lab. Invest.* 61, 20-26.

Fernandez, P.L., Merino, M.J., Nogales, F.F., Charonis, A.S., Stetler-Stevenson, W. and Liotta, L.A. (1992) Immunohistochemical profile of basement membrane proteins and 72 kilodalton type IV collagenase in the implantation placental site. *Lab. Invest.* 66, 572-579.

Fisher, S.J., Cui, T.-Y., Zhang, L., Hartman, L., Grahl, K., Guo-Yang, Z., Tarpey, J. and Damsky, C.H. (1989) Adhesive and degradative properties of human placental cytotrophoblast cells in vitro. *J. Cell Biol.* 109, 891-902.

Graham, C.H. and Lala, P.K. (1992) Mechanisms of placental invasion of the uterus and their control. *Biochem. Cell Biol.* 70, 867-874.

Hall, D.E., Reichardt, L.F., Crowley, E., Holley, B., Moezzi, H., Sonnenberg, A. and Damsky, C.H. (1990) The $\alpha 1/\beta 1$ and $\alpha 6/\beta 1$ integrin heterodimers mediate cell attachment to distinct sites on laminin. *J. Cell Biol.* 110, 2175-2184.

Hofmann, G.E, Glatstein, I., Schatz, F., Heller, D. and Deligdish, L. (1994) Immunohistochemical localization of urokinase-type plasminogen activator and plasminogen activator inhibitors 1 and 2 in early human implantation sites. *Am. J. Obstet. Gynecol.* 170, 671-676.

Hynes, R.O. (1992) Integrins: Versatility, modulation, and signalling in cell adhesion. *Cell* 69, 11-25.

Kato, N., Nawa, A., Tamakoshi, K., Kikkawa, F., Suganuma, N., Okamoto, T., Goto, S., Tomoda, Y., Hamaguchi, M. and Nakajima, M. (1995) Suppression of gelatinase production with decreased invasiveness of choriocarcinoma cells by human recombinant interferon beta. *Am. J. Obstet. Gynecol.* 172, 601-606.

Kirby, D.R. (1960) The development of mouse eggs beneath the kidney capsule. *Nature* 187, 707-708.

Kirby, D.R. (1965) The invasiveness of the trophoblast. In: *The Early Conceptus Normal And Abnormal*, (ed.) W.W. Park, pp. 68-74.

Kisalus, L.L., Herr, J.C. and Little, C.D. (1987) Immunolocalization of extracellular matrix proteins and collagen synthesis in first-trimester human decidua. *Anat. Rec.* 218, 402-415.

Korhonen, M., Ylänne, J., Laitinen, L., Cooper, H.M., Quaranta, V., and Virtanen, I. (1991) Distribution of the α1-α6 integrin subunits in human developing and term placenta. *Lab. Invest.* 65, 347-355.

Kurman, R.J., Main, C.S. and Chen, H.C. (1984) Intermediate trophoblast: A distinctive form of trophoblast with specific morphological, biochemical and functional features. *Placenta* 5, 349-370.

Lala, P.K. and Graham, C.H. (1990) Mechanisms of trophoblast invasiveness and their control: The role of proteases and proteases inhibitors. *Cancer Metast. Rev.* 9, 369-379.

Librach, C.L., Werb, Z., Fitzgerald, M.L., Chiu, K., Corwin, N.M., Esteves, R., Grobelny, D., Galardy, R. and Damsky, C.H. (1991) 92-kD type IV collagenase mediates invasion of human cytotrophoblasts. *J. Cell Biol.* 113, 437-449.

Liotta, L.A., Rao, C.N. and Wewer, U.M. (1986) Biochemical interactions of tumor cells with the basement membrane. *Ann. Rev. Biochem.* 55, 1037-1057.

Loke, Y.W. and King, A. (1995) Trophoblast interaction with extracellular matrix. In: *Human Implantation: Cell Biology And Immunology*, (eds.) Y.W. Loke and A. King, pp. 151-179.

Maquoi, E., Polette, M., Nawrocki, B., Bischof, P., Noël, A., Pintiaux, A., Santavicca, M., Schaaps, J.-P., Pijnenberg, R., Birembaut, P. and Foidart, J.-M. (1997) Expression of stromelysin-3 in the human placenta and placental bed. *Placenta* 18, in press.

Martelli, M., Campana, A. and Bischof, P. (1993) Secretion of matrix metalloproteinases by human endometrial cells *in vitro*. *J. Reprod. Fertil.* 98, 67-76.

Moll, U.M. and Lane, B.L. (1990) Proteolytic activity of first trimester human placenta: localization of interstitial collagenase in villous and extravillous trophoblast. *Histochemistry* 94, 555-590.

Monsky, W.L., Kelly, T., Lin, C.-Y., Yeh, Y., Stetler-Stevenson, W.G., Mueller, S.C. and Chen, W.-T. (1993) Binding and localization of Mr 72,000 matrix metalloproteinase at cell surface invapodia. *Cancer Res.* 53, 3159-3164.

Mylona, P., Kielty, C.M., Hoyland, J. and Aplin, J.D. (1995) Expression of type VI collagen in human endometrium and decidua. *J. Reprod. Fert.* 103, 159-167.

Nawrocki, B., Polette, M., Marchand, V., Maquoi, E., Beorcha, A., Tournier, J.M., Foidart, J.M., and Birembaut, P. (1996) Membrane-type matrix metalloproteinase-1 expression at the site of human placentation. *Placenta* 17, 565-572.

Pignatelli, M. and Stamp, G. (1995) Integrins in tumour development and spread. *Canc. Sur.* 24, 113-127.

Polette, M., Nawrocki, B., Pintiaux, A., Massenat, C., Maquoi, E., Volders, L., Schaaps, J.P., Birembaut, P. and Foidart, J.-M. (1994) Expression of gelatinases A and B and their tissue inhibitors by cells of early and term placenta and gestational endometrium. *Lab. Invest.* 71, 838-846.

Rawdanowicz, T.J., Hampton, A.L., Nagase, H., Woolley, D.E. and Salamonsen, L.A. (1992) Matrix metalloproteinase production by cultured human endometrial stromal cells: identification of interstitial collagenase, gelatinase A, gelatinase B and stromelysin 1 and their differential regulation by interleukin-1α and tumor necrosis factor α. *Proc. Aust. Soc. Med. Res.* p. 18.

Remacle, A.G., Baramova, E.N., Weidle, U.H., Krell, H.W. and Foidart, J.M. (1995) Purification of progelatinases A and B by continuous-elution electrophoresis. *Prot. Expr. Purif.* 6, 417-422.

Ries, C. and Petrides, P.E. (1995) Cytokine regulation of matrix metalloproteinase activity and its regulatory dysfunction in disease. *Biol. Chem. Hoppe-Seyler* 376, 345-355.

Salamonsen, L.A. (1994) Matrix metalloproteinases and endometrial remodelling. *Cell Biol. Intern.* 18, 1139-1144.

Sang, Q.-X., Thompson, E.W., Grant, D., Stetler-Stevenson, W.G. and Byers, S.W. (1991) Soluble laminin and arginine-glycine-aspartic acid containing peptides differentially regulate type IV collagenase messenger RNA, activation, and localization in testicular cell culture. *Biol. Reprod.* 45, 387-394.

Sato, H., Takino, T., Okada, Y., Cao, J., Shinagawa, A., Yamamoto, E. and Seiki, M (1994) A matrix metalloproteinase expressed on the surface of invasive tumor cells. *Nature* 370, 61-65.

Seltzer, J.L., Lee, A.Y., Akers, K.T., Sudbeck, B., Southon, E.A., Wayner, E.A. and Eisen, A.Z. (1994) Activation of 72-kDa type IV collagenase/gelatinase by normal fibroblasts in collagen lattices is mediated by integrin receptors but is not related to lattice contraction. *Exp. Cell Res.* 213, 365-374.

Shimonovitz, S., Hurwitz, A., Dushnik, M., Anteby, E., Geva-Eldar, T. and Yagel, S. (1994) Developmental regulation of the expression of 72 and 92 kd type IV collagenases in human trophoblasts: a possible mechanism for control of trophoblast invasion. *Am. J. Obstet. Gynecol.* 171, 832-838.

Sonnenberg, A. (1993) Integrins and their ligands. *Curr. Topics Microbiol. Immunol.* 184, 7-33.

Stetler-Stevenson, W.G. (1990) Type IV collagenases in tumor invasion and metastasis. *Cancer Metast. Rev.* 9, 289-303.

Stetler-Stevenson, W.G. (1995) Progelatinase A activation during tumor cell invasion. *Invasion Metast.* 14, 259-268.

Stetler-Stevenson, W.G., Aznavoorian, S. and Liotta, L.A. (1993) Tumor cell interactions with the extracellular matrix during invasion and metastasis. *Annu. Rev. Cell Biol.* 9, 541-573.

Strongin, A.Y., Collier, I., Bannikov, G., Marmer, B.L., Grants, G.A. and Goldberg, G.I. (1995) Mechanism of cell surface activation of 72-kDa type IV collagenase. *J. Biol. Chem.* 270, 5331-5338.

Strongin, A.Y., Marmer, B.L., Grant, G.A. and Goldberg, G.I. (1993) Plasma membrane-dependent activation of the 72-kDa type IV collagenase is prevented by complex formation with TIMP-2. *J. Biol. Chem.* 268, 14033-14039.

Takino, T., Sato, H., Shinagawa, A. and Seiki, M. (1995) Identification of the second membrane-type matrix metalloproteinase (MT-MMP-2) gene from a human placenta cDNA library. *J. Biol. Chem.* 270, 23013-23020.

Thompson, E.W., Yu, M., Bueno, J., Jin, L., Maiti, S.N., Palao-Marco, F.L., Pulyaeva, H., Tamborlane, J.W., Tirgari, R., Wapnir, I. and Azzam, H. (1994) Collagen induced MMP-2 activation in human breast cancer. *Breast Canc. Res. Treat.* 31, 357-370.

Timpl, R., Bruckner, P. and Fietzek, P. (1979) Characterization of pepsin fragments of basement membrane collagen obtained from a mouse tumor. *Eur. J. Biochem.* 95, 255-263.

Vassalli, J.D. and Pepper, M.S. (1994) Membrane proteases in focus. *Nature* 370, 14-15.

Wewer, U.M., Damjanov A,.Weiss, J., Liotta, L.A. and Damjanov, I. (1986) Mouse endometrial stromal cells produce basement-membrane components. *Differentiation* 32, 49-58.

Wewer, U.M., Faber, M., Liotta, L.A. and Albrechtsen, R. (1985) Immunochemical and ultrastructural assessment of the nature of the pericellular basement membrane of human decidual cells. *Lab. Invest.* 53, 624-633.

Wilhem, S.M., Collier, I.E., Marmer, B.L., Eisen, A.Z., Grant, G.A. and Goldberg, G.I. (1989) SV40-transformed human lung fibroblasts secrete a 92-kDa type IV collagenase which is identical to that secreted by normal human macrophages. *J. Biol. Chem.* 264, 17213-17221.

Will, H. and Hinzmann, B. (1995) cDNA sequence and mRNA tissue distribution of a novel human matrix metalloproteinase with a potential transmembrane segment *Eur. J. Biochem.* 231, 602-608.

Yagel, S., Parhar, R.S., Jeffrey, J.J. and Lala, P.K. (1988) Normal non-metastatic trophoblast cells share in vitro invasive properties of malignant cells. *J. Cell Physiol.* 136, 455-464.

Yamada, K.M. (1983) Isolation of fibronectin from plasma and cells. In: *Immunochemistry Of The Extracellular Matrix,* (ed.) H. Furthmayr, pp. 111-123.

Trophoblast Research 10:143-162, 1997

LAMININS IN DECIDUA, PLACENTA AND CHORIOCARCINOMA CELLS

Heather J. Church[1], Allan J. Richards[2], and John D. Aplin[1,3]

[1]Department of Obstetrics and Gynaecology
and School of Biological Sciences
University of Manchester
Manchester, United Kingdom

[2]MRC Connective Tissue Genetics Group
Strangeways Laboratory and Department of Pathology
University of Cambridge
Cambridge, United Kingdom

INTRODUCTION

The Laminin Family

Laminins are a family of extracellular matrix glycoproteins (Engel, 1992, 1993; Engvall, 1993; Burgeson et al., 1994; Timpl and Brown, 1994). The prototype laminin (laminin 1; M_r~850,000) was originally described in the matrix of the Engelbreth-Holm-Swarm (EHS) mouse tumor and on isolation was found to be a large heterotrimer (Timpl et al., 1979). The three subunits α1 (MWt~400kDa), β1 (~220kDa), and γ1 (~200kDa) assemble to yield a cruciform molecule with one long and three short arms (Figure 1). The long arm of the cross comprises a linear assembly of α-helical domains associated in a triple helix. Further subunits have been identified that show degrees of sequence identity to the subunits of laminin 1, each homologous subunit being a product of a distinct gene (Burgeson et al., 1994). The assembly of various possible heterotrimeric combinations has been described, giving rise to a family of related laminin isoforms. To date eleven different laminin isoforms are predicted and these are shown in Table 1 and Figure 1.

Laminin 2 (merosin) was originally isolated from human placenta (Leivo and Engvall, 1988) and mouse heart (Paulsson and Saladin, 1989), and found to contain a novel α chain homologue (α2) along with a β1 and γ1 subunit (Ehrig et al., 1990; Paulsson et al., 1991). The α2 subunit is unique in having two non-covalently associated chains of ~300 and 80kDa; these are proteolytic products derived from one gene product (LAMA2). The light chain contributes the last two of five globular domains at the carboxy terminus of the long arm (Vuolteenaho et al 1994; Figure 1). Laminin 3 (s-laminin) contains a β chain homologue (β2; ~190kDa; Hunter et al., 1989a) associated with an α1 and γ1 chain. Laminin 4 (s-merosin) is a combination of the α2 and β2 subunits along with a γ1 chain (Engvall et al., 1990). More recently identified members of the laminin family are laminin 5 (kalinin/nicein; α3β3γ2; Rousselle et al., 1991; Verrando et al., 1988), laminin 6 (k-laminin; α3β1γ1; Marinkovich et al., 1992a) and laminin 7 (k-s-

[3]To Whom Correspondence Should Be Addressed: Research Floor, St. Mary's Hospital, Manchester, M13 0JH, United Kingdom

Table 1

Summary of the laminin family to date showing current and previous names and heterotrimeric composition. Laminins 8-11 are predicated but yet to be confirmed (Timpl, 1996)

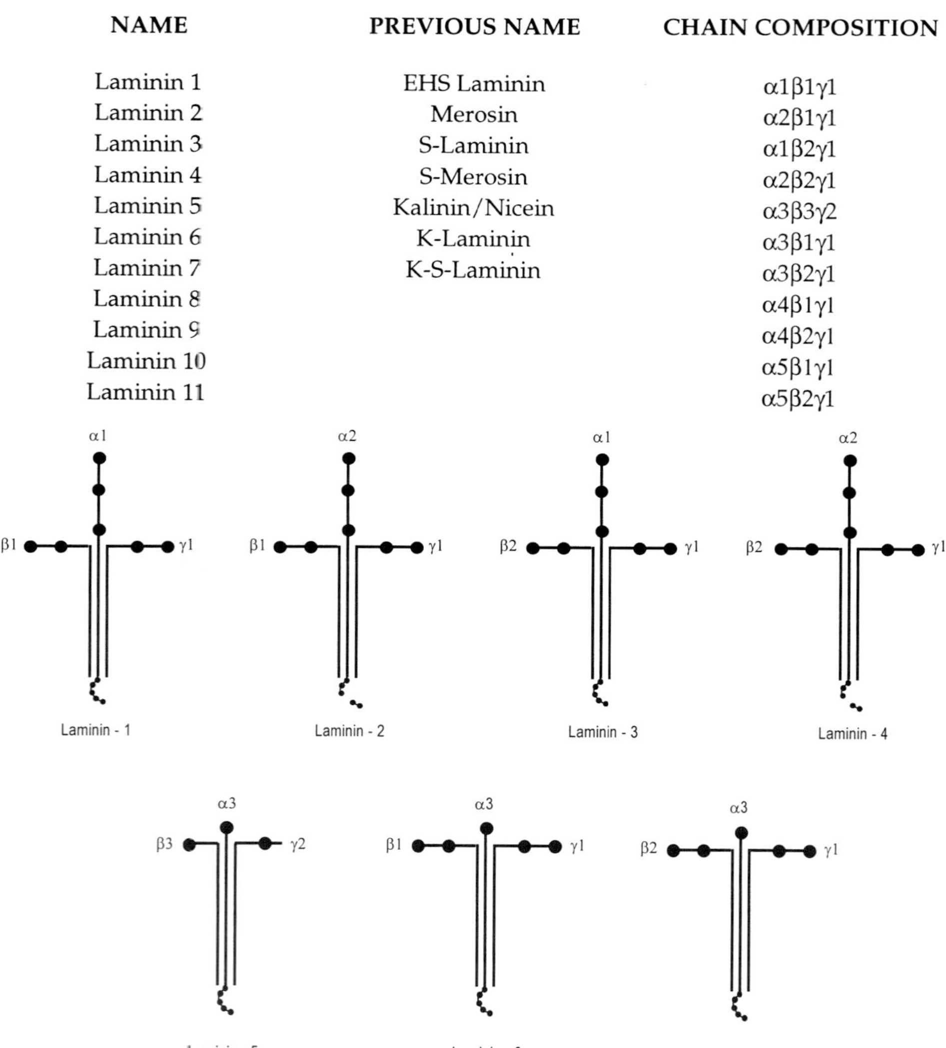

NAME	PREVIOUS NAME	CHAIN COMPOSITION
Laminin 1	EHS Laminin	α1β1γ1
Laminin 2	Merosin	α2β1γ1
Laminin 3	S-Laminin	α1β2γ1
Laminin 4	S-Merosin	α2β2γ1
Laminin 5	Kalinin/Nicein	α3β3γ2
Laminin 6	K-Laminin	α3β1γ1
Laminin 7	K-S-Laminin	α3β2γ1
Laminin 8		α4β1γ1
Laminin 9		α4β2γ1
Laminin 10		α5β1γ1
Laminin 11		α5β2γ1

Figure 1. Overview of laminin heterotrimer assembly. The α4 subunit (not shown) has structural similarities to α3, most notably in its N-terminally truncated short arm. The α5 subunit (not shown) has a longer short arm than α1 or α2. Note that each α chain has a long arm that interacts with the β and γ chain short arms, C-terminal of which in each case is a region of 5 globular domains. In α2 this is cleaved so that the two C-terminal globular domains form a light chain of ~80kDa (adapted from Timpl and Brown 1994).

laminin; α3β2γ1; Burgeson 1993; Champliaud et al., 1996). Splice variants of the α3 subunit are also observed with distinct N-terminal domains (Vidal et al., 1995). Genomic and cDNA clones have been isolated that encode subunit α4 (Richards et al., 1994; Iivanainen et al., 1995), but it has not been continued which heterotrimeric combinations it forms. The α4 protein is more homologous to α3 than α1 and α2 (55, 50, and 45% respectively; Iivanainen et al., 1995). Laminin α5 is slightly larger than α1, is homologous to drosophila laminin and is widely expressed in mouse (Miner et al., 1995); again its repertoire of subunit interactions is uncharacterized. Other novel laminin-like molecules have been reported (Mizayaki et al., 1993; Lindblom et al., 1994; Sorokin et al., 1994). Thus the laminin family is not yet fully characterized.

In general it appears that a specific subset of laminin isoforms is present in each basement membrane. Thus for example, the human placental villous basement membrane has been reported to contain laminins 1-4 (Engvall et al 1990). Certain tumor cells produce a complex repertoire of laminin subunits, only some of which are assembled for secretion (Wewer et al., 1994a). Laminin 1 expression is widespread in embryonic epithelia (Ekblom, 1996), while laminin 2 is found in muscle, peripheral nerve and trophoblast. The α2 subunit appears to be expressed in several mesenchymal locations in the mid trimester fetus (Vuolteenaho et al., 1994). Laminin 3 was originally identified in glomerular basement membrane and neuromuscular junction where it is restricted to the synapse (Hunter et al., 1989a) but it is also found in trophoblast, and laminin 4 is present in trophoblast and myotendinous junction (Engvall et al., 1990). Laminins 5, 6, and 7 are found in certain epithelial basement membranes (Verrando et al., 1987; Verrando et al., 1988; Rousselle et al., 1991; Aplin and Church, 1994). The α4 subunit has been found to be expressed by mesenchymal tissues at the RNA level, although no protein data is currently available (Iivanainen et al., 1995).

Laminin Functions

Laminin 1 has been more widely studied than other members of the family. It binds to many of the other components of basement membranes including collagen IV, perlecan, entactin, and also to itself (cf., Beck and Gruber, 1995). It is the first extracellular matrix component to have been detected in the developing embryo, the α1 subunit being first expressed in mouse at the 16-cell stage, and the β1 and γ1 chains even earlier (Cooper and MacQueen, 1983). In this context laminin may play a role both in the organization of early matrix deposition and the polarization of the trophectodermal epithelium. Laminin-cell interactions are important in mediating polarization, adhesion, and migration in a wide variety of cell types (Mercurio, 1995); mouse trophoblast outgrows efficiently on laminin 1-containing substrates (Armant et al., 1986) and normal human cytotrophoblast (Burrows et al., 1995) and choriocarcinoma cells (Aplin and Charlton, 1990) attach to such surfaces.

Laminin binding to the cell surface can be mediated by several members of the integrin family: α1β1, α2β1, α3β1, α6β1, α7β1, α9β1, α6β4, αvβ3 and αvβ8 (Mecham, 1991; Hynes, 1992; Mercurio, 1995). These interactions have in some cases been localized to discrete domains of the laminin molecule; for example, α6β1 has been shown to bind to a site in the C-terminal region on the long arm of laminin 1 (Gehlsen et al., 1989; Hall et al., 1990) while α1β1 binds to a site at the N-terminal globular domain in the short arm of the α1 chain. The binding site for integrin α2β1 is located on the laminin β1 subunit in the central cross region (Underwood et al., 1995). The muscle glycoprotein α-dystroglycan has been shown to bind the c-terminal globular domains of Laminin 2 (Mercurio, 1995). Functional differences between the isoforms are less clear. Laminin 3 in muscle is concentrated at the sites of synapses; after injury, it can mediate the regeneration of motor axons to these sites and this activity is associated with a 16 amino acid sequence in the $\beta 2$ subunit (Hunter et al., 1989b; Martin et al., 1995). This has further been confirmed by the observation that mutant mice lacking $\beta 2$ show defects in neuromuscular junctions (Noakes et al., 1995). Abnormal expression of the laminin $\alpha 2$ chain is associated with some forms of muscular dystrophy (Hayashi et al., 1993; Tome et al., 1994; Hebling-Leclerk et al., 1995) while other forms show abnormal expression of $\beta 1$ and $\beta 2$ (Yamada et al., 1995). Mutations in the subunits of laminin 5 are associated with the blistering disease epidermolysis bullosa (Aberdam et al., 1994; Pulkkinen et al., 1994; Vailly et al., 1995).

Laminin 1 has been shown to modulate other aspects of cell phenotype. It can stimulate the production of matrix metalloproteinases and invasive behavior in malignant cells (Royce et al., 1992; Bresalier et al., 1995) as well as promoting angiogenesis (Kibbey et al., 1994). This activity is mediated by the peptide sequence IKVAV from the distal part of the long arm of the $\alpha 1$ subunit. Laminin can trigger secretory differentiation in mammary epithelial cells, and this activity is associated with the globular domains at the end of the $\alpha 1$ long arm (Streuli et al., 1995). The peptide YIGSR from the short arm of the $\beta 1$ subunit can induce apoptosis if presented in a multimeric form (Kim et al., 1994).

Laminin is highly antigenic and autoantibodies are associated with a number of pathologies including pre-eclampsia (Foidart et al., 1986) and systemic lupus erythematosus (Garcia Lerma et al., 1995; Church, unpublished data). Interestingly, administration of laminin antibodies to pregnant mice can cause spontaneous abortions (Foidart et al., 1983). Furthermore, monkeys immunized with YIGSR abort spontaneously, and their immune sera are embryotoxic (Chambers et al., 1995).

Laminins in the Endometrium and Placenta

Undifferentiated endometrial stroma contains a mesenchymal-type extracellular matrix with collagens I, III, V, VI, and fibronectin (Aplin et al., 1988; Aplin, 1989; Aplin and Jones, 1989) and periglandular deposits of tenascin (Vollmer et al., 1990). The glandular and endothelial basement membranes contain type IV collagen, laminin, and heparan sulphate proteoglycan (Aplin et al., 1988; Aplin, 1989; Aplin and Jones, 1989). On decidualization the stromal cells produce a pericellular basement membrane containing collagen IV, heparan sulphate proteoglycan, BM-40, and laminin (cf., Aplin, 1989). Its function is unknown but it is possible that it plays a role in controlling the adhesion, migration, and differentiation of invading trophoblast cells and therefore in the successful implantation of the blastocyst. In placenta, cytotrophoblastic columns contain rather little extracellular matrix in the proximal regions, but as the distance increases from the villous basement membrane, increasing amounts appear of an intercellular deposit that lacks banded collagen fibrils (Enders, 1968) but contains fibronectin and laminin (Frank et al., 1994). This has been described as 'matrix-type fibrinoid' to distinguish it from the 'fibrin-type fibrinoid' that is abundant at sites of trophoblastic denudation, mainly on villi and the basal plate, and in Rohr's and Nitabuch's striae (Frank et al., 1994). In this study we have extended existing data on the distribution of different laminin isoforms throughout the maternal decidua, placenta, and also in choriocarcinoma cell lines.

MATERIALS AND METHODS

Antibodies

Monoclonal antibodies 4C7, 3E5, 2E8 (Engvall et al., 1986) and 5H2 (Leivo and Engvall, 1988; Leivo et al., 1989) against laminin subunits α5, β1, γ1, and α2 (light) respectively, were a generous gift from Dr. Eva Engvall, La Jolla Cancer Research Foundation, La Jolla, California, USA. 4C7 was originally reported as an anti-α1 antibody, but recent data suggest it recognizes α5. Monoclonal antibody C4 against laminin β2 (Hunter et al., 1989a) was obtained from the Developmental Studies Hybridoma Bank, University of Iowa. Monoclonal antibody BM165 (Rousselle et al., 1991) against laminin α3 was kindly donated by Dr. Patricia Rousselle, Institut de Biologie et Chimie des Protéines, Lyon, France. Monoclonal antibody K140 (Marinkovich et al., 1992b) against laminin β3 was kindly donated by Professor Robert Burgeson, Harvard Medical School, Massachusetts, USA. A polyclonal antiserum against laminin α2 (heavy) subunit (Paulsson et al., 1991) was kindly provided by Dr. Mats Paulsson, Biocenter, University of Basel, Switzerland and polyclonal antiserum R9 against laminin 1 (EHS) by Professor David Garrod, School of Biological Sciences, University of Manchester, UK. Polyclonal antisera P3 and P4 against laminin α4 were raised to the following peptide sequences: P3: KPPVKRPELT; P4: LSSTAEEKFI. These were conjugated to keyhole limpet hemocyanin and injected subcutaneously into rabbits. P3 was found to react more strongly with the tissues under study. Polyclonal anti-α1 antiserum YY4 was kindly provided by Dr. H. Kleinman, NIH, Bethesda, Maryland, USA.

Tissue

First trimester placenta and parietal decidua was obtained with local ethical committee approval from patients undergoing elective termination of pregnancy at 6-12 weeks of gestation as estimated from the date of the last menstrual period. Tissue was collected in sterile PBS, examined under a dissecting microscope, embedded in OCT, and snap frozen in liquid nitrogen for cryosectioning.

Cells

The choriocarcinoma cell lines BeWo, JAr, and JEG-3 were cultured in DMEM/Hams F12 (1:1) containing 10% FCS as previously described (Aplin et al., 1992). For immunofluorescence the cells were plated on coverslips and cultured for 48 hours.

Immunofluorescence

Indirect immunofluorescence was performed on cryosections (5 mm) of first trimester placenta and decidua. Sections were air dried and fixed in ice cold acetone for 10 minutes. They were then incubated in primary antibody dilutions for 1 hour at room temperature, washed in PBS, incubated in fluorescein isothiocyanate-conjugated secondary antibody (DAKO, High Wycombe, UK) for 1 hour, washed in PBS, and mounted in non-fade aqueous mountant (Immumount, Life Sciences International,

Basingstoke, UK). Immunofluorescence was performed on acetone-fixed confluent cells after two days in culture. When the primary antibody was omitted, there was no staining. Other primary monoclonal antibodies provided negative controls as appropriate - anti-keratin for decidua, anti-vimentin for trophoblast cell lines and columns.

RESULTS

Expression of Laminin Subunits in Decidua

Figure 2 shows immunolocalization data using a panel of monoclonal antibodies specific for laminin subunits in cryosections of human first trimester decidua. The decidual stromal cells show clear pericellular staining for subunits α2(80) and α2(300) (Figures 2a and 2b), which corresponds to the stromal cell basement membrane. The α4 subunit is expressed uniformly and shows the same pericellular distribution (data not shown). The α5 subunit shows very weak and variable expression through the stroma (Figure 2c) and in some areas is completely negative. The α3 subunit was not detected.

Localization data for β1, β2, and γ1 are shown in Figures 2d, e, and f, respectively. The β1 staining is very strong and shows the same pericellular distribution within the stroma. Staining for β2 is also strong and pericellular in distribution. Immunoblotting data indicate that β2 is present at a level comparable with β1. The γ1 subunit shows very strong staining in the pericellular basement membranes.

In addition to the stromal cells, laminins occur in the glandular and vascular basement membranes of decidua where they also show a distinct pattern of subunit expression. The α5 subunit is expressed in the vessels (Figure 2c) and the glands. The α2 (300) chain occurs in the vessels appearing to stain smooth muscle cells surrounding the spiral arteries, while the α2(80) is present in both the glands (Figure 2a) and the vessels. Neither the α3 nor α4 subunit is expressed by the glands or vessels. The β1 subunit is strongly expressed by both the glands and the vessels, the vascular staining apparently localizing to smooth muscle cells surrounding the vessels as seen in Figure 2d. The β2 subunit is expressed by both the glands and vessels. The γ1 subunit is strongly expressed throughout the tissue and a brightly staining vessel can be seen in Figure 2f. These immunolocalization data are summarized in Table 2.

Laminin Expression and Distribution in Placenta

Laminins are expressed by placenta at the villous basement membrane, in the fetal vessels and also by the extravillous cytotrophoblast cells. This is demonstrated in Figure 3a which shows a column originating from a first trimester villous, stained for laminin 1 (α1β1γ1). There is very strong staining in the villous basement membrane, and there is also a clearly stained blood vessel within the villous. Punctate deposits of extracellular laminin are apparent in the distal region of a small cytotrophoblast column. The villous basement membrane contains subunits α5, α2, β1, β2, and γ1 as seen in Figures 3b-f, respectively. This pattern is also observed at term (Engvall et al 1990). Laminin α3 was not detected in first trimester placenta while the anti-α4 antibody shows weak staining in a basolateral distribution around the villous cytotrophoblast cells (data not shown).

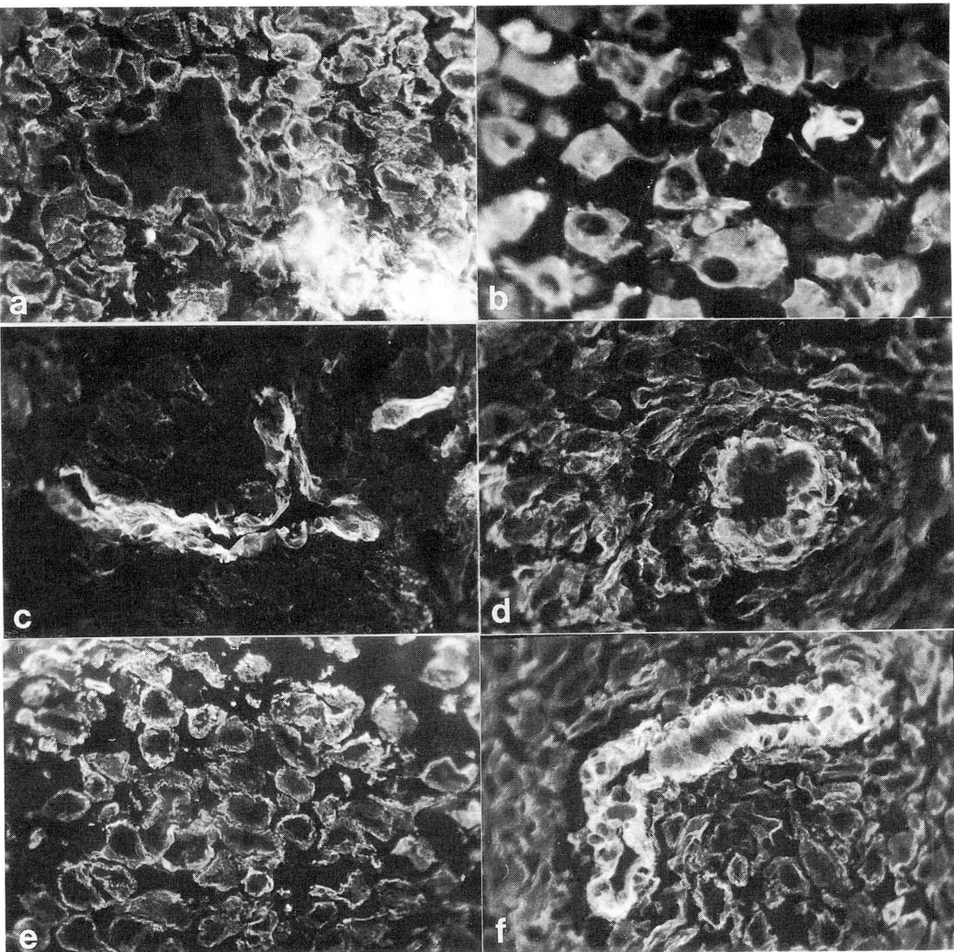

Figure 2. Immunofluorescence of laminin subunits in first trimester decidua. (a) α2(80); (b) α2 (300); (c) α5; (d) β1; (e) β2; (f) γ1. Note the pericellular basement membrane of decidual stromal cells is immunopositive in a, b, d, e, and f. Vessel walls are stained in c, d, and f. A glandular basement membrane is immunopositive in a (X400).

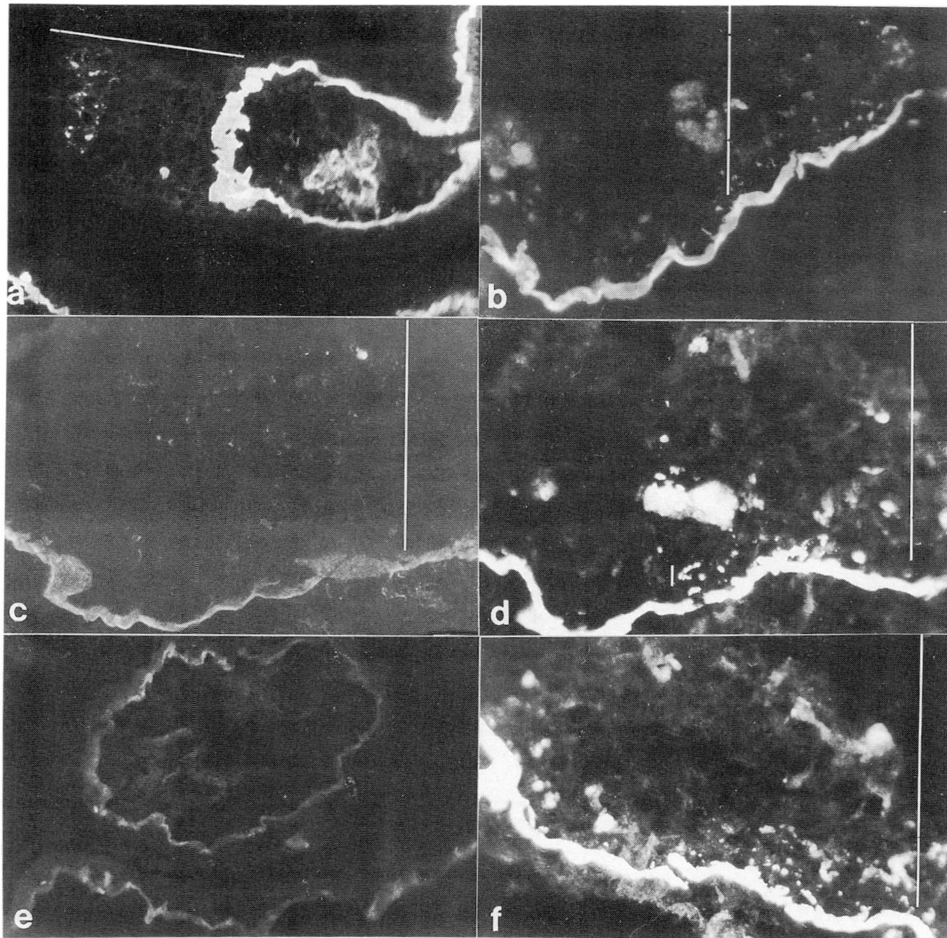

Figure 3. Immunofluorescence of laminin subunits in first trimester placenta. (a) Staining with a polyclonal antiserum to laminin 1 that recognizes α1, β1, and γ1 subunits; (b) α5; (c) α2 (80); (d) β1; (e) β2; (f) γ1. In a, a small villous is present with strong staining in the trophoblast basement membrane and in a blood vessel in the mesenchyme. A small column is present with punctate staining in the more distal layers of cytotrophoblast. b, c, d, and f all show serial sections of the same villous basement membrane with an associated column. This contains weak α5 and α2 immunoreactivity together with stronger β1 and γ1 staining. In e, a villous is shown with β2 reactivity. Lines indicate the position of columns in a, b, c, d, and f (X400).

Table 2

Summary Of The Distribution Of Laminin Subunits In The First Trimester Decidua

ANTIGEN	ANTIBODY	GLAND/VESSEL	STROMA
Laminin 1	R9 (polyclonal)	+++	+++
α5	4C7	+++	+/-
β1	3E5	+++	++/-
γ1	2E8	+++	+++
α2 (300)	polyclonal	++/-vessels	+++
α2 (80)	5H2	++	+++
β2	C4	++/-	++/-
α3	BM165	+/-	-
α4	P3 + P4	+	++
β3	K140	+/-	-

+++ Strong staining; ++/- Positive variable staining; +/- Weak and variable staining

Figures 3b, c, d, and f show serial sections through a small column and confirm that extravillous cytotrophoblast cells also produce laminins. The α5 subunit can be seen in association with cells just adjacent to the villous BM as shown in Figure 3b. Figure 3c shows staining for the α2(80) subunit which is present only in trace amounts. The β1 and γ1 subunits are clearly present as extracellular matrix within the columns as seen in Figures 3d and 3f and both subunits show very bright staining. We did not see staining for the β2 subunit in the columns. The staining pattern for first trimester placenta is summarized in Table 3.

Table 3

Distribution Of Laminin Subunits Within First Trimester Placenta

SUBUNIT	VILLOUS BM	PROXIMAL COLUMNS	DISTAL COLUMNS
α5	+++	+++	+++
α2(300)	ND	ND	ND
α2(80)	+++	+/-	+/-
α3	-	-	-
α4	++/-	-	-
β1	+++	+++	+++
β2	++/-	-	-
γ1	+++	+++	+++

+++ Strong staining; ++/- Positive variable staining;
+/- Weak and variable staining; ND Not done

Table 4

Laminin Expression By Choriocarcinoma Cell Lines

ANTIGEN	ANTIBODY	BeWo	JAr	JEG-3
Laminin 1 (α1β1γ1)	R9 polyclonal	++E	++1	++1*
α1	YY4 Polyclonal	+E	ND	ND
α5	4C7	+/-	+/-	+1
α2 (300)	M4	-	ND	ND
α2(80)	5H2	-	-	-
α3	BM165	+/-	ND	ND
α4	P3 polyclonal	-	ND	ND
β2	C4	+/-	ND	+/-
β1	3E5	++E	+/-	++1
γ1	2E8	++E	+/-	++1

++E strong extracellular staining
+E positive extracellular staining
++1 strong intracellular staining
+1 positive intracellular staining
+/- weak staining
ND, not done

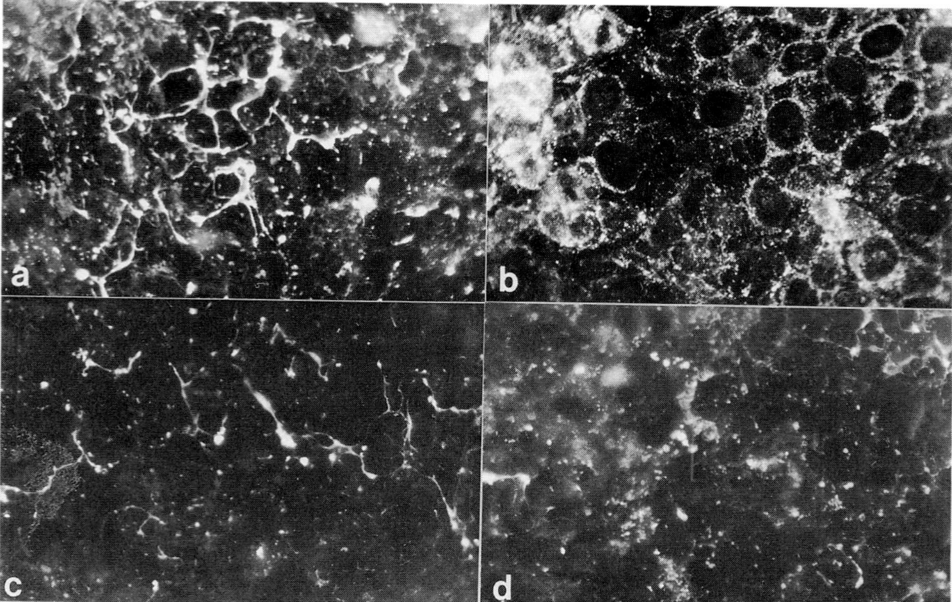

Figure 4. Immunofluorescence of BeWo (a, c, d) and JEG cells (b) with anti-laminin 1 (a, b) and antibodies to β1 (c) and γ1 (d) (X400).

Production of Laminins by Choriocarcinoma Cell Lines

We have studied the production of laminin isoforms by three trophoblast cell lines BeWo, JAr, and JEG-3 and the data are summarized in Table 4. BeWo cells produce high levels of laminin which is exported and assembled into an extracellular matrix. Figure 4a shows a BeWo monolayer stained for laminin 1 (α1β1γ1). Staining for the β1 and γ1 subunits is shown in Figures 4c and 4d, respectively and shows a similar extracellular pattern of staining. The α5 subunit isonly expressed weakly by BeWo cells The α3, α4, and β2 subunits are not present in BeWo extracellular matrix.

Laminin production by JEG-3 cells as identified by staining for laminin 1 is shown in Figure 4b. Strong intracellular staining is present, but there is also extracellular matrix staining in some areas. Staining using subunit-specific antibodies indicates that JEG-3 cells produce α5, β1 and γ1 with trace amounts of β2 (data not shown). JAr cells were found to produce subunits α5, β1, and γ1. This is consistent with previous studies (Peters et al., 1985).

DISCUSSION

Laminin subunits α2 (300 and 80kDa), β1, β2, and γ1 colocalize in the decidual stromal cell basement membrane, suggesting that laminins 2 (α2β1γ1) and 4 (α2β2γ1) are present. This has been confirmed biochemically (Church et al., 1996). The α4 subunit is also present, and biochemical analysis will be required to define the β and γ chains that associate with this novel α subunit. Limited and variable expression of the α5 subunit suggests that in some areas of stroma the cells may also be producing laminins 10 (α5β1γ1) and/or 11 (α5β2γ1). The vascular and glandular basement membranes contain subunits α1, β1, β2, and γ1, suggesting the major isoforms to be laminins 2, 4, 10 and 11. The γ1 subunit is common to 6 of the 7 laminin isoforms currently described and this is reflected in its high staining intensity throughout the tissue. These data indicate that different laminin isoforms are expressed in different cellular compartments of decidua as previously observed in kidney (Hunter et al., 1989a) and heart (Engvall et al 1990). Laminin β2 has previously been observed in vascular tissue (Wewer et al., 1994b).

Subunits α2, α5, β1, β2, and γ1 colocalize in the villous basement membrane. This is consistent with previous studies which have demonstrated the villous basement membrane to contain laminins 1-4 (Engvall et al., 1990). Laminins 8-11 are also probably present. Laminin α4 subunit is also expressed by the cytotrophoblast stem cells. Our results demonstrate that extravillous cytotrophoblast cells also produce laminin, with subunits α5, β1, γ1, and some α2 within the columns of first trimester tissue. We have not observed the α4 or the β2 subunits in the columns. This suggests that as the cytotrophoblast cells migrate away from the villous basement membrane, they continue to secrete laminin into a matrix that may in turn contribute to the development and stability of the column. It also suggests that there is a selective secretion of laminins by the extravillous cytotrophoblast cells, with an apparent down-regulation of α4 and β2. The data are consistent with studies of trophoblast cell columns in first trimester placental bed (Damsky et al., 1992).

It has been demonstrated that first trimester villi retain the ability to develop into trophoblast cell columns in coculture with parietal decidua (Vicovac et al., 1993; Vicovac et al., 1995a) and when cultured on gels of collagen type 1 or matrigel (Vicovac, 1995b) . These *in vitro* column structures produce the same phenotypic changes observed *in vivo* such as upregulation of integrins α5β1 and α1β1 (Damsky et al., 1992). They also show intercellular accumulations of laminin in the medial and distal columns (Vicovac et al., 1995a), confirming that this material is of trophoblastic origin. This collagen gel culture system offers the potential to study the composition and function of laminin in extravillous trophoblast without the added complexity of maternally-derived extracellular matrix.

Choriocarcinoma cell lines offer a model system in which to characterize laminin production by trophoblast. However, our data demonstrate that their repertoire of subunit production is more restricted than that of normal trophoblast. BeWo cells produce the β1 and γ1 subunits along with smaller amounts of α1 and α5 We have detected little α2, α3, or α4 subunit production. JAr and JEG-3 cell lines express laminin 10 (α5, β1, γ1) (Peters et al., 1985). It appears that on becoming transformed all three cell lines have a greatly reduced ability to make α2 and β2.

The relationship between expression of the individual subunits and secretion of laminin is not clearly understood. Current evidence suggests that laminin secretion requires assembly into a trimeric species containing an α, β, and γ subunit. When only β and γ subunits are produced, they remain intracellular (Wewer et al., 1994b). When β and γ chains are present in excess they accumulate both as single subunits and as dimers within the cytoplasm, but are rapidly exported into matrix or medium on assembly into a trimer, demonstrating synthesis of α chain to be a rate-limiting factor (Peters et al., 1985; Ecay and Valentich, 1992). Consistent with this, BeWo cells contain an excess of β1 and γ1 over α subunits (Church and Aplin, unpublished). Selective assembly has been described in the human adenocarcinoma cell line HU-1, where although the presence of α1, α2, β1, β2, and γ1 was demonstrated both at the mRNA and the protein level, assembly was found to be restricted to the α2, β1, and γ1 yielding only laminin 2 (Wewer et al., 1994a). Similar results were obtained when the cells were grown as tumors in mice, with only laminin 2 being assembled into a basement membrane and the β2 subunit remaining within the cells. Thus there appears not to be a general hierarchy of preferential chain assembly in which α1 is preferred to α2. The β2 subunit has been shown not to compete well with β1 when both are present; secretion of trimer was observed in the pheochromocytoma cell line PC12 only after the transfection of the β1 subunit into the cells, despite the presence of mRNA for the β2 subunit (Reing et al., 1992). Therefore, although the choriocarcinoma cell lines all produce little α2 or β2 polypeptide, it remains to be determined whether down-regulation occurs at a transcriptional or post-transcriptional stage.

The function of decidual laminin remains unknown. It is possible that the specific isoforms expressed in the decidual stromal pericellular basement membrane may play a role in regulating the invasion of trophoblast into maternal decidua (Burrows et al., 1995), a process that leads to the remodeling of maternal arteries to allow increased blood flow to the placenta (Pijnenborg, 1994). There are also adhesive interactions occurring between the decidual basement membranes and bone marrow-derived cell populations (Aplin and Jones, 1989). In addition, the basement membrane may play a role in stabilizing the three dimensional architecture of decidual tissue, which is remodeled in pregnancy (Aplin et al., 1995). Similarly the production of laminin isoforms by extravillous cytotrophoblast may aid the growth and development of anchoring villi. Laminins within the columns could act as cell adhesion ligands in the initial phases of extravillous cell anchorage until the decidual matrix is available for interaction with the invading cytotrophoblast population. These processes are all likely to be important for pregnancy success.

SUMMARY

A panel of subunit-specific antibodies has been used to characterize the expression of laminin polypeptides in decidua, placenta, and choriocarcinoma cells. Decidual stromal cells express laminins 2 and 4 together with isoform(s) containing the α4 chain, and these are all present in the pericellular basement membranes. The vascular and epithelial basement membranes of decidual tissue are predicted to contain laminins 2, 4, 10 and 11. First trimester villous basement membrane contains subunits consistent with the presence of laminins 1-4 and 8-11. Extravillous trophoblast in columns probably produces laminins 2 and 10. Amongst the choriocarcinoma cell lines studied, BeWo produce the largest amounts of basement membrane material. This contains the β1 and γ1 subunits together with α1 and α5, suggesting the presence of laminins 1 and 10.

ACKNOWLEDGEMENTS

We thank Wellbeing and the Medical Research Council for funding (to HJC).

REFERENCES

Aberdam, D., Galliano, M.-F., Vailly, J., Pulkkinen, L., Bonifas, J., Christiano, A.M., Tryggvason, K., Uitto, J., Epstein, E.H., Ortonne, J.-P. and Meneguzzi, G. (1994) Herlitz's junctional epidermolysis bullosa is linked to mutations in the gene (LAMC2) for the g2 subunit of nicein/kalinin (Laminin-5). *Nature Genetics* 6, 299-304.

Aplin, J.D., Charlton, A.K. and Ayad, S. (1988) An immunohistochemical study of human endometrial extracellular matrix during the menstrual cycle and first trimester of pregnancy. *Cell Tissue Res.* 253, 231-240.

Aplin, J.D. (1989) Cellular biochemistry of the endometrium. In: *Biology of the Uterus,* (eds.) R.M. Wynn and W.P. Jollie, Plenum Medical, New York, pp. 89-129.

Aplin, J.D. and Jones, C.J.P. (1989) Extracellular matrix in endometrium and decidua. In: *Placenta as a Model and a Source,* (eds.) O. Genbacev, A. Klopper and R. Beaconsfield, Plenum Medical, New York, pp. 115-128.

Aplin, J.D. and Charlton A.K. (1990) The role of matrix macromolecules in the invasion of decidua by trophoblast. Trophoblast Res. 4, 139-158.

Aplin, J.D., Sattar, A. and Mould, P. (1992) Variant choriocarcinoma (BeWo) cells that differ in adhesion and migration on fibronectin display conserved patterns of integrin expression. *J. Cell Sci.* 103, 435-444.

Aplin, J.D. and Church, H.J. (1994) Basement membrane - hemidesmosome interactions. In: *Molecular Biology of Desmosomes and Hemidesmosomes,* (eds.) D.R. Garrod and J.E. Collins, R.G. Landes Company, Austin, pp. 87-106.

Aplin, J.D., Mylona, P., Kielty, C.M., Ball, S., Church, H.J., Williams, J.D.L. and Jones, C.J.P. (1995) Collagen VI and laminin as markers of decidualisation in human endometrial stroma. In: *Molecular and Cellular Aspects of Peri-implantation Processe,* (ed.) S.K. Dey, Serono Symposium Series. Springer-Verlag, New York, pp. 331-351.

Armant, D.R., Kaplan, H.A. and Lennarz, W.J. (1986) Fibronectin and laminin promote in vitro attachment and outgrowth of mouse blastocysts. *Dev. Biol.* 116, 519-523.

Beck, K. and Gruber, T. (1995) Structure and assembly of basement membrane and related extracellular matrix protein, In: *Principles of cell adhesion* (ed.) P.D. Richardson, CRC Press, Boca Raton, pp. 219-252.

Bresalier, R.S., Schwarz, B., Kim, Y.S., Duh, Q.Y., Kleinman, H.K. and Sullam, P.M. (1995) The laminin α1 chain IKVAV-containing peptide promotes liver colonisation by human colon cancer cells. *Cancer Res.* 55, 2476-2480.

Burgeson, R. (1993) Dermal-epidermal adhesion in skin. In: *Molecular and Cellular Aspects of Basement Membranes,* (eds.) D.H. Rohrbach and R. Timpl, Academic Press, New York, pp. 49-66.

Burgeson, R.E., Chiquet, M., Deutzmann, R., Ekblom, P., Engel, J., Kleinman, H., Martin, G.R., Ortonne, J.-P., Paulsson, M., Sanes, J., Timpl, R., Tryggvason K., Yamada, Y. and Yurchenco, P.D. (1994) A new nomenclature for laminins. *Matrix Biol.* 14, 209-211.

Burrows, T.D., King, A., Smith, S.K. and Loke, Y.W. (1995) Human trophoblast adhesion to matrix proteins: Inhibition and signal transduction. *Mol. Hum. Reprodn.* 10, 2489-2500.

Chambers, B.J., Klein, N.W., Conrad, S.H., Ruppenthal, G.C., Sackett, G.P., Weeks, B.S. and Kleinman H.K. (1995) Reproduction and sera embryotoxicity after immunization of monkeys with the laminin peptides YIGSR, RGD, and IKVAV. *Proc Natl Acad Sci USA* 92, 6818-6822.

Champliaud, M.F., Lunstrum, G.P., Rousselle, P., Nishyama, T., Keene, D.R. and Burgeson, R.E. (1996) Human amnion contains a novel laminin variant, laminin 7, which like laminin 6, covalently associates with laminin 5 to promote stable epithelial-stromal attachment. *J. Cell. Biol.* 132, 1189-1198.

Church, H.J., Vicovac, Lj., Williams, J.D.L., Hey, N.A. and Aplin, J.D. (1996) Laminins 2 and 4 are expressed by human decidual cells. *Lab. Invest.* 74, 21-32.

Cooper, A.R. and Macqueen, H.A. (1983) Subunits of laminin are differentially synthesised in mouse eggs and early embryos. *Devl. Biol.* 96, 467-471.

Damsky, C.H., Fitzgerald, M.L. and Fisher, S.J. (1992) Distribution patterns of extracellular matrix components and adhesion receptors are intricately modulated during first trimester trophoblast differentiation along the invasive pathway in vivo. . *Clin. Invest.* 89, 210-222.

Ecay, T.W. and Valentich, J.D. (1992) Basal lamina formation by epithelial cell lines correlates with laminin A chain synthesis and secretion. *Exp. Cell Res.* 203, 32-38.

Ehrig, K., Leivo, I., Argraves, W.S., Ruoslahti, E. and Engvall, E. (1990) Merosin, a tissue-specific basement membrane protein, is a laminin-like protein. *Proc. Nat. Acad. Sci. USA* 86, 3264-3268.

Ekblom, P. (1996) Receptors for laminins during epithelial morphogenesis. *Curr. Opin. Cell. Biol.* 8, 700-706.

Enders, A.C. (1968) Fine structure of anchoring villi of the human placenta. *Am. J. Anat.* 122, 419-452.

Engel, J. (1992) Laminins and other strange proteins. *Biochemistry* 31, 10643-10651.

Engel, J. (1993) Structure and function of Laminin. In: *Molecular and Cellular Aspects of Basement Membranes,* (eds.) D.H. Rhorbach, and R. Timpl, Academic Press, New York, pp. 147-176.

Engvall, E. (1993) Laminin variants: why, where and when? *Kidney Int.* 43, 2-6.

Engvall, E., Davis, G.E., Dickerson, K., Rouslahtim E., Varonm, S. and Manthorpe, M. (1986) Mapping of domains in human laminins using monoclonal antibodies: Localisation of the neurite-promoting site. *J. Cell Biol.* 103, 2457-2465.

Engvall, E., Earwicker, D., Haaparanta, T., Ruoslahti, E. and Sanes, J.R. (1990) Distribution and isolation of four laminin variants; tissue restricted distribution of heterotrimers assembled from five different subunits. *Cell Regulation* 1, 731-740.

Foidart, J.M., Yaar, M., Figueroa, A., Wilk, A., Brown, K.S. and Liotta, L.M. (1983) Abortion in mice induced by intravenous injections of antibodies to type IV collagen or laminin. *Am. J. Pathol.* 110, 346-357.

Foidart, J.M., Hunt, J., Lapiere, C.-M., Nusgens, B., de Rycker C., Bruwier, M., Lambotte, R., Bernard, A. and Mahieu, P. (1986) Antibodies to laminin in pre-eclampsia. *Kidney International* 29, 1050-1057.

Frank, H.-G., Malekzadeh, F., Kertschanska, S., Crescimanno, C., Castellucci, M., Lang, I., Desoye, G. and Kaufmann, P. (1994) Immunohistochemistry of two different types of placental fibrinoid. *Acta Anat.* 150, 55-68.

Garcia Lerma, J., Moneo, I., Ortiz de Landazuri, M. and Sequi Navarro, J. (1995) Comparison of the anti-laminin antibody response in patients with systemic lupus erythematosus (SLE) and parasitic diseases (filariasis). *Clin. Immunol. Immunopathol.* 76, 19-31

Gehlsen, K.R., Dickerson, K., Argraves, W.S., Engvall, E. and Ruoslahti, E. (1989) Subunit structure of a laminin-binding integrin and localisation of its binding site on laminin. *J. Biol. Chem.* 264, 19034-19038.

Hall, D.E., Reichardt, L.F., Crowley, E., Holley, B., Moezzi, H., Sonnenberg, A. and Damsky, C.H. (1990) The $\alpha 1\beta 1$ and $\alpha 6\beta 1$ integrin heterodimers mediate cell attachment to distinct sites on laminin. *J. Cell Biol.* 110, 2175-2184.

Hayashi, Y.K., Engvall, E., Arikawa-Hirasawa, E., Goto, K., Koga, R., Nonake, I., Sugita, H., and Arahata, K. (1993) Abnormal localisation of laminin subunits in muscular dystrophies. *J. Neurol. Sci.* 119, 53-64.

Helbling-Leclerc, A., Zhang, H., Topaloglu, H., Cruaud, C., Tesson, F., Weissenbach, J., Tome, F.M., Schwartz, K., Fardeau, M. and Tyrggvason, K. (1995) Mutations in the laminin alpha 2-chain gene (LAMA2) cause merosin-deficient congenital muscular dystrophy. *Nat. Genet.* 11, 216-218.

Hunter, D.D., Shah V., Merlie, J.P. and Sanes, J.R. (1989a) A laminin-like adhesive protein concentrated in the synaptic cleft of the neuromuscular junction. *Nature* 338, 229-233.

Hunter, D.D., Porter, B.E., Bulock, J.W., Adams, A.P., Merlie, J.P. and Sanes, J.P. (1989b) Primary sequence of a motor neuron-selective adhesive site in the synaptic basal lamina protein s-laminin. *Cell* 59, 905-913.

Hynes, R.O. (1992) Integrins: Versatility, modulation, and signaling in cell adhesion. *Cell* 69, 11-25

Iivanainen, A., Sainio, K., Sariola, H. and Tryggvason, K. (1995) Primary structure and expression of a novel human laminin $\alpha 4$ chain. *FEBS Letters* 365, 183-188.

Kibbey, M.C., Corcoran, M.L., Wahl, L.M. and Kleinman, H.K. (1994) Laminin SIKVAV peptide-induced angiogenesis in vivo is potentiated by neutrophils. *J. Cell Physiol.* 160, 185-193.

Kim W.H., Schnaper, H.W., Nomizu, M., Yamada, Y. and Kleinman, H.K. (1994) Apoptosis in human fibrosarcoma cells is induced by a multimeric synthetic Tyr-Ile-Gly-Ser-Arg (YIGSR)-containing polypeptide from laminin. *Cancer Res.* 54, 5005-5010

Leivo, I. and Engvall, E. (1988) Merosin, a protein specific for basement membranes of Schwann cells, striated muscle and trophoblast, is expressed late in nerve and muscle development. *Proc. Natl. Acad. Sci. USA* 85, 1544-1548.

Leivo, I., Laurila, P., Wahlstrom, T. and Engvall, E. (1989) Expression of merosin, a tissue-specific basement membrane protein, in the intermediate trophoblast cells of choriocarcinoma and placenta. *Lab. Invest.* 60, 783-790.

Lindblom, A., Marsh, T., Fauser, C., Engel, J. and Paulsson, M. (1994) Characterisation of native laminin from bovine kidney and comparison with other laminin variants. *Eur. J. Biochem.* 219, 383-392.

Marinkovich, M.P., Lunstrum, G.P., Keene, D.R. and Burgeson, R.E. (1992a) The dermal-epidermal junction of human skin contains a novel laminin variant. *J. Cell Biol.* 119, 695-703.

Marinkovich, M.P., Lunstrum, P. and Burgeson, R.E. The anchoring filament protein kalinin is synthesized and secreted as a high molecular weight precursor. (1992b) *J. Biol. Chem.* 267, 17900-17906.

Martin, P.T., Ettinger, A.J. and Sanes JR. (1995) A synaptic localisation domain in the synaptic cleft protein laminin β2 (s-laminin). *Science* 269, 413-416.

Mecham, R.P. (1991) Receptors for laminin in mammalian cells. *FASEB J.* 5, 2538-2546.

Mercurio A.M. (1995) Laminin receptors: achieving specificity through cooperation. *Trends in Cell Biology* 5, 419-423.

Miner, J.H., Lewis, R.M. and Sanes, J.R. Molecular cloning of a novel laminin chain, α5, and widespread expression in adult mouse tissues. *J. Biol. Chem.* 270, 28523-28526.

Mizayaki, K., Kikkawa, Y., Nakamura, A., Yasumitsu, H. and Umeda, M. (1993) A large cell-adhesive scatter factor secreted by human gastric carcinoma cells. *Proc. Nat. Acad. Sci. USA*, 90, 11767-11771.

Noakes, P.G., Gautam, M., Mudd, J., Sanes, J.R. and Merlie, J.P. (1995) Aberrent differentiation of neuromuscular junctions in mice lacking s-laminin/laminin β2. *Nature* 374, 258-262

Paulsson, M. and Saladin, K. (1989) Mouse heart laminin. Purification of the native protein and structural comparison with Engelbreth-Holm-Swarm tumor laminin. *J. Biol. Chem.* 264, 18726-18732.

Paulsson, M., Saladin, K. and Engvall E. (1991) Structure of laminin variants. The 300-kDa chains of murine and bovine heart laminin are related to the human placenta merosin heavy chain and replace the A chain in some laminin variants. *J. Biol. Chem.* 266, 17545-17551.

Peters, B.P., Hartle, R.J., Krzesicki, R.F., Kroll, T.G., Perini, F., Balun, J.E., Goldstein, I.J. and Ruddon, R.W. (1985) The biosynthesis and secretion of laminin by human choriocarcinoma cells. *J. Biol. Chem.* 260, 14732-14742.

Pijnenborg, R. (1994) Trophoblast invasion. *Reprod. Med. Rev.* 3, 53-73.

Pulkkinen, L., Christiano, A.M., Airenne, T., Haakana, H., Tryggvason, K. and Uitto, J. (1994) Mutations in the γ2 chain gene (LAMC2) of kalinin/laminin 5 in the junctional forms of epidermolysis bullosa. *Nature Genetics* 6, 293-297.

Reing, J., Durkin, M.E. and Chung, A.E. (1992) Laminin B1 expression is required for laminin deposition into the extracellular matrix of PC12 cells. *J. Biol. Chem.* 267, 23143-23150.

Richards, A., Al-Imara, L. and Pope, F.M. (1996) The complete cDNA sequence of laminin alpha 4 and its relationship to the other human laminin alpha chains. *Eur. J. Biochem.* 238, 813-821.

Richards, A.J., Al-Imara, L., Carter, N.P., Lloyd, J.C., Leversha, M.A. and Pope F.M. (1994) Localisation of the gene LAMA4 to chromosome 6q21 and isolation of a partial cDNA encoding a variant laminin A chain. *Genomics* 22, 237-239.

Rousselle, P., Lunstrum, G.P., Keene, D.R. and Burgeson, R.E. (1991) Kalinin: an epithelium-specific basement membrane adhesion molecule that is a component of anchoring filaments. *J. Cell Biol.* 114, 567-576.

Royce, L..S., Martin, G.R. and Kleinman H.K. (1992) Induction of an invasive phenotype in benign tumor cells with a laminin A-chain synthetic peptide. *Invasion Metastasis* 12, 149-155.

Sorokin, L., Girg, W., Gopfert, T., Hallman, R. and Deutzmann, R. (1994) Expression of novel 400-kDa laminin chains by mouse and bovine endothelial cells. *Eur. J. Biochem.* 223, 603-610.

Streuli, C.H., Schmidhauser, C., Bailey, N., Yurchenco, P., Skubitz, A.P.N., Roskelly, C. and Bissell, M. (1995) Laminin mediates tissue-specific gene expression in mammary epithelia. *J. Cell Biol.* 129, 591-693.

Timpl, R. (1996) Macromolecular organization of basement membranes. *Curr. Opin. Cell. Biol.* 8, 618-624.

Timpl, R., Rohde, H., Gehron Robey, P., Rennard, S.I., Foidart, J.M. and Martin, G.R. (1979) Laminin - A glycoprotein from basement membranes. *J. Biol. Chem.* 254, 9933-9937.

Timpl, R. and Brown, J.C. (1994) The Laminins. *Matrix Biol.* 14, 275-281.

Tome, F.M.S., Evangelista, T., Leclerc, A., Sunada, Y., Manole, E., Estournet, B., Barois, A., Campbell, K.P. and Fardeau, M. (1994) Congenital muscular dystrophy with merosin deficiency. *C.R. Acad. Sci., Paris* 317, 351-357.

Underwood, P.A., Bennett, F.A., Kirkpatrick, A., Bean, P.A. and Moss, B.A. (1995) Evidence for the location of a binding sequence for the α2β1 integrin of endothelial cells in the β1 subunit of laminin. *Biochem. J.* 309, 765-771.

Vailly, J., Pulkkinen, L., Miquel, C., Christiano, A.M., Gerecke, D., Burgeson, R.E., Uitto, J., Ortonne, J.-P. and Meneguzzi G. (1995) Identification of a Homozygous one-basepair deletion in exon 14 of the LAMB3 gene in a patient with herlitz junctional epidermolysis bullosa and prenatal diagnosis in a family at risk for recurrence. *J. Invest. Dermatol.* 104, 462-466.

Verrando, P., Hsi, B.L., Yeh, C.J., Pisani, A., Serieys, N. and Ortonne, J.-P. (1987) Monoclonal antibody GB3, a new probe for the study of human basement membranes and hemidesmosomes. *Exp. Cell Res.* 170, 116-128.

Verrando, P., Pisani, A. and Ortonne, J.-P. (1988) The new basement membrane antigen recognized by the monoclonal antibody GB3 is a large size glycoprotein: Modulation of its expression by retinoic acid. *Biochim. Biophys. Acta* 942, 45-56.

Vicovac, Lj., Papic, N. and Aplin, J.D. (1993) Tissue interactions in first trimester trophoblast-decidua cocultures. *Trophoblast Res.* 7, 223-236.

Vicovac, Lj., Jones, C.J.P. and Aplin, J.D. (1995a) Trophoblast differentiation during formation of anchoring villi in model of the early human placenta in vitro. *Placenta* 16, 46-56.

Vicovac, Lj., Jones, C.J.P., Church, H.J. and Aplin, J.D. (1995b) Factors controlling trophoblast entry into the extravillous lineage. *Placenta* 16, A74.

Vidal, F., Baudoin, C., Miquel, C., Galliano, M.F., Christiano, A.M., Uitto, J., Ortonne, J.P. and Meneguzzi, G. (1995) Cloning of the Lam α3 chain gene (LAMA3) and identification of a homozygous deletion in a patient with herlitz junctional epidermolysis bullosa. *Genomics* 30, 273-280.

Vollmer, G., Siegal, G.P., Chiquet-Ehrismann, R., Lightner, V.A., Arnholdt, H. and Knuppen, R. (1990) Tenascin expression in the human endometrium and in endometrial adenocarcinomas. *Lab Invest.* 62, 725-730.

Vuolteenaho, R., Nissinen, R., Sainio, K., Byers, M., Eddy, R., Hirvonen, H., Shows, T.B., Sariola, H., Engvall, E. and Tryggvason, K. (1994) Human laminin M chain (merosin): Complete primary structure, chromosomal assignment, and expression of the A and M chain in human fetal tissues. *J. Cell Biol.* 124, 381-394.

Wewer, U.M., Wayner, E.A., Hoffstrom, B.G., Lan, F., Meyer-Neilsen, B., Engvall, E. and Albrechtsen, R. (1994a) Selective assembly of laminin variants by human carcinoma cells. *Lab. Invest.* 71, 719-730.

Wewer, U.M., Gerecke, D.R., Durkin, M.E., Kurtz, K.S., Mattei, M.-G., Champliaud, M.-F., Burgeson, R. and Albrechtsen, R. (1994b) Human β2 chain of laminin (formerly S chain): cDNA cloning, chromosomal localization and expression in carcinomas. *Genomics* 24, 243-252.

Yamada, H., Tome, F.M.S., Higuchi, I., Kawai, H., Azibi, K., Chaouch, M., Roberds, S.L., Tanaka, T., Fujita, S., Mitsui, T., Fukunaga, H., Miyoshi, K., Osame, M., Fardeau, M., Kaplan, J.C., Shimizu, T., Campbell, K.P. and Matsumura, K. (1995) Laminin abnormality in severe childhood autosomal recessive muscular dystrophy. *Lab. Invest.* 72, 715-722.

Trophoblast Research 10:163-172, 1997

TROPHOBLAST - MATRIX INTERACTIONS IN HUMAN IMPLANTATION
-A Review-

Tanya D. Burrows, Ashley King., S.K. Smith, and Y.W. Loke

Research Group in Human Reproductive Immunobiology
Department of Pathology
University of Cambridge
Cambridge, United Kingdom CB2 1QP

The process of implantation in humans during the first trimester of pregnancy requires a degree of tissue remodeling. Trophoblast undergoes a series of transformations from a single polarized epithelial layer resting on a basement membrane (villous trophoblast) to cellular aggregates (trophoblast columns), which ultimately disperse and invade the decidua as individual cells (interstitial trophoblast) (Aplin, 1991; Loke and King, 1995). Thus trophoblast is transformed from a sessile to motile phenotype. Cellular migration ceases at the superficial layer of the myometrium where placental bed giant cells form (Pijnenborg, 1994). During this migration, many cells also invade the maternal spiral arteries, forming endovascular trophoblast.

In many biological systems, cells use a platform of extracellular matrix (ECM) for anchorage and migration. For example, keratinocytes at the periphery of a wound site migrate across a substrate of ECM rich in fibronectin (FN) during healing. Invasion of host tissues by cancer cells is also dependent upon the specific interaction of tumor cell surface receptors with ECM (Ruoslahti, 1992).

The ECM proteins laminin (LM) and FN are particularly abundant in the decidua during the first trimester of pregnancy and are distributed pericellularly around individual stromal cells (Charpin et al., 1985; Faber et al., 1986; Aplin et al., 1988; Loke et al., 1989). At present seven isoforms of LM (1-7) have been characterized. The expression of different LM isoforms in the placental bed is still unclear, although recent studies have shown that the predominant isoforms secreted by decidualized stromal cells are LM 2 and 4 (Church et al., 1995). Both adult and embryonic FN variants have been detected in the placental bed (Feinberg et al., 1991; Burrows et al., 1993).

The main family of receptors that mediate adhesion to the ECM are the integrins (Hynes, 1994). These cell adhesion molecules are heterodimers of non-covalently associated α and β subunits. The ligand specificity of these heterodimers is determined by the specific combination of α and β subunits (Hemler, 1991) (Table 1). Integrins can be polyspecific, binding several different ligands and in addition, individual ligands may be recognized by more than one integrin. A number of matrix receptors recognize the tri-amino acid sequence, arginine-glycine-aspartic acid (RGD), which is present in a number of ECM proteins such as FN and type I collagen (Ruoslahti and Pierschbacher, 1987). Adhesion to the RGD binding site of a ligand can be demonstrated by blocking cell-

Table 1

Integrins And Their Ligands

Integrin heterodimer	Extracellular matrix ligand
$\alpha_1\beta_1$	Collagen, laminin (E1 fragment)
$\alpha_2\beta_1$	Collagen, laminin
$\alpha_3\beta_1$	Collagen, laminin (E3 fragment), fibronectin
$\alpha_4\beta_1$	Fibronectin (CS-1 binding site)
$\alpha_5\beta_1$	Fibronectin (RGD binding site)
$\alpha_6\beta_1$	Laminin (E8 fragment)
$\alpha_6\beta_4$	Laminin (E8 fragment), epiligrin

ligand adhesion with synthetic peptides containing an RGD sequence. Similarly, the functional activity of each integrin matrix receptor expressed by a cell can also be demonstrated *in vitro* with function-perturbing monoclonal antibodies (mAbs), which block integrin-mediated adhesion to a substrate.

Cellular adhesion to LM mainly involves the $\alpha_6\beta_1$ integrin heterodimer (Sonnenberg et al., 1988). The $\alpha_6\beta_4$ heterodimer, localized at hemidesmosomes, has also been shown to bind LM in some cells (Dedhar, 1990; Lee et al., 1992). Two β_1 integrins that are important in interactions with FN are the $\alpha_4\beta_1$ and $\alpha_5\beta_1$ heterodimers, which utilize the cell surface binding domain-1 (CS-1) and RGD binding domains, respectively (Ruoslahti and Pierschbacher, 1987; Yamada, 1989).

It has been demonstrated by immunohistology on frozen sections of the implantation site, that as trophoblast cells migrate off the villous basement membrane they downregulate the β_4 integrin subunit and upregulate the β_1, α_5, and α_1 integrin subunits (Korhonen et al., 1991; Damsky et al., 1992; Burrows et al., 1993). Therefore there is a switch from the $\alpha_6\beta_4$ heterodimer to the $\alpha_5\beta_1$ and $\alpha_1\beta_1$ heterodimer as the cytotrophoblast cells migrate from the villi into the decidua. A proportion of extravillous trophoblast may also express the $\alpha_6\beta_1$ heterodimer. Thus villous trophoblast cells exclusively express LM receptors, but the extravillous trophoblast cells switch to express FN receptors and different LM receptors. This suggests that differential adhesion to different matrix proteins may be important in the transformation of trophoblast cells from a sessile to a motile phenotype, a conclusion which is supported by the *in vitro* studies of Damsky et al. (1994).

Interestingly, this switch in integrin expression by trophoblast does not occur when trophoblast migration is inadequate such as in the pathological condition of pregnancy, pre-eclampsia, although there are investigators who believe that this affects only endovascular trophoblast and not interstitial trophoblast (Pijnenborg, 1994). The invasive interstitial trophoblast of pre-eclamptic placentae continues to express the $\alpha_6\beta_4$ heterodimer, but fails to upregulate the $\alpha_1\beta_1$ heterodimer (Zhou et al., 1993). Therefore, changes in these two integrins, $\alpha_6\beta_4$ and $\alpha_1\beta_1$, may be important in determining trophoblast behavior in normal pregnancy. However, a subsequent study by Divers et al. (1995) failed to detect any differences in integrin phenotype of third trimester uteroplacental tissues between normal and pre-eclamptic pregnancies. It may be premature, therefore, to conclude that anomalies in the pattern of integrin expression and matrix interaction of trophoblast may explain a number of pregnancy disorders such as pre-eclampsia and placenta accreta.

The ECM proteins LM and FN are known to promote the attachment and outgrowth of mouse blastocysts *in vitro* (Armant, 1991; Romagnano and Babiarz, 1993). In addition, the expression and function of cell surface receptors in mouse blastocyst cells is well documented (Sutherland et al., 1988). The role of integrins in human trophoblast migration is well documented by the studies of Fisher et al. (1989), Aplin et al. (1992), and Loke et al. (1995). Human trophoblast cells are known to adhere to both LM (Loke et al., 1989) and FN *in vitro* (Burrows et al., 1993). We have demonstrated recently that these trophoblast-matrix interactions are mediated by specific functional integrin receptors (Burrows et al., 1995). Binding of trophoblast cells to purified plasma FN is mediated by the $\alpha_5\beta_1$ FN receptor. Inhibition of trophoblast cell adhesion to intact FN was observed when trophoblast cells were cultured in the presence of mAb to the α_5 and β_1 integrin subunits (Figure 1). As expected, peptides containing the tri-amino acid sequence RGD totally inhibited trophoblast adhesion to FN, whereas control peptides had no effect. This indicates that the recognition site of the $\alpha_5\beta_1$ FN receptor is the RGD sequence of the FN protein.

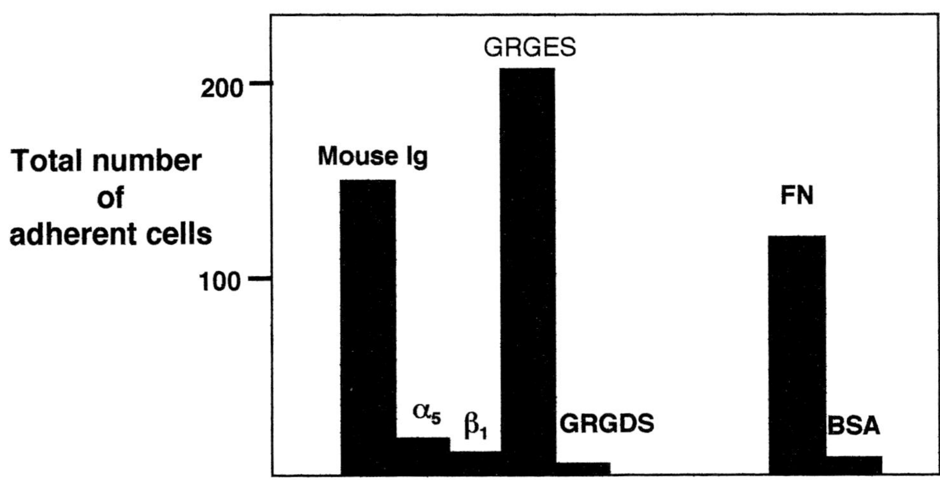

Figure 1. Inhibition of trophoblast adhesion to fibronectin by mAbs to specific integrin subunits and synthetic peptides. Trophoblast adhesion to fibronectin was significantly inhibited with mAbs to the α_5 and β_1 integrin subunits and with a GRGDS peptide. Control mouse immunoglobulin (Ig) or GRGES peptide had no effect. FN = Fibronectin, BSA = Bovine Serum Albumin.

Adhesion of isolated trophoblast to LM was significantly inhibited with mAbs to the α_6 and $\beta 1$ integrin subunits, whereas mAbs to the α_1 and β_4 had little or no effect (Figure 2). Therefore adhesion of trophoblast to LM *in vitro* is predominantly mediated by the α_6 LM receptor. The $\alpha_1\beta_1$ and $\alpha_6\beta_4$ integrins appear not to be involved in adhesion of trophoblast to LM *in vitro*. This study only examined the binding characteristics of trophoblast on one isoform of LM (LM 1), derived from Engelbreth Holm Swarm tumor cells. The $\alpha_1\beta_1$ and $\alpha_6\beta_4$ LM receptors may be important for cell adhesion to other isoforms of LM present in the placental bed.

Monolayers of decidual stromal cells were constructed *in vitro* and used as a substrate layer for trophoblast adhesion. This method was used to simulate the *in vivo* situation more closely and to determine which integrins mediate adhesion of trophoblast to the decidual stromal cell matrix (Burrows et al., 1995). Decidual stromal cells continue to produce the ECM proteins LM, FN, and collagen *in vitro*. Trophoblast cells bind strongly to decidual stromal cells *in vitro*, and extend pseudopodia across the surface of the stroma. This interaction could be blocked with mAbs to the α_5, α_6, β_1, and β_4 integrin subunits (Figure 3), which indicates that a number of matrix receptors mediate the trophoblast - stromal cell interaction, namely, the $\alpha_5\beta_1$, $\alpha_6\beta_1$ and $\alpha_6\beta_4$. Thus it appears that trophoblast express a number of functionally active matrix receptors which mediate adhesion to the decidual stromal cell pericellular matrix proteins LM and FN. These interactions may enable trophoblast to derive the necessary traction for migration through the decidua at the time of implantation. The regulation of these adhesive interactions and subsequent migration would therefore be crucial to successful implantation.

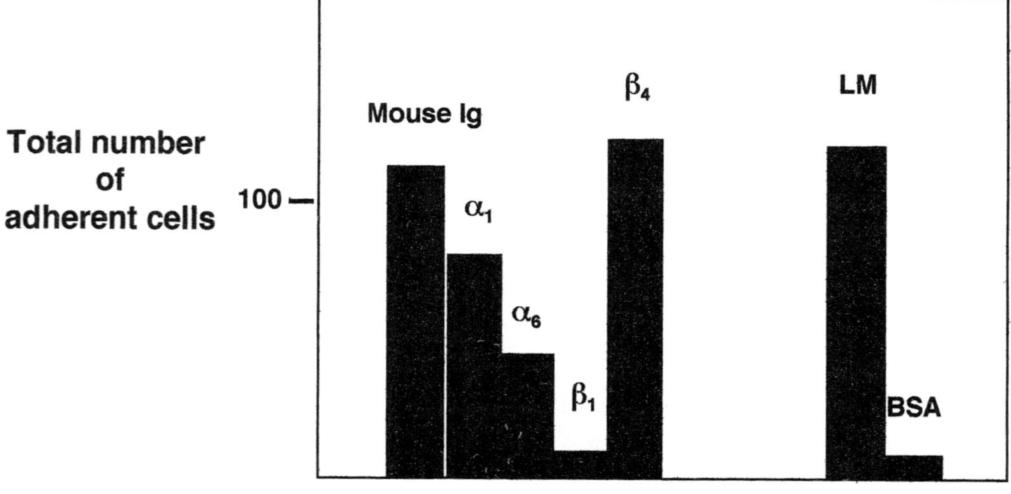

Figure 2. Inhibition of trophoblast adhesion to laminin by mAbs to specific integrin subunits and synthetic peptides. Trophoblast adhesion to laminin was significantly inhibited with mAbs to the α_1 and β_4 integrin subunits and control mouse immunoglobulin (Ig) had little or no effect. LM = Laminin, BSA = Bovine Serum Albumin.

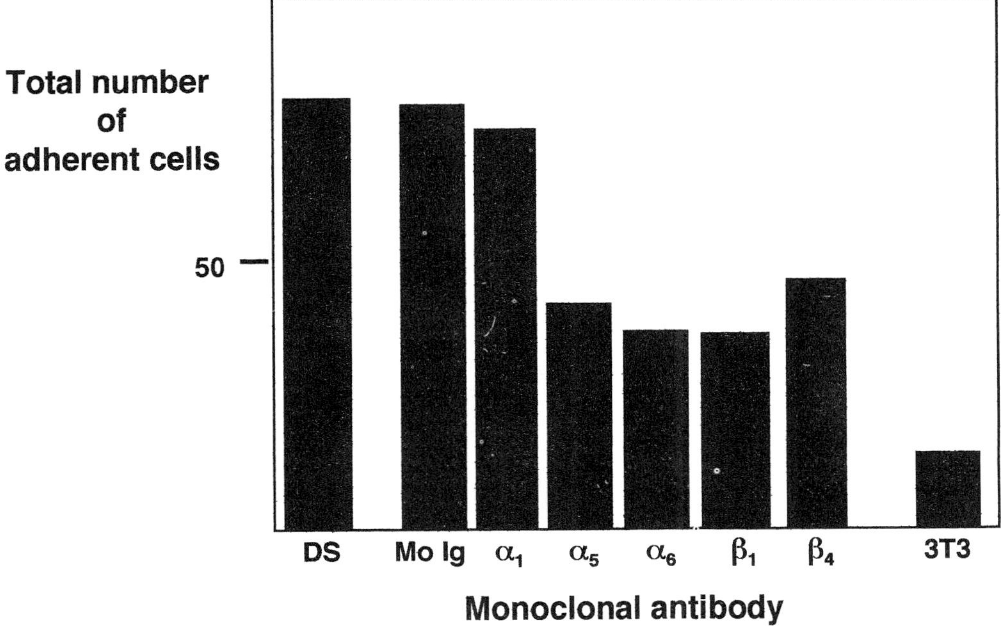

Figure 3. Adhesion of trophoblast cells to decidual stromal cells in the presence of various blocking agents and adhesion of trophoblast to NIH/3T3 fibroblast cells *in vitro*. Numerous trophoblast bound to decidual stromal cells (DS) in the absence of mAb. mAbs to the α_5, α_6, β_1, and β_4 integrin subunits significantly inhibit binding of trophoblast cells to decidual stromal cells in vitro. Anti-α_1 integrin subunit mAb and mouse immunoglobulin (MoIg) had no effect. Few trophoblast cells bind to the fibroblast monolayer (3T3). (Burrows et al., 1995).

In addition to acting as a substrate for cell adhesion, the ECM sends signals to the cell interior via the integrin matrix receptors (Hynes, 1994; Sastry and Horwitz, 1993; Schaller and Parsons, 1994). Transduction of molecular signals to the cell interior are now believed to play a pivotal role in cell behavior. Integrins induce a number of physiological changes within the cell, for example, calcium flux, changes in intracellular pH, and tyrosine phosphorylation of proteins (Figure 4). The initial event in integrin-mediated signal transduction is enhanced phosphorylation of intracellular proteins, in particular the focal adhesion kinase pp125^{FAK} (Guan et al., 1991; Nojima et al., 1992). Increases in tyrosine phosphorylation have also been shown to coincide with increased tumor cell invasion of adjacent ECM (Mueller et al., 1992). Therefore, adhesion of trophoblast cells to uterine extracellular matrix may result in the transmission of the necessary signals for migration and invasion, as well as for the subsequent differentiation of trophoblast within the decidual microenvironment.

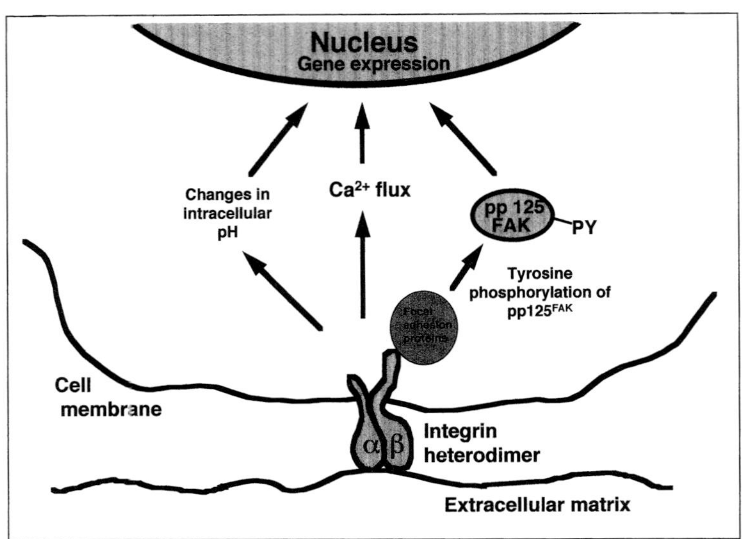

Figure 4. Integrin Signaling Pathways.

It has been demonstrated recently by Western blotting, that tyrosine phosphorylation of intracellular proteins is also used by both normal and malignant trophoblast cells as a mechanism of integrin-mediated signal transduction (Burrows et al., 1995). Adhesion of normal trophoblast cells to LM or FN *in vitro* can induce the phosphorylation of tyrosine residues of a 115 kDa protein. The intensity of the signal was greater in those cells plated on LM than FN. In addition, a 125 kDa protein band was detected but was of weaker intensity. Normal and malignant trophoblast exhibit different profiles of phosphorylated proteins after adhesion to either LM or FN. Tyrosine phosphorylation of three protein bands (125, 115, and 105 kDa) was detected in the malignant trophoblast cell line JEG-3 choriocarcinoma.

The identity and function of the different phosphorylated proteins has not been determined to date. However the 125 kDa protein is probably the focal adhesion kinase protein pp125^{FAK}. This kinase co-localizes with focal adhesions that form at the point of interaction of integrins with the ECM (Schaller and Parsons, 1994). pp125^{FAK} is one of the major substrates of integrin-mediated tyrosine phosphorylation and a number of integrin heterodimers have been shown to trigger tyrosine phosphorylation of this kinase (Kornberg et al., 1991; Vuori and Ruoslahti, 1993). This focal adhesion kinase has been found in most tissues examined (Hanks et al., 1992; Turner et al., 1993), and therefore the 125 kDa protein phosphorylated in trophoblast may well be this kinase. Andre and Becker-Andre (1993) have identified a tissue-specific isomer of pp125^{FAK} in the brain which lacks a portion of the N-terminus of this molecule. Thus it will be interesting to see whether the phosphorylated protein of smaller molecular weight detected in trophoblast represents a different isoform of pp125^{FAK} and whether this transduces a different signal to the cell.

At present we can only speculate about the downstream effect of integrin-mediated signal transduction on trophoblast cell behavior. Evidence from previous morphological studies (Burrows et al., 1993) suggests that laminin has a negative effect on trophoblast migration whereas FN may induce the opposite effect, stimulating trophoblast migration. Isolated first trimester trophoblast bind readily to both LM and FN in short-term culture *in vitro*, but only flatten and spread when plated on FN. After long-term culture trophoblast form large polygonal shaped cells on LM but remain irregular on FN. Trophoblast on FN possess many pseudopodia and under time-lapse photography are seen to actively move across the FN substrate (Burrows et al., 1993). Therefore, LM and FN send different signals to trophoblast, which leads to the differential behavioral patterns observed. It is tempting to suggest that signals from FN may be more important for trophoblast migration while LM gives opposing signals for trophoblast to stop migration.

Thus trophoblast invasion into the decidua requires both physical and biochemical cellular interactions. Transmission of the correct signal to invading trophoblast is likely to be important for successful implantation. In addition, the factors involved in controlling each stage of the integrin-mediated pathway will be important for the regulation of trophoblast invasion *in vivo*.

ACKNOWLEDGEMENTS

Tanya Dee Burrows is supported by Wellbeing. Ashley King is the Meres Senior Student for Medical Research at St John's College, Cambridge. Dr. Y.W. Loke is funded by the Medical Research Council and by the Special Programme of Research, Development and Research Training in Human reproduction, World Health Organization.

REFERENCES

Andre, E. and Becker-Andre, M. (1993) Expression of an N-terminally truncated form of human focal adhesion kinase in brain. *Biochem.Biophys.Res. Commun.* 190, 140-147.

Aplin, J.D. (1991) Implantation, trophoblast differentiation and haemochorial placentation, mechanistic evidence *in vivo* and *in vitro*. *J. Cell Sci.* 99, 681-692.

Aplin, J.D., Charlton, A.K. and Ayad, S. (1988) An immunohistochemical study of human endometrial extracellular matrix during the menstrual cycle and first trimester of pregnancy. *Cell Tissue Res.* 253, 231-240.

Aplin, J.D., Sattar, A., and Mould, A.P. (1992) Variant choriocarcinoma (BeWo) cells that differ in adhesion and migration on fibronectin display conserved patterns of integrin expression. *J.Cell Sci.* 103, 435-444.

Armant, D.R. (1991) Cell interactions with laminin and its proteolytic fragments during outgrowth of mouse primary trophoblast cells. *Biol. Reprod.* 45, 664-673.

Burrows, T.D., King, A., and Loke, Y.W. (1993) Expression of integrins by human trophoblast and differential adhesion to laminin and fibronectin. *Hum. Reprod.* 8, 475-484.

Burrows, T.D., King, A., Smith, S.K., and Y.W. Loke. (1995) Human trophoblast adhesion to matrix proteins : inhibition and signal transduction. *Hum. Reprod.* 10, 2489-2500.

Charpin, C., Kopp, F., Pourreau-Schneider, N., Lissitzky, J.C., Lavant, M.N., Martin, P.M., and Toga, M. (1985) Laminin distribution in human decidua and immature placenta. *Am. J. Obstet. Gynecol.* 151, 822-826.

Church, H.J., Vivovac, L.M., Williams, J.D.L., and Aplin, J.D. (1995) Decidualized endometrium expresses laminin isoforms 2 and 4. *Placenta* 16, A.12.

Damsky, C H., Fitzgerald, M.L., and Fisher, S.J. (1992) Distribution of extracellular matrix components and adhesion receptors are intricately modulated during first trimester cytotrophoblast differentiation along the invasive pathway in vivo. *J. Clin. Invest.* 89, 210-222.

Damsky, C. H., Librach, C., Lim, K-H., Fitzgerald, M. L., McMaster, M. T., Janatpour, M., Zhou, Y., Logan, S.K., and Fisher, S.J. (1994) Integrin switching regulates normal trophoblast invasion. *Development* 120, 3657-3666.

Dedhar, S. (1990) Integrins and tumor invasion. *BioEssays* 12, 583-590.

Divers, M.J., Bulmer, J.N., Miller, D., and Lilford, R.J. (1995) Beta 1 integrins in third trimester human placentae: No differential expression in pathological pregnancy. *Placenta* 16, 245-260.

Faber, M., Wewer, V.M., Berthelsen, J.G., Liotta, L.A. and Albrechtsen, R. (1986) Laminin production by human endometrial stromal cells relates to the cyclic and pathological state of the endometrium. *Am. J. Pathol.* 124, 384-398.

Feinberg, R.E.F., Kliman, H.J. and Lockwood, C.S. (1991) Is oncofetal fibronectin a trophoblast glue for human implantation? *Am. J. Pathol.* 138, 537-543.

Fisher, S.J., Cui, T.-Y., Zhang, L., Hartman, L., Grahl, K., Zhang, G.-Y., Tarpey, J., and Damsky, C.H. (1989) Adhesive and degradative properties of human placental cytotrophoblast cells in vitro. *J.Cell Biol.* 109, 891-902.

Guan, J-L., Trevithick, J.E., and Hynes, R.O. (1991) Fibronectin/integrin interaction induces tyrosine phosphorylation of a 120 kDa protein. *Cell Regul.* 2, 951-964.

Hanks, S.K., Calalb, M.B., Harper, M.C., and Patel, S.K. (1992) Focal adhesion protein tyrosine kinase phosphorylated in response to cell spreading on fibronectin. *Proc. Natl. Acad. Sci. USA* 89, 8487-8489.

Hemler, M.E. (1991) Structures and functions of VLA proteins and related integrins. In: *Receptors for Extracellular Matrix*, (eds.) J.A. McDonald and R.P. Mecham, Academic Press, San Diego, pp. 256-287.

Hynes, R.O. (1994) The impact of molecular biology on models for cell adhesion. *BioEssays* 16, 663-669.

Korhonen, M., Ylänne, J., Laitinen, L., Cooper, H.M., Quaranta, V., and Virtanen, I. (1991) Distribution of α_1-α_6 integrin subunits in human developing and term placenta. *Lab. Invest.* 65, 347-356.

Kornberg, L.J., Earp, H.S., Turner, C.E., Prockop, C., and Juliano, R.L. (1991) Signal transduction by integrins: Increased protein tyrosine phosphorylation caused by clustering of β_1 integrins. *Proc. Natl. Acad. Sci. USA* 88, 8392-8396.

Lee, E.C., Lotz, M.M., Steele Jr., G.D., and Mercurio, A.M. (1992) The integrin $\alpha_6\beta_4$ is a laminin receptor. *J.Cell Biol.* 117, 671-678.

Loke, Y.W., Gardner, L., Burland, K., and King, A. (1989) Laminin in human trophoblast-decidua interaction. *Hum. Reprod.* 4, 457-463.

Loke, Y.W. and King, A. (1995) *Human Implantation. Cell Biology and Immunology.* Cambridge University Press.

Loke, Y.W., King, A. and Burrows, T.D. (1995) Decidua in human implantation. In: *Regulators of Human Implantation,* (eds.) C. Simón and A. Pellicer. *Hum.Reprod.* 10 Supplement 2.

Mueller, S.C., Yeh, Y., and Chen, W.T. (1992) Tyrosine phosphorylation of membrane proteins mediates cellular invasion by transformed cells. *J.Cell Biol* 119, 1309-1325.

Nojima, Y., Rothstein, D.M., Sugita, K., Schlossman, S.F., and Morimoto, C. (1992) Ligation of VLA-4 on T cells stimulates tyrosine phosphorylation of a 105-kD protein. *J. Exp. Med.* 175, 1045-1053.

Pijnenborg, R. (1994) Trophoblast invasion. *Reprod. Med. Rev.* **3**, 53-73.

Romagnano, L. and Babiarz, B. (1993) Mechanisms of murine trophoblast interaction with laminin. *Biol. Reprod.* 49, 374-380.

Ruoslahti, E. (1992) Control of cell motility and tumour invasion by extracellular matrix interactions. *Br. J. Cancer* 66, 239-242.

Ruoslahti, E. and Pierschbacher, M.D. (1987) New perspectives in cell adhesion: RGD and integrins. *Science* 238, 491-497.

Sastry, S.K. and Horwitz, A.F. (1993) Integrin cytoplasmic domains: Mediators of cytoskeletal linkages and extra and intracellular initiated transmembrane signaling. *Curr. Opin. Cell Biol.* 5, 819-831.

Schaller, M.D. and Parsons, J.T. (1994) Focal adhesion kinase and associated proteins. *Curr. Opin. Cell Biol.* 6, 705-710.

Sonnenberg, A., Modderman, P.N., and Hogervorst, F. (1988) Laminin receptor on platelets is the integrin VLA-6. *Nature* 336, 487-489.

Sutherland, A.E., Calaro, P.G. and Damsky, C.H. (1988) Expression and function of cell surface extracellular matrix receptors in mouse blastocyst attachment and outgrowth. *J. Cell. Biol.* 106, 1331-1348.

Turner, C.E., Schaller, M.D., and Parsons, J.T. (1993) Tyrosine phosphorylation of the focal adhesion kinase $pp125^{FAK}$ during development: Relation to paxillin. *J.Cell Sci.* 105, 637-645.

Vuori, K. and Ruoslahti, E. (1993) Activation of protein kinase C precedes $\alpha 5\beta 1$ integrin-mediated cell spreading on fibronectin. *J. Biol. Chem.* 268, 21459-21462.

Yamada, K.M. (1989) Fibronectins: Structure, functions and receptors. *Curr. Opin. Cell Biol.* 1, 956-963.

Zhou, Y., Damsky, C.H., Chio, K., Roberts, J.M., and Fisher, S.J. (1993). Preeclampsia is associated with abnormal expression of adhesion molecules by invasive cytotrophoblast. *J. Clin. Invest.* 91, 950-960.

Trophoblast Research 10:173-180, 1997

TROPHOBLAST INTERACTION WITH DECIDUAL NK CELLS IN HUMAN IMPLANTATION
- A Review -

Ashley King and Y.W. Loke

Research Group in Human Reproductive Immunobiology
Department of Pathology
University of Cambridge
Cambridge CB2 1QP United Kingdom

INTRODUCTION

Implantation of the human placenta is accompanied by migration of trophoblast cells from the tips of the anchoring villi into the decidua. The extent of this migration is tightly controlled so that trophoblast cells do not normally penetrate beyond the inner third of the myometrium where they differentiate into multinucleated placental bed giant cells. The mechanisms of this control are not known. Because these trophoblast cells are fetally-derived and therefore foreign to the mother, it is expected that there may be some immunological recognition by the mother which could influence the migration. For this reason, reproductive immunologists are very interested in the type of immune response which may be generated locally in the uterus against trophoblast. Intriguingly, evidence to date has revealed that this uterine immune response is not the same as that encountered in classical transplantation immunology (Loke and King, 1995). There are fundamental differences occurring in both the trophoblast and the uterus. On the trophoblast side, it is now established that the HLA antigens expressed by trophoblast are not the same as those expressed by other somatic cells. On the uterine side, the mucosa of this organ is populated by cells of the innate immune system, such as Natural Killer (NK) cells rather than T and B lymphocytes characteristic of the specific acquired immune response. Thus, the immunological interaction between human trophoblast and the uterus is likely to be unusual.

Trophoblast Expression Of HLA Class I Antigens

It is now generally accepted that extravillous trophoblast (EVT), the population that migrates into decidua during implantation, expresses the non-classical HLA-G. There is evidence for the presence of both HLA-G mRNA (Yelavarthi et al., 1991; Chumbley et al., 1993) and protein (Kovats et al., 1990; Chumbley et al., 1994) in these cells. Although low levels of HLA-G message is found in a variety of cell types besides trophoblast, such as fetal eye, fetal thymus (Shukla et al., 1990), fetal liver (Houlihan et al., 1992), circulating T and B cells (Kirszenbaum et al., 1994) and even adult skin biopsies (Ulbrecht et al., 1994), recent studies with an HLA-G specific antibody has localized the protein only in EVT (Chumbley et al., 1994). Thus, it seems that expression of the HLA-G antigen is restricted to this trophoblast population. The HLA-G protein exists in several isoforms probably resulting from differential splicing (Ishitani and Geraghty, 1992). In addition, soluble as well as membrane forms have been observed (Fujii et al., 1994).

The function of HLA-G is unknown. It is relatively non-polymorphic at the genetic and protein level. Although a number of sequence variations in the HLA-G gene which could result in amino-acid substitutions have been reported in different individuals (van der Ven and Ober, 1994), only one of these variations is actually at a functional residue of the peptide binding groove (Parham, 1995). Therefore, unlike the classical class I genes, polymorphism of HLA-G does not appear to be selected for peptide-binding diversity which implies that it has probably not evolved for classical T cell interaction. Furthermore, the finding that HLA-G antigen is not expressed in the fetal thymus where T cell education takes place further supports the conclusion that HLA-G is not involved in influencing the T cell repertoire. We believe that HLA-G expressed by EVT functions as a target molecule for decidual Natural Killer (NK) cells rather than T cells (King and Loke, 1991; Loke and King, 1991). However, the HLA-G molecule is capable of binding endogenous nonapeptides in the same way as classical class I molecules. The peptide may merely enhance the stability of the class I molecule at the cell surface, but the possibility that it can also interact with T cells should not be entirely dismissed.

The potential role of HLA class I antigens in trophoblast-decidual NK interaction is further complicated by the finding that EVT appears to express another class I molecule besides HLA-G (Grabowska et al., 1990). The identity of this molecule has been much debated, but the accumulative evidence indicates that it is probably HLA-C (King et al., 1996). For a classical class I antigen, HLA-C is rather unusual and differs from HLA-A and -B in having a relatively low surface expression and being less polymorphic. Because of these characteristics, the importance of HLA-C in interacting with T cells has been questioned. Again, as for HLA-G, we believe that HLA-C expressed by EVT is recognized by decidual NK cells (Loke and King, 1995).

Decidual Natural Killer (NK) Cells

The most abundant lymphoid cells in decidua are NK cells which comprise about 70% of the total population with macrophages making up 20% and T lymphocytes 10% (King et al., 1989). There are insignificant numbers of B lymphocytes and granulocytes are absent. The majority of the NK cells have prominent cytoplasmic granules which have led to their being called Large Granular Lymphocytes (LGL). The number of these NK cells is low in the proliferative stage of the menstrual cycle, gradually increases during the mid-luteal phase and reaches a peak in the late secretory phase. They will show signs of apoptosis a few days before menstruation. However, if pregnancy occurs, their number increases during the early stages of pregnancy, particularly in the decidua basalis and then declines in the second trimester. Immunohistological studies have demonstrated these NK cells to be in close proximity to the invading trophoblast cells at the implantation site. This temporal and spatial association between decidual NK cells and EVT suggests a functional interaction between these two cell types.

Phenotypic analysis has revealed major differences between decidual NK cells and NK cells from peripheral blood, such as the expression of the NK cell marker, CD56 and the FcγRIII, CD16. In blood, 90% of the NK cells are $CD56^{dim}$ $CD16^{bright}$ and only 10% are $CD56^{bright}$ and $CD16^{dim}$. In contrast, in the uterus the majority of NK cells are $CD56^{bright}$ $CD16^{-}$. In addition, CD57, another marker for adult NK cells, is also not expressed by decidual NK cells (Table 1). It would seem, therefore, decidual NK cells probably represent a sub-population of the NK cell family which is restricted to the uterus.

Table 1

Characteristics Of NK Cells From Blood, Decidua And Fetal Liver
(From: Loke and King, 1995)

	Blood	Decidua	Fetal Liver
NK markers	$CD56^{dim}$ $CD16^{bright}$ $CD57^{+/-}$	$CD56^{bright+++}$ $CD16^{-}$ $CD57^{-}$	$CD56^{bright}$ $CD16^{-/dim}$ $CD57^{-}$
Cytoplasmic	$cyCD3^{-}$	$cyCD3^{+}$	$cyCD3^{+}$
Activation markers	$CD69^{-}$	$CD69^{+}$	$CD69^{+}$
NK activity	High Respond to high dose IL-2	Low Respond to low dose IL-2	Low Respond to low dose IL-2
Cytokine Production	$GM\text{-}CSF^{-}$ $CSF\text{-}1^{-}$	$GM\text{-}CSF^{+}$ $CSF\text{-}1^{+}$	$GM\text{-}CSF^{+}$ CSF-1?

The ontogeny of decidual NK cells is unclear. Two possible differentiation pathways can be proposed (Loke and King, 1995). Decidual NK cells could have arisen from a common NK cell progenitor as those in blood, but have subsequently undergone tissue-specific differentiation under the influence of some uterine signal resulting in a unique phenotype. Alternatively, decidual NK may represent a distinct NK cell population which has split off early in its differentiation pathway from that in blood to settle in the uterus.

The variation in number of NK cells over the menstrual cycle and in pregnancy suggests their recruitment/maintenance is likely to be under hormonal control, but the identity of the stimulus is not known. Decidual NK cells do not express estrogen, progesterone or prolactin receptors. They do express the intermediate affinity p75 and the high affinity p55 IL-2R (Nishikawa et al., 1991), and they have been observed to proliferate in the presence of IL-2 *in vitro* (King et al., 1992). However, IL-2 has not been localized in decidua *in vivo* (Loke and King 1995), and also IL-2 transforms decidual NK cells into powerful lymphokine-activated killer (LAK) cells which are capable of killing trophoblast cells *in vitro* (King and Loke, 1990), so this cytokine is unlikely to have a physiological role to play. The search is now on for a cytokine which can induce proliferation in decidual NK cells without transforming them into LAK cells *in vitro*, and which is demonstrable in uterine tissues *in vivo*.

Trophoblast-Decidual NK Cell Interaction

Although we have proposed that trophoblast invasion of the uterus during the process of implantation is controlled by decidual NK cells interacting with trophoblast HLA-G/HLA-C, elucidation of the exact mechanism of this interaction has proved elusive. This is because compared to T cells little is known about NK cell biology, NK target molecules, the NK cell receptors and the outcome of any interaction between NK cells and target cells (Figure 1).

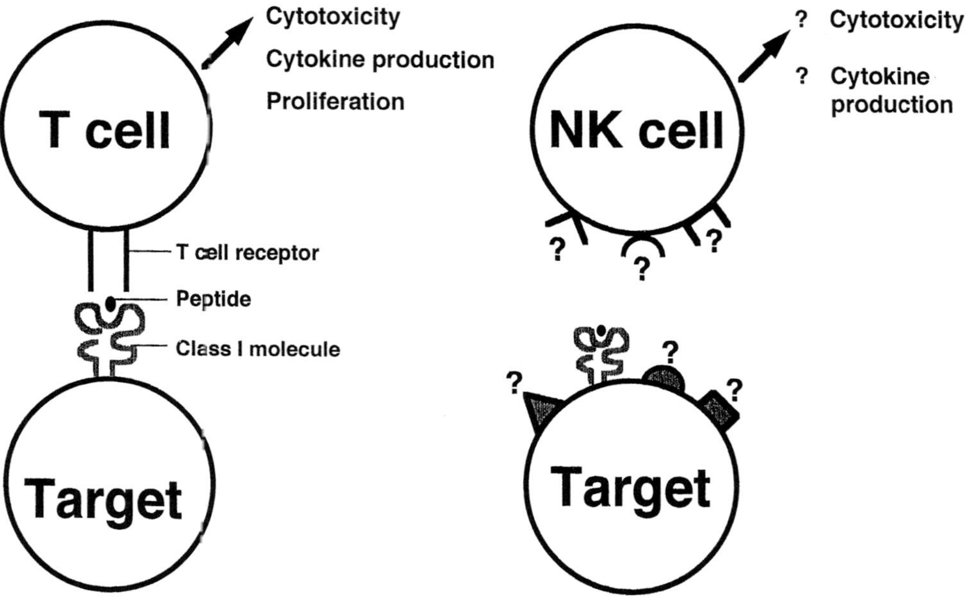

Figure 1. Diagram showing that the T cell receptors and its associated target ligand are known, but the equivalent structures for NK cells have not been established. (From: Loke and King, 1995)

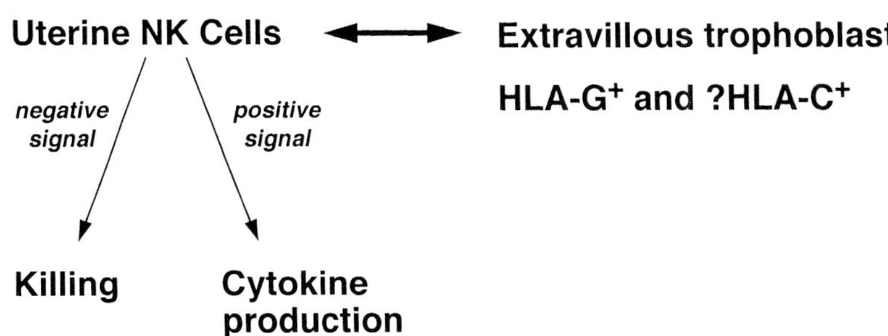

Figure 2. A hypothetical model of the possible outcome of interaction between uterine NK cells and extravillous trophoblast at the implantation site. (From: Loke and King, 1995)

However, it is now generally accepted that NK cells preferentially kill target cells which have low or absent HLA class I antigen expression. This observation had led to the formulation of the 'missing-self' hypothesis which postulated that NK cells eliminated cells lacking class I molecules in contrast to T cells which kill cells bearing foreign class I antigens. Thus the presence of class I molecules on target cells prevents NK-mediated lysis (Ljunggren and Kärre, 1990). The situation has since become more complex as it is now apparent that NK cells can discriminate between different class I alleles, although they do not detect fine polymorphic differences like T cells. Instead, NK cells appear to recognize a public, perhaps primordial, polymorphism in class I molecules. A family of human NK receptors which do recognize HLA class I antigens have recently been identified, the p58 (NKAT) family of receptors for Cw3/Cw4 and related alleles (Gumperz and Parham, 1995). There are now designated Killer Inhibitory Receptors (KIR). Each NK cell can express several different receptors, both KIR which recognize class I molecules, and other receptors of different molecular structure which possibly use target ligands such as oligosaccharides. It seems that this is the way receptor diversity is generated in NK cells rather than by somatic recombination of a single receptor as used by T and B cells. Expression of different combinations of these receptors on each NK cell will determine its repertoire (Yokoyama, 1995). Furthermore, the NK repertoire in blood NK cells varies between different individuals with the intriguing observation that KIR which have specificity for non-self class I molecules may occur (Gumperz and Parham, 1995). These recent findings raise the fascinating possibility that maternal allorecognition of the fetus does occur, but it is mediated in a completely novel manner by NK cells rather than a classical allograft reaction mediated by T cells.

In the context of interaction with trophoblast, the expression of HLA-G/HLA-C by trophoblast could influence their susceptibility to decidual NK cell lysis. Exposure of human trophoblast cells to exogenous IFN-γ, a cytokine which upregulates class I expression, has been observed to protect trophoblast cells against lysis by IL-2-stimulated decidual NK cells (King and Loke, 1993). Similarly, HLA-G transfected into class I-deficient cell lines was found to provide some protection from lysis by freshly isolated decidual NK effectors compared to the parental cell line (Chumbley et al., 1994).

The additional possibility that class I signals transmitted to decidual NK cells can affect not only cytotoxicity but also influence other NK cell functions such as cytokine production would need to be explored (Figure 2). Decidual NK cells are known to produce a variety of cytokines and trophoblast expresses appropriate receptors for many of these cytokines. Thus a potential cytokine network may be in place at the implantation site by which decidual NK cells influence trophoblast behavior.

CONCLUSION

From this brief review, it can be seen that the immunological relationship between the implanting placenta and the maternal uterus is not governed by the laws of classical transplantation immunology. Instead, it seems to involve a more primitive defense system whose mechanism of 'self' and 'non-self' recognition and the resultant reactions invoked, are more akin to that observed between unrelated invertebrates than between vertebrate allograft and host. This observation has completely altered our conceptual view of the immunology of reproduction.

ACKNOWLEDGEMENTS

AK is in receipt of the Meres Senior Studentship for Medical Research of St. John's College, Cambridge. YWL is funded by the Medical Research Council and by the Special Programme of Research, Development and Research Training in Human Reproduction, World Health Organization.

REFERENCES

Chumbley, G., King, A., Gardner, L., Howlett, S., Holmes, N. and Loke, Y.W. (1994) Generation of an antibody to HLA-G in transgenic mice and demonstration of the tissue reactivity of this antibody. *J. Reprod. Immunol.* 27, 173-186.

Chumbley, G., King, A., Holmes, N., and Loke, Y.W. (1993) In situ hybridization and northern blot demonstration of HLA-G mRNA in human trophoblast populations by locus-specific oligonucleotide. *Hum. Immunol.* 37, 17-22.

Fujii, T., Ishitani, A. and Geraghty, D.E. (1994) A soluble form of the HLA-G antigen is encoded by a messenger ribonucleic acid containing intron 4. *J. Immunol.* 153, 5516-5524.

Grabowska, A., Carter, N. and Loke, Y.W. (1990) Human trophoblast cells in culture express an unusual major histocompatibility complex class I-like antigen. *Am. J. Reprod. Immunol.* 23, 10-18.

Gumperz, J.E. and Parham, P. (1995) The enigma of the natural killer cell. *Nature* 378, 245-248.

Houlihan, J.M., Biro, P.A., Fergar-Payne, A., Simpson, K.L. and Holmes, C.H. (1992) Evidence for the expression of non-HLA,-A,-B,-C class I genes in the human fetal liver. *J. Immunol.* 149, 668-675.

Ishitani, A. and Geraghty, D.E. (1992) Alternative splicing of HLA-G transcripts yields proteins with primary structures resembling both class I and class II antigens. *Proc. Natl. Acad. Sci. USA* 89, 1-5.

King, A. and Loke, Y.W. (1990) Human trophoblast and JEG choriocarcinoma cells are sensitive to lysis by IL-2 stimulated decidual NK cells. *Cell. Immunol.* 129, 435-448.

King, A. and Loke, Y.W. (1991) On the nature and function of human uterine granular lymphocytes. *Immunol. Today* 12, 432-435.

King, A. and Loke, Y.W. (1993) Effect of IFN-γ and IFN-α on killing of human trophoblast by decidual LAK cells. *J. Reprod. Immunol.* 23, 51-62.

King, A., Wellings, V., Gardner, L. and Loke, Y.W. (1989) Immunocytochemical characterisation of the unusual large granular lymphocytes in human endometrium throughout the menstrual cycle. *Human Immunol.* 24,195-205.

King, A., Wheeler, R., Carter, N.P., Francis, D.P. and Loke, Y.W. (1992) The response of human decidual leukocytes to IL-2. *Cell. Immunol.* 140, 409-421.

King, A., Boocock, C., Sharkey, A., Gardner, L. and Loke, Y.W. (1996) Evidence for the expression of HLA-C class I mRNA and protein by human first trimester trophoblast. *J. Immunol.* 156, 2068-2076.

Kirszenbaum, M., Moreau, P., Gluckman, E., Dausset, J. and Carosella, E. (1994) An alternatively spliced form of HLA-G mRNA in human trophoblasts and evidence for the presence of HLA-G transcript in adult lymphocytes. *Proc. Natl. Acad. Sci. USA* 91, 4209-4213.

Kovats, S., Main, E.K., Librach, C., Stubblebine, M., Fisher, S.J. and DeMars, R. (1990) A class I antigen, HLA-G, expressed in human trophoblasts. *Science* 248, 220-223.

Ljunggren, H-G. and Kärre, K. (1990) In search of the 'missing self': MHC molecules and NK cell recognition. *Immunol. Today* 11, 237-244.

Loke, Y.W. and King, A. (1991) Recent developments in the human maternal-fetal immune interaction. *Curr. Opin. Immunol.* 3, 762-766.

Loke, Y.W. and King, A. (1995) Human Implantation: Cell Biology and Immunology. Cambridge University Press, Cambridge, England.

Nishikawa, K., Saito, S., Morii, T., Hamada, K., Ako, H., Narita, N., Ichijo, M., Kurahayashi, M. and Sugamura, K. (1991) Accumulation of $CD16^-$ $CD56^+$ natural killer cells with high affinity interleukin 2 receptors in human early pregnancy decidua. *Int. Immunol.* 3, 743-750.

Parham, P. (1995) Antigen presentation by class I major histocompatibility complex molecules: a context for thinking about HLA-G. *Am. J. Reprod. Immunol.* 34, 10-19.

Shukla, H. Swarocp, A., Srivastava, R. and Weissman, S.M. (1990) The mRNA of a human class I gene HLA G/HLA 6.0 exhibits a restricted pattern of expression. *Nucleic Acids Res.* 18, 2189.

Ulbrecht, M., Rehberger, B., Strobel, I., Messer, G., Kind, P., Degitz, K., Bieber, T., and Weiss, E.H. (1994) HLA-G: Expression in human keratinocytes in vitro and in human skin in vivo. *Eur. J. Immunol.* 24, 176-180.

van der Ven, K. and Ober, C. (1994) HLA-G polymorphisms in African Americans. *J. Immunol.* 153, 5628-5633.

Yelavarthi, K.K., Fishback, J.L. and Hunt, J.S. (1991) Analysis of HLA-G mRNA in human placental and extraplacental membrane cells by in-situ hybridization. *J. Immunol.* 146, 2847-2854.

Yokoyama, W.M. (1995) Natural killer cell receptors specific for major histocompatibility complex class I molecules. *Proc. Natl. Acad. Sci. USA* 92, 3081-3085.

Trophoblast Research 10:181-190, 1997

PARENT CELLS FOR TROPHOBLAST HYBRIDIZATION I: ISOLATION OF EXTRAVILLOUS TROPHOBLAST CELLS FROM HUMAN TERM CHORION LAEVE

Gabriele Gaus[1], Hitoshi Funayama[1,2], Berthold Huppertz[1], Peter Kaufmann[1] and Hans-G. Frank[1,3]

[1]Department of Anatomy
Technical University
D-52057 Aachen, Germany

[2]Department of Obstetrics and Gynecology
Medical College Hospital
6-7-1 Nishishinyuku Shinyuku-ku
160 Tokyo, Japan

INTRODUCTION

Various approaches have been used to isolate trophoblast cells from the human placenta for *in vitro* studies. These comprise methods developed by groups in Philadelphia/New Haven (Feinman et al., 1986; Kliman et al., 1986; Kliman and Feinberg, 1990), in Cambridge (Loke, 1983, 1990; Loke et al., 1989; Burrows et al., 1993), in Davis (Douglas and King 1989, 1990), and in Aarhus (Jie et al., 1990; Toth et al., 1990; Aboagye-Mathiesen et al., 1993). Improvements of the various methods and their combinations meanwhile guarantee high purities of the resulting trophoblast cells (comparison of these methods was subject of a workshop on "Trophoblast Cell Culture" organized and chaired by P.K. Lala and L. Guilbert, as a part of the XIII Rochester Trophoblast Conference in Banff, 1996). However, in most preparations there are still uncertainties as to the villous or extravillous origin of the cells.

Because of several reasons, for our purposes less the absolute trophoblastic purity but rather the exclusive extravillous origin of trophoblast is important: Anchoring of the placenta to the uterine wall is related to invasive extravillous trophoblast cells and their oncofetal extracellular matrix, matrix-type fibrinoid (Feinberg et al., 1991; Damsky et al., 1992; Frank et al., 1994, 1995; King and Blankenship 1994; Lang et al., 1994; Huppertz et al., 1995). This special matrix is secreted during invasion by differentiated, invasive extravillous trophoblast cells, which are no longer proliferative. The composition of this special invasion-related extracellular matrix has been analyzed in detail (Frank et al., 1995; Huppertz et al., 1995, 1996) and is clearly different from the extracellular matrix present in the subtrophoblastic basement membrane of the villous trees. Since biochemical isolation of complete matrix-type fibrinoid is impossible and molecular biological approaches are hampered by the oncofetal and predominantly posttransscriptional modifications characterizing this matrix, we started adapting the hybridoma technique known from monoclonal antibody technology to this problem.

[3]To Whom Correspondence Should Be Addressed.

To gain analytical and productive access to the complex mixture of invasion-related extracellular matrix-molecules we started to immortalize differentiated extravillous trophoblast cells by fusion with their malignant trophoblastic counterparts (Figure 1a-c). Already existing alternatives would have been to use SV40-T-antigen transformed trophoblast cells (Lei et al., 1992; Graham et al., 1993; Lewis et al., 1996) or extravillous trophoblast cells derived from cultured villous explants (Yagel et al., 1989; Irving et al., 1995). Both methods seemed to be less suitable for our specific purposes; transformation is connected with uncontrolled amplification and splicing of amplified viral DNA into the cellular genome (Old and Primrose, 1992) leading to heterogenous genotypes with limited stability. For our purposes, a stable proliferative population, which is not based on uncontrolled transformation by non-human regulative genes or gene products is advantageous. Culture systems starting from villous explants (Yagel et al., 1989) end up in trophoblast cells with a somehow extended proliferative phenotype. However, these cells obviously correspond to the stem cells of the extravillous pathway (Benirschke and Kaufmann, 1995) and not the differentiated, matrix-secreting, non-proliferative phenotype.

Because of these reasons, we decided for the hybridoma approach, which is based on two major prerequisites:

- Generation and selection of suitable malignant counterparts for fusion. We recently have produced, selected and cloned such cell lines, mutants of the choriocarcinoma cell line JEG-3 and deficient of the enzyme hypoxanthine-guanine-phosphoribosyltransferase (HGPRT) (Frank et al., 1996; Funayama et al., this volume).

- Isolation of viable differentiated extravillous trophoblast cells. In order to avoid hidden admixtures of villous trophoblast which produces a quite different extracellular matrix (Duance and Bailey, 1983; Amenta et al., 1986; Castellucci et al., 1993; cf., Benirschke and Kaufmann, 1995), and to establish a quick isolation method with a high yield of viable cells, we have chosen term chorion laeve as starting material. Chorion laeve is primarily free of villous trophoblast and therefore a suitable source of extravillous trophoblast. As compared to extravillous trophoblast from cell columns of early pregnancy, term membranes are largely void of proliferating trophoblastic stem cells (Kaufmann and Castellucci, 1997), but rather predominantly contain the differentiated, matrix-secreting subset. Since *in vitro* proliferation shall be achieved only by hybridization, the differentiated phenotype is even advantageous for our purposes.

Preparation of tissues and enzymatic digestion was adapted following the method of Perkins and Linton (1995). As a positive marker suitable for selection of extravillous trophoblast cells we have chosen human placental alkaline phosphatase. Within chorion laeve, this enzyme is expressed only by extravillous trophoblast cells. Subsequent separation of antibody-labeled extravillous trophoblast cells is achieved using magnetic particles coated with secondary antibody.

Based on these methods we developed a protocol for a rapid, gentle and selective isolation of term extravillous trophoblast . In addition to our HGPRT-negative JEG-3 subclones, these cells represent the second parental cell population necessary for fusion experiments. The suitability of for production of hybridomas has already been proven in pilot experiments.

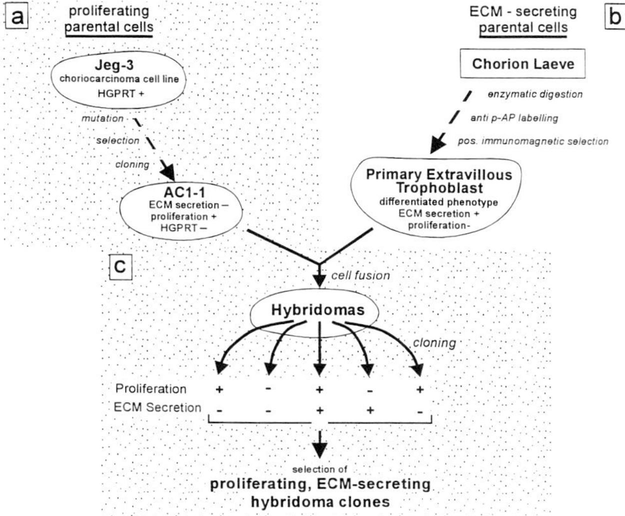

Figure 1. Methodological approach for generation of extravillous trophoblast hybridomas combining the proliferative activity of choriocarcinoma cells with the capability of differentiated extravillous trophoblast cells to secrete extracellular matrix (ECM). (a) Mutation, selection and cloning of a HGPRT-negative, proliferating choriocarcinoma cell line (Funayama et al., this volume). (b) Isolation of primary extravillous trophoblast cells which secret oncofetal extracellular matrix, but do not proliferate, as described in this paper. (c) Fusion of both parent cell lines to form hybridomas; subsequent cloning and selection of ECM-secreting, proliferating clones. This part of the project is in progress.

MATERIALS AND METHODS

Preparation Of Immunoreagents

Mouse-anti-human placental alkaline phosphatase (8B6, DAKO) was desalted using a DG-10 desalting column (Bio-Rad) equilibrated with 30 ml RPMI 1640. One hundred µl of the antibody solution were diluted in 3 ml RPMI 1640, loaded onto the column and subsequentlty eluted with 4 ml RPMI 1640. Magnetic particles (Dynabeads M-280 Sheep anti-Mouse IgG, Dynal) were washed according to manufacturers instructions.

Tissues And Cell Isolation

Placentae were obtained from either normal term deliveries or cesarean sections at term. Samples of complete membranes were snap frozen in liquid nitrogen. Binding patterns of mouse anti-human placental alkaline phosphatase (dilution 1:25; see above) and anti-cytokeratin (dilution 1:500; MNF 116, pan-cytokeratin, Dako) were analyzed in cryostat sections of this material (Figure 2a, b) using the AC-Kit protocol (Frank et al., 1994). Omitting first antibody or replacing it with normal mouse serum (dilution 1:20) were used as control reactions.

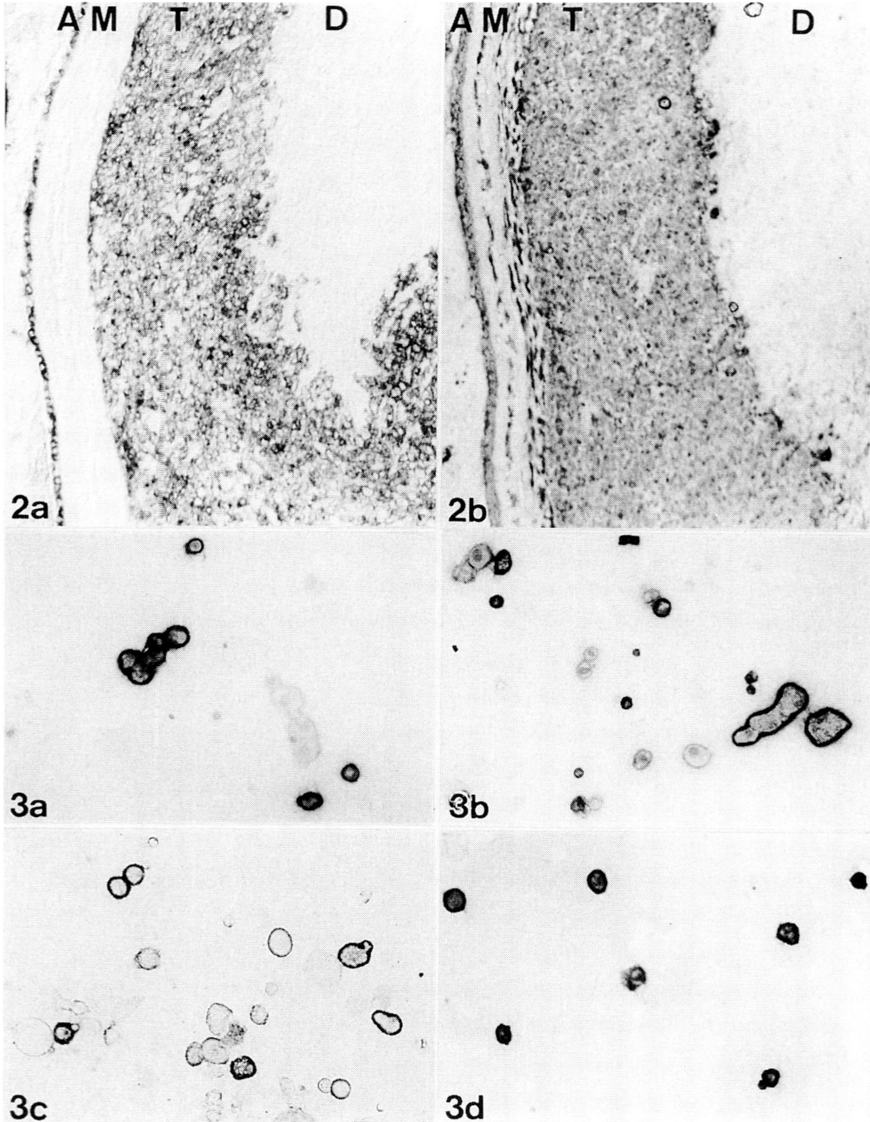

Figure 2. Serial cryosections of embryonic membranes. A, amnionic epithelium; M, amnionic and chorionic mesenchyme; T, trophoblast layer; D, decidua. Immunohistochemical detection of placental alkaline phosphatase (a) and cytokeratin (b). Mesenchymal cells in both amnionic and chorionic mesenchyme stain positive for cytokeratin but not for placental alkaline phosphatase. Magnification bar, 100 μm.

Figure 3. Smears of the cellular suspension after enzymatic digestion and washing but before addition of anti-alkaline phosphatase (a, b, c) and after immunomagnetic separation (d). Immunohistochemical detection of cytokeratin (a), vimentin (b) and placental alkaline phosphatase (c, d). Magnification bar, 100 μm.

For isolation of extravillous trophoblast, amnionic epithelium and amnionic mesenchyme were peeled off the chorionic parts (Figure 2a, b). The remaining layers of the membranes which make up the chorion laeve (chorionic mesenchyme, trophoblast and decidua (cf., Figures 2a, b) were washed three times rapidly in cold (4°C) Hank's balanced salt solution (HBSS) containing 0.37 % EDTA and three times in phosphate buffered saline, PBS, until the washing fluid was macroscopically free of blood. Approximately 30-35 g of chorion laeve can be obtained from one term placenta.

After the washing steps, the tissues were incubated with 25 U of type XIV protease (Sigma) per gram wet weight in 20 ml RPMI 1640 (Gibco), gently rotating for 75 minutes at 37°C. Thereafter the membranes were washed three times with PBS and placed in 30 ml RPMI 1640 containing 5% fetal calf serum (FCS, Gibco), 1450 U of collagenase type IV (Sigma) and 2950 U of hyaluronidase type I-S (Sigma) per gram wet weight, and incubated gently rotating for 90 minutes at 37°C. The digested material was then stirred for 5 minutes at room temperature and centrifuged at 4°C and 800 rpm for 10 minutes. The supernatant was transferred, diluted with a minimum of 20 ml RPMI 1640 and centrifuged again under the same conditions. Both pellets were combined and washed once with Ham´s F12 containing antibiotics (100 U/ml penicillin; 100 µg/ml streptomycin; 2.5 µg/ml amphotericin B [PSF]). The pellets were resuspended in 10 ml RPMI 1640 and an aliquot of the cells was counted in a Neubauer hemocytometer. Viability of the cells was estimated by dye exclusion after incubation with Trypan Blue (Lindl and Bauer, 1989). Smears of this cell suspension were used for immunohistochemical detection of cytokeratin (dilution 1:500; MNF 116, pan-cytokeratin, Dako), vimentin (dilution 1:10; V9 anti-vimentin, Dako) and placental alkaline phosphatase (dilution 1:25; 8B6, anti-human placental alkaline phosphatase, Dako), see Figures 3 a-c.

For immunomagnetic separation 4 ml RPMI 1640 containing 100 µl of desalted monoclonal anti-human alkaline phosphatase were added and the cell suspension was incubated for 60 minutes at room temperature by gentle end to end mixing. The suspension was washed three times with Ham´s F12/PSF and the pellet was resuspended in 10 ml RPMI 1640. A prewashed aliquot of magnetic particles coated with a secondary antibody was added and incubated with the cell suspension for 30 minutes at room temperature by gentle end to end mixing. The optimum concentration of magnetic particles was tested in a range from 5 µl to 15 µl per gram wet weight of tissue. Fifteen µl beads per gram wet weight of tissue proved to be the optimal working concentration . Cells labeled with magnetic particles were separated with a magnetic particle concentrator (Dynal MPC-1). To avoid problems by highly viscous or clumpy cell suspensions affecting the mobility of labeled cells in the magnetic field, the suspension was diluted with 10 ml RPMI 1640 prior to separation. After separation of the cells the supernatant was discarded. Using the magnetic particle concentrator, the labeled cells were washed twice and resuspended in Ham´s F12/PSF.

Aliquots of this cell suspension were taken for counting of viable cells as described above and for immunohistochemical detection of placental alkaline phosphatase in smears. Detection of alkaline phosphatase was performed without newly adding primary antibody (Figure 3d); rather only secondary antibody was used according to the AC-kit protocol (Frank et al., 1994).

RESULTS AND DISCUSSION

Within the fetal membranes, the antibody against human placental alkaline phosphatase labels only amnionic epithelial cells and extravillous trophoblast cells (Figure 2a), while anti-cytokeratin labels trophoblast cells, amnionic epithelium and mesenchymal cells in both chorionic and amnionic mesenchyme (Figure 2b). Decidua did not react with anti-cytokeratin. Since amnionic epithelium and mesenchyme can be removed easily and completely by mechanical separation along the spongy layer of the membranes, the remaining chorion laeve contains extravillous trophoblast cells as the only alkaline phosphatase-positive cell population. Due to the more trophoblast specific reaction pattern of anti-alkaline phosphatase as compared to anti-cytokeratin this antibody was used as purity control.

The enzymatic digestion we have used, leads to a complete disintegration of extracellular compounds and leaves the cells freely moving in the viscous digestion mixture. If necessary, disintegration can be completed by repeatedly pipetting up and down the mixture. An average yield of 10^7 living cells per gram wet weight can be obtained after the washing steps.

Immunohistochemical detection of cytokeratin, vimentin and human placental alkaline phosphatase in smears from such preparations is shown Figures 3a-c. Generally and reproducibly, half of the cells were vimentin-positive and half cytokeratin-positive, while only 40% of all cells expressed human placental alkaline phosphatase. This result implies that about 20% of the cytokeratin-positive cells do not express placental alkaline phosphatase. With regard to our immunohistochemical findings, cyotokeratin-positivity does not prove the trophoblastic nature of a cell. In particular, in stem villi, chorionic plate and membranes cytokeratin-positive mesenchymal cells are well-known (Kasper et al., 1988 a,b). Because of this, we have chosen a positive selection method based on the cell surface marker alkaline phosphatase. This approach is in analogy to another positive selection strategy for trophoblast cells using laminin coated magnetic beads (Loke et al., 1989a, b; Loke, 1990). Any isolation based on negative selection has to guarantee complete removal of undesired cells without saturation of selective capacity. Although successfully performed (Douglas and King, 1989; Graham et al., 1993), it is more difficult to meet the objective of negative selection than that of positive selection; also in the latter case selectivity is necessary, but without substantial need for completeness. Our experience shows that most of the beads are shed from the cell surface within short. Eventually remaining beads do not hinder hybridization. After immunomagnetic separation, the percentage of selected cells being effectively labeled with anti-placental alkaline phosphatase was about 95% (Figure 3d). The remaining about 5% consisted of cellular detritus, dead cells and aggregated cells. Even though the latter consisted at least mostly of alkaline phosphatase-positive cells, an accidental admixture of non-trophoblastic cells to these aggregates could not be excluded. It is unclear, whether the aggregates result from incomplete digestion or bridging by secondary antibody coated beads during the isolation procedure.

The overall time needed to perform the whole protocol takes between 8-10 hours. The 95%-purity obtained with this method is sufficient for our hybridoma project and the procedure is reliable and reproducible. Access to isolated extravillous trophoblast, which so far was a tricky and time consuming task endangering viability of the isolated cells, can be achieved within one day using this protocol.

Extravillous trophoblast of term membranes, which is purified according to this protocol, should not be regarded to be identical in every aspect with extravillous trophoblast of first trimester cell columns, the usual target of cell biologists studying extravillous trophoblast. This is particularly valid for the composition of the population of cells. Whereas in first trimester columns all stages of the invasive pathway are present, in term membranes the differentiated, matrix-secreting subset dominates How far the results obtained with such a cell preparation derived from the membranes can be transferred to other populations of extravillous trophoblast, needs to be established.

SUMMARY

Extravillous trophoblast cells of human placenta produce matrix-type fibrinoid, an oncofetally modified extracellular matrix. Matrix-type fibrinoid is linked to differentiation of extravillous trophoblast cells. It is involved in anchoring of the placenta to the uterine wall, at the same time enabling the trophoblast cells to migrate and to invade maternal tissues. Biochemical isolation of matrix-type fibrinoid is impossible. *In vitro* production is hampered by the lack of proliferation of differentiated extravillous trophoblast cells. Therefore, we want to immortalize the capability for production of matrix-type fibrinoid. In order to avoid any introduction of non human regulative genes into the trophoblastic genome, we have chosen the hybridoma technique for immortalization. Negatively selectable choriocarcinoma cell lines are available as trophoblast-related malignant parent line for fusion. Here we describe a rapid isolation protocol for differentiated term extravillous trophoblast cells. With chorion laeve as source we avoid any accidental admixture of trophoblast cells of villous origin. Immunomagnetic isolation of extravillous trophoblast cells was performed using placental alkaline phosphatase as highly trophoblast specific positive marker molecule. The relatively simple method yields preparations of differentiated extravillous trophoblast cells of high purity, which are used as second parental cell population for generation of hybridomas.

ACKNOWLEDGEMENTS

We thank Prof. L. Zardi, Genoa and Dr. D. Blanchard, Nantes for their generous gift of antibodies. Skillful technical and photographical assistance was provided by U. Zahn, M.v. Bentheim, H. Kriegel and G. Bock. In part, this project was supported by the Deutsche Forschungsgemeinschaft (Ka 360/7-3).

REFERENCES

Aboagye-Mathiesen, G., Toth, F.D., Petersen, P.M., Gildberg, A., Norskov Lauritsen, N., Zachar, V. and Ebbesen, P. (1993) Differential interferon production in human first and third trimester trophoblast cultures stimulated with viruses. *Placenta* 14, 225-234.

Amenta, P.S., Gay, S., Vaheri, A. and Martinez Hernandez, A. (1986) The extracellular matrix is an integrated unit: Ultrastructural localization of collagen types I, III, IV, V, VI, fibronectin, and laminin in human term placenta. *Collagen and Related Research* 6, 125-152.

Benirschke, K. and Kaufmann, P. (1995) *The Pathology Of The Human Placenta*. Heidelberg, New York: Springer. Ed. 3.

Burrows, T.D., King, A. and Loke, Y.W. (1993) Expression of integrins by human trophoblast and differential adhesion to laminin or fibronectin. *Human Reproduction* 8, 475-484.

Castellucci, M., Crescimanno, C., Schroeter, C.A., Kaufmann, P. and Mühlhauser, J. (1993) Extravillous trophoblast: immunohistochemical localization of extracellular matrix molecules. In: *Frontiers In Gynecologic And Obstetric Investigation,* (eds.) A.R. Genazzani, F. Petraglia, and A.D. Genazzani, New York: The Parthenon Publishing Group, pp. 19-25.

Damsky, C.H., Fitzgerald, M.L. and Fisher, S.J. (1992) Distribution patterns of extracellular matrix components and adhesion receptors are intricately modulated during first trimester catotrophoblast differentiation along the invasive pathway, in vivo. *J. Clin. Invest.* 89, 210-222.

Douglas, G.C. and King, B.F. (1989) Isolation of pure villous cytotrophoblast from term human placenta using immunomagnetic microspheres. *J. Immunol. Meth.* 119, 259-268.

Douglas, G.C. and King, B.F. (1990) Isolation and morphologic differentiation in vitro of villous cytotrophoblast cells from rhesus monkey placenta. *In Vitro Cell. Dev. Biol.* 26, 754-758.

Duance, V.C. and Bailey, A.J. (1983) Structure of the trophoblast basement membrane. In: *Biology of Trophoblast,* (eds.) Y.W. Loke and A. Whyte, Amsterdam: Elsevier Science Publisher, B.V, pp. 597-625.

Feinberg, R.F., Kliman, H.J. and Lockwood, C.J. (1991) Is oncofetal fibronectin a trophoblast glue for human implantation? *Am. J. Pathol.* 138, 537-543.

Feinman, M.A., Kliman, H.J., Caltabiano, S. and Strauss, J.F. (1986) 8-Bromo-3',5'-adenosine monophosphate stimulates the endocrine activity of human cytotrophoblasts in culture. *J. Clin. Endocrinol. Metab.* 63, 1211-1217.

Frank, H.-G., Malekzadeh, F., Kertschanska, S., Crescimanno, C., Castellucci, M., Lang, I., Desoye, G. and Kaufmann, P. (1994) Immunohistochemistry of two different types of placental fibrinoid. *Acta Anat.* 150, 55-68.

Frank, H.-G., Huppertz, B., Kertschanska, S., Blanchard, D., Roelcke, D. and Kaufmann, P. (1995) Anti-adhesive glycosylation of fibronectin-like molecules in human placental matrix-type fibrinoid. *Histochem. Cell Biol.* 104, 317-329.

Frank, H.-G., Funayama, H., Gaus, G., Ebeling, I., Huppertz, B. and Kaufmann, P. (1996) Hybridisation of extravillous trophoblast I: Selection of a family of HGPRT-negative Jeg-3 mutants as fusion partners. *Placenta,* 17, A.33 (Abstract).

Funayama, H., Gaus, G., Ebeling, I., Takayama, M., Fuezesi, L., Huppertz, B., Kaufmann, P. and Frank, H.-G. Parent cells for trophoblast hybridisation II: AC1 and related trophoblast cell lines, a family of HGPRT-negative mutants of the choriocarcinoma cell line JEG-3. *Trophoblast Res.* 10, 191-201, this volume.

Graham, C.H., Hawley, T.S., Hawley, R.G., MacDougall, J.R., Kerbel, R.S., Khoo, N.K.S. and Lala, P.K. (1993) Establishment and characterization of first trimester human trophoblast cells with extended lifespan. *Exp. Cell Res.* 206, 204-211.

Huppertz, B., Kertschanska, S., Frank, H.-G. and Kaufmann, P. (1995) Immuncytochemische und immunbiochemische Charakteristika des Matrix-Typ Fibrinoid der menschlichen Placenta. *Ann. Anat.* 177 (Suppl), 26. (Abstract).

Huppertz, B., Kertschanska, S., Frank, H.-G., Gaus, G., Funayama, H. and Kaufmann, P. (1996) Extracellular matrix components of placental extravillous trophoblast: Immunocytochemistry and ultrastructural distribution. *Histochem. Cell Biol.* 106, 291-301.

Irving, J.A., Lysiak, J.J., Graham, C.H., Hearn, S., Han, V.K.M. and Lala, P.K. (1995) Characteristics of trophoblast cells migrating from first trimester chorionic villous explants and propagated in culture. *Placenta* 16, 413-433.

Jie, Z., Fey, S.J., Hager, H., Hollsberg, P., Ebbesen, P. and Larsen, P.M. (1990) Markers for human placental trophoblasts in two-dimensional gel electrophoresis. *In Vitro Cell. Dev. Biol.* 26, 937-943.

Kasper, M., Moll, R. and Stosiek, P. (1988a) Distribution of intermediate filaments in human umbilical cord: Unusual triple expression of cytokeratins, vimentin, and desmin. *Zoologisches Jahrbuch der Anatomie* 117, 227-233.

Kasper, M., Stosiek, P. and Karsten, U. (1988b) Coexpression of cytokeratins and vimentin in hyaluronic acid-rich tissues. *Acta Histochem.* 84, 107-108.

Kaufmann, P. and Castellucci, M. (1997) Extravillous trophoblast in the human placenta. *Trophoblast Res.* 10, 21-65, this volume.

King, B.F. and Blankenship, T.N. (1994) Ultrastructure and development of a thick basement membrane-like layer in the anchoring villi of macaque placentas. *Anat. Rec.* 238, 498-506.

Kliman, H.J., Nestler, J.E., Sermasi, E., Sanger, J.M. and Strauss, J.F. (1986) Purification, characterization, and in vitro differentiation of cytotrophoblasts from human term placentae. *Endocrinology* 118, 1567-1582.

Kliman, H.J. and Feinberg, R.F. (1990) Human trophoblast-extracellular matrix (ECM) interactions in vitro: ECM thickness modulates morphology and proteolytic activity. *Proc. Natl. Acad. Sci. USA* 87, 3057-3061.

Lang, I., Hartmann, M., Blaschitz, A., Dohr, G., Kaufmann, P., Frank, H.-G., Hahn, T., Skofitsch, G. and Desoye, G. (1994) Differential lectin binding to the fibrinoid of human full- term placenta: Correlation with a fibrin antibody and the PAF-Halmi method. *Acta Anat.* 150, 170-177.

Lei, K.J., Gluzman, Y., Pan, C.J. and Chou, J.Y. (1992) Immortalization of virus-free human placental cells that express tissue-specific functions. *Mol. Endocrinol.* 6, 703-712.

Lewis, M.P., Clements, M., Takeda, S., Kirby, P.L., Seki, H., Lonsdale, L.B., Sullivan, M.H.F., Elder, M.G. and White, J.O. (1996) Partial characterization of an immortalized human trophoblast cell-line, TCL-1, which possesses a CSF-1 autocrine loop. *Placenta* 17(2-3), 137-146.

Lindl, T. and Bauer, J. (1989) *Zell- und Gewebekultur.* Stuttgart New York: Gustav Fischer Verlag. Ed. 2,

Loke, Y.W. (1983) Human trophoblast in culture. In: *Biology of Trophoblast,* (eds.) Y.W. Loke and A. Whyte, Amsterdam: Elsevier Scientific Publishers, B.V., pp. 663-701.

Loke, Y.W. and Burland, K. (1988) Human trophoblast cells cultured in modified medium and supported by extracellular matrix. *Placenta* 9, 173-182.

Loke, Y.W., Gardner, L. and Grabowska, A. (1989a) Isolation of human extravillous trophoblast cells by attachment to laminin-coated magnetic beads. *Placenta* 10, 407-415.

Loke, Y.W., Gardner, L., Burland, K. and King, A. (1989b) Laminin in human trophoblast-decidua interaction. *Hum. Reprod.* 4, 457-463.

Loke, Y.W. (1990) Experimenting with human extravillous trophoblast: A personal view. *Am. J. Reprod. Immunol.* 24, 22-28.

Old, R.W. and Primrose, S.B. (1992) *Gentechnologie.* Stuttgart: Georg Thieme Verlag.

Perkins, A.V. and Linton, E.A. (1995) Identification and isolation of Corticotrophin-Releasing Hormone-positive cells from the human placenta. *Placenta* 16, 233-244.

Toth, F.D., Juhl, C., Norskov Lauritsen, N., Mosborg Petersen, P. and Ebbesen, P. (1990) Interferon production by cultured human trophoblast induced with double stranded polyribonucleotide. *J. Reprod. Immunol.* 17, 217-227.

Yagel, S., Casper, R.F., Powell, W., Parhar, R.S. and Lala, P.K. (1989) Characterization of pure human first-trimester cytotrophoblast cells in long-term culture: Growth pattern, markers, and hormone production. *Am. J. Obstet. Gynecol.* 160, 938-945.

Trophoblast Research 10:191-201, 1997

PARENT CELLS FOR TROPHOBLAST HYBRIDIZATION II: AC1 AND RELATED TROPHOBLAST CELL LINES, A FAMILY OF HGPRT-NEGATIVE MUTANTS OF THE CHORIOCARCINOMA CELL LINE JEG-3

Hitoshi Funayama[1,2], Gabriele Gaus[1], Ingke Ebeling[1], Masaomi Takayama[2], Laszlo Füzesi[3], Berthold Huppertz[1], Peter Kaufmann[1] and Hans G. Frank[1,4]

[1]Department of Anatomy
Technical University
Wendlingweg 2
D-52057 Aachen, Germany

[2]Department of Obstetrics and Gynecology
Medical College Hospital
6-7-1 Nishishinyuku Shinyuku-ku
160 Tokyo, Japan

[3]Department of Pathology
Technical University
Pauwelsstrasse 30
D-52057 Aachen, Germany

INTRODUCTION

This manuscript deals with the second prerequisite, which is necessary for immortalization of extravillous trophoblast by fusion (see Gaus et al., this volume).

Since the matrix-secreting extravillous trophoblast cells do not proliferate at all, immortalization preferably by hybridization with suitable malignant tumor cells, i.e. choriocarcinoma cells, is presumed to offer a suitable alternative. Successful examples for this technology comprise the classical hybridoma technique for production of monoclonal antibodies (cf., Winter and Milstein, 1991) and the fusion of trophoblastic cells (JEG-3) for analysis of endocrine secretion patterns in human:mouse and human:human hybridomas (Kohler et al., 1981; Bordelon et al., 1976; Hardin et al., 1983).

A major requirement for any hybridoma culture is the specific elimination of all unfused tumor cells which due to their high proliferative capacity would rapidly overgrow the slower proliferating hybridomas. This can successfully be done with tumor cell mutants which are deficient for the enzyme hypoxanthine-guanine-phosphoribosyl transferase (HGPRT-negative mutants). Unfused mutant cells can easily be eliminated after hybridization by azaserine treatment whereas the successful hybridomas compensate the HGPRT deficiency by the normal activity of the HGPRT enzyme of the wild type fusion partner and thus are resistant to azaserine. The HGPRT gene is located

[4]To Whom Correspondence Should Be Addressed.

on the X chromosome and functional defects are often associated with X-chromosomal aberrations (Cox and Masson, 1978).

As a consequence, for our hybridization experiments as fusion partners (a) primary invasive, non proliferating extravillous trophoblast cells (see Gaus et al., this volume) and (b) HGPRT-negative choriocarcinoma cells are needed. In preparation for fusion, we describe herein the production, selection and cloning of a family of HGPRT-negative JEG-3 mutants. In one of the cloned lines no cells did show signs of production of marker molecules of Matrix-Type Fibrinoid (MTF). Due to their homogenous negativity for MTF markers these cells seem to be particularly suitable fusion partners for the generation of hybridomas secreting placental oncofetal matrix molecules. Cytogenetic analysis was focused on this cell line.

MATERIALS AND METHODS

Mutagenesis

JEG-3 cells (American Type Culture Collection; ATCC, Rockville, Maryland, USA) were propagated for five passages in normal culture medium (Table 1) in an atmosphere containing 5% CO_2, 95% air and 100% humidity at 37°C. In order to increase the number of mutants, the cells were treated with the chemical mutagen methane-sulfonic acid methyl ester (Sigma M 4016) at a concentration of 1.25 µM for 1 hour. This concentration was determined empirically in pilot experiments as the dosage showing a 50% survival rate of cells 24 hours after treatment was stopped.

Selection

After treatment with the mutagen, the cells were allowed to stabilize for two passages and then cultivated in an atmosphere containing 5% CO_2, 90% N_2, a controlled level of 5% O_2, at 100% humidity and at 37°C. Selection was started in selection medium 1 (Table 1) with an initial concentration of 5 µM 6-thioguanine. This agent is a cytostatic and cytotoxic guanine analogon, which is able to substitute guanine or hypoxanthine as intermediates for RNA and DNA synthesis. Since effective substitution depends on functionally active HGPRT enzyme, it can only be incorporated into nucleic acids by the wild type cells, which are HGPRT positive. The 6-thioguanine incorporating cells decease soon so that only the HGPRT-negative mutants survive treatment. Concentration of 6-thioguanine was raised stepwise from passage to passage during 25 passages (during the first ten passages 3-5 µM per step, at higher passage numbers up to 15 µM per step) to reach a final concentration of 150 µM. If the death rate at a new dosage level of 6-thioguanin raised to a critical level (at least 70% of cells dead after 12 hours), the exposure to 6-thioguanine was interrupted for one passage in normal culture medium to allow recovery of the cells. After recovery the cells were reexposed to the same dosage until stable growth was achieved.

JEG-3 cells express alkaline phosphatases (Edlow et al., 1975; Ogata et al., 1988; Takami et al., 1988; Wada and Chou 1993; Takami et al., 1992) which are known to cause resistance to mercaptopurines like 6-thioguanine (Efferth and Volm, 1993). Three strategies were used to avoid such a resistance as well as any other type of resistance to guanine analoga.

Table 1

The different media and their compositions. Normal medium was composed of Ham´s F12 as basic solution with the addition of 100 U/ml Penicillin G, 100 µg/ml Streptomycin, 2.5 µg/ml Amphotericin B, 25 mM N-2-hydroxyethyl-piperazine-N´-2-ethane-sulphonic acid (HEPES; pH 7.2) and 10% (v/v) fetal calf serum.

Media	Basic Solution	Additives
Selection medium 1	normal medium	6-thioguanine (150 µM) levamisol (10 µM)
Selection medium 2	normal medium	8-azaguanine (130 µM) levamisol (10 µM)
Selection medium 3	normal medium	azaserine (5.7 µM) hypoxanthine (100 µM)
Selection medium 4	normal medium	azaserine (5.7 µM) hypoxanthine (100 µM) bromodeoxyuridine (100 µM)

a) The inhibitor of alkaline phosphatase, levamisole, was added to selection medium 1.

b) At the end of the selection procedure, the cells were cultured for three passages in selection medium 2 containing 8-azaguanine (Table 1). Since alkaline phosphatases do lead to resistance to mercaptopurines, but not to aza-analoga of purines, this treatment was designed to eliminate mutants with overexpression of alkaline phosphatase as main mechanism of resistance.

c) We have cultured the cells for five days in selection medium 3 (Table 1). HGPRT-negative cells are not able to produce substantial amounts of DNA under such conditions. To remove all cells still escaping the suppressive effects of selection medium 3, we have cultured the cells for 24 hours in selection medium 4. Selection medium 4 (Table 1) differs from selection medium 3 only by the addition of bromodeoxyuridine as a toxic substance analogue to thymidine. Prolonged treatment with this medium kills all cells still synthesizing relevant amounts of DNA, i.e. all cells that escape the suppression by selection medium 3.

After this treatment the cells were cultured again in selection medium 1. The resulting population of cells was designated as AC1 and cultured routinely in selection medium 1. The treatment as described above under b and c is repeated every 15th to 20th passage.

Cloning

AC1 cells were seeded at a concentration of 2-3 cells per well diluted in 100 µl selection medium 1 per well in 96-well plates. The cells were incubated for seven days in the low oxygen atmosphere (see above) before first inspection. Inspection of all wells was performed in weekly intervals. Fifty µl selection medium 1 was added each week. Every two weeks the medium was replaced by 100 µl fresh selection medium 1. Growing clones were detected by inspection. Only wells showing rapid growth of a single colony were selected for further evaluation. Three clones (AC1-1, AC1-5 and AC1-9) resulted from this step. The treatment for removal of undesired cell types (see above under 2.2 b and c) was done with each of the clones as soon as enough cells were available. Throughout the processes of selection (2.2.) and cloning (2.3.) sample aliquots of the cells were frozen in liquid nitrogen.

Suppression Of Proliferation In Selection Medium 3

JEG-3, AC1, AC1-1, AC1-5 and AC1-9 cells to be tested for suppression of proliferation by selection medium 3 were seeded in 8-chamber slides (Nunc, Germany). For the first 24 h they received normal culture medium (Table 1). After 24 hours half of the chambers was exposed to selection medium 3. After five days of culture all media were spiked and pulse labeled with bromodeoxyuridine (BrdU, 100 μM) for one hour. After this labeling the slides were washed with Ham´s F12 without serum, fixed in neutrally buffered formaldehyde (4%; 0.1M phosphate buffer) for 15 minutes, washed with PBS and air dried. The slides were either used directly for immunohistochemical detection or stored for maximally 4 weeks at -20°C. Before start of the immunohistochemical analysis, the slides were pretreated with microwaves in aqua dest. (600 W 3 x 5 minutes) to denature DNA. BrdU was detected with a monoclonal antibody (clone Bu20a; dilution 1:75; Dako, Denmark) and using a standardized histochemical sequence based on the streptavidin-biotin procedure (AC-Kit, for detailed description see Frank et al., 1994, with streptavidine-texas red as fluorochrome). In a second series of experiments the suppressive effect of selection medium 3 was monitored for the cell line AC1-1. After 24 hours of culture in normal medium (Table 1), a series of culture bottles containing AC1-1 cells were exposed to selection medium 3. Samples of the cells were inspected and passaged in 48 hour intervals, the cells were counted and viability was estimated by trypan blue exclusion (Lindl and Bauer, 1989).

Detection Of Matrix Proteins

To detect in parallel intracellular occurrence of MTF-markers (whether secreted or not), insoluble matrix deposits as well as inhomogeneities of cells within a cell line, we have preferred immunohistochemical detection. Cells (JEG-3, AC1, AC1-1, AC1-5 and AC1-9) were seeded in 8-chamber slides (Nunc, Germany). Shortly before the culture reached confluence (at about 60 - 70% of culture area covered by cells) the slides were washed in Ham´s F 12 (without serum). The slides were fixed for 15 minutes in neutrally buffered formaldehyde (4%; 0.1M phosphate buffer). Finally, all slides were treated with acetone for 10 minutes at -20°C.

As marker molecules for the production of MTF, the presence of fibronectins (antibody IST-9, dilution 1:4, Prof. L. Zardi, Genoa, Italy, (Carnemolla et al., 1987) and oncofetal fibronectins (antibody BC-1, dilution 1:4, Prof. L. Zardi, Genoa, Italy, (Carnemolla et al., 1989; Carnemolla et al., 1987); and antibody FDC-6 as culture supernatant of clone FDC-6, ATCC, Rockville, Maryland, USA, dilution 1:5, (Feinberg et al., 1995) was examined. Anti-i (D. Blanchard, Nantes, [Blanchard et al., 1992]) was used as an additional antibody recognizing oncofetally glycosylated fibronectin-like molecules (Frank et al., 1995). We have used a standardized histochemical protocol based on the biotin-streptavidin procedure (AC-Kit; description see Frank et al., 1994, with streptavidine-texas red as fluorochrome).

Table 2

Overview Cell Lines

Cell lines	Proliferation		Immunoreactivity			
	Incorporation of BrdU in:		Antibodies used for screening of MTF marker molecules			
	Selection medium 1	Selection medium 3	IST-9	BC-1	FDC-6	anti-i
JEG-3	+ [(1)]	+	-	-	-	-
AC-1	+	-	-	-	-	-
AC1-1	+	-	-	-	-	-
AC1-5	+	-	+ [(2)]	-	+ [(2)]	-
AC1-9	+	-	+ [(2)]	-	+ [(2)]	-

[(1)] The cells proliferate, but die within 48 hours by incorporation of 6-thioguanine
[(2)] Weak, spot like reactivity could be seen in a very small minority of cells (below 5% of total cells)

Cytogenetic Analysis

From the original choriocarcinoma cell line (JEG-3) and the clone AC1-1, chromosome preparations were made. Cells were treated during logarithmic growth for 1-2 h with 100 ng/ml colcemid. After treatment the cells were harvested by 0.25% trypsin in Hank's balanced salt solution and resuspended in hypotonic KCl (75 mM). The cells were sedimented at 400 g for 5 minutes. The supernatant was removed and the sediment was fixed (methanol/glacial acetic acid 3:1) for 24 hours at 4°C. Thereafter the cells were suspended in fresh fixative and 1-2 drops of suspension given onto slides wetted with a film of aqua dest. The slides were air dried, trypsinized (2 seconds) with 0.2% trypsin in physiological saline, stained with Giemsa´s solution (5%, 5 min., room temperature). Chromosomal aberrations were classified according to ISCN (1995).

RESULTS AND DISCUSSION

The experiments described above resulted in the new choriocarcinoma cell line AC1 and its subclones AC1-1, AC1-5 and AC1-9. These cell lines showed normal proliferative behavior in culture media containing 6-thioguanine (BrdU incorporation rate was 68%). Their proliferation was completely stopped when cultured in media containing azaserine and hypoxanthine (Figure 1a-d; Table 2). AC1-1 was chosen for further hybridization experiments, since with AC1-1 we have a homogenously MTF-negative and cloned line available (Table 2). In the cell lines AC1-5 and AC1-9 two antibodies against MTF marker molecules were focally positive both intracellularly and

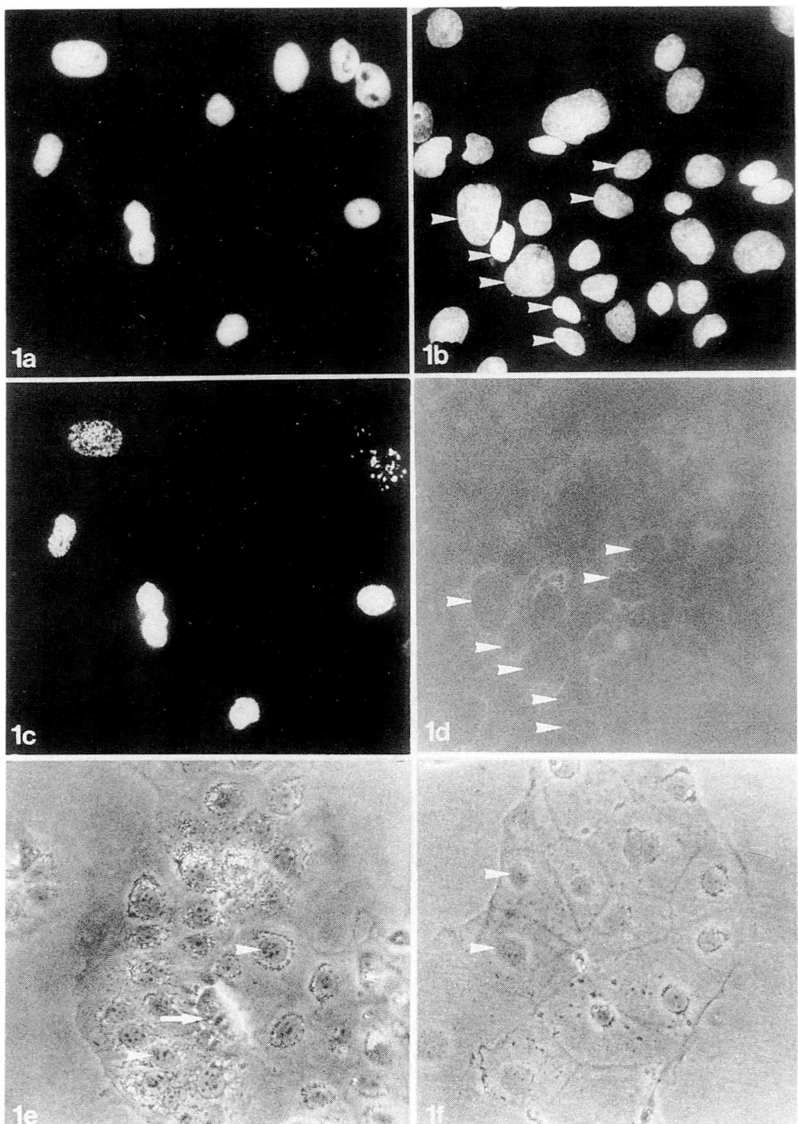

Figure 1. Cell line AC1-1. (a, c, e) AC1-1 cultured in selection medium 1. (b, d, f) AC1-1 after five days of culture in selection medium 3. (a, b) Fluorescent staining of all nuclei with bisbenzimide. (c, d) Immunohistochemical detection of incorporated bromodeoxyuridine in the same nuclei as shown in a and b. Exposure time was 62 seconds (c) and 320 seconds (d). The very long exposure time in d was necessary to demonstrate the nuclei as negative in the surrounding background fluorescence. Arrowheads mark some corresponding nuclei in b and d. Comparison of c and d demonstrates suppression of DNA-synthesis in selection medium 3. (e, f) Phase contrast morphology. (e) The cells show nuclei with prominent nucleoli (arrowhead). A mitotic cell (arrow) can be seen. (f) Flattened cells with pronounced appearance of the cellular border. Nuclei (arrowheads) without prominent nucleoli are fading out. (X 325)

in surrounding extracellular spots (below 5% of total cells positive; Table 2). Interpreting this reactivity as a sign for some kind of differentiation, it may be speculated that such expression can be due to the action of activated tumor suppressor genes or repressed oncogene products (Levine, 1993). However, independent from any search for reasons such a heterogenous reactivity correlates with a somehow unstable phenotype or even genotype. There is some danger, that the cells import such instability into fusion products. Therefore, we prefer the homogenously reacting monoclonal cell line AC1-1 for fusion experiments.

During prolonged exposure of all HGPRT negative cell lines to selection medium 3 we found severe changes in phase contrast morphology. Although the cells remained attached, they flattened. The number of cells showing nuclei with clearly visible nucleoli was decreasing constantly (Figure 1 e, f). The border between nuclei and cytoplasm was fading out (Figure 1 e, f). Number of cells as well as viability were decreasing constantly. After twenty days of culture in medium containing azaserine and hypoxanthine no viable cells were present. These time dependent changes clearly document the action of selection medium 3. Their slowly progressive nature is in accordance with the substantial lack of nucleic acid precursors, which is the main effect of selection medium 3 on HGPRT negative cells. The cells can be negatively selected by azaserine and hypoxanthine and thus are suitable for hybridization experiments, in which only the HGPRT-positive extravillous trophoblast cells (although not proliferating) as well as hybridomas (AC1-1/extravillous trophoblast) will survive in selection medium 3.

Since the complete cytogenetical analysis is beyond the scope of this short report we will focus only on those aspects, which are a) essential for the hybridoma project and b) which could be detected in all cells.

In this sense, the cytogenetical analysis (Figure 2a, b) confirms the close relationship between JEG-3 and AC1-1. The 5 marker chromosomes mar1-5 were constantly observed in JEG-3 as well as in AC1-1. The most significant differences between the two lines were the loss of the translocation t(4;11) (p16; q13) and the presence of a new marker chromosome (mar6) in AC1-1 presumably involving X chromosome (Figure 2b). The latter information is important since the HGPRT-gene is known to be located on the X chromosome (Cox and Masson, 1978).

The differences make cells of AC1-1 particularly suitable as fusion partners since the presence of a normal X-chromosome in hybridomas proves successful hybridization with a wild type cell i.e. an extravillous trophoblast cell. Vice versa, the large number of marker chromosomes of AC1-1 opens a good opportunity to prove the participation of at least one AC1-1 cell in a fusion by conventional cytogenetic analysis of the fusion product.

Meanwhile, our group has refined a technique to isolate primary invasive extravillous trophoblast cells (Gaus et al., this volume). Using these pure extravillous trophoblast preparations and the above choriocarcinoma cell line AC1-1, we have performed pilot hybridization studies yielding viable hybridomas.

Further steps include (a) optimization of the fusion protocols, (b) selection of long-term *in vitro* proliferating hybridomas, (c) ELISA- based screening of supernatants and selection of a variety of monoclonal MTF-secreting hybridomas, (d) detailed

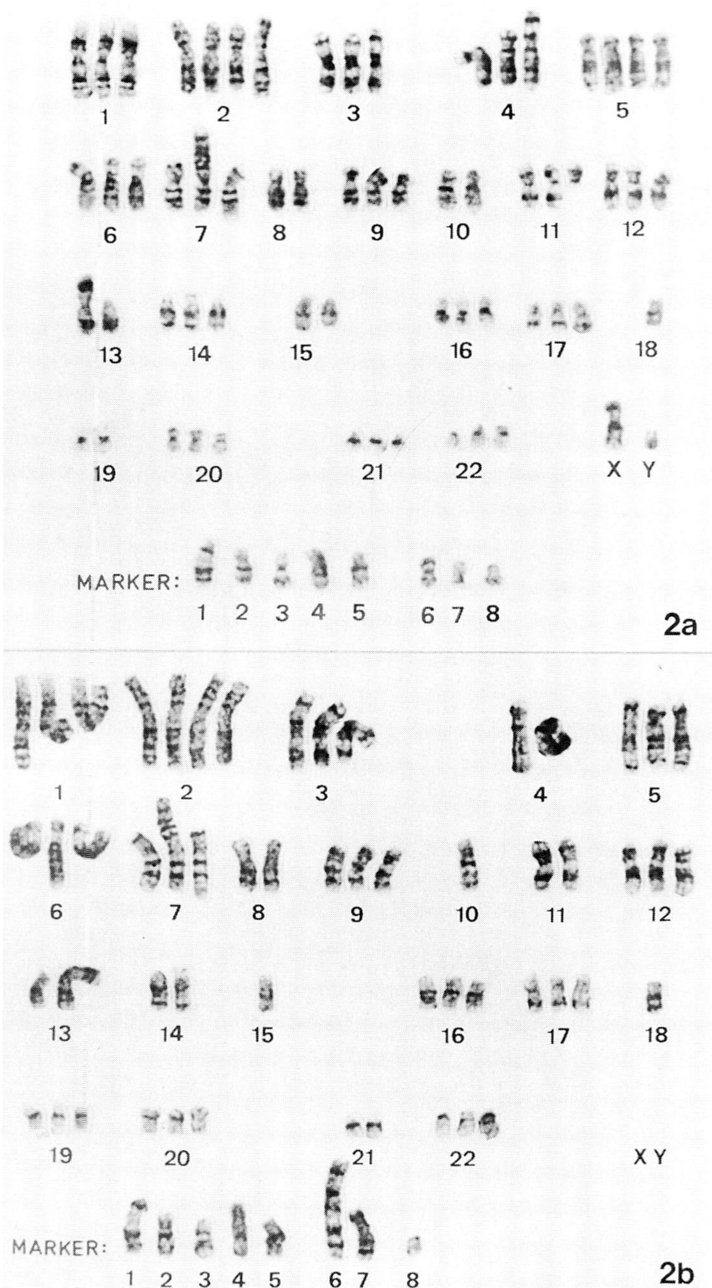

Figure 2. Representative karyotypes of JEG-3 and AC1-1.
(a) JEG-3. 71,X,Y,+1,+2,+2,+3,+t(4;11)(p16;q13),+5,+5,+6,add(7)(q36),+add(7)(p22),+9, -10,+12,i(13)(p10),+14,+16,+17,-18,+20,+21,+22,+1~8mar
(b) AC1-1. 63,-X,-Y,+1,+2,+2,+3,+5,+6,add(7)(q36),+add(7)(p22),+9,-10,+12,i(13)(p10), -15,+16,+17,-18,+19,+20,+22,+1~8mar

characterization and mapping of supernatants of monoclonal MTF-secreting hybridomas for other so far unknown proteins of MTF by appropriate 2D-electrophoretical techniques and (e) mapping of tumorigenicity in nude mice experiments with parallel cytogenetic analysis in order to identify locations of putative trophoblast-relevant tumor suppressor genes.

In conclusion, AC1-1 is a sufficiently characterized fusion partner which fulfills all requirements for the immortalization of human invasive extravillous trophoblast cells in order to study their extracellular matrix production *in vitro*.

SUMMARY

To gain analytical and productive access to the oncofetally modified molecules in the extracellular matrix of human non-proliferative invasive trophoblast cells, we intend to immortalize these cells together with their productive capacity by fusion with their malignant trophoblast counterpart. To generate a suitable malignant fusion partner we have selected hypoxanthine-guanosin-phosphoribosyl-transferase-negative mutants of the choriocarcinoma cell line JEG-3. We obtained the cell line AC1 and its subclones AC1-1, AC1-5 and AC1-9. AC1-1 is our preferred candidate for fusion with matrix-producing invasive trophoblast cells since this subclone homogenously did not react with antibodies recognizing trophoblast matrix molecules.

ACKNOWLEDGEMENTS

We thank Prof. L. Zardi, Genoa and Dr. D. Blanchard, Nantes for their generous gift of antibodies. Skillful technical and photographical assistance was provided by U. Zahn, M. v. Bentheim and G. Bock. For fruitful discussions and corrections of the manuscript we thank Dr. G. Kosanke and B. Gunawan. We gratefully acknowledge the language editing by M. Franssen, M.A. In part, this project was granted by the German Research Foundation (Ka 360/7-3).

REFERENCES

Blanchard, D., Bernard, D., Loirat, M.J., Frioux, Y., Guimbretière, J. and Guimbretière, L. (1992) Caractérisation d'anticorps monoclonaux murins dirigés contre les érythrocytes foetaux. *Rev. Fr. Transfus. Hémobiol.*, 35, 239-254.

Bordelon, M.R., Coon, H.G. and Kohler, P.O. (1976) Human glycoprotein hormone production in human-human and human-mouse somatic cell hybrids. *Exp. Cell Res.*, 103, 303-310.

Carnemolla, B., Borsi, L., Zardi, L., Owens, R.J. and Baralle, F.E. (1987) Localization of the cellular-fibronectin-specific epitope recognized by the monoclonal antibody IST-9 using fusion proteins expressed in E. coli. *FEBS Lett.* 215(2), 269-273.

Carnemolla, B., Balza, E., Siri, A., Zardi, L., Nicotra, M.R., Bigotti, A. and Natali, P.G. (1989) A tumor-associated fibronectin isoform generated by alternative splicing of messenger RNA precursors. *J. Cell Biol.* 108, 1139-1148.

Cox, R. and Masson, W.K. (1978) Do radiation-induced thioguanine resistant mutants of cultured mammalian cells arise by HGPRT gene mutation or X-chromosome rearrangement? *Nature* 276, 629-630.

Edlow, J.B., Ota, T., Relacion, J.R., Kohler, P.O. and Robinson, J.C. (1975) Enzymes of normal and malignant trophoblast: Phosphoglucose isomerase, phosphoglucomutase, hexokinase, lactate dehydrogenase, and alkaline phosphatase. *Am. J. Obstet. Gynecol.* 121, 674-681.

Efferth, T. and Volm, M. (1993) Increase of alkaline phosphatase in multidrug resistant tumor cells and their cross-resistance to 6-thioguanine. *Drug Res.* 43(II), 1118-1121.

Feinberg, R.F., Kliman, H.J., Bedian, V., Monzon-Bordonaba, F., Menzin, A.W. and Wang, C.L. (1995) Oncofetal fibronectin: Antibodies old and new. *J. Soc. Gynecol. Invest.* 2, 180. (Abstract)

Frank, H.-G., Malekzadeh, F., Kertschanska, S., Crescimanno, C., Castellucci, M., Lang, I., Desoye, G. and Kaufmann, P. (1994) Immunohistochemistry of two different types of placental fibrinoid. *Acta Anat.* 150, 55-68.

Frank, H.-G., Huppertz, B., Kertschanska, S., Blanchard, D., Roelcke, D. and Kaufmann, P. (1995) Anti-adhesive glycosylation of fibronectin-like molecules in human placental matrix-type fibrinoid. *Histochem. Cell Biol.* 104, 317-329.

Gaus, G., Funayama, H., Huppertz, B., Kaufmann, P. and Frank, H.-G.(1997) Parent cells for trophoblast hybridisation I: Isolation of extravillous trophoblast cells from human term chorion laeve. *Trophoblast Res.*10, 181-190, this volume.

Hardin, J.W., Riser, M.E., Trent, J.M. and Kohler, P.O. (1983) The chorionic gonadotropin alpha-subunit gene is on human chromosome 18 in JEG cells. *Proc. Natl. Acad. Sci. USA* 80, 6282-6285.

ISCN (1995) *An International System For Human Cytogenetic Nomenclature.* Basel: S. Karger.

Kohler, P.O., Riser, M., Hardin, J., Boothby, M., Boime, I., Norris, J. and Siciliano, M.J. (1981) Chorionic gonadotropin synthesis and gene assignment in human: mouse hybrid cells. *Adv. Exp. Med. Biol.* 138, 405-418.

Levine, A.J. (1993) The tumor suppressor genes. *Ann. Rev. Biochem.* 62, 623-651.

Lindl, T. and Bauer, J. (1989) *Zell- und Gewebekultur.* Stuttgart New York: Gustav Fischer Verlag. Ed. 2.

Ogata, S., Hayashi, Y., Takami, N. and Ikehara, Y. (1988) Chemical characterization of the membrane-anchoring domain of human placental alkaline phosphatase. *J. Biol. Chem.* 263, 10489-10494.

Takami, N., Ogata, S., Oda, K., Misumi, Y. and Ikehara, Y. (1988) Biosynthesis of placental alkaline phosphatase and its post-translational modification by glycophospholipid for membrane-anchoring. *J. Biol. Chem.* 263, 3016-3021.

Takami, N., Oda, K. and Ikehara, Y. (1992) Aberrant processing of alkaline phosphatase precursor caused by blocking the synthesis of glycosylphosphatidylinositol. *J. Biol. Chem.* 267, 1042-1047.

Wada, N. and Chou, J.Y. (1993) Characterization of upstream activation elements essential for the expression of germ cell alkaline phosphatase in human choriocarcinoma cells. *J. Biol. Chem.* 268, 14003-14010.

Winter, G. and Milstein, C. (1991) Man-made antibodies. *Nature* 349, 293-299.

ANGIOGENESIS AND BLOOD VESSELS

Trophoblast Research 10:205-213, 1997

STRUCTURE AND PERMEABILITY IN HUMAN PLACENTAL CAPILLARIES
-A Review-

J. Anthony Firth[1] and Lopa Leach[2]

[1]Department of Anatomy and Cell Biology
Imperial College School of Medicine at St. Mary's
London W2 1PG, United Kingdom

[2]Department of Human Morphology
University of Nottingham
Queen's Medical Centre
Nottingham NG7 2UH, United Kingdom

INTRODUCTION

The human placental barrier at term consists of two continuous layers, the syncytiotrophoblast and the endothelium. Both of these are of embryonic origin, and both lie in the path of any molecule crossing the placenta in either direction between the maternal and fetal circulations. However, these two barriers have not attracted equal interest from investigators of transplacental transport and permeability. Indeed, most studies tend to assume that any interesting contributions to total placental transport depend on the syncytiotrophoblast. It is easy to understand the steps in the argument through which this view has become established. The syncytiotrophoblast is by definition a syncytium and therefore presumed devoid of paracellular channels; all syncytial transport must therefore be transcellular, dependent on specific pathways and subject to regulation. The endothelium is very thin, endothelial cells are separated by short paracellular clefts which lack proper junctional complexes, and in addition are packed with components of a vesicular-caveolar system which is well established as a major transendothelial transport shuttle.

It is now evident that at least two elements of this argument are seriously defective. There is good physiological evidence for the existence of extracellular, water-filled pathways across the human and other hemochorial placentae (cf., Stulc, 1989), though it is not yet clear whether these depend mainly on fibrin-filled syncytial deficits (Nelson et al., 1990) or on transtrophoblastic channels (Kaufmann et al., 1982; Kertschanska and Kaufmann, 1992) (Figure 1). The long-standing assumption that the endothelial caveolae and vesicles form the predominant pathway for transport of water and solutes across non-fenestrated (continuous) endothelia has now succumbed to a rising tide of both physiological and morphological evidence that this is unlikely (Bundgaard, 1980; Crone, 1984; Frøkjær-Jensen, 1990) and of evidence for a quite different function for plasma membrane caveolae (Anderson, 1993).

The aim of this review is to draw attention to the growing evidence that the permeability of continuous capillaries in the placenta and elsewhere depends largely on the detailed organization of the intercellular junctions in the endothelial paracellular clefts, and to point out that this pathway may be significant in placental permeability as a whole.

Some Tight Junctions Are Tighter Than Others

Viewed from the standpoint of organization rather than cell lineage, endothelia are a variation on the epithelial theme. The cells form a closely adherent monolayer on a basement membrane, their adjoining plasma membranes bounding paracellular clefts with a fairly uniform width of just under 20 nm. As with epithelia, the clefts are the sites of specialized intercellular junctions of the zonular type, known as junctional complexes, which surround the entire apical part of the cells and attach them to all of their neighbors; the topology of these zonular junctions has been aptly compared by Diamond (1978) to the plastic band which holds together a six-pack of beer. Within the epithelial junctional complex there is a clear distinction between a luminal band, the zonula occludens or tight junction, in which there appear in thin sections to be multiple contacts between the plasma membranes, and a more abluminal band, the zonula adherens or adhesion belt, in which the membranes maintain station at a near-20 nm separation but show both intercellular and intracellular specializations (Farquhar and Palade, 1963). In endothelia the paracellular clefts tend to be very short so that the junctional complex occupies a large proportion of the intercellular cleft, and the adhesion belt is intermingled with the tight junctions rather than entirely abluminal to them (Franke et al., 1988; Leach and Firth, 1992; Schulze and Firth, 1992a).

Studies on epithelia have uniformly identified restriction of paracellular permeability with the tight junctions. Epithelial tight junctions consist of anastomozing linear contacts between adjacent plasma membranes which effectively form zippers joining the subapical region of each cell to its neighbors; an excellent recent review is provided by Citi (1993). The electrical resistance and permeability to small molecules of such junctions depends among other things on the number of linear contacts separating the luminal and abluminal parts of the paracellular cleft (Claude, 1978; Claude and Goodenough, 1973; Martinez-Palomo and Erlij, 1975; Stevenson et al., 1988), and there is still scope for debate as to whether the membrane junctional molecules are proteins or lipids (Furuse et al., 1993; Kachar and Reese, 1982; Kan, 1993), but there is no doubt that undamaged, mature epithelial tight junctions are highly selective to crystalloid solutes and impermeable to macromolecules (Gumbiner, 1993). The epithelial adhesion belt appears to be a cytoskeleton-linked region, rich in adhesion molecules of the cadherin family, which determines the assembly and maintenance of the tight junctions but which of itself is not a limiting factor in the permeability of the epithelial paracellular cleft (Geiger et al., 1985; Grunwald, 1993).

Most Endothelial Tight Junctions Are Leakier

The only endothelial tight junctions which appear comparable in structure and permeability to those of epithelia are found in the capillaries of the blood-brain barrier. These endothelia have high electrical resistances and are impermeable to macromolecules and selectively permeable to crystalloids; they have continuous, multistranded tight junctions with close contact between the membranes along the length of each junctional strand (Brightman and Reese, 1969; Schulze and Firth, 1992b). With this exception, the organization and permeability of endothelial tight junctions differ considerably from those of their epithelial counterparts.

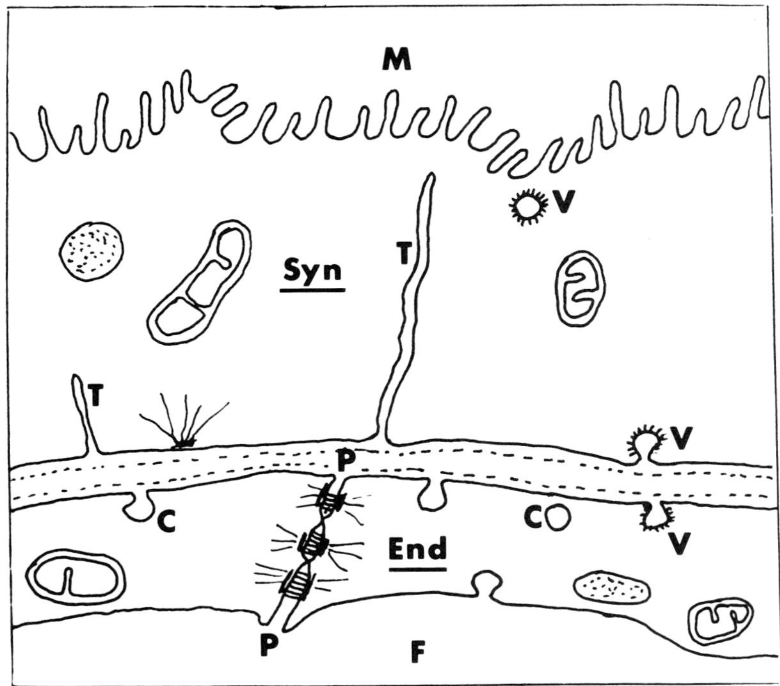

Figure 1. Structures likely to be related to permeability in the term placental barrier. The maternal intervillous space (M) is separated from the fetal capillary lumen (F) by the syncytiotrophoblast (Syn) and the endothelium (End), which are separated by their basal laminae and a variable but usually very thin layer of connective tissue matrix. Receptor-mediated transport across the syncytiotrophoblast involves endocytosis by clathrin-coated vesicles (V), whereas the tubules (T) may not reach the maternal surface normally but may do so under conditions of bulk flow (such as raised fetal venous pressure). In the endothelium the clathrin-coated vesicles may transcytose certain specific solutes (such as Immunoglobulin G), but are greatly outnumbered by caveolae (C) which may have a limited role in transfer of very large solutes but are still poorly understood. The bulk of transendothelial solute and water transfer is probably through the paracellular clefts (P-P); the organization of their tight junctions and adhesion belts is shown in Figure 2.

Perfusion of placental, cardiac or muscle capillaries with electron microscopical tracers of various sizes, such as hemepeptides, shows that the paracellular clefts are permeable to these macromolecules up to a molecular mass of 20-40 KDa, corresponding to a diameter of about 6 nm (Sibley et al., 1982; Ward et al., 1988; Wissig, 1979). Measurement of the permeability-surface area product of human placental capillaries shows restriction of cyanocobalamin (diameter 1.7 nm), and coupled with morphometic estimation of capillary area gives absolute permeabilities for Cr-EDTA and cyanocobalamin slightly lower than those for muscle or cardiac capillaries (Eaton et al., 1993).

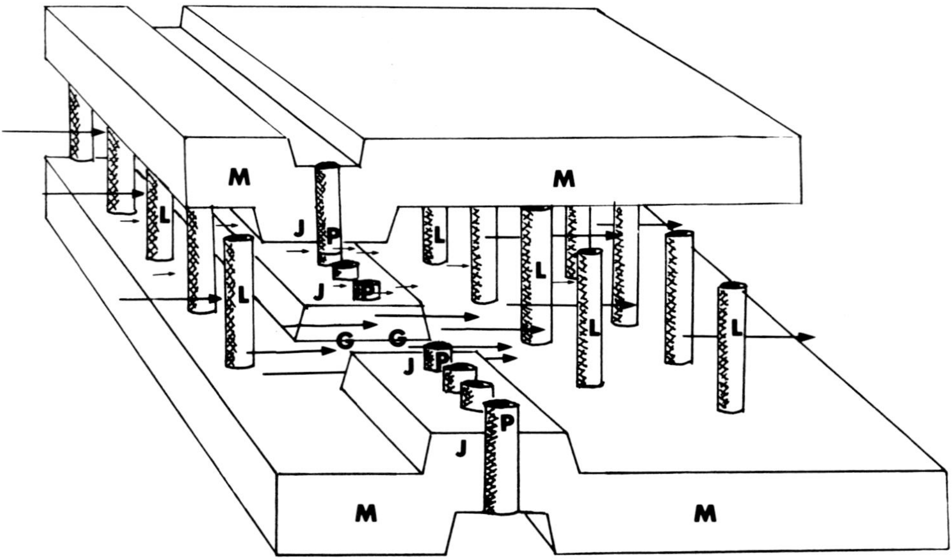

Figure 2. Model of part of a tight junctional strand and adjacent adhesion belt from a human placental capillary paracellular junction. Membranes (M) show linear close approach at tight junctional strand (J) where they are held together by intramembrane particles (P). Junctional strands are interrupted by gaps (G) which provide pathways for water and solute flow. More widely spaced membranes in adhesion belt regions are connected by linkers (L), thought to be rich in VE-cadherin, which hold cells together and may serve as a fiber matrix restricting size of permeant molecules. Large arrows indicate pathways for water and solutes, small arrows a possible water-only pathway across the tight junction.

Structural studies on endothelial tight junctions show two principal differences from the leakier end of the epithelial junctional spectrum (Figure 2). Goniometric tilting of suitably stained thin sections of mesenteric, placental, and cardiac capillaries shows that the tight junctions are less intimate than in epithelia, with a 4 nm separation between the membranes rather than the tight contact characteristic of epithelia (Adamson and Michel, 1993; Firth et al., 1983; Leach and Firth, 1992; Ward et al., 1988). In addition serial sectioning in these capillaries reveals large discontinuities in the junctional strands themselves (Adamson and Michel, 1993; Bundgaard, 1984; Leach and Firth, 1992). Adamson and Michel (1993) pointed out that neither of these differences provide an obvious basis for the pore properties of capillaries; the strand discontinuities are too large, while the 4 nm membrane separation is almost big enough but provides too much total pore area. It seems likely that the discontinuities may account for total pore area by restricting the proportion of the cleft available as a filter, but that the structural basis for pore size restriction needs to be sought elsewhere.

Whatever the function of the 4 nm membrane separation in endothelial tight junctions, it implies that there may be interesting molecular differences between endothelial and epithelial tight junctions. To date this prediction has proved unfounded, as tight junctional plaque and membrane marker molecules such as ZO-1, cingulin, and occludin are all found in endothelial as well as epithelial tight junctions (Anderson et al., 1993; Citi, 1993; Furuse et al., 1993). The tight junctional plaque molecule ZO-1, which is probably involved in linking actin to the junctional membrane, is expressed as a different isoform in all endothelia and in a few epithelia, but the endothelial isoform is found in brain as well as non-brain capillaries and seems to be a feature of labile junctions rather than of leaky ones (Balda and Anderson, 1993; Willott et al., 1992).

Fiber Matrix Models

The fiber matrix model of capillary permeability (Curry and Michel, 1980) presented an alternative concept of capillary pores. Its central idea is that paracellular clefts and any other water-filled transendothelial channels may contain a three-dimensional network of fibers which can act as a molecular sieve; furthermore, by undergoing a conformational change on binding serum albumin, the matrix may offer an explanation for the well-established ability of albumin and some other proteins to reduce capillary permeability (Curry et al., 1984). The model was conceived in terms of generic glycocalyx elements, but more recently has been interpreted by some in terms of specific adhesion molecules and of specific subcellular structures. Silberberg (1988) suggested on theoretical grounds that the stability of dimensions of endothelial paracellular clefts required a system of "posts", presumably glycoprotein aggregates, which could hold adjacent cells together. Structures spanning endothelial clefts in placental capillaries in the guinea pig which resemble Silberberg's conceptual posts are seen in placental capillaries of the guinea pig after tannic acid-osmium and similar mordanting procedures (Firth et al., 1983) and in cardiac muscle capillaries after freeze substitution (Frøkjær-Jensen, 1990), and have now been described in a wide range of endothelia including that of human placental terminal villi (Leach and Firth, 1992). Immunogold histochemistry on ultrathin frozen sections shows that two adhesion molecules are strongly concentrated in the adhesion belt region of the endothelial paracellular cleft in which the linkers (the cleft-spanning structures found by electron microscopy) are found. One is the familiar immunoglobulin superfamily adhesion molecule PECAM-1, mainly associated with heterophilic cellular adhesion, while the second is a recently characterized and endothelium-specific homophilic adhesion molecule termed VE-cadherin (Leach et al., 1993). This study raises the interesting possibility that the fiber matrix and the linkers are both based on adhesion molecule assemblies in the adhesion belt and may even be the same structures. Using adhesion molecules both for cellular adhesion and for setting of paracellular permeability would have the merit of economy.

Change In Capillary Permeability

Acute inflammation involves a large increase in microvascular permeability and may be expected to involve reorganization of any structures which normally limit permeability. Stimulation of endothelial cells *in vitro* with inflammatory agents such as thrombin and cytokines, or chelation of Ca^{2+} with EGTA, leads to rapid redistribution away from intercellular contacts of VE-cadherin but not of PECAM-1 (Ayalon et al., 1994; Dejana et al., 1995; Lampugnani et al., 1992). In the perfused delivered term human placenta, inclusion of 100 μM histamine in the perfusate produces a significant increase in permeability to cyanocobalamin and Cr-EDTA accompanied by redistribution of VE-cadherin and, to a lesser extent, of PECAM-1 away from the paracellular clefts to a more diffuse distribution (Leach et al., 1995).

SUMMARY

Capillaries in the human and guinea pig placenta closely resemble those of cardiac and skeletal muscle both in their continuous endothelium and in their relatively high restrictiveness to permeation by macromolecules. Their detailed paracellular junctional architecture not only closely resembles other continuous capillaries, but has served as an important model for the investigation of major questions about the subcellular and molecular basis of capillary permeability. In such capillaries the main site for transendothelial water and solute flows and for imposition of molecular size restriction appears to be the paracellular clefts of the endothelium. Within these clefts the discontinuous tight junctions, which show extensive molecular homology with epithelial junctions, may be rate-limiting for hydraulic conductivity but are not size-limiting for permeability of hydrophilic solutes. It is likely that the main sites of size restriction are adhesion belts of the paracellular cleft, where a likely filtering structure is the linker array which is rich in the endothelium-specific adhesion molecule VE-cadherin. An important area for experiments in the next few years will be the identification of the cellular signals involved in the regulation of endothelial permeability which lead to repositioning of the adhesion molecules of the paracellular cleft.

REFERENCES

Adamson, R.H. and Michel, C.C. (1993) Pathways through the intercellular clefts of frog mesenteric capillaries. *J. Physiol.* 466, 303-327.

Anderson, J.M., Balda, M.S. and Fanning, A.S. (1993) The structure and regulation of tight junctions. *Current Opinion Cell Biol.* 5, 772-778.

Anderson, R.G.W. (1993) Caveolae: Where incoming and outgoing messengers meet. *Proc. Natl. Acad. Sci. USA* 90, 10909-10913.

Ayalon, O., Sabanai, H., Lampugnani, M.G., Dejana, E. and Geiger, B. (1994) Spatial and temporal relationships between cadherins and PECAM-1 in cell-cell junctions of human endothelial cells. *J. Cell Biol.* 126, 247-258.

Balda, M.S. and Anderson, J.M. (1993) Two classes of tight junctions are revealed by ZO-1 isoforms. *Am. J. Physiol.* 264, C918-C924.

Brightman, M.W. and Reese, T.S. (1969) Junctions between intimately apposed cell membranes in the vertebrate brain. *J. Cell Biol.* 40, 648-677.

Bundgaard, M. (1980) Transport pathways in capillaries - in search of pores. *Ann. Rev. Physiol.* 42, 325-336.

Bundgaard, M. (1984) The three dimensional organization of tight junctions in a capillary endothelium revealed by serial-section electron microscopy. *J. Ultrastruct. Res.* 88, 1-17.

Citi, S. (1993) The molecular organization of tight junctions. *J. Cell Biol.* 121, 485-489.

Claude, P. (1978) Morphological factors influencing transepithelial permeability: A model for the resistance of the zonula occludens. *J. Membr. Biol.* 39, 219-232.

Claude, P. and Goodenough, D.A. (1973) Fracture faces of zonulae occludentes from "tight" and "leaky" epithelia. *J. Cell Biol.* 58,390-400.

Crone, C. (1984) The function of capillaries. In: *Recent Advances in Physiology 10*, (Ed.) P.F. Baker, Churchill Livingstone: Edinburgh, pp. 125-162.

Curry, F.E. and Michel, C.C. (1980) A fiber matrix model of capillary permeability. *Microvasc. Res.* 20, 96-99.

Curry, F.E., Michel, C.C. and Phillips, M.E. (1984) The effects of serum albumin on the effective osmotic pressures exerted by myoglobin solutions across the walls of single frog mesenteric capillaries. *J. Physiol.* 358, 116P.

Dejana, A., Corada, M. and Lampugnani, M.G. (1995) Endothelial cell-to-cell junctions. *FASEB J.* 9, 910-918.

Diamond, J.M. (1978) Channels in epithelial cell membranes and junctions. *Federation Proc.* 37, 2639-2644.

Eaton, B.M., Leach, L. and Firth, J.A. (1993) Permeability of the fetal villous microvasculature in the isolated perfused human term placenta. *J. Physiol.* 463, 141-155.

Farquhar, M.G. and Palade, G.E. (1963) Junctional complexes in various epithelia. *J. Cell Biol.* 17, 375-409.

Firth, J.A., Bauman, K.F. and Sibley, C.P. (1983) The intercellular junctions of guinea-pig placental capillaries: A possible structural basis for endothelial solute permeability. *J. Ultrastruct. Res.* 85, 45-57.

Franke, W.W., Cowin, P., Grund, C., Kuhn, C. and Kapprell, H.-C. (1988) The endothelial junction. The plaque and its components. In: *Endothelial Cell Biology in Health and Disease,* (Eds.) N. Simionescu and M. Simionescu, Plenum: New York, pp. 147-166.

Frøkjær-Jensen, J. (1990) Anatomical correlates of capillary permeability. In: *Endothelial Cell Function in Diabetic Microangiopathy: Problems in Methodology and Clinical Aspects* Vol. 9, (Eds.) G.M. Molinatti, R.S. Bar, F. Belfiore, and M. Porta, Karger: Basel, pp. 28-42.

Furuse, M., Hirase, T., Itoh, M., Nagafuchi, A., Yonemura, S., Tsukita, S. and Tsukita, S. (1993) Occludin: A novel integral membrane protein localizing at tight junctions. *J. Cell Biol.* 123, 1777-1788.

Geiger, B., Avnur, Z., Volberg, T. and Volk, T. (1985) Molecular domains of adhaerens junctions. In: *The cells in Contact: Adhesions and Junctions as Morphogenetic Determinants*, (Eds.) G.M. Edelman and J.-P. Thiéry, Wiley, New York, pp. 461-489.

Grunwald, G.B. (1993) The structural and functional analysis of cadherin calcium-dependent cell adhesion molecules. *Current Opinion Cell Biol.* 5, 797-805.

Gumbiner, B.M. (1993) Breaking through the tight junction barrier. *J. Cell Biol.* 123, 1631-1633.

Kachar, B. and Reese, T.S. (1982) Evidence for the lipidic nature of tight junctional strands. *Nature* 296, 464-466.

Kan, F.W.K. (1993) Cytochemical evidence for the presence of phospholipids in epithelial tight junction strands. *J. Histochem. Cytochem.* 41, 649-656.

Kaufmann, P., Schröder, H. and Leichtweiss, H.-P. (1982) Fluid shift across the placenta: II. Feto-maternal transfer of horseradish peroxidase in the guinea pig. *Placenta* 3, 339-348.

Kertschanska, S. and Kaufmann, P. (1992) Morphological evidence for the existence of transtrophoblastic channels in human placental villi. *Placenta* 13, A33.

Lampugnani, M.G., Resnati, M., Raiteri, M., Pigott, R., Pisacane, A., Houen, G., Ruco, L.P. and Dejana, E. (1992) A novel endothelial cell specific membrane protein is a marker of cell-cell contacts. *J. Cell Biol.* 118, 1511-1522.

Leach, L., Clark, P., Lampugnani, M.G., Arroyo, A.G., Dejana, E. and Firth, J.A. (1993) Immunoelectron characterization of the inter-endothelial junctions of human term placenta. *J. Cell Sci.* 104, 1073-1081.

Leach, L., Eaton, B.M., Westcott, E.D.A and Firth, J.A. (1995) Effects of histamine on endothelial permeability and structure and adhesion molecules of the paracellular junctions of perfused human placental microvessels. *Microvasc. Res.* 50, 323-337.

Leach, L. and Firth, J.A. (1992) Fine structure of the paracellular junctions of terminal villous capillaries in the perfused human placenta. *Cell Tissue Res.* 268, 447-452.

Martinez-Palomo, A. and Erlij, D. (1975) Structure of tight junctions in epithelia with different permeability. *Proc. Natl. Acad. Sci. USA* 72, 4487-4491.

Nelson, D.M., Crouch, E.C., Curran, E.M. and Farmer, D.R. (1990) Trophoblast interaction with fibrin matrix. Epithelialization of perivillous fibrin deposits as a mechanism for villous repair in the human placenta. *Am. J. Pathol.* 136, 885-865.

Schulze, C. and Firth, J.A. (1992a) The interendothelial junction in myocardial capillaries: Evidence for the existence of regularly spaced, cleft-spanning structures. *J. Cell Sci.* 101, 647-655.

Schulze, C. and Firth, J.A. (1992b) Interendothelial junctions during blood-brain barrier development in the rat: Morphological changes at the level of individual tight junctional contacts. *Dev. Brain Res. 69, 85-95.*

Sibley, C.P., Bauman, K.F. and Firth, J.A. (1982) Permeability of the foetal capillary endothelium of the guinea-pig placenta to haem proteins of various molecular sizes. *Cell Tissue Res.* 223, 165-178.

Silberberg, A. (1988) Structure of the interendothelial cell cleft. *Biorheology* 25, 303-318.

Stevenson, B.R., Anderson, J.M., Goodenough, D.A. and Mooseker, M.S. (1988) Tight junction structure and ZO-1 content are identical in two strains of Madin-Darby canine kidney cells which differ in transepithelial resistance. *J. Cell Biol.* 107, 2401-2408.

Stulc, J. (1989) Extracellular transport pathways in the haemochorial placenta. *Placenta* 10,113-119.

Ward, B.J., Bauman, K.F. and Firth, J.A. (1988) Interendothelial junctions of cardiac capillaries in rats: their structure and permeability properties. *Cell Tissue Res.* 252, 57-66.

Willott, E., Balda, M.S., Heinzelman, M., Jameson, B. and Anderson, J.M. (1992) Localization and differential expression of two isoforms of the tight junctional protein ZO-1. *Am. J. Physiol.* 262, C1119-C1124.

Wissig, S.L. (1979) Identification of the small pore in muscle capillaries. *Acta Physiol. Scand. Suppl.* 463, 33-44.

Trophoblast Research 10:215-258, 1997

HEPARIN-BINDING ANGIOGENIC GROWTH FACTORS IN PREGNANCY
-A Review-

Asif Ahmed

Department of Obstetrics and Gynaecology
Birmingham Maternity Hospital
University of Birmingham
Edgbaston, Birmingham, B15 2TG, United Kingdom

INTRODUCTION

The vasculature conveys the components of blood that are essential for cell life. The importance of vessel growth during normal development and during the formation of collateral circulation in wound healing was realized over two hundred years ago. In 1794, John Hunter, the founder of experimental surgery and pathology, wrote about the development of vessels in the chick embryo "Blood is formed before the vessels, and when coagulated, the vessels appear to arise. When new vessels are produced in part they are not always elongations from the original ones, but newly formed". Although his ideas of formation of new vessels from coagulated blood were not correct, he came close to the concept of the development of blood and endothelial cells from a primitive mesoderm. Correct interpretation of the growth of new capillaries from pre-existing vessels came from Meyer (1853), who wrote, "thin fibers formed from spindle-like cells which later become hollow" (i.e., capillary sprouts), and it was almost a hundred years after Hunter that the role of endothelial cell migration in the formation of new capillaries was first realized (Bobritzky, 1885).

Today, it is well established that two different mechanisms are involved in the formation of blood vessels, namely, vasculogenesis and angiogenesis. Vasculogenesis is the development of blood vessels from *in situ* differentiating endothelial cells and may be unique to embryonic development (Risau et al., 1988). In contrast, angiogenesis can occur during the entire life span as it is a basic requirement for normal growth of tissue and involves the formation of new blood vessels by the sprouting of capillaries from the venous end of pre-existing blood vessels (Ausprunk and Folkman, 1977; Folkman and Shing, 1992). Angiogenesis is a multistep phenomenon which sequentially involves the degradation of vascular basement membrane and interstitial matrix by endothelial cells, the co-ordinated migration and proliferation of endothelial cells and tubulogenesis and the formation of capillary loops. In male adult life angiogenesis seldom occurs and the turnover of endothelial cells is very low, measured in years. The process occurs normally as part of the body's repair processes, as in wound healing and bone fracture and is under rigorous control. Unrestrained angiogenesis can lead to pathological conditions such as atherosclerosis, diabetic retinopathy, rheumatoid arthritis, and solid tumor growth.

In the female reproductive system, however, angiogenesis occurs in monthly cycles. In the ovary, angiogenesis occurs during the development of the follicle, its rupture, and the subsequent formation of the corpus luteum (cf., Reynolds et al, 1992). It also occurs during the regeneration of endometrium at the end of menstruation for the preparation of a receptive endometrium for implantation (cf., Smith, 1995). In pregnancy, the establishment of the vascular structures involved in transplacental exchange requires extensive angiogenesis as does placental vascular growth. Only in the last few years has the identification and cellular localization of angiogenic growth factors been described in the placenta and investigations are now under way to determine their function in trophoblast-endothelial cell interaction. In this review I will focus on the expression and the roles of the heparin-binding angiogenic growth factors in pregnancy.

Placental Angiogenesis

The term "angiogenesis" was coined to describe the formation of new blood vessels in the placenta (Hertig, 1935). Extensive angiogenesis is required to establish the vascular structures both in the fetal villi and in the maternal decidua necessary for transplacental exchange (Findlay, 1986). This involves proliferation, migration, and maturation of maternal and fetal vascular endothelial cells (Cross et al., 1994).

The earliest morphological step in the development of new capillary outgrowth is the local degradation of the basement membrane of the parent venule (Ausprunk and Folkman, 1977). During this process, the basement membrane is broken down by proteolytic enzymes produced by endothelial cells in response to angiogenic growth factors (Figure 1). Proteolytic enzymes include plasminogen activators (PA) consisting of urokinase plasminogen activator (u-PA) and tissue-type plasminogen activator (t-PA) together with their natural inhibitors, plasminogen activator inhibitor (PAI). PA converts the latent protease plasminogen to plasmin (Presta et al., 1986). Plasmin is a wide spectrum protease which degrades the extracellular matrix, including laminin and fibronectin (Rifkin et al., 1982) and activates the collagenases produced by the endothelial cells (Figure 1). Collagenase, stromeolysin, and gelatinase are other matrix proteases collectively called tissue metalloproteinases (Docherty and Murphy, 1990). The action of these matrix proteases is kept in balance by their inhibitors, namely the tissue inhibitors of metalloproteinases (TIMP's). These proteolytic enzymes and their inhibitors are produced not only by endothelial cells (Rifkin et al., 1982) but also by trophoblast (Queenan et al., 1987; Feinberg et al., 1989; Emonard et al., 1990). Trophoblast cells may limit their own invasion since cultured human cytotrophoblast up-regulate TIMP-3 in parallel with gelatinase B (Cross et al., 1994).

The degradation of the basement membrane causes gaps to develop in the capillary structure through which vascular endothelial cells migrate. Migrating endothelial cells accumulate at definite focal points on the parent capillary wall. The bleb so formed develops into a bud which in turn forms a sprout and becomes the leading tip of the nascent vessel. Accompanying this migratory process is the organized formation of solid strands of cells behind the leading tip as well as extensive proliferation of endothelial cells proximal to the leading tip of the newly emerging blood vessels. Elongation of these sprouts is via recruitment of endothelial cells from the store in the parent vessel which are intercalated into the growing solid strands (Cliff, 1965; Ausprunk and Folkman, 1977). New vessel formation is complete by the development of a lumen. Intracellular lumina are thought to be initiated by the appearance of vacuoles within the cytoplasm of endothelial cells. Subsequent enlargement and fusion of vacuoles within neighboring endothelial cells leads to the development of a tube-like structure (Folkman and Haudenschild, 1980).

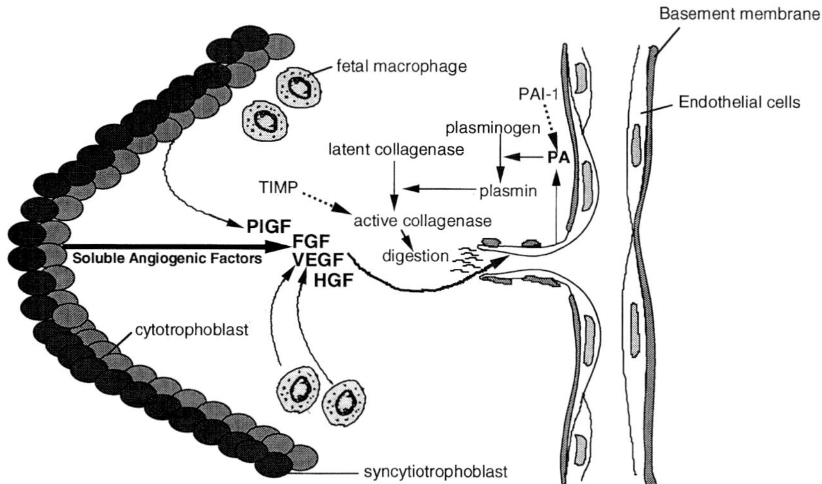

Figure 1. Role of angiogenic growth factors in placental angiogenesis. The structural integrity of the basement membrane is conserved by preserving a fine balance between the proteolytic enzymes that are produced locally and their inhibitors. Plasminogen activators (PA) and metalloproteinases (e.g., collagenase) together with their natural inhibitors, plasminogen activator inhibitor (PAI), and tissue inhibitors of metalloproteinases (TIMP's) respectively, regulate the degrade the extracellular matrix in response to angiogenic growth factors.

In the human placenta, angiogenesis is important both for the development of the villous vasculature and the formation of terminal villi. Placental vascular growth begins early in pregnancy and continues throughout gestation in association with a continual and dramatic increase in rates of uterine and umbilical blood flows (Rosenfeld et al., 1974; Ferrell, 1989). Increased blood flow to placental tissues satisfies the steadily increasing metabolic demands of the growing fetus (Ferrell, 1989). The development of terminal villi is considered to be regulated by the relative rates of longitudinal growth of mature intermediate villi and their contained capillaries (Kaufmann et al., 1985). When placental capillary growth exceeds that of the mature intermediate villi, the capillaries coil and form loops (Benirschke and Kaufmann, 1990). The capillary coils obtrude from the surface of the mature intermediate villi, raising arcades of trophoblast and forming new terminal villi. Specific angiogenic factors and inhibitors must regulate these processes.

Angiogenic Factors

The development of both *in vitro* and *in vivo* angiogenesis models within the last 20 years has greatly increased the number of angiogenic factors (Table I) and has contributed to the understanding of the complex phenomenon of angiogenesis. The assays most commonly used to evaluate the entire process of neovascularization are the corneal pocket assay, which permits linear measurement of capillary growth induced by a substance implanted in the cornea of the rabbit and the chorio-allantoic membrane (CAM) assay, which evaluates the ability of a substance to stimulate or inhibit angiogenesis.

Table I

Angiogenic Factors

Angiogenic Factors	Heparin-binding	Endothelial cell mitogenic or migratory activity	Angiogenesis	In vitro trophoblast proliferation
FGF-1	yes	+	+	+
FGF-2	yes	+	+	+
VEGF	yes	+	+	0
PIGF	yes	±	0	0
HGF	±	+	+	±
PD-ECGF	0	+	+	unk
EGF	0	+	+	+
TGFα	0	+	+	unk
TGFβ	0	-	+	-
TNFα	0	-	+	unk
GM-CSF	0	+	unk	+
hr-E-selectin	unk	+	+	unk
hr-VCAM-1	unk	+	+	unk
Angiogenin	0	0	+	unk
Angiotropin		+	+	unk
Angiotensin II	0	+	+	±
Bradykinin	0	+	+	±
Adenosine	0	±	+	0
PAF	unk	+	+	unk
Okadaaic acid	0	±	+	unk
PGE_1	0	±	+	unk
PGE_2	0	±	+	unk
1-butyryl-glycerol	0	±	+	unk
Nicotinamide	0	±	+	unk

+ = strong stimulation; ± = weak stimulation; 0 = no effect; - = inhibition; unk = unknown

Many of these factors are described in detail in the review by Klagsbrun and D'Amore, 1991.

Angiogenic growth factors have the ability to induce vascularization, either by direct or indirect action. In microgram quantities, they can induce vascular proliferation in soft tissue. The direct-acting growth factors specifically stimulate endothelial cell proliferation, migration, and tube formation in soft agar gel by binding to specific membrane surface receptors on endothelial cells. The indirect-acting growth factors have no direct stimulatory effect on vascular endothelial *cells in vitro*, but induce angiogenesis by an indirect means *in vivo*, by acting on cells which secrete a direct-acting growth factor. The focus of this review will be on the heparin binding polypeptide growth factors which have a direct effect on cultured endothelial cell growth and also may act on trophoblast.

Fibroblast Growth Factors

Fibroblast growth factors (FGFs) are a family of at least nine genetically distinct but structurally related heparin-binding polypeptide growth factors. They generally encode proteins with a molecular mass of 18-30 kDa, most are secreted constitutively *in vitro*, and act on many cell type. *In vitro* actions of FGFs include the modulation of cell motility, differentiation, cell survival, and extension of neurites as well as proliferation. *In vivo*, FGFs have been implicated in normal physiological processes such as embryonic and fetal development, neovascularization, nerve regeneration and wound healing (for a review on FGF see Klagsbrun and D'Amore, 1991; Basilico and Moscatelli, 1992; Mason, 1994).

It was proposed that this growth factor was primary released from dead or damaged cells as acidic (FGF-1) and basic (FGF-2) fibroblast growth factors lack a consensus signal peptide found in most exported proteins. However, experiments with single cells, where release due to cell death can be ruled out, show that FGF-2 is released, and that inhibitors of endo- and exocytosis block its release, while ER-Golgi inhibitors do not (Mignatt et al., 1992). Thus, it is clear that a mechanism exists for export of these FGF types, which does not involve the classical protein secretion pathway. The FGF-1 and FGF-2 are single copy genes, located on chromosomes 3 and 4 respectively and share a 55% overall sequence homology with proteins of molecular weights of 18 kDa. Although both FGF-1 and FGF-2 bind to the same cellular receptors and display similar biological activity, FGF-2 is 30 times more potent than FGF-1 and heparin potentiates FGF-2 activity but not FGF-1 (Klagsbrun and D'Amore, 1991; Basilico and Moscatelli, 1992). FGF-2 is highly mitogenic for capillary endothelial (Schweigerer et al., 1987) and human umbilical vascular endothelial cells (Ahmed et al., 1994) in vitro, and both FGF-1 and FGF-2 are potent angiogenic factors at picomolar quantities in the CAM and cornea bioassays (Lobb et al., 1985; Folkman et al., 1987; Broadley et al., 1989). Four of the FGFs have been identified and isolated from various tumors as oncogene products such as FGF-3 or *int-2* (Dickson et al., 1990), FGF-4 or *hst-1/K-fgf* (Delli-Bovi et al., 1987) and *FGF-5* (Zhan et al., 1988). Molecular cloning of genes related to FGF-4 led to the discovery of FGF-6 [hst-2, a molecule sharing 80% identity in amino acid sequence with FGF-4] (Marics et al., 1989). They all encode for proteins which share 40-50% sequence homology with FGF-1 or FGF-2. In addition, FGF-like growth factors include FGF-7, also called keratinocyte growth factor, which stimulates epithelial cell proliferation in addition to endothelial cell growth (Finch et al., 1989), FGF-8, a FGF isolated from bovine uterus (Milner et al., 1989) and FGF-9, also called glia-activating factor, which like FGF-1 and FGF-2 lacks a consensus signal peptide. Only FGF-1 and FGF-2 have been described in the placenta.

FGF Receptors

FGFs exert their actions through specific cell-surface receptors which have intrinsic tyrosine kinase activity (Asai et al., 1993). Five distinct receptors for FGFs cloned and characterized include FGFR1 or [also known as flg (fms-like gene 1) or cek-1 (chicken embryo kinase-1)], FGFR2 [also known as bek (bacterial-expressed kinase gene product) or cek-3], FGF-R3 [or cek-2], FGF-R4 and FGF-R5 or flg-2 (cf., Basilico and Moscatelli, 1992; Mason, 1994). The FGFR are highly homologous receptor consisting of a extracellular domain composed of three immunoglobulin-like repeats and an acidic region, a transmembrane domain, and a cytoplasmic region containing a tyrosine kinase domain. The human FGFR1 receptor binds both FGF-1 and FGF-2 with high affinities and FGF-4 with low affinity while FGFR2 binds FGF-1, FGF-2 and FGF-4 with equal high affinity, but does not bind FGF-7 or FGF-5. FGFR3 is rather specific to FGF-1 and has low affinity for FGF-2 while FGFR4 binds strongly to FGF-1 and FGF-4 and FGF-2 with lower affinity (Hughes and Hall, 1993). Expression of FGFR1 mRNA is widespread compared with FGFR2 mRNA which is expressed primarily in the fiber tracts of the rat nervous system (Asai et al., 1993). The expression of these receptors has not been described in human placenta although FGFR1 is expressed in human endometrium (Shams et al., 1996).

Cellular Distribution And Function In Placenta

Human placenta is a rich source of FGF-2 as this growth factor was purified from it and shown to stimulate capillary endothelial cell protease production, DNA synthesis, and cellular migration (Moscatelli et al., 1986; Uhlrich et al., 1991). FGF-2 has been identified in dendritic and Hofbauer cells in the placental villous mesenchyme (Schulze-Osthoff et al., 1990) and indirect evidence suggests that this factor is associated with actively dividing cells in the human placenta (Cattini et al., 1991).

Immunocytochemistry indicates that extravillous trophoblast and endothelial cells in the decidua stain positively for FGF-2 and are detected in or around cytotrophoblast but not in the syncytiotrophoblast in first trimester villae (Ferriani et al., 1994). Identical staining patterns is produced by FGF-1 antibody (Figure 2) indicating co-localization of FGF-1 and FGF-2. By the third trimester of pregnancy, weaker and more diffuse staining, seen in the syncytiotrophoblast surrounding the placenta villi, suggests that the principal function of FGFs may be trophoblast proliferation and placental angiogenesis.

Immunolocalization of FGF-2 gives little information regarding the *in vivo* cellular source of FGF-2. The first description of the pattern of FGF-2 mRNA expression during pregnancy has further elucidated the potential role of FGF-2 in promoting placental development and neovascularization within the placenta (Shams and Ahmed, 1994). These expression studies reveal that FGF-2 mRNA signal is present in both cytotrophoblast and syncytiotrophoblast surrounding the first trimester villi and even by the third trimester of pregnancy, the hybridization signal persists in vasculosyncytial membrane, although the Hofbauer cells within the villous mesenchyme show little or no expression for FGF-2 mRNA.

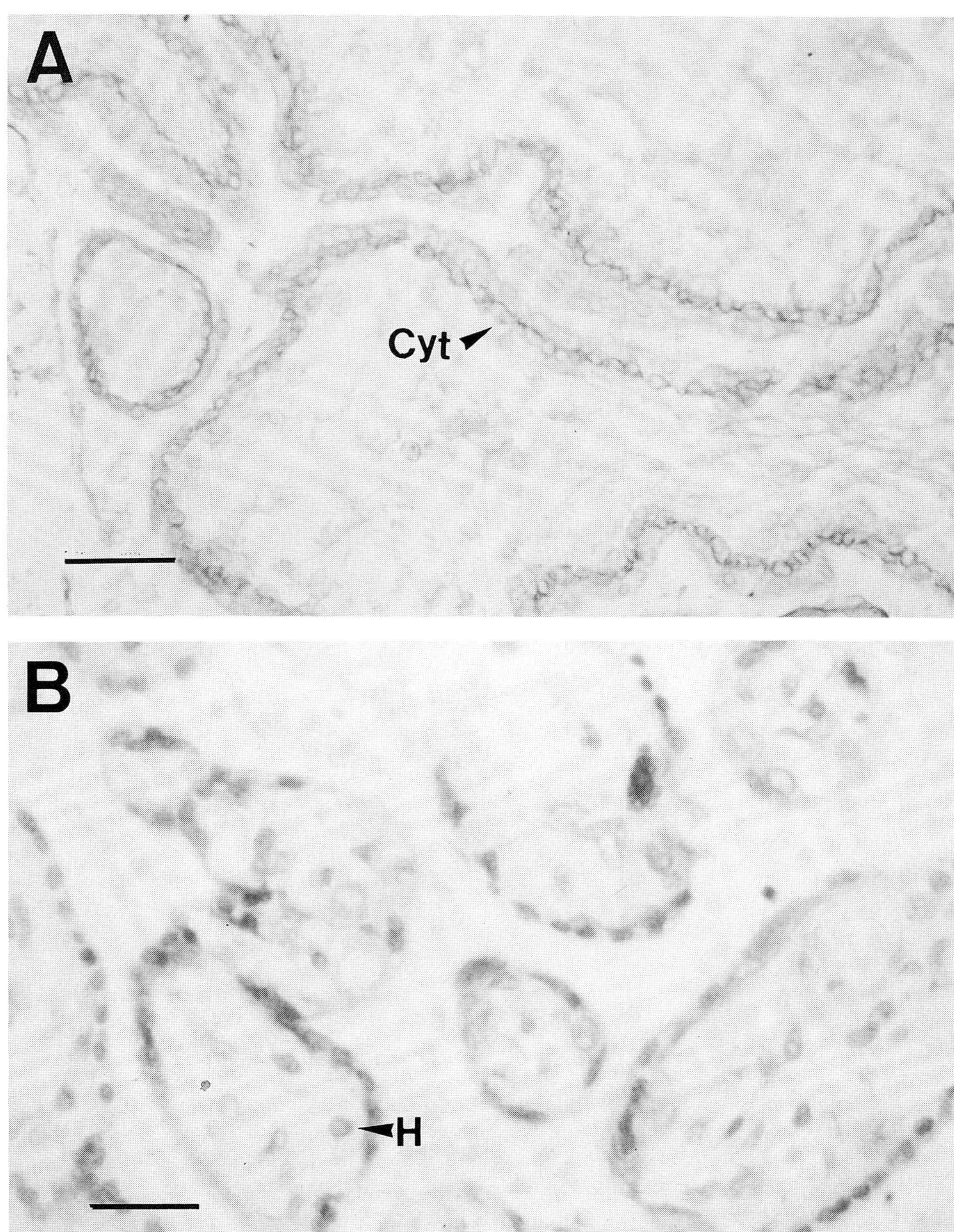

Figure 2. Localization of FGF-1 immunoreactivity in villi of human placenta. Panel A show first trimester villi (11 weeks) and panel B shows term villi. There is weak immunoreactivity for FGF-1 in villous trophoblast term placenta, but stronger staining of Hofbauer cells in the villous mesencyme (labeled H in panel B). Bar corresponds to 25 μm. (Reproduced from *Growth Factors* 10, 259-268, 1994.)

FGF-2 increases nitric oxide synthase (NOS) production in bovine endothelial cells (Kostyk et al., 1995) and trophoblast (A. Ahmed, unpublished data). Strong FGF immunostaining (Figure 3 A, B, C, and D) and intense FGF-2 mRNA expression (Figure 3E and F) seen in the smooth muscle cells around mid and large sized placental vessels demonstrates efficient translation of mRNA into the protein which probably acts on endothelial cells to produce nitric oxide for regulating placental vascular tone as well neocapillary formation. In fetal membrane, the expression of FGF-2 mRNA is strongest in the surface epithelium of the amnion (Shams and Ahmed, 1994) suggesting that amnion produces FGF-2. Circulating levels of FGF-2 in amniotic fluid have been reported (Hill et al., 1995). This group showed that levels of FGF-2 were elevated (2 to 4 fold) in cord serum and amniotic fluid in pregnancy complicated with diabetes probably due to metabolic stress. The FGF-2 levels also positively correlated with fetal and placental size.

Few functional studies relating to the effect of FGFs on trophoblast have been reported. The nuclear localization of FGF-2 immunoreactivity in dividing but not non-dividing placental cells (Cattini et al., 1991) together with the ability of FGF-2 to stimulate DNA synthesis in choriocarcinoma cell line JEG-3 (Ferriani et al., 1994) suggests a role for FGF-2 in cytotrophoblast proliferation *in vivo*. FGF-2 also causes an increase inositol phosphate accumulation in JEG-3 cell pre-labeled with myo-[^{3}H]inositol (Figure 4) at concentrations at which it stimulates DNA synthesis suggesting a role of phospholipase C in JEG-3 cell proliferation. Although FGF-2 stimulates phospholipase D activity in human umbilical vascular endothelial cells in the absence of inositol phosphate turnover (Ahmed et al., 1994), FGF-2 does not activate phospholipase D in JEG-3 cells (Ferriani et al., 1994). These studies clearly suggest that FGF-2 in stimulating growth is acting via different mechanisms and that it may also be acting via different FGF receptors in the two cell types.

This polypeptide growth factor clearly plays an important role the development and maintenance of placental vasculature. FGF-2 stimulates production of plasminogen activator proteases in rat granulosa cells (Lapolt et al., 1990), bovine capillary endothelial cells (Moscatelli et al., 1986), breast tumor cells (Mira et al., 1986) and porcine sertoli cells (Jaillard et al., 1987) capable of degrading basement membrane. It also induces capillary endothelial cells to migrate into three-dimensional collagen matrices to form capillary-like tubes (Montesano et al., 1986). As trophoblast invasion is essential for successful implantation it is probable that FGF-2 may increase trophoblast invasive properties by stimulating the activity of plasminogen activators. Cultured human trophoblast secrete urokinase plasminogen activator (u-PA) which plays a role in the degradation of the extracellular matrix (Queenan et al., 1987). u-PA activity is regulated by plasminogen activator inhibitors, PAI-1 primarily in the intermediate trophoblast and PAI-2 in the villous trophoblast (Feinberg et al., 1989). As extracellular proteolysis is a component of the angiogenic process and FGF-2 stimulates plasminogen activator activity, FGF-2 may play a role in the establishment of pregnancy by initiating the proteolytic degradation of the extracellular matrix.

A

B

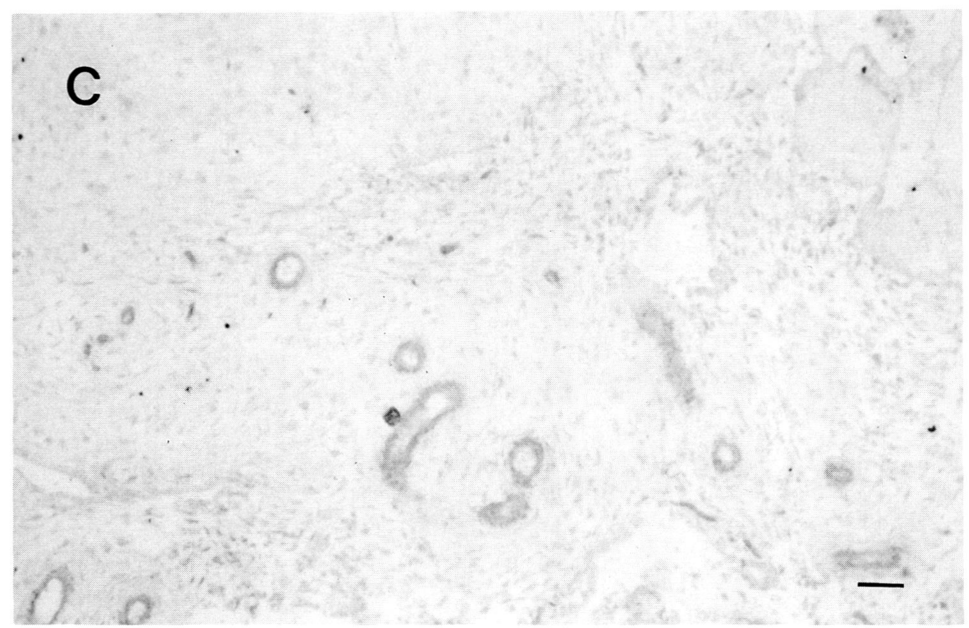
C

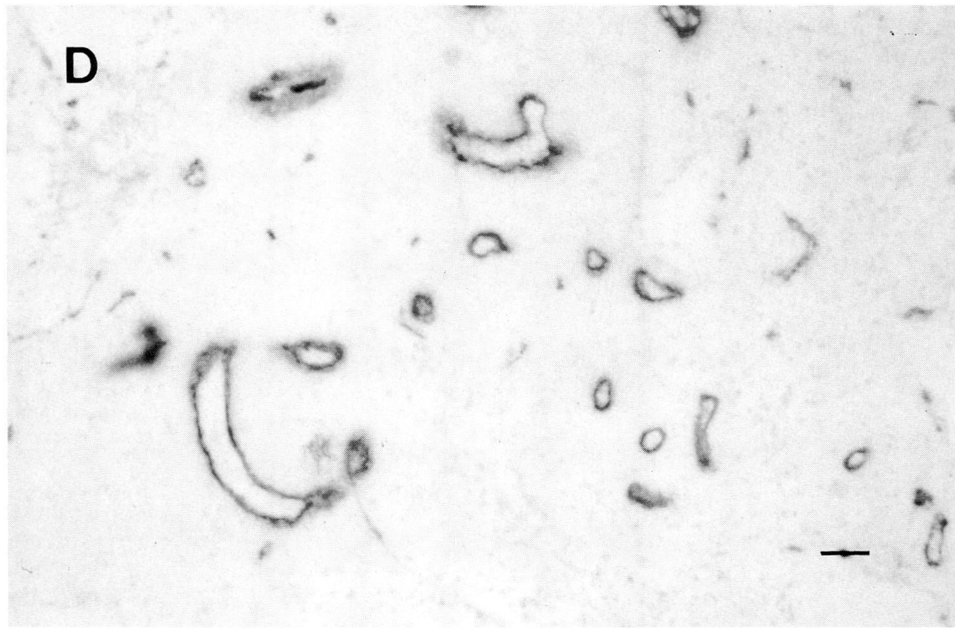
D

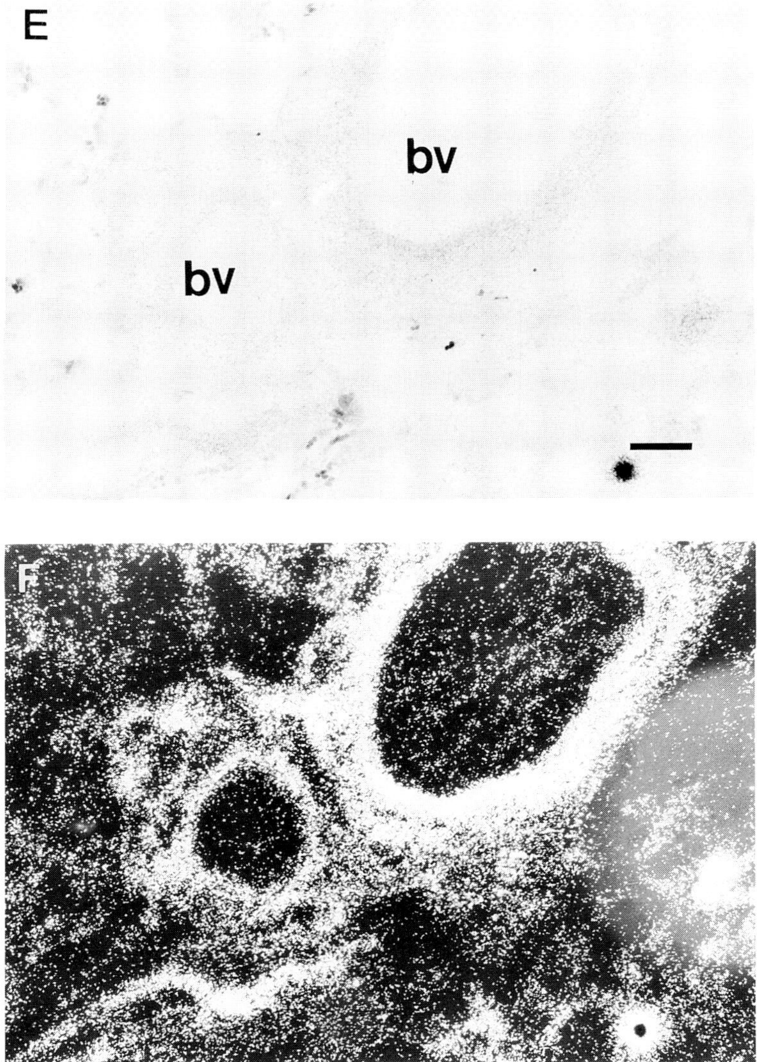

Figure 3. FGF expression in blood vessels in human placenta. Panels A and B (page 223), blood vessel in term placental villous, stained with antibody against the synthetic peptide sequence of FGF-1 (panel A), or non-immune serum (panel B). Staining is seen in the smooth muscle cells of a large size placental vessel. Panels C and D (page 224) show maternal decidua at 9 weeks gestation, with staining of FGF-1 in endothelial cells of mid-size vessels (C). Endothelial cells were identified by staining a semi-serial section with antibody to factor VIII (panel D). Localization of FGF-2 mRNA expression bFGF mRNA of blood vessel in term placental villous is seen in panel E (this page) (bright-field) and panel F (this page) (dark-field). ^{35}S-labeled bFGF cRNA antisense probe was used for hybridization. bv = Blood vessels. Bars in A and B corresponds to 25 μm, in C and D to 70 μm and E and F represents 50 μm. (Reproduced from *Growth Factors* 10, 259-268 and 11, 105-111, 1994.)

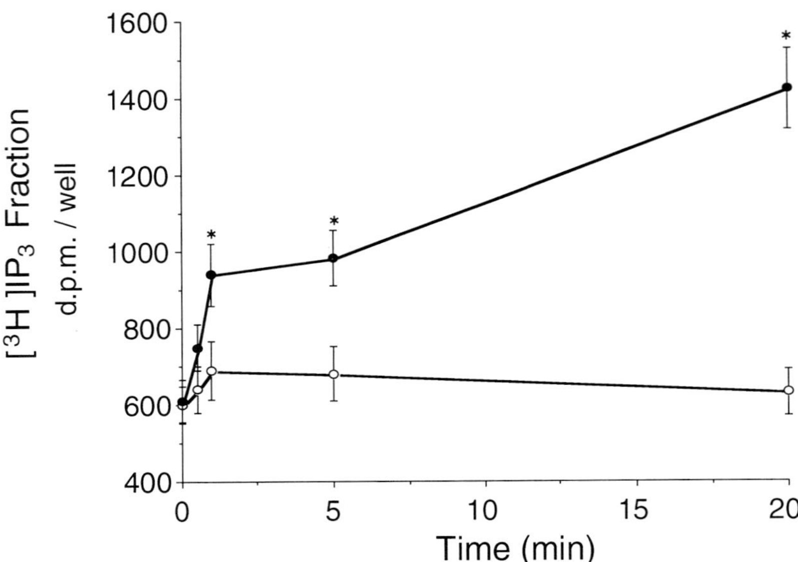

Figure 4. Effect of FGF-2 on inositol phosphate accumulation in JEG-3 cells. Cells were pre-labeled with myo-[^{3}H]inositol at 37°C for 48 hours. Cells were stimulated at 37°C for the indicated times (in the presence of 10 mM LiCl) following addition of 10 ng/ml recombinant FGF-2 (closed circle), or no addition (open circle). accumulation of inositol trisphosphate ([^{3}H]IP$_3$) fraction, after stimulation at 37°C for 20 minutes. After extraction of cells, accumulation of inositol trisphosphate ([^{3}H]IP$_3$) fraction was isolated by Dowex ion-exchange chromatography as described previously (Ahmed and Smith, 1992). Each point represents the mean ± sem of triplicate determination of three separate experiments. * $p \leq 0.05$. (Reproduced from *Growth Factors* 10, 259-268, 1994.)

Vascular Endothelial Growth Factor

Another heparin binding growth factor identified is vascular endothelial growth factor (VEGF), less commonly known and originally discovered as vascular permeability factor (Sanger et al., 1983; Ferrara and Henzel, 1989). It is composed of two identical subunits covalently linked by disulphide bonds, with a combined molecular weight of 36 to 46k (depending on degree of glycosylation), and exhibiting low (18%) but significant overall sequence homology with platelet derived growth factor (Ferrara et al., 1992) and closer (53%) homology with placenta growth factor (Maglione et al., 1991). Five different VEGF transcripts encoding polypeptides of 206, 189, 145, 165, and 121 amino acids are expressed by human cells (Tischer et al., 1991; Charnock-Jones et al., 1993). All isoforms arise from alternative splicing of a signal gene whose coding region is divided among eight exons. Exon 1 contains a 26 amino acid hydrophobic consensus secretory signal common to all peptides (Tischer et al., 1991). VEGF$_{165}$ lacks the product of exon 6 and the residues encoded by exons 6 and 7 are missing from VEGF$_{121}$. VEGF$_{189}$ contains all 8 exons and VEGF$_{206}$ has an additional 17 codons after the 24 amino acids insertion in VEGF$_{189}$. Little is known about VEGF$_{145}$ except that its expressed in endometrium. The VEGF transcripts differ in their efficiency of secretion, affinity for heparin and relative potency for vascular permeabilizing and mitogenic activities (Houck et al., 1991; Houck et al., 1992). $VEGF_{165}$ is expressed predominantly in most instances and $VEGF_{121}$ and $VEGF_{165}$ are secreted in soluble form whereas the $VEGF_{189}$ and $VEGF_{206}$ remain cell-associated (Houck et al., 1992). Although $VEGF_{189}$ is tightly bound to proteoglycans in the cell membrane or in the extracellular matrix, it is released into culture media following treatment with suramin and heparinases I and III (Houck et al., 1992). The affinity of VEGF for heparin is substantially lower than that of other typical heparin-binding growth factors such as FGF-2 (Sanger et al., 1983) and the smaller $VEGF_{121}$ isoform does not bind to heparin (Houck et al., 1992). More recently cDNAs for two new factors structurally homologous to VEGF, designated as VEGF-B and VEGF-C have been cloned (Olofsson et al., 1996; Joukov et al., 1996).

VEGF is a powerful multifunctional polypeptide that is a specific potent mitogen for endothelial cells, but also mediates a number of other endothelial (Sanger et al., 1983; cf., Dvorak et al., 1995) and non-endothelial (Ahmed et al., 1995, 1997) effects. VEGF was reported to stimulate the secretion of collagenase IV, u-PA, t-PA and PAI-1 in microvascular endothelial cells (Pepper et al., 1991; Knoll et al., 1992). VEGF evokes transient accumulation of cytoplasmic calcium and activates both phospholipases C and D in human umbilical vascular endothelial cells (Seymour et al., 1996) unlike FGF-2 which only stimulates phospholipase D activity in these cells (Ahmed et al., 1994). $VEGF_{121}$ is more potent than $VEGF_{165}$ in stimulating endothelial cell proliferation *in vitro* and angiogenesis *in vivo* (Birkenhager et al., 1996). The expression of VEGF in endometrial epithelial cell line (Charnock-Jones et al., 1993) and human cervix (Sisi et al., 1994) is up-regulated by estrogen, and hypoxia increases VEGF mRNA expression in the heart (Ladoux and Frelin, 1993) and tumors (Shweiki et al., 1994). In addition to the endothelial mitogenic capacity of the VEGF family, $VEGF_{189}$ is known to induce fluid and protein extravasation from blood vessels (cf., Dvorak et al., 1995).

VEGF Receptors

Several receptor tyrosine kinases related to the PDGF receptor family have been identified as VEGF receptors (Matthews et al., 1991). Two receptor that bind VEGF with high affinity are Flt-1 [fms-like tyrosine kinase] (De Vries et al., 1992) and KDR [kinase-insert-domain-containing receptor] (Terman et al., 1992). The Flt gene family belongs to the class III of the superfamily of receptor tyrosine kinases, which also include two protooncogenes *c-fms* and *c-kit*, [the receptors for M-CSF and stem cell factor respectively], the α and β chains of platelet-derived growth factor receptors and the product of the Flt3/Flk2 gene. The fms-like tyrosine kinase-4 (Flt4) cDNA cloned from a human placental cDNA library (Galland et al., 1993) is related to Flt-1 and KDR and is the receptor for VEGF-C, but does not bind VEGF (Pajusola et al., 1994). The Flt-1, Flt-4, and KDR receptor tyrosine kinases differ from other members of class III receptor tyrosine kinases by having seven instead of five Ig-like loops in their extracellular domains and constitute a novel subfamily of class III tyrosine kinases.

Both Flt-1 and KDR have been shown to undergo autophosphorylation upon binding of VEGF in transfected cells (Waltenburger et al., 1994). These transfection studies on the KDR and Flt-1 receptors showed that KDR has a functional role in regulating cell proliferation, while Flt-1 underwent a unique phosphorylation sequence on binding of VEGF but its function was not elucidated (Waltenburger et al., 1994). Inactivation of the mouse *flk-1* gene (equivalent to human KDR) is important in yolk-sac blood island formation and vasculogenesis in the mouse embryo (Shalaby et al., 1995).

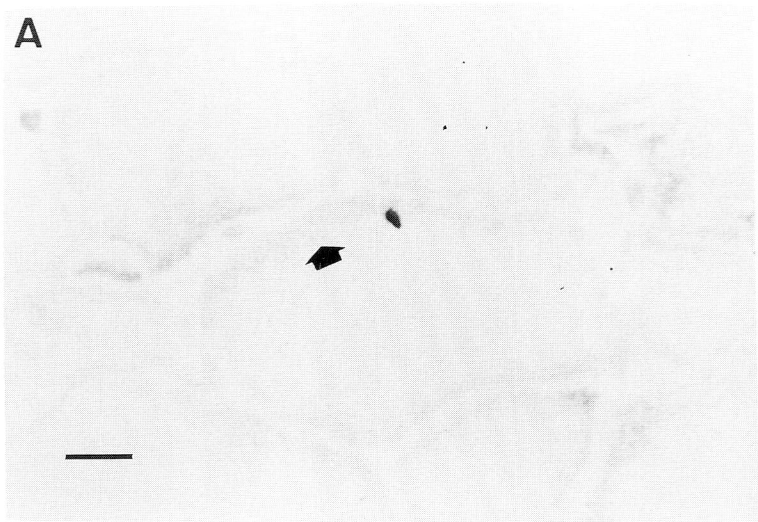
A

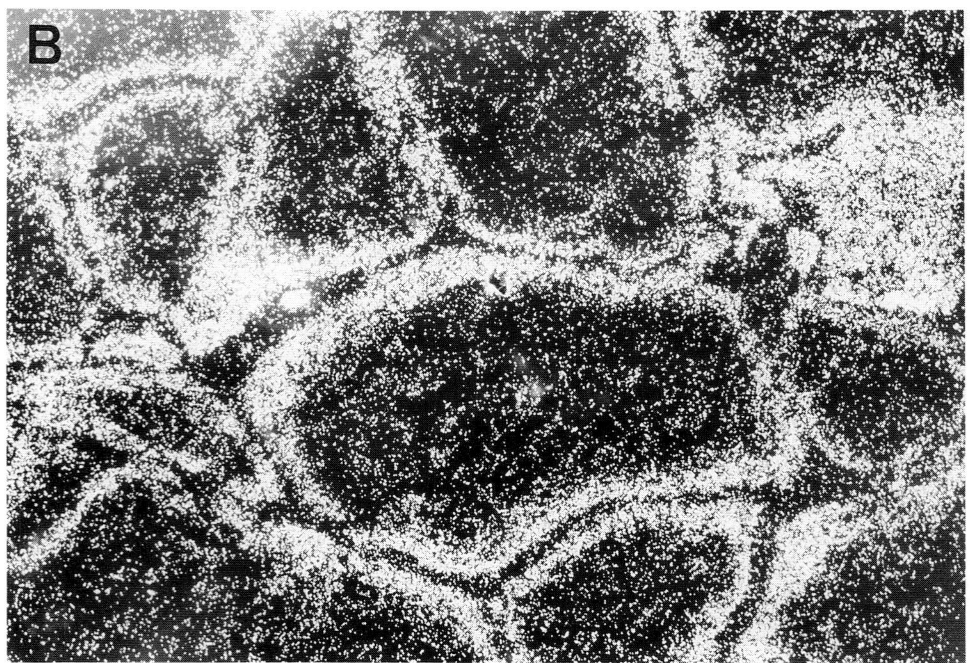
B

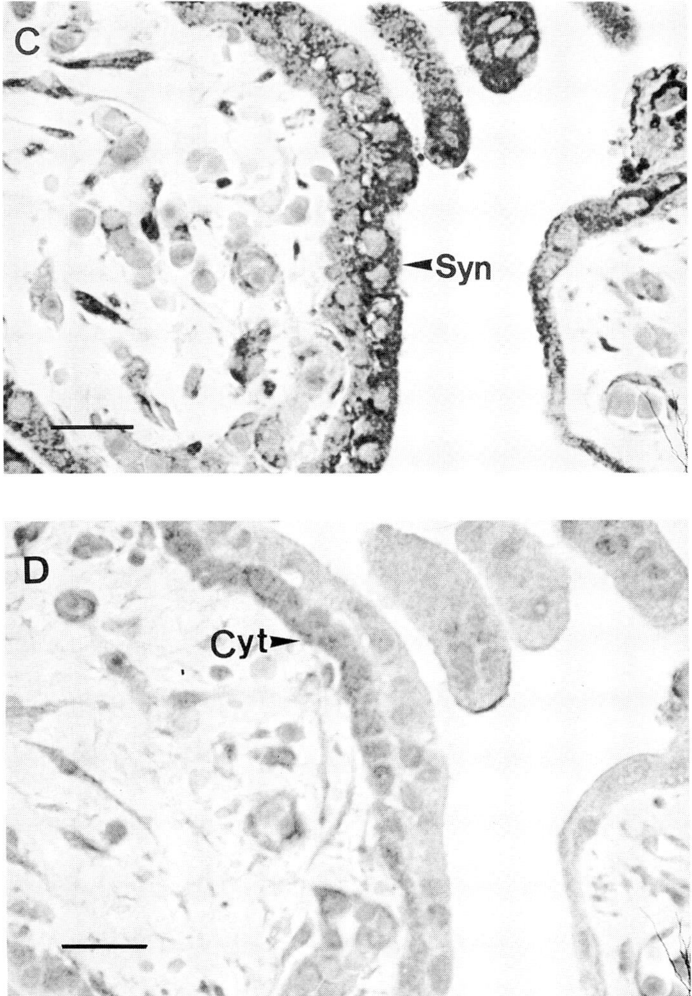

Figure 5. Expression of VEGF mRNA and co-localization of VEGF and Flt-1 in early gestational placental villi. The synthetic oligonucleotide probes (21 bases) directed against base pair 874 to 853 of human VEGF cDNA were tail-labeled with [^{35}S] dATP for in situ hybridization by 3' terminal deoxynucleotidyl transferase using a commercial available kit. Photomicrographs of the tissue hybridized with an antisense probe is shown in bright field (A, facing page). The antisense signal is seen as silver grains in the same field visualized with dark field optics (B, facing page). Panel C show first trimester villi stained with anti-VEGF antibody raised against the synthetic peptide sequence of VEGF 1-20 and panel D shows the same section stained with anti-Flt-1 antibody raised against a peptide corresponding to amino acids 1312-1328 mapping at the carboxy terminus of flt of human cell origin. Strong VEGF staining is seen in or around the syncytiotrophoblast (labeled "syn" and arrowhead in panel C, this page) and cytotrophoblast layer (labeled "cyto"). Positive Flt-1 immunostaining (Panel D, this page) was detected in the bilayer of syncytiotrophoblast and cytotrophoblast and in Hofbauer cells. Bars in A, B, C, and D corresponds to 100 μm.

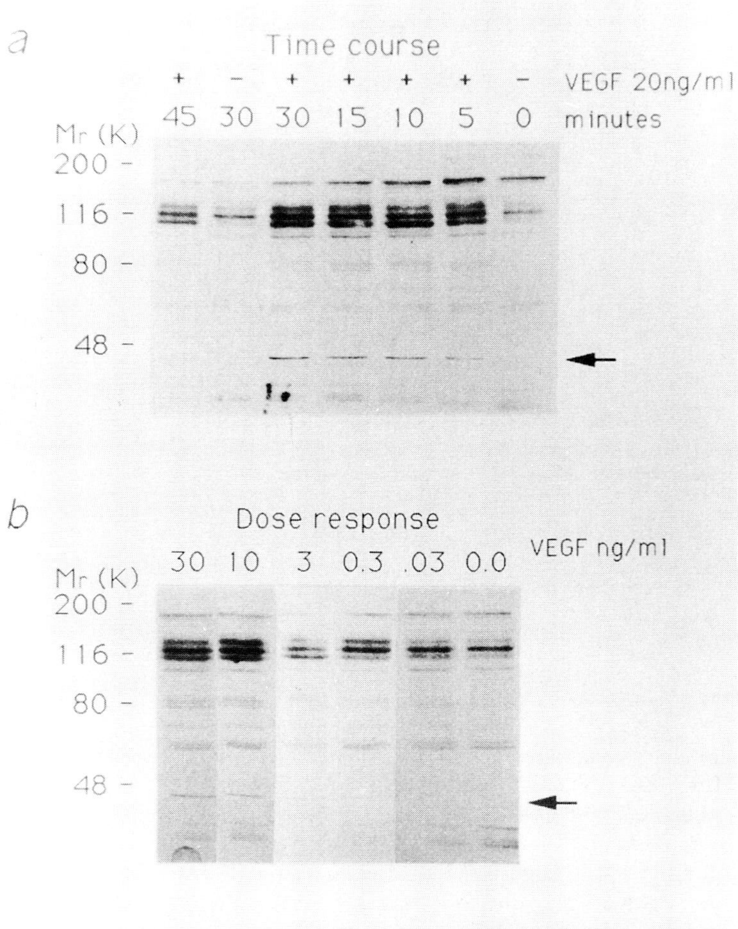

Figure 6. Effect of recombinant $VEGF_{165}$ on protein tyrosine phosphorylation and DNA synthesis in BeWo, a human trophoblast cell line. Tyrosine phosphorylation was measured by lysates immunoblotted using a rabbit polyclonal antiphosphotyrosine antibody (see a and b). Cells were stimulated with VEGF (20 ng/ml) in a time-dependent fashion (a) and exposed to increasing concentration of VEGF (b). DNA synthesis was measured by $[^3H]$-thymidine incorporation (see c, opposite page) and each point represents the mean ± sem of quintuplicate determination of three separate experiments. * $p \leq 0.05$. (Reproduced from *Biol. Reprod.* 51, 524-530, 1994.)

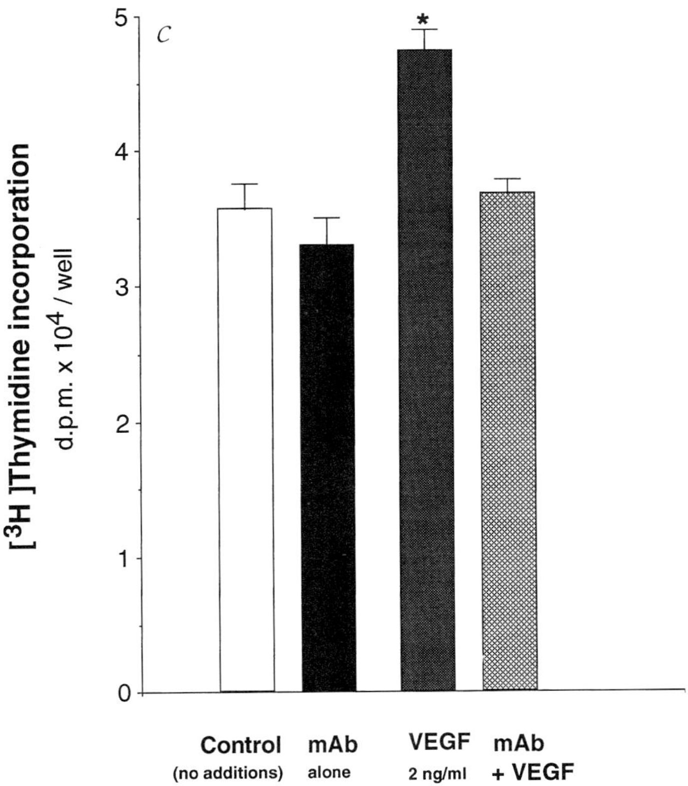

Mutation in the Flt gene leads to disorganized vascular endothelium from the earliest stages of development and affects the development of embryonic and extra-embryonic vessels, endocardium, and the microvasculature (Fong et al., 1995). We have recently demonstrated that one of the functions of Flt-1 receptor is its ability to stimulate NO release in response to VEGF in endothelial cells (Ahmed et al., 1997). Although not eluded to by the authors, a recent *in vivo* study on the effects VEGF in inducing the development of brush-like vessels in the pre-capillary region of chorioallantoic membrane of chicken embryos, also shows an increased vessel diameter (Birkenhager et al., 1996). The only other biological function attributed to the Flt-1 receptor is the ability of peripheral blood monocytes which express Flt-1 and not KDR receptors to migrate in response to VEGF (Clauss et al., 1996).

In addition to the membrane-spanning Flt-1, a cDNA encoding a soluble truncated form of Flt-1 was cloned from human vascular endothelial cell library. The recombinant soluble human Flt receptor (sFlt) binds VEGF with high affinity and inhibits VEGF its mitogenic activity for human umbilical vascular endothelial cells (Kendall and Thomas, 1993). It is possible that sFlt may be an endogenous VEGF antagonist and exogenous sFlt could serve as therapeutic agent for inhibiting excessive angiogenesis.

Cellular Distribution and Function in Placenta

VEGF has been studied in the genital tract and it is known to have a role in the ovary, producing neovascularization of the corpus luteum in the luteal phase of the menstrual cycle (Clark, 1990). In the endometrium, RT-PCR analysis revealed that five different molecular species of VEGF are generated by alternative splicing of mRNA (Charnock-Jones et al., 1993) and immunocytochemistry showed the localization of VEGF immunoreactive protein in human endometrium (Li et al., 1995).

The mRNA encoding VEGF (Sharkey et al., 1993) and its protein (Ahmed et al., 1995) are expressed by human villous and extravillous trophoblast and maternal macrophages while mRNA encoding the Flt-1 is highly expressed in cytotrophoblast shell and columns and also in trophoblast-like choriocarcinoma cell line BeWo (Charnock-Jones et al., 1994). Flt-1 protein is localized predominantly to the extravillous trophoblast and to the syncytio-cytotrophoblast bilayer and endothelial cells of the capillaries within the placental villi and its expression changes throughout gestation (Ahmed et al., 1995). VEGF acts directly on trophoblast (Ahmed et al., 1997, Figure 7) and mRNA encoding KDR, Flt-1 and Flt-4 are expressed in human placenta (Ahmed, unpublished data).

In situ hybridization and immunocytochemcial studies reveal that both VEGF mRNA (Figure 5A and 5B) and VEGF protein (Figure 5C) as well as Flt-1 (Figure 5D) are co-localized in the syncytio-cytotrophoblast bilayer of the first trimester placental villi. This clearly suggests that the VEGF produced by trophoblast acts on trophoblast in an autocrine fashion, which is unlike the paracrine effect of VEGF on endothelial cells. Extensive angiogenesis and extravillous trophoblast migration are required to establish adequate utero-placental vascular structures for exchange. Intense Flt-1 staining and mRNA expression of extravillous trophoblast cells at eight weeks and the localization of VEGF protein in maternal macrophages and decidual cells has led to the hypothesis that VEGF secreted by the maternal cells may act as a chemoattractant for the invading trophoblast. This view was further supported by the observation that Flt immunoreactivity in extravillous trophoblast diminished at twenty-four weeks of gestation when the migration of extravillous trophoblast is completed.

In contrast, staining for Flt-1 in decidual stromal cells increase at twenty-four week gestation and these cells also express strong VEGF immunoreactivity suggesting that VEGF has functions other than cell proliferation and migration during pregnancy (Ahmed et al., 1995). Moreover, Cooper et al. (1996) showed that both VEGF and Flt-1 mRNA expression in the placenta decreased with increasing gestation. These descriptive localization and quantitative expression studies have been challenged by functional studies which show that extravillous trophoblast do not migrate in response to VEGF (P.K. Lala, personal communication). Clearly, descriptive studies may provide a working hypothesis but alone are of limited value, regulation and functional activity are essential if we are to understand the role of VEGF in the placenta.

VEGF immunoreactive protein (Ahmed et al., 1995) and mRNA (Sharkey et al., 1993) are expressed in the Hofbauer cells within the placental villous mesenchyme throughout gestation demonstrating that these cells produce VEGF protein. As VEGF is a secreted protein, its localization in Hofbauer cells which are adjacent to fetal capillaries suggests that VEGF may promote new capillary growth within the placenta itself during placental development (Figure 1). A recent study showed that VEGF mRNA is expressed in cells surrounding the expanding vasculature and was predominantly produced in tissues that acquire new capillary networks (Shweiki et al., 1992). VEGF may also act on Hofbauer cells as they too expressed Flt-1 immunoreactive protein (Ahmed et al., 1995a). This novel finding further demonstrates that Flt-1 receptor is not specific to endothelial cells and that VEGF can act on non-endothelial cells, in particular, trophoblast, and macrophages. As fetal Hofbauer cells within the villi are the key site of VEGF and Flt-1 expression, VEGF may induce production of other angiogenic growth factors from these cells in an autocrine manner. In contrast, KDR receptor has been reported to be localized only on endothelial cells in human placenta (Vuckovic et al., 1996; Clark et al., 1996; Holt et al., 1997). It is important to note that Flk-1, the mouse homolog of human KDR, is highly expressed in trophoblast of midgestation mice (Millauer et al., 1993) and RT-PCR has shown the presence of KDR in the human choriocarcinoma cell line, BeWo (Charnock-Jones et al., 1994) and in non-transformed first trimester cytotrophoblast-like cell line, ED27 (Ahmed et al., 1977). Thus, it may be that this receptor exhibits temporal shifts in expression in trophoblast during human pregnancy that are below the detection limits of *in situ* techniques. Indeed, using immunocytochemistry with microwave procedure, we have localized KDR on trophoblast cells in the placenta and on epithelial cells in the cervix (unpublished data).

Intensive research is under way to establish the functions of VEGF in trophoblast. In BeWo cells, VEGF stimulates tyrosine phosphorylation of 42-44 kDa protein corresponding to mitogen-activated protein kinase [MAP kinase] (Figure 6a and b). This enzyme family plays a pivotal role in the regulation of cell division in a number of cell types (Pulvener et al., 1991; Wood et al., 1992). VEGF is a potent mitogen for endothelial cells but only a relatively weaker mitogen for BeWo cell proliferation (Figure 6C) and may be one of many growth factors involved as regulators of cytotrophoblast growth. The weak mitogenic activity may be due to the expression of KDR receptor in BeWo cells. A recent study reported that VEGF mRNA levels are closely associated to the process of cellular differentiation and that VEGF protein secretion is also induced in adipocyte differentiation (Claffey et al., 1994). One of the putative functions of VEGF in the placenta may be to induce trophoblast differentiation probably via Flt-1 receptor.

Flt-1 is essential for endothelial organization during the development of the mouse embryonic vasculature (Fong et al., 1995). VEGF is likely to play a role in endothelial cell maintenance as non proliferating adult endothelium also express Flt (Jakeman et al, 1992; Peters et al., 1993). Since the trophoblast replace the maternal endothelial cells of blood vessels during trophoblast invasion of the maternal decidua, it is likely that trophoblast have endothelial cell-like properties. At term, when extensive angiogenesis has diminished, sustained localization of VEGF in the syncytiotrophoblast shell and Hofbauer cells indicates that VEGF is also involved in regulating functions of the trophoblast other that migration. VEGF stimulates release of vasorelaxants from non-transformed first trimester trophoblast cells. Trophoblast cells exposed to $VEGF_{165}$ for 20 hours releases $PTHrP_{1-86}$ into the culture medium (Ahmed et al., 1995a). This potent vasorelaxant is localized in the brush border of the syncytiotrophoblast and fetal vessels of placental villi (Emly et al., 1994) and human placenta possesses bioactive PTHrP (Bowden et al., 1994). As PTHrP causes a concentration-dependent vasodilator effect in the dual perfused placental cotyledons that has been preconstricted with a thromboxane mimetic U46619 (Mandsager et al., 1994), it is proposed that the role of VEGF is also to maintain vascular smooth muscle compliance. Moreover, VEGF stimulates nitric oxide (NO) release via Flt-1 receptor in human trophoblast and endothelial cells suggesting that the Flt-1 receptor is likely be to important for maintaining endothelial cell integrity and vessel patency (Ahmed et al., 1997). Although not eluded to by the authors, the recent in vivo studies on the effects of VEGF in inducing the development of brush-like vessels in the pre-capillary region on chorioallantoic membrane of chicken embryos, also shows an increased vessel diameter (Birkenhager et al., 1996).

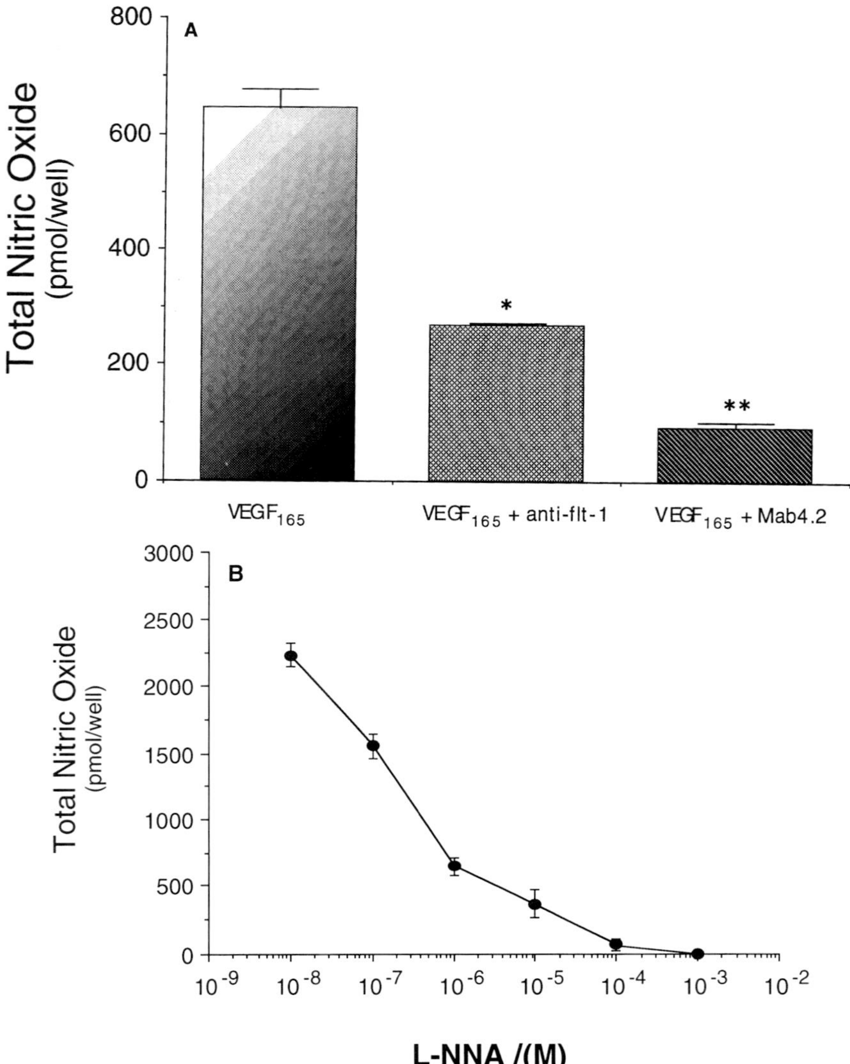

Figure 7. Human recombinant $VEGF_{165}$-evoked nitric oxide release via the Flt-1 receptor in a non-transformed trophoblast cell line. a) Inhibition of NO release by anti-Flt-1 (crossed column) and anti-VEGF (hatched column) antibodies and b) inhibition of NO release by N^G-monomethyl-L-arginine (L-NNA). Cells were grown to confluency in 15% FCS DMEM/Ham-F12 media. Confluent cells were incubated in serum free media for 24 hours. Cells were stimulated with 10 ng/ml $VEGF_{165}$ alone or in the presence of five-fold greater concentration of a polyclonal IgG anti-Flt-1 antibody. Media was collected for assay by the Seivers Chemiluminescence Analyzer which analyses gaseous NO purged from nitrites and nitrates in solution. All results shown have been corrected for background levels of nitric oxide present in the media. Data is expressed in pmoles of nitric oxide per well. * $p \leq 0.02$ and ** $p \leq 0.01$. (Reproduced from *Lab. Invest.* 76(1), 779-791, 1997.)

Human trophoblast contains the high-affinity Flt-1 receptor which triggers the synthesis and release of NO by the activation of constitutive NOS (cNOS). Immunoreactive endothelial isoform of NOS is expressed in human villous trophoblast (Myatt et al., 1993) and extravillous trophoblast (Nanaev et al., 1995). Our studies demonstrates that the Flt-1 receptor is functionally coupled to cNOS activation and NO release (Figure 7). A recent report showed that increased dilation of the uteroplacental arteries in the guinea pig was a prerequisite for trophoblast invasion and remodeling of the endothelium (Nanaev et al., 1995). It is interesting to note that NOS activity is highest during the first trimester (Rutherford et al., 1995), during which time, the placental VEGF levels are also maximal (Khaliq et al., 1996a). Moreover, VEGF does not stimulate the migration of extravillous trophoblast (P.K. Lala, personal communication). We have proposed that it is VEGF and/or other autocoid-mediated trophoblast-derived NO release that promotes the increased dilatation of the spiral arteries required for trophoblast to penetrate into the spiral arteries to replace the smooth muscle and endothelium that is required for the formation of low resistance blood flow seen in the uteroplacenta circulation (Ahmed et al., 1997). Failure to generate adequate amounts of trophoblast-derived NO may predispose women to preeclampsia.

Development of pregnancy-induced hypertension is associated with impaired blood flow to the placenta and coagulopathy leading to severe maternal hypertension and fetoplacental compromise. As VEGF (Sharkey et al., 1994; Ahmed et al., 1995), Flt-1 (Charnock-Jones et al., 1994; Ahmed et al., 1995) and endothelial cNOS (Myatt et al., 1993; Rutherford et al., 1995) are all localized to the syncytium, the layer of trophoblast that line the intervillous space occupying the analogous position of the endothelium in a blood vessel. Since NO at the levels produced by endothelial cNOS has been shown to prevent platelet aggregation and is highly vasodilatory, it is proposed that NO generated following Flt-1 activation by the villous trophoblast may serve to regulate the hemodynamics of the maternal-fetal interface. Decrease in maternal endothelial cNOS activity as seen in pregnancy-induced hypertension (Rutherford et al., 1995) may be a result of reduced expression of VEGF or its Flt-1 receptor.

Although high levels of NO are cytotoxic (Hibbs et al., 1988), lower levels of NO may inhibit growth of mesangial cells (Appel, 1990) and vascular smooth muscle (Garg and Hassid, 1989) without toxic effects. Addition of increasing concentration of sodium nitroprusside (SNP) to trophoblast causes an increase in DNA synthesis at picomolar concentrations, while at nano- and micromolar concentrations, SNP produces marked inhibition in [^{3}H]thymidine incorporation (Ahmed et al., 1997). Although VEGF at 1-2 ng/ml is a weak mitogen for trophoblast (Charnock-Jones et al., 1994). VEGF at 10 ng/ml, when it generates maximal NO production has no mitogenic activity (Ahmed et al., ,1997). More importantly, endogenous basal trophoblast proliferation as measured by DNA synthesis was reported to be markedly potentiated in the presence of anti-Flt-1 neutralizing antibody (Ahmed et al., 1997). The increased DNA synthesis in the presence of anti-Flt-1 antibody may be due to inhibition of NO production. I have proposed that over production of trophoblast-derived NO production via the Flt-1 receptor may negatively regulate DNA synthesis in these cells. Intrauterine growth restriction is characterized not only by the abnormal Doppler wave forms in the umbilical artery but also by the relatively small placenta and depleted numbers of arteries and arterioles in the tertiary stem villi (Jackson et al., 1995). Rutherford et al. (1995) recently reported that NOS activity was elevated in placenta from intrauterine growth restricted (IUGR) pregnancies and my group has shown that the intensity of localization of immunoreactive endothelial cNOS is markedly increased in IUGR compared to normal placentae (Ahmed et al., 1997). These results support my proposition that excess trophoblast-derived NO may limit the growth of the placenta and that one of the functions of Flt-1 receptor is to suppress VEGF-induced proliferation mediated via the KDR receptor both in endothelial cells as well as in trophoblast.

Clearly, so far, the only biological function of Flt-1 receptor following activation by VEGF is the autocrine regulator of NO biosynthesis within the trophoblast. Moreover, Flt-1 receptor is clearly growth suppressive and this appears to be mediated by trophoblast-derived NO. The detection of endogenous soluble Flt suggests that the soluble Flt may regulate the VEGF-mediated effects both in endothelial and trophoblast cells.

Placenta Growth Factor

Following identification and characterization of VEGF as potent angiogenic growth factor, another putative angiogenic growth factor was identified using a placental cDNA library (Maglione et al., 1991). Placenta growth factor (PlGF) is a recently described secreted dimeric growth factor which shares biochemical and structural features with VEGF. In particular, PlGF and VEGF both contain very similar (53% homology) platelet-derived growth factor-like domains. Owing to this similarity with VEGF, it was proposed that PlGF may also act as an angiogenic factor (Maglione et al., 1991; Maglione et al., 1993a). The single copy gene for human PlGF is organized into seven exons and has been mapped to chromosome 14 (Maglione et al., 1993a; 1993b). There are at least two isoforms of PlGF (PlGF-1 and PlGF-2) resulting from alternative splicing of the mRNA encoding monomeric PlGF precursors containing 149 and 170 amino acid residues (Maglione et al., 1993a; Hauser and Weich, 1993). Both forms of PlGF showed very similar competing activity when ^{125}I-VEGF$_{165}$ was used for endothelial cell receptor binding (Hauser and Weich, 1993). PlGF-2 is now known to bind with high affinity to Flt-1 (Park et al., 1994; Kendall et al., 1994) while PlGF-1 appears to be heparin independent.

Cellular Distribution and Function in Placenta

PlGF mRNA was shown to be abundantly expressed in placenta by Northern blot analysis (Maglione et al., 1991, 1993a: Hauser and Weich, 1993) and is detected in a limited number of tissues including the placenta and the kidney (Takahashi et al., 1994). The cellular localization of PlGF in any human tissue until now was not known. Using immunocytochemistry and *in situ* hybridization we have reported the distribution of PlGF mRNA and protein in placenta (Khaliq et al., 1996).

The predominant localization of PlGF mRNA and protein in the vasculosyncytial membrane of the placental villi (Figure 8A and 8B) indicates that PlGF may act as a paracrine mediator of blood vessel formation during placental angiogenesis. This is supported by the fact that PlGF mRNA is also markedly elevated in the hypervascular renal cell carcinoma tissues (Takahashi et al., 1994). The detection of PlGF protein in the placenta demonstrates that the mRNA for PlGF is efficiently translated into the protein.

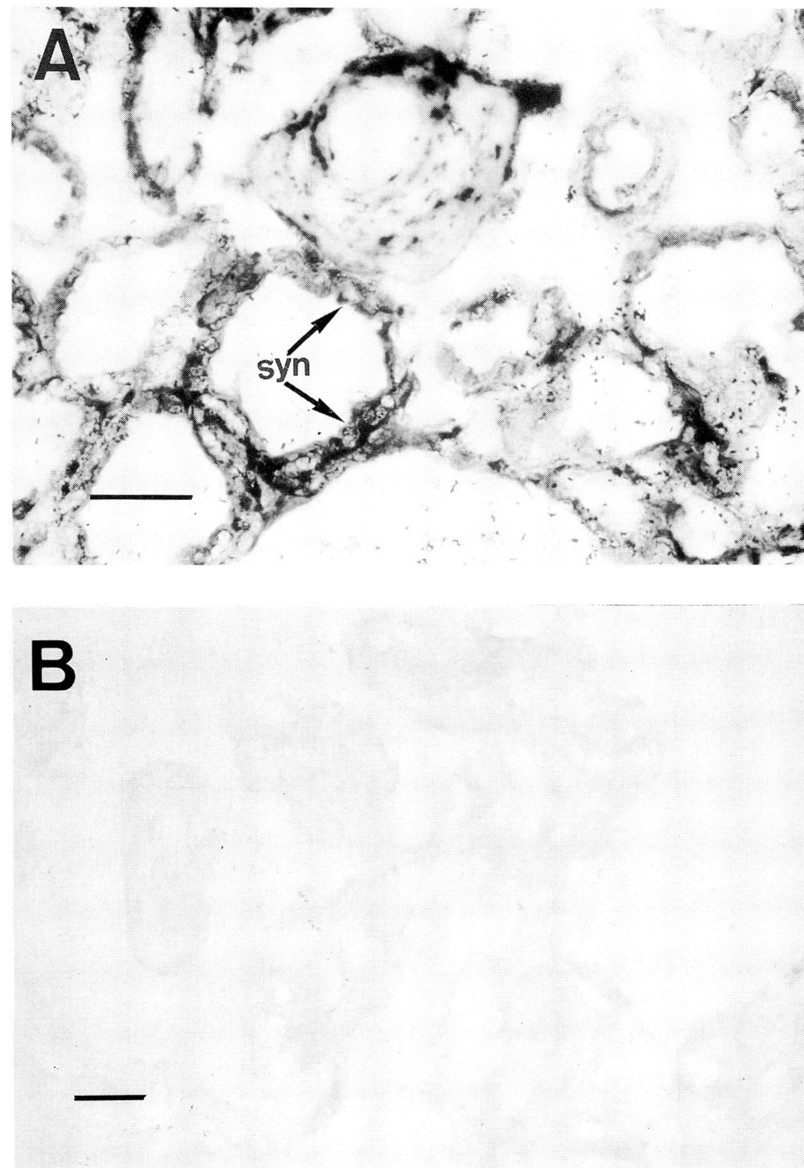

Figure 8. Localization of PlGF mRNA expression in term placenta. Panel A is a photomicrograph of placental section after hybridization with fluorescein-labeled PlGF-2 anti-sense riboprobe and shows intense staining for PlGF mRNA in the vasculosyncytial membrane of villous trophoblast (syn). Weak staining is observed within the villous mesenchymal tissue. Panel B is a photomicrograph of placental section after hybridization with the control fluorescein-labeled PlGF-2 sense riboprobe and shows negligible hybridization with PlGF mRNA. Bars correspond to 50 μm for plate A and B. (Reproduced from *Growth Factors* 13, 243-250, 1996.)

PlGF was initially reported to stimulate endothelial cell proliferation. Data showing that PlGF stimulates the proliferation of the CPA endothelial cell line (Maglione et al., 1991) and bovine aortic endothelial cells (Hauser and Weich, 1993) may be misleading as they were obtained from crude conditioned medium after transfection of insect cells with PlGF cDNA. Recently, Di Salvo et al. (1995) have shown that pure PlGF monodimers are mitogenic for cultured human umbilical vein endothelial (HUVE) cells only at non-physiological concentration (Di Salvo et al., 1995). The likely explanation for the biological activity of crude conditioned medium may be that the recombinant human PlGF formed a more active heterodimer with the host cell VEGF. Cao and colleagues (1996) have shown that the VEGF:PlGF heterodimer is less potent than VEGF in stimulating DNA synthesis in HUVE cells, but is more potent than PlGF alone (Cao et al., 1996). In the CAM assay, PlGF did not induce angiogenesis and PlGF (<100 ng/ml) has no effect on human vascular endothelial cell DNA synthesis or proliferation (Birkenhager et al., 1996).

Unlike VEGF which induces vascular permeability in the Miles Assay (Connolly et al., 1989; Keck et al., 1989), PlGF has no effect on vascular permeability (Park et al., 1994). However, PlGF was demonstrated to potentate the permeability action of low doses of VEGF, that alone were ineffective at inducing vascular permeability. This type of potentiation of VEGF activity was also demonstrated on the mitogenesis of adrenal cortex capillary endothelial cells (Park et al., 1994). It is proposed that PlGF may potentate the actions of VEGF on vascular endothelial cells and therefore activate KDR.

PlGF may also act in an autocrine fashion on trophoblast differentiation but at physiological concentration $\leq$ 100 ng/ml does not stimulate trophoblast DNA synthesis or nitric oxide release (unpublished data from my laboratory). I would like to propose that PlGF may be important in controlling the degree of neovascularization via the Flt-1 receptor by regulating the interaction of VEGF with Flt-1. The biological actions of PlGF in the trophoblast remains to be elucidated and is under investigation in this laboratory.

Semi-quantitation of both PIGF and VEGF by western blotting has demonstrated that levels of PIGF increase from first trimester to term placenta while in contrast, the levels of VEGF decrease from first trimester to term placenta (Khaliq et al., 1996b). This observation suggests there is a fine balance between the levels of PIGF and VEGF during the development of the placenta. However, the factors regulating this balance are as yet undefined. One plausible factor may be the local oxygen environment within the placenta during gestation. The first trimester is characterized as hypoxic (lacking a blood circulation) where the partial pressure of oxygen (pO_2) in the intervillous space is around 18 mm Hg. True intravillous blood flow is established at about 12 weeks gestation. As the placenta progresses through towards the second trimester and maternal blood flow starts, the pO2 increases to around 61 mm Hg and probably increases further towards the end of the third trimester (Rodesch et al., 1992). Oxygen is know to be an important mediator of cell proliferation, growth factors production and cell invasion (Boulton et al., 1994; Khaliq et al., 1995; Genbacev et al., 1996). *In vitro* studies have demonstrated that hypoxia decreases the steady state of mRNA levels of PIGF and increases VEGF mRNA, while higher oxygen concentrations are know to elevate the levels of PIGF mRNA and decrease VEGF mRNA (Gleadle et al., 1995a, b; Holt et al., 1997). These observations support our hypothesis that oxygen is a key regulator of the balance between PIGF and VEGF in the placenta during gestation. This may explain some of the changes observed under pathological conditions which are thought to be related to intrauterine hypoxia, such as an increase in the mitotic index of cytotrophoblast cells (Fox, 1970; Arnholdt et al.,

1991), reduced thickness in the syncytiotrophoblast and increased syncytial knots (Alvarex et al., 1969, 1970). Consequently, these changes may arise due to an alteration in the homeostasis of local oxygen-sensitive growth factors such as PIGF and VEGF. Genbacev and colleagues also demonstrated that hypoxia increased the incorporation of [3H]-thymidine and 5-bromo-2'-deoxyuridine in trophoblast cells and inhibited their invasive capacity by altering their integrin profile (Genbacev et al., 1996). Furthermore, since hypoxia is known to modulate the expression of Flt-1 and KDR (Thieme et al., 1995; Nomura et al., 1995) it is also possible that direct regulation of Flt-1 and KDR expression by oxygen may also be important in the development of the placenta.

VEGF and PIGF share a common receptor, namely Flt-1, while PIGF does not bind the other VEGF receptor, KDR (Park et al., 1994; Terman et al., 1994; Sawano et al., 1996). Recent studies in our laboratory have demonstrated that stimulation of Flt-1 receptor by VEGF increases the production of NO in cultured trophoblast cells (Ahmed et al., 1997). In contrast we demonstrated that low doses of PIGF downregulate basal production of NO, and that VEGF could eliminate this down regulation (Khaliq et al., 1996b). This is not surprising since VEGF has a higher affinity for the Flt-1 receptor (Park et al., 1994). High doses of PIGF potentiate the permeability action of ineffective doses of VEGF. This type of potentiation of VEGF activity was also demonstrated on the mitogenesis of adrenal cortex capillary endothelial cells (Park et al., 1994). The fact that PIGF can induce the phosphorylation of a 38 kDa protein while VEGF does not (Cao et al., 1996) supports the view that PIGF and VEGF can exert different biological actions through the same receptor. This is further supported by our observation that VEGF stimulates NO release, while PIGF appears to inhibit it. Furthermore, naturally occurring VEGF:PIGF hetrodimers have recently been identified, and have less affinity for the Flt-1 and KDR receptors than VEGF, but have a higher affinity for the Flt-1 receptor than PIGF (DiSalvo et al., 1995; Cao et al., 1996; Birkenhager et al., 1996). These studies suggest that regulation of both VEGF and PIGF activity may also be regulated by VEGF:PIGF heterodimer formation. We propose that the changes observed in cytotrophoblast in preeclampsia and IUGR are mediated by a fine balance between VEGF and PIGF, which in turn is regulated by the changes in the oxygen environment within the placenta under pathological conditions.

Hepatocyte Growth Factor

Hepatocyte growth factor (HGF) is a pleiotrophic growth factor with angiogenic properties which regulates cell growth, motility, and morphogenesis (Bussolino et al., 1992) all of which are important in placental development. HGF is a heterodimeric molecule composed of a 69 kDa α-chain and a 34 kDa β-chain, linked by one disulfide bridge (34 kDa). The α-chain of HGF has four kringle domains and β-chain has a serine-protease like domain (38% homology with plasminogen) without catalytic activity (Boros and Miller, 1995). The functions attributed to the α-chain include both receptor and heparin binding (Lyon et al., 1994). Single chain HGF has no apparent biological activity and the activation of HGF is coupled with conversion from a single- to a two-chain HGF (Boros and Miller, 1995). This growth factor is predominantly expressed in cells of mesenchymal origin, including fibroblast cells (Stoker et al., 1987), smooth muscle (Rosen et al., 1990a), placenta (Rosen et al., 1990b), and endothelial cells (Noji et al., 1990). Experiments in mice using a targeting construct to delete the first kringle domain for HGF required for biological activity indicates that production of this cytokine within the embryonic period appears to be essential for fetal growth and development of the liver and the placenta (Schmitdt et al., 1995; Uehara et al., 1995).

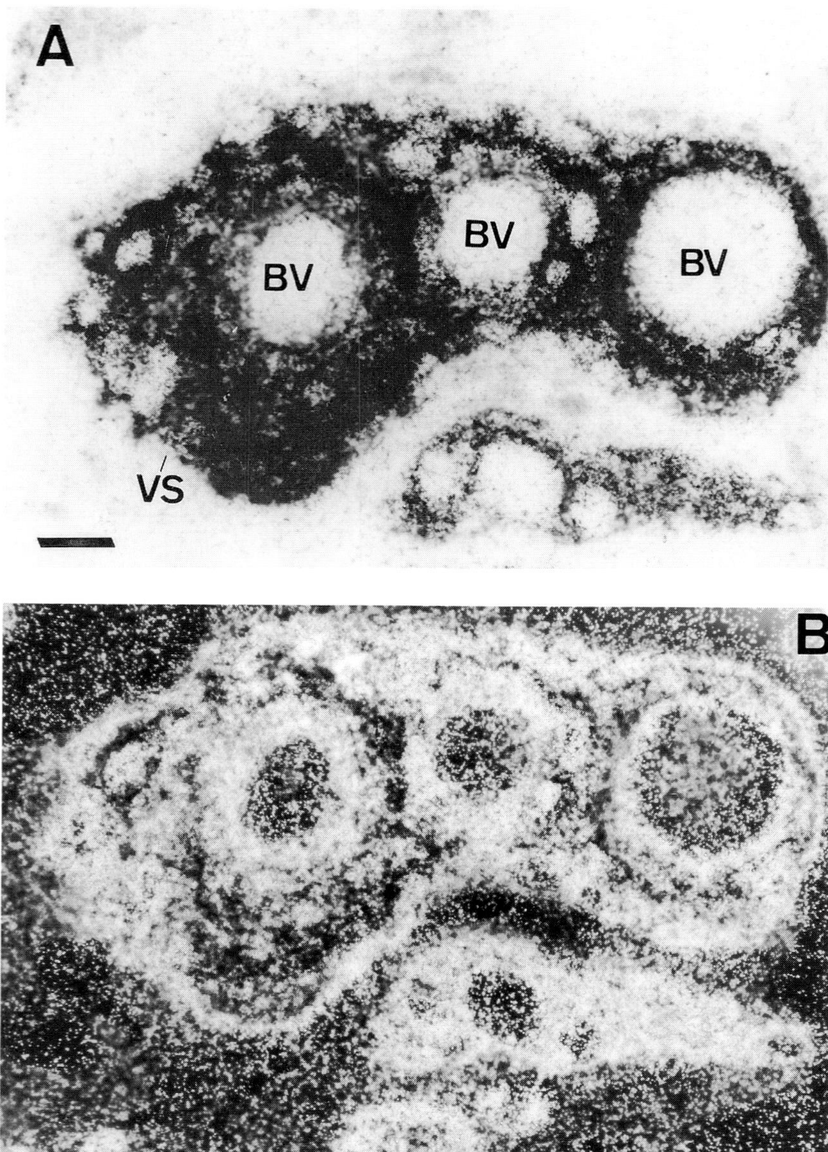

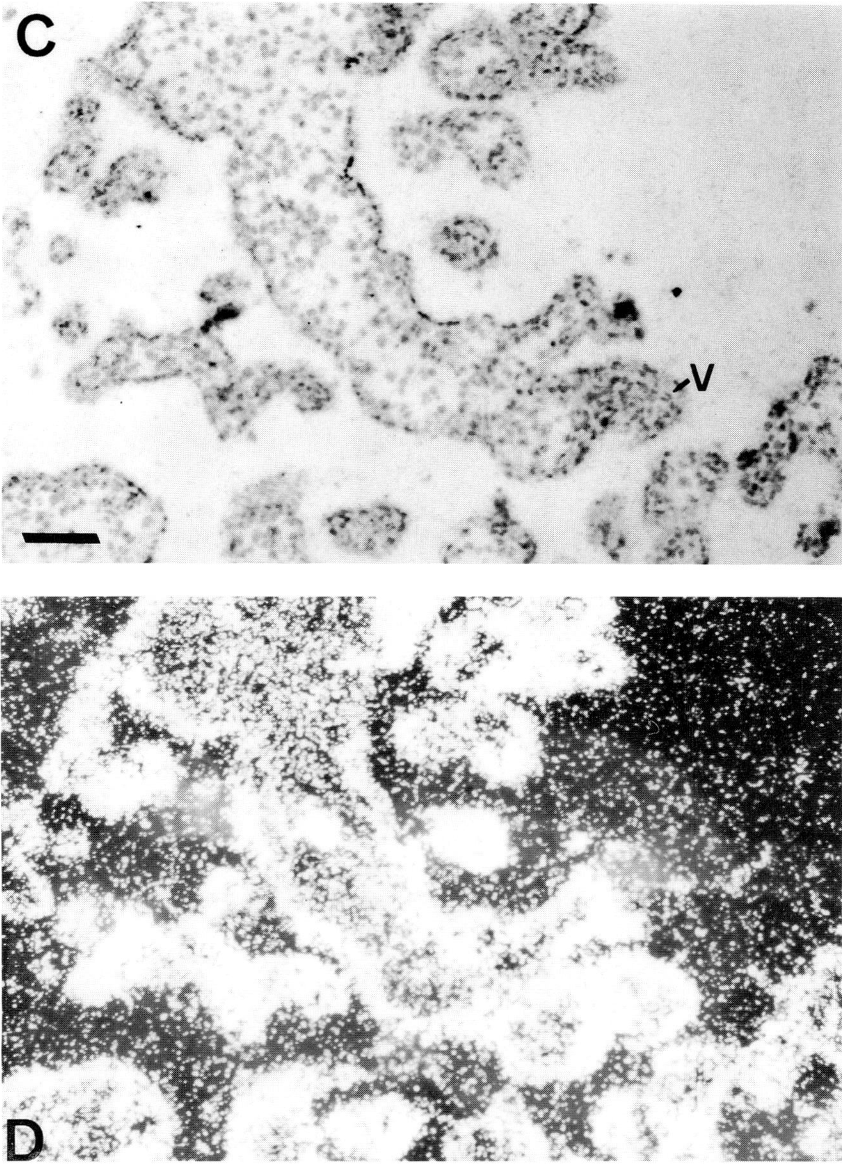

Figure 9. Cellular expression of HGF and c-met mRNA in human term placenta. The brightfield (A and C, upper panels, both pages) and darkfield (B and D, lower panels, both pages) photomicrographs of tissue hybridized with the antisense ^{35}S-labeled cRNA probe for HGF (A and B, opposite page) and c-met (C and D, this page) show intense expression of HGF mRNA in the perivascular stroma of the villous mesenchyma, including the smooth muscle within the arterioles (BV). The hybridization signal for c-Met is most intense in the vasculosyncytial (vs.) membrane. Bars correspond to 50 μm for A and B and 100 μm for C and D. (Reproduced from *Growth Factors* 13, 133-139, 1996.)

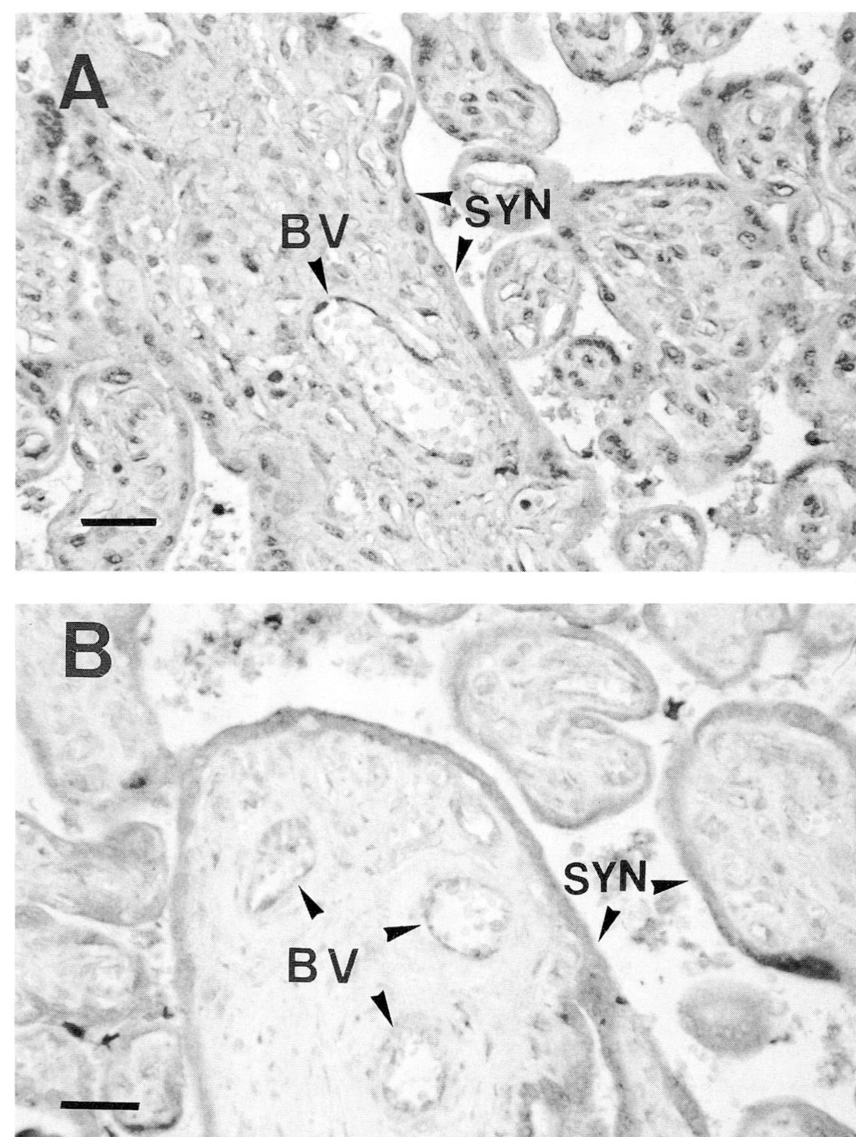
A
SYN
BV
B
SYN
BV

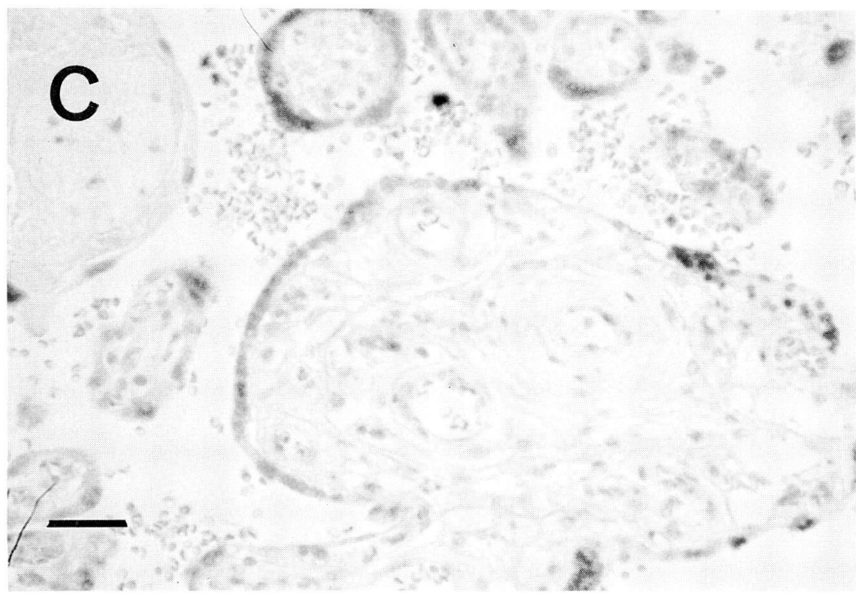

Figure 10. Localization of immunoreactive HGF and c-Met in human term placenta. Positive immunoreactivity to the HGF polyclonal antibody (A, opposite page) and c-Met polyclonal antibody (B, opposite page) was noted in the villus syncytium (syn) and the vascular endothelial cells of blood vessels (Bv). No c-met staining of the mesenchymal stroma of the villous core was noted. In control section (C, this page), no positive staining was found when primary c-Met antibody was omitted. Bars correspond to 50 μm for A, B, and C. (Reproduced from *Growth Factors* 13, 133-139, 1996.)

HGF Receptor c-Met

High-affinity receptor for HGF is a trans-membrane protein encoded by the proto-oncogene *c-met*. The c-Met/HGF receptor is a heterodimeric molecule composed of the 50 kDa α-subunit and the 145 kDa β-subunit. The intracellular domain of the β subunit has tyrosine kinase activity and transduces the majority of the effects of HGF (Boros and Miller, 1995). Binding of HGF to the low-affinity binding sites is completely displaced by heparin suggesting that heparan sulphate proteoglycans in the cell surface and extracellular matrix offer low-affinity binding sites to HGF (Mizuno et al., 1994).

Cellular Distribution and Function in Placenta

HGF protein extracted from human (Rosen et al., 1990) and rat (Jin et al., 1993) placentae has 'scatter factor' activity to cultured Madin-Darby canine kidney cells. The expression of HGF and c-met mRNA as measured by Northern blot analysis was described during development of the rat placentae (Jin et al., 1993).

The rat HGF and c-met mRNA seemed to correlate with HGF *in vitro* activity, as measured by 'cell scatter' with both HGF and its receptor mRNA reaching a maximal level in six days (first trimester) and then declining (Jin et al., 1993). Immunolocalization of HGF was previously described both in human (Wolf et al., 1991a) and rat placentae (Wolf et al., 1991b) which reported that HGF protein was predominantly localized in the villous syncytium and endothelial cells. Studies from this laboratory have described the cellular distribution of HGF and c-met mRNA by *in situ* hybridization in human placenta and fetal membranes (Kilby et al., 1996). HGF mRNA is intensely localized in the villous mesenchymal core, particularly the perivascular stroma and the vascular smooth muscle (Figure 9A and B). No HGF mRNA is expressed in the villous syncytium while strong immunoreactive HGF is detected in the villous syncytium (Figure 10A). The mesenchymal core of the villi shows a weak generalized immunostaining to HGF (Figure 10A). The mRNA encoding c-Met (Figure 9C and D) and immunoreactive c-Met protein (Figure 10B and C) are expressed in the vasculosyncytial membrane and the endothelial cells lining the villous vasculature indicating that HGF has a paracrine action on trophoblast.

Immunostaining of HGF in the absence of hybridization signal for HGF mRNA in trophoblast suggests that immunoreactivity may be due to binding of secreted HGF to its receptor. HGF also binds to heparan sulphate protoglycans which are present in the extracellular matrix between these cell types and may partially explain the diffused immunostaining of HGF in placental tissue.

Unlike VEGF and PlGF, HGF is a potent stimulator of both cell proliferation and DNA synthesis in a variety of cell types (Boros and Miller, 1995) including trophoblast (Kilby et al., 1996). *In vitro,* HGF stimulates proliferation and migration of vascular endothelial cells (Morimoto et al., 1991; Bussolino et al., 1992) and when HGF is implanted into mice or into the acascular cornea, HGF induced neovascularization *in vivo,* without inducing significant inflammation reaction (Grant et al., 1993). The presence of HGF mRNA in the peri-vascular stromal, vascular smooth muscle cells of the placenta villi and the placental membranes indicates that HGF synthesized in these tissues may act to promote trophoblast growth in a paracrine fashion and on the vascular endothelial cells lining the villous vasculature to initiate neocapillary formation, probably in concert with other angiogenic factors, in an autocrine manner.

Recent 'knock out' studies in mice using homologous recombination to one parental gene encoding HGF, indicates that this pleiotrophic growth hormone is essential for fetoplacental growth and development (Uehara et al., 1995). The spatial distribution of HGF and the c-Met within the human placenta suggests potential physiological actions, including angiogenesis and trophoblast cell proliferation. Since such functions are an integral component of placental development, and consequently fetal well-being, it remains to be determined whether alterations in HGF or its receptor may compromise human pregnancy.

Other Angiogenic Factors

This review has focused on the structure and functions of the angiogenic polypeptide growth factors with heparin binding activity in the placenta.

There are a wide range of other factors with the ability to stimulate new blood vessel formation(Table I) which may also play an important role in the placenta and need to be investigated. Among the non-heparin binding agents include platelet-derived endothelial cell growth factor (PD-ECGF), epidermal growth factor (EGF) and transforming growth factor-alpha (TGF-α), all of which stimulate proliferation of endothelial cells in vitro and induces angiogenesis *in vivo*, although EGF is less potent than TGF-α. In contrast, transforming growth factor-beta (TGF-β) and tumor necrosis factor alpha (TNF-α) inhibit endothelial cell proliferation *in vitro* but stimulate angiogenesis *in vivo* and promote tube formation *in vitro* (cf., Klagsburn and d'Amore, 1991). Both these factors may influence angiogenesis by promoting endothelial cell differentiation such as tube formation and matrix production, rather than proliferation. Alternatively, TGF-β and TNF-a produce angiogenesis by stimulating other cells such as trophoblast and macrophages to produce other angiogenic factors. The expression of FGF-2 is increased by TNF-α. All these growth factors have been described in pregnancy (Cross et al., 1994). In addition, there are low molecular weight peptides such as angiotensin II (Le Noble et al., 1993) and bradykinin (Hu and Fan, 1993) and non-peptides such as prostaglandins (PGE) and platelet-activating factor (PAF) which promote neovascularization. More recently, soluble forms of adhesion molecules such as vascular cell adhesion molecule-1 (VCAM-1) were described as inducers of angiogenesis (Koch et al., 1995).

CONCLUSION AND CLINICAL SIGNIFICANCE

Successful pregnancy depends upon placental growth and fetal development which relies on a series of complex coordinated interactions between maternal and fetal tissues that is facilitated by the trophoblast. It has been postulated that capillary loop elongation within intermediate villi may have a key role to play in the placental remodeling that occurs with gestation (Kaufmann et al., 1988) and this has been confirmed both stereologically and morphologically (Jackson et al., 1992). The growth factors described in this review may play a key role in these processes. Some of the key growth factors with angiogenic properties but equally important are anti-angiogenic factors and soluble receptors which control the activity of angiogenic factors have not been described in this review. Regulation of angiogenesis in pregnancy is tightly controlled and probably involves not one but an orchestra of these growth factors acting in a co-ordinated fashion under the influence of endocrine factors such as ovarian steroid hormones, and oxygen.

Failure of trophoblast to invade the maternal spiral arterioles is associated with defective vascular development and growth retardation. The intense expression of VEGF protein in the maternal decidua and the high NOS activity in first trimester together with the localization of Flt-1 and cNOS in extravillous trophoblast suggests one of the primary roles of VEGF in early pregnancy may be to promote trophoblast to penetrate the spiral arteries. Low levels of VEGF secretion in maternal decidua and/or poor expression of Flt-1 on extravillous trophoblast may be a contributory factor in intrauterine growth retardation and preeclampsia. It is also possible that increased levels of endogenous soluble Flt may antagonize the beneficial effects of VEGF.

The spatial distribution of HGF mRNA surrounding the villous vasculature may indicate that, although HGF may have a site of action on the placental vasculature, it may also be transported in the 'portal' fetoplacental circulation producing proliferation and growth of liver cells. HGF has also been noted to be important in embryonic and fetal liver development (Schmidt et al., 1995). Indeed, in severe fetal growth restriction, secondary to uteroplacental insufficiency in which perinatal death occurs, reduced hepatic size and hemilateral degeneration have been commonly described (Daamen and Schaberg, 1968). Based on the promising finding that exogenously injected HGF enhances regeneration of the liver, kidney, and lung in experimental animals, it is possible that HGF may become an effective treatment to limit hepatic and renal dysfunction.

Increasing evidence suggests that the angiogenic factors are involved in the maintenance of vessel patency of the utero-placental and fetoplacental circulation by releasing vasorelaxants. In both endothelial and trophoblast cells FGF-2 and VEGF have been reported to stimulate NO biosynthesis. The potential implication of VEGF-mediated NO synthesis and release from human trophoblast are far reaching. Pre eclampsia is a typical disease of pregnancy and is associated with endothelial cell dysfunction (Roberts et al., 1989). Its symptoms are characterized by a progressive hypertension, pathologic edema, proteinuria and abnormal clotting, and failure of liver and renal function. Pre eclampsia is associated with impaired blood flow to the placenta and since NO at the levels produced by endothelial cNOS has been shown to prevent platelet aggregation and is highly vasodilatory, it is highly probable that NO produced by VEGF from the trophoblast may serve to regulate the hemodynamics at the maternal-trophoblast interface. In the next few years, we will have a greater understanding of the role of placental angiogenic factors as their mechanisms of actions are elucidated.

ACKNOWLEDGEMENTS

Most of the work described here was supported by a grant from the Wellcome Trust (grant no. 042930/z/94).

REFERENCES

Ahmed, A. and Smith, S.K. (1992) Platelet activating factor stimulates phospholipase C activity in human endometrium. *J. Cell.Physiol.* 152, 207-214.

Ahmed, A., Plevin, R., Shoaibi, M. A., Fountain, S. A., Ferriani, R.A. and Smith, S.K. (1994) Basic FGF activates phospholipase D in endothelial cells in the absence of inositol-lipid hydrolysis. *Am. J. Physiol.* 266, C206-C212.

Ahmed, A., Li, X.F., Dunk, C., Whittle, M.J., Rushton, I.D., and Rollason, T. (1995) Colocalisation of vascular endothelial growth factor and its Flt-1 receptor in human placenta. *Growth Factors* 12, 235-243.

Ahmed, A., Dunk, C., Kniss, D., and Wilkes, M. (1997) Role of VEGFR-1 (Flt-1) receptor in mediating calcium-dependent nitric oxide release and limiting DNA synthesis in human trophoblast. Lab. Invest. 76(1), 779-791.

Alvarez, H., Benedetti, W.L., Morel, R.L., and Scavarelli, M. (1970) Trophoblast development gradient and its relationship to placental hemodynamics. *Am. J. Obstet. Gynecol.* 10, 416-420.

Alvarez, H., Morel, R.L., Benedetti, W.L., and Scavarelli, M. (1969) Trophoblast hyperplasia and maternal arterial pressure at term. *Am. J. Obstet. Gynecol.* 105, 1015-1021.

Appel, R.G. (1990) Mechanism of atrial natriuretic factor-induced inhibition of rat mesangial cell mitogenesis. *Am. J. Physiol.* 259, E312-E318.

Arnholdt, H., Mesiel, F., Fandrey, K. and Lohrs, U. (1991) Proliferation of villous trophoblast of the human placenta in normal and abnormal pregnancies. *Virchows. Arch. B. Cell. Pathol.* 60, 365-372.

Ausprunk, D.H. and Folkman, J. (1977) Migration and proliferation of endothelial cells in preformed and newly formed blood vessels during tumor angiogenisis. *Microvas. Res.* 14, 53-56.

Basilico, C. and Moscatelli, D. (1992) The FGF family of growth factors and oncogenes. *Adv. Cancer Res.* 59, 115-165.

Benirschke, K. and Kaufmann, P. (1990) *Pathology Of The Human Placenta* (2nd Editon), Springer-Verlag:Berlin, pp. 98.

Birkenhager, R., Schneppe, B., Rockl, W., Wilting, J., Weich, H.A. and McCarthy, J.E.G. (1996) Synthesis and physiological activity of heterodimers comprising different splice forms of vascular endothelial growth factor and placenta growth factor. *Biochem. J.* 316, 703-707.

Boros, P. and Miller, C.M. (1995). Hepatocyte growth factor: A multifunctional cytokine. *Lancet* 345, 293-295.

Boulton, M.E., Khaliq, A., McLeod, D. and Moriarty, P. (1994) Effect of 'Hypoxia' on the proliferation of retinal microvascular cells in vitro. *Exp. Eye Res.* 59, 243-246.

Bowden, S.J., Emly, J.F., Hughes, S.V., Powell G., Ahmed. A., Whittle, M.J., Ratcliffe, J.G., and Ratcliffe, W.A. (1994) Parathyroid hormone-related protein (PTHrP) in term placenta and membrane. *J. Endocrinol.* 142, 217-224.

Broadley, K.N., Aquino, A.M., Woodward, S.C., Buckley, S.A., and Sato, Y. (1989) Monospecific antibodies implicate basic fibroblast growth factor in normal wound repair. *Lab. Invest.* 61, 571-75.

Burgess, W.H. and Maciag, T. (1989) Heparin binding growth factor family of proteins. *Annu. Rev. Biochem.* 58, 575-606.

Bussolino, F., Di Renzo, M.F., Ziche, M., Bocchietto, E., Olivero, M., Naldini, L., Gaudino, G., Tamagnone L., Coffer, A., and Comoglio, P.M. (1992) Heptocyte growth factor is a potent angiogenic factor which stimulates endothelial cell motility and growth. *J. Cell. Biol.* 119, 629-641.

Cao, Y., Chen, H., Zhou, L., Chiang, M-K., Anand-Apte, B., Weatherbee, J.A., Wang, Y., Fang, F., Flanagan, J.G., and Tsang M.L-S. (1996) Heterodimers of placenta growth factor/vascular endothelial growth factor: Endothelial activity, tumor cell expression, and high affinity binding to flk-1/KDR. *J. Biol. Chem.*, 271, 3154-3162.

Cattini, P.A., Nickel, B., Bock, M. and Kardami, E. (1991) Immunolocalization of basic fibroblast growth factor (bFGF) in growing and growth-arrested placental cells: A possible role for bFGF in placental cell development. *Placenta* 12, 341-352.

Charnock-Jones, D.S., Sharkey, A., Rajput-Williams, J., Burch, D., Schofield, J.P., Fountain, S.A., Boocock, C.A. and Smith, S.K. (1993) Identification and localisation of alternately spliced mRNAs for vascular endothelial growth factor in human uterus and estrogen regulation in endometrial carcinoma cell lines. *Biol. Reprod.* 48, 1120-1128.

Charnock-Jones, D.S., Sharkey, A.M., Boocock, C.A., Ahmed, A., Plevin, R., Ferrara, N. and Smith, S.K. (1994) Localisation and activation of the receptor for vascular endothelial growth factor on human trophoblast and choriocarcinoma cells. *Biol. Reprod.* 51, 524-530.

Claffey, K.P., Wilkison, W.O. and Spiegelman, B.M. (1994) Vascular endothelial growth factor-regulation by cell differentiation and activated second messenger pathways. *J. Biol. Chem.* 267, 16317-16322.

Clark, J.G. (1990) The origin, development and degeneration of the blood vessels of the human ovary. *Johns Hopkins Hosp. Rep.* 9, 593-676

Claus, M., Weich, H., Breier, G., Knies, U., Rock, W., Waltenberger, J. and Risau, W. (1996) The vascular endothelial growth factor receptor Flt-1 mediated biological activities. *J. Biol. Chem.* 271, 17629-17634.

Cliff, W.J. (1965) Kinetics of wound healing in rabbit ear chamber, a time lapse cinemicroscopic study. *Q. J. Exp. Physiol.* 50, 79-89.

Connolly, D.T., Olander, J.V., Heuvelman, D., Nelson, R., Monsell, R., Siegel, N., Haymore, B.L., Neimgruber, R. and Feder, J. (1989) Human vascular permeability factor. Isolation from U937 cells. *J. Biol. Chem.* 264, 20017-20024.

Cooper, J.C., Sharkey, A.M, Charnock-Jones, D.S., Palmer, C.R. and Smith, S.K. (1996) VEGF mRNA levels in placentae from pregnancies complicated by pre-eclampsia. *Br. J. Obstet. Gynaecol.* 103, 1191-1196.

Cross, J.C., Werb, Z. and Fisher, S.J. (1994) Implantation and the placenta: Key pieces of the developmental puzzle. *Science* 266, 1508-1518.

Daamen, C.B.F. and Scaberg, A. (1968). Hemilateral liver degeneration in perinatal death. *J. Pathol.* 97, 29-34.

De Vries, C., Escobedo, J.A., Ueno, H., Houck, K., Ferrara, N. and Williams, L.T. (1992) The fms-like tyrosine kinase, a receptor for vascular endothelial growth factor. *Science* 255, 989-991.

Delli-Bovi, P., Curatola, A.M., Kern, F.G., Greco, A., Ittmann, M. and Basili-Co, C. (1987) An oncogene isolated by transfection of Kaposi's sarcoma DNA encodes a growth factor that is a member of the FGF family. *Cell* 50, 729-737.

Di Salvo, J., Bayne, M.L., Conn, G., Kwok, P.W., Prashant, G.T., Soderman, D.D., Palisi, M.T., Sullivan, K.A. and Thomas, K.A. (1995) Purification and characterisation of a naturally ocurring vascular endothelial growth factor placenta growth factor heterodimer. *J. Biol. Chem.* 270, 7717-7723.

Dickson, C., Smith, R., Brookes, S. and Peters, G. (1990) Proviral insertions within the int-2 can generate multiple anomalous transcript but leave the protein-coding domain intact. *J. Virol.* 64, 784-93.

Docherty, A.J.P. and Murphy, G. (1990) The tissue metalloproteinase family and inhibitor TIMP: A study using cDNAs and recombinant proteins. *Ann. Rheum. Dis.* 49, 469-479.

Dvorak, H.F., Brown, L.F., Detmar, M. and Dvorak, A.M. (1995) Vascular permeability factor/vascular endothelial growth factor, microvascular hyperpermeability, and angiogenesis. *Am. J. Path.* 146, 1029-1039.

Emly, J.F., Gregory, J., Bowden, S.J., Ahmed, A., Whittle, M.J,. and Ratcliffe, W.A. (1994) Immunohistochemical localisation of parathyroid hormone-related protein (PTHrP) in human term placenta and membranes. *Placenta* 15, 653-660.

Emonard, H., Christiane, Y., Smet, M., Grimaud, J.A., and Foidart, J.M. (1990) Type IV and interstitial collagenolytic activities in normal and malignant trophoblast cells are specifically regulated by the extracellular matrix. *Inv. Met.* 10, 170-177.

Feinberg, R.F., Kao, L., Haimowitz, J.E., Queenan, J.T., Wun, T., Strauss, J.F., and Kliman, H.J. (1989) Plasminogenogen activator inhibitor types 1 and 2 in human trophoblasts. PAI-1 is an immunocytochemical marker of invading trophoblasts. *Lab. Invest.* 61, 20-26.

Ferrara, N. and Henzel, W.J. (1989) Pituitary follicular cells secrete a novel heparin-binding growth factor specific for vascular endothelial cells. *Biochem. Biophys. Res. Commun.* 161, 851-858.

Ferrara, N., Houck, K., Jakeman, L., and Leung, D.W. (1992) Molecular and biological properties of the vascular endothelial growth factor family of proteins. *Endocrin. Rev.* 13, 18-32.

Ferrell, C. L. (1989) Placental regulation of fetal growth. In: *Animal Growth Regulation*, (eds.), D.R. Campion, G.J. Hausman, and R.J. Martin. Plenum:New York, pp. 1-19.

Ferriani, R.A., Ahmed, A., Sharkey, A., and Smith, S.K. (1994) Colocalization of acidic and basic fibroblast growth factor (FGF) in human placenta and the cellular effects of bFGF in trophoblast cell line JEG-3. *Growth Factors* 10, 259-268.

Findlay, J.K. (1986) Angiogenesis in reproductive tissues. *J. Endocrinol.* 111, 357-366.

Folkman, J. and Haudenschild, C. (1980) Angiogenesis in vitro. *Nature* 228, 551-556.

Folkman, J. and Shing Y. (1992) Angiogenesis. *J. Biol. Chem.* 267, 10931-10934.

Fong, G.H., Rossant, J., Gertsenstein, M., and Breitman, M.L. (1995). Role of the Flt-1 receptor tyrosine kinase in regulating the assembly of vascular endothelium. *Nature* 376, 66-70.

Fox, H. (1970) Effect of hypoxia on trophoblast in organ culture: A morphologic and autoradiographic study. *Am. J. Obstet. Gynecol.* 107, 1058-1064.

Galland, F., Karamysheva, A., Pebusque, M.-G., Borg, J.-P., Rottapel, R., Dubreuil, P., Rosnet, O., and Birnbaum, D. (1993) The Flt-4 gene encodes a transmembrane tyrosine kinase related to the vascular endothelial growth factor receptor. *Oncogene* 8, 1233-1240.

Garg, U.C. and Hassia, A. (1989) Nitric oxide-generating vasodilators and 8-bromo-cyclic guanosine monophosphate inhibit mitogenesis and proliferation of cultured rat vascular smooth muscle cells. *J. Clin. Invest.* 83, 1774-1777.

Genbacev, O., Joslin, R., Damsky, C.H., Polliotti, B.M. and Fisher, S.J. (1996) Hypoxia alters early gestation human cytotrophoblast differentiation/invasion in vitro and models placental defects that occur in preeclampsia. *J. Clin. Invest.* 97, 540-550.

Hauser, S. and Weich, H.A. (1993) A heparin-binding form of placenta growth factor (PlGF-2) is expressed in human umbilical vein endothelial cells and in placenta. *Growth Factors* 9, 259-268.

Hertig, A.T. (1935) Angiogenesis in the early human chorion and in the primary placenta of the macaque monkey. *Contrib. Embryol.* 25, 37-82.

Hibbs, J.B., Taintor, R.R., Vavrin, Z. and Rachlin, E.M. (1988) Nitric oxide: A cytotoxic activated macrophage effector molecule. *Biochem. Biophys. Res. Commun.* 157, 87-94.

Hill, D.J., Tevaarwerk, G.J.M., Caddell, C. Arany, E., Kilkenny, D. and Gregory, M. (1995) Fibroblast growth factor 2 is elevated in term maternal and cord serum and amniotic fluid in pregnancies complicated by diabetes: Relationship to fetal and placental size. *J. Clin. Endocrinol. Metab.* 80, 2626-2632.

Holt, V.J., Wang, T-H., Wang, C-L., Torry, R.J., Caudle, M.R., Desai, M. and Torry, D. (1997) Normal human trophoblast express and respond to angiogenic growth factors. *Placenta*, in press.

Houck, K.A., Leung, D.W., Rowland, A.M., Winer, J. and Ferrara, N. (1992) Dual regulation of vascular endothelial growth factor availability by genetic and proteolytic mechanisms. *J. Biol. Chem.* 267, 26031-26037.

Houck, K.A., Ferrara, N., Winer, J., Cachianes, G., Li, B. and Leung, D.W. (1991) The vascular endothelial growth factor family: Identification of a fourth species and characterisation of alternative splicing of RNA. *Mol. Endocrinol.* 5, 1806-1814.

Hu, De-E. and Fan, T-P.D. (1993) [Leu^{8}]des-Arg^{9}-bradykinin inhibits the angiogenic effects of bradykinin and interleukin-1 in rats. *Br. J. Pharmacol.* 109, 14-17.

Hughes, S.E. and Hall, P.A. (1993) Overview of the fibroblast growth factor and receptor families: Complexity, functional diversity, and implications for future cardiovascular research. *Cardiovas. Res.* 27, 1199-1203.

Jackson, M.R., Mayhew, T.M. and Boyd, P.A. (1992) Quantitative description of the elaboration maturation of villi from 10 weeks to term. *Placenta* 13, 357-370.

Jackson, M.R., Allen, L., Morrow, R.J., Lye, S.L. and Ritchie, J.R. (1995) Reduced placental villous tree elaboration in SGA pregnancies: Umbilical artery Doppler waveforms. *Am. J. Obstet. Gynecol.* 1722, 518-522.

Jaillard, C., Chatelain, P.G. and Saez, J.M. (1987) In vitro regulation of pig sertoli cell growth and function: Effects of fibroblast growth factor and somatomedin-C. *Biol. Reprod.* 37, 665-674.

Jakeman, L.B., Wuner, J., Bennett, G.L., Altar, C.A., and Ferrara, N. (1992) Binding sites for vascular endothelial growth factor are localized on endothelial cells in adult rat tissue. *J. Clin. Invest.* 89, 244-253.

Jin, L., Pang, Y.S., Joseph, A., Li, Y., Rosen, E., Bhargava, M. and Goldberg, I.T. (1993) Rat placental hepatocyte growth factor/Scatter factor: Purification, characterisation and developmental regulation. *P.S.E.B.M.* 204, 75-80.

Joukov, V., Pajusola, K., Kaipainen, A., Chilov, D., Lahtinen, I., Kukk, E., Saksela, O., Kalkkinen, N. and Alitalo, K. (1996) A novel vascular endothelial growth factor, VEGF-C, is a ligand for the Flt-4 (VEGFR-3) and KDR (VEGFR-2) receptor tyrosine kinase. *EMBO J.* 15, 290-298.

Kaufmann, P., Bruns, U., Leiser, R., Luckhardt, M. and Winterhanger, E. (1985) The fetal vascularisation of term human placental villi. II Intermediate and terminal villi. *Anat. Embryol.* 173, 203-214.

Keck, P.J., Hauser, S.D., Krivi, G., Sanzo, K., Warren, T., Feder, J. and Connolly, D.T. (1989) Vascular permeability factor, an endothelial mitogen related to PDGF. *Science* 246, 1309-1312.

Kendall, R.L. and Thomas, K.A. (1993) Inhibition of vascular endothelial cell growth factor activity by an endogenously encoded soluble receptor. *Proc. Natl. Acad. Sci. USA* 90, 10705-10709.

Kendall, R.L., Wang, G., DiSalvo, J. and Thomas, K.A. (1994). Specificity of vascular endothelial growth factor receptor ligand binding domains. *Biochem. Biophys. Res. Comm.* 201, 326-330.

Khaliq, A., Shams, M, Li, X-F., Sisi, P., Acevedo, C.A., Weich, H., Whittle, M.J. and Ahmed, A. (1996a) Localisation of Placenta Growth Factor (PlGF) in Human Term Placenta. *Growth Factors* 13, 243-250.

Khaliq, A., Lix, X-F., Dunk, C., Shams, M., Whittle, M.J. and Ahmed, A. (1996b) Placental growth factor (PIGF) expression and function in human placenta. *Placenta* 17, A43.

Khaliq, A., Patel, B., Jarvis-Evans, J., McLeod, D. and Boulton, M.E. (1995) The effect of oxygen on the production of bFGF and TGF-β by retinal cells in vitro. *Exp. Eye Res.* 60, 414-424.

Kilby, M.D., Afford, S., LI, X-F., Strain, A.J., Ahmed, A. and Whittle, M.J. (1996) Localisation of hepatocyte growth factor and its receptor (c-met) protein and mRNA in human term placenta. *Growth Factor* 13, 133-139.

Klagsburn, M. and D'Amore, P.D. (1991) Regulators of angiogensis. *Annu. Rev. Physiol.* 53, 217-239.

Knoll, R., Westernacher, D., and Weich, H.A. (1992) Enhanced transcription of the plasminogen/plasmin system in macrovascular endothelial cells treated with vascular endothelial growth factor, *Circulation* (*Suppl.* 1) 86, I-740-741 .

Koch, A.E., Halloran, M.M., Haskell, C.J., Shah, M.R., and Polverini, P.J. (1995) Angiogenesis mediated by soluble forms of E-selectin and vascular cell adhesion molecule-1. *Nature* 376, 517-519.

Kostyk, S.K., Kourembanas, S., Wheeler, E.L., Medeiros, D., McQuillan, L.P., D'Amore, P.D. and Braunhut, S.J. (1995) Basic fibroblast growth factor increases nitric oxide synthase production in bovine endothelial cells. *Am. J. Physiol.* 269, H1583-1589.

Ladoux, A. and Frelin, C. (1993) Hypoxia is a strong inducer of vascular endothelial growth factor mRNA expression in the heart. *Biochem. Biophys. Res. Commun.* 195, 1005-1010.

Lapolt, S.P., Yamoto, M., Veljkovic, M., Sincich, C., Tor., N.Y., Tsafriri, A., and Hsueh, J.A. (1990) Basic fibroblast growth factor induction of granulosa cell tissue type plasminogen activator expression and oocyte maturation: Potential role as a paracrine ovarian hormone. *Endocrinology* 127, 2357-2363.

Le Noble, F.A.C., Schreurs, N.H.J.S., Van Straaten, H.W.M., Slaaf, D.W., Smits, J.F.M., Rogg, H., and Struijker-Boudier, A.J. (1993) Evidence for a novel angiotensin II receptor involved in angiogenesis in chick embryo chorioallantoic membrane. *Am. J. Physiol.* 264, R460-R465.

Li, X.F., Gregory, J., and Ahmed, A. (1994) Immunohistochemical localization of vascular endothelial growth factor in human endometrium. *Growth Factors* 11, 277-282.

Lobb, R.R., Alderman, E.M., and Fett, J.W. (1985) Induction of angiogenesis by bovine brain derived class I heparin-binding growth factor. *Biochemistry* 24, 4969-4973.

Lyon, M., Deakin, J.A., Mizuno K., Nakamura, T., and Gallagher, J.T. (1994) Interaction between hepatocyte growth factor with heparin sulphate. Elucidation of the major heparin sulphate structural determinants. *J. Biol. Chem.* 269, 11216-11223.

Maglione, D., Guerriero, V., Rambaldi, M., Russo, G., and Persico, M.G. (1993b) Translation of the placenta growth factor mRNA is severely affected by a small open reading frame localized in the 5' untranslated region. *Growth Factors* 8, 141-152.

Maglione, D., Guerriero, V., Viglietto, G., Delli-Bovi, P., and Persico M.G. (1991) Isolation of a human placenta cDNA coding for a protein related to the vascular permeability factor. *Proc. Natl. Acad. Sci.* USA 88, 9267-9271.

Maglione, D., Guerriero, V., Viglietto, G., Ferraro M.G., Aprelikova, O., Alitalo, K., Del Vecchio, S., Lei, K-J., Chou, J.C., and Persico M.G. (1993a) Two alternative mRNAs coding for the angiogenic factor PlGF are transcribed from a single gene of chromosome 14. *Oncogene* 8, 173-179.

Mandsager, N.T., Brewer, A.S. and Myatt, L. (1994) Vasodilator effects of parathyroid hormone, parathyroid hormone-related protein, and calcitonin gene-related peptide in the human fetal-placental circulation. *J. Soc. Gynecol. Invest.* 1, 19-24.

Marics, I., Adelaide, J., Raybaud, F., Mattei, M.G., Coulier, F., Planche, J., de Lapeyriere, O., and Birnbaum, D. (1989) Characterization of the HST-related FGF 6 gene, a new member of the fibroblast growth factor gene family. *Oncogene* 4(3), 334-340.

Mason, I.J. (1994) The ins and outs of fibroblast growth factors. *Cell* 78, 547-552.

Matthews, W., Jordon, C.T., Gavin, M., Jenkins, N.A., Copeland, N.G. and Lemischka, K. (1991) A receptor tyrosine kinase cDNA isolated from a population of enriched primitive hematopoietic cells and exhibiting close genetic linkage to c-kit. *Proc. Natl. Acad. Sci.* USA. 88, 9026-9030.

Mignatti, P., Morimoto, T. and Rifkin, B. D. (1992) Basic fibroblast growth factor, a protein devoid of secretory signal sequence, is released by cells via a pathway independent of the endoplasmic reticulum -Golgi complex. *J. Cell. Physiol.* 151, 81-93.

Millauer, B., Wizigmann-Voos, Schnurch, H., Martinez, R., Moller, N.P., Risau, W. and Ullrich, A. (1993) High affinity VEGF binding and developmental expression suggest Flk-1 as a major regulatory of vasculogenesis and angiogenesis. *Cell* 72, 835-846.

Mira, Y., Lopez, R., Joseph-Silverstein, J., Rifkin, D.B., and Ossonski, L. (1986) Identification of a pituitary factor responsible for enhancement of plasminogen activator activity in breast tumor cells. *Proc. Natl. Acad. Sci. USA* 83, 7780-7784.

Mizuno, K., Inoue, H., Hagiya, M., Shimizu, S., Nose, T., Shimohigashi, Y., and Nakamura, T. (1994) Hapirpin loop and second kringle domain are essential sites for heparin binding and biological activity of hepatocyte growth factor. *J. Biol. Chem.* 269, 1131-1136.

Morimoto, A.K., Okamura, K., Hamanaka, T., Sato, Y., Shima, N., Higashio, K., and Kunano, M. (1991) Hepatocyte growth factor modulates migration and proliferation of human microvascular endothelial cells in culture. *Biochem. Biophys. Res. Commun.* 179, 1042-1049.

Moscatelli, D., Presta, M., and Rifkin, D.B. (1986) Purification of a factor from human placenta that stimulates capillary endothelial cell protease production, DNA synthesis, and migration. *Proc. Nat. Acad. Sci. USA* 83, 2091-2095.

Myatt, L., Brockman, D.E., Langdon, G., and Pollock, J.S. (1993) Constitutive calcium-dependent isoform of nitric oxide synthase in the human Placental villous vascular tree. *Placenta* 14, 373-383.

Nanaev, A., Chwalisz, K., Frank, H., Kohnen, G., Hegele-Hartung, C., and Kaufmann, P. (1995) Physiological dilation of uteroplacental arteries in the guinea pig depends on nitric oxide synthase activity of extravillous trophoblast. *Cell Tissue Res.* 282, 407-421.

Noji, S., Tashiro, K., Koyama, E., Nohno, T., Ohyama, K., Taniguchi, S., and Nakamura, T. (1990) Expression of hepatocyte growth factor gene in endothelial and Kupffer cells of damaged rat liver, as revealed by in situ hybridization. *Biochem Biophys. Res. Commun.* 173, 42-47.

Olofsson, B., Pajusoal, K., Kaipainen, A., von Euler, G., Joukov, V., Petterson, R., Alitalo, K. and Eriksson, U. (1997) Vascular endothelial growth factor B (VEGF-B), a novel growth factor for endothelial cells. *Proc. Natl. Acad. Sci. USA*, in press.

Pajusola, K., Aprelikova, O., Pelicci, G., Weich, H., Claesson-Welsh, L., and Alitalo, K. (1994) Signalling properties of Flt-4, a portellytically processed receptor tyrosine kinase related to two VEGF receptors. *Oncogene* 9, 3545-3555.

Park, J.E., Chen, H.H., Winer, .J, Houck, K.A., and Ferrara, N. (1994) Placenta growth factor. Potentiation of vascular endothelial growth factor bioactivity, *in vitro* and *in vivo*, and high affinity binding to flt-1 but not flk-1/KDR. *J. Biol. Chem.* 269, 25646-25654.

Pepper, M.S., Ferrara, N., Orci, L., and Montesano, R. (1991) Vascular endothelial growth factor (VEGF) induces plasminogen activators and plasminogen activator inhibitor-1 in microvascular endothelial cells. *Biochim. Biophys. Acta.* 181, 902-906.

Peters, K.G., De Vries, C., and Williams, L.T. (1993) Vascular endothelial growth factor receptor expression during embryogenesis and tissue repair suggests a role in endothelial differentiation and blood vessel growth. *Proc. Natl. Acad. Sci. USA* 90, 8915-8919.

Plate, K.H., Breier, G., Weich, H.A., and Risau, W. (1992) Vascular endothelial growth factor is a potential tumour angiogenesis factor in human gliomas in vivo. *Nature* 359, 845- 848

Presta, M., Moscatelli, D., Joseph, S.J., and Rifkin, D.B. (1986) Purification from a human hepatoma cell line of a basic fibroblast growth factor-like molecule that stimulates capillary endothelial cell plasminogen activator production, DNA synthesis, and migration. *Mol. Cell. Biol.* 6, 4060-4066.

Pulvener, B.J., Kymakis, J.M., Avruch, J., Nikolakaki, E., and Woodgeti, J.R. (1991) Phosphorylation of c-jun by MAP kinase. *Nature* 353, 670-674.

Queenan, J.T., Kao, L.C., Arboleda, C.E., Ullo-Aguirre, A., Golos, T.G., Cines, D.B., and Strauss, J.F. (1987) Regulation of urokinase-type plasminogen activator production by cultured human trophoblasts. *J. Biol. Chem.* 262, 10903-10906.

Reynolds, P.R., Killilea, S.D., and Redmer D.A. (1992) Amgiogenesis in the female reproductive system. *FASEB J.* 6, 886-892.

Rifkin, D.B., Gross, J.L., Moscatelli, D., and Jaffe, E. (1982) Proteases and angiogenesis: production of plasminogen activator and collagenase by endothelial cells. In: *Pathobiology of the Endothelial Cell,* (eds.), H.L. Nossel and H.J. Vogel, Academic Press: New York, pp. 419-472.

Roberts, J.M., Taylor, R.N., Musci, T.J., Rodgers, G.M., Hubel, C.A. and McLaughlin, M.K. (1989) Pre-eclampsia: An endothelial cell disorder. *Am. J. Obstet. Gynecol.* 161, 1200-1204.

Rodesch, F., Simon, P., Donner, C. and Jauniaux, E. (1992) Oxygen measurements in endometrial and trophoblastic tissues during early pregnancy. *Obstet. Gynecol.* 80, 283-285.

Rosen, E.M., Meromsky, L., Setter, E., Vinter, D.W., and Goldberg, I.D. (1990a) Smooth muscle derived factor stimulates mobility of human tumor cells. *Invasion Metastatsis* 10, 49-64.

Rosen, E.M., Meromsky, L., Romero, R., Setter E., and Goldberg, I.D. (1990b) Human placenta contains epithelial scatter factor protein. *Biochem. Biophys. Res. Commun.* 176, 1082-1088.

Rosenfeld, C.R., Morriss, F.H., Jr., Makowski, E.L., Meschia, G., and Battaglia, F.C. (1974) Circulatory changes in the reproductive tissues of ewes during pregnancy. *Gynecol. Invest.* 5, 252-268.

Rutherford, R.A.D., McCarthy A., Sullivan, M.H.F., Elder, M.G., Polak, J.M. and Wharton, J. (1995) Nitric oxide synthase in human placenta and umbilical cord from normal, intrauterine growth-retarded and pre-eclamptic pregnancies. *Br. J. Pharmacol.* 116, 3099-3109.

Sanger, D.R., Perruzzi, C.A., Feder, J,. and Dvorak, H.F. (1983) Tumor cells secrete a vascular permeability factor that promotes accumulation of ascites fluid. *Science* 219, 983-985.

Sawano, A., Takahashi, T., Yamaguchi, S., Aonuma, M. and Shibuya, M. (1996) Flt-1 but not KDR/Flk-1 tyrosine kinase is a receptor for placental growth factor, which is related to vascular endothelial growth factor. *Cell Growth Differ.* 7, 213-221.

Schmidt, C., Bladt, F., Goedecke S., Brinkman, V., Zschiesche, W., Sharpe, M., Gherardi E. and Birchmeier, C. (1995) Scatter factor/Hepatocyte Growth factor is essential for liver development. *Nature* 373, 699-702.

Schulze-Osthoff, K., Risau, W., Vollmer, E. and Sorg, C. (1990) In situ detection of basic fibroblast growth factor by highly specific antibodies. *Am. J. Pathol.* 137, 85-92.

Seymour, L.W., Shoaibi, M.A., Martin, A., Ahmed, A., Elvin, P., Kerr, D.J., and Wakelam, M.J.O. (1996) Vascular endothelial growth factor stimulates protein kinase C dependent phospholipase D activity in endothelial cells. *Lab. Invest.* 75, 427-437.

Shalaby, F., Rossant, J., Yamaguchi, T.P., Gertsenstein, M., Wu, X.F., Breitman, M.L., and Schuh, A.C. (1995). Failure of blood-island formation and vasculogenesis in Flk-1-deficient mice. *Nature* 376, 62-66.

Shams, M. and Ahmed, A. (1994) Localization of mRNA for basic fibroblast growth factor in human placenta. *Growth Factors* 11, 105-111.

Shams, M., Li, X-F., and Ahmed, A. (1996) Expression of basic fibroblast growth factor and its receptor (Flg) in human endometrium. *J. Soc. Gynecol. Invest. (Suppl.)* 3(2), 265A.

Sharkey, A.M., Charnock-Jones, D.S., Boocock, C.A., Brown, K.D., and Smith, S.K. (1994) Expression of mRNA for vascular endothelial growth factor in human placenta. *J. Reprod. Fertil.* 99, 609-615.

Shifren, J.L., Doldi, N., Ferrara, N., Mesiano, S., and Jaffe, R.B. (1994) In the human fetus, vascular endothelial growth factor is expressed in epithelial cells and myocytes, but not vascular endothelium: Implications for mode of action. *J. Clin. Endocrinol. Metab.* 79, 316-322.

Shweiki, D., Itan, A., Soffer, D., and Keshet, E. (1992) Vascular endothelial growth factor induced by hypoxia may mediate hypoxias-initiated angiogenesis, *Nature* 359, 843-845.

Sisi, P., Li, X-F, Rollason, T., Gee, H., and Ahmed A. (1994) Expression and localisation of vascular permeability factor in human cervix. *J. Reprod. Fertil.* (Abs. Series) 14, 9.

Smith, S.K. (1995) Angiogenic growth factor expression in the uterus. *Hum. Reprod. Update* 1, 162-172.

Stoker, M., Gherardi, E., Perryman, M., and Gray, J. (1987) Scatter factor is a fibroblast derived modulator of epithelial cell mobility. *Nature* 327, 239-242.

Takahashi, A., Sasaki, H., Kim, S.J., Tobisu, K., Kakizoe, T., Tsukamoto, T., Sagimura, T., and Terada, M. (1994) Markedly increased amounts of messenger RNAs for vascular endothelial growth factor and placenta growth factor in renal cell carcinoma associated with angiogenesis. *Cancer Res.* 54, 4233-4237.

Terman, B.I., Dougher-Vermazen, M., Carrion, M.E., Dimitrov, D., Armellino, D.C., Gospodarowicz, D., and Bohlen, P. (1992) Identification of the KDR tyrosine kinase as a receptor for vascular endothelial cell growth factor. *Biochem. Biophys. Res. Commun.* 187, 1579-1586.

Terman, B.T., Khandke, L., Douger-Vermazan, M., Maglione, D., Lassam, N.J., Gospodarowicz, D., Persico, M.G., Bohlen, P. and Eisinger, A. (1994) VEGF receptor subtypes KDR and Flt-1 show different sensitivities to heparin and placenta growth factor. *Growth Factors* 11, 187-195.

Thieme, H., Aiello, L.P., Takagi, H., Ferrara, N. and King, G.L. (1995) Comparative analysis of vascular endothelial growth factor receptors on retinal and aortic endothelial cells. *Diabetes* 44, 98-103.

Tischer, E., Mitchell, R., Hartman, T., Silva, M., Gospodarowicz, D., Fiddes, J.C., and Abraham, J.A. (1991) The human gene for vascular endothelial growth factor. *J. Biol. Chem.* 266, 11947-11954.

Uehara, Y., Minowa, O., Mori, C., Shiota, K., Kuno, J., Noda, T., and Kitamura, N. (1995) Placental defect and embryonic lethality in mice lacking hepatocyte growth factor/scatter factor. *Nature* 373, 702-705.

Uhlrich, S., Tiollier, J., Tardy, M., and Tayout, J.L. (1991) Isolation and characterization of two different molecular forms of basic fibroblast growth factor extracted from human placental tissue. *J. Chromato.* 539, 393-403.

Vuckovic, M., Ponting, J., Terman, B.I., Niketic, V., Seif, M.W., and Kumar, S. (1996) Expression of the vascular endothelial growth factor receptor, KDR, in human placenta. *J. Anat.* 188, 361-366.

Waltenburger, J., Claesson-Welsh, L. Siegbahn, A., Shibuya, M., and Heldin, C.H. (1994) Different signal transduction properties of KDR and Flt1, two receptors for vascular endothelial growth factor. *J. Biol. Chem.* 269, 26988-26995.

Wolf, H., Zarnegar, R., Oliver, L. and Micholopoulos, G.K. (1991a) hepatocyte growth factor in human placenta and trophoblastic disease. *Am. J. Pathol.* 138, 1035-1043.

Wolf, H., Zarnegar R. and Micholopoulos, G.K. (1991b) Localization of hepatocyte growth factor in human and rat tissues. *Hepatology* 14, 488-494.

Wood, K.W., Sarnecki, C., Roberts, T.M., and Blenis, J. (1992) Ras mediates nerve growth factor receptor modulation of three signal-transducing protein kinases: MAP kinase, Raf-1 and RSK. *Cell* 68, 1041-1050.

Yagel, S., Parha, R.S., Jeffrey, J.J., and Lala, P.K. (1988) Normal nonmetastatic human trophoblast cells share in vitro invasive properties of malignant cells. *J. Cell. Physiol.* 136, 455-462.

Zhan, X., Bates, B. Hu, X.G., and Gold-farb, M. (1988) The human FGF-5 oncogenes encodes a novel protein related to fibroblast growth factors. *Mol. Cell. Biol.* 8, 3487-95.

Trophoblast Research 10:259-277, 1997

ENDOTHELINS AND HUMAN PLACENTAL GROWTH
-A Review-

Christelle Bourgeois[1], Thérèse-Marie Mignot[1], Bruno Carbonne[1,2], and Françoise Ferré[1]

[1]U. 361 INSERM
Reproduction et Physiopathologie Obstétricale
[2]Maternité Port-Royal - Baudelocque
Université René Descartes, Paris V
123, bd de Port-Royal
75014 Paris, France

INTRODUCTION

Endothelin (ET) was first identified by Yanagisawa et al. (1988) on medium from cultured aortic endothelial cells. It belongs to a super family of twenty-one amino acid peptides with two disulfide bonds linking the two pairs of cysteine residues.

Three isoforms, ET-1, 2, and 3, were discovered, cloned, and sequenced in the cardiovascular system and then in other mammalian tissues; ET-1, which is widely distributed, has been the most thoroughly investigated. Other related compounds, including vasoactive intestinal contractor (VIC), described as a particular ET-2 isoform in the intestinal tract of mice and rats, and sarafotoxin, the isoforms of which are present in the venom of the viper (Atractaspis engaddensis), have also been isolated.

ET-1 is an important mediator of vascular function. It appears to be the most potent natural agent involved in contraction of vessels and elevation of blood pressure. It is also a potent contractile agent in all types of smooth muscle studied (respiratory tract, uterus, vas deferens, stomach, etc.). Today, it is evident that, in a variety of target cells, endothelins are pleiotropic factors controlling a number of biological effects such as cellular ion fluxes, cell to cell communication, cell migration, neural transmission and hormone and cytokine release in neuroendocrine and immune systems. In addition, like several other vasoactive substances, endothelins exhibit growth factor properties (Rubanyi and Polokoff, 1994; Naruse et al., 1994).

The field of endothelins has greatly expanded and at the same time, our means of investigation have considerably diversified. Bearing this in mind, the most salient research aspects currently being explored will be briefly pointed out.

We will then attempt to examine these aspects within the framework of current knowledge and the study of these ubiquitous peptides in the human placenta, which is the prototype of a developmental organ.

Control Of Endothelin Biosynthesis And Production

Major attention is being focused on the sites and factors controlling ET isoform expression.

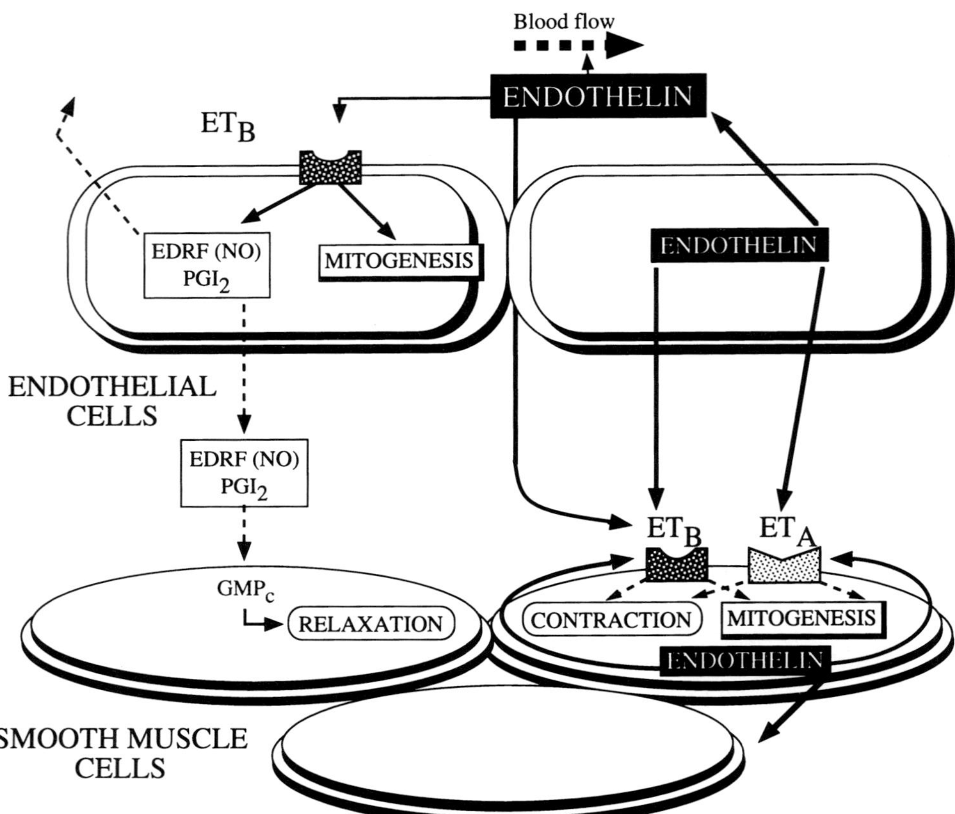

Figure 1. Role of ET-1 in the vascular system. ET-1 synthesized and secreted by endothelial cells activates ET_A and ET_B receptors of underlying smooth muscle cells (SMC) and initiates contraction and mitogenesis in a paracrine fashion. In an autocrine signaling mode, ET-1 also acts on endothelial ET_B receptors, stimulating EDRF (including NO) and PGI_2 synthesis and mitogenesis. Endothelium-released PGI_2 and EDRF induce SMC relaxation by stimulating GMPc production. ET-1 is also produced by SMC and acts in a paracrine/autocrine fashion on SMC ET_A and ET_B receptors to produce contraction and mitogenesis. NO = nitric oxide; PGI_2 = prostacyclin; EDRF = endothelium-derived relaxing factor; GMPc : cyclic guanosine monophosphate.

Each isoform is encoded by a specific gene (Inoue et al., 1989). ET-1 was previously considered to be produced by epithelial cells (for example vascular endothelium), thus acting in a paracrine fashion upon the underlying mesenchymal structures (e.g., vascular smooth muscle). While there is evidence that epithelia are one of the main sources of ET-1, autocrine mechanisms are also implicated due to the fact that epithelial cells express ET receptors, and that many types of cells, including smooth muscle, themselves produce ET-1 (Figure 1).

At present, substantial data on control of expression of each isoform are available mainly for ET-1 in endothelial cells. ET-1 biosynthesis can be modulated by different physical and biochemical stimuli: fluid shear stress, mechanical strain, hypoxia/anoxia, nitric oxide, thrombin, insulin, TGFβ, TNFα, and IFNγ, IL-1, angiotensin II, vasopressin, bradykinin, and natriuretic peptides. The strongest regulation appears to occur at the transcriptional level (Kurihara et al., 1989; Oliver et al., 1991; Marsden and Brenner, 1992; Benatti et al., 1994; Fujisaki et al., 1995).

Response elements linked to transcription of preproET-1 mRNA were identified in the 5'-flanking region of the human gene and, notably, an AP1 cis-acting DNA sequence which responds to phorbol esters, stimulated by the Fos/Jun complex. A GATA motif controlled by the transacting factor GATA2 involved in down-regulation of gene expression by retinoid acid is also present in other promoter region (Rubanyi and Polokoff, 1994). An additional upstream cis-element was found to be implicated in down-regulation of endothelial ET-1 synthesis in response to shear stress (Malek et al., 1993). Post-transcriptional regulation was observed with ANP, which affects the stability of preproET-1 mRNA and inhibits its translation, representing a negative feedback mechanism. An ET-1 autoinduction phenomenon was previously observed in vascular endothelial and smooth muscle cells and, more recently, in rat mesangial cells. ET-1 is able to induce its own synthesis by both transcriptional and post-transcriptional mechanisms (Iwasaki et al., 1995).

Post-translational preproET processing is essential for the synthesis of bioactive endothelins (Figure 2). After removal of the signal peptide in the lumen of the rough endoplasmic reticulum, the propeptide is cleaved by endoproteases called substilisin-like-proprotein convertases (SPCs). Following this, big ET in turn generates the mature peptide. Metalloendopeptidases of the enkephalinase family are implicated in this process as well as in the endothelin degradation (Vijayaraghavan et al., 1990; McMahon et al., 1991; Opgenorth et al., 1992; Emoto and Yanagisawa, 1995). Each of these successive enzymatic systems consists of isoforms, differently expressed, depending on the cell type and species. Research focused on endothelin-converting-enzymes (ECEs). Some of them have already been cloned and extensive studies are now in progress to develop selective non-peptidic inhibitors, in addition to phosphoramidon analogues, in view of potential therapeutic use in altering ET-1 production (Xu et al., 1994; Emoto and Yanagisawa, 1995).

ET-1 appears to be released according to a classical Golgi-mediated secretory pathway and there is no conclusive report of storage granules in cells containing endothelins (Huggins et al., 1993). It must be borne in mind, particularly in epithelial cells which present polarity in their plasmatic membrane, that ET-1 is secreted in an asymmetrical fashion. In vascular endothelial cells, a major fraction of ET-1 produced is released in the basolateral direction, reaching local concentrations which are higher than in circulation (Wagner et al., 1992).

Endothelin Receptors And Transduction Pathways Mediating Biological Actions

Three different receptors, all of the seven-transmembrane domain group coupled to guanyl nucleotide-binding regulatory proteins (G-proteins), have been cloned and sequenced. They differ in their respective affinities for natural isopeptides (Masaki et al., 1994). Only ET_A (ET-1 = ET-2 > ET-3) and ET_B (ET-1 = ET-2 = ET-3) receptors were found in mammalian tissues, exhibiting about 65% homology in their amino-acid sequence in humans. Cis-elements and a nuclear factor that regulate their gene transcription have been identified. A receptor specific for ET-3 : ET_C (ET-3 > ET-1) has been reported in dermal melanophores from amphibians (Karne et al., 1993).

ET_B receptors were found to be predominant in epithelial and neural tissues and are probably the only receptors found in the vascular endothelium, where they mediate relaxation of the neighboring muscular layer through the release of relaxing and antiproliferative factors such as nitric oxide and prostacyclin (Masaki, 1995). ET-3, via ET_B receptors, is able to promote ET-1 production in endothelial and mesangial cells (Yokohawa et al., 1991; Iwasaki et al., 1995). In addition, these receptors might be involved in the clearance of circulating endothelins (Frelin and Guedin, 1994). In smooth muscles, ET_A and ET_B receptors coexist, mediating contraction and, as in other cell types, proliferation (Wang et al., 1994; Smith et al., 1995; Douglas et al., 1995) (Figure 1). The density of ET_A receptors is often higher than that of ET_B. Compared to ET_A receptors, those of ET_B are more easily down-regulated by ET-1-prolonged activation (Sudjarwo et al., 1994).

The recent availability of large number of peptidic and non-peptidic agonists or antagonists with high selectivity enables the description of atypical ET receptors (Douglas et al., 1994; Bax and Saxena, 1994), as well as the existence of subtypes within the ET_A and ET_B groups (Nishiyama et al., 1995). These subtypes could be generated through divergent genes or by alternative splicing from a single gene, as was demonstrated for ET_A receptors and, more recently, for ET_B receptors in human tissues including placenta (Shyamala et al., 1994). It is clear that the nature and expression rate of these different kinds of receptors is highly dependent on subtle modifications in the cell phenotype.

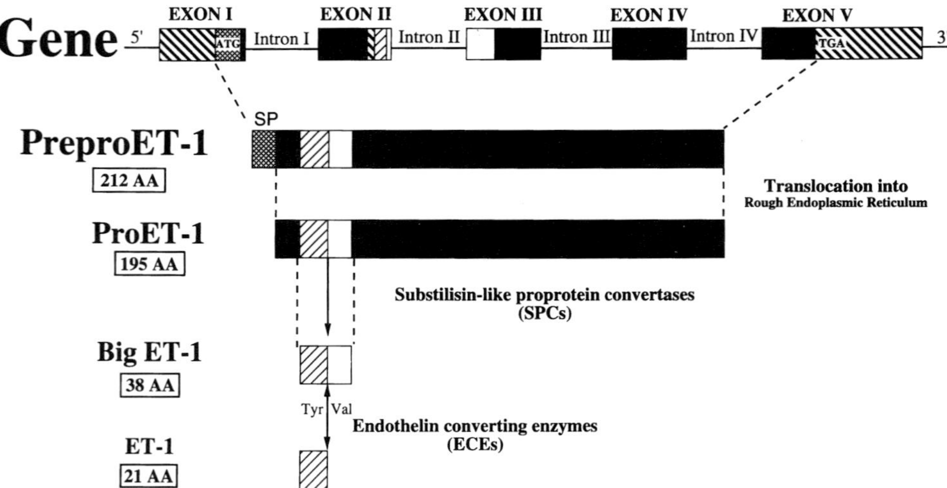

Figure 2. Post-translational processing of preproET-1. From the preproET-1 mRNA, a 212 amino acid preproET-1 arises which is subsequently cleaved in proET-1, then big ET-1. The latter is next cleaved by endothelin-converting-enzyme (ECE) to yield the final 21-amino acid product ET-1.

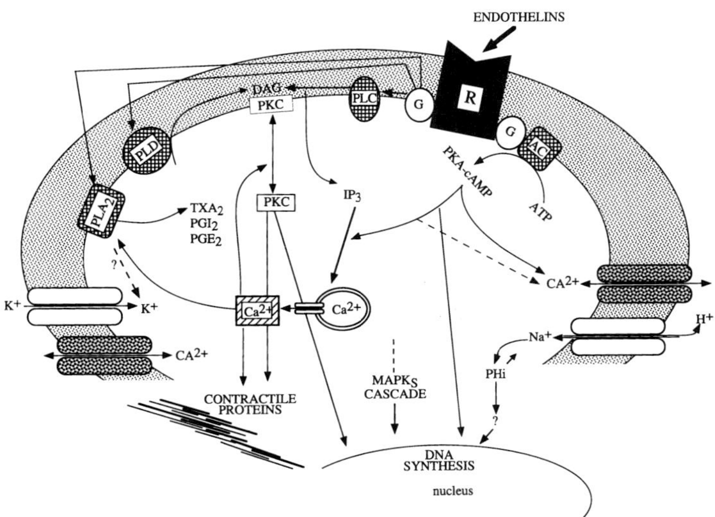

Figure 3. ET signal transduction pathways in smooth muscle cell. ETs bind to their receptors which stimulate the GTP turnover of G proteins. This results in modulation of intracellular enzymes (PLC, PLD, PLA_2, AC), sodium proton exchange, ion channels, etc. G: GTP binding proteins; PLC: phospholipase C; PLD: phospholipase D; PLA_2: phospholipase A_2; AC: adenylyl cyclase; PKA: protein kinase A; PKC: protein kinase C; MAPK: mitogen-activated protein kinase; IP_3: inositol triphosphate; DAG: diacylglycerol; TXA_2: thromboxane A_2; PGI_2: prostacyclin; PG: prostaglandin; cAMP: cyclic adenosine monophosphate.

Such receptor diversity, in addition to their coupling to different G proteins and transduction pathways can explain differential or even opposite endothelin biological effects. Indeed, in smooth muscle, endothelins are linked to phospholipases C, A2, D, adenylate cyclase, sodium proton exchange, ion channels, and other transport mechanisms (Rae et al., 1995; La and Reid, 1995). Consequently, kinases such as several protein kinase C isoforms and mitogen-activated protein kinases are involved in the modulation of contractile elements and nuclear events by endothelins (short- and long-term effects) (Figure 3).

Endothelins As Growth Factors

The relevance of ET-1 in cell proliferation/differentiation processes is now well documented. The peptide could induce cell hyperplasia and/or hypertrophy and is implicated in the control of extracellular matrix components. ET-1 elicits mitogenic and comitogenic effects in various cell types in culture (endothelial, smooth muscle, fibroblast, mesangial cells) increasing DNA, cell number and protein synthesis, and particularly muscle-specific genes in cardiomyocytes (Ito et al., 1993; Chollet et al., 1993; Weber et al.,

1994; Stewart et al., 1994; Tsuboi et al., 1994). ET-1 has been shown to stimulate the expression of several other vasoactive and growth factors, and that of immediate-early genes (c-fos, c-jun, Egr-1, c-myc, c-zif), considered to be important in the signaling of growth responses (Komoro et al., 1988; Simonson et al., 1992; Jones et al., 1992; Bruneau and de Bold, 1994). However, ET-1 mitogenic effects are sometimes weaker than those induced by fetal calf serum alone or by other growth factors (Swope et al., 1995). The ET-1 effect could also become antiproliferative depending on the medium supplement. Such an opposite effect might represent an inhibitory feedback mechanism meant to reduce an excessive proliferative cell phenotype (Fujitani et al., 1995).

Significant expression of ET-1 has been reported in normal embryonic and fetal tissues (Schiff et al., 1993b). The importance of endothelins in mammalian development is now supported by gene knock-out experiments (cf., Battistini et al., 1995). Disruption of the ET-1 gene caused severe morphological cranio-facial abnormalities in homozygous ET-1 $^{-}/^{-}$ mice which died soon after birth as a result of respiratory failure. On the contrary, ET-1 $^{+}/^{-}$ heterozygous mice were viable. They presented lower tissue and plasma levels of ET-1 than wild type mice but higher blood pressure (Kurihara et al., 1994). Knock-out of ET_A receptor gene in mice, mimicked the ET-1 $^{-}/^{-}$ phenotype (Hosoda et al., 1994). Interestingly, homozygous ET3 $^{-}/^{-}$ mice (Baynash et al., 1994) or ET_B receptor $^{-}/^{-}$ mice (Hosada et al., 1994) were viable at birth. However, they developed pigmentary disorders and toxic megacolon and died at three to four weeks of age. No rescue mechanism by other ETs seems to occur within the cells of knocked out mice as reported by Maemura et al. (1994) in microvascular endothelial cells isolated from ET-1 $^{-}/^{-}$ mice.

Together, these results indicate that ET-1 and ET-3 genes as well as ET_A and ET_B receptors genes are essential regulators of mammalian neural crest derived tissue development. The ET-1 isoform or the ET_A receptor controls cardiomyocyte lineage, and the ET-3 isoform or the ET_B receptor controls epidermal and choroidal melanocytes and enteric ganglion neurons (Baynash et al., 1994). Recent use of a pluripotential embryonic carcinoma cell line which constitutes a model of differentiation for cardiomyocytes and neural lineage confirms these results (Monge et al., 1995). In this context, it is interesting to note that a defect in human ET-3 isoform and ET_B receptor genes was found to be related to Hirschsprung's disease (Puffenberger et al., 1994; Hosoda et al., 1994). Transgenic techniques in animals will significantly contribute to our understanding of endothelin regulatory functions in health and disease (Harats et al., 1995).

Increased plasmatic levels of immunoreactive (ir) ET-1, about 2-to 3-fold the normal values, are associated with various human cardiovascular disorders such as vasospasm and hypertension, septic shock, diabetes, and proliferative processes including atherosclerosis, carcinogenesis and neointima formation after balloon artery angioplasty (Rubanyi and Polokoff, 1994; Rae et al., 1995). In contrast to ET-1, little is known about the specific roles of ET-2 and ET-3 isoforms, although the three peptides originate from three different genes and show differential patterns of expression and pharmacological activities. Considering the potential roles of endothelins in control of physiological and pathophysiological states, it is not surprising that improvement of our knowledge of the endothelin system is becoming one of the most attractive areas of medical research.

Endothelins In Human Placenta

irET-1 levels have been determined by numerous authors in maternal plasma and in various fluids of the feto-placental unit obtained during cesarean section performed at term, before the onset of labor (cf., Ferré, 1995).

The very high levels found in amniotic fluid were largely explained by the fact that amnion is a rich source of endothelin. As in previous works, it has been confirmed that irET-1 levels in umbilical cord vessels are also significantly higher than in maternal peripheral plasma. Data concerning endothelin levels in umbilical arteries compared to vein remain controversial. In our study, no difference was shown in umbilical arteriovenous concentrations (Figure 4).

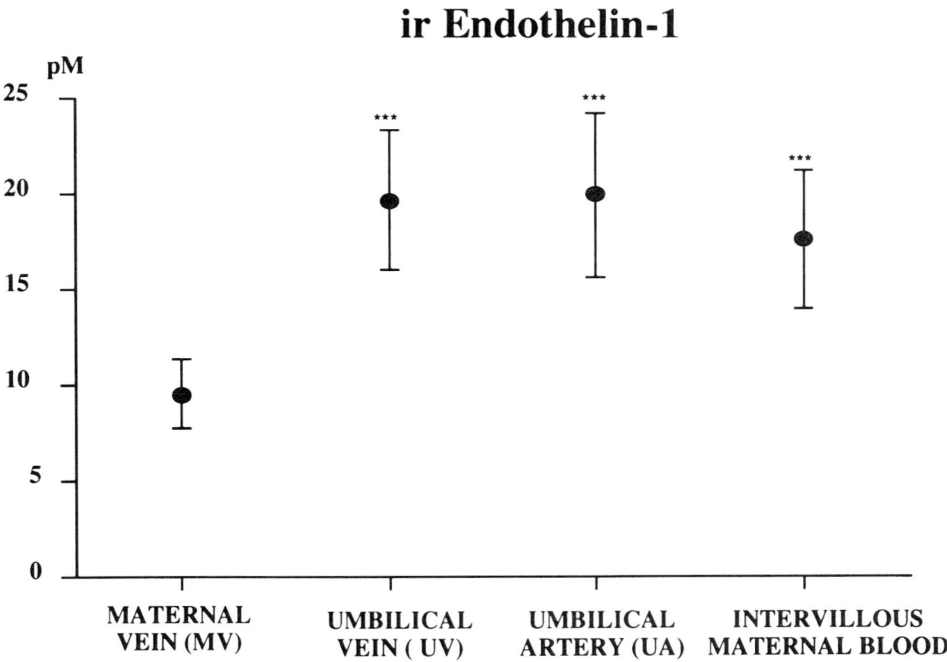

Figure 4. Concentrations of ir ET-1 in the plasma of the maternal vein (MV), umbilical vein (UV), umbilical artery (UA), and intervillous maternal blood from healthy women at the end of pregnancy. After extraction as previously described (Malassiné et al., 1993a), irET-1 was quantified using an ET 1-21 specific [^{125} I] assay system (Amersham RPA 555). Recovery of 54% was afforded by this method. ET-1, ET-2, ET-3 and big ET-1 (100%, 144%, 52%, and 0.4%, respectively) were immunoreactive. Ninety-five percent confidence intervals of ir ET-1 concentrations in MV, UV, UA, and intervillous maternal blood of pregnant healthy women (n=22). Statistical significance was determined by a nonparametrical test (Mann-Whitney U test) *** $p< 0.001$ versus MV.

These observations raise the question of the origin of endothelins present in feto-placental blood. In addition to the fetus, the placenta appears to be an important source of endothelins. The preproET-1 and ET-3 genes were initially isolated from a human placental cDNA library (Itoh et al., 1988; Onda et al., 1990) and then were found to be expressed in placental villi (Benigni et al., 1991; McMahon et al., 1993). irET-1 has been localized in various placental cell types (Hemsén et al., 1991; Van Papendorp et al., 1991; Svane et al., 1993; Malassiné et al., 1993a; Wilkes et al., 1993). Cultured umbilical endothelial cells (Clozel and Fischli, 1989), trophoblast cells (Malassiné et al, 1993a; Ferré et al., 1993 Fant and Nanu, 1995) and placental fibroblast cells (Fant and Nanu, 1995) were shown to release ir ET-1 in the medium or to express preproET-1 mRNA. By contrast, in the rat placenta, ET-1 was detected in fetal vessels but not in trophoblastic cells (Horwitz et al., 1995).

In human term placenta, the villous syncytiotrophoblast which lines the intervillous space containing maternal blood can be compared to an endothelial layer. As is the case for endothelial cells, one feature of syncytiotrophoblast cells is their capacity to be polarized, and the manner in which endothelin synthesized by the different trophoblast cell populations (villous and extravillous trophoblast cells) is released remains to be clarified. Indeed, endothelin secreted by trophoblast cells could reach other proximal placental structures, but could also directly enter into the intervillous space and contribute, along with the endothelin produced by decidual cells (Kubota et al., 1992), to the high amount of ir ET-1 observed in maternal blood at the feto-maternal interface (Figure 4). Furthermore, it will be interesting to elucidate whether vascular smooth muscle and particularly inflammatory cells (macrophages and mast cells) which are present in placenta, are another sources of endothelins as is the case for these cell types in other organs.

Studies are in progress in an attempt to fully clarify the relationships between the sites of endothelin isoform production and those of their biological actions in human placenta. The expression of endothelins and of their specific binding sites in placental vessels and trophoblast cells, in addition to the frequent proximity of these different cell types, argues for a paracrine/autocrine role of endothelins in controlling placental fibroblast growth (Fant et al., 1992; Fant and Nanu, 1995), trophoblast steroidogenic function (Mignot et al., 1995), and feto-placental blood flow (Wilkes et al., 1990; Gude et al., 1991; MacLean et al., 1992; Myatt et al., 1992; Le et al., 1993; Mombouli et al., 1993; Svane et al., 1993; Sabry et al., 1995a). In stem villi vessels, considered as the major site of placental vascular resistance, the action of ET-1 is dual: in addition to its classical vasoconstrictor effect, ET-1 at low concentrations causes vasodilatation by inducing endothelial release of relaxing factors (Sabry et al., 1995b).

The human placenta displays a high density of endothelin receptors initially identified in the whole organ. Mixed populations of ET_A/ET_B receptors, which have been detected on both placental vessels and trophoblast cells, are not distributed in a homogeneous manner. The relative proportion of these subtypes varies along the placental vascular tree (Robaut et al., 1991; Rutherford et al., 1993), and the kinds of receptors expressed by vascular endothelial and smooth muscle cells, respectively, remain to be established. In the media of stem villi vessels, both ET_A and ET_B receptors, but above all, ET_A receptors, are coupled to the phospholipase C transducing system (Mondon et al., 1995). In syncytiotrophoblast, endothelin receptors on the basal plasma membrane facing fetal circulation appear to be mainly of the ET_A subtype, whereas only an ET_B-like subtype is found in the microvillous membrane facing the maternal blood space (Figure 5) (Mondon et al., 1993; Malassiné et al., 1993b).

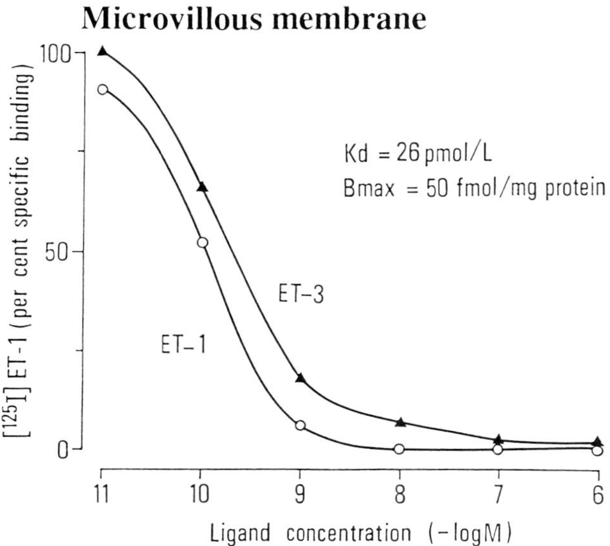

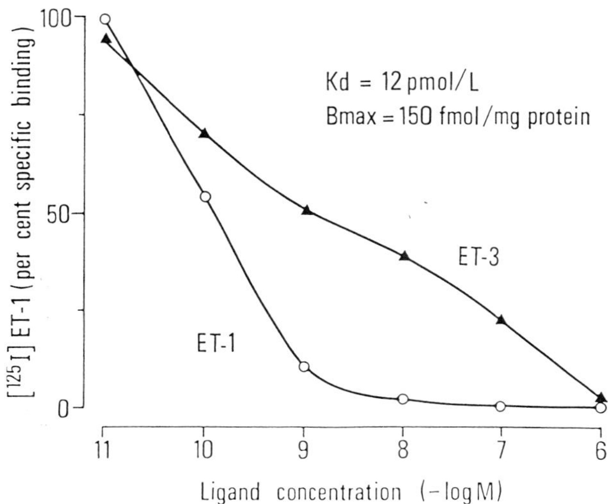

Figure 5. Effects of ET-1 and ET-3 on [^{125}I] ET-1 specific binding to microvillous and to basal plasma membranes. Syncytiotrophoblast membranes were incubated with 20 pmol/L [^{125}I]ET-1 for 120 minutes in the presence of increasing concentrations of the displacing ligands. Each point represents the mean of duplicate determinations obtained from a single representative experiment [Mondon et al., (1993) *J. Clin. Endo Metab.* 76, 237-244].

Potential Role Of Endothelins In Placental Growth

A large majority of studies have been performed in term placenta. Up to now, little information is available on the ontogenesis of endothelins and their receptors at earlier stages of pregnancy or on how these receptors are coupled to multiple transduction pathways when activated in a specific placental cell.

PreproET-1 mRNA appears to be expressed in villous placenta in a developmentally regulated manner, increasing throughout pregnancy (Fant et al., 1992) while ET receptor density, which is highest in the first trimester placenta, decreases progressively toward term, possibly due to down-regulation (Kilpatrick et al., 1993). Nevertheless, when preterm and term periods are compared, opposite changes in the densities of ET-1 binding sites are observed between placental villous tissue and cultured trophoblast cells (Cervar et al., 1995).These findings suggest differential evolution of the endothelin system in each placental cell type during gestation.

Considering that endothelin cells elicit mitogenic effects toward a variety of cell types, we are able to postulate that these peptides are potent growth factors in placenta. It was reported that ET-1, in part due to an autocrine mechanism, regulates cell growth in the stromal compartment. Indeed, ET-1 is mitogenic *per se* and in synergy with IGF1. In addition, ET-1 stimulates the secretion of IGF2 and IGF binding proteins in placental fibroblasts in culture (Fant et al., 1992). A potential role of endothelin cells at the time of embryo implantation, in the establishment of placental cell lineage, in physiological angiogenesis and in early trophoblast proliferation and invasion, which are all essential processes for normal placentation, should be explored. Interestingly, as previously mentioned, among the other growth factors expressed by the placenta in the first trimester of pregnancy, none exhibits as great a receptor concentration as endothelin cells (Kilpatrick et al., 1993). The fact that, under certain conditions, endothelin cells have been shown to possess antiproliferative properties should lead to clarification of their importance starting from mid-gestation, when the critical phases of morphological and functional differentiation are taking place in the placenta. These different responses to endothelin cells are generally considered to be related to the phenotypic state of the cell when exposed to the peptide. Alternatively, it remains to be determined to what extent endothelins contribute forward promoting a particular phenotype. It is likely that culture cell models, associated with molecular biology techniques, will greatly contribute in determining whether modifications in gene expression of the various components of the endothelin system are involved, for example when cytotrophoblast cells develop to form specialized syncytiotrophoblast cells (Knoll et al., 1992). Uncovering the factors or, more likely, the combinations of factors related to specific transcriptional and/or post-transcriptional control of the endothelin genes is a preliminary step in understanding how their expression is spatio-temporally regulated during the complex dynamic processes underlying placental development. While it is evident that ET-1 and ET-3 play an important role in fetal development in mice, no information is available concerning the placenta after disruption of ET genes in embryonic stem cells. In any case, studies conducted on other species, which have revealed different situations with regard to morphology, structure, and function of placenta and in its interactions with embryo and fetus are difficult to extend, in their entirety, to humans.

Endothelins in Pathology

Elevated circulating ir ET-1 levels have been detected by some authors in maternal plasma and umbilical cord vessels during pre-eclampsia, whereas in other studies, no elevation was observed. Pre-eclampsia often provokes intrauterine fetal growth retardation, and premature labor. The etiology of pre-eclampsia is still unknown, but the placenta appears to play a major role since normal trophoblast invasion of maternal uterine tissues is reduced, leading to placental ischemia and altered endothelial cell function which in turn increases endothelin production (Roberts and Redman, 1993; Walker, 1994). Indeed, pre-eclampsia is associated with modifications in preproET-1 mRNA level (McMahon et al., 1993), expression of an endothelin-converting-enzyme (ECE) (Li et al., 1995) and in density of ET-1 specific binding sites (Schiff et al., 1993a; Cervar et al., 1995). However, other substances and other regulatory mechanisms are also affected in pre-eclampsia. Consequently, to date, a role for endothelin cells in the initiation of this disease and/or its development at later stages, has not been confirmed (Handwerger, 1995). Elevated tissue expression and circulating levels of endothelin cells have also been reported in human tumors (Kusuhara et al., 1990; Ishibashi et al., 1993). An interesting future question which must be considered concerns a possible link between these peptides, and the formation of human trophoblastic tumors and phenotypic transitions involved in hydatidiform moles and choriocarcinoma cells.

The fact that endothelin cells appear to be more important in development than originally thought could lead to a reassessment of their role in placental biology. However, additional studies are needed to verify such a hypothesis.

SUMMARY

Endothelins (ETs) were first discovered in medium culture of endothelial cells and were reported to be potent vasoconstrictors of smooth muscle cells. They currently appear to demonstrate a broad range of biological actions on a wide spectrum of target cells.

The three isoforms of ET (ET-1, 2, 3) are encoded by distinct genes from which a peptidic precursor is raised and subsequently proteolyzed to give biologically active ETs. On target cells, ETs bind to seven transmembrane domain receptors coupled through G proteins to various intracellular signaling effectors. At least three ET receptors, each encoded by a distinct gene, have been cloned. The combination of ETs and ET receptor isoforms could account for the variety of bioactivities observed.

It is well established that ETs exert mitogenic effects alone or in synergy with growth factors. ETs are also involved in the human fetal development of neural crest derived tissues. At the end of pregnancy, high ir ET-1 levels have been described in feto-placental blood. The human placenta contributes to this production, as demonstrated in trophoblast, endothelial, and fibroblast cells by molecular and immunohistochemical studies. The presence of specific ET-1 binding sites on placental vasculature and trophoblast suggests autocrine/paracrine roles of this peptide in the control of placental blood flow and endocrine function. Considering the growth factor properties of endothelins, their role in normal placental development as well as in the pathophysiology of severe pregnancy disorders are discussed in this review.

ACKNOWLEDGEMENTS

Determination of irET-1 levels was supported by grants from MGEN (Mutuelle Générale de l'Education Nationale). The authors wish to thank Ph. Zamia for statistical advice and F. Mondon and M. Verger for their kind assistance in the preparation of this manuscript.

REFERENCES

Battistini, B., Botting, R. and Warner, T.D. (1995) Endothelin: A knockout in London. *TiPS* 16, 217-222.

Bax, W.A. and Saxena, P.R. (1994) The current endothelin receptor classification: Time for reconsideration? *TiPS* 15, 379-386.

Baynash, A.G., Hosoda, K., Giaid, A., Richardson, J.A., Emoto, N., Hammer, R.E. and Yanagisawa, M. (1994) Interaction of endothelin-3 with endothelin-B receptor is essential for development of epidermal melanocytes and enteric neurons. *Cell* 79, 1277-1285.

Benatti, L., Fabbrini, M.S. and Patrono, C. (1994) Regulation of endothelin-1 biosynthesis. In: *Platelet-dependent Vascular Occlusion*, vol. 714 of the Annals of the New York Academy of Sciences, pp. 109-115.

Benigni, A., Gaspari, F., Orisio, S., Bellizzi, L., Amuso, G., Frusca, T. and Remuzzi, G. (1991) Human placenta expresses endothelin gene and corresponding protein is excreted in urine in increasing amounts during normal pregnancy. *Am. J. Obstet. Gynecol.* 164, 844-848.

Bruneau B.G. and de Bold, A.J. (1994) Selective changes in natriuretic peptide and early response gene expression in isolated rat atria following stimulation by stretch or endothelin-1. *Cardiov. Res.* 28, 1519-1525.

Cervar, M., Kainer, F. and Desoye, G. (1995) Pre-eclampsia and gestational age differently alter binding of endothelin-1 to placental and trophoblast membrane preparations. *Mol. Cell. Endoc.* 110, 65-71.

Chollet, P., Malecaze, F., Gouzi, L., Arne, J.L. and Plouet, J. (1993) Endothelin-1 is a growth factor for corneal endothelium. *Exp. Eye Res.* 57, 595-600.

Clozel, M. and Fischli, W. (1989) Human cultured endothelial cells do secrete endothelin-1. *J. Cardiovasc. Pharmacol.* 13, S229-S231.

Douglas, S.A., Meek, T.D. and Ohlstein, E.H. (1994) Novel receptor antagonists welcome a new area in endothelin biology. *TiPS* 15, 313-316.

Douglas, S.A., Vickery-Clark, L.M., Louden, C. and Ohlstein, E.H. (1995) Selective ET_A receptor antagonism with BQ-123 is insufficient to inhibit angioplasty induced neointima formation in the rat. *Cardiovasc. Res.* 29, 641-646.

Emoto, N. and Yanagisawa, M. (1995) Endothelin-converting enzyme-2 is a membrane-bound, phosphoramidon-sensitive metalloprotease with acidic pH optimum. *J. Biol. Chem.* 270, 15262-15268.

Fant, M.E. and Nanu, L. (1995) Human placental endothelin: Expression of endothelin-1 mRNA by human placental fibroblasts in culture. *Mol. Cell. Endoc.* 109, 119-123.

Fant, M.E., Nanu, L. and Word, R.A. (1992) A potential role for endothelin-1 in human placental growth: Interactions with the insulin-like growth factor family of peptides. *J. Clin. Endoc. Metab.* 74, 1158-1163.

Ferré, F. (1995) Endothelin - its possible role during pregnancy. *Eur. J. Obstet. Gynecol. Repr. Biol.* 59, 1-4.

Ferré, F., Mondon, F., Mignot, T.M., Cronier, L., Cavero, I., Rostène, W. and Malassiné, A. (1993) Endothelin-1 binding sites and immunoreactivity in the cultured human placental trophoblast: Evidence for an autocrine and paracrine role for endothelin-1. *J. Cardiov. Pharmacol.* 22, S214-S218.

Frelin, C. and Guedin, D. (1994) Why are circulating concentrations of endothelin-1 so low? *Cardiov. Res.* 28, 1613-1622.

Fujitani, Y., Ninomiya, H., Okada, T., Urade, Y. and Masaki, T. (1995) Suppression of endothelin-1-induced mitogenic responses of human aortic smooth muscle cells by interleukin-1 beta. *J. Clin. Invest.* 95, 2474-2482.

Fujisaki, H., Ito, H., Hirata, Y., Tanaka, M., Hata, M., Lin, M., Adachi, S., Akimoto, H., Marumo, F. and Hiroe, M. (1995) Natriuretic peptides inhibit angiotensin II-induced proliferation of rat cardiac fibroblasts by blocking endothelin-1 gene expression. *J. Clin. Invest.* 96, 1059-1065.

Gude, N.M., King, R.G. and Brennecke, S.P. (1991) Endothelin: Release by and potent constrictor effect on the fetal vessels of human perfused placental lobules. *Reprod. Fertil. Devel.* 3, 495-500.

Handwerger, S. (1995) Endothelins and the placenta. *J. Lab. Clin. Med.* 125, 679-681.

Harats, D., Kurihara, H., Belloni, P., Oakley, H., Ziober, A., Ackley, D., Cain, G., Kurihara, Y., Lawn, R. and Sigal, E. (1995) Targeting gene expression to the vascular wall in transgenic mice using the murine preproendothelin-1 promoter. *J. Clin. Invest.* 95, 1335-1344.

Hemsén, A., Gillis, C., Larsson, O., Haegerstrand, A. and Lundberg, J.M. (1991) Characterization, localization and actions of endothelins in umbilical vessels and placenta of man. *Acta Physiol. Scand.* 143, 395-404.

Horwitz, M.J., Clarke, M.R., Shakir, A.K. and Amico, J.A. (1995) Developmental expression and anatomical localization of endothelin-1 messenger ribonucleic acid and immunoreactivity in the rat placenta: A Northern analysis and immunohistochemistry study. *J. Lab. Clin. Med.* 125, 713-718.

Hosoda, K., Hammer, R.E., Richardson, J.A., Baynash, A.G., Cheung, J.G., Giaid, A. and Yanagisawa, M. (1994) Targeted and natural (Piebald-Lethal) mutations of endothelin-B receptor gene produce megacolon associated with spotted coat color mice. *Cell* 79, 1267-1276.

Huggins, J.P., Pelton, J.T. and Miller, R.C. (1993) The structure and specificity of endothelin receptors: Their importance in physiology and medicine. *Pharmac. Ther.* 59, 55-123.

Inoue, A., Yanagisawa, M., Kimura, S., Kasuya, Y., Miyauchi, T., Goto, K. and Masaki, T. (1989) The human endothelin family: Three structurally and pharmacologically distinct isopeptides predicted by three separate genes. *Proc. Natl. Acad. Sci. USA* 86, 2863-2867.

Ishibashi, M., Fujita, M., Nagai, K., Kako, M., Furue, H., Haku, E., Osamura, Y. and Yamaji, T. (1993) Production and secretion of endothelin by hepatocellular carcinoma. *J. Clin. Endocrinol. Metab.* 76, 378- 383.

Ito, H., Hirata, Y., Adachi, S., Tanaka, M., Tsujino, M., Koike, A., Nogami, A., Marumo, F. and Hiroe, M. (1993) Endothelin-1 is an autocrine/paracrine factor in the mechanism of angiotensin II-induced hypertrophy in cultured rat cardiomyocytes. *J. Clin. Invest.* 92, 398-403.

Itoh, Y., Yanagisawa, M., Ohkubo, S., Kimura, C., Kosaka, T., Inoue, A., Ishida, N., Mitsui, Y., Onda, H., Fujino, M. and Masaki, T. (1988) Cloning and sequence analysis of cDNA encoding the precursor of a human endothelium-derived vasoconstrictor peptide, endothelin: Identity of human and porcine endothelin. *FEBS Lett.* 231, 440-444.

Iwasaki, S., Homma, T., Matsuda, Y. and Kon, V. (1995) Endothelin receptor subtype beta mediates autoinduction of endothelin-1 in rat mesangial cells. *J. Biol. Chem.* 270, 6997-7003.

Jones, L.G., Rozich, J.D., Tsutsui, H. and Cooper IV G. (1992) Endothelin stimulates multiple responses in isolated adult ventricular cardiac myocytes. *Am. J. Physiol.* 263, H1447-H1454.

Karne, S., Jayawickreme, C.K. and Lerner, M.R. (1993) Cloning and characterization of an endothelin-3 specific receptor (ETc receptor) from Xenopus laevis dermal melanophores. *J. Biol. Chem.* 268, 19126-19133.

Kilpatrick, S.J., Roberts, J.M., Lykins, D.L. and Taylor, R.N. (1993) Characterization and ontogeny of endothelin receptors in human placenta. *Am. J. Physiol.* 264, E367-E372.

Knoll, B.J. (1992) Gene expression in the human placental trophoblast: A model for developmental gene regulation. *Placenta* 13, 311-327.

Komoro, I., Kurihara, H., Sugiyama, T., Takafu, F. and Yazaki, Y. (1988) Endothelin stimulates c-fos and c-myc expression and proliferation of vascular smooth muscle cells. *FEBS Lett.* 238, 249-252.

Kubota, T., Kamata, S., Hirata, Y., Eguchi, S., Imami, T., Marumo, F. and Aso, T. (1992) Synthesis and release of endothelin-1 by human decidual cells. *J. Clin. Endocrinol. Metab.* 75, 1230-1234.

Kurihara, H., Yoshizumi, M., Sugiyama, T., Takaku, F., Yanagisawa, M., Masaki, T., Hamaoki, M., Kato, H. and Yazaki, Y. (1989) Transforming growth factor-beta stimulates the expression of endothelin mRNA by vascular endothelial cells. *Bioch. Biophys. Res. Comm.* 159, 1435-1440.

Kurihara, Y., Kurihara, H., Suzuki, H., Kodama, T., Maemura, K., Nagai, R., Oda, H., Kuwaki, T., Cao, W.H., Kamada, N., Jishage, K., Ouchi, Y., Azuma, S., Toyoda, Y., Ishikawa, T., Kumada, M. and Yazaki, Y. (1994) Elevated blood pressure and craniofacial abnormalities in mice deficient in endothelin-1. *Nature* 368, 703-710.

Kusuhara, M., Yamaguchi, K., Nagasaki, Y., Hayashi, C., Suzuki, A., Hori, S., Handa, S., Nakamura, Y. and Abe, K. (1990) Production of endothelin in human cancer lines. *Cancer Res.* 50, 3257-3261.

La, M. and Reid, J.J. (1995) Endothelin-1 and the regulation of vascular tone. *Clin. Exper. Pharmacol. Physiol.* 22, 315-323.

Le, S.Q., Wasserstrum, N., Mombouli, J.V. and Vanhoutte, P.M. (1993) Contractile effect of endothelin in human placental veins : role of endothelium prostaglandins and thromboxane. *Am. J. Obstet. Gynecol.* 169, 919-924.

Li, X.M., Moutquin, J.M., Deschênes, J., Bourque, L., Marois, M. and Forest, J.C. (1995) Increased immunohistochemical expression of neutral metalloendopeptidase (enkephalinase ; EC 3.4.24.11) in villi of the human placenta with pre-eclampsia. *Placenta* 16, 435-445.

MacLean, M.R., Templeton, A.G.B. and McGrath, J.C. (1992) The influence of endothelin-1 on human foeto-placental blood vessels : a comparison with 5-hydroxytryptamine. *Br. J. Pharmacol.* 106, 937-941.

Maemura, K., Kurihara, H., Kurihara, Y., Nagai, R. and Yazaki, Y. (1994) Isolation and characterization of vascular endothelial cells derived from mice lacking Endothelin-1. *Bioch. Biophys. Res. Com.* 201, 538-545.

Malassiné, A., Cronier, L., Mondon, F., Mignot, T.M. and Ferré, F. (1993a) Localization and production of immunoreactive endothelin-1 in the trophoblast of human placenta. *Cell Tissue Res.* 271, 491-497.

Malassiné, A., Mondon, F., Robaut, C., Vial, M., Bandet, J., Tanguy, G., Rostène, W., Cavero, I. and Ferré, F. (1993b) Trophoblastic localization of [^{125}I] endothelin-1 binding sites in human placenta. *Trophoblast Research* 7, 271-285.

Malek, A.M., Greene, A.L. and Izumo, S. (1993) Regulation of endothelin-1 gene by fluid shear stress is transcriptionally mediated and independent of protein kinase C and cAMP. *Proc. Natl. Acad. Sci. USA* 90, 5999-6003.

Marsden, P.A. and Brenner, B.M. (1992) Transcriptional regulation of the endothelin-1 gene by TFN-alpha. *Am. J. Physiol.* 262, C854-C861.

Masaki, T. (1995) Possible role of endothelin in endothelial regulation of vascular tone. *Annu. Rev. Pharmacol. Toxicol.* 35, 235-255.

Masaki, T. Vane, J.R. and Vanhoutte M.V. (1994) International union of pharmacology nomenclature of endothelin receptors. *Pharmacol. Rev.* 46, 137-142.

McMahon, E.G., Palomo, M.A., Moore, W.M., McDonald, F.J. and Stern, M.K. (1991) Phosphoramidon blocks the pressor activity of porcine big endothelin-1-(1-39) in vivo and conversion of big endothelin-1-(1-39) to endothelin-1-(1-21) in vitro. *Proc. Natl. Acad. Sci.* USA., 88, 703-707.

McMahon, L.P., Redman, C.W.G. and Firth, J.D. (1993) Expression of the three endothelin genes and plasma levels of endothelin in pre-eclamptic and normal gestations. *Clin. Science* 85, 417-424.

Mignot, T.M., Malassiné, A., Cronier, L. and Ferré, F. (1995) Endothelin increases progesterone release by human trophoblast in culture. In: *Endothelin in Endocrinology : New Advances,* (eds.) E. Baldi, M. Maggi, I.T. Cameron, and M.J. Dunn, vol. 15, Ares-Serono Symposia Publications, pp. 277-283.

Mombouli, J.V., Le, S.Q., Wasserstrum, N. and Vanhoutte, P.M. (1993) Endothelins 1 and 3 and big endothelin-1 contract isolated human placental veins. *J. Cardiov. Pharmacol.* 22, S278-S281.

Mondon, F., Doualla-Bell Kotto Maka, F., Sabry, S. and Ferré, F. (1995) Endothelin-induced phosphoinositide hydrolysis in the muscular layer of stem villi vessels of human term placenta. *Eur. J. Endoc.* 133,.606-612.

Mondon, F., Malassiné, A., Robaut, C., Vial, M., Bandet, J., Tanguy, G., Rostène, W., Cavero, I. and Ferré, F. (1993) Biochemical characterization and autoradiographic localization of [^{125}I]endothelin-1 binding sites on trophoblast and blood vessels of human placenta. *J. Clin. Endoc. Metab.* 76, 237-244.

Monge, J.C., Stewart, D.J. and Cernacek, P. (1995) Differentiation of embryonal carcinoma cells to a neural or cardiomyocyte lineage is associated with selective expression of endothelin receptors. *J. Biol. Chem.* 270, 15385-15390.

Myatt, L., Bewer, A.S. and Brockman, D.E. (1992) The comparative effects of big endothelin-1, endothelin-1, and endothelin-3 in the human placental circulation. *Am. J. Obstet. Gynecol.* 167, 1651-1656.

Naruse, M., Naruse, K. and Demura, H. (1994) Recent advances in endothelin research on cardiovascular and endocrine systems. *Endocrine J.* 41, 491-507.

Nishiyama, M., Moroi, K., Shaw, L.H., Yamamoto, M., Takasaki, C., Masaki, T. and Kimura, S. (1995) Two different endothelin-B receptor subtypes mediate contraction of the rabbit saphenous vein. *Jpn. J. Pharmacol.* 68, 235-243.

Oliver, F.J., de la Rubia, G., Feener, E.P., Lee, M.E., Loeken, M.R., Shiba, T., Quertermous, T. and King, G.L. (1991) Stimulation of endothelin-1 gene expression by insulin in endothelial cells. *J. Biol. Chem.* 266, 23251-23256.

Onda, H., Ohkubo, S., Ogi, K., Kosaka, T., Kimura, C., Matsumoto, H., Suzuki, N. and Fujino, M. (1990) One of the endothelin gene family, endothelin 3 gene, is expressed in the placenta. *FEBS Lett.* 261, 327-330.

Opgenorth, T.J., Wu-Wong, J.R. and Shiosaki, K. (1992) Endothelin-converting enzymes. *FASEB J.* 6, 2653-2659.

Puffenberger, E.G., Hosoda, K., Washington, S.S., Nakao, K., de Wit, D., Yanagisawa, M. and Chakravarti, A. (1994) A missense mutation of the endothelin-B receptor gene in multigenic Hirschsprung's disease. *Cell* 79, 1257-1266.

Rae, G.A., Calixto, J.B. and D'Orleans-Juste, P. (1995) Effects and mechanisms of action of endothelins on non-vascular smooth muscle of the respiratory, gastrointestinal and urogenital tracts. *Regulatory Peptides* 55, 1-46.

Robaut, C., Mondon, F., Bandet, J., Ferré, F. and Cavero, I. (1991) Regional distribution and pharmacological characterization of [^{125}I]endothelin-1 binding sites in human fetal placental vessels. *Placenta* 12, 55-67.

Roberts, J.M. and Redman, C.W.G. (1993) Pre-eclampsia: More than pregnancy-induced hypertension. *Lancet* 341, 1447-1451.

Rubanyi, G.M. and Polokoff, M.A. (1994) Endothelins: Molecular biology, biochemistry, pharmacology, physiology, and pathophysiology. *Pharmacological Reviews* 46, 325-415.

Rutherford, R.A.D., Wharton, J., McCarthy, A., Gordon, L., Sullivan, M.H.F., Elder, M.G. and Polak, J.M. (1993) Differential localization of endothelin ET_A and ET_B binding sites in human placenta. *Br. J. Pharmacol.* 109, 544-552.

Sabry, S., Mondon, F., Ferré, F. and Dinh-Xuan, A.T. (1995a) *In vitro* contractile and relaxant responses of human resistance placental stem villi arteries of healthy parturients: Role of endothelium. *Fundam. Clin. Pharmacol* 9, 46-51.

Sabry, S., Mondon, F., Levy, M., Ferré, F. and Dinh-Xuan, A.T. (1995b) Endothelial modulation of vasoconstrictor responses to endothelin-1 in human placental stem villi small arteries. *Br. J. Pharmacol.* 115, 1038-1042.

Schiff, E., Galron, R., Ben-Baruch, G., Mashiach, S. and Sokolovsky, M. (1993a) Endothelin-1 receptors on the human placenta and fetal membranes: Evidence for different binding properties in pre-eclamptic pregnancies. *Gynecol. Endocrinol.* 7, 67-72.

Schiff, E., Zael, Y., Friedman, S.A. and Shalev, E. (1993b) Fetal circulatory endothelin-1,2 in the mid-trimester. *Gynecol. Obstet. Invest.* 35, 185-186.

Shyamala, V., Moulthrop, T.H.M.B., Strattow-Thomas, J. and Tekamp-Olson, P. (1994) Two distinct human endothelin beta-receptors generated by alternative splicing from a single gene. *Cell Molec. Biol. Res.* 40, 285-296.

Simonson, M.S., Jones, J.M. and Dunn, M.J. (1992) Differential regulation of *fos* and *jun* gene expression and AP-1 *cis*-element activity by endothelin isopeptides. *J. Biol. Chem.* 267, 8643-8649.

Smith, P.L., Lee, C.P., Pullen, M., Ohlstein, E.H., Beck, G., Eddy, E.P. and Nambi, P. (1995) Non peptide endothelin receptor antagonists. IV. Identification of receptors in rabbit colonic mucosa and smooth muscle and correlation with physiological effects. *J. Pharmacol. Exp. Ther,* 272, 1204-1210.

Stewart, A.G., Grigoriadis, G. and Harris, T. (1994) Mitogenic actions of endothelin-1 and epidermal growth factor in cultured airway smooth muscle. *Clin. Exp. Pharmacol. Physiol.* 21, 277-285.

Sudjarwo, S.A., Hori, M., Tanaka, T., Matsuda, Y., Okada, T. and Karaki, H. (1994) Subtypes of endothelin ET_A and ET_B receptors mediating venous smooth muscle contraction. *Biochem. Biophys. Res. Comm.* 200, 627-633.

Svane, D., Larsson, B., Alm, P., Andersson, K.E. and Forman, A. (1993) Endothelin-1: Immunocytochemistry, localization of binding sites, and contractile effects in human uteroplacental smooth muscle. *Am. J. Obstet. Gynecol.* 168, 233-241.

Swope, V.B., Medrano, E.E., Smalara, D. and Abdel-Malek, Z.A. (1995) Long-term proliferation of human melanocytes is supported by the physiologic mitogen alpha-melanotropin, endothelin-1, and basic fibroblast growth factor. *Exp. Cell Res.* 217, 453-459.

Tsuboi, R., Sato, C., Shi, C.M., Nakamura, T., Sakurai, T. and Ogawa, H. (1994) Endothelin-1 acts as an autocrine growth factor for normal human keratinocytes. *J. Cell. Physiol.* 159, 213-220.

Van Papendorp, C.L., Cameron, I.T., Davenport, A.P., King, A., Barker, P.J. Huskisson, N.S., Gilmour, R.S., Brown, M J. and Smith, J.K. (1991) Localization and endogenous concentrations of endothelin-like immunoreactivity in human placenta. *J. Endocrinology* 131, 507-511.

Vijayaraghavan, J., Scicli, A.G., Carretero, O.A., Slaughter, C., Moomaw, C. and Hersh, L.B. (1990) The hydrolysis of endothelins by neutral endopeptidase 24-11 (enkephalinase). *J. Biol. Chem.* 265, 14150-14155.

Wagner, O.F., Christ, G., Wojta, J., Vierhapper, H., Parzer, S., Nowotny, P.J., Schneider, B., Waldhäusl, W. and Binder, B.R. (1992) Polar secretion of endothelin-1 by cultured endothelial cells. *J. Biol. Chem.* 267, 16066-16068.

Walker, J.J. (1994) Hypertension in pregnancy. *Br. J. Obstet. Gynaecol.* 101, 639-644.

Wang, Y., Rose, P.M., Webb, M.L. and Dunn, M.J. (1994) Endothelins stimulate mitogen-activated protein kinase cascade through either ET_A or ET_B. *Am. Physiol.* 267, C.1130-C.1135.

Weber, H., Webb, M.L., Serafino, R., Taylor, D.S., Moreland, S., Norman, J. and Molloy, C.J. (1994) Endothelin-1 and angiotensin-II stimulate delayed mitogenesis in cultured rat aortic smooth muscle cells : evidence for common signaling mechanisms. *Molec. Endocr.* 8, 148-158.

Wilkes, B.M., Mento, P.F., Hollander, A.M., Maita, M.E., Sung, S. and Girardi, E.P. (1990) Endothelin receptors in human placenta: Relationship to vascular resistance and thromboxane release. *Am. J. Physiol.* 258, E864-E870.

Wilkes, B.M., Susin, M. and Mento, P.F. (1993) Localization of endothelin-1-like immunoreactivity in human placenta. *J. Histochem. Cytochem.* 41, 535-541.

Xu, D., Emoto, N., Giaid, A., Slaughter, C., Kaw, S., de Wit, D. and Yanagisawa, M.(1994) ECE-1: A membrane-bound metalloprotease that catalyzes the proteolytic activation of big endothelin-1. *Cell* 78, 473-485.

Yanagisawa, M., Kurihara, H., Kimura, S., Tomobe, Y., Kobayashi, M., Mitsui, Y., Yazaki, Y., Goto, K. and Masaki, T. (1988) A novel potent vasoconstrictor peptide produced by vascular endothelial cells. *Nature* 332, 411-414.

Yokohawa, K., Kohno, M., Yasunari, K., Marakawa, K. and Takeda, T. (1991) Endothelin-3 regulates endothelin-1 production in cultured human endothelial cells. *Hypertension* 18, 304-315.

Trophoblast Research 10:279-284, 1997

EARLY DETECTION OF ENDOMETRIAL VESSELS BY POWER DOPPLER

Jean-Pierre Schaaps

Department of Obstetrics and Gynecology
University of Liège
CHR Citadelle
Bd du 12e de Ligne, 1
4000 Liège, Belgium

INTRODUCTION

The regulation of invasion and destruction is involved in the implantation process. The basal membranes of the endometrial superficial layer are destroyed. The interstitial tissue is invaded by the trophoblastic columns. After that first step, the growth of the trophoblast does not disrupt either the compact layer of the endometrium or the capsularis even if, during the intra-endometrial progression some vascular and glandular basal membranes are destroyed. A local, extremely specific regulation of the invasion should exist (Pijnenborg et al., 1981; Pijnenborg, 1994; Ramsey et al., 1976).

As in the metastatic process, the first intravascular invasion is performed from the outside towards the internal part of the capillaries (Hustin et al., 1988).

A secondary effect, after colonization of the lumen by the trophoblastic cells, is that the invasion will progress deeper in the vessels, replacing the endothelial cells, intruding into the wall and stopping at the external part without any vascular effraction.

As in the metastatic invasion, angiogenic or vasoactive agents should modify the vascular network around the implantation area (Liotta, 1986). A large amount of anastomic and dilated vessels can be detected by ultrasound all around the trophoblastic ring from its formation. This vascular network, associated to the specific parietal modifications of the uteroplacental arteries, induces an increase of the diastolic component of the flow (Wladimiroff et al., 1995).

This vascular angiogenesis can be observed, *in vivo* by ultrasound, from the first stages of human embryonic development.

MATERIALS AND METHODS

One of the most recent developments of Color Doppler is the Power Doppler. Using such technology, vessels with a inner diameter of less than 1 mm perfused with a very slow flow (<2 cm/sec) can be detected and the vascular mapping of the uterus (from the uterine to the basal arteries) can be obtained (Figures 1 and 2).

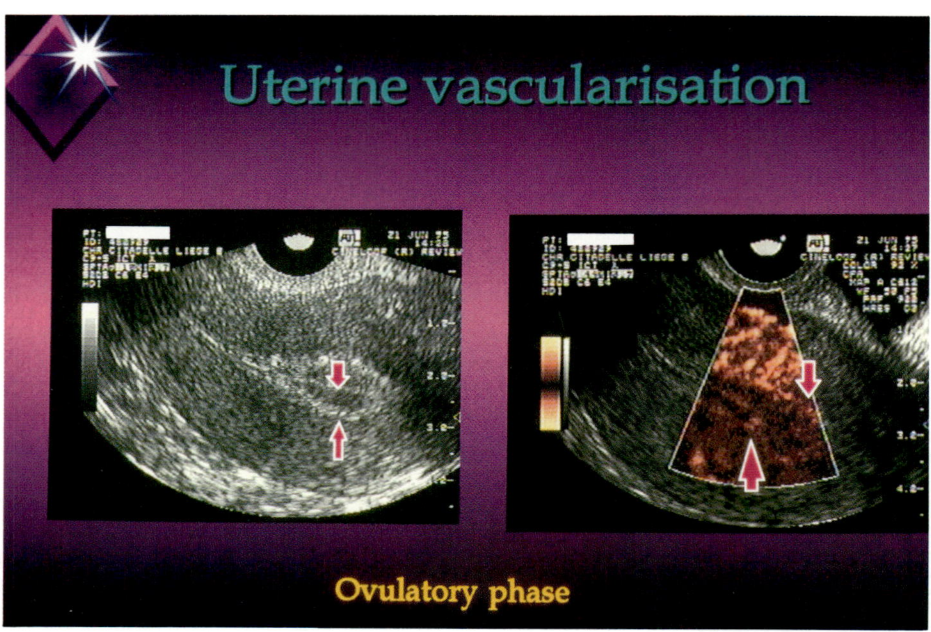
Uterine vascularisation
Ovulatory phase

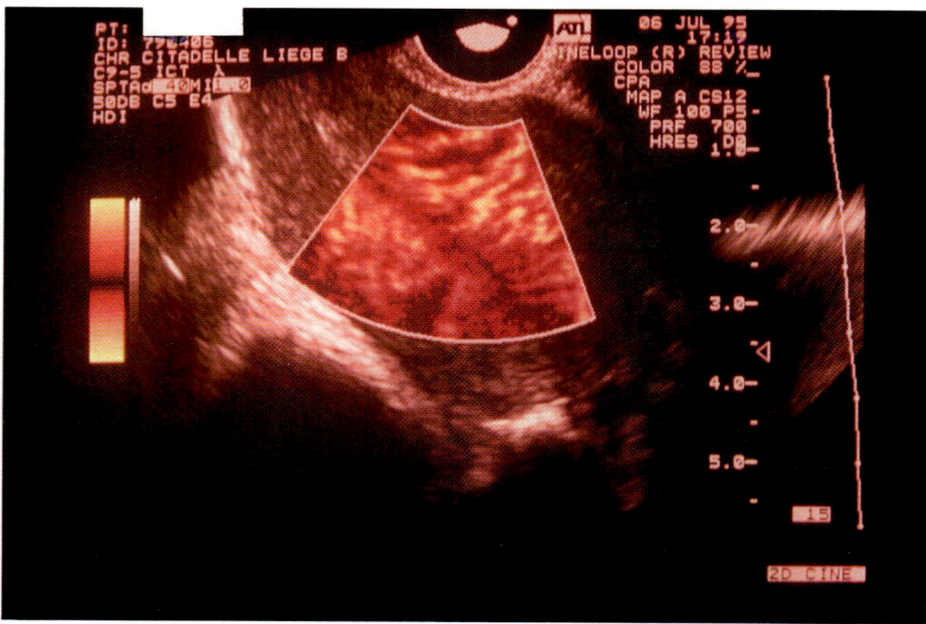
PT:
CHR CITADELLE LIEGE B
C9-5 ICT
50DB C5 E4
HDI
06 JUL 95
17:19
CINELOOP (R) REVIEW
COLOR 88 %
CPA
MAP A CS12
WF 100 PS
PRF 700
HRES D0
1.0
2.0
3.0
4.0
5.0
15
2D CINE

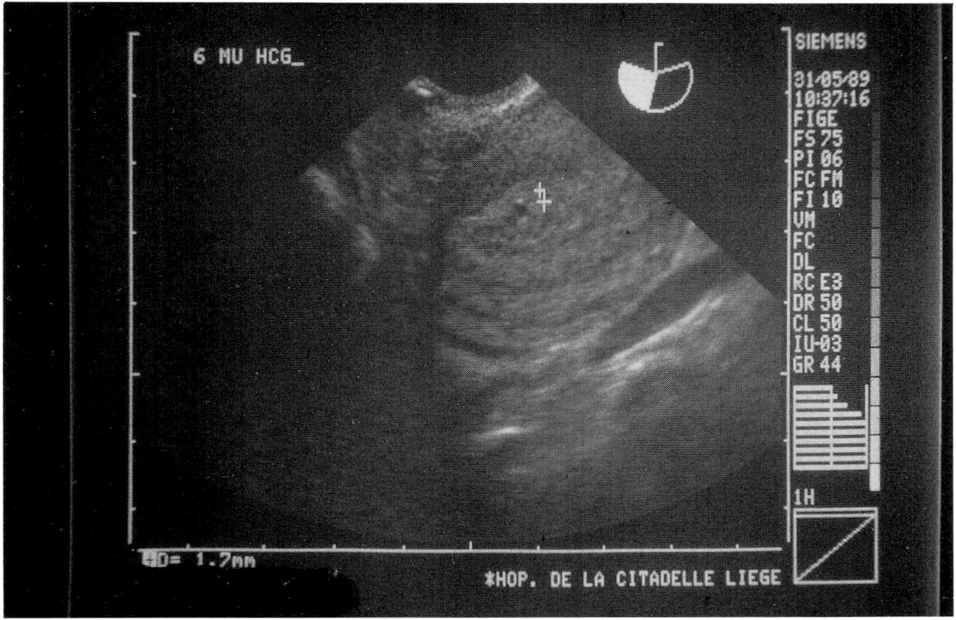

Figure 3. Small echo-free lacuna detected in the endometrium (< 2 mm) at the 27th day of the cycle.

The intra-endometrial spiral arteries were not detected by these highly sensitive systems at any time during the menstrual cycle. In abnormal cases, like the presence of fibroids or polyps (organic pathologies) an intra-endometrial blood flow can be found.

Functional disturbances, able to lead to a mucosal hypertrophy, do not induce such a vascularization.

The casual diagnosis of pregnancy has been performed in 11 patients undergoing a vaginal ultrasonic exploration for other reasons (cycle irregularities or statement of infertility problem).

No prospective study has been performed. For ethical reasons, even in an IVF program, it was out of the question to insonate such very young pregnancies.

Figure 1. Endometrium pattern observed around the ovulation. Clear visualization of the radial arteries.

Figure 2. Late secretory phase: Increasing in number and size of the intra myometrial circulation. The endometrium remains silent.

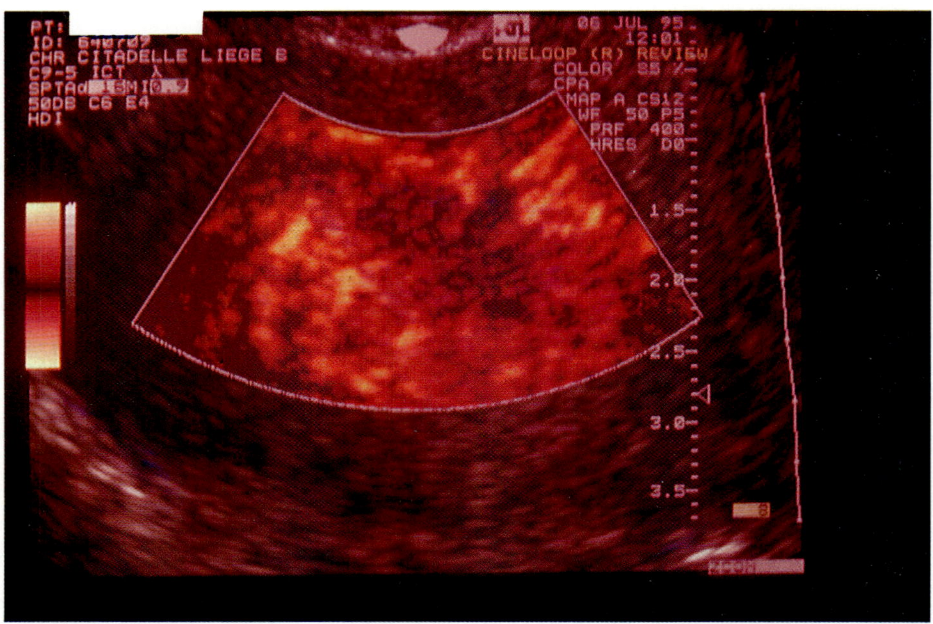
PT:
CHR CITADELLE LIEGE B
C9-5 ICT
50DB C6 E4
HDI
06 JUL 95
12:01
CINELOOP (R) REVIEW
COLOR 85 %
CPA
MAP A CS12
WF 50 P5
PRF 400
HRES D0
1.5
2.0
2.5
3.0
3.5

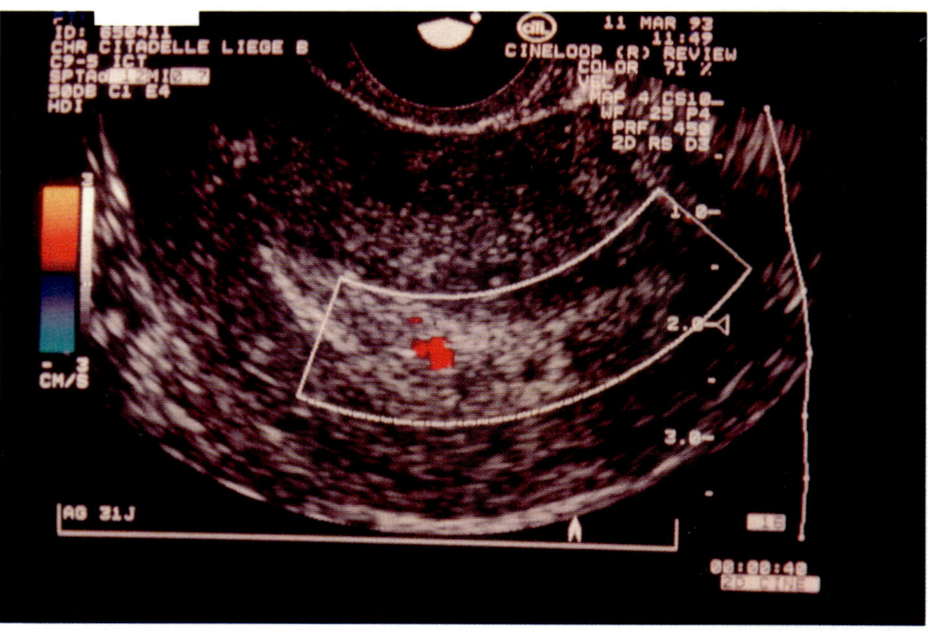
CHR CITADELLE LIEGE B
C9-5 ICT
50DB C1 E4
HDI
11 MAR 93
11:49
CINELOOP (R) REVIEW
COLOR 71 %
VEL
MAP 4 CS10
WF 25 P4
PRF 450
2D RS D3
1.0
2.0
3.0
CM/S
AG 31J
00:00:40
2D CINE

The precise estimation of the age of the pregnancy was performed later in the first trimester using the usual biometric data.

RESULTS

Before the detection of a typical gestational sac, an intraendometrial flow was detected in the non invaded endometrium.

At this period (29th, 30th day of the cycle), ultrasound examination can provide only a non specific shape of the gestational sac (Figure 3). It appears as a 2 mm intraendometrial echo-free lacuna which can be confused with a glandular dilatation. The flow was detected close to the lacuna and at several millimeters distance (Figures 4 and 5).

In each case, an HCG titration was requested after the ultrasound examination with a range of between 450 and 1000 mUI/ml.

DISCUSSION

The early modifications of the vascular network assessed by ultrasound seem to be closely related to the trophoblastic activity. This vascular appearance is similar to a major aspecific inflammatory response, but the morphology is different. The inflamed endometrium does not show the same pattern as the normal late luteal phase (homogenicity, regularity, echo density).

The development of this network will continue throughout the first trimester and concern the deepest part of the myometrium. The size and number of detected vessels will constitute the dense peritrophoblastic network.

CONCLUSION

The ultrasonic data provided by the highly sensitive Doppler systems suggest:

1. the existence, from the first stages of the embryonic implantation (less than 10 days after hatching), of angiotrophic signals diffusing out of the trophoblastic area and modifying the endometrial vasculature;

2. the angiotrophic stimulation appears before the trophoblastic shell organization.

Figure 4. Intraendometrial circulation observed the day before the expected period.

Figure 5. Thirty-five days of amenorrhea, clear detection of vessels located at distance from the gestational sac.

REFERENCES

Hustin, J., Schaaps, J.P. and Lambotte, R. (1988) Anatomical studies of the utero-placental vascularization in the first trimester of pregnancy. *Trophoblast Res.* 3, 49-60.

Liotta, L.A. (1986) Tumor invasion and metastases-role of the extracellular matrix: Rhoads Memorial Award Lecture. *Cancer Res.* 46, 1-7.

Pijnenborg, R., Bland, J.M., Robertson, W.B., Dixon, G. and Brosens, I. (1981) The pattern of interstitial trophoblastic invasion of the myometrium in early human pregnancy. *Placenta* 2, 303-316.

Pijnenborg, R. (1994) Trophoblast invasion. *Reprod. Med. Rev.* 3, 53-73.

Ramsey, E.M., Houston, M.L. and Harris, J.W.S. (1976) Interactions of the trophoblast and maternal tissues in three closely related primate species. *Am. J. Obstet. Gynecol.* 124, 647-652.

Wladimiroff, J.W. and Van Splunder I.P. (1995) Assessment of the early foetal circulation. In: *Transvaginal Colour Doppler*, (eds.) T.H. Bourne , E. Jauniaux and D. Jurkovic, Springer-Verlag Heidelberg, pp. 175-183

Trophoblast Research 10:285-290, 1997

UNEXPECTED PERSISTENCE OF AN INTRAMYOMETRIAL HIGH FLOW AFTER MOLA DESTRUENS

Jean-Pierre Schaaps

Department of Obstetrics and Gynecology
University of Liège
CHR Citadelle
Bd du 12e de Ligne, 1
4000 Liège, Belgium

INTRODUCTION

Trophoblastic tissues induce an increase in the number and the size of the uterine vessels located in the deepest area of the myometium. These modifications are due to angiogenic signals, acting from the first stages of the implantation (Hustin et al., 1988; Schaaps et al., 1988; Hustin, 1992, 1995; Wladimiroff et al., 1995).

The functional and anatomical changes are dependent on the vitality or the presence of active trophoblastic cells. Soon after abortion or delivery, the uterine vascularization is seen returning to a normal status (Dickey et al., 1995).

In cases of persistent trophoblastic tissue in activity, this phenomenon does not appear in this area (Figure 1). The vascular evolution after an hydatiform mole is the same as after a normal abortion. The clinical follow up of such a pathology is based on the circulating HCG levels titration and do not request the use of any imaging technique.

In cases of uterine choriocarcinoma, typical iliac arteriographic patterns have been described as a bunch of grapes due to the presence of arterio-venous shunts.

The vascular evolution of the uterus in cases of chorioadenoma requiring a chemotherapy has been followed by ultrasound.

MATERIALS AND METHODS

Eight patients suffering from chorioadenoma have requested chemotherapy. The abnormal decreasing or persistence of circulating HCG, after curettage or aspiration of the molar tissue, has indicated the necessity for this treatment.

One of them received a monochemotherapy, six others a polychemotherapy and the last one underwent an hysterectomy because of advanced AIDS with a T cell count of less than 200 per ml. She died eight months later from immunodeficiency syndrome after complete negativation of the HCG.

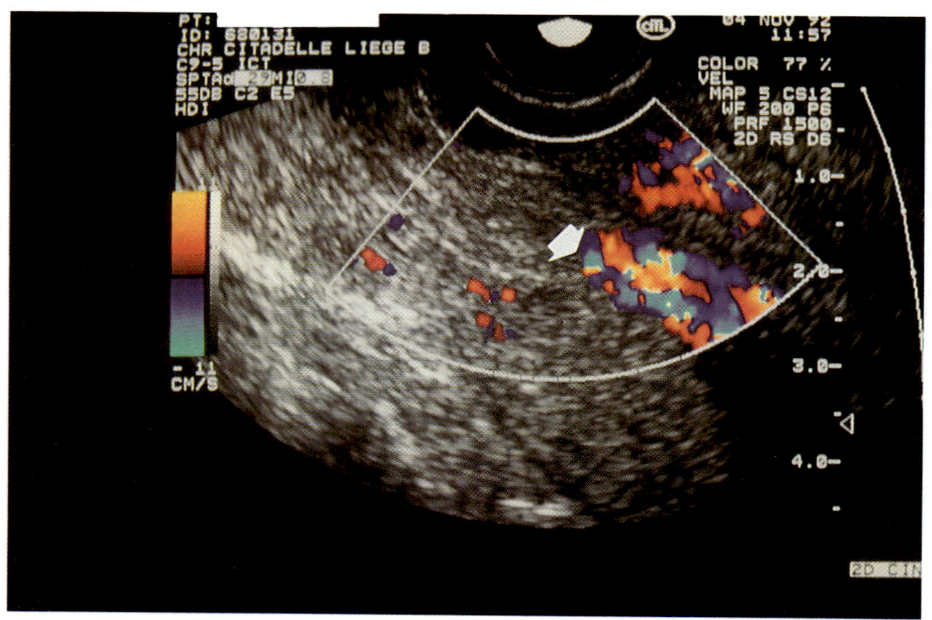
PT:
ID: 680131
CHR CITADELLE LIEGE B
C9-5 ICT
55DB C2 E5
HDI
04 NOV 92
11:57
COLOR 77 %
VEL
MAP 5 CS12
WF 200 P6
PRF 1500
2D RS D6
1.0
2.0
3.0
4.0
- 11
CM/S

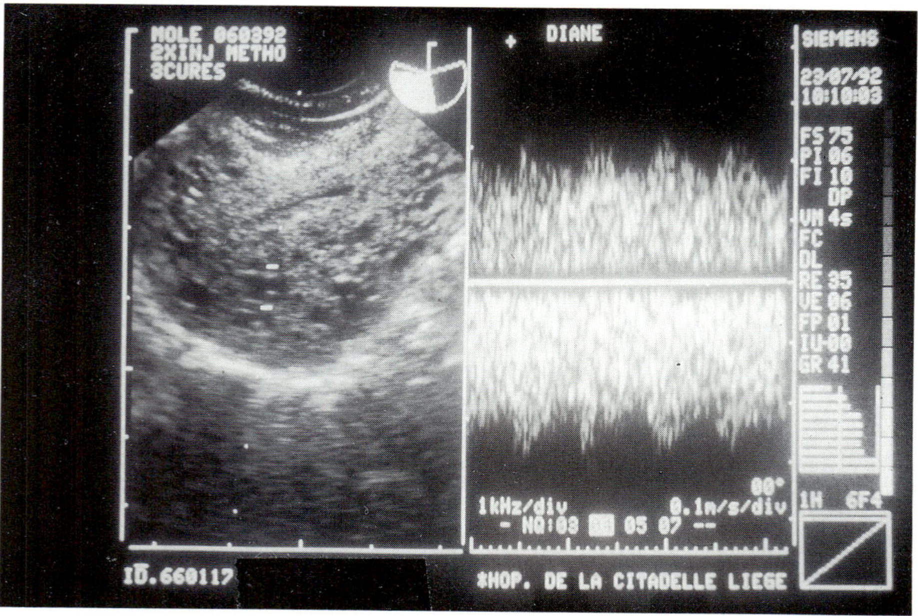
MOLE 060392
2XINJ METHO
3CURES
DIANE
SIEMENS
23/07/92
10:10:03
1kHz/div
0.1m/s/div
ID.660117
*HOP. DE LA CITADELLE LIEGE

RESULTS

The seven patients were followed by Doppler ultrasound before their antimitotic treatment for a period of between 16 and 72 months. All of them showed a complete biological recovery after the treatment.

In one patient, the evolution of the uterine circulation was as expected: a quick reduction of the flow and the disappearance of the pregnant uterine vascular network.

In the six others, a persistence of very prominent arterio-venous shunts located in the endometrium was observed (Figures 2 and 3).

Those shunts have been considered as high flow rate, blood speed were detected at more than 120 cm/sec: more than in the aortic arch (± 100 cm/sec) (Figure 4).

In two patients, a selective arteriography was performed: they show a typical pattern of shunts described in the choriocarcinoma (Figure 5).

DISCUSSION

Vessel stimulation is known to be dependent on the trophoblastic tissue. The long term follow-up of the uterine vascular network we performed indicates that the angiographic signals issued from the trophoblastic cells can be different in normal or abnormal tissues.

In the first situation the trophic effect is closely associated with the presence of trophoblastic cells. The induced anatomical disturbances disappear as the pregnant process is removed.

The chorioadenoma is characterized by a more aggressive behavior in comparison with the hydatiform mola. The reversibility of the vascular effects do not seem to be the rule.

Even with the biological evidence of the complete destruction of the abnormal cells, the vascular imprinting persists for a long time (72 months) and is probably permanent.

Data provided by angiographic explorations of the uterus in case of suspicion of trophoblastic diseases must be considered differently. The detection of such vascular structure is not a specific indicator of a malignant process. It can be a persistent modification of the vascular network.

Figure 1. Persistence of a local abnormal circulation 10 days after curettage and persistence of blood losses. Hysteroscopy and biopsy : persistence of small amount of trophoblastic tissue.

Figure 2. Two months after negativation of HCG - 3 courses of methotrexate have been needful. Intra myometrial blood flow with high diastolic component.

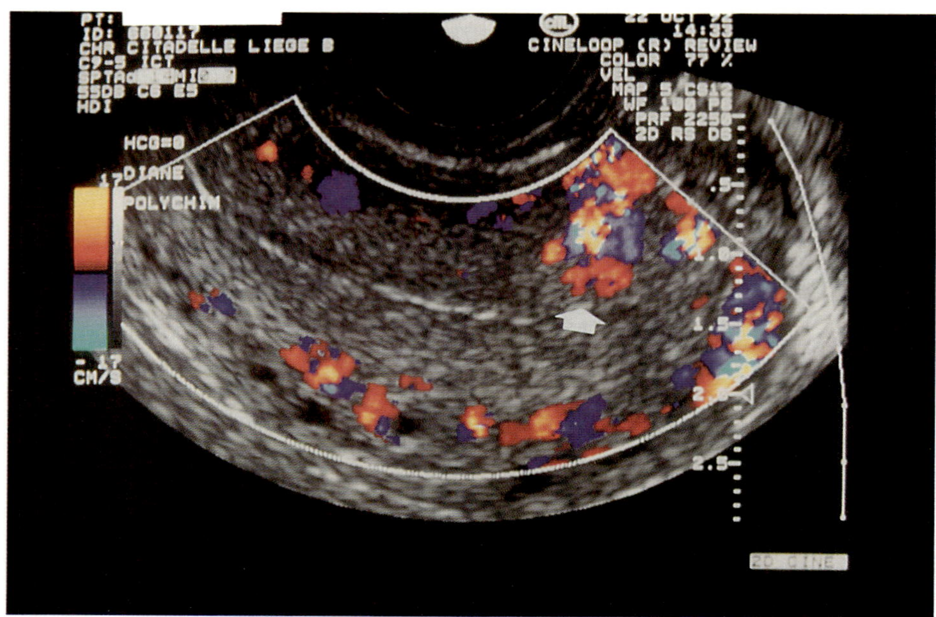

CHR CITADELLE LIEGE B
C9-5 ICT
14:33
CINELOOP (R) REVIEW
COLOR 77 %
VEL
HCG=0
DIANE
POLYCHIM
CM/S
2D CINE

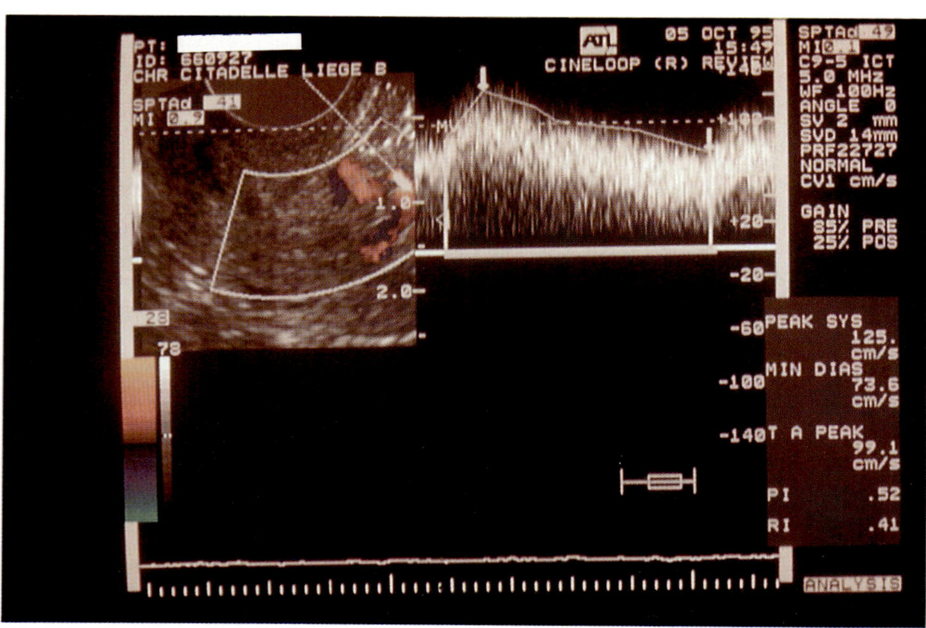

ID: 660927
CHR CITADELLE LIEGE B
05 OCT 95
15:49
CINELOOP (R) REVIEW
SPTAd 49
C9-5 ICT
5.0 MHz
WF 100Hz
ANGLE 0
SV 2 mm
SVD 14mm
PRF22727
NORMAL
CV1 cm/s
GAIN
85% PRE
25% POS
PEAK SYS
125.
cm/s
MIN DIAS
73.6
cm/s
T A PEAK
99.1
cm/s
PI .52
RI .41
ANALYSIS

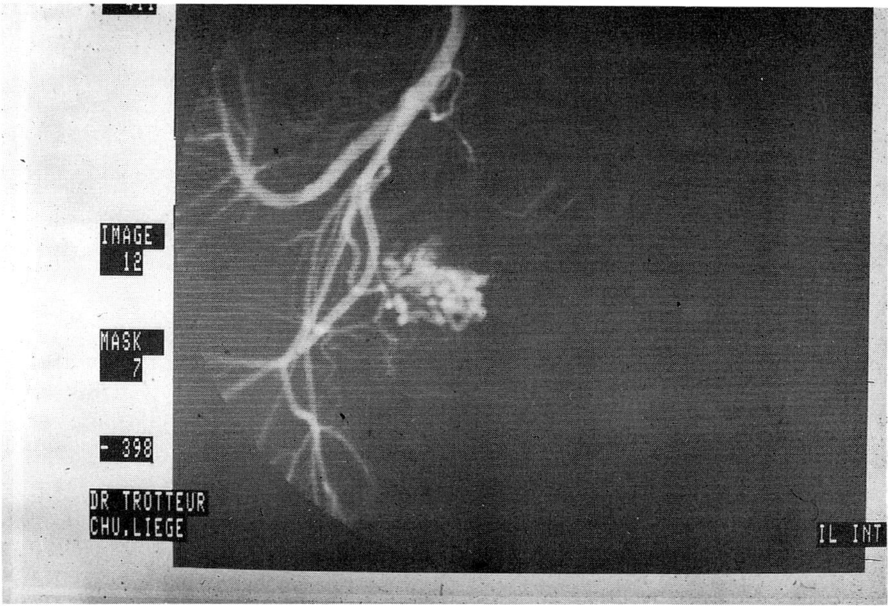

Figure 5. Arteriography performed the same week than the Figure 2. Aspect of bunch of grapefruit indicating a choriocarcinoma.

CONCLUSION

Eight chorioadenomas have been observed by power Doppler for a long time after complete biological recovery.

The persistence of a particular vascular network (high velocity shunts) does not indicate the persistence of a malignant tissue. Rather it suggests the secretion of other angiogenic signals inducing a non reversible vessel modification.

Figure 3 (opposite page, upper). HCG = 0, the shunt area is clearly located by Color Doppler.

Figure 4 (opposite page, lower). Approximation of the flow speed (125 cm/sec), 5 years after chorioadenoma.

REFERENCES

Dickey, R.P. and Hower, J.F. (1995) Ultrasonographic features of uterine blood flows during the first 16 weeks of pregnancy. *Human Reproduction* 10, 2448-2452.

Hustin, J., Schaaps, J.P. and Lambotte, R. (1988) Anatomical studies of the utero-placental vascularization in the first trimester of pregnancy. *Trophoblast Res.* 3, 49-60.

Hustin, J. (1992) The materno-trophoblastic interface: Uteroplacental blood flow. In: *The First Twelve Weeks Of Gestation*, (eds.) E.R. Barnea, J. Hustin and E. Jauniaux Springer-Berlin: Heidelberg New York, pp. 97-110.

Hustin, J. (1995) Vascular physiology and pathophysiology of early pregnancy. In: *Transvaginal Colour Doppler*, (eds.) T.H. Bourne T.H., E. Jauniaux and D. Jurkovic, Springer-Verlag, Heidelberg, pp. 47-56.

Schaaps, J.P. and Hustin, J. (1988) In vivo aspects of the materno-trophoblastic border during the first trimester of gestation. *Trophoblast Res.* 3, 39-48

Wladimiroff, J.W. and Van Splunder I.P. (1995) Assessment of the early foetal circulation. In: *Transvaginal Colour Doppler*, (eds.), T.H. Bourne T.H., E. Jauniaux and D. Jurkovic, Springer-Verlag: Heidelberg, pp. 175-183.

Trophoblast Research 10:291-309, 1997

PATHOLOGICAL BASIS FOR ABNORMAL UMBILICAL ARTERY DOPPLER WAVEFORMS IN PREGNANCIES COMPLICATED BY INTRAUTERINE GROWTH RESTRICTION
- A Review -

John C.P. Kingdom[1,5], Lena Macara[2], Christiane Krebs[3], Rudolph Leiser[3] and Peter Kaufmann[4]

[1]Department of Obstetrics and Gynaecology
University College London
London, United Kingdom

[2]Department of Obstetrics and Gynaecology
University of Glasgow
Glasgow, Scotland

[3]Department of Veterinary Anatomy
University of Giessen
Giessen, Germany

[4]Department of Anatomy
Technical University of Aachen
Aachen, Germany

INTRODUCTION

Doppler ultrasound assessment of the umbilical arteries is an important diagnostic tool for identifying the small preterm fetus with intrauterine growth restriction (IUGR) due to placental disease. The latter is recognized by absent end-diastolic flow velocity (AEDFV) in the umbilical artery, and is associated with a perinatal death rate of 40% (Karsdorp et al., 1994). Randomized trials of Doppler in high-risk preterm pregnancies have demonstrated that, as a result of subsequent intensive fetal monitoring and timely elective delivery, this risk can be reduced by about one third (Alfirevic and Neilson, 1995). No other tests of fetal well-being, such as biometry or cardiotocography, have been shown through randomized controlled trials to influence perinatal mortality. It is thus extremely important to understand the underlying pathology of AEDFV.

Loss of end-diastolic frequencies in the umbilical arterial waveform is due to a substantial rise in fetoplacental vascular impedance - demonstrated by embolization studies in the sheep fetus (Morrow et al., 1989). The human placenta has a completely different structure from that of the sheep and human placental IUGR is a chronic process. For these reasons we must be careful when considering the possible mechanism by which AEDFV is generated in the human fetoplacental circulation. We have previously summarized the arguments for and against the possible anatomical and vasomotor mechanisms as shown in Table 1. (Macara et al., 1993). In this chapter we will discuss our

[5]To Whom Correspondence Should Be Addressed: Department of Obstetrics and Gynaecology, University College London Medical School, 86-96 Chenies Mews, London WC1E 6HX United Kingdom

collaborative research since 1993 and relate these to developments in the literature during this time.

Clinical Background

Samples of fetal umbilical venous blood from pregnancies complicated by AEDFV frequently demonstrate evidence of chronic hypoxia and metabolic acidosis (Nicolaides et al., 1988; Nicolini et al., 1990) and radioisotope studies indicate a >50% reduction in uteroplacental blood flow (Nylund et al., 1983). The latter can be detected non-invasively in the human by Doppler ultrasound assessment of the uteroplacental circulation (Bower et al., 1993). The basis for uterine ischemia is thought to be an inadequate trophoblast invasion of spiral arteries (de Wolf, Robertson and Brosens, 1986) and the currently prevailing view is that the uteroplacental ischemia results in placental, and therefore fetal, hypoxia and acidosis in IUGR.

The Embolization Theory

The first histological studies performed on placentae from IUGR pregnancies with AEDFV in the umbilical artery were conducted on immersion-fixed paraffin-embedded sections viewed by light microscopy. Giles and colleagues made the original observation of a reduction in the density of small stem villous arteries in the 20-90 μM diameter range, proposing that an obliterative process was responsible for these findings (Giles et al., 1985). Other groups confirmed these findings (McCowan et al., 1987; Bracero et al., 1989), and experiments conducted in the fetal sheep model were used to advance the obliteration/embolization theory (Trudinger et al., 1987; Morrow et al., 1989). In 1993 Jauniaux and Burton summarized the technical limitations of these studies, reminding those of us involved in this field to pay careful attention to clinical selection (need for preterm control placentae for comparison with AEDFV cases) and appropriate techniques for tissue sampling, fixation and imaging (Jauniaux and Burton, 1993).

Table 1

Possible Mechanisms Leading To Increased Fetoplacental Vascular Impedance In Preterm IUGR With Absent End-Diastolic Flow Velocity In the Umbilical Artery

ANATOMICAL
Loss of previously-formed small stem villous arterioles
Failure of normal development of terminal villi
VASOMOTOR
Enhanced action of local or endocrine vasoconstrictor mechanism
Reduced action of local or endocrine vasodilator mechanisms

Morphometric studies, subsequently performed on systematic randomly-sampled immersion-fixed placental villous tissue from AEDFV pregnancies, indicated a reduction in the sectional area of placenta occupied by terminal villi. No direct evidence of stem vessel obliteration was found (Hitschold et al., 1993; Jackson et al., 1995). Terminal villi develop from mature intermediate villi during the second half of gestation (Kaufmann, 1982; Castellucci et al., 1990) in tandem with the known appearance of (Fisk et al., 1988), and progressive rise in (Hendricks et al., 1989), diastolic velocities in the umbilical artery as gestation advances. The capillary network of the mature placenta confers both a low fetoplacental vascular impedance and a large surface area for gas and nutrient transfer; necessary for the high fetal metabolic demands in late pregnancy (Kaufmann and Burton, 1994).

The importance of the terminal capillaries for fetoplacental vascular impedance has been stressed previously (Kaufmann et al., 1985, 1988 ; Leiser et al., 1991). Moreover the same authors have reported the existence of different kinds of maldevelopment of terminal capillary convolutes which were likely to affect vascular impedance (c.f., Benirschke and Kaufmann, 1995). Terminal villi could therefore be the principal anatomical site of functional pathology in the most severe forms of IUGR, characterized by AEDFV in the umbilical artery and elective preterm delivery by cesarean section in the fetal interest. Our group began collaborative work in 1991 to determine the major site within the placenta responsible for the phenomenon of AEDFV. This review covers the material and discussion from the three main papers arising from this work (Macara et al., 1995; Macara et al., 1996 ; Krebs et al., 1996).

METHODS

Clinical Details

We studied singleton preterm IUGR fetuses with AEDFV in the umbilical artery, delivered by planned cesarean section prior to labor. The cord was clamped immediately at delivery; all neonates were structurally normal, displaying clinical evidence of intrauterine starvation. All women were healthy and normotensive in early pregnancy and gestational age was verified by ultrasound prior to 16 weeks. Our criteria for IUGR were: estimated fetal weight <5th centile for gestation, or 5th-10th centile where serial measurements crossed below the 10th centile; AEDFV in the umbilical artery; reduced amniotic fluid volume (vertical depth of cord-free amniotic fluid <30 mm). In this way severe preterm IUGR, of obstetrical and pediatric significance, was clearly separated from the more common healthy small-for-gestational age term fetus (typically identified following a spontaneous vaginal birth, when an erroneous diagnosis of IUGR is easily made).

Since preterm IUGR with AEDFV frequently coexists with pre-eclampsia (Karsdorp et al., 1994), we studied normotensive and preeclamptic pregnancies, the latter identified by; blood pressure recordings >140/100 mmHg on two or more occasions, with a rise in phase IV diastolic blood pressure >20 mmHg occurring after 20 weeks of gestation, together with proteinuria (>0.3g/24 hours or >+2 on dipstick testing on two occasions where a 24 hour urine collection was not possible due to the need for delivery) to define preeclampsia (Redman and Jeffries 1988; Perry and Beevers, 1994). Local birthweight centiles were used (Information and Statistics Division, 1990). Preterm gestational age-matched deliveries with birthweight >25th were selected as controls to allow for the expected change in placental structure with gestational age (Kaufmann and

Burton, 1994); most were delivered vaginally as a result of unexplained spontaneous preterm labor, but again we were careful to ensure the cord was clamped immediately following delivery in order to prevent vessel collapse. Individual cases are summarized in Tables in our publications and we would encourage wider use of this approach in order that the reader can judge the clinical severity of cases studied.

Tissue Sampling

The placenta was inspected from its chorionic surface to identify peripheral artery-vein pairs suitable for perfusion-fixation. Cotyledons were flushed with saline followed by 4% glutaraldehyde (for transmission [TEM] or scanning [SEM] electron microscopy), or by saline followed by Mercox plastic mixture to obtain casts of villous vascular trees (Lametschwandter et al., 1990; Leiser and Koob, 1992). Full-thickness vertical sections of placenta were randomly excised from the remaining placenta and immersion-fixed in formaldehyde prior to paraffin embedding (Macara et al., 1995).

Imaging Techniques

Concentric rings of α-smooth muscle actin-positive cells were identified immunohistochemically from 4 μM paraffin sections. This technique permits the identification of the smallest stem villous arterioles with a post-fixation vessel diameter 10-20 μM (Macara et al., 1995), which permitted us to construct histograms in 15 μM increments, representing the elaboration pattern of the vasculature down to the very smallest vessels, this is not possible using conventional staining techniques. The mean vessel diameter includes a contribution from every vessel counted, and thus this parameter will be very sensitive to any "obliteration", or vessel loss, originally proposed by Giles and colleagues in 1985.

Small cubes of perfusion-fixed villous tissues were randomly-prepared for both TEM and SEM studies, and the perfusion-fixed vessels casts were prepared by corrosion; the detailed methods and definitions of structures are outlined in Macara et al. (1996) and Krebs et al. (1996). Parallel immunohistochemical studies, for collagens I, III, IV, laminin and fibronectin, were performed on 4 μM paraffin sections from the immersion-fixed blocks.

RESULTS

Stem Villous Vessels

The immunohistochemical localization studies of stem villous arterioles using an antibody to α-smooth muscle actin failed to demonstrate evidence of any loss of small arterioles, even at the lowest possible diameter range (10-25 μM) (Figure 1). The mean vessel diameter in IUGR group and the control group were therefore very similar (31.8 μM [2.4] vs. 29.6 μM [2.3]; $p = 0.13$). However we did observe a reduction in the sectional area of placenta occupied by peripheral (non-muscularized) villi in the IUGR cases (Macara et al., 1995) consistent with the stereological data reported from Toronto (Jackson et al., 1995). Some α-SMA positive cells, identified by TEM as myofibroblasts (Macara et al., 1996), were identified in the stroma of mature intermediate and terminal villi in some of the IUGR cases but in none of the controls.

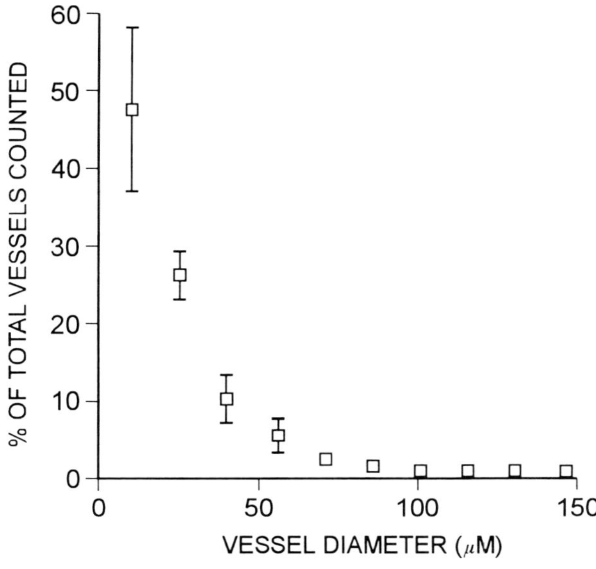

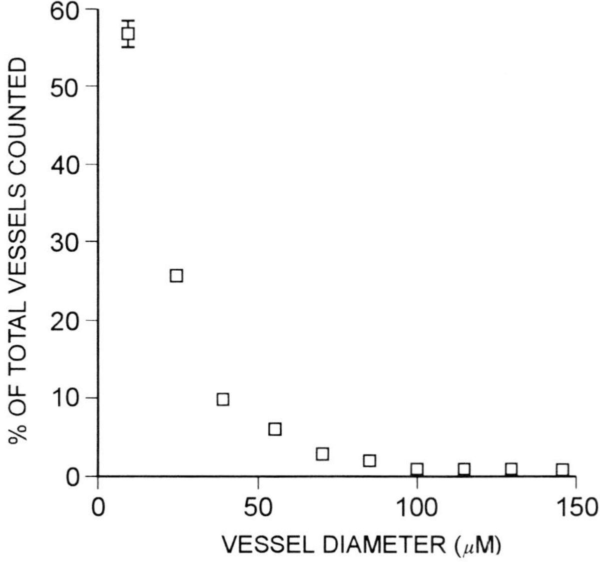

Figure 1. Mean percentage of vessel profiles within ten 15 μM diameter categories (error bar = standard deviation) from 10-160 μM for the IUGR (upper graph) and gestation age-matched control group (lower graph). The SD was tightly distributed around the mean in the majority of diameter categories hence error bars not visible. There was no significant difference in the distribution pattern of either group. Data from Macara et al., 1995, with permission.

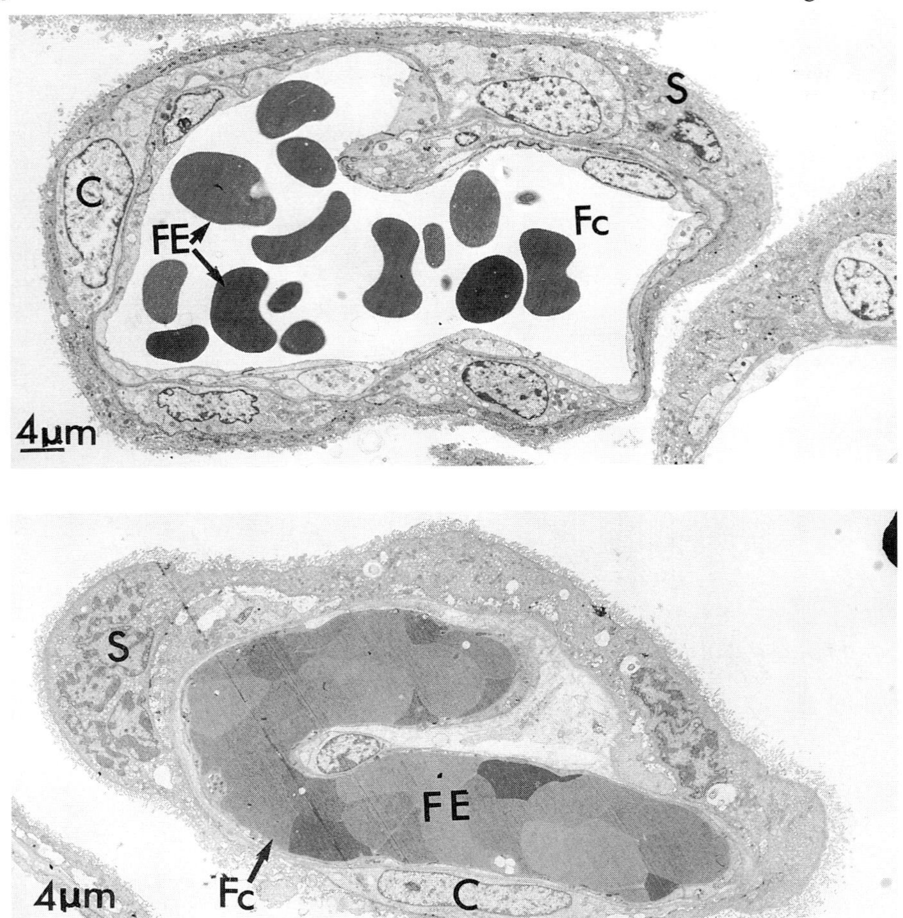

Figure 2. Transmission electron micrographs of a terminal villous from (2a) control and (2b) IUGR placentae. There was an increased number of syncytial nuclei in the IUGR villi which were aggregated into syncytial knots. Within the fetal capillaries of the IUGR terminal villi there was evidence of antenatal congestion with molding of the fetal erythrocytes and lysis of the erythrocyte cell membranes. (S) syncytiotrophoblast, (C) cytotrophoblast cells, (FC) fetal capillary, (FE) fetal erythrocyte. From Macara et al., 1996, with permission.

Two-Dimensional Structure Of Terminal Villi

Terminal villi from the IUGR group were smaller in diameter. They contained an increased ratio of syncytiotrophoblast nuclei (SN) to cytotrophoblast nuclei (CN), the former becoming aggregated into syncytial knots, demonstrating evidence of nuclear senescence (Figure 2). Figure 3 shows higher magnification views through the wall of representative terminal villi, demonstrating the thickened lamina densa and basal lamina-like material in the stroma in IUGR. These TEM sections demonstrate occlusion of the fetal capillaries by densely-packed erythrocytes with heterogeneous electron density and abnormal shapes. It is worth emphasizing that these are perfusion-fixed tissues, with few erythrocytes remaining in the normal samples. Dense compaction of fetal erythrocytes with local dissolution of their plasma membranes in the IUGR cases is indicative of antenatal capillary congestion. Examination of parallel paraffin sections stained with hematoxylin and eosin confirmed the generalized findings of capillary congestion and increased syncytial knotsin the preterm IUGR cases. The stroma of the terminal villi in the IUGR group contained a large number of fibroblast-like cells with ultrastructural features typical of myofibroblasts; winding bands of basal lamina-like material were noted to surround these cells (Figure 4). The myofibroblast-like nature of these cells was further verified according to Kohnen et al., (1995) on light microscopy sections stained with α-smooth muscle actin (see Macara et al., 1995). The stroma of terminal villi in the control cases demonstrated a minimal amount of background fibrillar material, occasional collagen fibers and no myofibroblast-like cells.

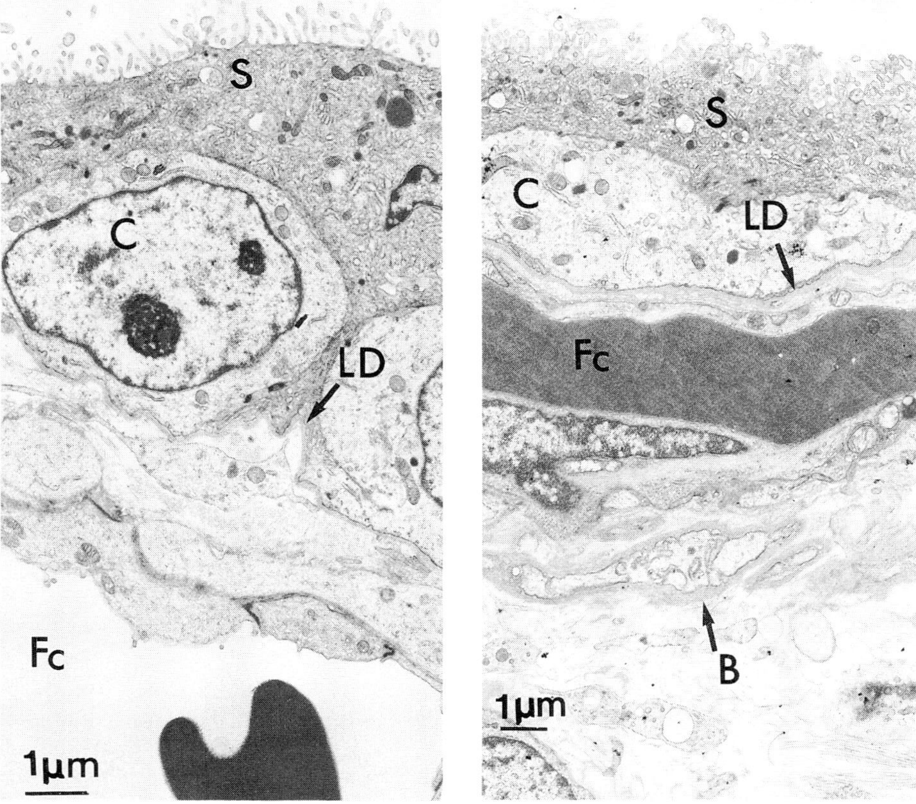

Figure 3. Transmission electron micrographs of a terminal villous from (3a, left panel) control and (3b, right panel) IUGR placentae. The basal lamina densa was twice as thick in the IUGR placentae. Basal lamina-like material (B) was also very evident within the stroma of the IUGR terminal villi. (S) syncytiotrophoblast, (C) cytotrophoblast, (FC) fetal capillary, (LD) basal lamina. From Macara et al., 1996, with permission.

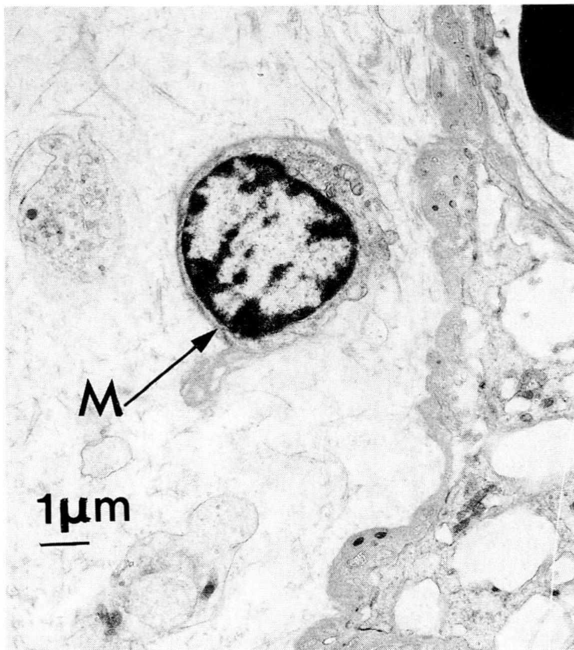

Figure 4. TEM of the stroma from an IUGR terminal villous demonstrating a myofibroblast-like cell in cross-section (M). From Macara et al., 1996, with permission.

Three-Dimensional Structure Of Terminal Villi

The typical SEM appearances of terminal capillary casts from preterm control placentae and IUGR cases are shown in Figure 5. The capillary loops were significantly longer in the IUGR group as shown in Figure 6 group (mean length 218 µM [SD72] vs. 137 [SD30] ; p <0.001). Terminal capillary loops were very sparse in number among the IUGR cases, since each convolute tended to have a single malformed loop. Figure 7 shows highly magnified SEM pictures demonstrating these appearances. The elongated loops in the IUGR cases had significantly fewer branches (4.0 per loop [SD 1.9] vs 6.1 [SD 2.2] ; p = 0.03) and the majority of IUGR capillary loops showed no evidence of coiling (79% vs 18% in the control group; p <0.05).

Corresponding views of the peripheral villi in the IUGR group confirmed the observations from the cast data, namely that terminal villi were much fewer in number, but considerably elongated, compared with the preterm controls. As can be seen from Figures 8 and 9, the syncytiotrophoblast surface of the intermediate and terminal villi in IUGR is very abnormal; the surface is wrinkled, with areas of fibrin plaque deposition identified in 75% of villous tissues examined in the IUGR group and only one specimen from the preterm controls.

The wrinkled appearance of the trophoblast gives the impression of an aged surface. Since the syncytiotrophoblast/cytotrophoblast nucleus ratio was found to be increased in IUGR, we speculated that these findings could be explained by a reduction in division of the cytotrophoblast. We assessed cytotrophoblast turnover using the antibody MIB-1 and found a reduction in the numbers of MIB-1 positive nuclei on villous cross-sections viewed by light microscopy (Macara et al., 1996).

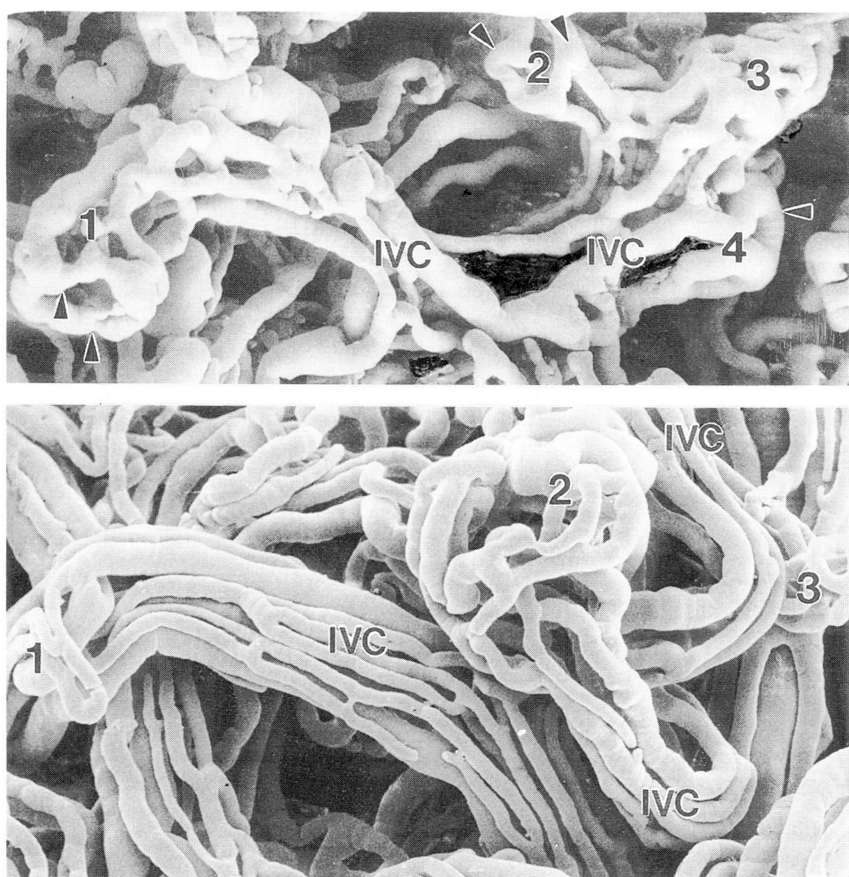

Figure 5. SEM of a vascular cast from a control (5a, top panel) and from an IUGR case (5b, bottom panel) . 5a: Overview of an intermediate villous capillaries (IVC) branching into 4 terminal capillary convolutions (1-4) with numerous loops (arrowheads). X1140. 5b: Bundles of IVC are extremely elongated with sparse convolutions and inconspicuous loops X1140.

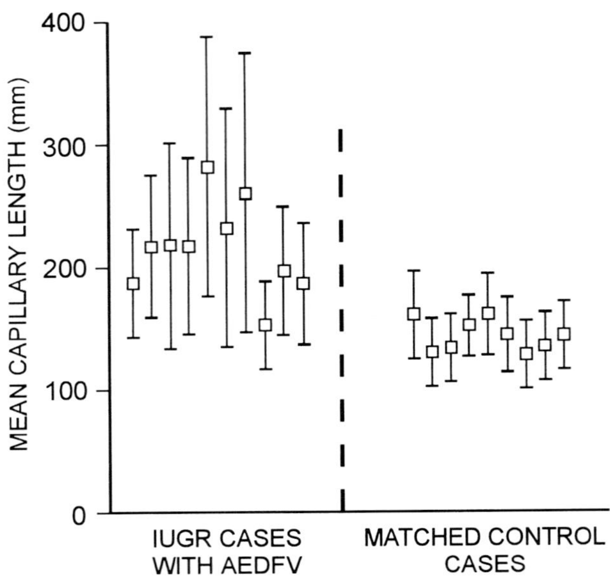

Figure 6. Mean capillary length of terminal capillary loops in each of the 10 IUGR and 9 gestation age-matched control placental vascular casts (indicate one standard deviation).

DISCUSSION

In these studies we have discovered a number of abnormalities of the peripheral non-muscularized villi of preterm IUGR placentae delivered preterm because of fetal compromise following absent end-diastolic flow velocity in the umbilical artery (AEDFV). The severity of IUGR was such that fetal death would have ensued unless cesarean section was carried out - several weeks prior to term. From a clinical perspective, these pregnancies are entirely different from those ending at term following spontaneous vaginal birth of a small healthy neonate. With this in mind, it is not surprising that detailed studies of the placentae from such pregnancies appear to cast new light on the underlying pathology of severe IUGR in comparison to studies conducted 15-20 years ago on principally term material (Sandstedt, 1979).

The application of immunohistochemistry, localizing concentric rings of α-smooth muscle actin-positive cells, allowed us to study the smallest arterioles surrounded by a monolayer of contractile cells. We were unable to detect any ***proportional*** loss of these smallest vessels in IUGR, reflected also by no difference in mean vessel diameter in comparison with controls. We did not find any difference in the proportion of sectional area of placenta occupied by stem villi in the IUGR cases, though this was the case for sectional area occupied by peripheral villi (Macara et al., 1995) and Jackson et al., (1995) came to similar conclusions from their stereological analysis. We can possibly reconcile these observations with those of earlier workers (e.g., Giles et al., 1985) since they may not have been able to visualize very small arterioles using conventional stains. We are currently evaluating both the density and lumen/media thickness of stem villous arterioles from perfusion-fixed systematic random-sampled placental blocks in order to resolve these differences of opinion. However, the reader needs to bear in mind the fact that AEDFV, from embolization studies carried out in Toronto (Morrow et al., 1989) implies up to a four-fold increase in vascular impedance! This could not entirely be accounted for by minor anatomical differences in the structure and number of small stem arterioles.

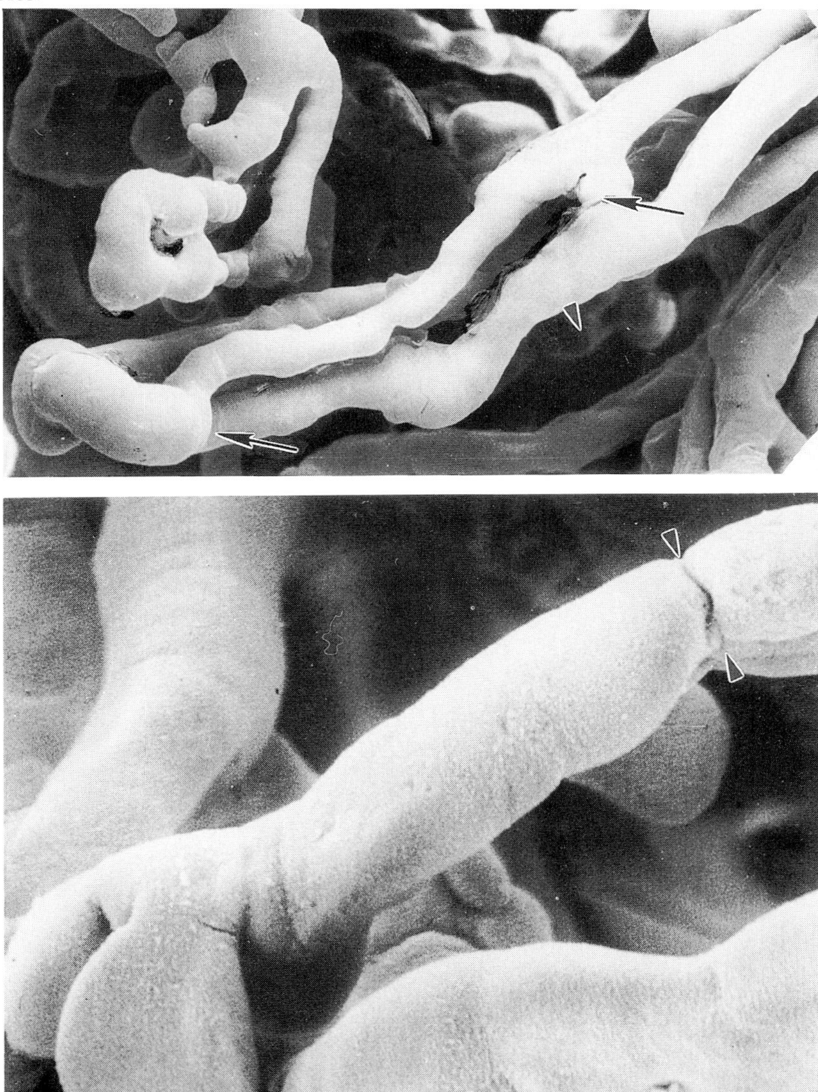

Figure 7. Highly-magnified SEM showing extremely elongated villi in IUGR. 7a (upper panel): Typical terminal capillary loop which is typically uncoiled and with few branches (arrows) X2700. 7b (lower panel): The shape of the terminal villous surface is likened to a drainpipe, with two abnormal-shaped buds located on the tip of the villous (lower left). Note the distinct constrictions (arrowheads). X2700.

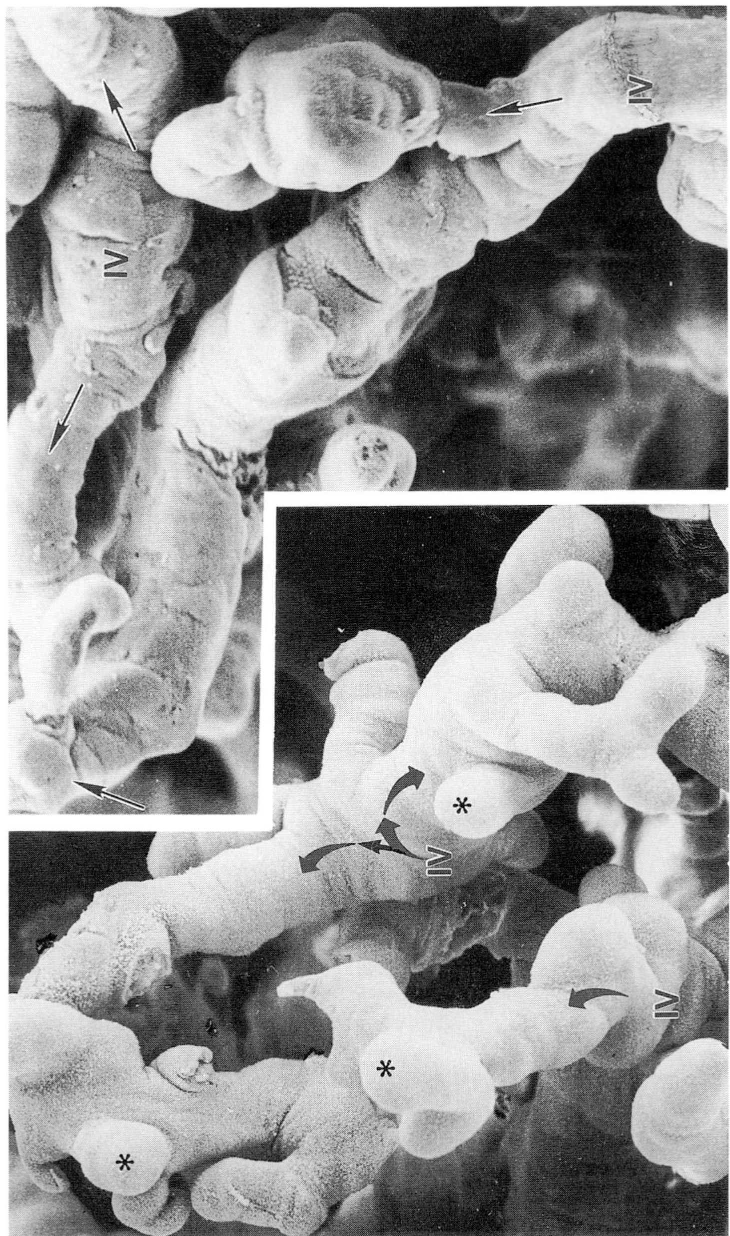

Figure 8. SEM of the terminal villous surface. 8a: A preterm control case demonstrating the smooth distal parts of mature intermediate villi (IV) which ramify (arrows) into multiple bud-like projections (asterisks) containing terminal capillary loops (visible in Figure 5a). X1500. 8b: Corresponding image from an IUGR case; IV with few elongated and characteristically twisted terminal villi (arrows). The surface is uneven, with bumps and constrictions. X1500.

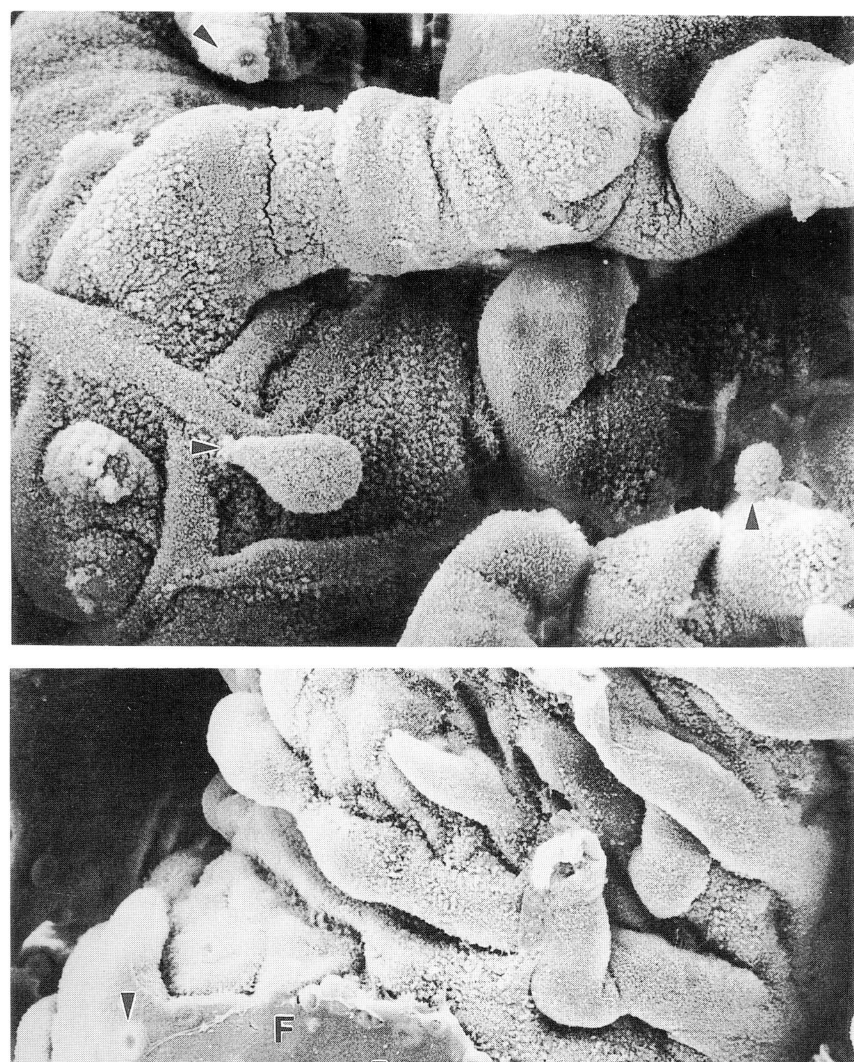

Figure 9. SEM of the villous surface from an IUGR case 9a (top panel): Demonstration of an intermediate villous (top left) giving off an elongated terminal villous whose surface is wrinkled. Several pedunculated projections are visible (arrowheads). The syncytiotrophoblast microvilli are shorter on the surface of the wrinkles than in the troughs between them. X2650. 9b (bottom panel): A fibrin plaque (F), containing maternal erythrocytes (arrowheads), is conspicuous on the beginning of an intermediate villous. The syncytiotrophoblast surface is typically irregular with wrinkles arranged circumferentially around the villous. These structures seem to be most recently grown, because of small and dense arrangement of microvilli. X2650.

The population of stem villi and their muscularized vessels increases rapidly during the first trimester due to proliferation of their distally-located precursors, the immature intermediate villi (IIV) (Castelluci et al., 1990). Ultimately 6-26 (mean 10) generations of muscularized stem villi will be formed such that by the early third trimester IIV have largely disappeared, differentiating into mature intermediate villi, from which exponential numbers of terminal villi form - thereby facilitating oxygen and nutrient transfer to the fetus (Kaufmann and Burton, 1994). The placental pathology of preterm IUGR must take account of the observation that fetus with AEDFV in the umbilical artery is usually chronically-hypoxic (Nicolaides et al., 1988; Nicolini et al., 1990). Furthermore, the reduced veno-arterial umbilical oxygen gradient in the presence of AEDFV (to 7 mmHg; Steiner et al., 1995) implies a reduced extraction of oxygen from the intervillous space by the villous structures.

Is The Placenta Hypoxic In IUGR?

Our detailed studies of terminal villi from IUGR pregnancies revealed an aged villous trophoblast, represented by reduced cytotrophoblast proliferation, increased numbers of syncytial knots, syncytial nuclear sensescence and a wrinkled surface. Beneath a thickened basal lamina was a fibrotic stroma, containing myofibroblast-like cells and capillaries congested with abnormally-shaped fetal erythrocytes. Local oxygen tension has a significant effect upon villous structure (Benirschke and Kaufmann, 1995). Under physiological circumstances the high-flow low-impedance fetoplacental circulation at term results in a high fractional extraction of oxygen from the intervillous space. In preterm IUGR, despite a reduction in uteroplacental blood flow (Nylund et al., 1983), samples of uteroplacental venous blood have a higher oxygen content than normal (Pardi et al., 1992) dismissing the myth of placental hypoxia in preterm IUGR. The differences in structure of terminal villi in preterm IUGR provides indirect evidence that local oxygen tension was elevated for some time prior to delivery (full discussion in Macara et al., 1996). During analysis of these studies we became aware of the seminal publication from the National Institutes of Health by Panigel and Myers (1972) who studied the structure of peripheral villi in the rhesus monkey following either fetectomy or ligation of the umbilical cord. Both maneuvers prevented fetal extraction of oxygen from the intervillous space, such that uteroplacental venous oxygen tension was elevated; after 7-14 days detailed analysis of the terminal villi indicated very similar findings to those we have observed in the presence of AEDFV, namely stromal fibrosis, trophoblast aging and reduced cytotrophoblast proliferation. In functional terms, AEDFV indicates a profound reduction in umbilical cord volume flow coupled to a similar reduction in fetal oxygen consumption - circumstances not far removed, in pathological terms, from the experimental conditions created by Panigel and Myers (1972).

Our scanning electron microscopic studies of the peripheral villi and their capillary structures indicated reduced density in IUGR, consistent with sectional data (Jackson et al., 1995 ; Macara et al., 1995). Furthermore these were abnormally elongated and their capillary structure, namely loss of branching and sinusoidal dailations, would be expected to confer a large increase in vascular impedance to the fetoplacental circulation. The reduced volume flow through these malformed capillaries in association with AEDFV would be predicted to reduce the extraction of oxygen even further; hence uteroplacental venous oxygen tension (and thus mean intervillous oxygen tension) will rise.

The appropriate response of a terminal villous network to intervillous space hypoxia, for example with maternal anemia or pregnancy at high altitude, is an increase in capillarization of the terminal villi and a reduction in diffusion distance (Bacon et al., 1984; Scheffen et al., 1990; Reshetnikova et al., 1995; Burton et al., 1996) i.e., the exact ***opposite*** of our findings. The observation of a reduced umbilical arterio-venous oxygen gradient in preterm IUGR with AEDFV (Steiner et al., 1995) confirms that the IUGR fetus is centrally hypoxic and unable to extract oxygen from the intervillous space.

These conclusions have important clinical implications, in particular the wisdom of maternal hyperoxygenation in IUGR (Battaglia et al., 1992; Johanson et al., 1995). In essence, the fetus becomes progressively hypoxic with AEDFV in tandem with peripheral villous fibrosis due to rising mean intervillous oxygen tension. Hyperoxygenation would, on the basis of our observations, be predicted to accelerate the process of villous fibrosis though to our knowledge the placentae from pregnant women given this treatment have not been examined in detail. Depending on gestation and degree of prior villous damage, this treatment may confer a short-term survival advantage, perhaps by facilitating corticosteroid treatment to improve neonatal lung function (Battaglia et al., 1992; Johanson et al., 1995).

Our studies offer a possible alternative to the platelet embolization theory advanced by Trudinger and colleagues (Wilcox and Trudinger, 1991). Hypoxia increases the fetal hematocrit and the passage of more densely-packed erythrocytes is compounded by an abnormal arrangement of the villous capillaries. It is perhaps not surprising that we observed capillary congestion at both the light and electron microscope level (Macara et al., 1996). This process, together with deposition of fibrin plaques on the trophoblast surface, provides an explanation for the deterioration in umbilical artery Doppler blood flow and fetal oxygenation observed in some cases of IUGR associated with pre-eclampsia (Arduini et al., 1993).

Our data may provide a pathological explanation for the phenomenon of AEDFV, principally based at the level of terminal villi. As a consequence the frequently perpetuated assumption of placental hypoxia in IUGR needs serious reconsideration and has important implications for those involved in *in vitro* studies of villous explants and cultured trophoblast cell lines. In addition our findings should help clinicians to appreciate the placental basis of fetal hypoxia, from which a more rational approach to fetal heath assessment should follow.

SUMMARY

The abnormal umbilical artery Doppler waveform represented by absent end-diastolic flow velocity (AEDFV) identifies a group of preterm fetuses with intrauterine growth restriction (IUGR) at high risk of perinatal death due chronic hypoxia from compromised umbilical artery blood flow. Initial studies of AEDFV placentae suggested a loss of small stem arterioles, though by immunohistochemical localization with an antibody to α-smooth muscle actin, we were unable to demonstrate any selective loss of even the smallest vessels in the 10-25 μM post-fixation diameter range. Further studies of villous ultrastructure and immunohistochemical localization of matrix molecules (laminin and collagens I, III and IV) demonstrated evidence of villous fibrosis and trophoblast aging. Parallel scanning electron microscopy studies demonstrated reduced numbers of malformed peripheral villi, whose capillaries were elongated and only poorly branched. Several of the structural differences in the terminal villi of this IUGR group, such as reduced cytotrophoblast proliferation and stromal fibrosis, are incompatible with the prevailing view of placental hypoxia in preterm IUGR with AEDFV. Reduced numbers of malformed terminal villi would be predicted to both increase capillary vascular impedance, and compromise placental gas exchange. Despite the known reduction in uteroplacental arterial blood flow in IUGR with AEDFV in the umbilical arteries, there is a greater fall in extraction of oxygen from the intervillous space: as such that maternal blood leaving the placenta has a higher content than under normal circumstances. Our data challenge the presumption of "placental hypoxia" in this subset of preterm IUGR pregnancies complicated by AEDFV in the umbilical arteries, which has both scientific and clinical implications.

ACKNOWLEDGEMENTS

Funded by; Scottish Hospitals Endowment Research Trust (grant 1163) to J.K. , Deutsche Forschungsgemeinschaft (grants Ka 360/7-1 and 7-2) to P.K., an Ethicon Foundation Fund travel award to L.M. and an E.C. Biomed-1 Concerted Action grant to J.K. and PK.

REFERENCES

Alfirevic, Z. and Neilson, J.P. (1995) Doppler ultrasonography in high-risk pregnancies: A systematic review with meta-analysis. *Am. J. Obstet. Gynecol.* 172, 1379-1387.

Arduini, D., Rizzo, G. and Romanini, C. (1993) The development of abnormal heart rate patterns after absent end-diastolic velocity in umbilical artery-analysis of risk factors. *Am. J. Obstet. Gynecol.* 168, 43-50.

Battaglia, C., Artini, P.G., D'Ambrogio, G., Galli, P.A., Segre, A. and Genazzani, A.R. (1992) Maternal hyperoxygenation in the treatment of intrauterine growth retardation. *Am. J. Obstet. Gynecol.* 167, 430-435.

Bacon, B.J., Gilbert, R.D., Kaufmann, P., Smith, A.D., Trevino, F.T. and Longo, L.D. (1984) Placental anatomy and diffusing capacity in guinea pigs following long-term maternal hypoxia. *Placenta* 5, 475-488.

Benirschke, K. and Kaufmann. P. (1995) *Pathology of the Human Placenta*, 3rd Ed. Springer:New York.

Bower, S., Bewley, S. and Campbell, S. (1993) Improved prediction of preeclampsia by two stage screening of uterine arteries using the early diastolic notch and color Doppler imaging. *Obstet. Gynecol.* 82, 78-83.

Bracero, L.A., Beneck, D., Kirshenbaum, N., Peiffer, M., Stalter, P. and Schulman, H. (1989) Doppler velocimetry and placental disease. *Am. J. Obstet. Gynecol.* 161, 388-393.

Burton, G.J., Reshetnikova, O.S., Milovanov, A.P. and Teleshova, O.V. (1996) Stereological evaluation of vascular adaptions in human placental villi to differing forms of hypoxic stress. *Placenta* 17, 49-55.

Castellucci, M., Scheper, M., Scheffen, I., Celona, A. and Kaufmann P (1990) The development of the human placental vascular tree. *Anat. Embryol.* 181, 117-128.

de Wolf, F., Robertson, W.B. and Brosens, I. (1986) Inadequate maternal vascular response to placentation in pregnancies complicated by pre-eclampsia and by small for gestational age infants. *Br. J. Obstet. Gynaecol.* 93, 1049-1059.

Fisk, N.M., MacLachlan, N., Ellis, C., Tannirandorn, Y., Tonge, H.M. and Rodeck, C.H. (1988) Absent end-diastolic flow in first trimester umbilical artery. *Lancet* ii, 1256-7.

Giles, W.B., Trudinger, B.J. and Baird, P. (1985) Fetal umbilical artery flow velocity waveforms and placental resistance: Pathological correlation. *Br. J. Obstet. Gynaecol.* 92, 31-38.

Hendricks, S.K., Sorensen, T.K., Wang, K.Y., Bushnell, J.M., Seguin, E.M. and Zingheim, R.W. (1989) Doppler umbilical artery waveform indices - normal values from fourteen to forty-two weeks. *Am. J. Obstet. Gynecol.* 161, 761-765.

Hitschold, T., Weis, E., Beck, T., Hunterfering, H. and Berle, P. (1993) Low target birth weight or growth retardation? Umbilical Doppler flow velocity waveforms and histometric analysis of the fetoplacental vascular tree. *Am. J. Obstet. Gynecol* 168, 1260-1264.

Information and Statistics Division. *Centile Values Of Birthweight For Gestational Age In Scottish Infants* 1982 - 1986. Scottish Health Service, Edinburgh, 1990.

Jackson, M.R., Walsh, A.J., Morrow, R.J., Mullen, B.J., Lye, S.J. and Ritchie, J.W.K. (1995) Reduced placental villous tree elaboration in small-for-gestational age newborn pregnancies: Relationship with umbilical artery Doppler waveforms. *Am. J. Obstet. Gynecol.* 172, 518-525.

Jauniaux, E. and Burton, G.J. (1993) Correlation of umbilical Doppler features and placental morphometry: the need for uniform methodology. *Ultrasound Obstet. Gynaecol.* 3, 233-235.

Johanson, R., Lindow, S.W., van der Elst, C., Jaquire, Z., van der Westhuizen, S. and Tucker, A. (1995) A prospective randomized comparison of the effect of continuous O_2 therapy and bedrest on fetuses with absent end-diastolic flow on umbilical artery Doppler waveform analysis. *Br. J. Obstet. Gynaecol.* 102, 662-665.

Karsdorp, V.H.M., van Vugt, J.M.G., van Geijn, H.P., Kostense, P.J., Arduini, D,, Montenegro, N. and Todros, T. (1994) Clinical significance of absent or reversed end diastolic velocity waveforms in umbilical artery. *Lancet* 344, 1664-1668.

Kaufmann, P. (1982) Development and differentiation of the human placental villous tree. *Bibl. Anat.* 22, 29-39.

Kaufmann, P., Bruns, U., Leiser, R., Luckhardt, M. and Winterhager. E. (1985) The fetal vascularisastion of term human placental villi. II. Intermediate and terminal villi. *Anat. Embryol.* 173, 203-214.

Kaufmann, P., Luckhardt, M. and Leiser, R. (1988) Three-dimensional representation of the fetal vessel system in the human placenta. *Tropho. Res.* 3, 113-137.

Kaufmann, P. and Burton, G. (1994) Anatomy and Genesis of the Placenta. In: *The Physiology of Reproduction,* (eds.) E. Knobil and J.D. Neill, Raven Press: New York, pp. 441-483.

Kohnen, G., Castellucci, M., Hsi, B.-L., Yeh, C.-J., and Kaufman, P. (1995) The monoclonal antibody GB42 - A useful marker for the differentiation of myofibroblasts. *Cell Tissue Res.* 281, 231-242.

Krebs, C., Macara, L.M., Leiser, R., Bowman, A.W., Greer, I.A. and Kingdom, J.C.P. (1996) Intrauterine growth restriction with absent end-diastolic flow velocity in the umbilical artery is associated with maldevelopment of the terminal placental villous tree. *Am. J. Obstet. Gynecol.* 175, 1534-1542.

Lametschwandtner, A., Lametschwandtner, U. and Weiger, T. (1990) Scanning electron microscopy of vascular corrosion cast - technique and applications: Updated review. *Scann. Micros.* 4, 889-941.

Leiser, R., Kosanke, G. and Kaufmann, P. (1991) Human placental vascularization: Structural and quantitative aspects. In: *Placenta - Basic Science For Clinical Application,* (ed.) H. Soma, Tokyo: Karger Publications, pp. 32-45.

Leiser, R. and Koob, B. (1992) Structural and functional aspects of placental microvasculature studied from corrosion casts. In: *Scanning Electron Microscopy of Vascular Casts: Methods and Applications,* (eds.) P.M.Motta, T. Murakami and H. Fujita, Kluwer Academic Publishers: Boston, chapter 20, pp. 261 - 277.

Macara, L., Kingdom, J.C.P. and Kaufmann, P. (1993) Control of the fetoplacental circulation. *Fet. Mat. Med. Rev.* 5, 167-179.

Macara, L.M., Kingdom, J.C.P., Kohnen, G., Bowman, A.W., Greer, I.A. and Kaufmann, P. (1995) Elaboration of stem villous vessels in growth-restricted pregnancies with abnormal umbilical artery Doppler waveforms. *Br. J. Obstet. Gynaecol.* 101, 807-812.

Macara, L.M., Kingdom, J.C.P., Kaufmann, P., Kohnen, G., Hair, J., More, I.A.R., Lyall, F. and Greer, I.A. (1996) Structural analysis of placental terminal villi growth-restricted pregnancies with abnormal umbilical artery Doppler waveforms. *Placenta* 17, 37-48.

McCowan, L.M., Mullen, B.M. and Ritchie, J.W.K. (1987) Umbilical artery flow velocity waveforms and the placental vascular bed. *Am. J. Obstet. Gynecol.* 157, 900-902.

Morrow, R.J., Adamson, S.L., Bull, S.B. and Ritchie, J.W.K. (1989) Effect of embolization on the umbilical arterial velocity waveform in fetal sheep. *Am. J. Obstet. Gynecol.* 161, 1055-1060.

Nicolaides, K.H., Bilardo, C.M., Soothill, P.W., and Campbell, S. (1988) Absence of end-diastolic frequencies in the umbilical artery: a sign of fetal hypoxia and acidosis. *Br. Med. J.* 297, 1026-1027.

Nicolini, U., Nicolaidis, P., Fisk, N.M., Vaughan, J.I., Fusi, L., Gleeson, R. and Rodeck, C.H. (1991) Limited role of fetal blood sampling in prediction of outcome in intrauterine growth retardation. *Lancet* 336, 768-772.

Nylund, L,, Lunell, N.O., Lewander, R., and Sarby, B. (1983) Uteroplacental blood flow index in intrauterine growth retardation of fetal or maternal origin. *Br. J. Obstet. Gynaecol.* 90,16-20.

Panigel, M. and Myers, R.E. (1972) Histological and ultrastructural changes in rhesus monkey placenta following interruption of fetal placental circulation by fetecomy or interplacental umbilical vessel ligation. *Acta. Anat.* 81,481-506.

Pardi, G., Cetin, I., Marconi, A.M., Bozzetti, P., Buscaglia, M., Makowski, E.L., and Battaglia, F.C. (1992) Venous drainage of the human uterus: Respiratory gas studies in normal and fetal growth retarded pregnancies. *Am. J. Obstet. Gynecol.* 166, 699-706.

Perry, I.J. and Beevers, D.G. (1994) The definition of pre-eclampsia. *Br. J. Obstet. Gynaecol.* 101, 587-591.

Redman, C.W.G. and Jefferies, M. (1988) Revised definition of pre-eclampsia. *Lancet* 1, 809-812.

Reshetnikova, O.S., Burton, G.J. and Milovanov, A.P. (1995) Effects of hypobaric hypoxia on the fetoplacental unit: The morphometric diffusing capacity of the villous membrane at high altitude. *Am. J. Obstet. Gynecol.* 171, 1560-1565.

Scheffen, I., Kaufmann, P., Philippens, L., Leiser, R., Geisen, C., and Mottaghy, K. (1990) Alterations of the fetal capillary bed in the guinea pig placenta following long-term hypoxia. In: *Oxygen Transport to Tissue XII*, (Ed) J. Piiper, T.K. Goldstick, and D. Meyer, Plenum Press: New York, pp. 77 -790.

Sandstedt, B. (1979) The placenta and low birthweight. *Curr. Top. Pathol.* 66, 1-55.

Steiner, H., Staudach, A., Spitzer, D., Schaffer, K.H., Gregg, A., and Weiner, C.P. (1995) Growth deficient fetuses with absent or reversed umbilical artery end-diastolic flow are metabolically compromised. *Early Human Dev.* 41, 1-9.

Trudinger, B.J., Stevens, D., Connelly A., Hales, J.R., Alexander, G., Bradley, L., Fawcett, A. and Thompson, R.S. (1987) Umbilical artery flow velocity waveforms and placental resistance: the effects of embolization of the umbilical circulation. *Am. J. Obstet. Gynecol.* 157, 1443-1448.

Wilcox, G.R. and Trudinger, B.J. (1991) Fetal platelet consumption: a feature of placental insufficiency. *Obstet. Gynecol.* 77, 616-621.

ENDOCRINE AND PARACRINE REGULATION

Trophoblast Research 10:313-328, 1997

PROBLEMS IN DEMONSTRATING PEPTIDE SECRETION FROM THE PLACENTA
- A Review-

Donald W. Morrish

Department of Medicine
362 Heritage Medical Research Center
University of Alberta
Edmonton, Alberta, Canada T6G 2S2

INTRODUCTION

The placenta is recognized as the source for a large number of substances, of which peptides constitute an important component. A diagram summarizing the sources of many of these illustrates their variety and different cellular location, both of which imply potential endocrine, paracrine, and autocrine regulatory loops (Figure 1). The methods used to establish these data comprise a spectrum from molecular characterization to whole animal studies. The principal question to be answered in these studies is: "Does the placental cell type in question produce a biologically active peptide at physiological concentrations?". In this review, these methods and their advantages and disadvantages will be discussed in reference to this question, paying particular attention to pitfalls encountered in studying the placenta. As examples of the value of comprehensive utilization of these approaches, studies of hCG, EGF, and TNFα are reviewed, representing respectively a classical placental hormone, the first described growth factor, and a more recently characterized placental-produced cytokine of likely great importance. Problems unique to the placenta will also be discussed. Placenta-specific variants also exist with some frequency and suggest the functional importance of fully characterizing a placenta-produced substance beyond preliminary localizing procedures.

Gene Expression

Northern blots, *in situ* hybridization and polymerase chain reaction (PCR) have been used to demonstrate peptide gene expression. Compared to immunostaining or radioimmunoassay (RIA) of tissue extracts, these methods have the advantage of not detecting a peptide that might be concentrated in blood or intercellular fluid, or bound to a membrane-inserted or solubilized receptor. Basic control studies for commonly used molecular techniques are shown in Table 1. Northern analysis has a somewhat low sensitivity that can be augmented about five to ten-fold by the use of poly A(+) mRNA. High stringency washes are needed to avoid nonspecific hybridization, a common problem under low stringency conditions. Some probes cross-hybridize with 18S ribosomal mRNA and their RNA transcript may be of a similar size, necessitating use of poly (A)+ mRNA, e.g. retinoblastoma (Rb) cDNA probes. Unless a purified cell population is used, northern analysis does not localize a product. RNAse protection

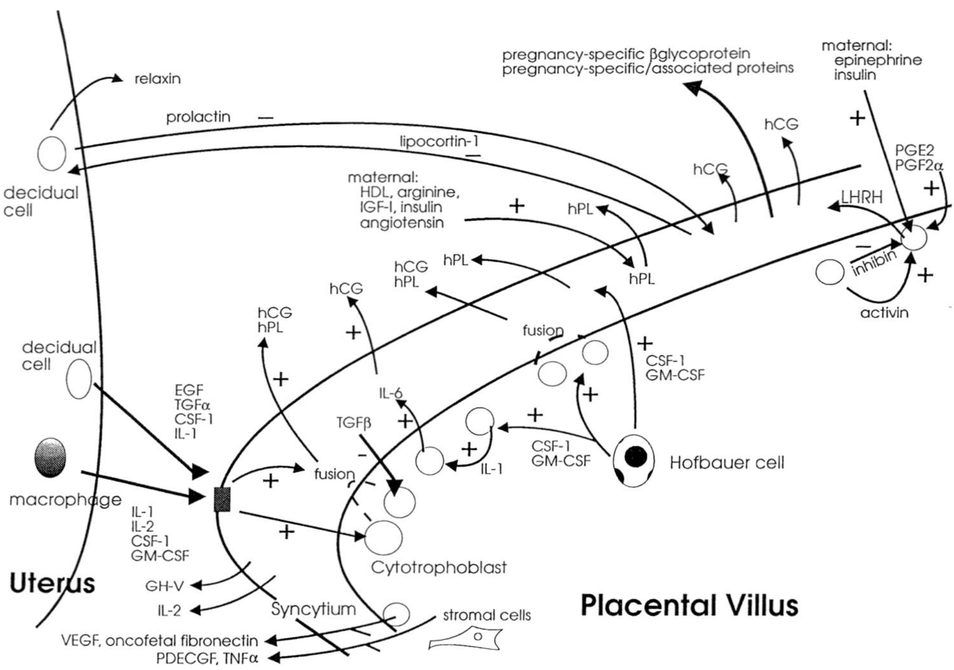

Figure 1. The uterus and a floating placental villous are shown. A large variety of substances can be shown to be localized in the uterus and placenta. Some of the implied and demonstrated functional relationships evolving from these studies are shown, including ones discussed in the text. Some of these relationships have been discussed elsewhere (Morrish et al., 1996).

assays are very sensitive and can demonstrate small changes in mRNA expression, but are more difficult technically to perform than northern blots. No studies using this technique in regard to a placental product have been reported to our knowledge.

In situ hybridization (ISH) is the technique of choice in localizing RNA expression. A refinement of this technique is to combine ISH with immunostaining to determine if the protein product is translated and where the majority resides (e.g., CRF binding protein localization in the placenta; (Petraglia et al., 1993). The information communicated by the two procedures is complementary: gene expression and peptide localization. ISH can be quantitated by grain counting and this can be important in determining if a low abundance signal is different from control. For example, McWilliams and Boime (McWilliams and Boime, 1980) found low signal over some non-cytotrophoblast cells in the placenta but statistical evaluation of grain counts compared to control showed no difference. Quantitative methods can also be used effectively in ISH in single cell mRNA expression in hemolytic plaque assays (Scarbrough et al., 1991). ISH has comparable sensitivity to that of immunostaining. PCR is the most sensitive technique for detecting genes of low abundance. This strength is also its weakness in that extreme care must be taken to avoid nonspecific amplification. Second rounds, or even first rounds of amplification, may produce artefactual products and "illegitimate transcription", i.e., non-specific amplification of a large variety of products with sufficient sequence homology with the primers to be amplified even though no such "real" gene sequences exist (Chelly et al., 1989). A modification of this technique is *in situ* PCR (Tsongalis et al., 1994; Hofler, 1993; Sarkar et al., 1991) which has yet to be applied to the placenta but would be useful in cellular localization of very low abundance genes. In this technique, multiple primer sets are needed to ensure specificity (Hofler, 1993). Use of primers spanning the intron-exon boundary also helps reduce the chances of non-specific amplification. Multiple primer sets and nested PCR are particularly helpful in eliminating non-specific amplification.

It is of value as well to actually clone and sequence the gene from the tissue of study because tissue-specific variants, presumably of biological significance, are known to exist. Such variants described to date in the placenta are as follows:

(1) The insulin receptor has two isoforms in placenta in equal amounts, compared to unequal quantities in other tissues (Moller et al., 1989; Ullrich et al., 1985; Ebina et al., 1985). As well there are atypical forms, appearing to be hybrid insulin/IGF-I receptors, which possess the binding and functional properties of the IGF-I receptor (Kasuya et al., 1993; Soos and Siddle, 1989)
(2) A growth hormone variant (GH-V) was predicted on the basis of the GH gene sequence (Seeburg et al., 1977) and first described in the syncytium by Hennen and colleagues (Frankenne et al., 1987; Hennen et al., 1985; Hennen et al., 1985). The mRNA for GH-V increases during gestation (MacLeod et al., 1992)
(3) A molecular variant of fibronectin found to be expressed only by tumor cells and by extravillous cytotrophoblast, named oncofetal fibronectin, exists. This variant may have a functional role as a trophoblast-uterine "glue" (Feinberg et al., 1991; Queenan et al., 1987)
(4) Placental isoforms for platelet-derived endothelial cell growth factor (PDECGF) and vascular endothelial growth factor (VEGF) exist (Maglione et al., 1991; Usuki et al., 1990).
(5) Biochemical evidence indicates there are separate growth hormone (GH) and placental lactogen (hPL) receptors (Freemark and Comer, 1989; Freemark and Handwerger, 1986). Cloning and sequencing both will be needed to ultimately demonstrate this finding.
(6) Kit-ligand has five isoforms with differing patterns of expression in amnion, chorion, and trophoblast (Sharkey et al., 1992)
(7) TNFα has a variant response element detectable in the placenta (Sarma et al., 1992).

Knowledge of these structural variants is important as the implied functional relationships and hypotheses generated from localization studies (e.g., Figure 1) may be quite different when specific identification of each variant is achieved. A model of this is GH-V, which increases during gestation, suppresses pituitary GH, and thus these data permit a new paradigm for GH during pregnancy, wherein the placenta assumes control of GH effects on the maternal environment during later gestation (Hennen et al., 1985a). Similarly, receptor isoforms and hybrids such as the insulin/IGF-I receptor are thought to be a mechanism for increasing receptor diversity. Localization and gestational changes in placenta of this particular receptor have not yet been performed but may have obvious significance for actions of insulin and IGF-I during pregnancy.

Table 1

Optimal Control Studies In Peptide Localization Methods

(a) Molecular Methods	
Northern blots	- high stringency washes - positive and negative control tissues - specific probe
In situ hybridization	- positive and negative control tissues - sense controls - quantitation of grains if low abundance message - pretreatment of slides with RNase A or H - substitution of tRNA for probes
Polymerase chain reaction	- positive and negative control tissues - two primer sets, over intron-exon border - nested PCR - omit reverse transcriptase - only expected number of bands on gel
(b) Immunostaining	- positive and negative tissue controls - absorb antibody with cross-reactivity substances - two monoclonal antibodies - preabsorb sections with IgG or serum (adequate time and concentration; pre-immune serum advantageous if available) - quench endogenous peroxidase - omit first antibody - IgG isotype controls - specific strong staining with low background - monoclonal antibodies or F(ab)' fragment of antibodies to avoid nonspecific binding to placental Fc receptors - absorb antibody with homologous antigen - if negative result on embedded sections, repeat using cryosections

Peptide Expression

The second essential step in demonstrating peptide production is to show that the protein is synthesized because some RNA's are not translated (Oestreicher and Scazzocchio, 1993; Lawrence et al., 1991). The most versatile method is immunostaining using avidin-biotin enhancement of sensitivity. Control studies for immunostaining are listed in Table 1. Possible artifacts and suggested control studies have been reviewed (van Leeuwen, 1986; Gosselin et al., 1986). False negative results may occur due to lack of antibody penetration, steric hindrance, low titre or affinity of the antibody, or modification of loss of antigenic sites during the staining procedure. Absence of a signal using wax or other embedded tissue methods must be confirmed using cryosections to avoid a false negative result. The placenta makes a large variety of substances; thus, it is essential to ensure that antisera used are absorbed for potentially cross-reacting substances, e.g., hPL antibody absorbed with growth hormone, or hCG absorbed with other gonadotropins and their subunits (Morrish et al., 1988; Morrish et al., 1987). An example of a probable artifact of localization is the study of Healy et al (Healy et al., 1977). At the time these experiments on prolactin localization in the placenta were done, it was not known that the placenta possessed a growth hormone receptor (Hill et al., 1992). The antibodies in this study were not absorbed with GH and so binding to GH receptors versus prolactin receptors cannot be avoided, resulting in likely false cellular localization of hPL. Similarly, a careful study of GH receptor localization in the placenta by Hill et al (Hill et al., 1992) used a monoclonal antibody that did not cross react with insulin or prolactin receptors. However, even this study may be partly inaccurate as it is not known if the antibody used cross-reacts with the more recently-described hPL receptors.

Most currently used antibodies are monoclonal, which provides excellent specificity. Nonetheless, if a new substance is being identified, it is most desirable to use two different antibodies against different epitopes, as monoclonal antibodies are directed against very small portions of the molecule ,i.e. about seven amino acids, and chance cross reaction with another similar or identical sequence is possible (Gosselin et al., 1986). For example, in localizing TNFα to the placenta for the first time, Chen et al. (Chen et al., 1991) used two different antibodies to avoid this potential problem. Unfortunately, this is rarely done due to cost and time involved, but results often in the necessity of a second study for confirmation. One of the most important negative control experiments, that of absorbing the antibody with the antigen, is not an absolute guarantee of specificity for this reason. Polyclonal antibodies may also show a lack of reactivity after homologous antigen absorption, but still show artifactual positivity due to Fc receptor determinants (Bass et al., 1994) as discussed below. Absorption with antigen does not guarantee against binding produced due to contaminating substances in the antigen used to raise the antibody (van Leeuwen, 1986).

A particular problem with polyclonal antiserum, and rarely with monoclonal antibodies, is that of placental Fc receptors (Bright and Ockleford, 1994; Koyama et al., 1991). Since the placenta expresses Fc receptors, intact polyclonal antisera will often show a positive reaction as they frequently include Fc antibodies. This problem can be overcome by using F(ab)' fragments or pre-absorbing the tissue section with adequate serum or IgG. For example, Fc receptor binding may explain the differences between positive localization of EGF to the placenta (Hofmann et al., 1991) and its lack of localization using F(ab)' antibodies (Bass et al., 1994). IgG isotype controls are necessary for all antisera types to ensure specificity (e.g., Yui et al., 1994). Many studies do not utilize positive or negative control tissues, but this can be of great value in validating the results, e.g. in EGF localization (Bass et al., 1994).

False-positive results may also arise due to tissue pseudoperoxidase, endoperoxidase, free radical groups, hydrophobic and ionic interactions, lipids, natural antibodies and contaminating antibodies (van Leeuwen, 1986). The standard procedure of pretreatment with hydrogen peroxidase prevents only some of these problems. Both thick sections and electron micrographs are susceptible to these artifacts. We have investigated some of these technical issues in EM sections elsewhere (Morrish and Marusyk, 1996).

RIA or ELISA of tissue culture medium or cell extracts is frequently performed. In these studies, it is necessary to ask the question whether the release of a substance is due to cell death, nonspecific release, or true secretion. Controls for cell viability and metabolic health using non-isotopic cell proliferation assays, e.g. MTT, WST, XTT assays, which are based on the reduction of tetrazolium salts by NADH/NADPH respectively, or acid phosphatase assays, or DNA content, should be done. An example of the MTT assay is in the studies of Yui et al (1994). Most studies omit these confirmatory assays which should be done at least the first time a new peptide is being studied even if the culture system is well defined. Medium lactate dehydrogenase (LDH) has also been used to ensure cell health: its absence in the medium shows the lack of leaky, damaged cell membranes. Vital dye exclusion , e.g., trypan blue or nigrosine, is probably inadequate as cells may be committed to apoptosis and so still appear "viable".

Protein synthesis studies can be done by demonstrating ^{3}H-amino acid incorporation into immunoprecipitable protein or cell free ribosomal protein synthesis. All these add important confirmation that a bioactive protein is actively secreted *in vitro*. These are often not done presently. It is also unfortunately now rare that bioassays of a secreted peptide are performed. An exception, due largely to the prior unavailability of immunoassays, has been the use of a bioassay for measuring TGFβ release from cultured trophoblast, wherein it was noted that no bioactive peptide was secreted (Matsuzaki et al., 1992). This study indicates the obvious pitfall of immunoassays.

Prior to the availability of gene cloning, amino acid sequencing was the final obligatory characterization of protein identity. Genetic methods are generally simpler, can be done on less tissue, and have the advantage of being able to definitively demonstrate the existence of isoforms of a peptide that might never be noticed otherwise due to small differences, as previously discussed.

Functional Studies

The major problem of *in vitro* studies is the lack of a definitive physiologic meaning of the phenomena shown in the study. Due to an abnormal *in vitro* environment lacking physiological substrates and regulatory factors, substances may be released in vitro that are not expressed *in vivo*. For example, syncytium contains only GH-V mRNA and protein in tissue section studies, but secretes only pituitary GH in vitro (Evain-Brion et al., 1990). This surely is an in vitro artifact. Another artifact is trypsin effects during cell dispersion, which may induce "secretion" due to enzyme effects (e.g., IL-1 release

Table 2. Application of Methodologies to Localization of hCG, EGF and TNFα

Method	hCG	EGF	TNFα
Northern blot (a) Molecular studies	Plouzek et al., 1993	Bissonnette et al., 1992	Hunt et al., 1993
	Fiddes and Goodman, 1979; 1980		
in situ hybridization	Hoshina et al., 1982; 1983	ND	Chen et al., 1991; Yelavarthi et al., 1991*
PCR	ND	Bass et al., 1994; Amemiya et al., 1994; Haining et al., 1996	Heinig et al., 1993
cloning gene from placenta	Fiddes and Goodman, 1979; 1980	ND	Sarma et al., 1992
(b) Peptide studies			
immunostaining	Tabarelli et al., 1983; Sasagawa et al., 1987	Hofmann et al., 1991; Amemiya et al., 1994; Bass et al., 1994	Chen et al., 1991
^{3}H-amino acid incorporation	Gitlin and Biasucci, 1969	ND	ND
in vitro translation	Landefeld et al., 1976; Chatterjee et al., 1976	ND	ND
immunoassays	Vaitukaitis et al., 1971	Kajikawa et al., 1991; Bass et al., 1994	Vince et al., 1995; Laham et al., 1994
bioassays	Aschheim and Zondek, 1927; Catt et al., 1972; Saxena and Rathnam, 1983	ND	De et al., 1992*
(c) Functional studies			
animal studies		Cohen, 1962* no placental data	Carbo et al., 1995*
transgenic mouse	Strauss et al., 1994*	ND	Giroir et al., 1992*
knockout mouse	ND	ND	Kolls et al., 1994*
in vivo human studies	oophorectomy causes failed pregnancy	ND	ND
disease models	none known	Fondacci et al., 1994	Kolls et al., 1994*; Satoh et al., 1989*

ND -not done yet in relation to placental physiology
Studies listed are in human unless otherwise indicated; * mouse studies

from trophoblast cells is induced by trypsin dispersion of the cells from the placenta, but probably is secreted very little if at all by intact tissue; (Kauma et al., 1992). Although blood concentrations can be measured of a peptide, these have relevance only to endocrine mechanisms. It is well recognized that all paracrine and autocrine mechanisms based on *in vitro* data can remain only theories until proven by some other means. We do not know the intercellular fluid concentrations of any cytokine due to the lack of sufficiently sensitive methods. Although single cell secretion and simultaneous peptide secretion can be measured by hemolytic plaque assays combined with *in situ* hybridization (Scarbrough et al., 1991; Smith et al., 1989), the data obtained are subject to the same limitations of in vitro culture.

Whole animal physiological studies are thus complementary and necessary to determine *in vivo* effects. Classical methods are administration of the peptide or ablation of the gland producing it. In the case of paracrine mechanisms, these studies become nearly impossible and results often are unexpected (e.g., EGF; Moore et al., 1986; Dembinski and Johnson, 1985). Researchers have thus resorted to mouse transgenic and knockout mouse models. These have the limitations of being valid only for the mouse, of being subject to strain variations (e.g., EGF receptor knockouts; Threadgill et al., 1995). These studies yield results only in a narrow range between degrees of lethality. Thus, subtle defects may be missed. Soares et al (1995) have summarized the important lessons learned about the placenta by knockout studies.

Ultimately, it is desirable to perform human studies. However, although *some in vivo* studies have been done during pregnancy (e.g., hypoglycemia stimulates hPL secretion *in vivo*; (Gaspard et al., 1974; Tyson et al., 1971), it is ethically impossible to do most such studies. Diseases representing excesses or ablations of a peptide have often proven important in determining the function of a substance in endocrinology. One puzzling natural mutation is the lack of effect of deletion of the hPL gene on human pregnancy (Simon et al., 1986).

Application Of Methodologies To hCG, EGF, and TNFα

Table 2 lists the techniques discussed and whether they have been applied to localizing these three substances in the placenta. hCG has been extensively tested by nearly all methods including assays of bioactivity. In regard to human "ablation" experiments, it is well recognized that loss of the ovaries in the first trimester leads to pregnancy failure (Osathanondh and Tulchinsky, 1980). Interestingly, no diseases of hCG absence or mutation have been described. Knockout models are not possible because rodents do not synthesize CG, although CG transgenesis of human genes into mouse has been done (Strauss et al., 1994). EGF has been well studied with immunoassay and molecular procedures. However, important studies of biological activity are lacking and none have been done to demonstrate placental or uterine production of a bioactive EGF. Radioreceptor assays of trophoblast cells do not show activity (Bass et al., 1994). TNFα has been well studied in the mouse, but again parallel human data are not available and have been generally limited to molecular and immunological localization studies. We thus lack final confirmation of a biologically active role of EGF and TNFα in human pregnancy.

SUMMARY

This review has sought to present the full range of studies that ideally should be done before a substance is considered to be a physiologically meaningful product. Such studies include molecular and immunological localization including cloning of the gene from the target tissue, bioassays, and whole animal experiments to demonstrate physiological activity *in vivo*. It is clear that few such placental products other that a classical hormone such as hCG have received such extensive testing in humans, though more complete data have been obtained in animals such as the mouse for technical and ethical reasons. The range of possible investigations should be kept in mind when interpreting new data on placental production of a peptide and its alleged significance. Furthermore, the placenta has several unique characteristics making technical interpretation of these studies difficult. One of these properties, placenta-specific isoforms, has yielded interesting new functional relationships between the placental and maternal environments.

REFERENCES

Amemiya, K., Kurachi, H., Morishige, K.-I., Adachi, K., Imai, T. and Miyake, A. (1994) Involvement of epidermal growth factor (EGF)/EGF receptor autocrine and paracrine mechanism in human trophoblast cells: Functional differentiation in vitro. *J. Endocrinol.* 143, 291-301.

Aschheim, S. and Zondek, B. (1927) Hypophysenvorderlappenhormon und ovarialhormon im Harn von Schwangeren (Anterior pituitary hormone and ovarian hormone in the urine of pregnant women. *Klinische Wochenschrift* 6, 1322

Bass, K.E., Morrish, D.W., Roth, I., Bhardwaj, D., Taylor, R., Zhou, Y. and Fisher, S.J. (1994) Human cytotrophoblast invasion is up-regulated by epidermal growth factor: evidence that paracrine factors modify this process. *Develop. Biol.* 164, 550-551.

Bissonnette, F., Cook, C., Geoghegan, T., Steffen, M., Henry, J., Yussman, M.A. and Schultz, G. (1992) Transforming growth factor-α and epidermal growth factor messenger ribonucleic acid and protein levels in human placentas from early, mid, and late gestation. *Am. J. Obstet. Gynecol.* 166, 192-199.

Bright, N.A. and Ockleford, C.D. (1994) Heterogeneity of FcΓ receptor-bearing cells in human term amniochorion. *Placenta* 15, 247-255.

Carbo, N., Lopez-Soriano, F.J. and Argiles, J.M. (1995) Administration of tumor necrosis factor-alpha results in a decreased placental transfer of amino acids in the rat. *Endocrinology* 136, 3579-3584.

Catt, K.J., Dufau, M.L. and Tsuruhara, T. (1972) Radioligand-receptor assay of luteinizing hormone and chorionic gonadotropin. *J. Clin. Endocrinol.* 34, 123-132.

Chatterjee, M., Munro, H.N. and Baliga, B.S. (1976) Synthesis of hPL and hCG by polyribosomes and messenger RNAs from early and full term placentas. *J. Biol. Chem.* 251, 2945-2951.

Chelly, J., Concordet, J.-P., Kaplan, J.-C. and Kahn, A. (1989) Illegitimate transcription: Transcription of any gene in any cell type. *Proc. Natl. Acad. Sci. (USA)* 86, 2617-2621.

Chen, H.-L., Yang, Y., Hu, X.-L., Yelavarthi, K.K., Fishback, J.L. and Hunt, J.S. (1991) Tumor necrosis factor alpha mRNA and protein are present in human placental and uterine cells at early and late stages of gestation. *Am. J. Pathol.* 139, 327-335.

Cohen, S. (1962) Isolation of a mouse submaxillary gland protein accelerating incisor eruption and eyelid opening in the new-born animal. *J. Biol .Chem.* 237, 1555-1562.

De, M., Sanford, T.H. and Wood, G.W. (1992) Detection of interleukin-1, interleukin-6, and tumor necrosis factor-alpha in the uterus during the second half of pregnancy in the mouse. *Endocrinology* 131, 14-20.

Dembinski, A.B. and Johnson, L.R. (1985) Effect of epidermal growth factor on the development of rat gastric mucosa. *Endocrinology* 116, 90-94.

Ebina, Y., Ellis, L., Jarnagin, K., Edery, M., Graf, L., Clauser, E., Ou, J., Masiarz, F., Kan, Y.W., Goldfine, I.D., Roth, R.A. and Rutter, W.J. (1985) The human insulin receptor cDNA: the structural basis for hormone-activated transmembrane signalling. *Cell* 40, 747-758.

Evain-Brion, D., Alsat, E., Mirlesse, V., Dodeur, M., Scippo, M., Hennen, G. and Frankenne, F. (1990) Regulation of growth hormone secretion in human trophoblastic cells in culture. *Horm. Res.* 33, 256-259.

Feinberg, R.F., Kliman, H.J. and Lockwood, C.J. (1991) Is oncofetal fibronectin a trophoblast glue for human implantation? *Am. J. Pathol.* 138, 537-543.

Fiddes, J.C. and Goodman, H.M. (1979) Isolation, cloning, and sequence analysis of the cDNA for the α subunit of hCG. *Nature* 281, 351-356.

Fiddes, J.C. and Goodman, H.M. (1980) The cDNA for the β-subunit of human chorionic gonadotropin suggests evolution of a gene by read-through into the 3' untranslated region. *Nature* 286, 684-687.

Fondacci, C., Alsat, E., Gabriel, R., Blot, P., Nessmann, C. and Evain-Brion, D. (1994) Alterations of human placental epidermal growth factor receptor in intrauterine growth retardation. *J. Clin. Invest.* 93, 1149-1155.

Frankenne, F., Rentier-Delrue, F., Scippo, M., Martial, J. and Hennen, G. (1987) Expression of the growth hormone variant gene in human placenta. *J. Clin. Endocrinol. Metab.* 64, 635-637.

Freemark, M. and Comer, M. (1989) Purification of a distinct placental lactogen receptor, a new member of the GH/prolactin receptor family. *J. Clin. Invest.* 83, 883-889.

Freemark, M. and Handwerger, S. (1986) The glycolytic effects of placental lactogen and growth hormone in ovine fetal liver are mediated through binding to specific fetal oPL receptors. *Endocrinology* 118, 613-618.

Gaspard, U., Sandcroft, H. and Luyckx, A. (1974) Glucose-insulin interaction and the modulation of human placental lactogen (HPL) secretion during pregnancy. *J. Obstet. Gynaecol. Brit. Cmwlth.* 81, 201-209.

Giroir, B.P., Peppel, K., Silva, M. and Beutler, B. (1992) The biosynthesis of tumor necrosis factor during pregnancy: Studies with a CAT reporter transgene and TNF inhibitors. *Eur. Cytok. Netw.* 3, 533-538.

Gitlin, D. and Biasucci, A. (1969) Ontogenesis of immunoreactive growth hormone, follicle-stimulating hormone, thyroid-stimulating hormone, luteinizing hormone, chorionic prolactin, and chorionic gonadotropin in the human conceptus. *J. Clin. Endocrinol. Metab.* 29, 926-935.

Gosselin, E.J., Cate, C.C., Pettengill, O.S. and Sorenson, G.D. (1986) Immunocytochemistry: its evolution and criteria for its application in the study of epon-embedded cells and tissue. *Amer. J. Anat.* 175, 135-160.

Haining, R.E.B., Schofield, J.P., Jones, D.S.C., Rajput-Williams, J. and Smith, S.K. (1996) Identification of mRNA for epidermal growth factor and transforming growth factor-α present in low copy number in human endometrium and decidua using reverse transcriptase-polymerase chain reaction. *J. Molec. Endocrinol.* 6, 207-214.

Healy, D.L., Muller, H.K. and Burger, H.G. (1977) Immunofluorescence shows localization of prolactin to human amnion. *Nature* 265, 642-643.

Heinig, J., Wilhelm, S., Bittorf, T., Brock, J. and Briese, V. (1993) Semiquantitative determination of IL-1 alpha, TNF-alpha, PDGF-A, PDGF-B, and PDGF-receptor in term human placenta using polymerase chain reaction (PCR). *Zentralblatt fur Gynakologie* 115, 317-322.

Hennen, G., Frankenne, F., Closset, J., Gomez, F., Pirens, G. and El Khayat, N. (1985a) A human placental GH: Increasing levels during second half of pregnancy with pituitary GH suppression as revealed by monoclonal antibody radioimmunoassays. *Intl. J. Fertil.* 30, 27-33.

Hennen, G., Frankenne, F., Pirens, G., Gomez, F., Closset, J., Schaus, C. and El Khayat, N. (1985b) New chorionic GH-like antigen revealed by monoclonal antibody radioimmunoassays. *Lancet* i, 399

Hill, D.J., Riley, S.C., Bassett, N.S. and Waters, M.J. (1992) Localization of the growth hormone receptor, identified by immunocytochemistry, in second trimester human fetal tissues and in placenta throughout gestation. *J. Clin. Endocrinol. Metab.* 75, 646-650.

Hofler, H. (1993) In situ polymerase chain reaction: Toy or tool? *Histochemistry* 99, 103-104.

Hofmann, G.E., Scott, R.T., Bergh, P.A. and Deligdisch, L. (1991) Immunohistochemical localization of epidermal growth factor in human endometrium, decidua, and placenta. *J. Clin. Endocrinol. Metab.* 73, 882-887.

Hoshina, M., Boothby, M. and Boime, I. (1982) Cytological localization of chorionic gonadotropin α and placental lactogen mRNA during development of the human placenta. *J. Cell Biol.* 93, 190-198.

Hoshina, M., Hussa, R., Patillo, R. and Boime, I. (1983) Cytological distribution of chorionic gonadotropin subunit and placental lactogen messenger RNA in neoplasms derived from the placenta. *J. Cell Biol.* 97, 1200-1206.

Hunt, J.S., Chen, H.L., Hu, X.L. and Pollard, J.W. (1993) Normal distribution of tumor necrosis factor-alpha messenger ribonucleic acid and protein in the uteri, placentas, and embryos of osteopetrotic (op/op) mice lacking colony-stimulating factor-1. *Biol. Reprod.* 49, 441-452.

Kajikawa, K., Yasui, W., Sumiyoshi, H., Yoshida, K., Nakayama, H., Ayhan, A., Yokozaki, H., Ito, H. and Tahara, E. (1991) Expression of epidermal growth factor in human tissues. *Virchows Arch. A.Patholog. Anat. Histopathol.* 418, 27-32.

Kasuya, J., Paz, B., Maddux, B.A., Goldfine, I.D., Hefta, S.A. and Fujita-Yamaguchi, Y. (1993) Characterization of human placental insulin-like growth factor-I/insulin hybrid receptors by protein microsequencing and purification. *Biochemistry* 32, 13531-13536.

Kauma, S.W., Walsh, S.W., Nestler, J.E. and Turner, T.T. (1992) Interleukin-1 is induced in the human placenta by endotoxin and isolation procedures for trophoblasts. *J. Clin. Endocrinol. Metab.* 75, 951-955.

Kolls, J., Peppel, K., Silva, M. and Beutler, B. (1994) Prolonged and effective blockade of tumor necrosis factor activity through adenovirus-mediated gene transfer. *Proc. Natl. Acad. Sci. (USA)* 91, 215-219.

Koyama, M., Saji, F., Kameda, T., Kimura, T., Nishikiori, N., Kikuchi, T. and Tanizawa, O. (1991) Differential mRNA expression of three distinct classes of FcΓ receptor at the fetomaternal interface. *J. Reprod. Immunol.* 20, 103-113.

Laham, N., Brennecke, S.P., Bendtzen, K. and Rice, G.E. (1994) Tumour necrosis factor alpha during human pregnancy and labour: Maternal plasma and amniotic fluid concentrations and release from intrauterine tissues. *Eur. J. Endocrinol.* 131, 607-614.

Landefeld, T.D., McWilliams, D.R. and Boime, I. (1976) The isolation of mRNA encoding the alpha subunit of human chorionic gonadotropin. *Biochem. Biophys. Res. Comm.* 72, 381-390.

Lawrence, J.B., Cochrane, A.W., Johnson, C.V., Perkins, A. and Rosen, C.A. (1991) The HIV-1 Rev protein: A model system for coupled RNA transport and translation. *New Biologist* 3, 1220-1232.

MacLeod, J.N., Lee, A.K., Liebhaber, S.A. and Cooke, S.A. (1992) Developmental control and alternative splicing of the placentally expressed transcripts from the human growth hormone gene cluster. *J. Biol. Chem.* 267, 14219-14226.

Maglione, D., Guerriero, V., Viglietto, G., Delli-Bovi, P. and Persico, M.G. (1991) Isolation of a human placenta cDNA coding for a protein related to the vascular permeability factor. *Proc. Natl. Acad. Sci. (USA)* 88, 9627-9271.

Matsuzaki, N., Li, Y., Masuhiro, K., Jo, T., Shimoya, K., Taniguchi, T., Saji, F. and Tanizawa, O. (1992) Trophoblast-derived transforming growth factor-β 1 suppresses cytokine-induced, but not gonadotropin-releasing hormone-induced, release of human chorionic gonadotropin by normal human trophoblasts. *J. Clin. Endocrinol. Metab.* 74, 211-216.

McWilliams, D. and Boime, I. (1980) Cytological localization of placental lactogen messenger ribonucleic acid in syncytiotrophoblast layers of human placenta. *Endocrinology* 107, 761-765.

Moller, D.E., Yokota, A., Caro, J.F. and Flier, J.S. (1989) Tissue-specific expression of two alternatively spliced insulin receptor mRNAs in man. *Mol. Endocrinol.* 3, 1263-1269.

Moore, G.P.M., Wilkinson, M., Panaretto, B.A., Delbridge, L.W. and Posen, S. (1986) Epidermal growth factor causes hypocalcemia in sheep. *Endocrinology* 118, 1525-1529.

Morrish, D.W., Marusyk, H. and Siy, O. (1987) Demonstration of specific secretory granules for human chorionic gonadotropin in placenta. *.J Histochem. Cytochem.* 35, 93-101.

Morrish, D.W., Marusyk, H. and Bhardwaj, D. (1988) Ultrastructural localization of human placental lactogen in distinctive granules in human term placenta: comparison with granules containing human chorionic gonadtotropin. *J. Histochem. Cytochem.* 36, 193-197.

Morrish, D.W., Dakour, J. and Li, H. (1996) The trophoblast as an active regulator of the pregnancy environment in health and disease: An emerging concept. In: *Pregnancy and Parturition,* (eds.) T. Zakar and C.T. Greenwich, JAI Press, *Advances in Organ Biology* 1, 121-152.

Morrish, D.W. and Marusyk, H. (1997) Localization of human chorionic gonadotropin (hCG) and placental lactogen (hPL) by immunogold labeling for electron microscopy: Technique and limitations. *Microscopic Research and Technique*, in press.

Oestreicher, N. and Scazzocchio, C. (1993) Sequence, regulation, and mutational analysis of the gene encoding urate oxidase in aspergillus nidulans. *J. Biol. Chem.* 268, 23382-23389.

Osathanondh, R. and Tulchinsky, D. (1980) Placental polypeptide hormones. In: *Maternal-Fetal Endocrinology,* (eds.) D. Tulchinsky and K.J. Ryan, Philadelphia: W.B.Saunders, p. 17.

Petraglia, F., Potter, E., Cameron, V.A., Sutton, S., Behan, D.P., Woods, R.J., Sawchenko, P.E., Lowry, P.J. and Vale, W. (1993) Corticotropin-releasing factor-binding protein is produced by human placenta and intrauterine tissues. *J. Clin. Endocrinol. Metab.* 77, 919-924.

Plouzek, C.A., Kimberly, K.L., Stephens, J.K. and Chou, J.Y. (1993) Differential gene expression in the amnion, chorion, and trophoblast of the human placenta. *Placenta* 14, 277-285.

Queenan, J.T., Kao, L.-C., Arboleda, C.E., Ulloa-Aguirre, A., Golos, T.G., Cines, D.B. and Strauss, J.F.I. (1987) Regulation of urokinase-type plasminogen activator production by cultured human cytotrophoblasts. *J. Biol. Chem.* 262, 10903-10906.

Sarkar, F.H., Ball, D.E., Li, Y.-W. and Crissman, J.D. (1991) Analysis of mRNA expression using a single 10μm frozen tissue section. *Clin. Biotech.* 3, 127-131.

Sarma, V., Wolf, F.W., Marks, R.M., Shows, T.B. and Dixit, V.M. (1992) Cloning of a novel tumor necrosis factor-alpha-inducible primary response gene that is differentially expressed in development and capillary tube-like formation in vitro. *J. Immunol.* 148, 3302-3312.

Sasagawa, M., Yamazaki, T., Endo, M., Kanazawa, K. and Takeuchi, S. (1987) Immunohistochemical localization of HLA antigens and placental proteins (hCG, hCG, CTP, hPL and SP1) in villous and extravillous trophoblast in normal human pregnancy: A distinctive pathway of differentiation of extravillous trophoblast. *Placenta* 8, 515-528.

Satoh, J., Seino, H., Abo, T., Tanaka, S., Shintani, S., Ohta, S., Tamura, K., Sawai, T., Nobunaga, T., Oteki, T., Kumagai, K. and Toyota, T. (1989) Recombinant human tumor necrosis factor suppresses autoimmune diabetes in nonobese diabetic mice. *J. Clin. Invest.* 84, 1345-1348.

Saxena, B.B. and Rathnam, P. (1983) Human chorionic gonadotropin in early pregnancy. In: *The Endocrinology of Pregnancy and Parturition,* (eds.) L. Martini and V.H.T. James, New York: Academic Press, p.97.

Scarbrough, K., Weiland, N.G., Larson, G.H., Sortino, M.A., Chiu, S., Hirshfield, A.N. and Wise, P.M. (1991) Measurement of peptide secretion and gene expression in the same cell. *Molec. Endocrinol.* 5, 134-142.

Seeburg, P.H., Shine, J., Martial, J.A., Ullrich, A., Baxter, J.D. and Goodman, H.M. (1977) Nucleotide sequence of part of the gene for human chorionic somatomammotropin: purification of DNA complementary to predominant mRNA species. *Cell* 12, 157-165.

Sharkey, A., Jones, D.S., Brown, K.D. and Smith, S.K. (1992) Expression of messenger RNA for kit-ligand in human placenta: localization by in situ hybridization and identification of alternatively spliced variants. *Molec. Endocrinol.* 6, 1235-1241.

Simon, P., Decoster, C., Brocas, H., Schwers, J. and Vassart, G. (1986) Absence of human chorionic somatomammotropin during pregnancy associated with two types of gene deletion. *Human Genetics* 74, 235-238.

Smith, P.F., Luque, E.H. and Neill, J.D. (1989) Detection and measurement of secretion from individual neuroendocrine cells using a reverse hemolytic plaque assay. In: *Neuroendocrine Peptide Methodology,* (eds.) P.M. Conn, New York: Academic Press, p. 205.

Soares, M.J., Chapman, B.M., Kamei, T. and Yamamoto, T. (1995) Control of trophoblast cell differentiation: Lessons from the genetics of early pregnancy loss and trophoblast neoplasia. *Develop. Growth Differ.* 37, 335-364.

Soos, M.A. and Siddle, K. (1989) Immunological relationships between receptors for insulin and insulin-like growth factor I. *Biochem. J.* 263, 553-563.

Strauss, B.L., Pittman, R., Pixley, M.R., Nilson, J.H. and Boime, I. (1994) Expression of the beta subunit of chorionic gonadotropin in transgenic mice. *J. Biol. Chem.* 269, 4968-4973.

Tabarelli, M., Kofler, R. and Wick, G. (1983) Placental hormones: I. Immunofluorescence studies of the localization of chorionic gonadotropin, placental lactogen and prolactin in human and rat placenta and in the endometrium of pregnant rats. *Placenta* 4, 379-388.

Threadgill, D.W., Dlugosz, A.A., Hansen, L.A., Tennenbaum, T., Lichti, U., Yee, D., LaMantia, C., Mourton, T., Herrup, K., Harris, R.C., Barnard, J.A., Yuspa, S.H., Coffey, R.J. and Magnuson, T. (1995) Targeted disruption of mouse EGF receptor: effect of genetic background on mutant phenotype. *Science* 269, 230-234.

Tsongalis, G.J., McPhail, A.H., Lodge-Rigal, R.D., Chapman, J.F. and Silverman, L.M. (1994) Localized in situ amplification (LISA): A novel approach to in situ PCR. *Clin. Chem.* 40, 381-384.

Tyson, J.E., Austin, K.L. and Farinholt, J.W. (1971) Prolonged nutritional deprivation in pregnancy: Changes in human chorionic somatomammotropin and growth hormone secretion. *Amer. J. Obstet. Gynecol.* 109, 1081-1082.

Ullrich, A., Bell, J.R., Chen, E.Y., Herrera, R., Petruzzelli, M., Dull, T.J., Gray, A., Coussens, L., Liao, Y., Tsubokawa, M., Mason, A., Seeburg, P.H., Grunfeld, C., Rosen, O.M. and Ramachandran, J. (1985) Human insulin receptor and its relationship to the tyrosine kinase family of oncogenes. *Nature* 313, 756-761.

Usuki, K., Norberg, L., Larsson, E., Miyazono, K., Hellman, U., Wernstedt, C., Rubin, K. and Heldin, C.-H. (1990) Localization of platelet-derived endothelial cell growth factor in human placenta and purification of an alternatively processed form. *Cell Regulation* 1, 577-596.

Vaitukaitis, J., Robbins, J.B., Nieschlag, E. and Ross, G.T. (1971) A method for producing specific antisera with small doses of immunogen. *J. Clin. Endocrinol.* 33, 988-991.

van Leeuwen, F. (1986) Pitfalls in immunocytochemistry with special reference to the specificity problems in the localization of neuropeptides. *Amer. J. Anat.* 175, 363-377.

Vince, G.S., Starkey, P.M., Austgulen, R., Kwiatkowski, D. and Redman, C.W. (1995) Interleukin-6, tumour necrosis factor and soluble tumour necrosis factor receptors in women with pre-eclampsia. *Br. J. Obstet. Gynaecol.* 102, 20-25.

Yelavarthi, K.K., Chen, H.L., Yang, Y.P., Cowley, B.D.J., Fishback, J.L. and Hunt, J.S. (1991) Tumor necrosis factor-alpha mRNA and protein in rat uterine and placental cells. *J. Immunol.* 146, 3840-3848.

Yui, J., Garcia-Lloret, M., Brown, A.J., Berdan, R.C., Morrish, D.W., Wegmann, T.G. and Guilbert, L.J. (1994) Functional, long-term cultures of human term trophoblasts purified by column-elimination of CD9 expressing cells. *Placenta* 15, 231-246.

Trophoblast Research 10:329-343, 1997

DECIDUAL SIGNALS IN THE ESTABLISHMENT OF PREGNANCY: THE PROLACTIN FAMILY
- A Review -

Kyle E. Orwig, Christine A. Rasmussen and Michael J. Soares[1]

Department of Molecular and Integrative Physiology
University of Kansas Medical Center
Kansas City, Kansas 66160 USA

INTRODUCTION

Pregnancy requires significant changes in the functioning of maternal tissues. Of primary importance is the redirection of resources and nutrients to the uterus, the site of embryonic development. In order to install this gestational conduit, the uterus must undergo dramatic reorganization. One of the earliest adaptations to pregnancy is the differentiation of uterine stromal cells, a process referred to as *decidualization*. Decidual cells are then responsible for controlling a cascade of other changes in maternal, extraembryonic, and embryonic tissues. Some of the functions of decidual cells are likely mediated by their secretion of a family of hormones related to pituitary prolactin (PRL).

The purpose of the present review is to provide a framework for understanding the biology of the decidual PRL family. The discussion will focus primarily on the rat and where applicable on other species for comparative purposes.

DECIDUA: MORPHOLOGICAL CONSIDERATIONS

Decidual tissue is a transient structure that first forms at the time of implantation, grows progressively larger until midgestation then begins to regress (Figure 1; Krehbiel, 1937; DeFeo, 1967; Enders and Schlafke, 1967; Bell, 1983; Welsh and Enders, 1985; Parr and Parr, 1989). The formation of decidual tissue requires prior hormonal stimulation of the uterus with progesterone and estrogen. An additional necessary stimulus in rodents is provided by either the implanting blastocyst or experimentally in the case of pseudopregnant animals via a non-specific irritation of the uterine lining (DeFeo, 1967). Experimentally induced decidual tissue is referred to as a *deciduoma* or *deciduomata* and is morphologically and biochemically similar to decidual tissue found in pregnancy (DeFeo, 1967; Bell, 1983; Parr and Parr, 1989). Cellular events during decidualization include a proliferation and differentiation of uterine stromal cells that begin in the *antimesometrial* endometrium (farthest from the incoming blood supply) at the site of implantation. Decidual tissue eventually surrounds the developing blastocyst providing a barrier between the blastocyst and the remainder of the uterus. Mesometrial and antimesometrial decidua are morphologically and functionally distinct (Bell, 1983; Gu and Gibori, 1995). As gestation advances, the mesometrial decidual compartment gives rise to the *decidua basalis*, whereas the antimesometrial decidual compartment forms the *decidua capsularis*. In addition to decidual cells, decidual tissue also contains endothelial

[1]To Whom Correspondence Should Be Addressed

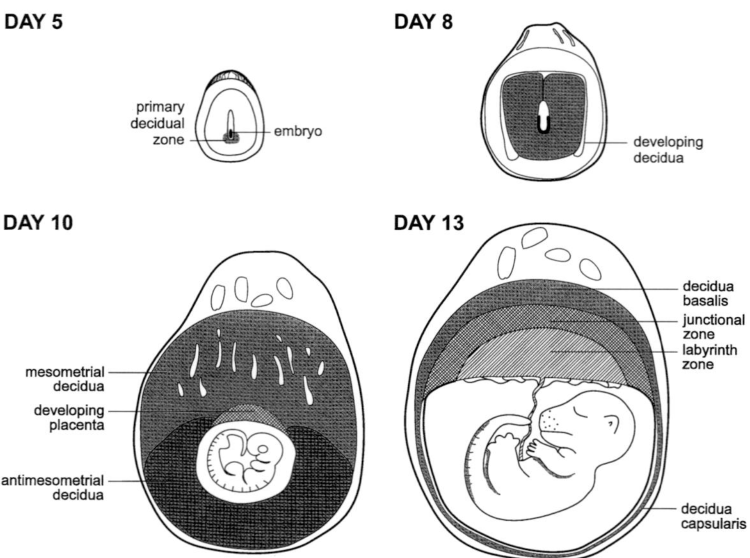

Figure 1. Schematic representations of developing decidua at different timepoints during gestation. Top left panel: A transverse view of the uterus at implantation, day 5 of pregnancy. The blastocyst implants on the antimesometrial side of the uterine lumen and decidualization begins in the stromal cells of the endometrium to form the primary decidual zone; Top right panel: A transverse view of the conceptus from day 8 of pregnancy. Decidualization continues forming the secondary decidual zone and engulfs the blastocyst forming a barrier between maternal and extraembryonic/embryonic environments. Bottom left panel: A transverse view of a conceptus at day 10 of pregnancy. Decidual structures have matured into the characteristic mesometrial and antimesometrial compartments. The development of the chorioallantoic placenta is at its beginning stages. Bottom right panel: A transverse view of a conceptus at day 13 of pregnancy. Mesometrial decidua is now referred to as the decidual basalis and the antimesometrial decidua is referred to as the decidual capsularis. Decidual regression is apparent at both locations. The chorioallantoic placenta has formed its distinctive junctional and labyrinth zones. Day 0 of pregnancy is defined by the presence of sperm in the vagina.

cells, an assortment of immune cells, and in the pregnant state, trophoblast cells (Parr and Parr, 1989). Each of these cell types may be targets for the actions of decidual cell secretory products.

DECIDUA: FUNCTIONAL CONSIDERATIONS

During gestation, decidual cells are located at the interface separating invading trophoblast cells from the maternal environment. A number of important functions have been attributed to decidua (cf., DeFeo, 1967; Bell, 1983; Parr and Parr, 1989): *i*) a protective role in controlling trophoblast cell invasion, *ii*) a nutritive role for the developing embryo, *iii*) a role in preventing immunological rejection of genetically disparate embryonic/fetal tissues, and *iv*) an endocrinological role in controlling maternal and fetal adaptations required for the establishment and maintenance of pregnancy. Pregnancy is dependent upon decidual cell acquisition of each of these specialized functions. Failure of normal decidual cell maturation may result in early pregnancy loss, uncontrolled trophoblast cell growth and invasion (gestational trophoblast disease/choriocarcinoma), immunologic rejection of the embryo, embryonic growth retardation or embryonic/fetal death. Progress has been limited in understanding any of the specialized decidual cell functions. Members of the PRL family may participate in many of the functions attributed to decidual cells.

DECIDUAL PROLACTIN FAMILY

The decidual PRL family consists of a group of proteins structurally related to pituitary PRL (Table I). Inclusion in the family does not imply functional relatedness (Soares et al., 1991).

Historical Evidence for the Existence of PRL-like Proteins in Decidua

Early identification of PRL-like hormones in decidual tissue was based principally on the trophic actions of PRL on the corpus luteum. An action referred to as *luteotrophic*. The concept of a decidual PRL-like luteotrophin grew from studies first reported by Ershoff and Deuel (1943). They demonstrated that the presence of deciduomal tissue extended the lifespan of the corpus luteum. Gibori and Rothchild and their colleagues further developed the concept of a *decidual luteotrophin* (dLTH; Gibori et al., 1974; Rothchild and Gibori, 1975). These researchers determined that luteal progesterone production could be maintained in pseudopregnant rats treated with potent inhibitors (ergot derivatives) of anterior pituitary PRL secretion, only if their uterine stroma was decidualized. Although, the rat dLTH has not yet been isolated, its initial characterization spurred related research in primates and work in the rat leading to the isolation and characterization of related gene products.

Primate Decidual PRL

A uterine PRL-like protein was first definitively isolated in the human (Golander et al., 1978; Riddick et al, 1978).

Structure

The primate decidual PRL protein is identical to primate pituitary PRL (Hwang et al., 1974; Golander et al., 1979); however, decidual PRL is encoded by a longer mRNA generated by alternative splicing resulting in the inclusion of a noncoding exon and the use of an upstream promoter (Gellersen et al., 1989; DiMattia et al., 1990; Brown and Bethea, 1994; Gellersen et al., 1994). PRL can be synthesized by decidual cells in glycosylated and nonglycosylated forms (Lee and Markoff, 1986).

Table I

The Decidual Prolactin Family

Hormone*	Mol wt (kDa)	Homology with Ant. Pit. PRL	Cysteine residues	Glyco-protein	Tissue Distribution**	PRL-like action
Primate						
PRL	23-25	100%	6	yes	ant. pit.	yes
Rat						
dLTH	28	?	?	?	?	yes
PLP-B	30-31	39	4	yes	placenta	no
dPRP	29	37	6	yes	placenta	no

* PRL, prolactin; dLTH, decidual luteotrophin; PLP-B, prolactin-like protein-B; dPRP, decidual prolactin-related protein
** Major alternative tissue sources of the decidual hormones

Expression

PRL expression in the human decidua is initiated prior to implantation and continues throughout pregnancy (Handwerger et al., 1991). Primate decidual cells do not respond to the same array of secretogogues as do lactotrophs of the anterior pituitary (Handwerger et al., 1991). Most of the modulators of decidual PRL production are produced in the uterus or extraembryonic tissues and thus form part of an intricate autocrine/paracrine regulatory network (Handwerger et al., 1989, 1991). There is considerable data indicating that progesterone, relaxin, and activation of the cAMP/protein kinase A pathway (possibly via prostaglandins) are involved in regulating decidual PRL gene expression (Zhu et al., 1990; Tang et al., 1993; Frank et al., 1994). Additional evidence suggests that a trophoblast cell secretory product, the α-subunit of chorionic gonadotropin, may be a specific positive regulator of decidual PRL secretion (Blithe et al., 1991). Whether these modulators of decidual PRL expression are secondary to their actions on the differentiation of decidual cells remains to be determined. Tissue-specific regulation of PRL gene expression can be accounted for, at least in part, through the utilization of different promoters (DiMattia et al., 1990; Berwaer et al, 1994; Gellersen et al., 1994). The decidual-specific promoter is located approximately 6 kb upstream of the pituitary-specific transcriptional start site (Gellersen et al., 1994).

Biological Actions

PRL is a multi-functional hormone with potentially a broad spectrum of biological activities (Nicoll et al., 1986). PRL of decidual origin is the principal source of amniotic fluid PRL (Andersen, 1990; Handwerger et al., 1991). Some experimentation has suggested a role for decidual PRL in water and electrolyte transport across extraembryonic membranes (Mulder, 1989; Andersen, 1990). Decidual PRL is capable of activating the PRL receptor signaling pathway (Handwerger et al., 1989; Andersen, 1990). In rodents, PRL receptors have a broad distribution throughout the uterus, placenta, and developing fetus (Freemark et al., 1993; Royster et al., 1995), suggesting an array of possible targets for decidual PRL. Although we have some insights regarding the biology of primate decidual PRL, limited progress has been made in understanding the physiological role of decidual PRL in the context of its expression within the primate uterus during pregnancy.

Rat Decidual PRL Family

In contrast, to the human where the PRL gene is expressed in both the anterior pituitary and decidua, rat decidual tissue does not express the PRL gene but instead expresses unique members of the PRL gene family (Gibori et al., 1987; Croze et al., 1990; Roby et al., 1993). There is evidence for a dLTH possessing characteristics resembling PRL and at least two proteins structurally related to PRL (Table I).

dLTH

Gibori and coworkers characterized the actions of the dLTH on the ovary and uterus and some aspects of its biochemical structure (Basuray et al., 1980; 1983; Gibori et al., 1984; Jayatilak et al., 1984, 1985, 1989; Herz et al., 1986; Gu et al., 1992). The dLTH interacts with luteal PRL receptors and mimics the actions of anterior pituitary PRL on the corpus luteum (Gibori et al., 1984). dLTH activities have been attributed to a 28-29 kDa protein produced by antimesometrial decidua (Jayatilak et al., 1989). This protein is recognized by an antiserum to human PRL but not by antisera to rat or ovine PRL (Gibori et al., 1974; Jayatilak et al., 1989). Expression of the rat dLTH is initiated shortly after implantation in the antimesometrial deciduum and terminates between days 13-14 of gestation (Jayatilak et al., 1989). As indicated above, the protein(s) responsible for dLTH actions is yet to be isolated.

PRL-like Protein-B

Croze *et al.* (1990) cloned a decidual cDNA encoding a protein homologous to pituitary PRL termed *PRL-like protein-B* (PLP-B).

Structure

PLP-B cDNA was originally identified from a rat placental cDNA library during a search for a placental lactogen cDNA (Duckworth et al., 1988). Both placental and decidual PLP-B mRNAs consist of two equally expressed transcripts estimated to be 0.9 and 1.2 kb in size and postulated to contain differing amounts of 5' untranslated sequence (Duckworth et al., 1988; Croze et al., 1990). Multiple transcription start sites have been identified in the PLP-B gene, possibly accounting for the two transcripts (Duckworth et al., 1988). The PLP-B mRNA encodes for a 201 amino acid mature protein with four cysteine residues, and a single putative *N*-linked glycosylation site (Duckworth et al., 1988). The PLP-B protein is secreted as a glycoprotein by both decidua and placental tissues (Cohick et al., 1996).

Expression

PLP-B is expressed at relatively low levels in the antimesometrial compartment of rat decidual tissue (Croze et al., 1990). Temporally, decidual PLP-B production follows the growth, development, and regression of the antimesometrial deciduum (Croze et al., 1990). After midgestation, PLP-B expression shifts to spongiotrophoblast cells of the rat chorioallantoic placenta where it is expressed at high levels until it declines during the final days of gestation (Duckworth et al., 1988; 1990; Ogilvie et al., 1990; Lu et al., 1994; Cohick et al., 1996). A unique feature of the PLP-B placental expression pattern is its absence in trophoblast giant cells. Consistent with these observations PLP-B is also not expressed in the trophoblast giant cell lineage-restricted Rcho-1 trophoblast cell line (Faria et al., 1990; Faria and Soares, 1991).

Biological Actions

The identification of PLP-B in decidual tissue prompted Croze and coworkers to speculate that it may represent the dLTH. In order to examine the actions of PLP-B, we expressed PLP-B in Chinese hamster ovary (CHO) cells using the pMSXND expression vector (Cohick et al., 1997). PLP-B does not effectively bind to PRL receptors or activate the PRL receptor signaling pathway (Cohick et al., 1997). Our observations to date have not supported a relationship between PLP-B and the dLTH. Although, specific biological actions for PLP-B have not been elucidated, the sustained pattern of expression of PLP-B throughout much of gestation suggests that its presence is likely beneficial for pregnancy.

Decidual PRL-Related Protein (dPRP)

A third member of the rat decidual PRL family was discovered in an attempt to characterize the PLP-B protein from rat decidual tissue (Roby et al., 1993).

Structure

A 29 kDa protein was isolated from medium conditioned by decidual explants. The protein possessed an affinity for Concanavalin A and crossreactivity with antibodies to amino acids 140-159 of PLP-B. *N*-terminal sequencing of the isolated decidual protein indicated it shared significant sequence identity with the *N*-terminus of a member of the rat placental PRL family, referred to as *PRL-like protein-C* (PLP-C), and limited similarity with PLP-B. In agreement with these observations, antibodies to PLP-C were shown to recognize the decidual protein. The decidual protein was termed *decidual PRL-related protein* (dPRP). A PLP-C cDNA was used to identify dPRP cDNAs from a rat decidual cDNA library (Deb et al., 1991; Roby et al., 1993). Nucleotide sequence analyses of the dPRP cDNAs predicted a mature protein of 239 amino acids, including a 28-amino acid signal sequence. The predicted dPRP amino acid sequence contains two putative *N*-linked glycosylation sites and six cysteine residues. The six cysteines are located in positions homologous to the cysteines of PLP-C and PRL. Overall, dPRP exhibits approximately 70% amino acid homology with PLP-C. Additional sequence similarities with members of the PRL family are evident. The dPRP gene was localized to rat chromosome 17, consistent with the chromosomal localization of all PRL family genes.

Distribution

Analysis of serum from pseudopregnant rats indicated that unlike other members of the PRL family, dPRP did not circulate at detectable levels (Rasmussen et al., 1996). dPRP associates with heparin containing molecules and resides, at least in part, within the decidual extracellular matrix (Rasmussen et al., 1996). Modulatory factors present in the decidual extracellular matrix would be well situated to influence the behavior of decidual cells and other cell types traversing the decidual extracellular matrix, such as trophoblast, endothelial, and various immune cells. Most interestingly, dPRP is expressed in the junctional zone of the chorioallantoic placenta during the latter half of pregnancy (Rasmussen et al., 1997), a region conspicuously devoid of heparan sulfate proteoglycan (Laurie, 1985). The distribution of dPRP within placental and extraplacental tissues during the latter part of gestation has not been determined.

Expression

Cell- and temporal-specific patterns of dPRP expression have been determined with cytochemical and biochemical procedures (Roby et al., 1993; Gu et al., 1994; Rasmussen et al., 1997). dPRP expression was first detected at day 6 of pregnancy. Expression increased with the growth of the deciduum and declined with decidual regression. Throughout the first half of pregnancy, dPRP protein and mRNA were predominantly localized to the antimesometrial deciduum of the developing conceptus. During the second half of gestation, dPRP expression shifts to the chorioallantoic placenta. Both trophoblast giant cells and spongiotrophoblast cells within the junctional zone of the chorioallantoic placenta express dPRP. The decidua-trophoblast shift in dPRP expression closely resembles that reported for PLP-B (see above). In contrast to PLP-B, dPRP is expressed at high levels in decidua and within trophoblast giant cells of the chorioallantoic placenta (Rasmussen et al., 1997). Factors and molecular mechanisms responsible for the control of PLP-B or dPRP expression in any tissue are yet to be elucidated.

Biological Actions

The coordinated pattern of dPRP expression from postimplantation until term involving both maternal and extraembryonic tissues implies some physiologic importance to dPRP. Similar to our approach in examining the actions of PLP-B, we generated recombinant dPRP protein (Rasmussen et al., 1996). Recombinant dPRP exhibited features that closely resembled native dPRP isolated from decidua. We next evaluated biological functions representing classical and nonclassical actions previously attributed to members of the PRL family. Classical actions involve utilization of the PRL receptor signaling pathway, whereas nonclassical actions utilize alternative mechanisms. dPRP failed to bind to PRL receptors and showed minimal abilities to promote the proliferation of the PRL-dependent Nb2 lymphoma cell line. Two members of the mouse placental PRL family influence angiogenesis through nonclassical mechanisms (Jackson et al., 1994). dPRP did not markedly influence the development of vascular structures as evaluated in both *in vitro* and *in vivo* assays (Rasmussen et al., 1996). Heterologous expression of dPRP in CHO cells, did however, significantly increase the ability of CHO cells to form tumors following transplantation into athymic mice (Rasmussen et al., 1996). We hypothesize that dPRP in some way alters the relationship between the tumor cells and the host, thereby increasing the success rate for establishing a tumor. dPRP may be mediating similar processes during the establishment of pregnancy.

CONCLUSIONS AND FUTURE DIRECTIONS

Production of members of the PRL family is part of the armamentarium through which decidual cells act to modify the uteroplacental environment. Experimentation on the decidual PRL family represents a window into mechanisms underlying decidualization and the initiation of pregnancy.

The decidual PRL family can be viewed as a powerful tool for the elucidation of regulatory mechanisms controlling uterine stromal cell differentiation and decidual cell-specific gene expression. An assortment of biochemical markers of decidualization have been presented, including: activin, follistatin, α1-acid glycoprotein, α2-macroglobulin, laminin, desmin, insulin-like growth factor binding proteins, and collagen type VI (Glasser and Julian, 1986; Wewer et al., 1986; Glasser et al., 1987; Bell et al., 1991; Giudice et al., 1991; Gu et al., 1992, 1995; Mulholland et al., 1992; Thomas, 1993); however, none of these indicies of decidual cell function possesses the combined specificity, fidelity, and abundance of members of the decidual PRL family. Identification of *cis*-elements and *trans*-acting factors responsible for decidual cell-specific expression of members of the decidual PRL family represents a strategy that may lead to understanding mechanisms controlling uterine stromal cell differentiation. Progesterone and prostaglandin E_2 are logical activators of pathways controlling decidualization (DeFeo, 1967; Kennedy, 1983, 1986; Yee and Kennedy, 1993; Frank et al., 1994; Gellersen et al., 1994). Decidual-specific expression patterns for a few DNA binding proteins, including the basic helix-loop-helix transcription factor Hed/Thing-2/d-HAND (Cross et al., 1995; Hollenberg et al., 1995; Srivastava et al., 1995), Wilms' tumor-1 (WT-1; Zhou et al., 1993), and the retinoid X receptor-α (Mangelsdorf et al., 1992), suggest the existence of additional candidate regulators controlling decidualization and decidual cell-specific gene expression. These processes are likely to be highly conserved across species similar to molecular mechanisms underlying the differentiation of other cell lineages (Weintraub et al., 1993). The recent establishment of rat uterine stromal cell lines (Cohen et al., 1993; Arslan et al., 1995; Srivastava et al., 1995) will hopefully facilitate investigations on uterine stromal cell differentiation.

At present, our knowledge of the biology of the decidual PRL family is incomplete. It is clear that decidual cells express members of the PRL family in cell- and temporal-specific patterns that likely reflect a key involvement in the initiation and maintenance of pregnancy. At present, we have only limited insights into possible functions of any member of the decidual PRL family. Advantages exist in studying primate decidual PRL. We know that primate decidual PRL activates the well-characterized PRL receptor signaling pathway. Ethical considerations will restrict investigations in humans but nonhuman primates may serve as an important animal model. Unfortunately, to date a nominal amount of information exists regarding decidual PRL and the distribution of PRL receptors in nonhuman primates. While primate decidual cells express a singular multi-functional PRL, rat decidual cells express a spectrum of more specialized ligands with differing biological responsibilities. An example of the evolution of two different strategies designed to accomplish the same tasks. A significant advancement of the field will require a concerted effort utilizing both primate and rodent models and array of *in vivo* and *in vitro* approaches.

SUMMARY

Decidual cells are responsible for creating a uterine environment supportive of the development of extraembryonic and embryonic tissues. A hormone family structurally related to pituitary PRL is a component of the efferent decidual cell response. Thus far, members of the PRL gene family have been identified in primates and in the rat. Patterns of expression during gestation suggest a role for members of the decidual PRL family in the establishment and maintenance of the gestational state. Precise physiological roles for each member of the decidual PRL family remain to be elucidated.

ACKNOWLEDGEMENTS

We would like to acknowledge current colleagues in our laboratory, Belinda M. Chapman, Christopher B. Cohick, Thomas J. Peters, Bing Liu, and Drs. Heiner Mueller, Guoli Dai, and Takayuki Kamei and also some previous trainees that made valuable contributions: Santanu Deb, Kazuyoshi Hashizume, and Katherine F. Roby. It is also important that we acknowledge a number of collaborators and contributors of valuable reagents: Drs. Geula Gibori, Frank Talamantes, Daniel I.H. Linzer, Henry G. Friesen, Claude Szpirer, and Simon C.M. Kwok. This work was supported by grants from the National Institutes of Health, HD 29036 and HD 29797.

REFERENCES

Andersen, J.R. (1990) Decidual prolactin. Studies of decidual and amniotic prolactin in normal and pathological pregnancy. *Dan. Med. Bull.* 37, 154-165.

Arslan, A., Almazan, G. and Zingg, H.H. (1995) Characterization and co-culture of novel nontransformed cell lines derived from rat endometrial epithelium and stroma. *In Vitro Cell. Dev. Biol.* 31, 140-148.

Basuray, R. and Gibori, G. (1980) Luteotropic action of the decidual tissue in the pregnant rat. *Biol. Reprod.* 23, 507-512.

Basuray, R., Jaffe, R.C., and Gibori, G. (1983) Role of decidual luteotropin and prolactin in the control of luteal cell receptors for estradiol. *Biol. Reprod.* 28, 551-556.

Bell, S.C. (1983) Decidualization: Regional differentiation and associated function. *Oxf. Rev. Reprod. Biol.* 5, 220-271.

Bell, S.C., Jackson, J.A., Ashmore, J., Zhu, H.H. and Tseng, L. (1991) Regulation of insulin-like growth factor binding protein-1 synthesis and secretion by progestin and relaxin in long term cultures of human endometrial stromal cells. *J. Clin. Endocrinol. Metab.* 72, 1014-1024.

Berwaer, M., Martial, J.A. and Davies, J.R.E. (1994) Characterization of an up-stream promoter directing extrapituitary expression of the human prolactin gene. *Mol. Endocrinol.* 8, 635-642.

Blithe, D.L., Richards, R.G. and Skarulis, M.C. (1991) Free alpha molecules from pregnancy stimulate secretion of prolactin from human decidual cells: A novel function for free alpha in pregnancy. *Endocrinology* 129, 2257-2259.

Brown, N.A. and Bethea, C.L. (1994) Cloning of decidual prolactin from Rhesus macaque. *Biol. Reprod.* 50, 543-552.

Cohen, H., Pageaux, J.-F., Melinand, C., Fayard, J.-M.,and Laugier, C. (1993) Normal rat uterine stromal cells in continuous culture: Characterization and progestin regulation of growth. *Eur. J. Cell Biol.* 61, 116-125.

Cohick, C.B., Xu, L. and Soares, M.J. (1997) Prolactin-like protein-B: Heterologous expression and characterization of placental and decidual species. *J. Endocrinol.* 152, 291-302.

Cross, J.C., Flannery, M.L., Blanar, M.A., Steingrimsson, E., Jenkins, N.A., Copeland, N.G., Rutter, W.J. and Werb, Z. (1995) *Hxt* encodes a basic helix-loop-helix transcription factor that regulates trophoblast cell development. *Development* 121, 2513-2523

Croze, F., Kennedy, T.G., Schroedter, I.C. and Friesen, H.G. (1990) Expression of rat prolactin-like protein-B in deciduoma of pseudopregnant rat and in decidua during early pregnancy. *Endocrinology* 127, 2665-2672.

Deb, S., Roby, K.F., Faria, T.N., Szpirer, C., Levan, G., Kwok, S.C.M. and Soares, M.J. (1991) Molecular cloning and characterization of prolactin-like protein-C complementary deoxyribonucleic acid. *J. Biol. Chem.* 266, 23027-23032

DeFeo, V.J. (1967) Decidualization. In: *Cellular Biology of the Uterus.* (ed.), R.M.Wynn Appleton-Century-Crofts; New York, pp. 191-290.

DiMattia, G.E., Gellersen, B., Duckworth, M.L. and Friesen, H.G. (1990) Human prolactin gene expression. The use of an alternative noncoding exon in decidua and the IM-9-P3 lymphoblast cell line. *J. Biol. Chem.* 265, 16412-16421.

Duckworth, M.L., Peden, L.M., and Friesen, H.G. (1988) A third prolactin-like protein expressed by the developing rat placenta: Complementary deoxyribonucleic acid sequence and partial structure of the gene. *Mol. Endocrinol.* 2, 912-920.

Duckworth, M.L., Schroedter, I.C. and Friesen, H.G. (1990) Cellular localization of rat placental lactogen-II and rat prolactin-like proteins A and B by in situ hybridization. *Placenta* 11, 143-155.

Enders, A.C. and Schlafke,. S. (1967) A morphological analysis of the early implantation stages in the rat. *Am. J. Anat.* 133, 291-316.

Ershoff, B.H. and Deuel, H.J. (1943) Prolongation of pseudopregnancy by induction of deciduomata in the rat. *Proc. Soc. Exp. Biol. Med.* 54, 167-168.

Faria, T.N., Deb, S., Vandeputte, M., Talamantes, F. and Soares, M.J. (1990) Transplantable rat choriocarcinoma cells express placental lactogen: Identification of Placental lactogen-I immunoreactive protein and messenger ribonucleic acid. *Endocrinology* 127, 3131-3137

Faria, T.N. and Soares, M.J. (1991) Trophoblast cell differentiation: Establishment, characterization, and modulation of a rat trophoblast cell line expressing members of the placental prolactin family. *Endocrinology* 129, 2895-2906

Frank, G.R., Brar, A.K., Cedars, M.I. and Handwerger, S. (1994) Prostaglandin E2 enhances human endometrial stromal cell differentiation. *Endocrinology* 134, 258-263.

Freemark, M., Kirk, K., Pihoker, C., Robertson, M.C., Shiu, R.P.C. and Driscoll, P. (1993) Pregnancy lactogens in the rat conceptus and fetus: Circulating levels, distribution of binding, and expression of receptor messenger ribonucleic acid. *Endocrinology* 133, 1830-1842.

Gellersen, B., DiMattia, G.E., Friesen, H.G. and Bohnet, H.G. (1989) Prolactin (PRL) mRNA from human decidua differs from pituitary PRL mRNA but resembles the IM-9-P3 lymphoblast PRL transcript. *Mol. Cell. Endocrinol.* 64, 127-130.

Gellersen, B., Kempf, R., Telgmann, R. and DiMattia, G.E. (1994) Nonpituitary human prolactin gene transcription is independent of Pit-1 and differentially controlled in lymphocytes and in endometrial stroma. *Mol. Endocrinol.* 8, 356-373.

Gibori, G., Jayatilak, P.G., Khan, I., Rigby, B., Puryear, T., Nelson, S. and Herz, Z. (1987) Decidual luteotropin secretion and action: Its role in pregnancy maintenance in the rat. In: *Regulation of Ovarian and Testicular Function*, (eds.) V.B. Mahesh, D.S. Dhindsa, E. Anderson and S.P Kalra SP, Plenum Press: New York, pp. 379-397.

Gibori, G., Kalison, B., Basuray, R., Rao, M.C. and Hunzicker-Dunn, M. (1984) Endocrine role of the decidual tissue: Decidual luteotropin regulation of luteal adenyl cyclase activity, luteinizing hormone receptors, and steroidogenesis. *Endocrinology* 115, 1157-1163.

Gibori, G., Rothchild, I., Pepe, G.I., Morishige, W.K. and Lam, P. (1974) Luteotrophic action of decidual tissue in the rat. *Endocrinology* 95, 1113-1118.

Giudice, L.C., Milkowski, D.A., Lamson, G., Rosenfeld, R.G. and Irwin, J.C. (1991) Insulin-like growth factor binding proteins in human endometrium: Steroid dependent messenger ribonucleic acid expression and protein synthesis. *Endocrinology* 72, 779-787.

Glasser, S.R. and Julian, J. (1986) Intermediate filament protein as a marker of uterine stromal cell decidualization. *Biol. Reprod.* 35, 463-474.

Glasser, S.R., Lampelo, S., Munir, M.I. and Julian, J. (1987) Expression of desmin, laminin, and fibronectin during in situ differentiation (decidualization) of rat uterine stromal cells. *Differentiation* 35, 132-142.

Golander, A., Hurley, T., Barrett, J. and Handwerger, S. (1979) Synthesis of prolactin by human decidua in vitro. *J. Endocrinol.* 82, 263-267.

Golander, A., Hurley, T., Barrett, J., Hize, A. and Handwerger, S. (1978) Prolactin synthesis by human chorion-decidual tissue: A possible source of amniotic fluid prolactin. *Science* 202, 311-312.

Gu, Y. and Gibori, G. (1995) Isolation, culture, and characterization of the two cell subpopulations forming the rat decidua: Differential gene expression for activin, follistatin, and decidual prolactin-related protein. *Endocrinology* 136, 2451-2458.

Gu, Y., Jayatilak, P.G., Parmer, T.G., Gauldie, J., Fey, G.H. and Gibori, G. (1992) α_2-macroglobulin expression in the mesometrial decidua and its regulation by decidual luteotropin and prolactin. *Endocrinology* 131, 1321-1328.

Gu, Y., Soares, M.J., Srivastava, R.K. and Gibori, G. (1994) Expression of decidual prolactin-related protein in the rat decidua. *Endocrinology* 135, 1422-1427.

Gu, Y., Srivastava, R.K., Ou, J., Krett, N.L., Mayo, K.E. and Gibori, G. (1995) Cell-specific expression of activin and its two binding proteins in the rat decidua: Role of α2-macroglobulin and follistatin. *Endocrinology* 136, 3815-3822.

Handwerger, S., Golander, A., Richards, R., Thrailkill, K., Jorgensen, V., Harman, I. and Grandis, A. (1989) Paracrine and autocrine factors involved in the regulation of the release of human decidual prolactin and human placental lactogen. In: *Placenta as a Model and a Source*, (eds.) O Genbacev, A Klopper, and R. Beaconsfield, Plenum Press: New York, pp 129-140.

Handwerger, S., Markoff, E. and Richards, R. (1991) Regulation of the synthesis and release of decidual prolactin by placental and autocrine/paracrine factors. *Placenta* 12, 121-130.

Herz, Z., Khan, I., Jayatilak, P.G. and Gibori, G. (1986) Evidence for the synthesis and secretion of decidual luteotropin: A prolactinlike hormone produced by rat decidual cells. *Endocrinology* 118, 2203-2209.

Hollenberg, S.M., Sternglanz, R., Cheng, P.F. and Weintraub, H. (1995) Identification of anew family of tissue-specific basic helix-loop-helix proteins with a two-hybrid system. *Mol. Cell. Biol.* 15, 3813-3822.

Hwang, P., Murray, J.B., Jacobs, J.W., Niall, H.D. and Friesen, H.G. (1974) Human amniotic fluid prolactin. Purification by affinity chromatography and amino-terminal sequence. *Biochemistry* 13, 2354-2358.

Jackson, D., Volpert, O.V., Bouck, N. and Linzer, D.I.H. (1994) Stimulation and inhibition of angiogenesis by placental proliferin and proliferin-related protein. *Science* 266, 1581-1584.

Jayatilak, P.G., Glaser, L.A., Basuray, R., Kelly, P.A. and Gibori, G. (1985) Identification and partial characterization of a prolactin-like hormone produced by the rat decidual tissue. *Proc. Natl. Acad. Sci. USA* 82, 217-221.

Jayatilak, P.G., Glaser, L.A., Warshaw, M.L., Herz, Z., Gruber, J.R. and Gibori, G. (1984) Relationship between luteinizing hormone and decidual luteotropin in the maintenance of luteal steroidogenesis. *Biol. Reprod.* 31, 556-564

Jayatilak, P.G., Puryear, T.K., Herz, Z., Fazleabas, A. and Gibori, G. (1989) Protein secretion by mesometrial and antimesometrial rat decidual tissue: Evidence for differential gene expression. *Endocrinology* 125, 659-666.

Kennedy, T.G. (1983) Prostaglandin E_2, adenosine 3':5'-cyclic monophosphate and changes in endometrial vascular permeability in rat uteri sensitized for the decidual cell reaction. *Biol. Reprod.* 29, 1069-1076.

Kennedy, T.G. (1986) Intrauterine infusion of prostaglandins and decidualization in rats with uteri differentially sensitized for the decidual cell reaction. *Biol. Reprod.* 34, 327-335.

Krehbiel, R.H. (1937) Cytological studies of the decidual reaction in the rat during early pregnancy and in the production of deciduomata. *Physiol. Zool.* 10, 212-234.

Laurie, G.W. (1985) Lack of heparan sulfate proteoglycan in a discontinuous and irregular placental basement membrane. *Dev. Biol.* 108, 299-309.

Lee, D.W. and Markoff, E. (1986) Synthesis and release of glycosylated prolactin by human decidua *in vitro*. *J. Clin. Endocrinol. Metab.* 62, 990-994.

Lu, X.-J., Deb, S. and Soares, M.J. (1994) Spontaneous differentiation of trophoblast cells along the spongiotrophoblast pathway: Expression of the placental prolactin gene family and modulation by retinoic acid. *Dev. Biol.* 163, 86-97.

Mulder, G.H. (1989) Effects of decidual prolactin on amnio-chorion water permeability. In: *Placenta as a Model and a Source,* (eds.) O. Genbacev, A. Klopper and R. Beaconsfield, Plenum Press: New York, pp. 103-114.

Mulholland, J., Aplin, J.D., Ayad, S., Hong, L. and Glasser, S.R. (1992) Loss of collagen type VI from rat endometrial stroma during decidualization. *Biol. Reprod.* 46, 1136-1143

Nicoll, C.S., Mayer, G.L. and Russell, S.M. (1986) Structural features of prolactins and growth hormones that can be related to their biological properties. *Endocr. Rev.* 7, 169-203.

Ogilvie, S., Duckworth, M.L., Larkin, L.H., Buhi, W.C. and Shiverick, K.T. (1990) *De novo* synthesis and secretion of prolactin-like protein-B by rat placental explants. *Endocrinology* 126, 2561-2566.

Parr, M.B. and Parr, E.L. (1989) The implantation reaction. In: *Biology of the Uterus.* (eds.), R.M. Wynn and W.P. Jollie, New York: Plenum Medical Book Company, pp. 233-278.

Rasmussen, C.A., Hashizume, K., Orwig, K.E., Xu, L. and Soares, M.J. (1996) Decidual prolactin-related protein: Heterologous expression and characterization. *Endocrinology* 137, 5558-5566.

Rasmussen, C.A., Orwig, K.E., Vellucci, S. and Soares, M.J. (1997) Dual expression of prolactin-related protein in decidua and trophoblast tissues during pregnancy in rats. *Biol. Reprod.* 56, 647-654.

Riddick, D., Luciano, A., Kusmik, W. and Maslar, I. (1978) De novo synthesis of prolactin by human decidua. *Life Sci.* 23, 1913-1929.

Roby, K.F., Deb, S., Gibori, G., Szpirer, C., Levan, G., Kwok, S.C.M. and Soares, M.J. (1993) Decidual prolactin related protein: Identification, molecular cloning, and characterization. *J. Biol. Chem.* 268, 3136-3142.

Rothchild, I. and Gibori, G. (1975) The luteotrophic effect of decidual tissue: The stimulating effect of decidualization on serum progesterone level of pseudopregnant rats. *Endocrinology* 97, 838-842.

Royster, M., Driscoll, P., Kelly, P.A. and Freemark, M. (1995) The prolactin receptor in the fetal rat: Cellular localization of messenger ribonucleic acid, immunoreactive protein, and ligand-binding activity and induction of expression in late gestation. *Endocrinology* 136, 3892-3900.

Soares, M.J., Faria, T.N., Roby, K.F. and Deb, S. (1991) Pregnancy and the prolactin family of hormones: Coordination of anterior pituitary, uterine, and placental expression. *Endocr. Rev.* 12, 402-423

Srivastava, D., Cserjesi, P. and Olson, E.N. (1995) A subclass of bHLH proteins required for cardiac morphogenesis. *Science* 270, 1995-1999.

Srivastava, R.K., Gu, Y., Zilberstein, M., Ou, J.S., Mayo, K.E., Chou, J.Y. and Gibori, G. (1995) Development and characterization of a Simian virus 40-transformed, temperature-sensitive rat antimesometrial decidual cell line. *Endocrinology* 136, 1913-1919.

Tang, B., Guller, S. and Gurpide, E. (1993) Cyclic adenosine 3',5'-monophosphate induces prolactin expression in stromal cells isolated from human proliferative endometrium. *Endocrinology* 133, 2197-2203.

Thomas, T. (1993) Distribution of α_2-macroglobulin and α_1-acid glycoprotein mRNA shows regional specialization in rat decidua. *Placenta* 14, 417-428.

Weintraub, H. (1993) The MyoD family and myogenesis: Redundancy, networks, and thresholds. *Cell* 75, 1241-1244.

Welsh, A.O. and Enders, A.C. (1985) Light and electron microscope examination of the mature decidual cells of the rat with emphasis on the antimesometrial decidual and its degeneration. *Am. J. Anat.* 172, 1-29.

Wewer, U.M., Damjanov, A., Weiss, J., Liotta, L.A. and Damjanov, I. (1986) Mouse endometrial stromal cells produce basement-membrane components. *Differentiation* 32, 49-58.

Yee, G.M. and Kennedy, T.G. (1993) Prostaglandin E2, cAMP and cAMP-dependent protein kinase isozymes during decidualization of rat endometrial stromal cells in vitro. *Prostaglandins* 46, 117-138.

Zhou, J., Rauscher, F.J. and Bondy, C. (1993) Wilms' tumor (WT1) gene expression in rat decidual differentiation. *Differentiation* 54, 109-114.

Zhu, H.H., Huang, J.R., Mazella, J., Rosenberg, M. and Tseng, L. (1990) Differential effects of progestin and relaxin on the synthesis and secretion of immunoreactive prolactin in long term culture of human endometrial stromal cells. *J. Clin. Endocrinol. Metabol.* 71, 889-899.

Trophoblast Research 10:345-352, 1997

THE HUMAN PLACENTAL GROWTH HORMONE VARIANT
-A Review-

Ahmed Igout and Georges Hennen

Biochimie et Laboratoire d'Endocrinologie
Université de Liège
Tour de Pathologie, B23
B-4000 Liège, Belgique

INTRODUCTION

Growth hormone (GH), prolactin (Prl), and choriosomato-mammotropin (CS) are peptide hormones which are structurally related, and which have arisen from a common ancestral gene (Niall et al., 1971; Miller and Eberhardt, 1983).

The human genome contains a cluster of five genes GH/CS localized on a segment of 48 kilobases of the long arm of chromosome 17 (Hirt et al., 1987; Chen et al., 1989). The prolactin gene is unique and located on chromosome 6 (Owerbach et al., 1981). The five GH/CS genes, for which the size is about 1.6 kbp for each, are separated by intergene regions of 6 to 13 kbp. They are oriented according to the scheme in Figure 1 (Hirt et al., 1987). A study of their sequence reveals a high degree of identity with a conservation of sequence varying from 91 to 99% (Miller and Eberhardt, 1983; Hirt et al., 1987; Chen et al., 1989). These five genes are all organized in the same manner with five exons interrupted by four introns as schematized in Figure 1 (DeNoto et al., 1981).

Each gene codes for a protein of approximately 200 amino acids, preceded by a signal peptide of 25 amino acids.

It has been known for a long time that three of these five genes i.e., hGH-N, hCS-A and hCS-B are functionally transcribed: hGH-N is selectively expressed in the pituitary giving human Growth Hormone (hGH) (Martial et al., 1979); while hCS-A and hCS-B are expressed during pregnancy in the placenta where they code for human Placental Lactogen (hPL) (also called chorio-somatomammotropin, hCS). Pituitary growth hormone is composed of two main variants, of 22 kDa and 20 kDa (Wallis, 1980), representing respectively 90 and 10% of the pituitary secretion of this hormone.

The hGH-N gene is transcribed to give one pre-mRNA, but alternative splicing of the transcript gives two mRNAs which are translated into the 22 kDa and 20 kDa forms. In contrast, the two distinct genes, hCS-A and hCS-B, code for the same protein (Barrera-Saldena et al., 1983).

Pituitary growth hormone and hCS possess similar biological properties (Niall et al., 1971). After birth, pituitary hGH-N maintains the staturo-ponderal linear growth of the infant by the stimulation of the production of IGF-1 (Daughaday and Rotwein, 1989). hGH-N has a direct effect on the metabolic activity of tissues, on nitrogen anabolism

(Henneman et al., 1960) and fat mobilization to the benefit of the muscular mass. hGH-N maintains the activity of gluconeogenesis and is responsible for the upkeep of glycemia during the post-absorptive and fasting periods. During fasting and starvation, hGH facilitates fat consumption (Manson and Wilmore, 1986; Ward et al., 1987). hGH also exerts a positive action on both humoral and cellular immunity (Church et al., 1989).

Recent Discoveries Concerning The Tissular Expression Of The hGH-V Gene

Even though the functions of the GH-N, CS-A and CS-B genes were well known the hGH-V gene was considered to be silent (Parks, 1984): no expression product of this gene had ever been detected in any tissue or biological fluid.

However, we detected by a specific immunological test in the serum of pregnant women a GH like factor (Figure 2) (Hennen et al., 1985a). With two specific monoclonal antibodies, each recognizing distinct epitopes of hGH, we have developed immunoassays for the measurement of GH in pregnant women. We were able therefore to observe that pituitary hGH progressively disappeared from the maternal circulation during pregnancy while a growth hormone-like factor appeared and increased in concentration until term (Hennen et al., 1985b; Hennen et al., 1985c). This factor, with antigenic properties distinct from pituitary growth hormone was detected in significant concentrations in the placenta. This suggested that the hormone was synthetized in the placenta before being secreted into the maternal circulation during pregnancy (Frankenne et al., 1988). We decided thus to call it human placental growth hormone (hPGH). Immunohistochemical studies showed that hPGH is synthetized by the syncytiotrophoblast (Jara et al., 1989).

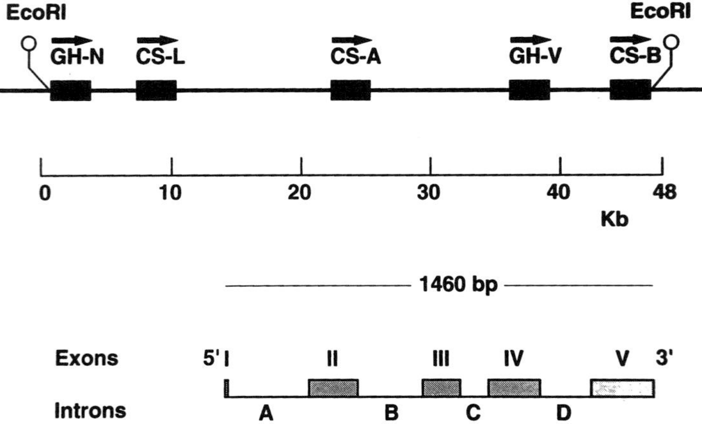

Figure 1. Locus and organization of the hGH/hCS gene family in chromosome 17. The upstream region before the start codon in exon I and downstream region after termination codon in exon V are not shown.

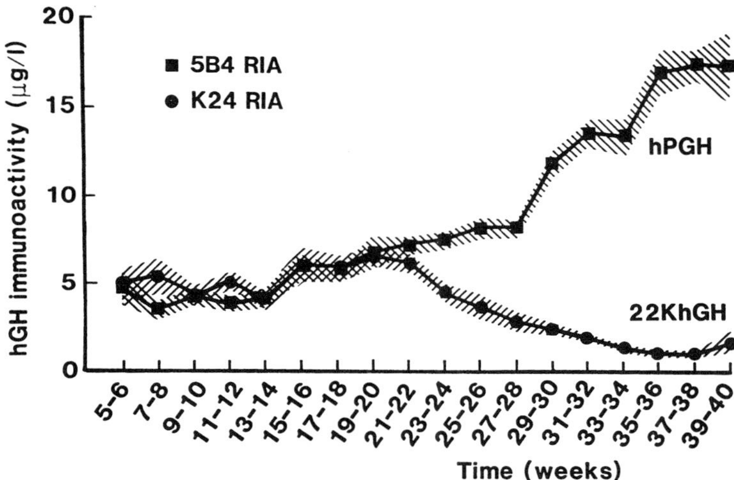

Figure 2. Serum hGH-N and hPGH (hGH-V) levels in women throughout pregnancy as measured by RIA using 5B4 and K24 MAbs.

hPGH was first purified from placental tissue recovered immediately after delivery. The biochemical characterization of purified hPGH showed that its isoelectric point (pI) was higher than that of the pituitary GH and was present in two forms which differed slightly in molecular weight (22 and 25 kDa); the 25 kDa form was shown to be glycosylated (Frankenne et al., 1990).

The determination of the N-terminal sequence demonstrated that the hormone was indeed the result of the expression of the hGH-V gene. The mode of expression of this gene *in vivo* was studied by several experiments. First evidence of the expression was shown by positive hybridization of the mRNA with a nucleotide probe specific to the hGH-V gene (Frankenne et al., 1987). By *in situ* hybridization we have shown that the hGH-V gene is expressed exclusively by the syncytiotrophoblast (Scippo et al., 1993). Finally the cDNA of hGH-V was cloned from a placental cDNA library (Igout et al., 1988). The sequence of this cDNA corresponded perfectly with what was known of the structure of the gene. All these elements confirmed that the splicing of the hGH-V pre-mRNA gives a mature mRNA coding for a protein of 191 amino acids of similar structure to other proteins of the GH/CS family. The level of expression of the hGH-V gene in the placenta is low and was estimated by dot blotting at 0.2 to 0.5 % of the level of the hCS gene and at 0.1 % by screening of a placental cDNA library (Chen et al., 1989; Scippo, 1992).

Biological Activities

We have produced the non-glycosylated form in *E coli* (Igout et al., 1993) and we have studied its biological activities (Igout et al., 1995). hPGH was found to be a potent pituitary GH agonist, through binding experiments in which plasma membranes from

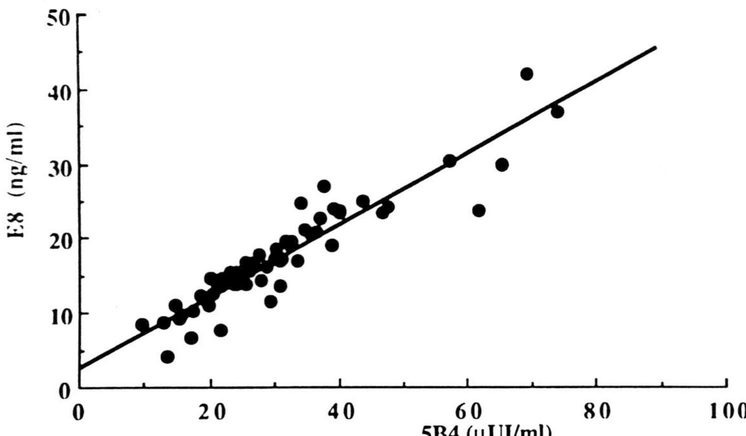

Figure 3. Correlation between the placental hGH values assayed with RIA in maternal sera (38 weeks of amenorrhea) using 5B4 and E8 MAbs, (r = 0.93).

pregnant rabbits were used and by Tibia test (Frankenne et al., 1990; Igout et al., 1995). hGH-V is twice as potent as hGH-N in its capacity for binding to the GH receptor (Igout et al., 1995) but its somatotrope effect is identical to that of hGH-N.

In the biological test using the NB2 cells its lactogenic activity was 6% that of the pituitary GH (Igout et al., 1995). These results suggest that the GH-V protein has high somatogenic activity but low lactogenic activity. hPGH possesses its specific receptor in the placenta, and certainly has both endocrine and paracrine/autocrine action via the GH hepatic receptor and its placental specific receptor (Igout, 1994). The IGF-I secretion in pregnant woman is well correlated with PGH secretion (Caufriez et al., 1990). Like hGH-N, hGH-V secretion is modulated by the variation of glycemia (Patel et al., 1995, data not shown).

Assay for hPGH

Using recombinant hPGH, we have produced two specific monoclonal antibodies, named 7C12 and E8 (Igout et al., 1993; Igout, 1994). The E8 Mab recognizes an epitope specific to hPGH but overlapping the 5B4 Mab epitope. Levels of placental GH measured with 5B4 and E8 Mabs are perfectly correlated (R=0.93) (Figure 3). The E8 and 7C12 Mabs recognize different epitopes and are presently used in an immunometric assay to study the physiopathology of the hormone.

CONCLUSIONS

The redundancy of the GH/CS genes with two GH genes and three CS genes is reflected in the expression of the corresponding proteins, in particular in the placenta. We have originally demonstrated the expression and studied in a systematic manner the placental growth hormone.

Placental GH is the only form of GH present in the maternal serum during the second half of pregnancy. During this period, the secretion of pituitary GH is suppressed. The hypothesis that hPGH could have a direct effect on fetal growth is improbable, since the hormone does not cross the maternal/fetal barrier and is not stimulated by GRF (DeZegher et al., 1990). However, fetal growth promotion is probably expressed via the modification of maternal metabolism induced by placenta GH. The feto-placental unit requires the transfer of high quantities of nutrients from the mother to the fetus. Placental GH having a somatotrophic activity, is well adapted to the stimulation of gluconeogenesis, lipolysis and anabolism thereby realizing optimal metabolic conditions for the fetus. The biological properties of hPGH, being close to those of pituitary GH, would indicate that hPGH takes over the role of pituitary GH during the second half of pregnancy. The non-pulsatile secretion of hPGH in the maternal compartment (Eriksson et al., 1989) could diminish its growth promoting activity while favoring the metabolic role this hormone plays.

The replacement of pituitary GH by placental GH could therefore indicate that the feto-placental unit takes control of the maternal metabolism during pregnancy.

SUMMARY

The hGH/CS genes are clustered in chromosome 17 in the 5' to 3' order GH-N, CSL, CSA, GH-V and CS-B. These genes show a high degree of sequence identity. During pregnancy we have demonstrated that the placenta secretes into the maternal circulation a placental variant of growth hormone. This is the only growth hormone secreted in pregnant women at the end of pregnancy. Since then we have purified this hormone from human placenta and have produced it by genetic engineering in *E.coli*. We have also studied its expression and its physiopathology.

ACKNOWLEDGEMENTS

This work was supported by a grant from the "Region Wallonne" (grant no 2640), from the EC (SCI-CT91-0759-TSTS) and "Fonds National de la Recherche Scientifique" (grant no FRSM-3.9004.91).

REFERENCES

Barrera-Saldena, H.A., Seeburg, P.H. and Saunders, G.F. (1983) Two structurally different genes produce the same secreted human placental lactogen hormone. *J. Biol. Chem.* 258, 3787.

Caufriez, A., Frankenne, F., Englert, Y., Golstein, J., Cantraine, F., Hennen, G. and Copinshi, G. (1990) Placental growth hormone as a potential regulator of maternal IGF-I during human pregnancy. *Am. J. Physiol. Endocrinol. Metab.* 258, E1014.

Chen, E.Y., Liao, Y.C., and Smith, D.H. (1989) The human growth hormone locus: nucleotide sequence, biology and evolution. *Genomics* 4, 479.

Church, J.A., Costin, G., and Brooks, J. (1989) Immune functions in children treated with biosynthetic growth hormone. *J. Pediatr.* 115, 420.

Daughaday, W.H. and Rotwein, P. (1989) Insulin-like growth factors I and II. Peptide, messenger ribonucleic acid and gene structures, serum and tissue concentrations. *Endocr. Rev.* 10, 68.

DeNoto, F.M., Moore, D.D., and Goodman, H.M. (1981) Human growth hormone DNA sequence and mRNA structure: Possible alternative splicing. *Nucleic Acids Res.* 9, 3719.

de Zegher, F., Vanderschueren-Lodeweyckx, M., Spitz, B., Faijerson, Y., Blomberg, F., Beckers, A., Hennen, G. and Frankenne, F. (1990) Perinatal growth hormone (GH) physiology: Effect of GH-releasing factor on maternal and fetal secretion of pituitary and placental GH. *J. Clin. Endocrinol. Metab.* 71, 520.

Eriksson, L., Frankenne, F., Eden, S., Hennen, G. and Von Schoultz, B. (1989) Growth hormone 24-hour serum profiles during pregnancy - lack of pulsatility for the secretion on the placental variant. *Br. J. Obstet. Gynaecol.* 96, 949.

Frankenne, F., Closset, J., Gomez, F., Scrippo, M.L., Smal. J. and Hennen, G. (1988) The physiology of growth hormones in pregnant women and partial characterization of the placental GH variant. *J. Clin. Endocrinol. Metab.* 66, 1171.

Frankenne, F., Rentier-Delrue, F., Scippo, M.L., Martial, J. and Hennen, G. (1987) Expression of the growth hormone variant gene in human placenta. *J. Clin. Endocrinol. Metab.* 64, 635.

Frankenne, F., Scippo, M.L., Van Beeumen, J., Igout, A. and Hennen, G. (1990) Identification of placental human growth hormone as the GH-V gene expression product. *J. Clin. Endocrinol. Metab.* 71, 15.

Henneman, P.H., Forbes, A.P., Moldawer, M., Dempsey, E.F. and Carroll, E.L. (1960) Effects of human growth hormone in man. *J. Clin. Invest.* 39, 1223.

Hennen, G., Frankenne, F., Pirens, G., Gomez, F., Closset, J. and El Khayat, N. (1985a) New chorionic GH-like antigen revealed by monoclonal antibody radioimmunoassays. *Lancet* Feb. 16, 399.

Hennen, G., Frankenne, F., Pirens, G., Gomez, F., El Khayat, N., Closset, J. and Schauss, Ch. (1985b) A chorionic GH-like antigen: increasing levels during second half of pregnancy with pituitary GH suppression as revealed by monoclonal antibody radioimmunoassays. *Int. J. Fertil.* 30, 27.

Hennen, G., Frankenne, F., and Closset, J. (1985c) Monoclonal antibodies: Basic principles, experimental and clinical applications in endocrinology. In: *Monoclonal Antibody To Growth Hormone: The Discovery Of A New Variant, Human Placental Growth Hormone*, (eds.) G. Forti, M.B. Lipsett, and M. Serio, Serono Symposia Publication: Raven Press, p. 29.

Hirt, H., Kimelman, J., Birnbaum, M.J., Chen, E.Y., Seeburg, P.H., Eberhardt, N.L. and Barta, A. (1987) The human growth hormone gene locus: Structure, evolution and allelic variations. *DNA* 6, 59.

Igout, A., Frankenne, F., L'Hermite-Baleriaux, M., Martin, A., and Hennen, G. (1995) Somatogenic and lactogenic activity of the recombinant 22 kDa isoform of human placental growth hormone. *Growth Regulation* 5, 60-65.

Igout, A., Scippo, M.L., Frankenne, F., and Hennen, G. (1988) Isolation and nucleotide sequence of hGH-V cDNA. *Arch. Int. Physiol. Biochim.* 96, 63-67.

Igout, A., Van Beeumen, J., Frankenne, F., Scippo, M.L., Devreese, B. and Hennen, G. (1993) Purification and biochemical characterization of recombinant human placental growth hormone produced in Escherichia coli. *Biochem. J.* 295, 719-724.

Igout, A. (1994) Le variant placentaire de l'hormone de croissance humaine: Clonage du cDNA, expression chez Escherichia coli et propriétés physicochimiques et biologiques de l'hormone recombinante. *Thèse de doctorat en Biochimie* (Ph.D.), Faculté des Sciences, Université de Liège.

Igout, A., Frankenne, F. and McNamara, M. (1995) Human placental growth hormone: A new monoclonal antibody for the study of its physiopathology. *75th Annual Meeting of the Endocrine Society, Las Vegas (Abstract #1127).*

Jara, C.S., Salud, A.A., Bryant-Greenwood, G.D., Pirens, G., Hennen, G. and Frankenne, F. (1989) Immunocytochemical localization of the human growth hormone variant in the human placenta. *J. Clin. Endocrinol. Metab.* 69, 1069.

Manson, J.M. and Wilmore, D.W. (1986) Positive nitrogen balance with human growth hormone and hypocaloric intravenous feedings. *Surgery* 100, 188.

Martial, J., Hallewell, R.A., Baxter, J.D., and Goodman, H.M. (1979) Human growth hormone: Complementary DNA cloning and expression in bacteria. *Science* 205, 602.

Miller, W.L. and Eberhardt, N.L. (1983) Structure and evolution of the growth hormone gene family. *Endocr. Rev.* 4, 97.

Niall, H.D., Hogan, M.L., Sauer, R., Rosenblum, I.Y. and Greenwood, F.C. (1971) Sequences of pituitary and placental lactogenic and growth hormones: Evolution from a primordial peptide by gene reduplication. *Proc. Natl. Acad. Sci. USA* 68, 866.

Owerbach, D., Rutter, W.J., Cooke, N.E., Martial, J.A. and Shows, T.B. (1981) The prolactin gene is located on chromosome 6 in humans. *Science* 212, 815.

Parks, J.S. (1984) Growth hormone and placental lactogen genes. In: *Endocrinology, 1st Edition*, (eds.) L. Labrie and L. Proulx, Elsevier Science Publishers BV: New York, p. 275.

Patel, N., Alsat, E., Igout, A., Baron, F., Hennen, G., Porquet, D. and Evain-Brion, D. (1995) Glucose inhibits human placental GH secretion, *in vitro*. *J. Clin. Endocrinol. Metab.* 80, 1743-1746.

Scippo, M.L. (1992) Le variant placentaire de l'hormone de croissance humaine: Purification et caractérisation partielles, étude de l'expression du gène. *Thèse de doctorat en Biochimie* (Ph.D.), Faculté des Sciences, Université de Liège.

Scippo, M.L., Frankenne, F., Hooghe-Peters, E.L., Igout, A., Velkeniers, B. and Hennen, G. (1993) Syncytiotrophoblastic localization of the human growth hormone variant mRNA in the placenta. *Mol. Cell. Endocrinol.* 92, R7-R13.

Wallis, M. (1980) Growth hormone: Deletions in the protein and introns in the gene. *Nature* 284, 512.

Ward, H.C., Halliday, D., Sim, A.J. (1987) Protein and energy metabolism with biosynthetic human growth hormone after gastrointestinal surgery. *Ann. Surg.* 206, 56.

Trophoblast Research 10:353-362, 1997

HUMAN CHORIONIC GONADOTROPHIN EARLY PREGNANCY LEVELS ARE MORE CLOSELY RELATED TO CHANGES IN β-SUBUNIT TROPHOBLAST PRODUCTION THAN TO VARIATIONS IN α-SUBUNIT PRODUCTION

Sylvain Meuris[1,4], Anne-Marie Nagy[1], Josiane Delogne-Desnoeck[1], Philippe Lebrun[1,2] and Eric Jauniaux[1,3]

[1]Research Laboratory on Reproduction
[2]Laboratory of Pharmacology
Faculty of Medicine
Free University of Brussels (ULB)
808 Route de Lennik
B-1070 Brussels, Belgium

[3]Academic Department of Obstetrics and Gynaecology
University College London Medical School
86-96 Chenies Mews
London WC1E 6HX, England

INTRODUCTION

Human chorionic gonadotrophin (hCG) is a glycoprotein composed of two non-covalently bound α and β subunits. These subunits are synthesized from separate mRNAs then combined in the rough endoplasmic reticulum of trophoblastic cells to form the heterodimer (Morgan et al., 1974; Peters et al., 1984). Lastly, hCG is secreted into the maternal circulation.

In addition to the dimer, uncombined subunits are also secreted leading to the polymorphism of the circulating hormone (Franchimont et al., 1972; Vaitukaitis et al., 1976). These circulating free subunits are not dissociation products since free α-subunits produced during pregnancy are unable to combine with β-subunit to form intact hCG (Cole et al., 1983; Blithe and Nisula, 1985). This inability to reassociate appears to be related to modifications in oligosaccharide structures of uncombined α molecules in the Golgi apparatus (Blithe, 1990).

The importance of the respective secretion of intact hCG and its free α- and β-subunits remains controversial. Indeed, it has recently been shown that, in early pregnancy, circulating free βhCG subunit is in excess over free αhCG subunit; whereas this balance is inverted in late gestation (Cole et al., 1984; Hay, 1985; Ozturk et al., 1987; Hay, 1988). These data, suggesting a two-phase regulation of hCG dimer formation, are in contrast with the excess of αhCG mRNA and of free αhCG subunit contents found in placental extracts from early and late pregnancy (Vaitukaitis, 1974; Ruddon et al., 1981; Boothby et al., 1983). Indeed, such data suggest that, throughout pregnancy, hCG dimer

[4]To Whom Correspondence Should Be Addressed.

formation is constantly dependent upon βhCG synthesis. This hypothesis is supported by our recent findings showing that the circulating levels of total α hCG and total βhCG subunits are equimolar in early pregnancy and that the ratio of α hCG to βhCG increased until term (Nagy et al., 1994a). The discrepancy between data obtained from placental extract and maternal serum may result from the greater metabolic clearance rate of circulating free αhCG subunit compared to that of the free βhCG subunit (Wehman and Nisula, 1979). However, it may also be caused by the use of different International Unit (IU) systems for the immunoassay of the circulating dimer and for the free subunits (Nagy et al., 1994a).

In early pregnancy, hCG is found in coelomic fluid in higher levels than in maternal circulation (Wathen et al., 1991; Iles et al., 1992; Jauniaux et al., 1993; Nagy et al., 1994b). From 5 to 12 weeks of gestation, the exocoelomic cavity separates the placenta from the amniotic cavity and contains the secondary yolk sac which floats in the coelomic fluid. Because of its close contact with placental tissue, biological investigations of coelomic fluid may give valuable information on the placental production of hCG and its subunits in early pregnancy. Indeed, in contrast with the maternal compartment, coelomic fluid may act as a transient reservoir with low clearance for proteins secreted by trophoblast.

The aim of the present study was to compare the relative amount of hCG and its free subunits directly secreted by trophoblast in coelomic fluid and maternal serum in order to evaluate the importance of their respective secretion by the trophoblast in early pregnancy.

MATERIALS AND METHODS

Patients

Blood and coelomic fluid samples collected after informed consent were obtained from 26 women with apparently normal gestation undergoing termination of pregnancy between 7-12 weeks gestation for psycho-social reasons. All patients were certain of the first day of their last menstrual period and pregnancy development was confirmed by ultrasonography. Coelomic fluid samples were retrieved by transvaginal puncture under ultrasonographic guidance; as previously described (Jauniaux et al., 1991c). Serum and coelomic fluid samples were stored at -20°C.

Radioimmunoassays

The assays for intact hCG and free α- and βhCG subunits were carried out using solid-phase two-site immunoradiometric assay (IRMA) kits kindly provided by bioMèrieux (Marcy-l'Etoile, France). These assays, using monoclonal antibodies, were calibrated against the International Reference Preparations (IRP) coded 75/537 for hCG, 75/569 for free α subunit and 75/551 for free β subunit. Cross reactivities were, respectively, <0.03% and <0.9% with α and β subunits in the assay for hCG, <0.3% and <0.001% respectively with intact hCG and β subunit in the assay for α subunit and <0.1% for hCG and <0.05% for α subunit in the assay for β subunit. All assays were performed in duplicate according to the manufacturer's instructions. Intra- and interassay coefficients of variations were respectively <7% and <10% for each assays. Sensitivities were 1 mIU/ml for hCG and 0.03 mIU/ml for both subunits.

Table 1

Comparisons between coelomic fluid and maternal serum median levels of hCG and its subunits in their respective International Unit (IU) systems, of total (free + combined) amounts of α- and βhCG subunits expressed in the same IU system (IRP 75/537) and of molar ratios between total α- and βhCG subunits. Samples were from 26 pregnant women at 7-12 weeks gestation. Comparisons between coelomic fluid and maternal serum were significant at $P<0.05$ using Kruskal-Wallis rank test for each variable; except for hCG (NS).

Parameter	Coelomic fluid	Maternal serum
hCG mIU/ml (IRP 75/537)		
Median	103515	90528
Interquartile range	90857-133250	67641-142580
Free αhCG mIU/ml (IRP 75/569)		
Median	12750	68.5
Interquartile range	10450-15650	53.6-98.6
Free βhCG mIU/ml (IRP 75/551)		
Median	1709	50.1
Interquartile range	1370-2045	33.3-90.2
Total α subunit mIU/ml (IRP 75/537)		
Median	1121760	98900
Interquartile range	894116-1334300	73001-146978
Total β subunit mIU/ml (IRP 75/537)		
Median	172002	93007
Interquartile range	155847-235847	70497-147985
α- to βhCG ratio		
Median	5.789	1.035
Interquartile range	4.759-7.187	1.007-1.073
Free αhCG (% of total)		
Median	89.9	5.45
Interquartile range	87.8-91.6	3.77-7.97
Free βhCG (% of total)		
Median	41.9	2.33
Interquartile range	37.8-46.3	1.75-3.38

Total amounts of each subunit were determined in each sample as the sum of hCG level and of subunit level expressed in the same IU system; as previously described (Nagy et al., 1994a). Briefly, conversion factors between the IU systems used for the α and β subunits and that used for the heterodimer were assessed using given amounts of intact and thermally dissociated purified hCG preparation. Using this method, we determined that the number of molecules of the hCG heterodimer contained in 1 IU (IRP 75/537) hCG corresponded to the same number of molecules of α and β subunits present in 0.0128 IU (IRP 75/569) α hCG subunit and 0.0215 IU (IRP 75/551) βhCG subunit, respectively (Nagy et al., 1994a).

Statistical Analysis

Regression analyses were calculated using individual values. Medians were expressed together with their interquartile range. A Kruskal-Wallis rank test was used to compare hormonal variables from the maternal serum and the coelomic fluid. Differences were considered statistically significant at $P < 0.05$.

RESULTS

Median level of hCG were similar in coelomic fluid and in maternal serum collected from the same women. By contrast, its free subunits were higher in coelomic fluid than in maternal serum: 186-fold for free αhCG subunit and 34-fold for free βhCG subunit (Table 1). There was no correlation for hCG or for its free subunits between coelomic fluid and maternal serum.

When hCG and its subunits were all expressed in the same IU system (IRP 75/537), free αhCG subunit was, on a molar basis, the most abundant form in coelomic fluid. Free αhCG subunit represented 90% of the total (free + combined) amount of αhCG subunit present in coelomic fluid but only 5.5% in maternal serum (Table 1). Free βhCG correlated with hCG in coelomic fluid ($r=0.725$, $p<0.001$) and in maternal serum ($r=0.687$, $p<0.001$).

Free βhCG subunit in coelomic fluid represented 42% of the total βhCG subunit . This last percentage was higher than that in maternal serum (Table 1). Free βhCG correlated with hCG in coelomic fluid (=0.457, $p<0.05$) and in maternal serum ($r=0.837$, $p<0.001$). The ratio between the amounts of total α- and total βhCG subunit was 5.8 in coelomic fluid but the subunits were equimolar in maternal serum (Table 1). There was a correlation between the total amounts of α- and βhCG subunits in coelomic fluid ($r=0.575$, $p<0.001$) as well as in maternal serum ($r = 0.999$, $p<0.001$).

DISCUSSION

Levels of intact hCG were relatively similar in matched samples of maternal serum and coelomic fluid. By contrast, free subunits were significantly higher in samples of coelomic fluid than in serum collected from the same women: 186-fold for free αhCG subunit and 34-fold for free βhCG subunit (Table 1). These enormous gradients are likely to be related to differences in the clearance rates of hCG and its subunits between maternal compartment and fetal fluid. Wehman and Nisula (1979) reported that the clearance rate of free αhCG is at least 3-fold more rapid than that for free βhCG and about 10-20 times more rapid than that reported for intact hCG. Moreover, these clearance rates do not depend on gestational age because there is no difference between men and women (Wehman et al., 1984).

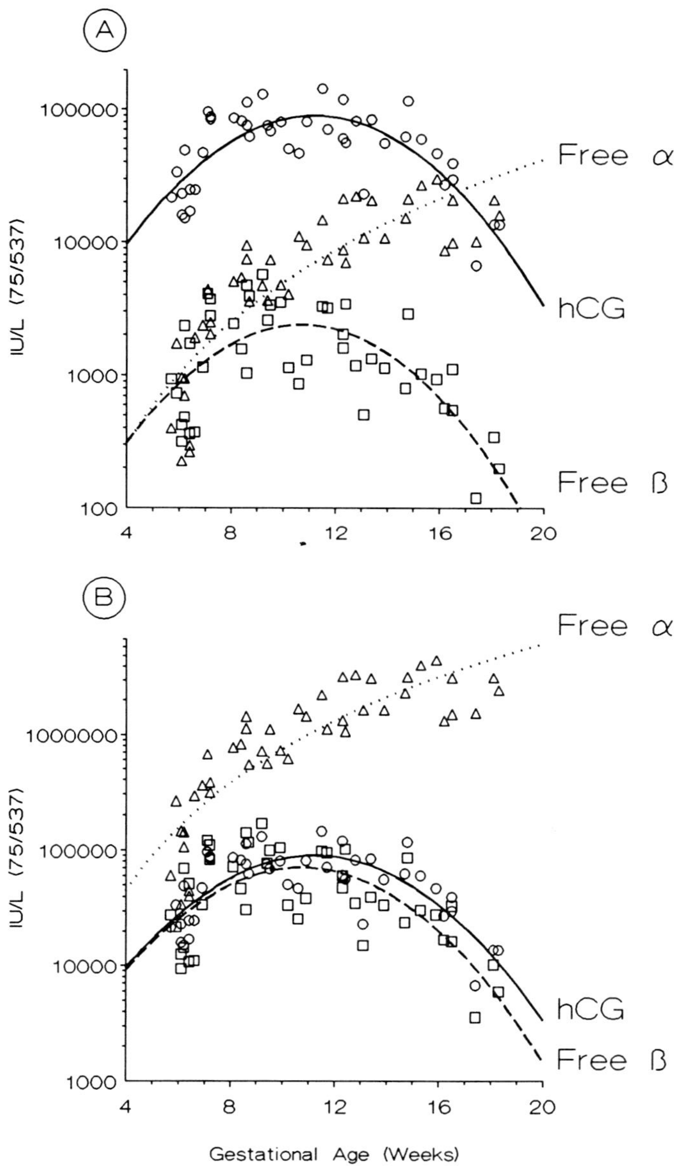

Figure 1. Circulating concentrations of hCG (O, solid line), free α- (Δ, dotted line) and βhCG (□, hatched line) subunits according to gestational age (Adaptation with permission from Meuris et al., *Hum. Reprod.* 10, 947-950, 1995). A) Circulating values of hCG and its subunits are expressed in the same International Unit System (IRP 75/537) after their conversion from their respective unit systems as described in Nagy et al. (*J. Endocrinol.* 140, 513-520, 1994). B) Circulating values of α and β subunits presented in A are corrected by x150 and x30, respectively, in order to reach the corresponding free to combined subunit molar ratio observed in the coelomic fluid.

Differences in levels of free subunits between coelomic fluid and maternal serum as well as the absence of any correlation for hCG or its free subunits between the two compartments strongly suggest independence between coelomic fluid and maternal serum. This may be explained by the anatomical relationships between the hCG secreting trophoblast and the two different compartments. HCG secreted into maternal circulation probably originates from the interface between trophoblast and the vascularized maternal decidua and from intervillous spaces though placental villi are bathed by a clear fluid apparently devoid of maternal red blood cells in early pregnancy (Hustin and Schaaps, 1987; Hustin, 1995). By contrast, hCG found in coelomic fluid is likely secreted from the internal layer of the trophoblastic shell directly into the extraembryonic coelom which is separated from the embryo by the amniotic cavity (Moore, 1982) in which little hCG is found (Wathen et al., 1991; Iles et al., 1992; Jauniaux et al., 1993). The coelomic cavity may therefore simply act as a "sump" for these products; as previously proposed (Wathen et al., 1991). Recent investigations have shown that the coelomic fluid contains high concentrations of specific placental and secondary yolk sac bioproducts, whereas their concentrations are comparatively low in samples of amniotic fluid and maternal serum (Jauniaux et al., 1991a, b; Wathen et al., 1991; Iles et al., 1992; Jauniaux et al., 1993, 1995). These gradients may be related to the absence of degradative process or, more likely, to very slow metabolic clearance rates in coelomic fluid. However, because the coelomic cavity is completely obliterated by the expanding amniotic cavity at the end of the first trimester, the biological significance of high levels of hCG and its subunits in coelomic fluid during early pregnancy remains unknown.

Considering coelomic fluid as a metabolic cul-de-sac into which hCG and its subunits accumulate and are slowly metabolized, their levels in this fluid may be considered as a more direct reflection of their trophoblastic production in early pregnancy than those measured in maternal serum.

Using the heat dissociation method for hCG to convert, on a molar basis, the International Unit systems used for the α- and βhCG subunits into that used for the heterodimer (Nagy et al., 1994a), our present data confirm the excess of total αhCG over βhCG subunits in coelomic fluid (Nagy et al., 1994b). This non equimolar ratio also confirms the much higher secretion on a molar basis of free than combined αhCG by tissue (Ruddon et al., 1981) and cell cultures (Kato and Braunstein, 1990) originating from first trimester placenta, as well as the excess of α- over βhCG subunits (Vaitukaitis, 1974) and of α- over βhCG mRNA (Boothby et al., 1983; Richards et al., 1995) in first trimester placental extracts. In addition, the enormous amount of hyperglycosylated free αhCG subunit in coelomic fluid which is not observed in maternal serum from early pregnancy (Nagy et al., 1994b; Blithe, 1995) suggests that coelomic fluid can be considered as a compartment where levels of hCG and its subunits reflect their secretion by the surrounding trophoblast.

It is therefore tempting to speculate that larger amounts of free α- and βhCG subunits are produced by the placenta from the onset of pregnancy than those inferred from their observed circulating levels (Figure 1A). When considering their respective concentrations in the coelomic fluid, we can easily adjust the circulating levels of the α (x150) and βhCG (x30) subunits in order to reach a similar molar ratio of free to combined subunit (Figure 1B). From this speculative figure, we clearly observed that the amount of free α subunit is in formidable excess when compared to free βhCG subunit and intact subunit (Figure 1B). From this speculative figure, we clearly observed that the amount of free α subunit is in formidable excess when compared to free βhCG subunit and intact hCG. Such an excess in α- to βhCG production by the trophoblast may explain why circulating levels of free α subunit regularly increase with gestational age when βhCG subunit and intact hCG levels decline. In other terms, we can deduce from this graph that only a small fraction (<10%) of the total amount of α subunit produced by the trophoblast is combined with a very large fraction (>90%) of ·hCG.

This proposal along with the above cited data are consistent with the early hypothesis that the control of glycoprotein hormone secretion (hCG, luteinizing hormone, follicle stimulating hormone and thyroid stimulating hormone) by the placenta and pituitary resides in the control of the production of the β subunits (Franchimont et al., 1972; Vaitukaitis et al., 1976). Finally, the close temporal relationship between the Doppler advent of intervillous maternal blood flow and the hCG peak (Meuris et al., 1995) suggests that the establishment of continuous intervillous blood flow is associated with an unexplained decline in β hCG subunit production and, as a consequence, with declining circulating hCG levels.

SUMMARY

Human chorionic gonadotrophin (hCG) is composed of two non-covalently bound α and β subunits synthesized from separate mRNAs. The hCG heterodimer and uncombined subunits are secreted during early gestation by trophoblast into the maternal bloodstream and into the exocoelomic cavity on the opposite side of the trophoblastic layer. The aim was to compare the relative amount of hCG and its free subunits in these compartments.

Levels of hCG were similar in coelomic fluid and in maternal serum collected from the same women. By contrast, levels of free subunits were higher in coelomic fluid than in maternal serum: 186-fold for free αhCG subunit and 34-fold for free βhCG subunit. These enormous gradients are likely to be related to differences in the clearance rates of hCG and its subunits between maternal and exocoelomic compartments. Considering coelomic fluid as a metabolic cul-de-sac into which hCG and its subunits accumulate and are slowly metabolized, their levels in this fluid may be reasonably considered as a direct reflection of their trophoblastic production. This hypothesis suggests that the amount of free α subunit is in formidable excess when compared to intact hCG and free βhCG subunit and that only a small fraction (<10%) of the total amount of α subunit produced by the trophoblast is combined with >90% of βhCG.

ACKNOWLEDGEMENTS

This work was granted by the National Fund for Medical Research (Belgium, FRSM grant Nr 3.4531.93) from which S.M. and P.L. are Senior Research Associate. A.-M. N. was supported by the Université Libre de Bruxelles, the Alice and David Van Buuren Foundation and the 'Institut pour l'Encouragement de la Recherche Scientifique dans l'Industrie et l'Agriculture'.

REFERENCES

Blithe, D.L. (1990) N-linked oligosaccharides on free α interfer with its ability to combine with human chorionic gonadotropin--βsubunit. *J. Biol. Chem.* 265, 21951-21956.

Blithe, D.L. and Iles, R.K. (1995) The role of glycosylation in regulating the glycoprotein hormone free α-subunit and free β-subunit combination in the extraembryonic coelomic fluid of early pregnancy. *Endocrinology* 136, 903-910.

Blithe, D.L. and Nisula, B.C. (1985) Variations in the oligosaccharides on free and combined α-subunits of human chorionic gonadotropin in pregnancy. *Endocrinology* 117, 2218-2228.

Boothby, M., Kukowska, J. and Boime, I. (1983) Imbalanced synthesis of human chorionic gonadotrophin α and β subunits reflects the steady state levels of the corresponding mRNAs. *J. Biol. Chem.* 258, 9250-9253.

Cole, L.A., Hartle, R.J., Laferla, J.J. and Ruddon, R.W. (1983) Detection of the free beta subunit of human chorionic gonadotropin (hCG) in cultures of normal and malignant trophoblast cells, pregnancy sera and sera of patients with choriocarcinoma. *Endocrinology* 113, 1176-1178.

Cole, L.A., Kroll, T.G., Ruddon, R.W. and Hussa, R.O. (1984) Differential occurrence of free beta and free alpha subunits of human chorionic gonadotropin (hCG) in pregnancy sera. *J. Clin. Endocrinol. Metab.* 58, 1200-1202.

Franchimont, P., Gaspard, U., Reuter, A. and Hennen, G. (1972) Polymorphism of protein and polypeptide hormones. *Clin. Endocrinol.* 1, 315-336.

Hay, D.L. (1985) Discordant and variable production of human chorionic gonadotropin and its free α- and β-subunits in early pregnancy. *J. Clin. Endocrinol. Metab.* 61, 1195-1200.

Hay, D.L. (1988) Placental histology and the production of human choriogonadotrophin and its subunits in pregnancy. *Br. J. Obstet. Gynaecol.* 95, 1268-1275.

Hustin, J. and Schaaps, J.P. (1987) Echographic and anatomic studies of the maternotrophoblastic border during the first trimester of pregnancy. *Am. J. Obstet. Gynecol.* 157, 162-168.

Hustin, J. (1995) Blood flow in the intervillous space in the first trimester. *Placenta* 16, 659-660.

Iles, R.K., Wathen, N.C., Campbell, D.J. and Chard, T. (1992) Human chorionic gonadotrophin and subunit composition of maternal serum and coelomic and amniotic fluids in the first trimester of pregnancy. *J. Endocrinol.* 135, 563-569.

Jauniaux, E., Jurkovic, D., Gulbis, B., Gervy, C., Ooms, H.A. and Campbell, S. (1991a) Biochemical composition of exocoelomic fluid in early human pregnancy. *Obstet. Gynecol.* 78, 1124-1128.

Jauniaux, E., Jurkovic, D., Henriet, Y., Rodesch, F. and Hustin, J. (1991b) Development of the secondary human yolk sac: correlation of sonographic and anatomic features. *Hum. Reprod.* 6, 1160-1166.

Jauniaux, E., Jurkovic, D. and Campbell, S. (1991c) In vivo investigations of anatomy and physiology of early human placental circulations. *Ultrasound Obstet. Gynaecol.* 1, 435-445.

Jauniaux, E., Gulbis, B., Jurkovic, D., Schaaps, J.P., Campbell, S. and Meuris, S. (1993) Protein and steroid levels in embryonic cavities in early human pregnancy. *Hum. Reprod.* 8, 782-787.

Jauniaux, E., Gulbis, B., Nagy, A.-M., Jurkovic, D., Campbell, S. and Meuris, S. (1995) Coelomic fluid chorionic gonadotrophin and protein concentrations in normal and complicated first trimester human pregnancies. *Hum. Reprod* 10, 214-220.

Kato, Y. and Braunstein, G.D. (1990) Purified first and third trimester placental trophoblasts differ in "in vitro" hormone secretion. *J. Clin. Endocrinol. Metab.* 70, 1187-1192.

Meuris, S., Nagy, A.-M., Delogne-Desnoeck, J., Jurkovic, D. and Jauniaux, E. (1995) Temporal relationship between the human chorionic gonadotrophin peak and the establishment of intervillous blood flow in early pregnancy. *Hum. Reprod* 10, 947-950.

Moore, K.L. (1982) *The Developing Human: Clinically Oriented Embryology*. Saunders, Philadelphia.

Morgan, F.J., Canfield, R.E., Vaitukaitis, J.L. and Ross, G.T. (1974) Properties of the subunits of human chorionic gonadotropin. *Endocrinology* 94, 1601-1606.

Nagy, A.-M., Glinoer, D., Picelli, G., Delogne-Desnoeck, J., Fleury, B., Courte, C., Kaufman, J.-M., Robyn, C. and Meuris, S. (1994a) Total amounts of circulating human chorionic gonadotrophin α and β subunits can be assessed throughout pregnancy using immunoradiometric assays calibrated with intact and thermally dissociated heterodimer. *J. Endocrinol.* 140, 513-520.

Nagy, A.-M., Jauniaux, E., Jurkovic, D. and Meuris S. (1994b) Placental production of human chorionic gonadotrophin α and β subunits in early pregnancy as evidenced in fluid from the exocoelomic cavity. *J. Endocrinol.* 142, 511-516.

Ozturk, M., Bellet, D., Manil, L., Hennen, G., Frydman, R. and Wands, J. (1987) Physiological studies of human chorionic gonadotropin (hCG), αhCG, and βhCG as measured by specific monoclonal immunoradiometric assays. *Endocrinology* 120, 549-558.

Peters, B.P., Krzesicki, R.F., Hartle, R.J., Perini, F. and Ruddon, R.W. (1984) A kinetic comparison of the processing and secretion of the αβ·dimer and the uncombined α and β subunits of chorionic gonadotropin synthesized by human choriocarcinoma cells. *J. Biol. Chem.* 259, 15123-15130.

Richards, R.G., Hartman, S.M. and Handwerger, S. (1994) Human cytotrophoblast cells in maternal serum progress to a differentiated syncytial phenotype expressing both human chorionic gonadotropin and human placental lactogen. *Endocrinology* 135, 321-329.

Ruddon, R.W., Hartle, R.J., Peters, B.P., Anderson, C., Huot, R.I. and Stromberg, K. (1981) Biosynthesis and secretion of chorionic gonadotropin subunits by organ cultures of first trimester human placenta. *J. Biol. Chem.* 256, 11389-11392.

Vaitukaitis, J. (1974) Changing placental concentrations of human chorionic gonadotropin and its subunits during gestation. *J. Clin. Endocrinol. Metab.* 38, 755-760.

Vaitukaitis, J.L., Ross, G.T., Braunstein, G.D. and Rayford, P.L. (1976) Gonadotropins and their subunits: Basic and clinical studies. *Rec. Prog. Horm. Res.* 32, 389-391.

Wathen, N.C., Cass, P.L., Kitau, M.J. and Chard, T. (1991) Human chorionic gonadotrophin and α-fetoprotein levels in matched samples of amniotic fluid, extraembryonic coelomic fluid, and maternal serum in the first trimester of pregnancy. *Pren. Diag.* 11, 145-151.

Wehmann, R.E. and Nisula, B.C. (1979) Metabolic clearance rates of the subunits of human chorionic gonadotropin in man. J. *Clin. Endocrinol. Metab.* 48, 753-759.

Wehman, R.E., Amr, S., Rosa, C. and Nisula, B.C. (1984) Metabolism, distribution and excretion of purified human chorionic gonadotropin and its subunits in man. *Annales d'Endocrinologie* (Paris) 45, 291-295.

Trophoblast Research 10:363-375, 1997

PROTEINS AND HORMONES OF THE PLACENTA AND EMBRYO: ADVANCES IN BIOCHEMICAL SCREENING FOR DOWN'S SYNDROME AND OTHER ANEUPLOIDIES IN THE FIRST TRIMESTER

-A Review-

J. Gedis Grudzinskas, Karen J. Powell and Pasquale Berlingieri

Academic Unit of Obstetrics and Gynaecology
Royal London Hospital
London, United Kingdom

INTRODUCTION

Down's syndrome occurs in Britain with a frequency of 1.2 to 1.3 per 1000 live births (Cuckle et al., 1991a), and in other countries with substantially similar frequencies dependent upon the age of reproducing women and the use of selective abortion of affected fetuses. It is the single most common cause of severe mental retardation. In addition, affected individuals show an excess of physical illnesses, such as leukemia and of malformations in the cardiac system (Fabia and Drolette, 1970), gastrointestinal tract and of the eyes and ears (Oster et al., 1975). Birth of a baby with Down's syndrome places enormous burdens upon the parents and immediate family. These burdens are not only emotional, physical and social, but also financial. In Britain the excess economic cost to society of each child with Down's syndrome, over and above the cost of a "normal" child has been estimated at £90,000 (Gill et al., 1987). Because of the severity and prevalence of the condition, it is considered acceptable and appropriate in most western developed countries to offer screening and diagnostic tests to pregnant women, followed by the facilities for selective abortion of affected fetuses. We review here placental and embryonic proteins and hormones which are being evaluated in screening for Down's syndrome and other aneuploidies in the first trimester.

SCREENING FOR DOWN'S SYNDROME

All the techniques for antenatal diagnosis of Down's syndrome currently available carry a risk of miscarriage being 1% for amniocentesis and a littler higher for chorionic villous sampling (CVS) (Rodeck and Johnson, 1995). Because of this, and the cost and limited availability of karyotyping, the tests are usually only offered to women considered to be at elevated risk of conceiving an affected fetus. Screening stratagems to identify these women have been based on maternal age but now include serum markers and ultrasound in the second and more recently the first trimester.

Screening By Age

In 1909 it was first noticed that the mothers of children with Down's syndrome tended to be older than average (Shuttleworth, 1909). Later the relationship between maternal age and Down's syndrome was investigated using population data, looking first at five year intervals of maternal age (Penrose and Smith, 1966) and later by yearly intervals. The confirmation that the risk of having a baby with Down's syndrome increased with increasing maternal age led the British government in 1976 to recommend that all pregnant women over the age of forty years should be offered amniocentesis. Today almost all hospitals offer amniocentesis or CVS to pregnant women who conceive after 36 or 37 years of age. However, this sort of screening has proved to be of only limited use in reducing the number of chromosomally abnormal babies born. The first reason for this is that the older mothers form a relatively small (although increasing) proportion of the population of pregnant women. In fact, 70% of affected babies are born to women under 36 years, who are individually at relatively low risk. Secondly, the women pregnant later on in their reproductive life are often those who have experienced infertility, or are aware that their fertility is declining. Because of this, not all the women selected for invasive testing on grounds of age accept the amniocentesis, (only 59% of women eligible to have amniocentesis) in one London teaching hospital in 1986 having the procedure (Knott et al., 1986).

Serum Markers In The Second Trimester

From the mid 1970's, all pregnant women in Britain were offered antenatal screening for neural tube defects based on the measurement of maternal serum alpha-fetoprotein (AFP). The first suggestion that it might also be possible to screen for aneuploidy using serum markers came in 1984 with the chance observation of a mother with a very low serum AFP who was later found to be carrying a fetus with Edward syndrome. Down's syndrome was then found to be similarly associated (Merkatz et al., 1984).

Following on from this, other pregnancy associated proteins and hormones were also investigated, and links made with elevated levels of human chorionic gonadotrophin (hCG) (Chard et al., 1984), and its subunits, and with low levels of unconjugated estradiol (uE_3) (Canick et al., 1988). None of these markers has yet been found to be associated with age. It is therefore possible to take a woman's risk based on her age and multiply it by a modifier based on the levels of the marker to produce a more accurate estimation of risk. It has proven to be convenient and effective to express serum marker values as a multiple of the normal median (MOM) at the relevant gestational age. This approach has allowed for systematic differences between laboratories and permits comparisons between different gestational ages.

Wald and his colleagues have combined use of several markers and maternal age into their "triple test", using AFP, uE_3 and hCG (Wald et al., 1988), claiming detection rates for this screening test of 67% for a false positive rate of 5%. This, and other similar combination biochemical screening tests have been widely used in hospitals throughout Britain. They have been widely acceptable to the population, with uptakes of 70 to 80% where they are offered, and with 75% of those found to be screen positive accepting the offer of diagnostic testing by amniocentesis (Wald et al., 1992). Very recently, the triple test has been further refined by replacing total hCG by free α and free β hCG subunit measurements, making it a quadruple test. Detection rates of 72% for a 5% false positive rate are claimed for this (Wald et al., 1994).

Although the use of second trimester screening for Down's syndrome represents a significant advantage over the use of age alone there remain some disadvantages. These tests are offered at 16 to 20 weeks of pregnancy, and positive results must be followed up with karyotyping. The implication is that a woman is well advanced in the second trimester before she knows that she is carrying a fetus affected by Down's syndrome, and if she opts for a termination of the abnormal pregnancy will face greater medical risks and possibly greater distress and long term psychological sequelae than if the procedure had been carried out earlier.

There is actually no very good reason why serum screening for Down's syndrome has to be at 16 weeks, other than the fact that the first observations of differential serum levels of pregnancy hormones in chromosomally normal and abnormal pregnancies arose from the biochemical tests used for neural tube defects. Now that screening for spina bifida and anencephaly has been largely replaced by the high resolution ultrasound scans available in almost all obstetric units, we do not need to be tied to 16 weeks. It is becoming clear that serum screening for Down's syndrome is possible and appropriate in the first trimester of pregnancy. In contrast, it is well known that chromosomally abnormal pregnancies are selectively miscarried. It has been calculated that at birth the maternal age-specific risk of trisomy 21 is 33% lower than at 15-20 weeks, and 54% lower than at 9-14 weeks (Snijders et al., 1994). This must be remembered when the performances of first and second trimester screens are compared.

Serum Markers In The First Trimester

Various stratagems have been tried to extend serum screening back into earlier weeks of pregnancy. Some groups have attempted merely using the same parameters as applied in the second trimester, but testing serum samples obtained progressively earlier. This does work to some extent, but results in losses of sensitivity and specificity. Others have chosen the same markers as those used in the second trimester, but have redefined the normal and abnormal ranges. Still others have looked at different proteins and hormones to find new first trimester markers. We will look at the various markers which have been proposed, comparing their efficiency in the first and second trimesters.

Pregnancy Associated Plasma Protein (PAPP-A)

Current observations suggest that PAPP-A is the single best biochemical marker for the detection of Down's syndrome in the first trimester (El Farra and Grudzinskas, 1995). PAPP-A is a pregnancy specific glycoprotein produced by trophoblast, and is detectable in maternal serum from 28 days after conception. A number of retrospective studies have defined a relationship between low levels of PAPP-A in the first trimester and Down's syndrome and other trisomies (Brambati et al., 1993 a, b; 1994; Wald et al., 1992; Spencer et al., 1992; Aitken et al., 1993). This difference is greatest at earliest gestations, and reduces rapidly towards the second trimester. Studies of levels in women attending for CVS have found PAPP-A in affected pregnancies at 6 to 11 weeks to be 0.27 MOM of levels in normal pregnancies (Brambati et al., 1993 a), and at 8 to 12 weeks to be 0.31 MOM. By the second trimester the difference is no longer evident (Aitken et al., 1993). On the basis of these results it has been suggested that using first trimester measurements of maternal serum PAPP-A and maternal age to define the 5% of women at highest risk of Down's syndrome, and offering karyotyping to this 5%, one could detect 71% of cases of Down's syndrome.

The studies carried out on PAPP-A so far have all been performed on small numbers of cases. Various groups have taken women at quite different gestations within the first trimester. Results have often been analyzed using large increments of gestation, with pregnancies dated to the nearest completed week, sometimes without ultrasonic confirmation of dates. The rapid changes in PAPP-A levels in the early weeks of pregnancy, and the changes in the relationship of levels between normal and abnormal pregnancies have made results from different studies very difficult to compare. Furthermore, a variety of PAPP-A assays have been used, and these have not been equally good at differentiating between normal and abnormal pregnancies. Because it is probable that PAPP-A will form an important component of any future first trimester serum screen for Down's syndrome, further investigations are warranted to develop the most effective assay, and using this, to measure PAPP-A in affected and normal pregnancies whose gestations have been carefully defined and confirmed using the crown rump length. A commercial assay is now available which will make it possible to resolve the outstanding issues.

Alpha Feto-Protein (AFP)

AFP is a protein synthesized by the fetal liver and yolk sac, which is believed to serve an oncotic function in the fetus similar to that of albumin in the adult. As already observed, maternal serum levels of AFP (MSAFP) in the second trimester tend to be lower in pregnancies affected by Down's syndrome. The median level in an affected pregnancy is 0.75 multiples of the median (MOM) of normal pregnancies. In the first trimester, the situation is very similar: a number of studies report MSAFP in Down's syndrome as 0.7 to 0.8 MOM of normal pregnancies (Brambati et al., 1986; Crandall et al., 1993). It would therefore be reasonable to expect MSAFP to perform a role in first trimester serum screening tests very much like its well established role in second trimester tests.

Human Chorionic Gonadotrophin (hCG)

HCG is a glycoprotein produced by trophoblast. It is composed of alpha and beta subunits which combine together before secretion but are actually synthesized separately. Most hCG circulates as the dimeric form, but 0.3 to 4% exists as free β hCG. In the second trimester, total MShCG is elevated in association with Down's syndrome (Bogart et al., 1987), and the small proportion circulating as free beta hCG is elevated by a greater extent.(Macri et al., 1990; Crossley et al., 1991; Spencer, 1991a; Ryall et al., 1992). Maternal serum free beta hCG is the single most efficient marker of Down's syndrome in the second trimester, and is used in almost all second trimester screening tests.

Unfortunately, levels of total hCG in chromosomally abnormal pregnancies in the first trimester are very similar to those in normal pregnancies, with a medians of around 0.97 MOM being quoted by several groups (Niebolo et al., 1990; Johnson et al., 1991; van Lith, 1992; Spencer et al., 1993; Aitken et al., 1993; Crandall et al., 1993; Macri et al., 1993). Nor is free beta hCG as useful as in the second trimester, apparently being elevated to only 1.9 MOM for normal pregnancies (Spencer et al., 1992). It is possible that free beta hCG could be used in future early screening tests for Down's syndrome, but it will not have the same power as in the second trimester, and may need to be used in combination with other markers.

Recently, interest has focused on measurements of urinary beta-core hCG. This is a breakdown product of hCG which is present in very small amounts in the serum (less than 0.2% of total hCG) but much larger proportions in urine (Cole et al., 1993). Cuckle and colleagues have found very high levels of urinary beta core-hCG in the second trimester in 24 pregnancies affected by Down's syndrome, with MOM greater than 2 compared to normal pregnancies (Cuckle et al., 1995). At present, these results await confirmation, and the conclusions have been contested.

Estriol

Estriol is synthesized by the syncytiotrophoblast from fetal precursors. In the maternal circulation approximately 90% is conjugated with glucuronide or sulphate. Low levels of estriol are found in Down's syndrome throughout pregnancy. This relationship was first observed as a reduction in levels of urinary estriol in pregnancies being screened for intrauterine growth retardation (Jorgensen et al., 1972) and was later confirmed to persist when serum levels were measured (Canick et al., 1988), and at earlier gestations.

In the first and second trimesters, maternal serum levels of unconjugated estriol (uE_3) are depressed to around 0.74 MOM of the levels in normal pregnancies. Maternal serum levels of uE_3 are used in most currently available second trimester biochemical screening tests for Down's syndrome, although there is continual debate about its utility. Concerns exist because be significantly correlated with AFP (Ryall et al., 1992; Davies et al., 1991) and because imprecision in assays for uE_3 may lead to imprecision in the final estimation of risk (Reynolds et al., 1992).

Pregnancy Specific Beta-1-Glycoprotein (SP1)

This is another pregnancy specific glycoprotein, and is detectable from the tenth day after conception. In the first trimester, levels of SP1 in affected pregnancies have been variously estimated as being between 0.79 and 0.4 MOM of normal (Brock et al., 1990; Macintosh et al., 1993). In the first study, serum was obtained relatively late (only one sample taken before 9 weeks) and the diagnosis made at term. In the second, the blood was taken much earlier (13 of the 14 samples reported were obtained before 9 weeks) and the diagnosis made by CVS. It cannot be known if the very different results obtained are due to the different gestation at which the measurements were made, or if some of the pregnancies in the second study had low levels because they were destined to miscarry. In the second trimester, high levels of SPI have been described in relation to Down's syndrome, but the findings are not consistent (Spencer et al., 1991b).

Inhibin

Comparison of levels of inhibin had, until recently, failed to show any difference between chromosomally normal and abnormal pregnancies However, four papers have now been published describing case control studies where levels of dimeric inhibin have been found to be elevated to between 1.59 and 2.6 MOM in affected pregnancies, both in the second trimester (Canick et al., 1994; Wallace et al., 1995a; Cuckle et al., 1995) and at 11 to 13 weeks (Wallace et al., 1995b). These results would suggest that addition of dimeric inhibin to existing serum screening tests should result in increased sensitivity.

Combinations Of Markers In The First Trimester

If a first trimester serum screen for Down's syndrome is eventually introduced, it is to be expected that, like the current second trimester tests, it will use multiple markers to increase its sensitivity. Such combination tests may be very efficient. It has been suggested that for a fixed 5% false positive rate, one could detect 22% of cases using maternal age alone, as compared with 71% by using maternal age and serum levels of PAPP-A, 33% by using age and free beta hCG, or 78-90% by using all three together (Brambati et al., 1994).

ULTRASOUND SCREENING FOR DOWN'S SYNDROME

Anomaly Scans

More than 95% of neonates with Down's syndrome are diagnosed by morphological criteria. Many have physical malformations or anomalies which, at least in theory, should allow for antenatal diagnosis by a detailed ultrasonographic survey of anatomy. These include brachycephaly, ventriculomegaly, cystic hygromata and nuchal edema, cardiac defects, duodenal atresia, hyperechogenic bowel, omphalocele, hydronephrosis, short femur, absent or anomalous middle phalanx of the little finger, wide gaps between the great and next toes, and intrauterine growth retardation. (Bronshtein and Blumenfeld, 1994). It is claimed that up to 80% of fetuses with Down's syndrome are potentially detectable by ultrasound scans based on the presence of one or more of these markers.

Nuchal Translucency

At 11 to 14 weeks of pregnancy an ultrasound scan performed can detect a small amount of fluid giving the appearance of an area of translucency behind the neck of the fetus, indicating an increased risk of fetal aneuploidy, cardiac defects or viral infection (Nicolaides et al., 1992). In a self selected group of volunteers attending for screening in a research setting, it has been found that increased nuchal translucency of 3 mm or greater is found in 86% of fetuses with trisomies 21, 18 or 13, and in 4.5% of chromosomally normal fetuses, independent of maternal age. (Nicolaides et al., 1994).

RECENT DEVELOPMENTS IN PRENATAL DIAGNOSIS

Detection Of Fetal Cells In Maternal Blood Or Cervical Secretions

The detection of XY metaphases in the circulation of women carrying male fetuses, suggest that it might be possible to provide antenatal diagnosis of aneuploidy in this way (Walkonowska et al., 1969). Various groups have looked in maternal blood for circulating fetal nucleated erythrocytes (Price et al., 1991; Simpson and Elias, 1992), or leukocytes (Iverson et al., 1979), or trophoblast cells (Covone et al., 1984) and in cervical secretions (Kingdom et al., 1995). The cells are separated out, usually using flow cytometry, on the basis of differential weight or labeling, and the chromosomal complement is analyzed using FISH analysis. Difficulties in obtaining sufficient numbers of a sufficiently pure population of fetal cells, combined with the limitations of fluorescence *in situ* hybridization (FISH) itself, have limited the success of these techniques so far (Elias and Simpson, 1994).

Proteins Of The Embryo Or Fetus

Two apparently hitherto unidentified proteins of fetal origin found in human amniotic fluid were first described by Fay and colleagues in 1988 and referred to as fetal antigens I (FA-I) and II (FA-II).

The fetal origin of FA1 was suggested in accordance with the quantitative distribution in fetal and maternal compartments and immunohistochemical analysis (Fay et al., 1988; Tornehave et al., 1989). High concentrations of FA I were found in second trimester amniotic fluid and in fetal serum, whereas the concentration in normal human serum was found to be approximately 20 ng/ml, which is 1000 times less than in second trimester fetal serum (Jensen, 1992). In preliminary tissue localization studies using the indirect immunoperoxidase technique on fetal tissues (week 7), FA I immunoreactivity was detected in the hepatocytes of the fetal liver (Tornehave et al., 1989). However, recent observations also revealed the presence of FA I within the cytoplasm of nearly all glandular cells of the first trimester anlage of the fetal pancreas (Tornehave et al., 1993). As the development of the fetal pancreas progresses, the relative number of FA 1 - positive glandular cells decreases and postnatally the expression of FA1 is restricted to the insulin-producing β cells of the islets of Langerhans (Tornehave et al., 1993; Jensen et al., 1993). It is suggested that FA1 is synthesized as a membrane anchored protein and released into the circulation after enzyme cleavage, and that circulating FAI represents the post-translationally modified gene product of human dlk which, in turn, is identical to human adrenal-specific mRNA pG2 (Jensen et al.., 1994). This protein is a putative mammalian homeotic protein differentially expressed in small cell lung carcinoma and neuroendocrine tumor cell lines.

Fetal antigen 2 (FA-2) first identified in amniotic fluid (Fay et al., 1988; Tornehave et al., 1989) has been shown to be identical to the aminopropeptide of the $\alpha 1$ chain of collagen type 1 (Teisner et al., 1992). Biochemical characteristics has revealed several molecular forms of the protein which differ in both size (Fay et al., 1988; Rasmussen et al., 1992a) and charge (Price et al., 1994a).

An inhibition ELISA has been developed and used to demonstrate the presence of FA2 in normal serum (Rasmussen et al., 1992b). A radioimmunoassay has been described for FA2 (Price et al., 1994b) and an assay standard isolated (Price et al., 1994a). Further analysis of amniotic fluid FA2 has revealed more extensive heterogeneity, namely two major high molecular mass FA2 types and several low molecular mass forms (Price et al. 1995a). Levels were seen to rise steeply from 10 to 14 weeks, peak at 17 weeks and then fall slightly by 23 weeks. A comparison of amniotic fluid FA2 levels between 10 and 23 weeks gestation in normal pregnancies and pregnancies affected by trisomy, showed significantly higher FA2 levels in trisomy 21 and significantly lower levels in trisomy 18 (Price et al., 1995b).

CONCLUSION

The ideal method of detecting fetal Down's syndrome may be karyotyping fetal cells harvested from maternal blood, avoiding all risk to the pregnancy. However, until this becomes available we continue to rely on invasive karyotyping offered to women identified at increased risk, calculated from maternal age and the levels of serum markers or the presence of signs seen by ultrasound.

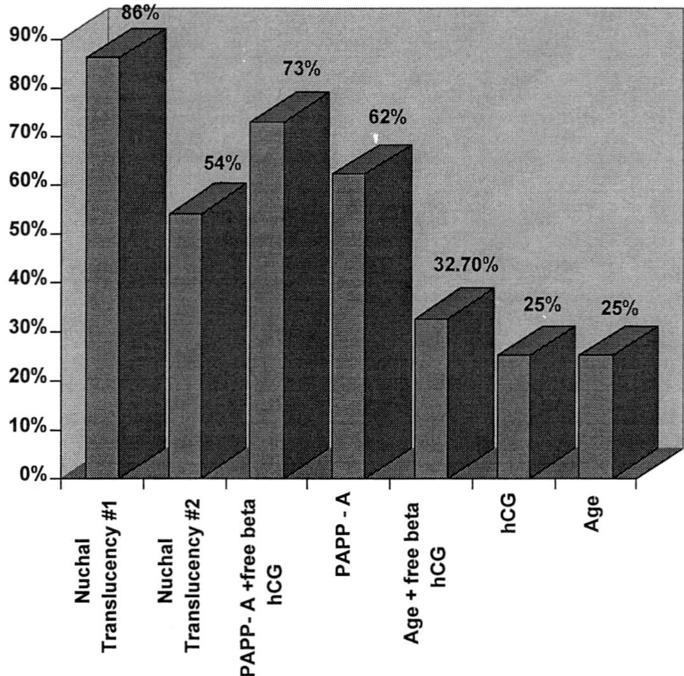

Figure 1. Detection rate of Down's syndrome using different markers in first trimester of pregnancy (El Farra and Grudzinskas, 1995). Nuchal translucency, Nicolaides et al., 1994; Nuchal translucency, Savoldelli et al., 1993; PAPP-A+ free beta-hCG, Brambati et al., 1994; PAPP-A, Brambati et al., 1993b; Age and free beta-hCG, Brambati et al., 1994; hCG, Van Lith, 1992; Age, Brizot et al., 1994.

There are now a variety of well established screening tests based on the levels of serum markers in the second trimester. However, these carry the substantial drawback that the diagnosis is made late.

A body of research now shows that first trimester screening for Down's syndrome should be possible, using nuchal translucency or the levels of serum markers including PAPP-A, or the two in combination (Figure 1) (El Farra and Grudzinskas, 1995). This may involve the use of resources to detect chromosomal abnormalities in pregnancies which were anyway doomed to miscarry, but there is a real demand from pregnant women for earlier screening tests, and this will drive antenatal screening for Down's syndrome into the first trimester in the future (Grudzinskas et al., 1997).

REFERENCES

Aitken D.A., McCaw. G. and Crossley, J.A. (1993) First trimester biochemical screening for fetal chromosome abnormalities and neural tube defects. *Pren. Diag.* 13, 68-69.

Bogart, M.H., Pandian, M.R. and Jones, O.W. (1987) Abnormal maternal serum chorionic gonadotrophin levels in pregnancies with fetal chromosome abnormalities. *Pren. Diag.* 7, 623-630.

Brambati, B., Simioni, G., Bonnachi, I. and Piceni, L. (1986). Fetal chromosomal aneuploidies and maternal serum alphafetoprotein levels in the first trimester. *Lancet* ii, 165-166.

Brambati, B., Macintosh, M.C.M., Teisner, B., Shrimanker, K., Lanzani, A., Maguiness, S., Bonacchi, I., Tului, L., Chard, T. and Grudzinskas, J.G.(1993a) Low maternal serum levels of pregnancy associated plasma protein A in the first trimester in association with abnormal karyotype. *Br. J. Obstet. Gynaecol.* 100, 324-326.

Brambati, B., Tului, L., Bonnachi, I., Shrimanker, K., Suzuki, Y. and Grudzinskas, J.G. (1993b) Serum PAPP-A and free beta hCG are first trimester screening markers for Down's syndrome. *Hum. Reprod.* 8 (Suppl. 1), 183 (Abstract).

Brambati, B., Tului, L., Bonnachi, I., Suzuki, Y., Shrimanker, K. and Grudzinskas, J.G. (1994) Biochemical screening for Down's syndrome in the first trimester. In: *Screening For Down's Syndrome*, (eds.) J.G. Grudzinskas, T. Chard, M. Chapman and T. Cuckle, Cambridge University Press:Cambridge, pp. 289-294.

Brock, D.J.H., Barron, L., Holloway, S., Liston, W.A., Hiller, S.G. and Seppala M. (1990) First trimester maternal serum biochemical indicators in Down's syndrome. *Pren. Diag.* 10, 245-251.

Cole, L.A., Kardana, A., Park, S-.Y. and Braunstein, G.D. (1993) The deactivation of hCG by nicking and dissociation. *J. Clin. Endocrinol. Metab.* 76, 704-710.

Covone, A.E., Johnson, P.M. and Mutton, D. (1984) Trophoblast cells in peripheral blood from pregnant women. *Lancet* I, 841-843.

Chard, T., Lowings, C. and Kitau, M.J. (1984) Alpha feto-protein and chorionic gonadotrophin in relation to Down's syndrome. *Lancet* 983, 750.

Canick, J., Knight, G., Palomaki, G., Haddow, J.E., Cuckle, H.S. and Wald N.J. (1988). Low second trimester serum unconjugated oestriol in pregnancies with Down's syndrome. *Br. J. Obstet. Gynaecol.* 95, 330-333.

Canick, J., Lambert-Messerlian, G.M., Palomaki G.E., Schneyer, A.L., Tumber, M.B., Knight, G.J. and Haddow, J.E. (1994) Maternal serum dimeric inhibin is elevated in Down's syndrome pregnancy. *Am. J. Hum. Genet.* 55(3), A9.

Crandall, B.F., Hansen, F.W., Hansen, F.W., Keener, M.S., Matsumo, B.S. and Miller, W. (1993) Maternal serum screening for alpha-fetoprotein, unconjugated oestriol and human chorionic gonadotrophin between 11 and 15 weeks of pregnancy to detect fetal chromosome abnormalities. *Am J. Obstet. Gynaecol.* 168, 1864-1869.

Crossley, J.A., Aitken, D.A. and Connor, J.M. (1991). Free beta human chorionic gonadotrophin and prenatal screening for chromosome abnormalities. *J. Med. Genet.* 28, 570.

Cuckle, H.C., Iles, R.K., Sehmi, I.K., Chard, T., Oakey, R.E., Davies, S. and Ind T. (1995) Urinary multiple marker screening for Down's syndrome. *Pren. Diag.* 15, 745-751.

Cuckle, H.S., Holding, S., Jones, R., Wallace, E.M. and Groome, N.P. (1995) Maternal serum dimeric inhibin A in second trimester Down's syndrome pregnancies. *Pren. Diag.* 15, 385-392.

Cuckle, H.S., Nanchahal, K. and Wald, N.J. (1991) Birth prevalence of Down's syndrome in England and Wales. *Pren. Diag.* 10, 643-651.

Davies, C.J., Selby, C. and Spencer, K. (1991) Multicentre retrospective clinical trial for the prenatal detection of Down's syndrome. *Clin. Chem.* 37, 943.

El Farra, K. and Grudzinskas, J.G. (1995) Will PAPP-A be a biochemical marker for screening of Down's syndrome? *Early Pregnancy: Biology and Medicine* I :47

Elias, S. and Simpson, J.L. (1994) Prenatal diagnosis using fetal cells isolated from maternal blood. In: *Screening For Down's Syndrome,* (eds.) J.G. Grudzinskas, T. Chard, M. Chapman and H.S. Cuckle, Cambridge University Press:Cambridge, pp. 193-210.

Fabia, J. and Drolette, M. (1970) Life tables up to age 10 for mongols with and without congenital heart defects. *J. Ment. Defic. Res.* 14, 235-242.

Fay, T., Jacobs, I., Teisner, B., Poulsen, O., Chapman, M., Stabile, I., Bohn, H., Westergaard, J.G. and Grudzinskas, J.G. (1988) Two fetal antigens (FAI and FA2) and endometrial proteins (PP12 and PP14) isolated from amniotic fluid. Preliminary observations in fetal and maternal tissues. *Eur. J. Gynaecol. Reprod. Biol.* 29, 73-85.

Gill, M., Murday, V. and Slack, J. (1987. An economic appraisal of screening for Down's syndrome in pregnancy using maternal age and serum alpha fetoprotein concentration. *Soc. Sci. Med.* 24, 725-731.

Grudzinskas, J.G. and Ward, R.H.J. (eds.) (1997) *Screening for Down's Syndrome in the First Trimester.* RCOG Press, London.

Iverson, G.M., Bianchi, D.W. and Cann, H.M (1981) Detection and isolation of fetal cells from maternal blood using the fluorescence-activated cell sorter (FACS). *Pren. Diag.* 1, 61-73.

Jensen, C.H., Krogh, T.N., Hoirup, P., Clausen, P.P., Skjodt, K., Larsson, L-I., Enghild, J.J. and Teisner, B. (1994) Protein structure of fetal antigen I (FAI). *Eur. J. Biochem.* 225, 83-92.

Johnson, A., Cowchock, F.S., Darby, M. and Jackson, L.G. (1991) First trimester maternal serum alpha-fetoprotein and human chorionic gonadotrophin in aneuploid pregnancies. *Pren. Diag.* 11, 377-380.

Jorgensen P.I. and Trolle D. (1972) Low urinary oestriol excretion during pregnancy in women giving birth to infants with Down's syndrome. *Lancet* ii, 782-784.

Kingdom, J., Sherlock, J., Rodeck, C.H. and Adinolfi, M. (1995) Selection of trophoblast cells in transcervical samples collected by ravage or cytobrush. *Obstet. Gynecol.* 86, 282-288.

Knott, P.D., Penketh, R.J.A. and Lucas, M.K. (1986) Uptake of amniocentesis in women aged 38 years or more by the time of the expected date of delivery: A two-year retrospective study. *Br. J. Obstet. Gynaecol.* 93, 1246-1250.

Macintosh, M.C.M., Brambati, B., Chard, T. and Grudzinskas, J.G. (1993) First trimester maternal serum Schwangerschafts protein I (SPI), in pregnancies associated with chromosomal abnormalities. *Pren. Diag.* 13, 567-568.

Macri, J.N., Kasturi, R.V. and Krantz, D.A. (1990). Maternal serum Down's syndrome screening: Free beta protein is a more effective marker than hCG. *Am. J. Obstet. Gynecol.* 163, 1248-1253.

Macri, J.N., Spencer, K, Aitken, D, Garver, K., Buchanan, P.D., Muller, F. and Boue, A. (1993) First trimester screening for Down's syndrome. *Pren. Diag.* 13, 557-562.

Merkatz, L., Nitowsk,i H., Macri, J. and Johnson, W.(1984) An association between low maternal serum alpha fetoprotein and fetal chromosomal abnormalities. *Am. J. Obstet. Gynecol.* 148, 886-894.

Nicolaides, K.H, Azar, G., Byrne, D., Mansur, C. and Marks, K (1992) Fetal nuchal translucency: Ultrasound screening for chromosomal defects in the first trimester of pregnancy. *Brit. Med. J.* 304, 867-869.

Nicolaides, K.H., Brizot, M.L. and Snijders, R.J.M. (1994) Fetal nuchal translucency: ultrasound screening for fetal trisomy in the first trimester of pregnancy. *Br. J. Obstet. Gynecol.* 101(9), 782-786.

Niebolo, L., Ozturk, M., Brambati, B., Miller, S.L., Wands, J. and Milunsky, A. (1990) First trimester maternal serum in alpha-fetoprotein and human chorionic gonadotrophin screening for chromosome defects. *Fetal Diag. Ther.* 7, 123-131.

Oster, J., Mikkelsen, M. and Nielsen, A. (1975) Mortality and life table in Down's Syndrome. *Acta Paed. Scand.* 64, 322-326.

Penrose, C.S. and Smith, G.F.(1966) *Down's Anomaly*, Churchill, London, p. 151.

Price, J., Elias, S. and Wachtel, S.S. (1991) Prenatal diagnosis using fetal cells isolated from maternal blood by multiparameter flow cytometry. *Am. J. Obstet. Gynecol.* 165, 1731-1737.

Price, K., Silman, R., Armstrong, P. and Grudzinskas, J.G (1995b) Abnormal amniotic fetal antigen 2 levels in trisomy 18 and 21. *Hum. Reprod.* 10,9, 2438-2440

Price, K.M., Silman, R. and Grudzinskas, J.G (1994a) Isolation of fetal antigen 2 assay standard. *Clin. Chim. Acta* 226, 83-88.

Price, K.M., Silman, R., Armstrong, P. and Grudzinskas, J.G. (1994b) Development of radioimmunoassay for fetal antigen 2. *Clin. Chim. Acta* 224, 95-102.

Price, K.M., Silman, R., Armstrong, P., Teisner, B. and Grudzinskas, J.G (1995) The typing of fetal antigen 2 in human amniotic fluid. *Clin. Chim. Acta* 236, 181- 194

Rassmussen, H.B., Teisner, B., Andersen, J.A., Yde-Andersen, E. and Leigh, I. (1992) Fetal antigen 2 (FA2) in relation to wound healing and fibroblast proliferation. *Brit. J. Dermatol.* 126, 148-153

Rassmussen, H.B., Teisner, B., Bangsgaard-Petersen, F., Yde-Andersen, E. and Kassem, M. (1992) Quantification of fetal antigen 2 (FA2) in supernatants of cultured osteoblasts, normal human serum, and serum from patients with chronic renal failure. *Nephrol. Dial. Transplant.* 7, 902-907.

Reynolds T. and John, R. (1992) A comparison of unconjugated oestriol assay kits shows that expression of the results as multiples of the mean causes unacceptable variation in calculated Down's syndrome risk factors. *Clin. Chem.* 38, 1888-1893.

Rodeck, C.H. and Johnson, P. (1995) Prenatal diagnosis. In: *Turnbull's Obstetrics*, (Eds.) G.V.P. Chamberlain, 2nd Edition, Churchill Livingstone, Edinburgh, pp. 253-282.

Ryall, R.G., Staples, A.J., Robertson, E.F. and Pollard, A.C. (1992) Improved performance in a prenatal screening programme for Down's syndrome incorporating serum free hCG subunit analysis. *Pren. Diag.* 12, 251-261.

Shuttleworth, G.E. (1909) Mongolian imbecility. *Br. Med. J.* 2, 661.

Simpson, J.L. and Elias, S. (1992) Isolating fetal erythroblasts from maternal blood with identification of fetal trisomy by fluorescent *in situ* hybridization. *Pren. Diag.* 12, S34.

Snijders, R.J.M., Holzgreve, W., Cuckle, H.S. and Nicolaides, K.H. (1994) Maternal age-specific risks for trisomies at 9-14 weeks gestation. *Pren. Diag.* 14, 543-552.

Spencer, K. (1991a). Evaluating an assay of the free beta subunit of choriogonadotrophin and its potential value in screening for Down's syndrome. *Clin. Chem.* 37, 809-814.

Spencer, K.(199lb) Pregnancy specific beta I glycoprotein in Down's syndrome screening: Does it have any value? Proc ACB Meeting, 1991. *Ann. Clin. Biochem.* C46, 108.

Spencer, K, Macri, J.N., Aitken, D. and Connor, J.M. (1992) Free beta hCG as a first trimester marker for fetal trisomy. *Lancet* 339, 1480.

Steele, M.W. and Bregg, W.R. Jr. (1966) Chromosome analysis of human amniotic fluid cells. *Lancet* i, 383-385.

Teisner, B., Rasmussen, H.B., Hojrup, P., Yde-Andersen, E. and Skjodt, K. (1992) Fetal antigen 2: Am amniotic protein identified as the aminopropeptide of the αl chain of human procollagen type I. *APMIS* 100, 1106-1114.

Van Lith, J.M.M. (1992) First trimester maternal serum human chorionic gonadotrophin as a marker for fetal chromosomal disorders. The Dutch working part on prenatal diagnosis. *Pren. Diag.* 12, 495-504.

Walknowska, J., Conte, F.A. and Grumbach, M.M. (1969) Practical and theroretical implications of fetal/maternal lymphocyte transfer. *Lancet* i, 1119-1122.

Wald, N.J., Cuckle, H.S., Densem, J.W., Nanchahal, K., Royston, P., Chard, T., Haddow, J.E., Knight, G.E., Palomaki, G.E. and Canick, J.A. (1988) Maternal serum screening for Down's syndrome in early pregnancy. *Br. Med. J.* 297, 883-887.

Wald, N.J., Wald, K. and Smith, D. (1992) The extent of Down's syndrome screening in Britain in 1991. *Lancet* 340, 494.

Wald, N.J., Densem, J.W., Smith, D. and Klee, G.G. (1994) Four-marker serum screening for Down's syndrome. *Pren. Diag.* 14, 707-716.

Wallace, E.M., Grant, V.E., Swanston, I.A., McNeilly, A.S., Blundell, G., Ashby, J.P. and Groome, N.P. (1995a) Second trimester screening for Down's syndrome using inhibin A. A. *J. Endoc.* 144 Suppl. p. 134.

Wallace, E.M., Grant, V.E., Swanston, I.A. and Groome, N.P. (1995b) Evaluation of maternal serum dimeric inhibin as a first trimester marker of Down's syndrome. *Pren. Diag.* 15, 359-362.

Trophoblast Research 10:377-391, 1997

TRANSFORMING GROWTH FACTOR β1 (TGFβ1) INHIBITS GAP JUNCTIONAL COMMUNICATION DURING HUMAN TROPHOBLAST DIFFERENTIATION

Laurent Cronier[1], Eliane Alsat[2], Jean-Claude Hervé[1], Jean Délèze[1] and André Malassiné[1,3]

[1]Laboratoire de Physiologie Cellulaire
CNRS URA 1869
40, Av du recteur Pineau
86022 Poitiers cedex, France

[2]INSERM U427
Faculté des sciences pharmaceutiques et biologiques de Paris
Université Paris V
4, Av. de l'observatoire
75270 Paris cedex 6, France

INTRODUCTION

The formation of the placenta is an important prerequisite for normal fetal development. In the human hemomonochorial placenta, the chorionic villi, in direct contact with the maternal blood, consist of trophoblast surrounding a core of connective tissue including fetal vessels, fibroblast cells and Hofbauer cells.

It is now well established that the villous trophoblast differentiates from the fusion of cytotrophoblastic cells into a syncytiotrophoblast, which expresses specific hormonal products (Hoshina et al., 1985; Aplin, 1991; Mason, 1993). This differentiation process is necessary for syncytiotrophoblast growth and also for regeneration of aging syncytiotrophoblast (Benirschke and Kaufmann, 1990). Therefore, the ability of trophoblast to differentiate is a major determinant of the structure and function of the chorionic villi.

The regulatory factors involved in villous trophoblast differentiation remain poorly understood. *In vitro* studies have shown that this phenomenon is modulated by interaction with extracellular matrix, soluble factors (mainly growth factors), and oxygen supply (Ong and Burton, 1990; Alsat et al., 1996). Thus, Epidermal Growth Factor (EGF), Colony Stimulating Factor (CSF), G-Macrophage-CSF (GM-CSF), human Chorionic Gonadotropin (hCG), and Leukemia Inhibitory Factor (LIF), all induce differentiation with concomitantly increased secretions of the syncytial hormones hCG and hCS. Based on the finding that TGFβ1 inhibits hCG and hCS production *in vitro*, Morrish et al. (1991) concluded that TGFβ1 inhibits trophoblast differentiation. Moreover, it was recently demonstrated that under hypoxia the formation of functional syncytiotrophoblast is impaired (Alsat et al., 1996).

[3]To Whom Correspondence Should Be Addressed: Laboratorie de Physiologie Cellulaire, CNRS URA 1869, Université de Poitiers, 40, Av du recteur Pineau, 86022 Poitiers cedex, France.

Gap junctions are clusters of transmembraneous channels composed of connexin proteins. It has been suggested that the exchange of molecules through gap junctions is involved in the control of cell and tissue differentiation. Ultrastructural studies have established the presence of gap junctions between the trophoblastic layers of hemochorial (Metz et al., 1976; De Virgiliis et al., 1982) and endotheliochorial placentae (Malassiné and Leiser, 1984). Moreover, in rat, different gap junction connexins are expressed during trophoblast invasion and placenta formation (Grümmer et al., 1995). Gap junctions were also reported to be present during trophoblastic cell fusion in the guinea pig placenta (Firth et al., 1980). Recently, it was demonstrated by means of the Fluorescence Recovery After Photobleaching method (gap-FRAP) that cell-to-cell communication through gap junctions precedes the formation of syncytiotrophoblast in culture (Cronier et al., 1994). This gap junctional communication (GJC) is required for trophoblast differentiation (Cronier et al., 1995). Moreover, trophoblastic GJC and the expression of connexin 43 are stimulated by hCG. This phenomenon is specific and involves the cAMP-Protein kinase A pathway (Cronier et al., 1997).

Transforming growth factor-β1 (TGFβ1) is composed of two identical subunits which are homologous to the inhibin/activin β-subunits. TGFβ1 is known as a multifunctional growth factor present in many cell types and can have either positive or negative effects on cell proliferation and differentiation (Sporn et al., 1986; Massague, 1987). It has been implicated in development and reproduction (Shull and Doetschman, 1994). The presence of mRNA for TGFβ1 and immunoreactive TGFβ1 has been demonstrated in human placenta throughout pregnancy (Kauma et al., 1990; Dungy et al., 1991). Moreover, TGFβ1 has been reported to be present at the human fetal/maternal interface, both in decidual and trophoblastic cells (Lysiak et al., 1995). In addition, receptors have been detected for TGFβ in the placenta (Mitchell and O'Connor-McCourt, 1991). Recently, the sites of TGFβ1 mRNA synthesis and immunoreactivity were localized in the cytoplasm of villous syncytiotrophoblast and extravillous trophoblast throughout gestation. Moreover, at term, TGFβ1 mRNA was also located in villous mesenchymal cells (Lysiak et al., 1995). In addition to its *in vivo* placental localization, this growth factor has been shown to modulate trophoblast differentiation. However, divergent results have been reported, either inhibition in term placenta (Morrish et al., 1991) or promotion of syncytial formation in first trimester and term placentae (Graham et al., 1992). Furthermore, the production of hCG is reported to be down regulated by TGFβ1 (Morrish et al., 1991; Matsusaki et al., 1992; Feinberg et al., 1994).

In this context, we initiated studies to determine the action of TGFβ1 on intercellular dye diffusion in cultured villous trophoblast of human term placenta, by means of the Fluorescence Recovery After Photobleaching (gap-FRAP) technique. In parallel, βhCG production, hCS expression, trophoblast differentiation, and desmoplakin immunostaining were assessed.

MATERIALS AND METHODS

Trophoblastic Cell Purification And Cell Culture

Human cytotrophoblastic cells were isolated from normal term placentae, obtained after uncomplicated cesarean section, using a method similar to that of Kliman et al. (1986). Briefly, after trypsin/DNase digestions followed by a Percoll gradient purification, the cell population was further purified by means of a monoclonal anti-

HLA-A, B, and C antibody (W_6-32HL, Sera Lab, Crawley Down, England). Cytotrophoblastic cells were diluted to a final concentration of 0.5×10^6 /ml in Minimum Essential Medium (MEM) containing 10% Fetal Calf Serum (FCS), 25 mM Glucose, 20 mM HEPES, 100 μg/ml Streptomycin, 100 IU/ml Penicillin, with or without the investigated factors. Cells were plated in 35 mm plastic dishes (Nunc, Nunclon). The medium was renewed every 24 hours for the duration of culture. These cell cultures have been previously characterized by transmission and scanning electron microscopy, immunohistochemistry for αhCG, βhCG, cytokeratin, and by βhCG and progesterone secretion (Malassiné et al., 1990).

Recombinant human TGFβ1 was purchased from Life Technologies Inc. (France). Purified hCG (3000 IU/mg; Pregnyl®) was purchased from Organon (Serifontaine, France). Rabbit anti-human Chorionic Gonadotropin from DAKO (Glostrup, Denmark) was dialyzed for 48 hour at 4°C in MEM before use.

Gap-FRAP Method

Communication among contacting trophoblastic cells in culture was determined by measuring the cell-to-cell diffusion of a fluorescent dye (Wade et al., 1986) using an interactive laser cytometer (ACAS 570, Meridian Instruments, Okemos, MI, USA). Briefly, the cells were internally loaded for 10 minutes at room temperature with the membrane permeant molecule 6-carboxyfluorescein diacetate (7 μg/ml in 0.25% DMSO). The highly fluorescent and membrane impermeant moiety 6-carboxyfluorescein is released and accumulates inside the cells. After washing off the excess extracellular fluorogenic ester to prevent further loading, a cell adjacent to other cells was selected, and its fluorescence was photobleached by strong laser pulses (488 nm). Digital images of the fluorescent emission excited by weak laser pulses were recorded at regular intervals (scanning period: 2 minutes before and after photobleaching) and stored for subsequent analysis. If the bleached cell is connected to its neighbors by permeable gap junctions, a fluorescence recovery occurs by diffusion of fluorescent molecules from adjacent unbleached cells.

GJC was analyzed two days after plating. Three different topologies of contacting cells were recognized, namely contiguous cytotrophoblastic cells, contiguous syncytiotrophoblast cells, and contiguous cyto- and syncytiotrophoblast. The percentage of coupled cells was analyzed whatever the topology of trophoblastic elements, during trophoblast differentiation and cell treatments.

Cell Staining

For desmoplakin immunostaining, cultured trophoblastic cells were rinsed with PBS, fixed, and permeabilized in methanol at -20°C for 25 minutes. A monoclonal antidesmoplakin antibody (Sigma), diluted at 1/400, was applied, followed by FITC-labelled goat antimouse immunoglobulin G (Sigma), diluted at 1/100, as described by Farmer and Nelson (1992).

Immunoblotting

Trophoblastic cells were washed twice with ice-cold PBS, scraped, and briefly centrifuged at 4°C. The cell pellet was extracted in RIPA buffer containing 150 mM NaCl,

1% Nonidet NP-40, 0.5% sodium deoxycholate, 0.1% sodium dodecyl sulfate, 50 mMTris-HCl (pH 8) and protease inhibitors, for 15 minutes at 4°C. After centrifugation, the supernatant was diluted with four-fold concentrated Laemmli sample buffer and boiled for 3 minutes. The samples were then aliquoted and stored at -20°C until use.

The indicated amounts of proteins were subjected to electrophoretic separation using 10% polyacrylamide-SDS gels and transferred to nitrocellulose membrane, as previously described (Alsat et al., 1993). Membranes were blocked by incubation in a saline solution (pH 7.4) containing 0.1% Tween 20 (PBS-T) and 5% non-fat dry milk, for 1 hour at room temperature. The blots were then incubated overnight at 4°C with a polyclonal antibody directed against human Placental Lactogen (hPL; 1/250, DAKO) diluted in PBS-T. The corresponding antigen was detected by a chemiluminescence system (Amersham, Les Ulis, France) after incubation with a peroxidase-coupled secondary antibody.

Densitometric analysis of hCS signal (22 kDa) was performed using an image-master-1D software from Pharmacia Biotech (Orsay, France).

Hormone Assay

hCG concentration was determined by a chemiluminescence immunoassay with a specific monoclonal antibody directed against the β subunit of hCG (Amerlite system, Amersham, Les Ulis, France) in cell culture media at various times. The values are the means ± SEM of triplicate determinations.

RESULTS

Differentiation

After adhesion and flattening, cells emitted protusions and pseudopodia to make initial contacts with neighboring cells. Later, groups of cytotrophoblastic cells in close apposition were observed, consistent with an aggregation stage. Finally, large cellular masses with central nuclei mount and expending cytoplasm were seen. These observations illustrate the process of interaction, aggregation and fusion of cytotrophoblastic cells (Douglas and King, 1990; Farmer and Nelson, 1992; Cronier et al., 1994).

In our experimental conditions (10% FCS, 0.5 x 10^6 cell/ml), 48 hours after plating, the syncytiotrophoblast cells were the dominant cellular elements observed; mononuclear cells and cellular aggregates were also present. The desmoplakin immunostaining revealed the inter cellular boundaries between aggregated cytotrophoblastic cells. Staining was also observed between cytotrophoblastic cells and syncytiotrophoblast and between contiguous syncytiotrophoblast cells. The staining disappeared in syncytiotrophoblast that was not in contact with other cells (Figure 1A).

In the presence of TGFβ1 (10 ng/ml) for the same period, the cytotrophoblastic cells remained essentially aggregated. This decrease or delay in the cytotrophoblastic cells fusion and syncytium formation was confirmed by the desmoplakin immunostaining at the intercellular boundaries (Figure 1B).

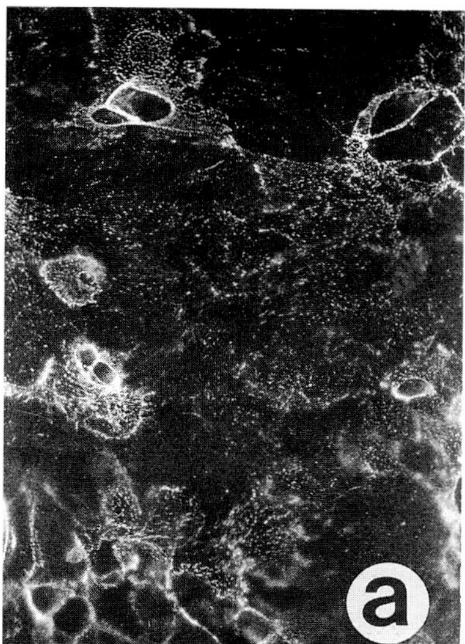

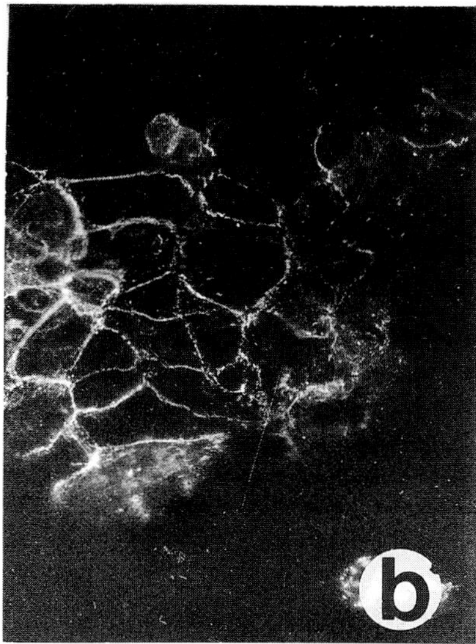

Figure 1. Effects of TGFβ1 on morphological differentiation of trophoblastic cells in culture. A) In control conditions, large syncytiotrophoblast cells were observed after two days of culture. Desmoplakin staining was noted at boundaries between cytotrophoblastic cells and syncytiotrophoblast. B) In the presence of 10 ng/ml TGFβ1, immunostaining was observed at intercellular borders, indicating that aggregates are the main trophoblastic configuration after 2 days in MEM + TGFβ1.

Gap FRAP

The development of gap junctional communication was previously demonstrated between trophoblastic cells in culture (Cronier et al., 1994, 1997). After two days of culture, fluorescence recovery never occurred in a majority of photobleached cells, demonstrating, in these cases, the absence of cell-to-cell communication. Fluorescence recovery was measured in the other trials. In some cases (34% of the records), the return of fluorescence followed a fast step-like course, reaching at least 90% of the final state in less than 30 second after photobleaching. This indicated that the diffusion of dye was neither prevented by cell membrane nor rate-limited by the presence of gap junctions. It was inferred that fusion of cell membranes had been completed and that the cellular elements were part of a true syncytium. In the other cases (6% of the tested cells), fluorescence recovery followed a much slower exponential time course, reflecting the presence of gap junctional communication. This slow fluorescence recovery can be reversibly interrupted by the exposure to a known junctional uncoupler (3 mM Heptanol). In contrast, exposure to heptanol did not prevent the fast step-like course of fluorescence recovery, confirming the cell membrane fusion of trophoblastic cells (Figure 2). Indeed, the specific block of junctional channels by heptanol is a mean to distinguish cell-to-cell communication by gap junctions or by cytoplasmic bridges (Bukauskas et al., 1992).

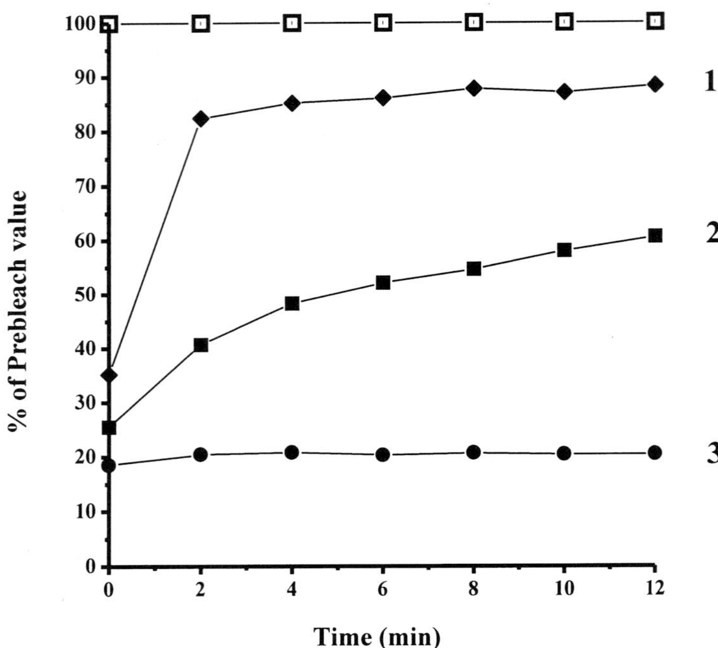

Figure 2. Evolution of fluorescence intensities with time after photobleaching of trophoblastic cells in contact. When trophoblastic cells were tested by the gap-FRAP technique, three different cases were demonstrated: 1) A very fast step-like fluorescence recovery, which was not affect by junctional uncoupler, corresponding to the diffusion of dye by cytoplasmic bridges during the process of membrane fusion. 2) A slower exponential time course, which was reversibly blocked by heptanol, reflecting the presence of gap junctional communication. 3) No return of fluorescence in the bleached cell, demonstrating the absence of cell-to-cell communication with neighboring cellular elements.

The presence of 1 ng/ml TGFβ1 for two days induced a slight increase of the frequency of coupled cells, while the presence of 5 to 10 ng/ml TGFβ1 strongly decreased the trophoblast coupling (Figure 3A). It has been previously demonstrated that, in the presence of 500 mIU hCG/ml, the proportion of coupled cells was increased and that a highest proportion of coupled cells was already observed after two days, instead of four days in control conditions (MEM, 10% FCS). Therefore, the effect of TGFβ1 was investigated in the presence of 500 mIU/ml hCG. In this condition, TGFβ1 (10 ng/ml) inhibited the stimulating action of hCG on GJC (Figure 3B). The role of endogenous hCG production on GJC during trophoblast differentiation was investigated by the study of the action of an hCG antibody. The addition of a polyclonal hCG antibody in excess decreased basal GJC. In the presence of hCG antibody, no significant additive inhibition by TGFβ1 was observed (2% of coupled cells, Figure 3B).

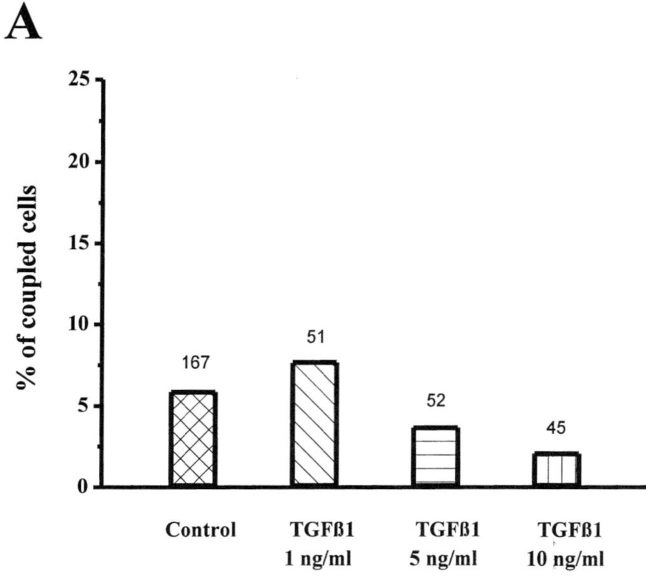

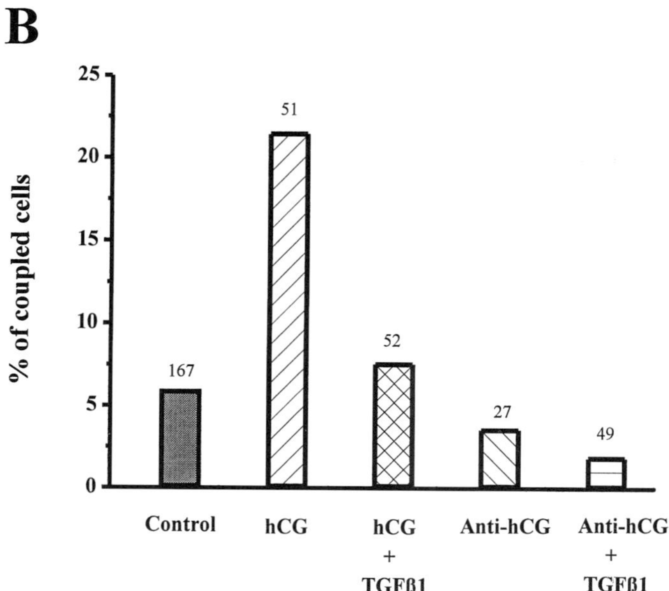

Figure 3. Effects of TGFβ1 on gap junctional intercellular communication after two days of culture. A) Percentage of coupled cells after two days of culture in the presence of TGFβ1 (1 to 10 ng/ml). B) Effect of hCG (500 mIU/ml) and of TGFβ1 (10 ng/ml) in the presence of hCG (500 mIU/ml) or hCG antibody (1/200). n values are indicated on top of the bars.

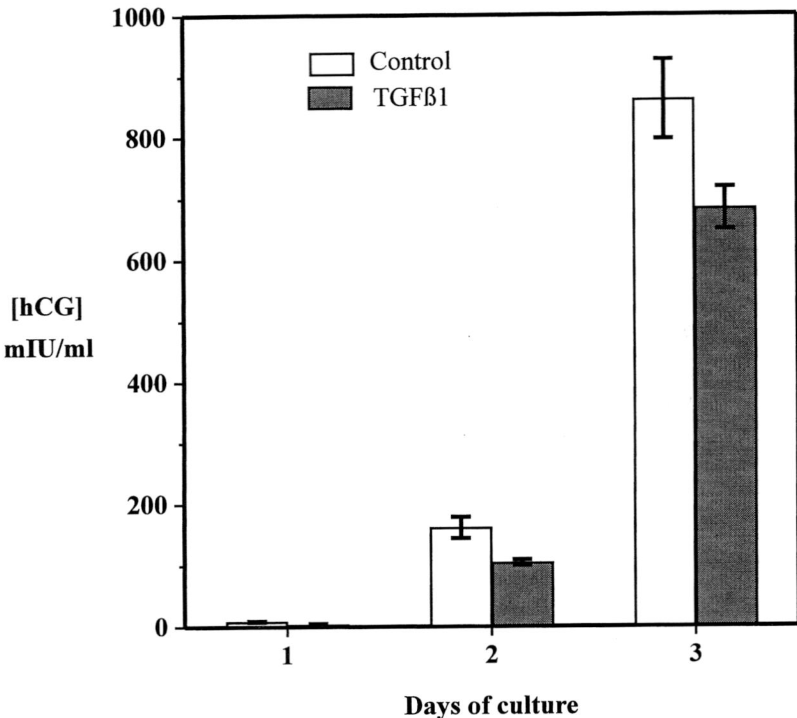

Figure 4. Effects of TGFβ1 (10 ng/ml) on βhCG secretion by trophoblastic cells in culture from one to three days. The culture media were renewed daily and assayed for hCG. Values represent the mean ± SEM of triplicate determinations.

Hormonal Production

The endocrine function of trophoblastic cells cultured in normal conditions or in the presence of 10 ng/ml TGFβ1 was assessed by determination of hCG release in the culture medium, and by immunoblotting of hCS in cell lysates. Increasing amounts of hCG with time were released by cell cultured in both normal condition and in the presence of TGFβ1 (10 ng/ml). However, the daily production of hCG by cells treated with TGFβ1 was significantly decreased when compared to the normal conditions (Figure 4).

The expression of hCS in cells cultured for two days in each of the experimental conditions is shown in Figure 5. After two days of culture, a 40% decrease of hCS immunodetection was measured in cells exposed to TGFβ1 compared to cells in control conditions. The reduction of the expression of this specific syncytial marker confirmed the decrease in the formation of syncytiotrophoblast in the presence of TGFβ1.

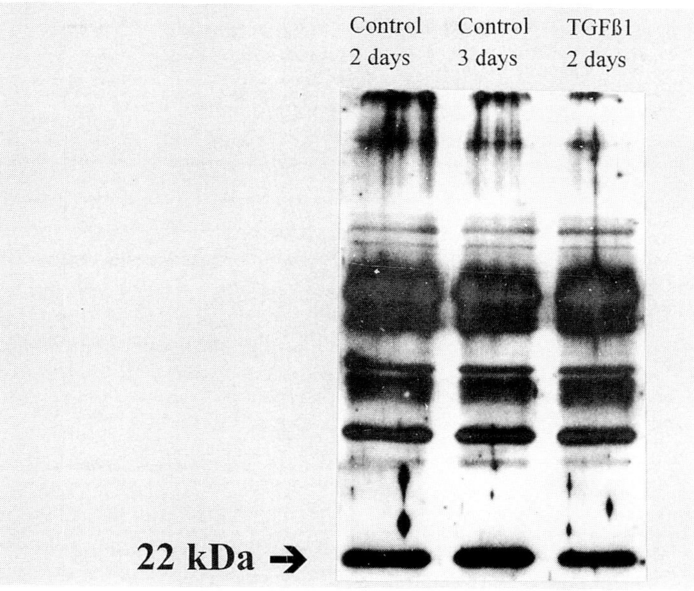

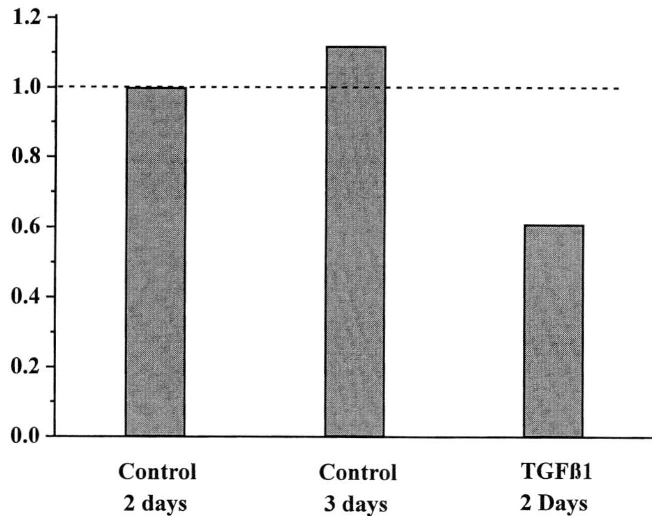

Figure 5. Effects of TGFβ1 (10 ng/ml) on hCS expression in trophoblastic cells revealed by immunoblot analysis. Cell lysates were prepared at the indicated periods of culture time and solubilized protein samples of 5 μg were used. In control conditions (MEM), there was an increase in hCS expression between two and three days. After two days of culture, the presence 10 ng/ml TGFβ1 induced a decrease of 40% of the hCS expression.

DISCUSSION

The presence of TGFβ1 (5 - 10 ng/ml) in the culture medium for two days induces a decrease of GJC. The main biological functions of gap junctions are control of intercellular electrical coupling and metabolic cooperation. Several observations indicate that the cell-to-cell exchange of metabolites and informative molecules through gap junctions would play an important role in the control of cell proliferation and differentiation and in embryonic development (Loewenstein and Rose, 1992; Hellmann et al., 1995). As previously reported, the gap-FRAP method has allowed us to demonstrate that a transfer of molecules between contiguous trophoblastic cells is established before the formation of the syncytiotrophoblast. In other words, a "physiological" syncytium, allowing the exchange of ions and small molecules, precedes the formation of a morphological syncytium. The end of GJC could then be considered as a physiological and objective criterion for the beginning of cellular fusion and the subsequent syncytium formation. Indeed, numerous authors have recognized the difficulty to distinguish aggregation of flattened cytotrophoblastic cells from syncytiotrophoblast (Aplin, 1991).

TGFβ1 inhibits basal GJC as well as the stimulatory effect of exogenous hCG on GJC. These results illustrate the importance of the stimulatory effect of hCG and the inhibitory effect of TGFβ1 on GJC. As previously reported (Cronier et al., 1994), trophoblastic GJC is specifically enhanced by hCG via the cAMP-Protein kinase A pathway. Furthermore, in the presence of an excess of hCG antibody, no significant additive inhibition by TGFβ1 is observed. Intercellular communication was not totally abolished in the presence of TGFβ1 and hCG antibody in excess. This fact suggests the presence in the culture medium of other regulators of GJC and trophoblast fusion (e.g., growth factors and hormones). Indeed, in a serum-free medium, cytotrophoblast spreading and fusion do not occur. In other cell types, TGFβ1 has been reported to stimulate, to inhibit or to have no effect on gap junctions, connexin expression or gap junctional communication (Albright et al., 1991; Chiba et al., 1994; Gibson et al., 1994). Its action seems to depend on cell type and culture conditions. It must be pointed out that the production of hCG is thought to be regulated by placental-derived cytokines, such as IL-1, IL-6, TNFβ, TGFβ1 and LIF (Ogren and Talamentes, 1994).

The mechanism by which TGFβ1 inhibits GJC is unknown. The growth factor could act directly on GJC (e.g., connexin expression) or indirectly by down-regulating trophoblastic hCG production.

The presence of TGFβ1 (10 ng/ml) largely inhibits or delays syncytium formation. The loss of desmoplakin immunoreactivity at intercellular boundaries of trophoblastic cells has been used to demonstrate complete syncytialization (Douglas and King, 1990), although the desmosomes persist after cellular fusion in the cytoplasm (Enders, 1965). In our study, in agreement with these reports, desmoplakin immunoreactivity was observed at intercellular boundaries between trophoblastic cells and was lacking in isolated syncytiotrophoblast cells. In the presence of TGFβ1 (10 ng/ml), desmoplakin immunostaining reveals a pavement-like pattern, indicative of cellular aggregates. Morrish et al. (1991) and Graham et al. (1992) have provided considerable evidence that TGFβ1 modulates villous trophoblast differentiation. Based on microscopic examination and hCG-hCS down-regulation, Morrish et al. (1991) concluded that TGFβ1 inhibited trophoblast differentiation. On the contrary, the presence of TGFβ1 for 72 hours has been reported by Graham et al. (1992) to increase the number of morphologically defined multinucleated cells. However, the disappearance of membranes was not assessed by desmoplakin immunostaining, and phenotypic markers of villous syncytiotrophoblast were not assayed.

It must be pointed out that divergent results have been reported on the effects of TGFβ1 on the invasiveness of the first trimester trophoblast *in vitro*, either an anti-invasive effect (Graham et al., 1992) or an absence of effects (Bass et al., 1994).

The presence of TGFβ1 (10 ng/ml) down-regulates hCG production and hCS expression by trophoblastic cells in culture, as previously demonstrated by Morrish et al. (1991). Furthermore, Matsuzaki et al. (1992) have reported that trophoblast-derived TGFβ1 suppresses cytokine-induced, but not gonadotropin-releasing hormone-induced release of hCG by human trophoblasts. Recently, Feinberg et al. (1994) demonstrated that TGFβ1 reduced βhCG production and up-regulated the expression of an anchoring trophoblast marker, oncofetal fibronectin. Thus, TGFβ1 could be critical for trophoblast-extracellular matrix interactions.

The decrease in hCG production could be due to an inhibitory effect of TGFβ1 on hCG gene expression, as suggested by Morrish et al. (1991) or to the TGFβ1 action on trophoblast differentiation. However, because the two phenomena are tightly linked (Hoshina et al., 1985; Merz, 1994), it would be difficult to evaluate their relative importance. Furthermore, due to its pleiotropic functions, TGFβ1 may act synergically with others growth factors and cytokines (IL-1, IL-6, TNFα, PDGF, IGF, and EGF).

TGFβ1 is known as a multifunctional growth factor present in many cell types. Recent reports, localizing TGFβ1 in the extravillous trophoblast, in villous trophoblast and at term in villous mesenchymal cells, favor the proposal of an autocrine/paracrine action of the growth factor (Lysiak et al., 1992; Selick et al., 1994). Moreover, in the human placenta, the level of expression of TGFβ1 mRNA increased at 17 and 34 weeks. These peaks in TGFβ1 expression were found to coincide with the overall cessation of trophoblast invasiveness and the later decline in absolute placental growth (Dungy et al., 1991). These results support the role of the growth factor in the differentiation process and in the physiology of the human trophoblast. It must be pointed out that the ability of trophoblast to differentiate is a major determinant for placental and fetal growth and development. Furthermore, because of the known antagonistic effects of TGFβ1 on systems associated with implantation and of the known immunosuppressive effects of TGFβ1, the growth factor has been proposed as a regulator of human implantation (Selick et al., 1994).

In conclusion, the inhibitory action of TGFβ1 further illustrates a clear correlation between inter-trophoblastic gap junctional communication, trophoblast differentiation and specific syncytial hormonal production.

SUMMARY

It was recently established that, during trophoblast differentiation, gap junctional communication (GJC) precedes the formation of the syncytiotrophoblast and is required for trophoblast cell fusion. Therefore, the end of cell-to-cell communication through gap junctions appears as a judicious criterion of cell fusion. Although GJC was seen to be stimulated by human Chorionic Gonadotropin (hCG), its regulation remains poorly understood. Transforming Growth Factor-β1 (TGF-β1), a multifunctional cytokine has been shown to modulate trophoblast differentiation. Therefore, the effects of TGF-β1 on intercellular dye diffusion have been investigated in cultured trophoblast of human term placenta, by means of the Fluorescence Recovery After Photobleaching (gap FRAP) technique. In parallel, trophoblast differentiation, hCG production and human Chorionic Somatomammotropin (hCS, a specific syncytiotrophoblast hormonal product) expression were assessed. The presence of TGF-β1 (5 or 10 ng/ml) in the culture medium for two days partially inhibited syncytium formation. The percentage of coupled cells was significantly decreased (2.8 times) after two days in presence of 10 ng/ml of TGF β1. Simultaneously, hCG release in culture medium was reduced (at this concentration, to 0.65 and 0.79 after respectively two and three days). In these conditions, Western blot analysis of trophoblast cellular proteins revealed that, after two days, hCS expression was reduced by 40% compared to control. Furthermore, the stimulation of trophoblastic GJC by exogenous hCG (500 mIU/ml) was considerably reduced by simultaneous exposure to TGF-β1 (10 mg/ml). The addition of a polyclonal hCG antibody in excess decreased basal GJC. In the presence of hCG antibody, no significant additive inhibition by TGFβ1 was observed. In conclusion, TGF-β1 was found to inhibit intercellular communication and, subsequently, differentiation and concomitant placental hormone secretions.

ACKNOWLEDGEMENTS

We are most grateful to Dr. J. Guibourdensche, Hôpital R. Debré, Paris, for hCG immunoassay. We thank Mrs. C. Bezagu for her photographic assistance. We would also like to thank the medical staff of the Clinique du Fief de Grimoire (Poitiers) for their kind cooperation in supplying us with the placentae. This work was supported in parts by grants from the Fondation de la Recherche Medicale, the Fondation Langlois, and the G.E.R.P. (Groupement d'Etude et de Recherche sur le Placenta).

REFERENCES

Albright, C.D., Grimley, P.M., Jones, R.T., Fontana, J.A., Keenan, K.P. and Resau, J.H. (1991) Cell-to-cell communication: a differential response to TGF-β in normal and transformed (BEAS-2B) human bronchial epithelial cells. *Carcinogenesis* 12, 1993-1999.

Alsat, E., Wyplosz, P., Malassiné, A., Guibourdenche, J., Porquet, D., Nessmann, C. and Evain-Brion, D. (1996) Hypoxia impairs cell fusion and differentiation process in human cytotrophoblast in vitro. *J. Cell Physiol.* 168, 346-353.

Alsat, E., Haziza, J. and Evain-Brion, D. (1993) Increase in epidermal growth factor receptor and its messenger ribonucleic acid levels with differentiation of human trophoblast cells in culture. *J. Cell. Physiol.* 154, 122-128.

Aplin, J.D. (1991) Implantation, trophoblast differentiation and haemochorial placentation: mechanistic evidence *in vivo* and *in vitro*. *J. Cell. Sci.* 99, 681-692.

Bass, K.E., Morrish, D.W., Roth, I., Bhardwaj, D., Taylor, R., Zhou, Y. and Fisher, S.J. (1994) Human cytotrophoblast invasion is up-regulated by epidermal growth factor: Evidence that paracrine factors modify this process. *Dev. Biol.* 164, 550-561.

Bernirschke, K. and Kaufmann, P. (1990) The Pathology of the Human Placenta. Springer Publ., New York.

Bukauskas, F., Kempf, C. and Weingart, R. (1992) Electrical coupling between cells of the insect *Aedes albopictus. J. Physiol.* 448, 321-337.

Chiba, H., Sawada, N., Oyamada, M., Kojima, T., Iba, K., Ishii, S. and Mori, M. (1994) Hormonal regulation of connexin 43 expression and gap junctional communication in human osteoblastic cells. *Cell Struct. Funct.* 19, 173-177.

Cronier, L., Bastide, B., Hervé, J.C., Délèze, J. and Malassiné, A. (1994) Gap junctional communication during human trophoblast differentiation: Influence of human Chorionic Gonadotropin. *Endocrinology* 135, 402-408.

Cronier, L., Bois, P., Hervé, J.C. and Malassiné, A. (1995) Effect of human chorionic gonadotropin (hCG) on chloride current in human syncytiotrophoblasts in culture. *Placenta* 16, 599-609.

Cronier, L., Hervé, J.C., Délèze, J. and Malassiné, A. (1997) Regulation of gap junctional communication during human trophoblast differentiation. *Microsc. Res. Tech.*, (in press).

De Virgiliis, G., Sideri, M., Fumagalli, G. and Remotti, G. (1982) The junctional pattern of the human villous trophoblast: A freeze fracture study. *Gynecol. Obstet. Invest.* 14, 263-272.

Douglas, G.C. and King, B.F. (1990) Differentiation of human trophoblast cells in vitro as revelated by immunocytochemical staining of desmoplakin and nuclei. *J. Cell. Sci.* 96, 131-141.

Dungy, L.J., Siddiqi, T.A. and Khan, S. (1991) Transforming growth factor-β1 expression during placental development. *Am. J. Obstet. Gynecol.* 165, 853-857.

Enders, A. (1965) Formation of syncytium from cytotrophoblast in the human placenta. *Obstet. Gynecol.* 25, 378-386.

Farmer, D.R. and Nelson, D.M. (1992) A fibrin matrix modulates the proliferation, hormone secretion and morphologic differentiation of cultured human placental trophoblast. *Placenta*, 13, 163-177.

Feinberg, R.F., Kliman, H.J. and Wang, C.L. (1994) Transforming growth factor-β stimulates trophoblast oncofetal fibronectin synthesis *in vitro*: Implications for trophoblast implantation *in vivo. J. Clin. Endocrinol. Metab.* 78, 1241-1248.

Firth, J.A., Farr, A. and Bauman, K. (1980) The role of gap junctions in trophoblastic cell fusion in the guinea-pig placenta. *Cell. Tissue Res.* 205, 311-318.

Gibson, D.F.C., Hossain, M.Z., Goldberg, G.S., Acevedo, P. and Bertram, J.S. (1994) The mitogenic effects of transforming growth factors β1 and β2 in C3H/10T½ cells occur in the presence of enhanced gap junctional communication. *Cell Growth Differ.* 5, 687-696.

Graham, C.H., Lysiak, J.J., McCrae, K.R. and Lala, P.K. (1992) Localization of transforming growth factor-β at the human fetal-maternal interface: Role in trophoblast growth and differentiation. *Biol. Reprod.* 46, 561-572.

Grümmer, R., Reuß, B., Hellmann, P., Soares, M.J. and Winterhager, E. (1995) Regulation of different connexin genes during differentiation of trophoblast cells in vivo and in vitro. *Placenta,* 16, A23.

Hellmann, P., Von Ostau, C., Grümmer, R. and Winterhager, E. (1995) Connexin40 expression in the human trophoblast: Implicator for proliferation and invasive properties. *Placenta,* 16, A26.

Hoshina, M., Boothby, M., Hussa, R.O., Patillo, R., Camel, H.M. and Boime, I. (1985) Linkage of human chorionic gonadotropin and placental lactogen to trophoblast differentiation and tumorigenesis. *Placenta* 6, 163-172.

Kauma, S., Matt, D., Strom, S., Eierman, D. and Turner, R. (1990) Interleukin-1β, human leucocyte antigen HLA-DRα, and transforming growth factor-β expression in endometrium, placenta, and placental membranes. *Am. J. Obstet. Gynecol.* 163, 1430-1437.

Kliman, H.J., Nestler, J.E., Sermasi, E., Sanger, J.M. and Strauss III, J.F. (1986) Purification, Characterization, and in vitro differentiation of cytotrophoblasts from human term placenta. *Endocrinology* 118, 1567-1582.

Loewenstein, W.R. and Rose, B. (1992) The cell-cell channel in the control of growth. *Semin. Cell. Biol.* 3, 59-79.

Lysiak, J.J., Hunt, J., Pringle, G.A. and Lala, P.K. (1995) Localization of transforming growth factor beta and its natural inhibitor decorin in the human placenta and decidua throughout gestation. *Placenta* 16, 221-231.

Malassiné, A. and Leiser, R. (1984) Morphogenesis and fine structure of near-term placenta of Talpa europea. I. Endotheliochorial labyrinth placenta. *Placenta,* 5, 145-158.

Malassiné, A., Alsat, E., Besse, C., Rebourcet, R. and Cedard, L. (1990) Acetylated low Density Lipoprotein endocytosis by human syncytiotrophoblast in culture. *Placenta* 11, 191-204.

Mason, J.I. (1993) 3β-hydrosteroid deshydrogenase and its regulation. In: *Progress in Endocrinology*, (ed) R. Mornex, C. Jaffiol and J. Leclère, Parthenon Publishing, Lancs, UK, pp. 509-513.

Massague, J. (1987) The TGF-β family of growth and differentiation factors. *Cell* 49, 437-438.

Matsusaki, N., Li, Y., Matsuhiro, K., Jo, T., Shimoya, K., Tanigushi, T., Saji, F. and Tanizawa, O. (1992) Trophoblast-derived transforming growth factor-β1 suppresses cytokine-induced, but not gonadotrophin-releasing hormone-induced, release of human chorionic gonadotropin by normal human trophoblasts. *J. Clin. Endocrinol. Metab.* 74, 211-216.

Merz, W.E. (1994) The primate placenta and human chorionic gonadotropin. *Exp. Clin. Endocrinol.* 102, 222-234.

Metz, J., Heinrich, D. and Forssmann, W.G. (1976) Gap junctions in hemodichorial and hemotrichorial placentae. *Cell. Tissue Res.* 171, 305-315.

Mitchell, E.J. and O'Connor-McCourt, M.D. (1991) A transforming growth factor β (TGFβ) receptor from human placenta exhibits a greater affinity for TGF-β2 than for TGFβ1. *Biochemistry* 30, 4350-4356.

Morrish, D.W., Bhardwaj, D. and Paras, M.T. (1991) Transforming growth factor β1 inhibits placental differentiation and human chorionic gonadotropin and human placental lactogen secretion. *Endocrinology* 129, 22-26.

Ogren, L. and Talamentes, F. (1994) The placenta as an endocrine organ. Polypeptides. In: *The Physiology of Reproduction,* second edition, (eds.) E. Knobil and J.D. Neill, Raven Press Ltd, New York, pp. 875-945.

Ong, P. and Burton, G.J. (1990) The effects of hypoxia and reoxygenation on barrier thickness of placental villi maintained in organ culture. In: *Placental Communications: Biochemical Morphological and Cellular Aspects,* (ed..) L. Cedard, E. Alsat, J.C. Challier, G. Chaouat and A. Malassiné, Colloque INSERM/John Libbey Eurotext Ltd, 199, p. 262.

Selick, C.E., Horowitz, G.M., Gratch, M., Scott, R.T., Navot, D. and Hofmann, G.E. (1994) Immunohistochemical localization of transforming growth factor-β in human implantation sites. *J. Clin. Endocrinol. Metab.* 78, 592-596.

Shull, M.M. and Doetschman, T. (1994) Transforming growth factor-β1 in reproduction and development. *Mol. Reprod. Dev.* 39, 239-246.

Sporn, M.B., Roberts, A.B., Wakefield, L.M. and Assoian, R.K. (1986) Transforming growth factor-β: Biological function and chemical structure. *Science* 233, 532-534.

Wade, M.H., Trosko, J.E. and Schindler, M. (1986) A fluorescence photobleaching assay of gap junction mediated communication between human cells. *Science* 232, 525-528.

LIST OF CONTRIBUTORS

Asif Ahmed
Department of Obstetrics and Gynaecology
Birmingham Maternity Hospital
University of Birmingham
Edgbaston, Birmingham, B15 2TG
United Kingdom

John D. Aplin
Department of Obstetrics and Gynaecology
School of Biological Sciences
University of Manchester
Manchester, M13 0JH, United Kingdom

Eliane Alsat
INSERM U427
Faculté des Sciences Pharmaceutiques et Biologiques de Paris
Université Paris V
4, Av. de l'observatoire
75270 Paris Cedex 6, France

Pasquale Berlingieri
Academic Unit of Obstetrics and Gynecology
Royal London Hospital
London, United Kingdom

Philippe Birembaut
INSERM U314
CHU Maison Blanche
45, rue Chognacq-Jay
51100 Reims, France

Paul Bischof
Department of Obstetrics and Gynaecology
Laboratoire d'Hormonologie Maternité
University of Geneva
1211 Genèva 14, Switzerland

Christelle Bourgeois
U.361 INSERM
Reproduction et Physiopathologie Obstétricale
Université René Descartes, Paris V
124, be de Port Royal
75014 Paris, France

Tanya D. Burrows
Research Group in Human Reproductive Immunobiology
Department of Pathology
University of Cambridge
Cambridge, BC2 1QP, United Kingdom

Bruno Carbonne
Maternitè Port-Royal - Baudelocque
Université René Descartes, Paris V
123, bd de Port Royal
75014 Paris, France

Mario Castellucci
Dipartimento di Scienze Mediche
Anatomia Umana
Via Solaroli, 17
I-28100 Novara, Italy

Heather J. Church
Department of Obstetrics and Gynaecology
School of Biological Sciences
University of Manchester
Manchester, M13 0JH, United Kingdom

Laurent Cronier
Laboratoire de Physiologie Cellulaire
CNRS URA 1869
Université de Poitiers
40, Av du recteur Pineau
86022 Poitiers Cedex, France

Jean Délèze
Laboratoire de Physiologie Cellulaire
CNRS URA 1869
Université de Poitiers
40, Av du recteur Pineau
86022 Poitiers Cedex, France

Josiane Delogne-Desnoeck
Research Laboratory on Reproduction
Free University of Brussels (ULB)
808 Route de Lennik
B-1070 Brussels, Belgium

Johannes Dietl
Department of Obstetrics and Gynecology
University of Tübingen
Liebermeisterstr. 8
72076 Tübingen, Germany

Ingke Ebeling
Department of Anatomy
Technical University
Wendlingweg 2
D-52057 Aachen, Germany

Allen C. Enders
Department of Cell Biology and Human Anatomy
University of California School of Medicine
Davis, California 95616 USA

Françoise Ferré
U.361 INSERM
Reproduction et Physiopathologie Obstétricale
Université René Descartes, Paris V
123, bd de Port Royal
75014 Paris, France

J. Anthony Firth
Department of Anatomy and Cell Biology
Imperial College School of Medicine at St. Mary's
London W2 1PG, United Kingdom

Jean-Michel Foidart
Laboratory of Biology
University of Liège
Tower of Pathology
B35 Sart Tilman
B-4000 Liège, Belgium

Hans-G. Frank
Department of Anatomy
Technical University
Wendlingweg 2
D-52057 Aachen, Germany

Hitoshi Funayama
Department of Obstetrics and Gynecology
Medical College Hospital
6-7-1 Nishishinyuku Shinyuuku-ku
160 Tokyo, Japan

Laszlo Füzesi
Department of Pathology
Technical University
Pauwelsstrasse 30
D-52057 Aachen, Germany

Gabriele Gaus
Department of Anatomy
Technical University
Wendlingweg 2
D-52057 Aachen, Germany

Joep P.M. Geraedts
Department of Molcular Cell Biology and Genetics
University of Limburg
PO Box 616
6200 MD Maastricht, The Netherlands

J. Gedis Grudzinskas
Academic Unit of Obstetrics and Gynecology
Royal London Hospital
London, United Kingdom

Georges Hennen
Biochemie et Laboratoire d'Endocrinologie
Université de Liège
Tour de Pathologie, B23
B-4000 Liège, Belgium

Jean-Claude Hervé
Laboratoire de Physiologie Cellulaire
CNRS URA 1869
Université de Poitiers
40, Av du recteur Pineau
86022 Poitiers Cedex, France

Hans-Peter Horny
Institute of Pathology
University of Tübingen
Liebermeisterstr. 8
72076 Tübingen, Germany

Berthold Huppertz
Department of Anatomy
Technical University
Wendlingweg 2
D-52057 Aachen, Germany

Ahmed Igout
Biochemie et Laboratoire
d'Endocrinologie
Université de Liège
Tour de Pathologie, B23
B-4000 Liège, Belgium

Eric Jauniaux
Research Laboratory on Reproduction
Free University of Brussels (ULB)
808 Route de Lennik
B-1070 Brussels, Belgium

Edwin Kaiserling
Institute of Pathology
University of Tübingen
Liebermeisterstr. 8
72076 Tübingen, Germany

Peter Kaufmann
Department of Anatomy
Technical University
Wendlingweg 2
D-52057 Aachen, Germany

Ashley King
Research Group in Human
Reproductive Immunobiology
Department of Pathology
University of Cambridge
Cambridge, BC2 1QP, United Kingdom

John C.P. Kingdom
Department of Obstetrics and
Gynaecology
University College London Medical
School
86-96 Chenies Mews
London, WC1E 6HX, United Kingdom

Stefan Kröber
Institute of Pathology
University of Tübingen
Liebermeisterstr. 8
72076 Tübingen, Germany

Christiane Krebs
Department of Veterinary Anatomy
University of Giessen
Giessen, Germany

Lopa Leach
Department of Human Morphology
University of Nottingham
Queen's Medical Centre
Nottingham NG7 2UH United Kingdom

Philippe Lebrun
Research Laboratory on Reproduction
Laboratory of Pharmacology
Free University of Brussels (ULB)
808 Route de Lennik
B-1070 Brussels, Belgium

Rudolph Leiser
Department of Veterinary Anatomy
University of Giessen
Giessen, Germany

Y.W. Loke
Research Group in Human
Reproductive Immunobiology
Department of Pathology
University of Cambridge
Cambridge, BC2 1QP, United Kingdom

Lena Macara
University of Glasgow
Glasgow, Scotland

André Malassiné
Laboratoire de Physiologie Cellulaire
CNRS URA 1869
Université de Poitiers
40, Av du recteur Pineau
86022 Poitiers Cedex, France

Erik Maquoi
Laboratory of Biology
University of Liège
Tower of Pathology
B23 Sart Tilman
B-4000 Liège, Belgium

Klaus Marzusch
Department of Obstetrics and Gynecology
University of Tübingen
Liebermeisterstr. 8
72076 Tübingen, Germany

Sylvain Meuris
Research Laboratory on Reproduction
Free University of Brussels (ULB)
808 Route de Lennik
B-1070 Brussels, Belgium

Thérèse-Marie Mignot
U.361 INSERM
Reproduction et Physiopathologie Obstétricale
Université René Descartes, Paris V
123, bd de Port Royal
75014 Paris, France

Donald W. Morrish
Department of Medicine
362 Heritage Medical Research Center
University of Alberta
Edmonton, Alberta, Canada T6G 2S2

Anne-Marie Nagy
Research Laboratory on Reproduction
Free University of Brussels (ULB)
808 Route de Lennik
B-1070 Brussels, Belgium

Béatrice Nawrocki
INSERM U314
CHU Maison Blanche
45, rue Chognacq-Jay
51100 Reims, France

Agnès Noël
Laboratory of Biology
University of Liège
Tower of Pathology
B23 Sart Tilman
B-4000 Liège, Belgium

Kyle E. Orwig
Department of Physiology
University of Kansas Medical Center
Kansas City, Kansas 66160 USA

Myriam Polette
INSERM U314
CHU Maison Blanche
45, rue Chognacq-Jay
51100 Reims, France

Karen J. Powell
Academic Unit of Obstetrics and Gynecology
Royal London Hospital
London, United Kingdom

Christine A. Rasmussen
Department of Physiology
University of Kansas Medical Center
Kansas City, Kansas 66160 USA

Allan J. Richards
MRC Connective Tissue Genetics Group
Strangeways Laboratory and
Department of Pathology
University of Cambridge
Cambridge, United Kingdom

Peter Ruck
Institute of Pathology
University of Tübingen
Liebermeisterstr. 8
72076 Tübingen, Germany

Jean-Pierre Schaaps
Department of Obstetrics and Gynecology
University of Liège
CHR Citadelle
Bd du 12e de Ligne, 1
4000 Liège, Belgium

S.K. Smith
Research Group in Human Reproductive Immunobiology
Department of Pathology
University of Cambridge
Cambridge, BC2 1QP, United Kingdom

Michael J. Soares
Department of Physiology
University of Kansas Medical Center
Kansas City, Kansas 66160 USA

Claude Sureau
Theramex Institute
Bioethics, Women's Health, and Society
38-40 Avenue de New-York
75016 Paris, France

Masaomi Takayama
Department of Obstetrics and
Gynecology
Medical College Hospital
6-7-1 Nishishinyuku Shinyuuku-ku
160 Tokyo, Japan

INDEX

All page numbers listed below represent the first page of each chapter where the subject is located.